Hsu
Fieldhouse
Pugliese
Loschiavo

essential
CHEMISTRY

PASCO
Education

Essential Chemistry
Copyright © 2018 PASCO Scientific
Digital Student Edition ISBN: 978-1-937492-21-2
Print Student Edition ISBN: 978-1-937492-17-5

1 2 3 4 5 6 7 8 9 - LSC - 22 21 20 19 18

All rights reserved. No part of this work may be reproduced or transmitted in any form or by any means, electronic or mechanical including photocopying and recording, or by any information storage or retrieval system, without written permission from the Publisher. For permission or other rights under this copyright, please contact:

PASCO Scientific
10101 Foothills Blvd.
Roseville, CA 95747

This book was written, illustrated, published, and printed in the United States of America.

Letter from the authors

Welcome to Essential Chemistry!

Chemistry is fundamentally important. Everyone ought to understand enough chemistry to make informed decisions in their lives. Everyday we face a range of questions, as simple as what to eat, and as complex as whether Earth's climate will change and why. Memorizing how to solve types of problems in a book will not prepare our students to answer these questions.

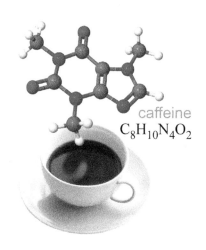

To use chemistry, our students need to understand chemistry, the underlying concepts as well as how to solve realistic problems. This is the goal of Essential Chemistry. You may find some of our teaching techniques different, but thousands of classroom hours have shown us that these hands-on, inquiry-based techniques create deeper learning that lasts.

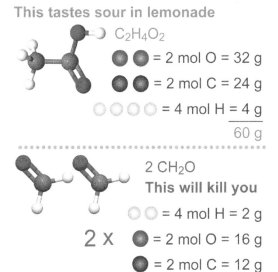

While the experience of chemistry is tangible and real, the explanation of chemistry is mostly microscopic and invisible. We can't sense an atom, even though we touch countless quintillions of them every second. The high level of abstraction in chemistry can easily lead to confusion. We based Essential Chemistry on both brain research and on common misconceptions.

For example, understanding subscripts and coefficients is at the heart of why many students have difficulty balancing chemical equations. We use molecular models right away to provide a concrete way to understand the chemical formula and why $2CH_2O$ is different from $C_2H_4O_2$. Almost like an abacus for atoms, the models give students something to put their hands on that helps them solidify their understanding of moles, molar masses and chemical formulas.

Of course, we are also committed to hands-on, quantitative chemistry! Essential Chemistry is accompanied by more than 70 investigations that use exceptional, easy-to-use, wireless sensors—measuring temperature, pH, pressure and color, to name a few. Students can do accurate experiments in which they use technology to make measurements that they can analyze and understand. More than five thousand illustrated slides support the lab work with detailed setup illustrations.

We hope you and your students enjoy learning chemistry!

Tom Hsu
Ronn Fieldhouse
Barbara Pugliese
Tom Loschiavo

About the authors

Dr. Tom Hsu is nationally known as an innovator in science education and has worked with more than 16,000 teachers over the past twenty-two years. The author of six science programs in physics, physical science, and chemistry, Dr. Hsu also invents unique lab apparatus for teaching and learning. As an educator, Dr. Hsu taught high school physics as well as graduate and undergraduate courses at MIT, where he was nominated for the 1991 Goodwin Medal for excellence in teaching. He holds a Ph.D. in applied plasma physics from MIT and previously worked as an engineer for Eastman Kodak and Xerox. Dr. Hsu was founder of CPO Science and cofounder of Ergopedia, Inc. He is Chief Product Officer at PASCO Scientific, a leader in science programs distributed to teachers and institutions worldwide.

Ronn Fieldhouse has over twenty years of classroom teaching experience in General Science, Chemistry, Honors Chemistry and AP® Chemistry. Prior to entering the field of education he worked as a chemist in the field of medical supplies, and brought this industry experience to the classroom. His passion for chemistry and teaching was the inspiration for many students to pursue science and engineering careers. Ronn is a recipient of the distinguished Golden Apple, awarded to outstanding teachers in northern Illinois. He holds a B.S. in Chemistry, M.S. in Curriculum and Instruction, and M.S. in Administration from Northern Illinois University. Ronn is a curriculum and training specialist at PASCO Scientific where he designs and develops curriculum and delivers professional development internationally.

Barbara Pugliese has a background in both chemistry and biology, and over 15 years of teaching experience at the high school and college levels. As an adjunct professor, Barbara taught Introductory Chemistry and Technology for Educators. Her focus at the high school level was mainly on Chemistry, Oceanography, and AP® Environmental Science. Throughout her teaching career Barbara made science meaningful for students providing them with opportunities to do science. Barbara earned a B.S. in Biology from California State Polytechnic University and M.A in Educational Technology from San Diego State University. Prior to joining PASCO Scientific, Barbara worked as an Instructional Designer specializing in traditional, blended, and directed study environments. At PASCO she applies those skills to delivering training to teachers, and designing and developing chemistry and biology curriculum.

Tom Loschiavo has over 12 years of teaching experience in high school that include AP® Chemistry, Honors Chemistry, Chemistry and Physics. At the outset of his teaching career Tom integrated technology in his science classes as a pedagogical tool to support student learning. Over time he adopted an inquiry-based learning approach where his instructional delivery ranged from teacher-guided to student-driven. Tom has a B.S. in Chemistry from the College of New Jersey and M.S. in Chemistry Education from the University of Pennsylvania. As a chemistry education manager at PASCO Scientific, Tom develops and advocates for chemistry products and curriculum that will help educators worldwide. He merges his chemistry knowledge and teaching skills to create innovative curriculum and to deliver outstanding trainings.

Contributors

Editors

Freda Husic, M.A., M.S.
Director of Curriculum
PASCO Scientific

Vivian Lemanowski, M.Ed.
Manager of Curriculum and Training
PASCO Scientific

Contributing Authors

Jared Lash
Product Specialist
PASCO Scientific

Sheila Nguyen
Assist. Professor, Chemistry Department
Cypress College

Amanda Zullo
Chemistry Teacher
Saranac Lake Senior High School

Susan Conant
Chemistry Teacher, retired

Patrick Escott
Chemistry Teacher
Charter H.S. of the Arts

Jason Lee
Chemistry Teacher
East Georgia College

Matthew D. Bannerman, M.A., M.Ed.
Curriculum specialist
PASCO Scientific

Technical support and development

Sean Morton, M.S.
Software Engineer
PASCO Scientific

Chris Murray
Software Engineer
PASCO Scientific

Phil Wong
Software Engineer
PASCO Scientific

Graphic Arts

Brennan Collins
Media Specialist
PASCO Scientific

Dan Cone
Graphic Artist

Jim Travers
Graphic Artist

Kevin Young
Marketing Operations
PASCO Scientific

Project Manager

Susan Watson
Project Administrator, Education Solutions
PASCO Scientific

Content Specialists

Cole Nichols
Sierra College

Tatyana Andriyenko
Sacramento State Univ.

Elisabeth Newman
Sacramento State Univ.

Anthony LaBarbera
Sacramento State Univ.

Beatrice Okokpujie
Sacramento State Univ.

Alastair Butterworth
Sacramento State Univ.

Reviewers

Lisa Tobias, M.Ed.
Science Teacher
St Peter Prince of the Apostles

Sally Mitchell, M.Ed.
Chemistry Teacher
Rye High School
Rye, NY

Fred Vital, M.A., M.S.
Chemistry Teacher
Darien High School
CT

John D. Kay, M. Ed.
Chemistry Teacher
Kingston High School, NY
Adjunct Instructor
SUNY New Paltz, New Paltz, NY

Helene, Marie-Afeli, Ph.D.
Assistant Professor of Chemistry
University of South Carolina
Palmetto College - Union

Kristen Drury
Chemistry Teacher
William Floyd High School
NY

Doug Ragan
Chemistry Teacher
Hudsonville High School
Michigan

Duane Swank, Ph.D.
Professor Emeritus
Pacific Lutheran University
Tacoma, WA

David M. Leal
Chemistry Teacher
Western Sierra
Collegiate Academy
CA

Michael Schaab
Professor Emeritus
Maine Maritime Academy

Jennifer Cook Gregory
Chemistry Teacher
Moline Sr. High School
Moline, IL

Louise Jakub-Cerro, M.A.
Chemistry Teacher
Salem High School
NJ

Brian Stagg, M.CLFS.
Chemistry Teacher
Eleanor Roosevelt
High School
MD

Van V. Truong
Science Teacher
Central High School
PA

Stephanie O'Brien, Ph.D.
Commack High School
Commack, NY

Kathie R. Alpert
Chemistry Teacher
Moorestown High School
NJ

Dale Jensen
Chemistry Teacher
St. Mary's School
Medford, OR

Cheri Smith
Chemistry Teacher
Yale Secondary, Canada

Todd Abronowitz, M.S.
Chemistry Teacher
Parish Episcopal School
Dallas, TX

Lisa McGaw
Chemistry faculty
Northern Oklahoma College
Tonkawa, Oklahoma

Rachel M. Harris, M.Ed., PA
Chemistry, Physics and Math Teacher
Ph.D. Graduate Student in Chemistry
The Johns Hopkins University
Baltimore, MD

Erica Posthuma-Adams, M.Ed.
Chemistry Teacher
University High School
of Indiana

Thomas Kunzleman, Ph.D.
Professor of Chemistry
Spring Arbor University
MI

Richard Schwenz
Chemistry Professor
University of
Northern Colorado

Francis P. Gasparro, Ph.D.
Hamden Hall Country
Day School
Hamden, CT

Features of the printed book

Essential Chemistry has been designed to be rigorous in its treatment of chemistry but also reader-friendly

1. One idea per page helps focus attention on content.
2. Paragraph outlining helps students key-in on the core idea of each paragraph.
3. Thousands of instructional graphics support learning, not just pretty pictures, but real chemistry content expressed in a visual language.
4. Hundreds of solved problems

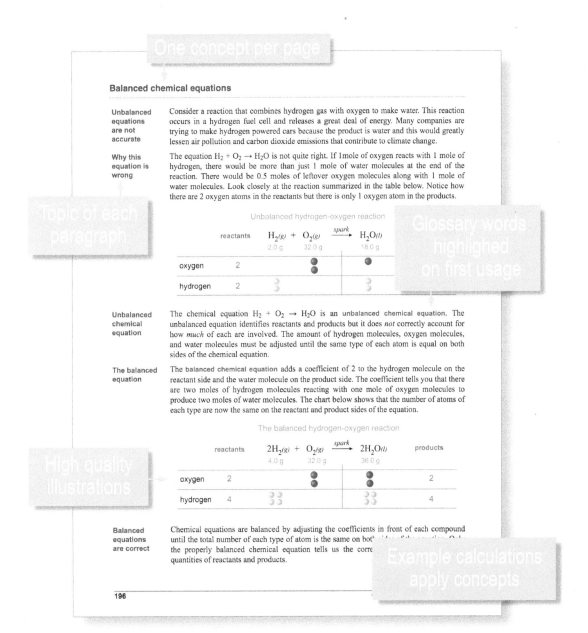

Features of the ebook

Essential Chemistry runs on any device in any browser. The navigation icons at the top of each page will help you move around and find what you are looking for.

The navigation toolbar on every page

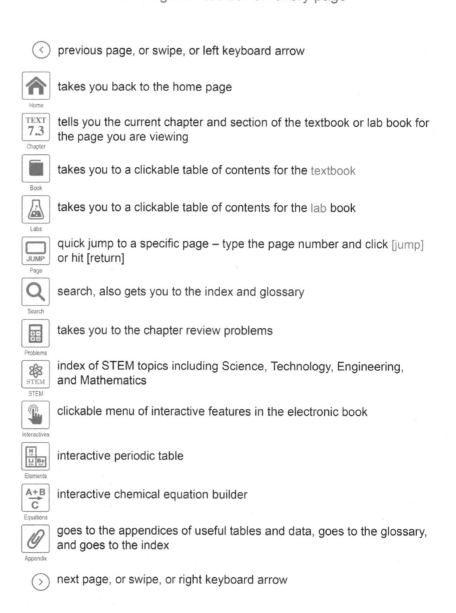

previous page, or swipe, or left keyboard arrow

takes you back to the home page

tells you the current chapter and section of the textbook or lab book for the page you are viewing

takes you to a clickable table of contents for the textbook

takes you to a clickable table of contents for the lab book

quick jump to a specific page – type the page number and click [jump] or hit [return]

search, also gets you to the index and glossary

takes you to the chapter review problems

index of STEM topics including Science, Technology, Engineering, and Mathematics

clickable menu of interactive features in the electronic book

interactive periodic table

interactive chemical equation builder

goes to the appendices of useful tables and data, goes to the glossary, and goes to the index

next page, or swipe, or right keyboard arrow

essential CHEMISTRY

Unit 1 Matter and Energy	Chapter 1	The Science of Chemistry	2
	Chapter 2	Measurement and Analysis	28
	Chapter 3	Classifying Matter	58
	Chapter 4	Temperature and Heat	92
Unit 2 Compounds and Reactions	Chapter 5	Chemical Compounds	126
	Chapter 6	Moles	162
	Chapter 7	Chemical Reactions	190
	Chapter 8	Stoichiometry	220
Unit 3 Atoms and Bonds	Chapter 9	Atomic Structure	252
	Chapter 10	Bonding and Valence	296
Unit 4 States and Solutions	Chapter 11	Energy and Change	344
	Chapter 12	Gases	378
	Chapter 13	Solutions	408
Unit 5 Rates and Equilibrium	Chapter 14	Reaction Rates	446
	Chapter 15	Equilibrium	476
	Chapter 16	Acids and Bases	506
Unit 6 Redox and Energy	Chapter 17	Oxidation and Reduction	542
	Chapter 18	Electrochemistry	572
	Chapter 19	Nuclear Chemistry	600
Unit 7 Organic and Biochemistry	Chapter 20	Organic Chemistry	640
	Chapter 21	Molecular Biology	678
	Chapter 22	Biochemistry	712
Unit 8 Earth and Beyond	Chapter 23	The Earth	746
	Chapter 24	The Universe	784
		Appendix	820
		Glossary	836
		Index	862

essential CHEMISTRY

Unit 1 Matter and Energy

Chapter 1
The Science of Chemistry 2
- 1.1 The Science of Matter and Change ... 4
- 1.2 Matter and Energy ... 14
- *essential CHEMISTRY* 1.3 Getting to Chemistry Class ... 23

Chapter 2
Measurement and Analysis 28
- 2.1 Measurement and Numbers ... 30
- 2.2 Representing and Analyzing Data ... 42
- *essential CHEMISTRY* 2.3 The Value of Measurement ... 51

Chapter 3
Classifying Matter 58
- 3.1 Atoms and the Periodic Table ... 60
- 3.2 Compounds and Molecules ... 68
- 3.3 Pure Substances and Mixtures ... 73
- 3.4 Physical and Chemical Changes ... 78
- *essential CHEMISTRY* 3.5 Magical Chemical Reactions ... 86

Chapter 4
Temperature and Heat 92
- 4.1 Temperature ... 94
- 4.2 Thermal Energy and Heat ... 101
- 4.3 Heat and Changes of State ... 113
- *essential CHEMISTRY* 4.4 Color Change Thermometers ... 120

Unit 2 Compounds and Reactions

Chapter 5
Chemical Compounds 126
- 5.1 The Past and Present Periodic Table ... 128
- 5.2 Ions ... 135
- 5.3 Naming and Writing Ionic Compounds ... 142
- 5.4 Covalent Compounds ... 149
- *essential CHEMISTRY* 5.5 Ban DHMO! ... 156

Chapter 6
Moles 162
- 6.1 The Mole ... 164
- 6.2 The Molar Mass of a Compound ... 168
- 6.3 Percent Composition ... 175
- *essential CHEMISTRY* 6.4 Nanotech - Molecular Machines ... 183

Chapter 7
Chemical Reactions 190
- 7.1 Chemical Equations ... 192
- 7.2 Types of Chemical Reactions ... 203
- *essential CHEMISTRY* 7.4 Accidental Discoveries ... 214

Chapter 8
Stoichiometry 220
- 8.1 Analyzing a Chemical Reaction ... 223
- 8.2 Percent Yield ... 234
- 8.3 Limiting Reactants ... 239
- *essential CHEMISTRY* 8.5 Engineering Green Chemistry ... 244

Unit 3 — Atoms and Bonds

Chapter 9
Atomic Structure — 252

- 9.1 Atoms Have Structure 254
- 9.2 The Quantum Atom 262
- 9.3 Electron Configurations 276

essential CHEMISTRY 9.4 Fireworks and Color 289

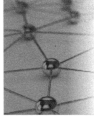

Chapter 10
Bonding and Valence — 296

- 10.1 Chemical Bonds 298
- 10.2 Bond Types 306
- 10.3 Molecular Geometry 318
- 10.4 Intermolecular Forces 330

essential CHEMISTRY 10.5 Stronger than Steel? 338

Unit 4 — States and Solutions

Chapter 11
Energy and Change — 344

- 11.1 Energy & Changes of State 346
- 11.2 Energy & Chemical Change 354
- 11.3 Entropy and Spontaneity 365

essential CHEMISTRY 9.4 The Water on Mars? 371

Chapter 12
Gasses — 378

- 12.1 Pressure and Gases 380
- 12.2 The Gas Laws 387
- 12.3 Kinetic Molecular Theory 396

essential CHEMISTRY 12.4 Deep Dives, High Altitudes 401

Chapter 13
Solutions — 408

- 13.1 What is a Solution? 410
- 13.2 Solubility 417
- 13.3 Concentration 424
- 13.4 Properties of Solutions 433

essential CHEMISTRY 13.5 Don't Drink the Water 439

Unit 5 — States and Solutions

Chapter 14
Reaction Rates — 446

- 14.1 Reaction Rates 448
- 14.2 Catalysis 458
- 14.3 Chemical Pathways 463

essential CHEMISTRY 13.5 Expiration Dates 469

Chapter 15
Equilibrium — 476

- 15.1 Chemical Equilibrium 478
- 15.2 The Equilibrium Expression 485
- 15.3 Le Chatelier's Principle 492
- 15.4 Solubility Product Constants 496

essential CHEMISTRY 12.4 Nitrogen & the Nobel Prize 500

Chapter 16
Acids and Bases — 506

- 13.1 Acids and Bases 508
- 13.2 The pH Scale 516
- 13.3 Acid - Base Equilibria 522
- 13.4 Acid - Base Reactions 526

essential CHEMISTRY 13.5 The *Basics* of Bicarbonate 535

essential CHEMISTRY

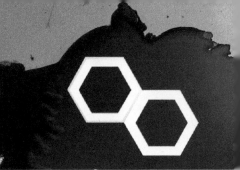

Unit 6 Redox and Energy

Chapter 17
Oxidation and Reduction — 542
- 17.1 Oxidation and Reduction 544
- 17.2 Oxidation Numbers 548
- 17.3 Redox Reactions 554
- 17.4 Half-Reactions 559

essential CHEMISTRY
- 17.5 Antioxidants 566

Chapter 18
Electrochemistry — 572
- 18.1 Fundamentals of Electricity .. 574
- 18.2 Electrochemical Cells and Electrolysis 578
- 18.3 Electricity from Electrochemical Cells 585

essential CHEMISTRY
- 18.5 Battery Shapes and Sizes ... 593

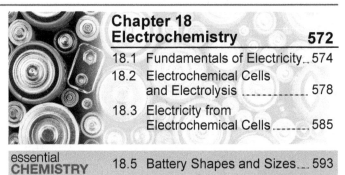

Chapter 19
Nuclear Chemistry — 600
- 18.1 Nuclear Chemistry 602
- 18.2 Radioactivity 610
- 18.3 Nuclear Energy 620
- 18.4 Radiation 627

essential CHEMISTRY
- 18.5 Radioactive Tracers 632

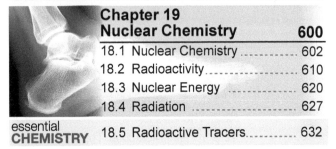

Unit 7 Organic and Biochemistry

Chapter 20
Organic Chemistry — 640
- 20.1 Carbon Chemistry 642
- 20.2 Organic Compounds 650
- 20.3 Functional Groups 657
- 20.4 Organic Reactions 665

essential CHEMISTRY
- 20.5 Recycling Plastic Numbers? 672

Chapter 21
Molecular Biology — 678
- 21.1 Carbohydrates 680
- 21.2 Lipids 686
- 21.3 Proteins 691
- 21.4 Nucleic Acids 701

essential CHEMISTRY
- 21.5 Semi-Synthetic Life 706

Chapter 22
Biochemistry — 712
- 22.1 Carbohydrate Synthesis 714
- 22.2 Cellular Respiration 721
- 22.3 Energy use in Cells 726
- 22.4 Maintaining Homeostasis 734

essential CHEMISTRY
- 22.5 Energy Transfer in Life 744

Unit 8 Earth and Beyond

Chapter 23
The Earth — 746
- 23.1 The Atmosphere 748
- 23.2 The Hydrosphere 758
- 23.3 The Geosphere 766

essential CHEMISTRY
- 23.4 Soil Health 777

Chapter 24
The Universe — 784
- 24.1 The Universe and Stars 786
- 24.2 The Solar System 795
- 24.3 Life Outside Earth 805

essential CHEMISTRY
- 24.4 Spectroscopic Astronomy 817

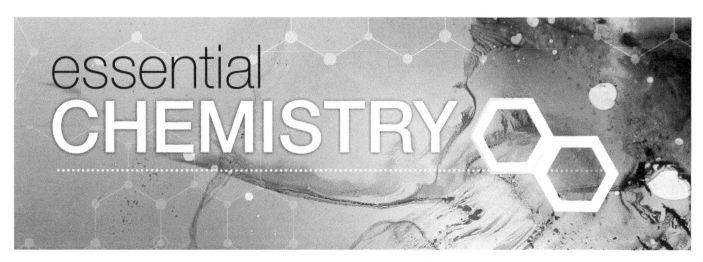

Tom Hsu
Ronn Fieldhouse
Barbara Pugliese
Tom Loschiavo

chapter 1: The Science of Chemistry

What is so essential about chemistry?

Have you ever seen a label on a product that touts it as being chemical free? The term "chemical free" can be used to denote something as being natural or safe, but can anything really be chemical free?

The answer, of course, is no. The word "chemical" literally means any substance made of atoms. The cleanest water contains a pure chemical of hydrogen and oxygen. Fresh fruits and vegetables are grown in soil mixed with potassium, nitrogen and phosphorous and can contain citric acid (lemons), beta-carotene (carrots), and capsanthin (red peppers). In fact, a bag with a chemical-free label should have nothing in it—not even air, because air contains chemicals too.

Everything you come in contact with is made of chemicals. The clothes you wear, the pen you write with, the air you breathe, the food you eat, and even YOU are a mixture of chemicals. The word chemical can be applied to you and every substance around you.

Not only is the world full of chemicals, but those chemicals are constantly changing. You see examples of this everyday such as when water evaporates on a warm day, plants grow in sunlight, batteries generate electricity, and food is digested. The study of the chemicals that make up the world and the changes that they can undergo is chemistry.

What does this have to do with you? As you gain more knowledge about chemistry, your understanding and appreciation of the world around you will increase. In the end, you will see that chemistry is not just a class you take in school. It is essential!

Chapter 1 study guide

Chapter Preview

Chemistry is the study of the basic structure and properties of matter as well as changes to matter and the energy used or released by those changes. In this chapter you will learn about the scientific method. A question is asked, a hypothesis is created and tested through experimentation with the evidence supporting or refuting that hypothesis. A well-designed experiment is repeatable and produces evidence that can be interpreted. It is key to control all the variables but the one being investigated. Accurate measurements are central to scientific investigations. The SI system is used in science for measurements of mass, weight, volume, and density. The Law of Conservation of Energy is a key concept in all of science.

Learning objectives

By the end of this chapter you should be able to:

- describe how the steps of the scientific method expand our understanding of chemistry;
- distinguish among a hypothesis, a theory and a scientific law;
- discuss reasons for using a standardized measurement system;
- explain how the mass-volume relationship of matter determines its density; and
- describe the law of conservation of energy and its applications in chemistry.

Investigations

1A: Experimental variables
1B: Investigating the temperature scale

Pages in this chapter

4	The Science of Matter and Change
5	Chemistry is the central science
6	The macroscopic and microscopic scale
7	Inquiry and the natural laws
8	Scientific knowledge and scientific evidence
9	What is not science
10	The scientific method
11	What is an experiment?
12	Experiment design and experimental variables
13	Section 1 Review
14	Matter and Energy
15	System International
16	Mass and weight
17	Volume
18	Density
19	Energy
20	Energy transformations
21	Conserving energy
22	Section 2 Review
23	The Chemistry of ... Getting to Chemistry Class
24	The Chemistry of ... Getting to Chemistry Class - 2
25	Chapter review

Vocabulary

matter	macroscopic scale	microscopic scale
natural laws	inquiry	objective
repeatable	hypothesis	theory
pseudoscience	scientific method	claim
experiment	procedure	conclusion
variable	control variables	experimental variable
measurement	System International	mass
weight	kilogram	gram
volume	liter	milliliter
graduated cylinder	meniscus	density
energy	joule	law of conservation of energy

1.1 - The Science of Matter and Change

The word chemistry comes from the word alchemy. As early as 1000 BCE, alchemists sought the Philosopher's Stone, which could turn any metal into gold. Today we know that there is no such thing as a Philosopher's Stone. Even though they were not successful, alchemists discovered some of the basic principles of science we still use today. They discovered how to extract metal from ores, make pottery and dyes, and other ways to manipulate matter. Modern chemistry is the study of matter.

What is Chemistry

Chemistry is the study of matter

Chemistry is the study of matter and the simple definition of **matter** is "stuff" that has mass and takes up space. In chemistry we need to dig deep. What is matter made of? How do we explain the infinite variety of matter around us? In your immediate vicinity you might reach out and touch wood, metal, orange juice, skin, hair, rubber, plastic ... the list has no end. Is every variety of matter unique, or is each kind of matter made from simpler constituents just as different words are built from the same twenty-six letters? This is the first question of chemistry - how do we explain the diversity of matter and its properties?

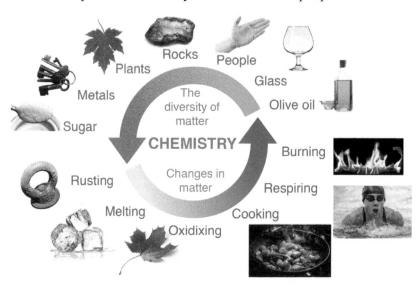

Chemistry explains how matter changes

The second question that motivates chemistry is how to explain changes in matter. Ice melts into liquid water. Gasoline burns into carbon dioxide, water, and trace gases. When food is digested, plant leaves are converted into energy, waste, and the material that makes up your body. Both the living and nonliving worlds are places of continuous changes in matter. Chemistry provides the explanation for how and why matter changes. Why does wood easily change to ashes and smoke but lead does not easily change to gold?

Technology and Chemistry

There is technological benefit to understanding matter and change. Once we understand the processes of chemistry we have the ability to shape both matter and change it to suit our needs and solve human problems. The creation of vitamins and medicines is done through chemistry. Advanced materials such as carbon fiber and super-strong metal alloys are achieved through chemistry. Everyday substances from dishwashing soap, to glue, to colored glass - these all owe their existence to chemistry. Virtually every human-made item you have ever used, or will ever use, owes some (or all) of its existence to chemistry.

Carbon fiber

Chemistry is the central science

How does chemistry fit?

While nature draws no boundaries, humans have divided up areas of science into the major areas of physics, chemistry, biology, and earth/space science. Where does chemistry fit into the general understanding of science?

The central science

The answer is that chemistry fits in the center. To understand biology, you need to know the chemical basis for living organisms. To understand chemistry you need to know about atoms, energy and systems, and these are the domain of physics. Chemistry is the bridge between the most fundamental properties of our universe - energy, mass, force, and atoms - and the most complex system in the universe - a living organism.

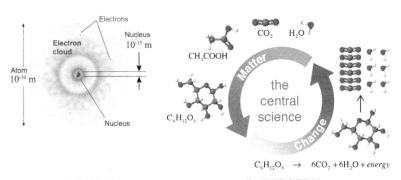

PHYSICS
Fundamental forces, atoms, systems and energy

CHEMISTRY
Atoms and forces interact in compounds and reactions

BIOLOGY
Exceeding complex systems of compounds and reactions

Physics

Physics is the most basic of the branches of science. Physics describes the laws that govern the behavior of objects and energy, such as the effect of gravity on the path of a ball flying through the air, or how light reflects from a mirror. Physics is concerned with fundamental forces, energy and particles, such as electrons, protons, and the attractive force between them that causes atoms to form. Because physics ultimately describes how atoms form and interact with each other it is often said that the explanations for chemistry are found in physics.

Biology

At the opposite end of the complexity spectrum are biology, ecology, and the life sciences. If you think about physics as explaining atoms, then chemistry explains how the atoms come together in millions of different chemicals that undergo billions of reactions to make other chemicals. Biology takes the millions of chemicals and billions of reactions and organizes them into extraordinarily complex interconnected systems that are living organisms and environments. Despite having sequenced the human genome we still understand very little of the details for how it all works together in our bodies.

Relationship between sciences

Just as the explanation for chemistry lies in physics, the fundamental processes of biology are explained by chemistry. This is why chemistry is the central science. Chemistry connects the inanimate world of physics to the living world of biology. Physics describes in detail how eight protons and eight electrons come together in an oxygen atom. Chemistry describes in detail how oxygen, carbon, phosphorus, and other atoms combine to make DNA and even how DNA replicates on the molecular level. But, chemistry cannot explain the details about how to get from DNA to a baby cow! However, every single process along the way from DNA to a cow obeys every law of physics and every principle of chemistry!

The macroscopic and microscopic scale

The macroscopic world

One of the most difficult aspects of chemistry is the different scale between what we actually observe and the understanding of what we observe. We observe the **macroscopic scale**. The prefix "macro" derives from the Greek makro, meaning "large." The macroscopic scale includes everything we can sense and measure directly such as temperature, color, size, weight, volume, and pressure. Rocks, shopping carts, dust specks, and planets are macroscopic objects. The macroscopic scale is the scale of ordinary life, from about 1/100th of a millimeter and larger.

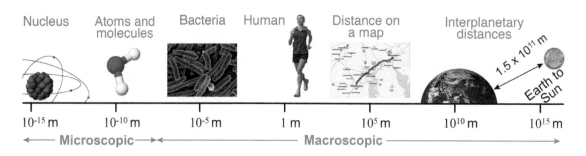

Distance scales in chemistry

The size of the microscopic world

Hidden inside the macroscopic world is the **microscopic scale** of atoms and molecules. In chemistry, the microscopic scale refers to the size of a molecule and smaller. Atoms and molecules are so small that even a dust speck contains trillions of atoms. A single atom has a diameter of around 10^{-10} m. In the context of chemistry, the word microscopic refers to things much, much smaller than are visible with an ordinary optical microscope. Bacteria are still macroscopic to a chemist!

Chemistry reactions are microscopic

Virtually all of the concepts in chemistry have their explanation on the microscopic scale! We feel macroscopic properties such as temperature and pressure, but both of these qualities are caused by the microscopic motion of trillions of atoms and molecules. We see the burning of wood into ashes and smoke but the explanation is that the atoms in cellulose are being rearranged into new compounds: carbon, water, carbon dioxide, and others. The battery in your cell phone supplies electricity but inside the battery, on the microscopic scale, trillions of chemical reactions are converting lithium ions and cobalt oxide into lithium cobalt oxide.

Chemistry is explained on the microscopic scale of atoms and molecules

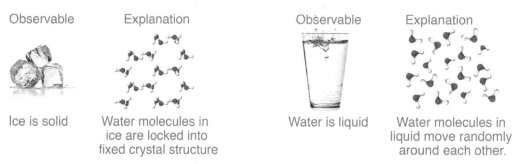

Inquiry and the natural laws

Chemistry and the natural laws

A central idea in science is that our universe obeys a set of natural laws. For example, a ball thrown upwards slows and eventually falls down again according to the law of gravity. With physical objects it is possible to see the action of the natural laws, such as gravity. However, the natural laws that chemistry obeys are often hidden on the microscopic scale! We observe that a fresh apple turns brown and a leaf changes color in the fall and neither change is reversible. On the other hand, water and ice are easily changed back and forth into each other. How do we explain this difference? Why are some changes in matter reversible and others are not?

Understanding chemical and physical changes

Fresh apples turn brown and don't change back.

Green leaves turn color and don't change back.

Water and ice can be changed back and forth into each other easily.

Inquiry

The primary goal of science is to discover the natural laws hidden in the behavior of the universe. The primary way we accomplish this is through the process of inquiry. The process of inquiry starts with asking questions, and proceeds by comparing possible answers to what is known and what is newly discovered. An inquiry is like a crime investigation in which we must unravel the clues to discover what really happened. In a scientific inquiry, the clues for natural laws are what we actually observe to happen and the mystery is the theory explaining why things happen the way they do.

Science is a process

Scientific inquiry is an ongoing process. The purpose of scientific research is to find new observations that cannot be explained by what we already know. An unexplained observation is often the first clue to a new discovery. For example, the kinetic theory of matter successfully explains how water freezes and the relationship between the solid, liquid, and gas phases of matter. Today new questions are being asked, like what happens if a material such as silicon freezes very fast? Does that change its properties? The answer is yes, the properties do change, and fast-frozen (amorphous) silicon has different electronic properties. New knowledge is added as we ask new questions that build on what we already know.

The process of scientific inquiry

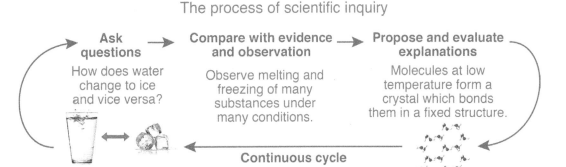

Continuous cycle

Scientific knowledge and scientific evidence

Scientific evidence

A major difference between science and other ways of learning is that science includes only knowledge that can be tested and verified by scientific evidence. Scientific evidence must be **objective** and **repeatable**. Objective means the evidence describes only what actually happens as exactly as possible. Opinions, past history, advice of friends, published materials, marketing claims, and internet references are not acceptable as scientific evidence. A repeatable experiment can be performed by anyone, and they will observe the same results. Scientific results must be confirmed independently before being accepted.

Examples of scientific evidence

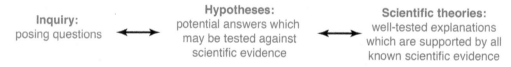

Darwin's observation of the beaks of Galapagos finches

Galileo's observations of Jupiter's moons

Measurements and data

Hypotheses

A **hypothesis** is a tentative explanation or conclusion about the physical world. Hypotheses do not have to be right! The logic of science assumes a hypothesis is correct and then seeks scientific evidence that either supports or refutes the hypothesis. Hypotheses are:

   a. tentative statements, which may or may not be correct,

   b. testable statements, that can be tested against scientific evidence, and

   c. necessary and fundamental steps in developing reliable scientific theories.

The development of scientific knowledge

Inquiry: posing questions ⟷ **Hypotheses:** potential answers which may be tested against scientific evidence ⟷ **Scientific theories:** well-tested explanations which are supported by all known scientific evidence

Theories and scientific knowledge

A scientific **theory** is a comprehensive and well-tested explanation of a natural or physical phenomenon. Unlike hypotheses, scientific theories are well established and highly reliable explanations that have been tested by multiple, independent researchers over a long period of time and under a wide range of conditions. In many ways the knowledge of science can be grouped roughly into the following categories.

1. Facts and data that describe what we know, such as the charge on the electron or the mass of a hydrogen atom.

2. Theories that provide explanations consistent with all known facts and data.

3. Ways of learning such as procedures, techniques, and tools that allow us to collect information about the universe.

4. The scientific method framework for reliably sorting truth from competing ideas.

What is not science

Ethics are outside of science

Science includes only a very specific kind of knowledge and thought. There is a great body of human thought that is valuable and important, but is not science. For example, consider questions about ethics - what is good and what is evil? Deciding between good and evil is extremely important but is not part of science. Ethics, philosophy, and morality are questions of human judgement and culture. Science includes only empirical knowledge that can be tested against objective and repeatable evidence. This is a very important limit. Science concerns the chemical and electrical workings of the brain, and even the structure of consciousness, but science cannot say anything about the feelings, decisions, or value judgements made by that very same brain. Whether a painting is beautiful or not is a great and worthy question, but not answerable by science because the answer is neither objective nor repeatable by others.

Misuse of science

To be "scientific" in the modern world conveys a certain authority and reliability. This is because true scientific knowledge has passed the most rigorous testing against accumulated facts and data over many years and many separate teams of people. Many claims are falsely advertised as "scientific" to take advantage of the well-earned scientific reputation for objectivity and reliability. You can ascertain whether a claim is backed by scientific evidence yourself.

a. Is there evidence that supports the claim?

b. Is the evidence objective and reproducible? Or, is it opinions of unnamed experts?

c. Are there other possible explanations that fit the same evidence?

d. Is the claim about a decision or value judgement that is not part of science?

Examples of *pseudoscience*

Alchemy   Phrenology   Astrology   Crystal healing

Pseudoscience

The term **pseudoscience** describes ideas that are often presented as scientific but are not supported by scientific evidence. For example, astrology claims that a person's characteristics and future are determined in part by the position of the planets and stars at the moment the person was born. Phrenology claimed that specific details of the shape of the head were linked to certain social and moral characteristics of a person.

Scientific words

Many everyday words are defined differently in science. For example, the word theory in science does not mean a hunch or a guess. In fact a scientific theory is almost the exact opposite of a guess! A scientific theory represents our best, most reliable knowledge about the natural or physical world. A closer synonym is "fact." When the "theory of evolution" is dismissed as "only a theory" this is a semantic misdirection of true meaning. In the ordinary sense of the word, evolution is a scientific fact that is broadly supported by an enormous body of scientific evidence collected by thousands of people from all parts of the world, in all cultures, and over hundreds of years.

The scientific method

The scientific method

The **scientific method** is the formal, logical thought process of science. It is also a natural cycle of learning you have used throughout your own life: question, try, remember, revise. The scientific method is a way of learning that is cautious, often slow, and accepting of only supportable, evidence-based conclusions. By the time a scientific explanation has reached the status of an accepted theory, the explanation has been compared to many, many observations, facts, and data, and found to be in agreement with every single one! Even **one** discrepancy means the explanation is incomplete or even incorrect. The high level of confidence we have in scientific knowledge comes from the continuous, relentless testing of every scientific claim against every new discovery AND everything already known.

Using the scientific method

The formal scientific method:

1. Scientists observe nature, then consider possible reasons that might explain the phenomenon;

2. Hypotheses are created, then tested against evidence collected from observations and experiments;

3. The hypotheses are modified and retested, and then a hypothesis which correctly accounts for ALL of the evidence from every experiment is a potentially correct theory;

4. A theory is put to the test when scientists collect new evidence. All it takes is one piece of evidence that disagrees with a theory to cause scientists to return to step one. If necessary, the theory will be revised.

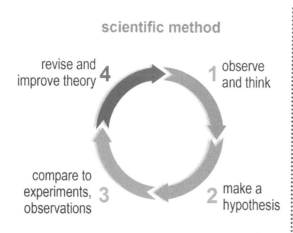
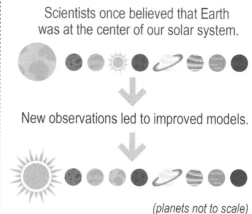

(planets not to scale)

Science is never finished!

There are no absolute truths in science because there are limits to everything we know. For example the gravitational laws are well-proven truths on the macroscopic scale but we still do not understand gravity on the microscopic scale. The scientific method requires that we continue to develop and refine theories to explain more and more different phenomena. As humans create better ways to observe, such as atomic force microscopes that can see single atoms, we make new observations that lead to deeper understanding in the way of theories that explain more aspects of nature.

What is an experiment?

Experiments

An **experiment** is a controlled situation created specifically to collect scientific evidence by observing what happens. Experiments are a core part of science because the evidence collected must be objective and must be confirmed by others who repeat the same experiment. Experiments may be exploratory, to see what happens without specifically testing any preconceived idea. However, experiments are often set up to test a specific hypothesis. The important criteria are that an experiment be (a) repeatable, (b) generate objective scientific evidence, and (c) provide a clear way to interpret the evidence, positive or negative. This last requirement (c) is the difference between a good experiment design and a poor one.

Claims and evidence

A **claim** is a statement that asserts something is true. A scientific claim must be supported by scientific evidence. For example, bleach is a common household cleaner which removes stains by chemical action. Advertisements for bleaches claim that they work equally well in hot or cold water. An experiment to see if the water temperature affects the rate at which bleach removes the color from "stains" made from food dye would be a good way to collect scientific evidence to test this claim.

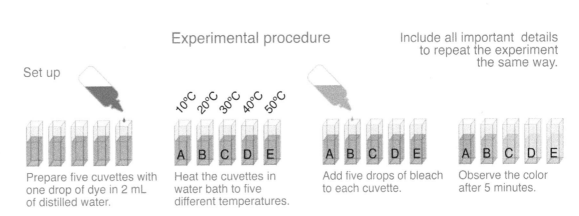

Procedures

A good experiment design starts with a **procedure** which describes the methods used to collect the evidence. The procedure should include all important set-up details and describe every important step and measurement. For example, set up the cuvettes by adding one drop of color per 2 mL of water. The instructions might include details such as heating the dyed water to 10 °C, 20 °C, 30 °C, 40 °C, and 50 °C, adding five drops of bleach, and measuring the remaining color after five minutes. The procedure should be clear and detailed enough that someone else can replicate the experiment the same way.

Keeping track of variables

A **conclusion** is a statement that the experimenter believes represents the results of an experiment or observation. For example, if the stain was removed in the higher temperature water and not the lower temperature water a conclusion might be that the bleach does not work at colder temperatures. However, what happens after 10 minutes or 20 minutes? If you do this experiment you will find that bleach takes time to act and it takes longer at colder temperatures. The "soaking time" effect is important and was not designed for in the experiment. This error in experiment design might cause an incorrect conclusion! A good experiment design should include controlling all variables, such as time and temperature, that affect the observed results. A **variable** is something that can be measured or changed in an experiment.

Experiment design and experimental variables

Planning experiments

In the previous bleach and dye example, some variables are the amount of dye, the amount of water, the type of water, the temperature, the amount of bleach, and others. Important variables are the ones which you believe may affect the experiment in any way. No experiment can keep track of all the variables that might affect the outcome - for example the time of day and the amount of sunlight in the room. One of the first steps in experiment design is to choose which variables will be tracked and which will be neglected.

Experimental variables

In a good experiment design only one variable is changed at a time. The variable that is changed is called the **experimental variable**. When only one variable is changed, any effect (such as color diminishing) can be attributed to the variable that was changed. If more than one variable is changed (bad design) then it is difficult to determine which variable caused any observed effects. The experiment still works, but you cannot easily understand the results. The experimental variable may also be described as the independent variable, because it is changed independently of other variables. Variables that are unchanged or kept constant during an experiment are called **control variables**. In the well-designed experiment below, wait time and the volumes of dye, bleach and water are control variables because they are kept constant during the experiment. Temperature is the experimental variable because it is changed in each trial.

Examples of good and bad experiment design

Variables that change   Variables that stay constant

Bad experiment design

Trial 1
1 drop dye
10 mL water
10 °C
1 drop bleach
10 minute wait

Trial 2
1 drop dye
20 mL water
20 °C
2 drops bleach
20 minute wait

Trial 3
3 drops dye
20 mL water
30 °C
2 drops bleach
15 minute wait

Good experiment design

Trial 1
1 drop dye
20 mL water
10 °C
1 drop bleach
10 minute wait

Trial 2
1 drop dye
20 mL water
20 °C
1 drop bleach
10 minute wait

Trial 3
1 drop dye
20 mL water
30 °C
1 drop bleach
10 minute wait

Solved problem

Identify the experimental variable in each testable question below:
  a. Which soil type will produce the tallest green bean plants?
  b. Does fuel economy change at different speeds?
  c. Does pasta cook faster in salty water or in pure water?

Asked Determine the experimental variable in each testable question.

Relationships The experimental variable is the one that is purposefully changed or defined to see if it causes some kind of response during the experiment. It is also called the independent variable.

Solve Decide which variable is set up to change or vary before the experiment begins. This is the experimental variable. Variables that change in response to another variable are <u>not</u> experimental variables.

Answer Each of the following is the experimental variable, because it will be changed or varied to see how another variable will respond:
a) soil type; b) speed; c) amount of salt

Section 1 review

Everything around you is made up of chemicals. Chemistry is the study of the fundamental nature of those chemicals, how they interact, and the changes they undergo. However, understanding why things happen comes from observing events and questioning why they occur. An inquiry can lead to a hypothesis, which is then tested by experiments and revised or rejected based on the evidence.

In a well-designed experiment all of the variables are controlled except for the one being investigated. This provides understandable results to compare to the hypothesis. A very well supported hypothesis or series of hypotheses help to create a potential scientific theory which itself is then tested and revised. This process is called the Scientific Method.

Vocabulary words: matter, macroscopic scale, microscopic scale, natural laws, inquiry, objective, repeatable, hypothesis, theory, pseudoscience, scientific method, claim, experiment, procedure, conclusion, variable, control variables, experimental variable

Key relationships

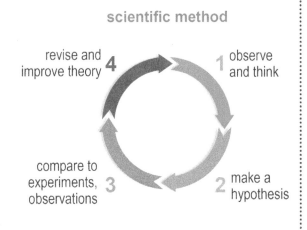
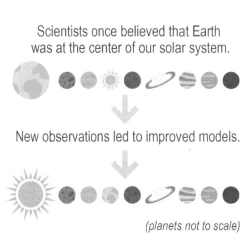

Review problems and questions

1. If you made an observation and wanted to know why it occurred, which is more useful: a law or a theory? Explain your answer.

2. Complete the sentence: If the evidence from an experiment shows a hypothesis is correct, and the hypothesis has 100% repeatability by scientists all over the world, it can become a(n) _____.

3. A scientist performs an experiment and finds evidence that does not agree with a theory. Is this enough evidence to change the theory?

4. Why is it important to only change one variable at a time during an experiment?

5. Is smoke matter? Why or why not? Use the definition of matter to support your answer.

Section 1.1: The Science of Matter and Change

1.2 - Matter and Energy

Chemistry is about matter and energy. Matter is familiar "stuff" that has mass and takes up space. You, wood, rocks, metal, glass, water, and air are examples of matter. Even though energy is a fundamental property of everything in our universe, it is more difficult to define. The best way to understand energy is through examples. What is the quantity that makes a hot cup of coffee different from a cold one? The answer is energy! Temperature is a measure of one kind of energy and higher temperatures mean more energy. Just as there are many kinds of matter, there are also many forms of energy.

Measurement

The purpose of units

To a scientist, saying "it's warm" is not an acceptable statement about temperature. Scientists want a numeric value that defines exactly how warm. The temperature 76 degrees Fahrenheit, 76 °F is such a value. The value 76 °F has two parts: a number, 76 and a unit, °F. The number is an amount and the unit gives you an idea of how to interpret the number. A number without a unit can cause a lot of confusion. If you say 76 degrees without the unit Celsius or Fahrenheit, anybody outside the United States might think you mean 76 degrees Celsius, 76 °C, which is nearly 170 °F. An outdoor temperature of 76 °C will kill you while 76 °F is a nice day! To understand chemistry, you need to speak the language of units. Almost all measurements have their own unit. Below are some of the most useful units for chemistry.

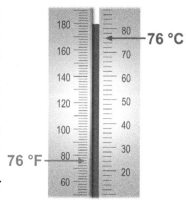

Measurement assigns value and units

You are measuring mass when you put a substance on a balance. A **measurement** provides both a number and a unit that describes a physical quantity. The value 32.10 grams is a measurement because it has a number (32.10) and a unit (grams) that provides information about a real, physical quantity.

chemistry units

UNIT	USE	UNIT	USE
Celsius degree (°C)	temperature	meter (m)	length
Fahrenheit degree (°F)	temperature	centimeter (cm)	length
kilogram (kg)	mass	mole (mol)	counting atoms
gram (g)	mass	joule (J)	energy
liter (L)	volume	watt (W)	power
milliliter (mL)	volume	pascal (Pa)	pressure
second (s)	time	atmosphere (atm)	pressure

Quantities may be expressed in different units

Notice that there is more than one unit for the same quantity. Temperature is commonly measured in both Celsius degrees and Fahrenheit degrees. Your laboratory balance will measure mass in grams but kilograms are another common mass unit. A graduated cylinder measures volume in milliliters. A stopwatch measures time in seconds. To make practical use of chemistry you need to become proficient in several different sets of units.

System International

A common measurement system

The measurement system used in the United States is the US Customary System (USCS) based on the English system of measurement units. The USCS is based on historical numbers such as 5,280 ft = 1 mile, 1 pound = 16 ounces, and 8 pints = 1 gallon. There is no common pattern to the numbers and you have to memorize all relationships. The world of science adopted a common system of measurement starting with the Treaty of the Meter in 1875. This system is called the **System International** and is abbreviated as SI.

Suffixes

Instead of having to memorize units like gallons and quarts, feet and inches, and pounds and ounces, the SI system provides a common unit for volume, mass and length. All volume measurements use the suffix liter, all length measurements use meter, and all mass measurements use gram.

MEASUREMENT	BASE UNIT
length	meter (m)
mass	gram (g)
volume	liter (L)

$1,000 = 10^3$	kilo (k)	kilogram (kg)
$100 = 10^2$	hecto (h)	hectogram (hg)
$10 = 10^1$	deka (da)	dekagram (dag)
base unit, for example gram		
$0.1 = 10^{-1}$	deci (d)	decigram (dg)
$0.01 = 10^{-2}$	centi (c)	centigram (cg)
$0.001 = 10^{-3}$	milli (m)	milligram (mg)

Prefixes

The SI system places prefixes in front of the suffix to indicate size. Although there are many prefixes, only the centi, milli, and kilo are commonly used. For example, 1000 meters is one kilometer and 10^{-2} (0.01) meters is one centimeter.

Solved problem

Express 1000 liters in kiloliters

Relationships The prefix kilo = 1,000

Solve The prefix kilo is placed in front of the suffix, liters.

Answer 1 kiloliter or 1 kL

Based on ten

The prefixes are based on the number ten. So instead of strange relationships like 5,280 feet to a mile or 12 inches to a foot, the SI system always uses the number ten. For example, 1000 milliliters is equal to one liter and 100 centimeters is equal to one meter.

Solved problem

How many millimeters are in 54.5 meters?

Given 54.5 m

Relationships 1000 mm = 1 m

Solve Start with the given amount of 54.5 m over the number one. Place the relationships so that the units cancel.

$$\frac{54.5 \text{ m}}{1} \times \frac{1000 \text{ mm}}{1 \text{ m}} = 54,500 \text{ mm}$$

Answer There are 54,500 mm in 54.5 m

Use the interactive equation titled Converting Units to select among numerous conversions between standard and non-standard units, such as meters to inches, kilograms to tons, and seconds to years.

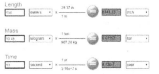

Mass and weight

Mass

Mass is a measure of how much matter is in an object. An eyelash has very little mass because there is very little matter in it, compared to your whole body. Mass uses the units, **kilogram** (kg) and **gram** (g). The mass of a single paperclip is about 1 gram, so 1,000 paperclips will have a mass of 1,000 grams. There are 1,000 grams in 1 kilogram and 0.001 kilograms in 1 gram. That means 1 paperclip has a mass of 0.001 kg and 1,000 paperclips have a mass of 1 kg.

common units of mass and weight

MASS		WEIGHT (force)		
gram (g)	0.001 kg	Newton (N)	0.225 lbs.	1 N
kilogram (kg)	1 kg	Pound (lb)	1 lb.	4.448 N
metric ton (t)	1000 kg	Ounce (oz)	1/16 lb.	0.278 N
		Ton (T)	2000 lbs.	8896 N

1 g   1 kg

Weight

Mass is acted upon by gravity and it is convenient to use an object's **weight** to measure mass. Weight is the force of gravity acting on mass. At Earth's surface each kilogram of mass has a weight of 9.81 newtons and each gram of mass has a weight of 0.00981 N. Newtons are a unit of force and a digital balance measures the weight force and calculates the mass by dividing the measured weight force by 0.00981 N/g. Gravity is different on different objects in the solar system. The moon is smaller than Earth and has a smaller force of gravity. On the moon each gram of mass has a weight only 1/6 of the amount on Earth, so you weigh less on the moon than on Earth. Divide your weight on Earth by 6 to calculate your weight on the moon. However, your mass does not change whether you are on Earth, on the moon, or anywhere else.

same size different mass   same mass different weight   same mass different size

Using mass to describe matter

Size, or volume does not tell you how much matter there is. Your smartphone and a sponge can be the same size, but one contains much more matter than the other does. The best way to communicate a quantity of matter is by its mass, not its volume.

The relationship between mass and matter

One kilogram of feathers has the same amount of matter as one kilogram of steel. Two kilograms of feathers has more matter than one kilogram of steel. We can even measure how much matter is in an unidentified object. Even though air is very light, one cubic meter of air at sea level has about 1 kg of mass. If you put your hand out of the window in a moving car, you can feel the mass of the air.

Volume

The description of volume

Volume is a three-dimensional section of space that has length, width and height. A large object like a house has a large volume because it takes up a lot of space. An object like a spoon is small and has a proportionally small volume.

Volume units

When you take volume measurements in the chemistry lab, you will use the unit **liter** (L), or more frequently, **milliliter** (mL). One liter is the volume of a cube that is 10 cm tall. One milliliter is the volume of a cube that is 1 cm tall. This means the unit "mL" and "cm^3" are totally interchangeable.

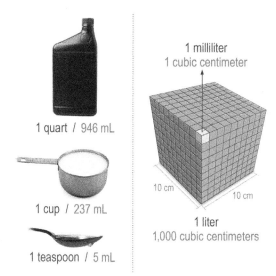

$$1 \text{ mL} = 1 \text{ cm}^3$$

The volume 15 mL also equals 15 cm^3. A milliliter is a small volume; there are five mL in a standard teaspoon. The images shown include several volume units and their equivalent in liters and milliliters.

Volume units are derived from length units

The unit for volume comes from the centimeter (cm), a unit for length. The mathematical formula for volume is: length × width × height. A cubic centimeter (cm^3) is a unit of volume because:

$$1 \text{ cm}^3 = 1 \text{ cm} \times 1 \text{ cm} \times 1 \text{ cm}.$$

Measuring volume with a graduated cylinder

An instrument or tool that has regularly marked intervals is said to be graduated. A ruler is graduated. The **graduated cylinder** is a tool used to measure the volume of liquids. Most graduated cylinders are marked in the milliliter unit. The example shows a volume of 68.4 mL, even though the sides show a slightly higher reading. Notice that you read the graduated cylinder at the lowest part of the liquid surface. You will notice that water tends to cling to the sides of a glass or plastic graduated cylinder. The concave shape that forms is called a **meniscus**. Always read a graduated cylinder with your eyes lined up exactly with the lowest point of the meniscus.

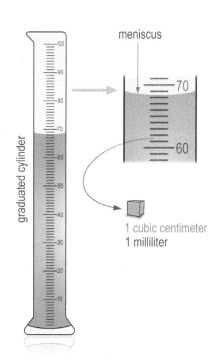

Density

Equal volume does not mean equal mass

Imagine having a piece of wood, a pile of sugar cubes, and some chunks of pencil lead (graphite) like those in the picture. If you had a 1 cm³ cube of each substance, all the cubes would be the same size. Each cube has the same volume, but does each cube have the same amount of matter? Does equal size imply equal mass?

Density is defined as mass per unit volume

Equal-sized cubes of graphite, sugar and wood have different amounts of mass. If you put the cubes in a glass of water, the mass difference would allow wood to float while graphite and sugar sink to the bottom. One factor that determines whether a substance will float or sink in water is its density. A substance's **density** is how much mass it packs in a given volume. Density is calculated by dividing mass in grams by volume in cm³ or mL. This gives density units of grams per cubic centimeter (g/cm³) for solids and grams per milliliter (g/mL) for liquids and gases. Graphite contains 2.66 grams of mass per cubic centimeter (density = 2.66 g/cm³). Wood is less dense than graphite because a 1 cm³ block only has 0.62 grams of matter (density = 0.62 g/cm³).

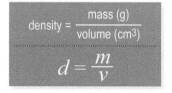

$$\text{density} = \frac{\text{mass (g)}}{\text{volume (cm}^3\text{)}}$$

$$d = \frac{m}{v}$$

key features of density

- Density is a property of matter, independent of size or shape.
- Density is mass per unit volume.

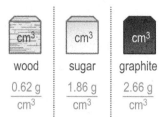

Water and air density

Solids range in density from cork, 0.12 g/cm³ to osmium, a precious metal with a density of 22.57 g/cm³. Water has a density of about 1.00 g/cm³. Air's density is much lower, only about 0.001 g/cm³. A common misconception is to think of air as "nothing." However, air is vital to life and is merely spread thinly over a large volume. Air has a low density at room temperature and normal pressure. Air's low density, caused by the distance between molecules, is the reason we can walk through air and feel almost no resistance.

Energy

Energy is a fundamental essence

Energy measures the capacity for change. Energy can create changes in temperature, pressure, height, or even turn matter from liquid into gas or back. Anything that changes over time by moving, turning on, or being alive, changes through the flow and transformation of energy. Without energy no change can occur—not color, pressure, or even temperature. Energy is a fundamental essence of our universe, and understanding the flow of energy through matter opens a powerful window into understanding chemistry.

Mechanical energy

Mechanical energy is energy that comes from position or motion. Gravitational potential energy, elastic potential energy, rotational energy, and ordinary kinetic energy are examples of mechanical energy.

Radiant energy

Radiant energy includes visible light, microwaves, radio waves, x-rays, and other forms of electromagnetic waves. Nearly all the energy on Earth ultimately comes from radiant energy that originates in the Sun.

Nuclear energy

Nuclear energy sometimes called atomic energy, is energy contained in matter itself. Nuclear energy can be released when atoms are changed from one element into another, such as in a nuclear reactor or in the core of the Sun.

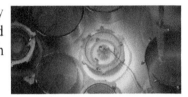

Electrical energy

Electrical energy moves in the form of electric currents that flow in response to electrical voltages, such as in a battery or wall socket. Electrical energy is used whenever we turn on an appliance such as a lamp or coffee maker.

Chemical energy

Chemical energy is stored in the bonds between the constituent atoms in molecules that make up most matter. Chemical energy can be released by rearranging the atoms into different molecules, such as by burning natural gas to produce water and carbon dioxide.

Thermal energy

Thermal energy is energy that comes from differences in temperature. Heat is one form of thermal energy. Thermal energy ultimately derives from the energy of moving atoms and molecules. Hot objects have more thermal energy than cold objects.

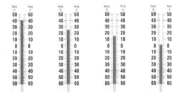

Energy of pressure

Pressure is a consequence of thermodynamic energy, a form of mechanical energy at the microscopic level that is important in gases and liquids. It takes work to inflate a tire; some of the work is stored as energy in the form of high-pressure air inside the tire.

Energy transformations

Energy

Energy is measured in a number of different units for historical reasons. Until just before the 20th century, it was thought that heat, mechanical energy, light, and electricity were completely different and unrelated phenomena. Each was assigned its own unit and only later did science discover that all were actually forms of the same thing, energy. The fundamental unit of energy is the **joule**. The table below shows some common units of energy and the equivalent in joules.

Energy units and equivalents in joules

joule (J)	1 J	*Primary SI unit of energy*
calorie (cal)	4.148 J	*Heat, raise 1 g water by 1 °C*
Calorie (kcal)	4,184 J	*Heat, raise 1 kg of water by 1 °C*
British thermal unit (Btu)	1,055 J	*Heat, raise 1 lb of water by 1 °F*
Kilowatt-hour (kwh)	3,600,000 J	*Electricity - 1000 watts for 1 hour.*

Energy changes forms

Almost every change that takes place involves some transformation of energy. For example, some of the chemical energy in fuel is converted into thermal (heat) energy when the fuel is burned. The heat energy might take the form of hot steam. Steam can be used to turn a turbine, transforming the energy to mechanical energy. A generator turns mechanical energy into electrical energy.

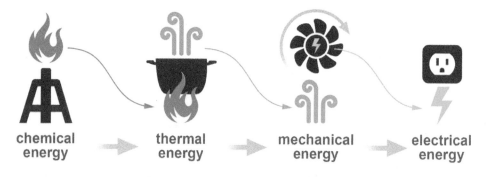

chemical energy → thermal energy → mechanical energy → electrical energy

Conservation of energy

The **law of conservation of energy** states that when energy converts from one form into another the total amount of energy stays the same. The law of conservation of energy is one of the most important laws in all of science and applies to all forms of energy. Energy can never be created or destroyed, but it can be converted from one form into another.

The flow from higher energy to lower energy

Systems in nature tend to move from higher energy to lower energy. Balls roll downhill because objects high up have more potential energy than objects lower down. Hot coffee cools off because warmer objects have more energy than cold ones. The same is true of chemical energy. We will see that many chemical reactions occur because the energy of the atoms and molecules is lower after the reaction than it was before the reaction.

Energy flow diagrams

Systems in nature tend to move in the direction of lower total energy. The energy flow diagram shows this process.

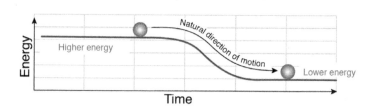

Conserving energy

Meaning of "conserve"

If energy can never be created or destroyed, how can it be "used up?" Why are some people worried about "running out" of energy? Messages are all around us about "not wasting energy" and "conserving energy." While correct and wise, these messages are using a different context for the word "conserve." In scientific context when a quantity is conserved that means the total amount of the quantity does not change over time.

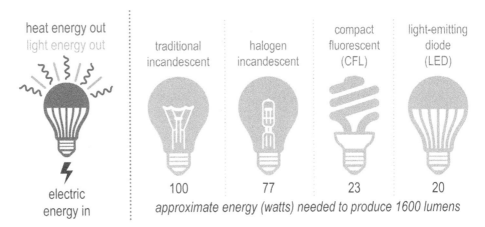

approximate energy (watts) needed to produce 1600 lumens

Using energy

An electric light bulb converts electrical energy to light. Notice the same amount of light (1600 lumens) is produced with 20 watts of electricity in an LED bulb compared to 100 watts of electricity in an incandescent bulb. The bulb doesn't really "use-up" energy. Every joule of electrical energy becomes light or heat. What gets "used up" is the electrical energy which is extremely convenient. It is easy to transmit energy in the form of electricity, but it must come from some other form of energy.

Chemical energy and fossil fuels

The majority of the electric power used in the US is converted from chemical energy in fossil fuels, mainly natural gas and coal. The major fossil fuels (natural gas, oil, coal) come from plant matter buried underground during the carboniferous era 300 to 360 million years ago. These sources are considered nonrenewable because they cannot be replenished once used.

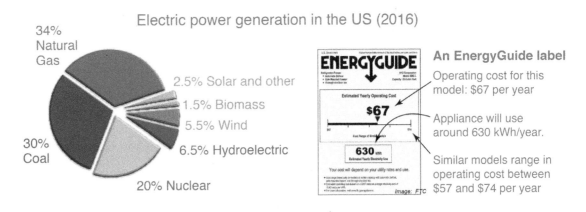

Home appliance energy ratings

Major home appliances are sold with an "Energy Guide" label that indicates their expected energy consumption per year. This standardized labeling compares the energy usage of a particular appliance with similar models produced by the same or other companies. In the example above, this refrigerator will cost approximately $67 to operate per year and consumes somewhat more energy than the average of other models.

Chapter 1 — Section 2 review

In science it is important to be able to measure things and to communicate those measurements clearly. A measurement describes a physical quantity using a number and a unit. The units used in chemistry come from the International System of Units (SI) and use prefixes based on powers of 10 when necessary. Mass and volume can be measured directly but density is calculated from the ratio of mass to volume. Mass is the amount of matter in an object, and the units often used are grams and kilograms. Volume is a measure of how much space something takes up, and the units often used are liters, milliliters, or cubic centimeters. For liquids volume is often measured with a graduated cylinder. Density is a measure of the amount of matter in a given volume of a substance. It is found by dividing mass by volume, and the units often used are g/cm^3 for solids and g/mL for liquids and gases. Fundamental to the measurement of quantities is the law of conservation of energy.

Vocabulary words: measurement, System International, mass, weight, kilogram, gram, volume, liter, milliliter, graduated cylinder, meniscus, density, energy, joule, law of conservation of energy

Key relationships:
- Volume unit equivalence: 1 mL = 1 cm^3
- Density = mass / volume

Measurement	Base unit					
Length	meter (m)	1,000	10^3	kilo	(k)	kilogram (kg)
mass	gram (g)	100	10^2	hecto	(h)	hectogram (hg)
volume	liter (L)	10	10^1	deka	(da)	dekagram (dag)
		0.1	10^{-1}	deci	(d)	decigram (dg)
		0.01	10^{-2}	centi	(c)	centigram (cg)
		0.001	10^{-3}	milli	(m)	milligram (mg)

Review problems and questions

1. Convert 15 mL to L.

2. Choose the correct letter that is another name for "size:"
    a. mass
    b. volume
    c. weight
    d. density

3. Choose the statement that correctly describes the density of Object A, a one cup system, compared to Object B, a two cup system:
    a. Object A is more dense than Object B.
    b. Object B is more dense than Object A.
    c. Object A is as dense as Object B.
    d. There is not enough information to answer this question.

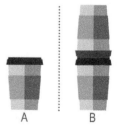

4. Explain the answer to the question above.

5. Suppose the mass of a 298-mL red cup equals 8.83 g. What is its density?

The Chemistry of ... Getting to Chemistry Class

Chemistry is called the central science because it is at the intersection of biology and physics. You may not have appreciated it before, but chemistry is also central to your life. As you go through this book you will start to see the everyday influences of chemistry.

We start the journey of the influences of chemistry by thinking about what you did today. This morning you woke up, opened your eyes and looked at the clock on your phone to see if you could snooze for a few more minutes. Even this simple act can be explained and understood with chemistry. The light that comes into your eyes causes a reversible change in a chemical called retinal that ultimately causes a cascade of other reactions and processes sending a nerve impulse to your brain to process an image.

After realizing that you cannot snooze anymore, you go to the kitchen to get breakfast. Looking around the kitchen, you see a variety of foods and drinks. If you look closer, you will actually see an even larger variety of … chemicals. Just to name a few, there is: protein in eggs, carbohydrate in breads, potassium in bananas, ascorbic acid in oranges, sugar in your drink, antioxidant in blueberries, iron in cereal, and calcium in milk.

You decide to take some of the items and make breakfast. A recipe for fluffy pancakes says you need to combine 1 cup flour, 2 tablespoons sugar, 2 tablespoons vinegar, ¾ cups of milk, 1 teaspoon baking powder, ½ teaspoon baking soda, ½ teaspoon salt, 1 egg, and 2 tablespoons butter to make the batter. Then you heat 1/3 cup of batter over medium heat for 2 minutes on each side. Looking at this from a chemist's perspective, the recipe is just like an experimental procedure you might perform in a lab. You are combining chemicals in measured amounts, adding energy for a known amount of time and testing the results. You could even change variables in a recipe to see how it affects the taste or texture of the pancakes. Of course, in the lab, we would probably convert those measurements to SI units and we would never taste the results!

The Chemistry of ... Getting to Chemistry Class - 2

Now you need to clean up the dishes. If you run the dishes under water, you will notice that they do not get completely clean. But, if you scrub with a little dish detergent, your silverware sparkles. How does dish detergent work? Dish detergent contains a cleaning agent, usually sodium lauryl sulfate. Part of the sodium lauryl sulfate molecule is attracted to the oil and grease, while another part of the molecule is attracted to water so it can be rinsed away. Using chemistry, we can explain the macroscopic observations of the dish detergent by looking at the microscopic properties of the molecules.

After cleaning up from breakfast, it's time to head off to school. You dress quickly and run outside to catch the bus. To get you from your house to your school, the bus relies on the conservation and transformation of chemical energy into mechanical energy. Chemical energy is stored in fuels, like gasoline or compressed natural gas. When fuels combust, they release energy that ultimately results in the movement of the bus. As we have become more concerned with carbon dioxide, a product of combustion, automakers have shifted to hybrid or electric vehicles. But, even electric vehicles rely on chemical energy. The purpose of a battery is to convert stored chemical energy into electrical energy. This electrical energy is then used to power the vehicle. The movement of the bus can be explained in physics terms, such as the distance traveled, velocity on the roads, and acceleration as it starts and stops. Yet under the hood, the underlying processes that make the bus move are explained with chemistry!

When you get to school you realize it is almost time for your chemistry class that is on the other side of the building. Since it is your favorite class and you don't want to be late, you jog down the hall. As you move quickly past other students you start breathing a little heavier and perspiring. In its simplest sense, when you breathe in, you are taking in oxygen molecules to react in your body. This reaction creates energy for your cells and produces carbon dioxide that you exhale. Perspiration is part of the thermoregulation of your body. When your body gets too hot, you sweat to cool down. From a molecular perspective, liquid water molecules in the sweat vaporize as they take energy from your body. Even the biological processes that occur as you move through the hallway are explained by chemistry on a fundamental level.

As you work through the topics in the book, you will have greater understanding of the central science and its relationship to the world around you. Even a trip from your home to chemistry class can be better appreciated with some essential chemistry concepts.

Chapter 1 review

Vocabulary
Match each word to the sentence where it best fits.

Section 1.1

claim	conclusion
control variables	experiment
experimental variable	hypothesis
inquiry	macroscopic scale
matter	microscopic scale
natural laws	objective
procedure	pseudoscience
repeatable	scientific method
theory	variable

1. The _____ is the formal thought process of science.

2. A(n) _____ is an unconfirmed explanation for an observation.

3. The basis for all scientific discoveries is _____ which is the process of learning through questioning.

4. A statement that asserts something is true is called a(n) _____.

5. Data is _____ when it describes what is actually observed or measured without opinion or interpretation.

6. Factors in an experiment that are kept constant are called _____.

7. The series of steps that outlines how to perform a particular experiment is called the _____.

8. An experiment or observation is _____ when others who do the same experiment or make the same observation get the same result.

9. A(n) _____ is a parameter or observable quantity which may change.

10. _____ is anything that takes up space and has mass.

11. The _____ includes things we can see and measure directly such as weight.

12. A situation specifically set up with certain conditions to make observations is called a(n) _____.

13. A central idea in science is that our universe obeys a set of _____.

14. Astrology is an example of a(n) _____.

15. A(n) scientific _____ is a statement that asks the reader or listener to believe something is true and supported by evidence provided in an experiment or observation.

16. Chemistry is explained on the _____ of atoms.

17. An explanation advances to become a(n) _____ when it has been tested by numerous, independent researchers over a long period of time.

18. The factor that is changed in an experiment is called the _____.

Section 1.2

density	energy
graduated cylinder	gram
joule	kilogram
law of conservation of energy	liter
mass	measurement
meniscus	milliliter
System International	volume
weight	

19. The world of science adopted the _____ as a common system of measurement.

20. A measure of the amount of matter in an object is given by the object's _____.

21. The _____ is the fundamental SI unit of mass.

22. A peanut has a mass of about one _____.

23. A(n) _____ is a specific kind of information that describes a physical quantity with both a number and a unit.

24. The SI unit for volume is the _____.

25. The fundamental unit of energy is the _____.

26. A(n) _____ is used to measure the volume of a liquid.

27. _____ is a quantity of space that may or may not be occupied by matter.

28. _____ is a force that results from gravity acting on mass.

29. One _____ is equivalent to one cm^3.

30. Graduated cylinders are read from the lowest point of the _____.

31. The _____ states that energy can never be created or destroyed.

32. A system's ability to change or create change is determined by the system's _____.

33. _____ is a ratio of the mass per unit volume.

Section 1.4: Chapter Review

Chapter 1 review

Conceptual questions

Section 1.1

34. What is a hypothesis?

35. What are the steps of the scientific method?

36. Give an example of how the scientific method could be used by:
    a. a runner
    b. an electrician

37. Design a simple experiment to test the temperature at which water freezes. Define your system, the variable that you are testing, and the variables you will hold constant.

38. When conducting an experiment, why is it important to have only one experimental variable?

39. Why is an "empty" water bottle sitting on your table not really empty?

40. What is an experiment?

41. What is a conclusion?

42. What are the steps in scientific inquiry?

43. What is the difference between an experimental variable and a control variable?

44. How are a theory and a hypothesis different?

45. Which of the following are examples of matter?
    a. the air
    b. gravity
    c. a cat
    d. you
    e. light beams

46. Good experiments often involve many measurements of the same parameter under the same conditions. Explain the benefits of making multiple measurements using the words "error" and "average."

47. What is the difference between the microscopic and macroscopic scales?

48. Write two sentences about differences between solid ice and liquid water.

49. Why is chemistry known as the central science?

50. What are two important characteristics of scientific evidence?

51. A person claims that deep dish pizza is better than thin crust pizza. Can this claim be supported by scientific evidence?

52. What are natural laws?

53. What is the difference between true science and pseudoscience?

Section 1.2

54. What is the difference between mass and weight?

55. Does a nuclear power plant create electrical energy? Explain your answer.

56. Describe an observation that is explained by air having mass.

Solid sphere   Hollow sphere

2 cm   2 cm

57. Two steel spheres have the same diameter and are made from identical kinds of steel. One sphere is solid and the other is hollow. Answer the following questions about the two spheres.
    a. Which sphere contains more matter?
    b. Which sphere has a larger volume?
    c. Which sphere has a greater density? Explain.

Is it tin, zinc, or silver?

58. An antique coin is given to a chemist who is asked to determine if the coin is tin, zinc, or silver. The chemist measures the following physical properties.
    1. Melting point = 431°C
    2. Boiling point = 900°C
    Of these three metals, which is the coin most likely to be?

59. What are the advantages of using the System International for measurements rather than the US Customary System?

60. Which unit represents a larger amount of energy, a joule or a calorie? Explain your reasoning.

61. Correct the following statement: Iron is heavier than cork that is why iron sinks in water and cork does not.

62. According to the law of conservation of energy, is the amount of energy in the universe increasing or decreasing?

Chapter 1 review

Quantitative problems

Section 1.3

63. Micah is a junior at Blue Valley High School. The table below shows the data for his study habits and grades.

Hours spent studying per week	Grades earned on exams
0	15%
4	40%
8	60%
12	78%
16	88%

  a. Propose a qualitative model that best describes how the length of time Micah spends studying per week relates to how well he does on his exams.
  b. If Micah wants to earn 98% on his next exam, should he study for about 4 hours, 8 hours, 16 hours, or 20 hours? Justify your answer.
  c. If Julie studies for 10 hours, do you think she will score 10%, 60%, 68%, or 80% on her exam? Justify your answer.

64. An object has a mass of 6×10^6 g, how much does it weigh:

  a. In kilograms?
  b. In pounds? (Hint: 1 kilogram = 2.20 pounds)
  c. In newtons?

65. Diamonds are measured in carats and 1 carat = 0.200 g. What is the volume of a 6.0-carat diamond ring? (Density of pure diamond = 3.52 g/cm^3.)

66. The maximum weight for a U.S. semi truck and full trailer is 80,000 pounds spread over 18 conventional wheels. What is its mass:

  a. In kilograms?
  b. On Earth's surface (in newtons)?
  c. On the moon (in pounds)?
  d. In grams?
  e. In metric ton?

67. Smaller cars generally have gas tanks that hold 12 gallons of gas, while larger cars can hold 15 or 16 gallons.

  a. How many liters of gas are in the small car?
  b. How many milliliters of gas are in the small car?
  c. If a large car holds 15 gallons of gas, how many liters of gas are in the large car?
  d. If a large car holds 15 gallons of gas, how many milliliters of gas are in the large car?

68. On average, a 42-gallon barrel of crude oil yields about 19 US gallons of gasoline when processed in an oil refinery.

  a. What is the volume of crude oil in liters?
  b. What is the volume of gasoline in liters?

69. Gasoline contains benzene and other known carcinogens and has a density 0.73 g/mL. How many liters of gasoline would be required to increase the mass of an automobile from 2142 kg to 2525 kg?

70. Isabella uses 75 cups of butter and 70 cups of sugar in baking Mia's wedding cake.

  a. How many liters of butter did Isabella use in baking?
  b. How many liters of sugar did Isabella use in baking?
  c. If the density of sugar is 1.59 g/mL, what is the mass of sugar used in baking Mia's wedding cake? (In kilograms).
  d. If the density of butter is 911 g/L, what is the mass of butter used in baking Mia's wedding cake? (In kilograms).

71. Ten cups of cream cheese, five cups of fresh strawberries and 2 teaspoons of vanilla extract are included in the recipe for Simply Bakers' Strawberry Cheesecake.

  a. How much cream cheese, in milliliters is used in this recipe?
  b. How much strawberries, in milliliters are used in this recipe?
  c. How much vanilla extract, in milliliters is used in this recipe?

72. Olivia is conducting an experiment. The table below shows her number of attempts and the results from each trial.

Experimental Results	
Trial 1	12.4 g
Trial 2	15.8 g
Trial 3	11.7 g
Trial 4	14.5 g
Trial 5	9.90 g
Trial 6	15.2 g
Trial 7	12.6 g
Trial 8	14.8 g

  a. What is the average mass of Olivia's yields in kilograms?
  b. If the substance Olivia measured is pure aluminum, what is the volume of aluminum used? The density of aluminum is 2.70 g/cm^3. (Use the average mass).
  c. If the substance Olivia measured is copper, what is the volume of copper used? The density of copper is 8.96 g/cm^3. (Use the average mass).
  d. If the substance Olivia measured is titanium, what is the volume of titanium used? The density of copper is 4.51 g/cm^3. (Use the average mass).
  e. If the substance Olivia measured is cesium, what is the volume of cesium used? The density of cesium is 1.93 g/cm^3. (Use the average mass).

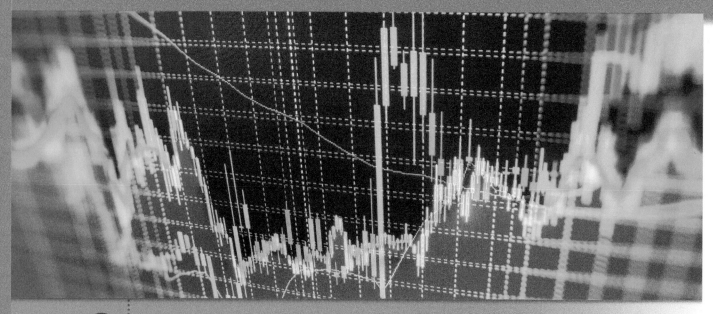

chapter 2: Measurement and Analysis

Performing experiments is an integral part of chemistry. During an experiment chemists make measurements, organize data, and analyze results. Another important part of science is communication. What would be the point of discovering a ground-breaking fuel source or a new medicine if you do not share that information with the rest of the world?

You might think of measurements and data as just numbers, but they are actually tools scientists use for communication! You probably know that a measurement needs to indicate the units of measure, such as grams, liters or meters. Reporting something as 1.0 kilogram instead of 1.0 gram may not seem like a big deal, but it is like the difference between $1,000 and $1! However, simply adding the correct unit to a number does not make a measurement valuable. The way the number is written tells you important information about the precision of the instruments being used and the error in the measurement.

As chemists perform experiments that generate more and more measurements, it becomes especially important to collect and visualize the data in a meaningful way. When you have 3 data points, it is easy to communicate them individually or to combine them in an easily understood table. However, when you have 3,000 data points, a table is not the most useful visualization tool. In this case a graph is much more useful to analyze the data and look for patterns and trends.

For a scientific experiment to be accepted into the larger community the results need to be reliable and repeatable. Once the results are published other scientists will repeat the experiment and verify the results. This is the importance of communication in science. The correct use of measurements helps scientists clearly express what is known and what is uncertain in a way that others can verify. Therefore, understanding the meaning of a measurement is not just important, it is essential.

Chapter 2 study guide

Chapter Preview

Proper measurement is a crucial part of scientific discovery. It includes being able to describe very large and very tiny things, understand which numbers in a measurement are significant, and convert between units. In this chapter you will learn about dimensional analysis, scientific notation, accuracy & precision, and uncertainty & error. You will also explore different ways of representing data and how to use data to determine when results are significant and what conclusions they can help you draw, as well as how certain you can be about those conclusions.

Learning objectives

By the end of this chapter you should be able to:
- determine the number of significant figures in a measurement;
- represent small or large numbers using scientific notation;
- use dimensional analysis and conversion factors to convert units of measurement;
- explain the difference between precision and accuracy;
- distinguish significant from insignificant experimental outcomes;
- use graphs of data to determine relationships between variables and determine rate of change; and
- draw the best-fit line for a collection of data points.

Investigations

2A: Density of a solid
2B: Density of a liquid

Pages in this chapter

30	Measurement and Numbers
31	Very large numbers
32	Very small numbers
33	Significant figures
34	Significant figures in practice
35	Zeros in significant figures
36	More zeros in significant figures
37	Rate, ratio, and proportionality
38	Converting between units
39	Dimensional analysis
40	Accurate and precise measurements
41	Section 1 Review
42	Representing and Analyzing Data
43	Uncertainty and error
44	When are differences significant?
45	Bar graphs and pie charts
46	X-Y or Line graphs
47	Slope and rate of change
48	Regression and curve fitting
49	Using data to draw conclusions
50	Section 2 Review
51	The Value of Measurements
52	The Value of Measurements - 2
53	Chapter review

Vocabulary

precision	accuracy	resolution
scientific notation	mantissa	exponent
significant figures	conversion factor	dimensional analysis
error	average	standard deviation
standard error	pie graph	bar graph
frequency table	independent variable	dependent variable
slope	error bars	box and whiskers plot
linear regression	correlation coefficient (r)	

2.1 - Measurement and Numbers

The question of quantity can have two kinds of answers. One kind of quantity, counting, can have an exact number. There can be 23 or 24 students but there cannot be 23.25 students. The second kind of quantity, a measurement, cannot have an exact answer! A box that is measured to be 1,000 grams on a balance that reads to 1 gram, might have any mass between 999.5 and 1,000.5 grams. No continuous real value, such as mass, length, or time, can ever be known exactly. The best we can ever do is measure a real value "plus or minus" some error. Knowing the possible error helps you interpret the value you are trying to measure.

Measurements in chemistry

Resolution

We can never determine that something has a mass of exactly 10 grams. The reason is because all real measurement instruments have limits to how small a variation they can sense and display. An ordinary lab balance has a **resolution** of 0.1 grams. This means 0.1 grams is the smallest mass difference that can be sensed and displayed. True masses of 10.02 grams and 9.96 grams will both read as 10.0 g. One way to show this is to write the measurement as 10.0 g +/- 0.05 g. This tells any reader the actual mass could have been anything between 9.95 g and 10.05 g.

Accuracy and precision

The **precision** of a measurement describes how close a sequence of identical events or measurements will be to each other. High precision means the event or measurement occurs almost in the same way every time. For example, if you go bowling and throw a gutter ball every time, you are highly precise! A balance that can read to 0.1 grams likely has a precision of 0.1 g. Precision is not the same as accuracy. **Accuracy** describes how close a measurement is to the true value being measured. A balance that has not been properly zeroed might make a measurement precise to 0.1 grams that is quite inaccurate because the zero is not set accurately! If you always throw gutter balls when bowling, you are highly precise because the outcome is repeated but your accuracy is low because you miss the goal of striking bowling pins.

Why isn't this mass exactly 5.2 g?

The mass could be between 5.15 and 5.24

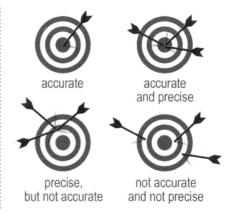

accurate

accurate and precise

precise, but not accurate

not accurate and not precise

The importance of accuracy and precision

To be practical, two masses of 10.01 g and 10.03 g are thought of as "equal" because they are only 0.02 grams apart. This difference is smaller than the balance shown above can measure. When analyzing your data in the lab, think about both the accuracy and the precision of your measurements before forming a conclusion. An inaccurate measurement could lead you to the wrong conclusion. A measurement that is not precise makes it impossible to know whether measured values agree or disagree with each other.

Very large numbers

Chemistry deals with large numbers

Atoms and molecules are so small that describing them requires proportionally small values that are difficult to write and understand. A single bacterium is so small you need a powerful microscope to see it. Yet one bacterium contains 1,000 billion atoms! This number written out the usual way is 1,000,000,000,000. This number is difficult to work with in ordinary form. Fortunately, **scientific notation** gives you a convenient way to perform calculations with extremely large or small numbers.

1 grain of sand contains 100,000,000,000,000,000 atoms

Dropping a penny on the moon would be like dropping a hydrogen atom on a penny.

large numbers	
10^1	10
10^2	100
10^3	1,000
10^4	10,000
10^5	100,000
10^6	1,000,000
10^7	10,000,000
10^8	100,000,000
10^9	1,000,000,000

Writing and reading scientific notation

Scientific notation expresses any number as the product of two numbers. The first number is called the **mantissa** and ranges from the numbers one to ten. The second number is a power of ten. Large or small numbers can be represented as a mantissa times a power of ten. Let's look at the number 3,400:

- 3,400 = 3.4 × 1000. The number 3.4 is the mantissa; the number 1000 is a power of 10. The mantissa is always written with a single number between 1 and 9 to the left of the decimal place.
- 1000 = 10 to the third power, and is usually written 10^3.
- So 3,400 = 3.4 x 10^3. The small superscript number 3 in 10^3 is called the **exponent**.

Scientific notation is useful

For the number 3,400, scientific notation does not seem to be very beneficial. However, 140,000,000,000 in scientific notation is 1.4 x 10^{11}. This number is much easier to work with when written in scientific notation. An important concept in chemistry is the mole, which uses a quantity of 6.02 x 10^{23}, also known as Avogadro's number. This is the number of atoms in one gram of hydrogen. A number of this size is virtually impossible to work with except in scientific notation.

Computers use scientific notation

All computers use scientific notation internally to represent numbers. Even a relatively simple number such as 3,400 would be represented as a floating point number in a computer as 3.4 x 10^3. To enter a number in a computer program such as a spreadsheet, you use the letter "E" to designate the next value as the exponent in a floating point number. For example, 3.4E3 is the same as 3,400 and 3.4 x 10^3.

Section 2.1: Measurement and Numbers

Very small numbers

Very small numbers

Chemistry requires very small numbers as well as very large ones. For example, a single atom is approximately 0.0000000001 m in diameter. The nucleus at the center of an atom is 10,000 times smaller! Fortunately, we can use scientific notation to easily perform calculations with either very large numbers or very small numbers.

Small numbers using scientific notation

Negative powers of ten are numbers smaller than one. Consider the number 0.0025 in scientific notation:

- $0.0025 = 2.5 \times 0.001$.
- The number $0.001 = 10^{-3}$.
- $0.0025 = 2.5 \times 10^{-3}$ in scientific notation.

A negative exponent means the number is less than 1, NOT that the number is negative.

On this calculator, to enter the number 2.5×10^{-3} press:

[2] [.] [5] [2nd] [EE] [(-)] [3]

numbers smaller than one

10^{-9}	0.000 000 001
10^{-8}	0.000 000 01
10^{-7}	0.000 000 1
10^{-6}	0.000 001
10^{-5}	0.000 01
10^{-4}	0.000 1
10^{-3}	0.001
10^{-2}	0.01
10^{-1}	0.1

Scientific notation using a calculator

The diagram shows how to enter a number in scientific notation on a calculator. On this calculator the EE or EXP key tells the calculator the next value is the exponent. The diagram shows the keystrokes to enter the number 2.5×10^{-3}. Keys may be different on other calculators, such as E instead of EE; however, one of these two keys is the one that allows you to enter the exponent. Do not try to enter [×] [1] [0] [-] [3]; this will not work on most calculators.

Use the interactive equation titled Scientific Notation Calculator to see how different tools convert and display small and large numbers in scientific notation. See which display matches your calculator.

Solved problem

Convert 0.000018 to scientific notation.

Asked Write the number with an exponent
Given The number 0.000018
Relationships $0.00001 = 10^{-5}$
Answer 1.8×10^{-5}

Significant figures

Why we use significant figures

Significant figures are a way of recording data that tells the reader how precise a measurement is. Every time a scientist writes a number down they use significant figures to communicate how precisely it was measured. They use specific rules that are universally agreed upon. The numbers 0.50 and 0.5 are mathematically the same value but 0.50 communicates that the number was read from a device that measures precisely to two decimal places, while 0.5 was measured from a device that precisely measures to one decimal place.

Recording significant figures

The two gold bars below are the same length. How the length is recorded depends on the device they are measured with. The ruler on top is marked every centimeter. To record the number correctly we list what we can read from the ruler, 3 cm. You also have to account for the part that goes past the 3 cm mark. At this point, you mentally divide the area between the 3 and 4 into ten imaginary sections and decide how many of the sections the gold bar covers. This last estimated number is up to the viewer's discretion. It appears to be 3 sections past the 3, and the correct way to record the number is 3.3 cm. But, if you thought it is was 3.2 cm or 3.4 cm you would also be correct.

The second ruler has finer graduations and can be read to greater precision. You can read 3.3 from the ruler and you can also see that the gold bar extends between 3.3 cm and 3.4 cm. Similar to the estimation with the previous ruler, you estimate the correct reading as 3.35 cm. Scientists will know that 3.35 cm was measured from a better measuring device than 3.3 cm.

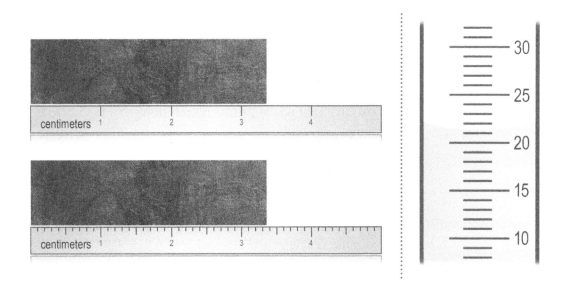

Solved problem

How much liquid is in the graduated cylinder above?

Asked Read the volume on the graduated cylinder.

Solve The fluid in a graduated cylinder is measured from the bottom of the meniscus. A meniscus is the curve of the liquid. From the graduated cylinder, we can see that the liquid is just over 21 mL. Mentally divide the section between 21 mL and 22 mL into ten pieces and determine how many of the ten pieces the liquid covers. In this case, it looks like 3 pieces are covered.

Answer 21.3 mL

Significant figures in practice

Scientific notation and sig figs

Scientific notation shows only significant figures and a power of ten. For example, 5.2×10^{-4} has two significant figures and 3.125×10^9 has four significant figures. In the first case the 2 is the estimated digit and in the second case the 5 is the estimated digit.

SIGNIFICANT FIGURES addition and subtraction		SIGNIFICANT FIGURES multiplication and division	
Round to the fewest number of decimal places in the problem.		Round to the fewest number of significant digits in the problem.	
one significant figure after decimal point	two significant figures after decimal point	two significant figures in problem	three significant figures in problem
89.353 + 2.5 ――― 91.853	3.72 − 1.9142 ――― 1.8058	32.563 × 2.1 ――― 68.3823	100.25 ÷ 3.54 ――― 28.31920
round to 91.9	round to 1.81	round to 68	round to 28.3

Addition and subtraction

Addition and subtraction follow the same rules for significant figures. The final answer is expressed to the precision of the least precise number. For example, 503.1 is measured to the tenths place and 346.87 is measured to the hundredths place. When you add or subtract these numbers, express the final answer to the tenths place because it is the least precise.

Solved problem

Add 5.03 to 23.051 and express the answer to the correct significant figure.

Asked   Add 5.03 to 23.051
Relationships   For addition, round the answer to the fewest number of decimal places in the given numbers.
Solve $5.03 + 23.051 = 28.081$ round 28.08

Your calculator will give you an answer of 28.081. However, you must round it to the fewest number of decimal places, in this case 28.08.

Answer   28.08

Multiplication and division

Multiplication and division follow the same rules for significant figures, however, they are different than the addition/subtraction rules. The final answer is expressed to the fewest number of significant figures. For example, 5.2×10^{-4} has two significant figures and 3.125×10^{-4} has four significant figures. Your final answer should have two significant figures.

Solved problem

Multiply 7.025×10^{-2} by 9.04×10^3 and express the answer to the correct sig figs.

Asked   Multiply 7.025×10^{-2} by 9.04×10^3
Relationships   For multiplication, round the answer to the fewest number of significant figures in the given numbers.
Solve $7.025 \times 10^{-2} \times 9.04 \times 10^3 = 6.3506 \times 10^2$ round 6.35×10^2

Answer 6.35×10^2

Zeros in significant figures

The problem with zeros
Is the zero in the number 510, a place holder or is it a measured number? The number zero represents a special case when counting significant figures. A zero can either be a measured number or a placeholder. You have to make a decision every time you calculate a number!

Zeros to the left
Some zeros never cause you problems. Zeros to the left of a decimal are placeholders and are not significant figures. For 0.00056, the four zeros in front of 5 and 6 are not significant figures, so 0.00056 has two significant figures.

Zeros in the middle
Zeros in the middle will not cause you problems either. All zeros between non-zero numbers are significant numbers. In the number 2003.5 both zeros are significant because they are between non-zero numbers. The decimal point does not affect the reasoning. For 80.34, 340.004 and 5.00089 the zeros are significant because the zeros are between non-zero numbers. The position of the decimal point has no effect.

zeros to the LEFT of a number are not significant	zeros in the MIDDLE of a number are significant	zeros to the RIGHT of a number less than one are significant
0.00234 3 sig figs	100.44 5 sig figs	0.2340 4 sig figs
0.000089 2 sig figs	5600.089 7 sig figs	0.890000 6 sig figs
0.000456 3 sig figs	1.00456 6 sig figs	0.004560 4 sig figs

sig figs is short for significant figures

Zeros to the right of a number
Zeros at the right end of numbers often cause questions. Zeros to the right can be either a placeholder or a significant figure. For the number 0.520, the zero at the end of the number is mathematically meaningless. The value does not change whether it is there or not. The zero tells you that 52 was read from a measuring device and 0 is the estimated digit.

Solved problem
Cross out the zeros that are not significant in each number: a) 60.20; b) 0.602; c) 0.00602

Asked   Put a line through non-significant zeros.

Solve
  a. 60.20: The first zero is in the middle of a number and the second zero is to the right of a number. Both zeros are significant.
  b. 0.602: The first zero is NOT significant because it is to the left of a number. The second zero is significant because it is in the middle of a number and also because it is to the right of a number.
  c. 0.00602: The first three zeros are NOT significant because they are to the left of a number. The fourth zero is significant because it is in the middle of a number and also because it is to the right of a number.

Answer a) 60.20; b) ~~0~~.602; c) ~~0.00~~602

Solved problem
How many significant figures are in the number 0.98050?

Given   0.98050

Solve   The red zero helps locate the decimal point and is not a significant figure. The green zero is between non-zero numbers and is a significant figure. The blue zero indicates that it is an estimated digit and is a significant number.

Answer The number 0.98050 has five significant digits.

Section 2.1: Measurement and Numbers

More zeros in significant figures

Zeros to the right

Zeros to the right of a number greater than or equal to 1 are placeholders unless they have a decimal point. The number 530 has two significant figures. The zero at the end is there to keep the value of the number.

Solved Problem

How many significant figures are in the number 67,030?

Given 67,030

Solve The red zero holds the value of the number and is not a significant figure. The green zero is between non-zero numbers and is a significant figure.

Answer The number 67,030 has four significant digits.

zeros to the right of a number bigger than one might be significant	adding a decimal point makes the zeros significant
2340......3 sig figs	2340.0......5 sig figs
89000......2 sig figs	89000.00......7 sig figs
456000......3 sig figs	4560.00......6 sig figs

Adding a decimal point

When a decimal point is added after a placeholder, the zeros become significant. The number 920.0 has four significant figures. The zero at the end of 920.0 is mathematically meaningless. It is there to express that the 9, 2, and the first 0 were read from a measuring device and the last 0 is the estimated digit. The number 340. has three significant figures. The decimal point at the end of the number indicates that the 3, 4, and 0 are read from a measuring device and the final zero is the estimated digit.

Solved problem

Cross out zeros that are not significant: a) 0.310; b) 31.0; c) 310; d) 310.; e) 0.03100

Asked Put a line through non-significant zeros.

Solve
  a. 0.310: The first zero is NOT significant because it is to the left of a number; the last zero is significant because it is to the right of a number and a decimal
  b. 31.0: The zero is significant because it is to the right of a number and a decimal
  c. 310: The zero is NOT significant because it is a placeholder
  d. 310.: The zero is significant because it is a placeholder with a decimal point
  e. 0.03100: The first two zeros are NOT significant because they are to the left of a number; the last two zeros are significant because they are to the right of a number and a decimal point

Answer a) 0̶.310; b) 31.0; c) 31̶0̶; d) 310.; e) 0̶.0̶3100

Solved problem

How many significant figures are in the number 45,010.?

Given 45,010.

Solve The green zero is between non-zero numbers and is a significant figure. The blue zero is followed by a decimal point, which indicates that it is an estimated digit, and is a significant number.

Answer The number 45,010. has five significant digits.

Rate, ratio, and proportionality

Rates

Rates are used by people every day, such as when they work 40 hours per week or buy apples for $4.25 per kilogram. The word "per" is the key to understanding the meaning of a rate. Literally, per translates to "for each" or "for every." So, the rate of $4.25 per kilogram of apples translates to $4.25 for each kilogram of apples. The concept of rates applies to many problems in chemistry. Consider a typical rate problem in the diagram below in which the rate of heating is used to make a prediction. You probably recognize the slope or rate of change equation from previous math classes.

Working with rates

Heat is added to a beaker of water and the temperature is observed to rise 20°C in 5 minutes. At this rate, when will the temperature reach 60°C?

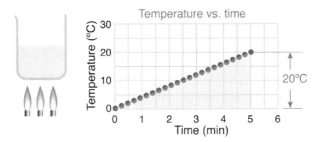

① Determine the rate

$$\text{slope} = \text{rate} = \frac{20°C}{5 \text{ min}} = \frac{4°C}{1 \text{ min}}$$

② Solve the problem

$$\frac{60°C}{1} \times \frac{1 \text{ min}}{4°C} = 15 \text{ min}$$

Using rates so the units must cancel

The problem asks for a time and you are given a temperature. The inverted rate is used because the units of temperature cancel leaving only the desired time unit.

- Rates are ratios and can be used as they are given or inverted.
- One-step rate problems (such as this one) can be solved using dimensional analysis similar to solving unit conversions.

Ratio and proportionality

Many chemistry problems use ratios and proportionality. Proportionality problems are similar to rate problems because ratios and proportions are used like rates in problem solving. The dimensional analysis rules for cancelling units tell you how to use the units in the numerator and denominator to solve the problem.

Proportionality problems

The combustion of 100 grams of gasoline in an engine requires 351 grams of oxygen. One liter of air contains 1.4 grams of oxygen.

How many liters of air are required to burn 1000g of gasoline (1 kg)?

① Determine the rates or ratios

$$\frac{100 \text{ g gas}}{351 \text{ g oxygen}} \qquad \frac{1.4 \text{ g oxygen}}{1 \text{ liter air}}$$

② Solve the problem

$$\frac{1000 \text{ g gas}}{1} \times \frac{351 \text{ g oxygen}}{100 \text{ g gas}} \times \frac{1 \text{ liter air}}{1.4 \text{ g oxygen}}$$

$$= \frac{1000 \times 351}{100 \times 1.4} \text{ liters}$$

$$= \boxed{2{,}507 \text{ liters}} \text{ Answer}$$

Section 2.1: Measurement and Numbers

Converting between units

Languages and units

There are just under 7,000 different languages spoken on Earth, of which 389 are spoken by more than one million people. It should be no surprise that the language of measurements is not universal. Fortunately, there are far fewer choices for units, and translation is a simple mathematical process that works for all of the choices in the same way. Consider a length of wood. The picture below shows different ways to say the same thing in different languages (left) and different units (right).

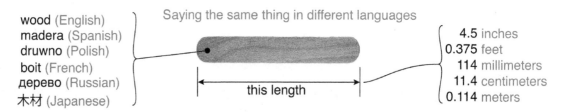

Conversion factors

A non-Spanish speaker needs a tool like a smartphone app or Spanish/English dictionary to translate Spanish. To translate between scientific units of measure you need a **conversion factor**. Conversion factors are "exchange rates" for different sets of units. The list of lengths for the stick above are all equal to one another because 1 inch = 1/12 foot = 25.4 mm = 2.54 cm; therefore, 4.5 inches = 0.375 feet = 114 mm = 11.4 cm. The chart below shows some common conversion factors.

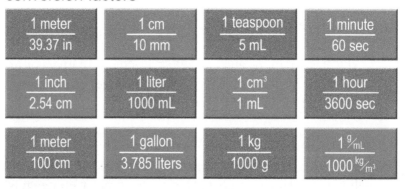

Using conversion factors

Each conversion factor has a value of exactly 1 because the quantities on the top and bottom of the fraction are the same. The key to using conversion factors is to know that multiplying anything by one leaves the real value of the thing unchanged, even if the numbers change. In the example below we convert 4.5 inches to centimeters using the conversion factor 1 inch = 2.54 cm. NOTE: conversion factors are the same value right-side up and upside down. This fact is necessary to use conversion factors correctly.

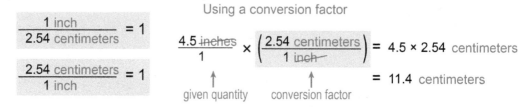

Notice that the conversion factor was chosen so the units of inches canceled on top and bottom, leaving the desired units of centimeters.

Dimensional analysis

Dimensional analysis
Units are also called dimensions. You perform **dimensional analysis** when you use conversion factors to convert one unit to another. Dimensional analysis is a very powerful tool used extensively by scientists, engineers, and everyday people to convert units. To perform dimensional analysis, you write a series of conversion factors as ratios and multiply them so all units cancel top and bottom except the units you want. Dimensional analysis often requires that conversion factors be inverted from how they are listed. For example 2.54 cm/inch is the same as 1 inch/2.54 cm. There are many more conversion factors in the appendix of this book.

Multiple conversion factors
There will be problems in which more than one conversion factor must be used. For example, water weighs 8.34 lbs/gallon. What is the density in g/mL (g/cm^3)? The illustration below shows how to set up the calculation. We are using the fact that 1 kilogram weighs 2.2 pounds as a conversion factor.

Using multiple conversion factors

$$\frac{8.34 \text{ lbs}}{\text{gal}} \left(\frac{1 \text{ gal}}{3.785 \text{ L}}\right)\left(\frac{1 \text{ L}}{1000 \text{ mL}}\right)\left(\frac{1 \text{ kg}}{2.2 \text{ lb}}\right)\left(\frac{1000 \text{ g}}{1 \text{ kg}}\right) = \frac{8.34 \times 1000}{3.785 \times 1000 \times 2.2} \frac{\text{g}}{\text{mL}}$$

$$= 1.00 \frac{\text{g}}{\text{mL}}$$

1000 g	2.2 lb	3.785 L	1 L
1 kg	1 kg	1 gal	1000 mL

conversion factors

Comparing values
When comparing two values it is important that they be in the same units. For example, what is bigger: 10 inches or 20 centimeters? The answer is not obvious unless one of the values is converted to the units of the other. Using the conversion factor from inches to centimeters we calculate that 10 inches = 25.4 centimeters. That means 10 inches is the larger distance, not 20 centimeters.

Compare quantities with the same units

10 inches
20 centimeters
Which is larger?

10 inches ⟷ 7.9 inches
25.4 centimeters ⟷ 20 centimeters

Doing calculations
When solving chemistry problems, units must be consistent for every value in the problem. For example, suppose you want the average density of a cake containing 1 kg of water and flour and 5 ounces of chocolate in a volume of 2 liters. You cannot add 1 kg and 5 ounces by simply adding 1.0 + 0.5 = 1.5. Both values have to be in the same units.

Solved problem
Gas is sold either by the liter or by the gallon. Convert $3.99/gallon to dollars per liter.

Given $3.99/gallon
Relationships 1 gallon = 3.785 liters
Solve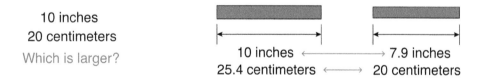

Answer $1.05 per liter

Section 2.1: Measurement and Numbers

Accurate and precise measurements

Methods to determine volume

From now on, apply the rules of accuracy and precision to all measurements and calculations. For example, to find the density of a material you can measure and record its mass and volume to the greatest level of precision offered by your lab instruments. Then you can calculate density by dividing mass by volume. Lastly, you need to round your answer to the correct number of significant figures. Mass is measured with a balance or a scale, but there are two common ways to determine volume.

- The volume of simple shapes is calculated by multiplying length × width × height.
- The volume of irregular shapes is measured using the water displacement method.

Determining density by water displacement

Suppose you want to know a colorful rock's density. You have five small samples. First, record the total mass of the rock samples with a balance. Next, correctly record the volume of water in a graduated cylinder by reading below the meniscus. Add the five rock samples to the water and record the volume again. The volume of the rocks is the amount of water displaced in the graduated cylinder, equal to the difference between the initial reading with water and the final reading with water and rocks. Calculate the density by dividing mass by volume. Lastly, round your answer to the correct number of significant figures.

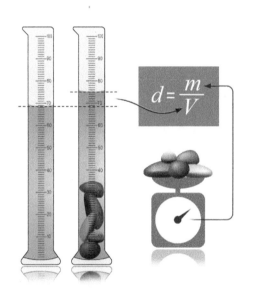

MATERIAL	DENSITY (g/cm³)
platinum	21.5
lead	11.3
iron	7.8
titanium	4.5
aluminum	2.7
granite	2.6
concrete	2.3
nylon plastic	2.3
rubber	1.2
liquid water	1.0
ice	0.92
oak (wood)	0.60
pine (wood)	0.44
cork	0.12

Solved problem

An engineer has samples of silvery metal of different sizes. One is 100.0 g of platinum and one is 50.0 g of titanium. The engineer wants to predict how much water each sample will displace in a graduated cylinder. Predict the change in volume when 50.0 g of titanium is added, then 100.0 g of platinum is added to the cylinder.

Asked Find the volume of water displaced by each sample.

Given 50.0 g titanium, from the table d = 4.5 g/cm³, 100.0 g platinum, from the table d = 21.5 g/cm³

Relationships d = m/V is rearranged as V = m/d and 1 mL = 1 cm³

Solve titanium: $\frac{50.0\ \cancel{g}}{1} \times \frac{mL}{4.5\ \cancel{g}} = 11.11111$ mL $\xrightarrow{\text{round}}$ 11 mL

platinum: $\frac{100.0\ \cancel{g}}{1} \times \frac{mL}{21.5\ \cancel{g}} = 4.65116$ mL $\xrightarrow{\text{round}}$ 4.65 mL

Answer The titanium displaces 11 mL and platinum displaces 4.65 mL.

Section 1 review

Chapter 2

Exact numbers are often used in math, but in science the measurements of real things can never be completely exact. When we take multiple measurements of the same object or phenomenon, how closely they match each other is the level of precision. If the true value is known then how far our measurements are from that true value is the level of accuracy. There is always a limit to how precise a measurement can be, usually due to the tools we use to make those measurements. The number of significant figures in a measurement is a way to indicate how confident or certain we are about that measurement. Very large or very small measurements are written using significant figures and multiplying by a positive or negative power of 10.

When two different measurements use different units we have to convert from one unit to another using a conversion factor. This process is called dimensional analysis. The change in a variable relative to another (often time) is known as rate. Proportions are comparisons of two ratios.

Vocabulary words: precision, accuracy, resolution, scientific notation, mantissa, exponent, significant figures, conversion factor, dimensional analysis

Review problems and questions

1. Sam measures the height of a single pea plant three times. His measurements are shown in the table. Describe Sam's accuracy and precision in measuring the height of the plant.

Pea Plant Data	
Attempt #	Height (cm)
1	13.5
2	10.1
3	14.9
Average:	12.8

2. Convert a-c to scientific notation, and convert d-f to standard form.
   a. 13,400
   b. 0.0032
   c. 780
   d. 5.7×10^{-3}
   e. 6.02×10^{1}
   f. 2×10^{3}

3. The graduated cylinder shows the volume of a liquid in mL.
   a. Read the graduated cylinder shown to the greatest accuracy possible.
   b. The density of the solution equals 1.21 g/mL. Calculate the mass of the solution. Report your answer with the correct number of significant figures.

   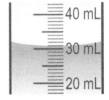

4. How many significant figures are in each of the following numbers?
   a. 200
   b. 200.
   c. 20.0×10^{2}
   d. 9,040
   e. 3.013
   f. 0.002100

5. Use dimensional analysis to solve the following, and round your answer to the correct number of sig figs: You have 1.5 gallons of milk. How many liters is this? There are 3.785 L per gallon.

Section 2.1: Measurement and Numbers

2.2 - Representing and Analyzing Data

A major difference between the scientific method and other ways of investigating questions is how to decide whether an explanation or idea is correct or not. Science is only concerned with explanations and ideas that can be tested and verified against scientific evidence, including observations and data. Humans understand pictures and patterns far better than columns of numbers. Therefore, the way data is represented is critical for supporting logical arguments that support a scientific claim or decision.

Scientific argument

The phlogiston theory

The story of phlogiston provides a good illustration for how data separates what we believe from what is actually true. For ages humans wondered about the nature of fire. What was fire and why do some things burn and not others? During the Renaissance and up to the emergence of modern chemistry around 1780, people believed wood and other combustibles contained a substance called phlogiston. Phlogiston was released into the air when wood burned. Air could only absorb so much phlogiston and that explains why a candle burns only a short time in a closed container. The phlogiston theory was the accepted scientific explanation for combustion at the time.

1. How can we tell whether the phlogiston theory is true or not?

2. A competing theory is proposed that offers a different explanation for combustion. How do we decide between the existing theory and the new theory?

Data and observations are *scientific* evidence of truth

Phlogiston theory

An object releases phlogiston during combustion and therefore mass should decrease.

Experimental data

Substance burned	Starting mass	Final mass	Air required
phosphorus	100 g	284 g	430 L
sulfur	100 g	200 g	333 L
iron	100 g	138 g	127 L

The discovery of oxygen

In 1772, Antoine Lavoisier demonstrated experimentally that the combustion products of phosphorus and sulfur had more mass instead of less. How could losing phlogiston increase the mass of a substance? Lavoisier was the first to measure chemical reactions in a sealed container. He proposed a new theory of combustion based on a new gas he named oxygen. The data was the deciding factor between the new "oxygen theory" of combustion and the accepted phlogiston theory.

- The data disagreed with the predictions of the phlogiston theory.
- The data agreed with the new oxygen theory of combustion.

Scientific logic

The if–then statement is a fundamental part of scientific logic. An if-then statement has the form "if ____ is true, then ____ must also be true." If the phlogiston theory were true then the combustion products should have less mass after burning. The data demonstrate the statement to be false, logically disproving the phlogiston theory. The same logic applied to the oxygen theory results in the statement: "if the oxygen theory is true the combustion products should have more mass because they incorporate oxygen from the air." This statement is in agreement with the data.

Uncertainty and error

"Error" is not the same as mistake!

In science, **error** is the difference between the true value of something being measured and its measured value. There is **always** unavoidable error in every measurement, no matter how careful the experimenter or how elaborate the equipment. This is why it is important to take multiple measurements. It is unfortunate that the common use of the word error means mistake but in the context of scientific data error is not a mistake!

Example of making a measurement

Consider a measurement of the number of heads in ten flips of a coin. The table below shows the results of thirty measurements. Intuitively, you know the most likely outcome is 5 heads out of 10 flips. However, the outcome of a single measurement could be any number between 0 and 10. The most likely number is 5 but the data shows 4 heads came up five times and 6 heads came up seven times.

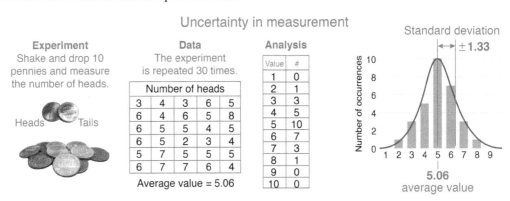

The average

All measurements have similar random variability to flipping a coin. One way scientists account for random error is to take the **average** of several measurements. To calculate the average, add all measurements together and divide by the number of measurements made. The average of the 30 coin-flip measurements is 5.06 - very close to 5. The average is a better estimate of the true value (5) because averaging smooths out random variability.

Standard deviation σ

Another reason for taking multiple measurements is they provide an estimate of uncertainty (or error). The solid red line on the graph shows the behavior of data that contains random variation. The difference between the average and the spread of the measurements is called the **standard deviation**. In the example the standard deviation is 1.33.

The standard error

The standard deviation is the uncertainty in a single measurement. The **standard error** in the average is smaller because many measurements are included. The standard error is the standard deviation divided by the square root of the number of measurements. A proper result states the average plus or minus the standard error.

Outliers

Outliers are "the exception," or data that are very different from the average. For example, a very warm winter day will shift the monthly average temperature higher than usual. An outlier is easier to spot when there are many other data points available to compare.

When are differences significant?

Different numbers can be "the same"

Scientific explanations make predictions that are tested by comparison with observational data. This brings up a **very** important question - how can a scientist tell if the data is the same as the prediction when all the data has uncertainties? In terms of experimental measurement, the word "same" does not mean "equal" or "identical." In an experiment, measurements are the same if their difference is less than or equal to the amount of uncertainty. This is important to remember.

> Differences are not *significant* unless they are larger than the uncertainties in measurement.

Significance

Any scientific use of data must consider the significance of differences between values. This rule has consequences for chemistry and all areas of science. The data below show the amount of carbon dioxide (CO_2) measured over 24 hours compared to the predictions of a plant growth model. The graph on the left shows just the data while the graph on the right includes the error bars based on the accuracy of the sensor making the measurements. If the error bars from the data overlap with the predicted results, the predicted results and actual results are not likely to be significantly different. If error bars do not overlap at all, there may be a significant difference between the data sets. In the example below, does the data support the predicted CO_2 or not? Can you tell without the error bars?

A carbon dioxide experiment on plant respiration

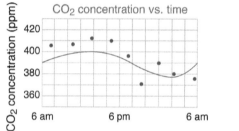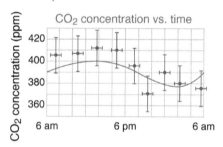

Estimating error

How can we know the error without knowing the true value of a measurement? The way scientists estimate the error is the following.

1. Assume the average is the true value.
2. Find the differences between each measurement and the average.
3. The standard deviation is the average difference between the individual measurements and the average.

Most calculators and spreadsheets have a function that calculates the standard deviation of a set of measurements. You can estimate the standard deviation as about 1/2 the difference between the average and the highest and lowest values.

Significant figures and results

The average gives a better result than individual measurements and may (or may not) have more significant figures than an individual measurement. A final result should have a number of significant figures consistent with the standard error. In the ten-coin-flip example the average of thirty flips is 5.0677, the standard deviation is 1.33 and the standard error is 0.24 for 30 trials. Since the standard error is in the tenth's place, the average result should be rounded to the nearest tenth and be stated as 5.1 +/- 0.2. In this case the average has one more significant figure than any of the individual measurements.

Bar graphs and pie charts

Pie charts

A **pie graph** is a way to show the relative proportions of individual categories in relation to a whole. The size of each slice of the "pie" is proportional to its associated quantity divided by the whole. In the example below the "whole" is the production of electricity in the US. The pie graph makes it easy to see that chemical energy in the form of coal and gas are the source of almost two-thirds (64%) of the electric power generated in the US, with nuclear being the third largest contributor at 20%.

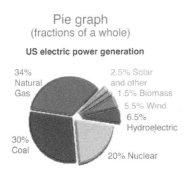

Pie graph
(fractions of a whole)

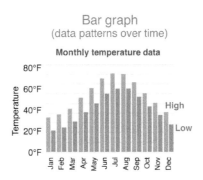

Bar graph
(data patterns over time)

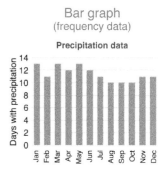

Bar graph
(frequency data)

Bar graphs

A **bar graph** represents each data point by the height of the bar. Bar graphs are effective at showing a pattern in relatively small sets of data, such as the seasonal variation with temperature. Bar graphs lose their visual effectiveness when they have more than a dozen or so bars. Notice the example has two bars for each month showing the high and low temperatures. It is common to use bar graphs to compare the patterns in two or more related data sets, such as the average high and low temperatures.

Frequency tables

Many areas of research generate a lot of data that may be highly variable from point-to-point but has an overall pattern that is important to show. A good example is the number of days it rains or snows over a year. Plotting all 365 days separately does not describe the overall pattern that the average precipitation is approximately constant in the region described by the graph. A **frequency table** is a way to group data into bins that are easier to represent on a graph. Combining many points into a bin helps smooth out fluctuations in the data to show the overall pattern.

Solved problem

The chemical phenylthiocarbamide (PTC) has a bitter taste to some people and no taste to others. The ability to taste PTC has been linked to a single gene TAS2R38 on chromosome 7. Since the test is harmless and easy to administer, PTC has become a standard marker for studying the genetics of populations. In a normally distributed population, 154 males could taste PTC while 78 could not; 165 females could taste PTC while 55 could not.

Asked: Choose a graph type and create graphs to demonstrate whether the data shows a difference between male and female ability to taste PTC in the population.

Solve: The answer must show if there is a difference in the percentage of males who can taste PTC compared to the percentage of females who can taste PTC.

Answer: A pie graph is a good choice for representing fractions of a population and a good solution would be to compare two pie graphs, one for males and one for females.

Population	tasters	Non-tasters
Male	154	78
Female	165	55

Section 2.2: Representing and Analyzing Data

X-Y or Line graphs

Dependent and independent variables

An x–y graph is a visual representation of the relationship between two variables. Just as words are spelled a certain way, graphs are drawn by following certain rules. In the example graph the temperature of a mixture of ice and water is shown while heat is being added. The experimental or **independent variable** (time) goes on the horizontal or x-axis, and the value that responds, called the **dependent variable** (temperature) goes on the vertical or y-axis. Time is often the independent variable when the graph shows how a measured quantity (the dependent variable) changes over time.

Examples of line (x-y) graphs

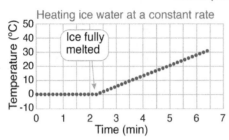
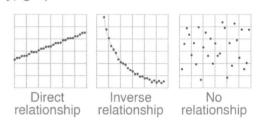

Direct and inverse relationships

Graphs are used to demonstrate the relationship between the variables on the x and y axes. When a change in one variable leads to a proportional change in the other we say there is a direct (linear) relationship. When an increase in one variable leads to a inverse (1/x) change in the other we say there is an inverse relationship. Other simple mathematical relationships that occur in chemistry are exponential (e^x), and logarithmic ($\log x$). A graph might also show that there is no relationship.

Slope and rate

In many experiments the rate at which something changes is an important parameter. In the heating experiment (above) the rate at which temperature increases depends on how quickly heat is added. The **slope** of a graph is equal to the rate of change. A steeper slope indicates a more rapid rate of change. A zero slope (flat) means no change.

Representing uncertainty on a graph

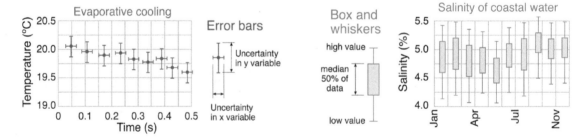

Showing uncertainty on a graph

To correctly interpret the results of data it is necessary to know the size of the uncertainties (error) in measurement. Uncertainties can be shown on a graph in many ways. **Error bars** on each data point are one method used to indicate the error on a graph. The size of the error bars are proportional to the error in each data point and may vary from one point to the next. A **box and whiskers plot** is another way to represent uncertainty when the data sets are too large to plot every point. The box represents the median (middle) fifty percent of the data in each group and the whiskers represent the upper and lower twenty five percent. The whiskers may also represent the largest and smallest measured values.

Slope and rate of change

Proportional relationships are everywhere

When a scientists says two things are in proportion they mean that a change in one thing results in a predictable and linear change in the other. For example if you double the amount of water in a graduated cylinder the height of the water will also double. The height of water in the graduated cylinder is proportional to the volume of water. Proportional relationships are all around you, in recipes, when you buy things, and countless other examples. The greater the slope, the greater the proportion or rate of change.

The straight line graph

Mathematically, a proportional relationship is a straight line on a graph. The rate of change is the slope of the graph. Straight lines have a constant slope - which means a constant rate of change. The example graphs below show the temperature of water as heat is added. The graph on the left increases by 25 °C every minute. The slope of the graph is 25 °C /min. The graph on the right shows a much lower rate of heating, as if the heater were set low. The rate of change for the graph on the right is only 5 °C every minute. Can you see the difference?

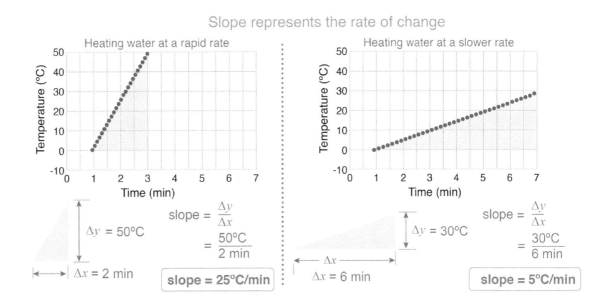

Slope

The slope of a line is defined as the change in the vertical variable divided by the change in the horizontal variable or "rise over run." Mathematically the slope is $\Delta y \div \Delta x$. The slope is the rate of change. The steeper the line, the greater the slope, the higher the rate of change. As long as the scales of the axes are the same, the graph with the steeper slope has the greater rate of change.

Visualizing patterns

Graphs of variables versus time are a visual history of what happened. A curved line indicates a varying rate of change. The amount of curve in a line tells you how fast the proportion is changing. The graph on the right shows the height of a 3-inch plant over 10 days. The plant had no growth (no slope) for the first 2 days, then grew at an increasing rate for four days, then a fairly constant rate for the next four days.

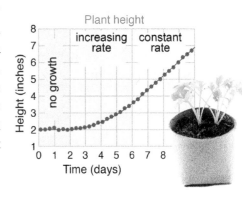

Section 2.2: Representing and Analyzing Data

Regression and curve fitting

Finding the "best" straight line

A theory or hypothesis often predicts a relationship between two variables. How do scientists evaluate whether data support or disprove a relationship? The answer is a form of "moving average" called a regression. The most common form of regression analysis is the **linear regression** in which the "best straight line" calculated for a set of x-y data is used to explain the relationship between them. For example, the ideal gas law predicts that the pressure of a gas increases linearly as the temperature changes. The data below show the results of an experiment to measure the pressure and temperature over a range.

Linear regression

Experiment
Measure pressure of fixed volume of air as temperature changes

Data

Temperature (°C)	Pressure (Pa)
20	111
22	111
25	106
33	112
44	117
47	122
59	123
70	128

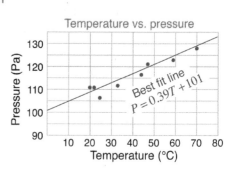

Temperature vs. pressure

Best fit line $P = 0.39T + 101$

Finding the best fit

The blue line represents the best straight line that fits the data. In a regression analysis, "best" means the unique line that results in the smallest average difference between the line and each data point. It is easy to "eyeball fit" the best straight line with a ruler but, most data software has a function to calculate the best straight line. The diagram shows the calculation on a spreadsheet. The averages of x and y have a line on top and the Greek letter sigma, Σ, which means "the sum of" so Σx^2 means "sum all the values of x^2."

Calculating the best fit straight line

x	y	$x-\bar{x}$	$(x-\bar{x})^2$	$y-\bar{y}$	$(x-\bar{x})(y-\bar{y})$
20	111	-20	400	-5.3	105.0
22	111	-18	324	-5.3	94.5
25	106	-15	225	-10.3	153.8
33	112	-7	49	-4.3	29.8
44	117	4	16	0.8	3.0
47	122	7	49	5.8	40.3
59	123	19	361	6.8	128.3
70	128	30	900	11.8	352.5
$\bar{x}$	$\bar{y}$		$\Sigma(x-\bar{x})^2$		$\Sigma(x-\bar{x})(y-\bar{y})$
40	116.3		2,324		907

Equation of a straight line $\quad y = mx + b$

Calculate best-fit slope, m

$$m = \frac{\Sigma(x-\bar{x})(y-\bar{y})}{\Sigma(x-\bar{x})^2} = \frac{907}{2324} = 0.39$$

Calculate best-fit intercept, b

$$b = \bar{y} - m\bar{x} = 116.3 - (0.39)(40) = 100.7$$

The coefficient of regression, r

The "goodness" of the fit is described by the **correlation coefficient (r)**, which varies from +1 to -1. A value of r = +1 indicates a perfect straight line with a positive slope (strong direct x-y relationship). A value of r = -1 describes a perfect straight line with a negative slope (strong inverse relationship). As r gets closer to zero, the linear relationship weakens. When r is equal to 0, the data do not have a linear relationship. The diagram below shows the patterns associated with different values for the correlation coefficient.

the correlation coefficient, r

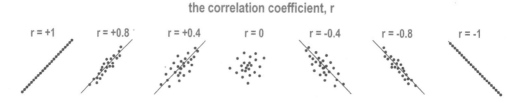

Using data to draw conclusions

The value of conclusions

A conclusion is a statement that is based on evidence and which describes the results or outcome of an experiment. The conclusion often states whether the data support or do not support a hypothesis or theory. For example, Lavoisier's data on masses of burned materials did not support the phlogiston theory but did support the oxygen theory of combustion.

Error and different results

A conclusion based on data cannot be made without considering the error. In 1811 Amadeo Avogadro proposed that a given volume of gas at the same temperature and pressure contained the same number of particles regardless of what kind of atoms or molecules they were. This was a very difficult hypothesis to test and it was not until 1909 that French physicist Jean Perrin made a reliable determination of Avogadro's number.

A hypothesis, the data, and conclusion?

Hypothesis: 22.4 liters of gas at room temperature and pressure contain 6.02×10^{23} molecules.

Data: 22.4 liters of air at room temperature and pressure are determined to contain 5.5×10^{23} +/- 0.7×10^{23} molecules.

Conclusion: Is the hypothesis supported or not?

Significant differences

The difference between the prediction and the data is 5.2×10^{22} which is less than the measurement uncertainty of 7×10^{22}. Therefore the data support the hypothesis even though the precise numbers are not the same. Most experimental conclusions use similar reasoning.

Conclusions about relationships

Conclusions involving relationships (not just numbers) require data that show the relationship agrees with the data to within the limits of error. When graphing data it is important not to "connect the dots!" Data are discrete measurements. Curve fits and theoretical relationships are the only elements that should show as a solid line because they give a value for every point.

Drawing conclusions about relationships

Experiment
Measure pressure of fixed volume of air as temperature changes

Theory
$$P = \frac{nR}{V}(T + 273)$$
$$P = 0.371\,T + 101.3$$

Data
$$P = 0.39\,T + 101$$
$$R = 0.95$$

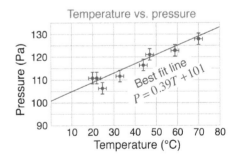

Causal relationships

The graph above shows as temperature increases, so does pressure. But it does not explain that an increase in temperature causes an increase in pressure. The graph on the right shows as cheese consumption increases, so do car thefts. Eating cheese does not cause car thefts! Sometimes data that appear to be related are not at all related. Graphs tell you the direction in which two variables co-occur, but they do not explain whether a change in one variable causes a change in the other.

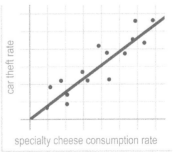

car theft vs. cheese consumption correlation

Section 2.2: Representing and Analyzing Data

Chapter 2

Section 2 review

Because our measurements always have some uncertainty or error, it is useful to take the average of our measurements. By assuming our average measurements are the true value for experiments we can see how large the level of error is based on how far individual measurements vary from the average. We can often draw a conclusion from the data of our experiment when the results are significant or substantially larger than our amount of error. Data can be presented in pie charts, bar graphs, frequency tables, and in graphs. Graphs allow us to note direct or inverse relationships and determine the rate of change. Regression analysis can be used to find the best fit line and determine the coefficient of regression.

Vocabulary words	error, average, standard deviation, standard error, pie graph, bar graph, frequency table, independent variable, dependent variable, slope, error bars, box and whiskers plot, linear regression, correlation coefficient (r)

Review problems and questions

1. The image to the right represents two data sets: A and B. Assume both data sets contain the same number of measurements.
   a. Which data set has a larger standard deviation, A or B? Explain how you arrived at your answer.
   b. Which data set has greater precision, A or B? Explain your answer.

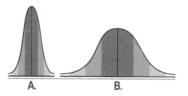

2. A student sets up an experiment to test how humidity affects the speed of mold growing on bread. What is the dependent variable in this experiment, and which axis would it appear on in a line graph once data is collected?

3. Which data set shown on the line graph has a greater rate of change, line 1 or line 2? How can you tell?

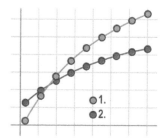

4. Which error bar shows greater uncertainty in the value measured on the y-axis: A or B? Explain your answer.

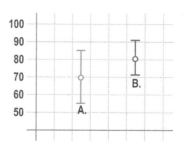

The Value of Measurements

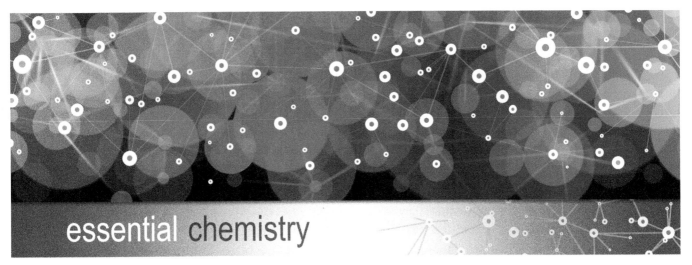

essential chemistry

As you begin your journey through chemistry, you will hopefully be inspired to ask questions, design experiments, collect data, and analyze results. While performing the experiments, you will make observations and measurements. In fact, making measurements are so vital to the scientific process that Nobel Prize winner Max Planck wrote, "An experiment is a question which science poses to nature and a measurement is the recording of nature's answer."

Scientists are always pushing the boundaries of precision and accuracy as they look for nature's answers. For example, scientists use scanning tunneling microscopes (STMs) to visualize and manipulate individual atoms on the 0.1 nanometer scale. That is a tenth of a billionth of a meter! Researchers have become so skilled that they were able to make a stop motion movie by taking pictures and moving atoms with a STM frame by frame.

simulated image of atoms from an STM

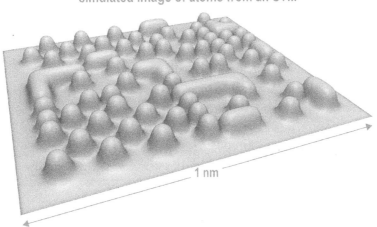

1 nm

(a human hair would be over 100,000 times more wide than this image!)

In science not only do we want to see atoms, but we want to know how they behave. The breaking and forming of bonds during reactions happens very, very quickly. In the late 1980s and 1990s Ahmed Zewail developed a technique of pulsing lasers that are a few femtoseconds (10^{-15} seconds) apart. To put that in perspective, 1 fs is to 1 second as 1 second is to 32 million years! As lasers pulse every few femtoseconds, images are taken, and atoms in motion can be studied. Observations on the femtosecond time-scale range anywhere from studying reactions that take place in the upper atmosphere to understanding the molecular basis of vision. For his contributions to science, Ahmed Zewail is considered the father of femtochemistry.

Section 2.3: Connections

The Value of Measurements - 2

Water quality is another set of measurable environmental factors that can have a very big impact on health. The source of your water has a lot to do with its initial quality. Generally surface water, like a river or stream, is more corrosive than groundwater. If the source water is too corrosive, then it can leach even more harmful chemicals, including lead and copper, from the pipes in your house. These chemicals can make the water look brown and cause some serious health issues. Surface water can also contain more particles, microorganisms, organic matter, and taste-and-odor-causing compounds than groundwater. This is where measurements come in! Scientists test and measure the water's contents, and engineers can design systems that will make the water safe for you and your community.

Measurements are taken to study the health of the environment, and doctors make measurements to monitor your health. When you go to the doctor, they will routinely check your height and weight to see if you are growing. They can also order blood tests to check your kidneys and liver, and to make sure your levels of sugar, protein and electrolytes are balanced. Glucose, a sugar, is a major source of energy for your body. Typical glucose levels should be in the 70-100 mg/dL range. Milligrams are thousandths of a gram. Deciliters are a tenth of a liter. One hundred mg/dL may seem like a small amount, but glucose measurements higher than this are indicative of diabetes.

Of course, measurements are not just used by people in white coats like scientists, doctors, and engineers. Just as chemistry is essential, measurements are essential to your everyday life. When you go to the store and look at a food label, you see nutritional measurements of Calories, milligrams of sodium and percent carbohydrates. When you are looking for a new smartphone or tablet, there are a lot of details about the size, weight and storage capacity of the device. Storage capacity is measured in GB (Gigabyte). In SI units, Giga is a prefix meaning a billion so a Gigabyte is approximately 10^9 bytes of information.

From the smallest atomic scales to the quality of your air and water, from your health to the products you use every day, measurements provide quantitative information about your world.

Chapter 2 review

Vocabulary
Match each word to the sentence where it best fits.

Section 2.1

accuracy	conversion factor
dimensional analysis	exponent
mantissa	precision
resolution	scientific notation
significant figures	

1. A(n) _____ is a ratio of two different units that has a value of 1 even though numbers are different.
2. _____ is a technique of using ratios that have a value of one to change the units of a quantity by cancelling out units.
3. The decimal number that multiplies the power of 10 in scientific notation is called the _____.
4. The _____ of a measurement describes how close the measured value is to the known value.
5. It is often more convenient to express very large or very small numbers using _____.
6. The _____ is the power of 10 for a number written in scientific notation.
7. A volume measured to be 15.0 mL has more _____ than the same volume measured to be 15 mL.
8. Using _____ is a way of recording data that tells the reader how precise a measurement is.
9. The _____ of an instrument is the smallest measured difference that can be sensed or displayed.

Section 2.2

average	bar graph
box and whiskers plot	correlation coefficient (r)
dependent variable	error
error bars	frequency table
independent variable	linear regression
pie graph	slope
standard deviation	standard error

10. The closer the _____ value is to zero, the weaker the relationship between x-y values; the closer to + or - 1, the stronger the relationship.
11. A best fit line is calculated with a mathematical technique called _____.
12. The _____ is the measured quantity that responds to a change in another quantity.
13. To minimize measurement _____, the data should be accurate and precise.
14. The _____ is the quantity in an experiment that is changed by design to see how another quantity responds, and is plotted on the x-axis.
15. In a graph, the _____ tells you the rate of change as a y-to-x value ratio.

16. The _____ is calculated by adding all values and dividing by the number of values.
17. The difference between the mean and the spread of the data in a graph is called the _____.
18. A(n) _____ is a good way to show the relative proportions of individual categories in relation to a whole.
19. A(n) _____ is a good choice to represent small sets of data because the size of each data point is represented with height.
20. _____ can be placed on each data point to indicate the error of the point.
21. A(n) _____ is a way to represent uncertainty when the data sets are too large to plot every point.
22. A(n) _____ combines many points into a single bin to help smooth out variations in data.
23. Error in the average of a set of measurements is called _____.

Conceptual questions

Section 2.1

24. What is the best measurement for the graduated cylinder in the figure above? Explain why.

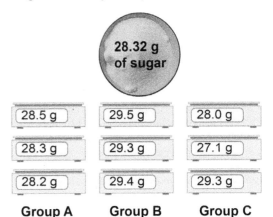

25. All real measurements are limited by the precision of the techniques and equipment used. In the experiment above, three groups of students make measurements of the same quantity of sugar. You will need to calculate the averages to answer part of the problem.

   a. Which group has the most precise measurement? Explain.
   b. Which group has the most accurate measurement? Explain.
   c. Is group C more or less accurate than group B? Explain.

Chapter 2 review

Section 2.1

26. How would you determine how many dozens of eggs are represented by 46 individual eggs. (1 dozen=12 eggs)

27. Which is a cheaper way to buy orange juice: one gallon for $3.00 or one liter for $3.00? Explain.

28. Why are there specific rules to record the number of significant figures a measurement has?

Section 2.2

29. Each item below represents the shape of a line on a line graph. What does each line shape tell you about the rate of change?
    a. Flat line
    b. Straight, but not flat line
    c. Curved line

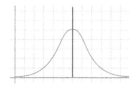

30. What does the vertical line through the highest part of the curve shown above represent?

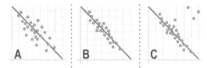

31. Use the data sets pictured above to answer the following questions:
    a. Which data set shows the least variation?
    b. Which data set shows the closest relationship among the data?
    c. Which data set has the most outliers?
    d. What is the blue line through the data called?
    e. What is the purpose of the blue line going through each data set?
    f. All three data sets show the blue line pointing in the same direction. Does that mean there is a positive, negative, or no relationship among the data?
    g. Based on the direction of the blue line, is there a direct, inverse, or no relationship among the data?

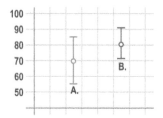

32. According to the graph above, are A and B significantly different? Why or why not?

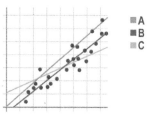

33. Use the graph above to answer the following:
    a. Identify the most appropriate best fit line (A, B or C) for the data shown in the graph above. Defend your choice.
    b. Does this graph show a positive slope or negative slope?
    c. What does the direction of the slope in this graph indicate about the relationship between the x- and y-variables?

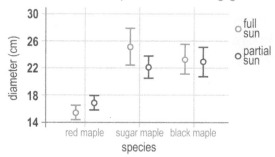

34. According to the graph above, are the diameters of red maple and black maple trees significantly different? Explain your answer.

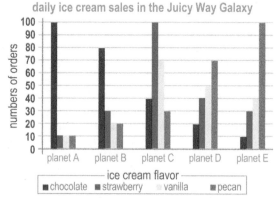

35. Aliens in the fictional Juicy Way galaxy have different ice cream preferences, as you can see on the graph above.
    a. What type of graph is used to represent ice cream sales on planets in the Juicy Way galaxy?
    b. If the data was in percentages, what type of graph would be more appropriate?
    c. Which planet purchases the most pecan ice cream?
    d. What information can you get from the graph that tells you why pecan ice cream has the best sales on that planet?

36. Explain how taking an average of multiple measurements helps reduce error.

Chapter 2 review

Section 2.2

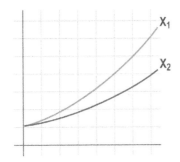

37. Which run shown in the graph above (X_1 or X_2) changes at a faster rate? How can you tell?

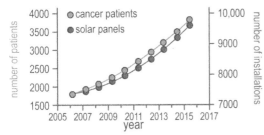

38. Your uncle posts a claim he found the cause of cancer on a social media website, and presents the graph above as evidence. Write a polite reply to his post that addresses your concerns about his claim.

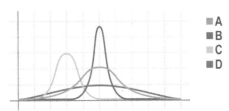

39. Assume all data sets (A, B, C and D) have the same amount of measurements but are measuring different quantities.

    a. Which data line would have the smallest standard deviation? Explain your answer.
    b. Which data set has the most random variation? Defend your answer.
    c. Which data set is more precise, A or C? How can you tell?

40. According to the table below, which student (W, X, Y or Z) would you consider as the outlier for time spent studying for chemistry quizzes?

Number of Hours Studying

Student:	W	X	Y	Z
Quiz #1	1.5	2.1	0	2.0
Quiz #2	2.0	2.7	0	1.8
Quiz #3	2.2	3.4	0.5	2.2
Quiz #4	2.5	3.9	0	2.9
Average:	2.1	3.0	0.1	2.2

41. Use the graphs above to answer the following questions.

    a. Which graph shows no relationship between the x and y axes? Explain your answer.
    b. Which graph shows an inverse relationship the between the x and y axes? Explain your answer.
    c. Which graph shows a direct relationship between the data of the x and y axes? Explain your answer.

42. A college student from the nearest university visits your school to collect data for her research project. She wants to know how much time on average a chemistry student spends answering homework problems. Suggest a dependent and independent variable for her to measure so she can complete her research.

Quantitative problems

Section 2.1

43. Write each of the following numbers in scientific notation.

    a. 650,000
    b. 1645
    c. 0.89
    d. 0.0000059

44. Write out the following numbers as decimal numbers.

    a. 8.6×10^{-2}
    b. 5.55×10^{-9}
    c. 4.7×10^4

45. Calculate the following. Be sure to report the correct number of significant figures.

    a. 9.65 meters × 0.35695 meters
    b. 9.8 meters ÷ 6.5 meters
    c. 8.90 meters × 0.5 meters

46. How many significant figures are in each of the following:

    a. 20. divers
    b. 11,300 kilograms
    c. 1.02340 grams
    d. 8.00 liters
    e. 0.86 N

47. Perform the following conversions:

    a. 3.50 quarts (qt) to mL
    b. 3.0 m to feet
    c. 65 km to inches
    d. 25 minutes to hours

Chapter 2 review

Section 2.1

48. Calculate the following to the appropriate number of significant figures.

    a. 36.4 liters - 5.21 liters + 14.3 liters
    b. 3.48 grams + 8.19 grams
    c. 2.61 meters + 15.68 meters

49. Prior to 1982 a penny was 95% copper. After 1982 the composition was changed to be mostly zinc with a small amount of copper plated on the surface. The official mass of a penny is 2.500 grams. Calculate the average density given the composition of 97.5% zinc and 2.5% copper. You will need to research the densities of copper and zinc.

50. An iceberg has a volume of 783 m^3. If the density of ice = 0.917 g/cm^3, how much mass does the iceberg have in kg?

51. A mineralogist receives three samples of yellow metal to identify. The easiest numerical test is to measure the density. The density of gold is 19.3 g/mL, brass is 8.73 g/mL and iron pyrite is 5.01 g/mL. Identify the samples from the data below.

    Mass and Volume Data

Sample	Mass	Volume
A	133 g	6.9 mL
B	206 g	41.2 mL
C	197 g	22.6 mL

52. A golden coin is measured to have a mass of 15.67 g, a radius of 2.27 cm, and a thickness of 0.514 cm. If the density of gold is 19.32 g/cm^3, is the coin made of pure gold? (Hint: $V = \pi r^2 h$)

53. If 21.0 grams of titanium (Ti) and 35.0 grams of zinc (Zn) are added to a graduated cylinder which contains 80.1 mL of water, where will the level of water be? (Density of Ti = 4.51 g/cm^3 and Density of Zn = 7.14 g/cm^3)

54. If a child weighs 98.5 lb, what is their mass in kilograms?
    (On Earth : 1 kg = 2.205 pound)

55. A bar of aluminum has a volume of 4.0 mL, and a mass of 10.78 g.

    a. Calculate its density.
    b. If 89.56 grams of aluminum are added to a graduated cylinder containing 60.0 mL of water, what will the level of water be in the cylinder?

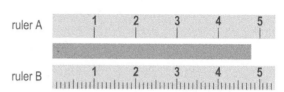

56. The rulers above are measured in centimeters.

    a. According to each ruler, what is the length of the object?
    b. How many significant figures are in ruler B's measured length?
    c. Which ruler is more precise?

57. A paperback book has what volume if it is 19 cm tall, 13 cm wide and 3.8 cm thick?

58. How many liters are in a 12 oz. can? (1 oz = 29.57 mL)

59. The density of silver is 10.5 g/mL.

    a. What volume would 3.47 g of silver occupy?
    b. What is the mass of a 24.6 cm^3 piece of silver?

60. An iceberg is floating in the ocean. A scientist measures the volume of the iceberg that is above the water surface and finds it to be 75 m^3. If 91.0% of the iceberg is underwater what is the total mass in kilograms of the iceberg? (density of ice = 0.917 g/cm^3)

61. Which number is smaller: 0.001 or 1 x 10^{-4}?

62. ⟪ If a recycling center collects 2558 aluminum cans and there are 22 aluminum cans in one pound, what volume in liters of aluminum was collected? (Density of aluminum = 2.70 g/mL.)

63. What is the volume of the liquid in the graduated cylinder shown above?

64. ⟪ How many aspirin tablets can be made from 100.0 grams of aspirin if each tablet contains 5.00 grains of aspirin? (7.00 x 10^3 grains = 1.00 lb; 1 kg = 2.20 lb).

Section 2.2

65. Research data for social media use by age group. Choose any type of graph or chart presented in this chapter. Display your findings in the graph you choose, then defend your graph choice.

66. ⟪ The following data are taken using different techniques to measure the mass of an object.

    a. Calculate the average mass and compare the three groups.
    b. Calculate the standard deviation and compare the three groups.

Group A	10.11 g	9.74 g	10.01 g	9.82 g	10.31 g
Group B	9.66 g	10.95 g	9.45 g	8.71 g	11.22 g
Group C	9.98 g	10.05 g	9.98 g	9.96 g	10.02

Chapter 2 review

Section 2.2

67. The data below are for a density experiment. Use the information to answer the following questions.

Mass vs. Volume

Trial	Volume (mL)	Mass (g)
1	55.34	25.23
2	56.38	26.53
3	55.86	26.06
4	52.01	26.52
5	54.99	24.23
6	57.01	27.22
Mean	55.27	25.97
Standard Deviation	1.7503	1.0735

a. Use a graphing program or calculator to graph the relationship between mass and volume.
b. Describe how the variables are related.
c. Which run is the greatest outlier?
d. What would happen to the standard deviation for each column with the outlier run removed?
e. What would happen to the mean for each column with the outlier run removed?
f. Describe why it is useful to identify and remove outliers.

68. ❮❮ The following data have an identical average and standard deviation. Graph them and discuss why a graph provides more information than statistics such as the standard deviation and average. These data sets were discovered in 1973 by the statistician Francis Anscombe.

Anscombe Quartet

x	y_1	y_2	y_3	y_4
10	8.04	9.14	7.46	6.58
8	6.95	8.24	6.77	5.76
13	7.58	8.74	12.74	7.71
9	8.81	8.77	7.11	8.84
11	8.33	9.26	7.81	8.47
14	9.96	8.1	8.84	7.04
6	7.24	6.13	6.08	5.25
4	4.26	3.1	5.39	12.5
12	10.84	9.13	8.15	5.56
7	4.82	7.26	6.42	7.91
5	5.68	4.74	5.73	6.89
Average	7.50	7.51	7.50	7.50
Standard Dev.	1.94	1.94	1.94	1.94

69. ❮❮ The data below show the average, high, and low temperatures for Palm Springs, California over two different ten-year periods. Use a spreadsheet, or Sparkvue to calculate the following for each of the two ten-year periods.

a. the ten-year mean
b. the standard deviation
c. the standard error for the ten year period.

After doing the calculations discuss whether the two ten-year periods are statistically the same or different and why you think so.

1975 - 1984

Year	Average (°F)	High (°F)	Low (°F)
1975	90	97	82
1976	82	96	69
1977	86	98	73
1978	78	91	66
1979	82	98	66
1980	72	82	63
1981	86	99	72
1982	76	88	64
1983	80	90	70
1984	90	106	75

2004 - 2015

Year	Average (°F)	High (°F)	Low (°F)
2005	84	97	70
2006	92	107	78
2007	85	102	68
2008	84	101	66
2009	84	99	68
2010	82	97	67
2011	77	90	64
2012	96	114	78
2013	93	109	77
2014	88	106	70

chapter 3 Classifying Matter

"What's the matter?" is a question you might hear when someone wants to know if everything is ok. But in chemistry, we can literally ask "What is the matter?" Matter can describe the "stuff" that makes up everything in the physical world – from the rocks and trees outside, to the air you breathe, and even you!

What makes up matter and how can the matter around you be so diverse? Normal matter is made up of atoms. Those atoms can be mixed, combined with, and rearranged to form all the matter around you. For example, carbon can be found as a pure substance in diamonds, graphite, or even nanotubes. When combined in the right conditions, carbon can form carbon dioxide in the air, carbonates that make up sea shells, plastics that hold your beverages, and even amino acids that make up you!

carbon can be found in many things, including you

Of course, the universe would be a pretty boring place if all the matter stayed exactly the same. Thankfully, this is not the case. Matter is constantly changing as energy is being added to or released from the matter. Sometimes it just changes state or shape as atoms are rearranged. This happens when water boils, or plastic is stretched. Other times, the matter is fundamentally changed as atoms chemically combine with different atoms. This happens when fuel is burned, or plants undergo photosynthesis.

So, as you look around the world with a chemist's eye, you should think to yourself, "What's the matter … and what's happening to it?"

Chapter 3 study guide

Chapter Preview

Atoms exist on the atomic scale, smaller than we have tools to see. They are classified into elements with distinct properties based on their number of protons or Atomic Number. The Periodic Table organizes all known elements based on their properties and relationships with other elements. Elements can combine to form compounds through the formation of chemical bonds. The chemical formula notes the number and type of elements in a compound. Pure substances are made up of one type of atom or molecule and cannot be separated. Mixtures contain more than one pure substance as either a homogenous or heterogeneous mixture. Substances may undergo physical changes that change physical properties or chemical changes that change the chemical composition of that substance.

Learning objectives

By the end of this chapter you should be able to:

- describe the scale of an atom;
- identify basic ways elements are organized on the periodic table;
- distinguish among atoms, elements, and compounds;
- identify parts of a chemical formula;
- compare the kinds of information different types of molecular models provide;
- distinguish between pure substances and mixtures; and
- use evidence to distinguish between physical and chemical changes.

Investigations

3A: Chemical formula

3B: Pure substances and mixtures

3C: Physical or chemical change

Pages in this chapter

60	Atoms and the Periodic Table
61	The microstructure of matter
62	The elements
63	The periodic table
64	Information on the periodic table
65	Elements and life on Earth
66	Macronutrients and trace elements
67	Section 1 Review
68	Compounds and Molecules
69	Chemical formulas
70	Representing molecules
71	Isomerism
72	Section 2 Review
73	Pure Substances and Mixtures
74	Homogeneous mixtures
75	Heterogeneous mixtures
76	Tap water vs. pure water
77	Section 3 Review
78	Physical and chemical changes
79	Thinking about physical changes
80	Chemical changes are not easily reversed
81	Chemical properties and chemical change
82	Clues for chemical change
83	Signs of a chemical change
84	Energy and change
85	Section 4 Review
86	Magical Chemical Reactions
87	Magical Chemical Reactions - 2
88	Chapter review

Vocabulary

scale	element	atomic number (Z)
element symbol	periodic table	atomic mass
period	group	trace amount
macronutrients	trace elements	compound
chemical bond	chemical formula	isomers
pure substance	mixture	homogeneous mixture
solution	heterogeneous mixture	tap water
distilled water	distillation	physical property
malleable	brittle	ductile
physical change	chemical change	chemical property
chemical reaction	reactivity	precipitate
intermolecular forces	intramolecular forces	

3.1 - Atoms and the Periodic Table

The variety we see in matter is truly staggering. Within ten meters of you there are examples of matter that is hard, soft, clear, plastic, metal, liquid, solid, gas, hot, cold, alive, ... and the list goes on. How do we explain all the different kinds of matter? After thousands of years of thinking and observing, we know the explanation is that matter is made of combinations of atoms of different elements. Just as 26 letters in the alphabet can spell an enormous number of words, the 92 different elements found on Earth can "spell" an enormous variety of matter.

The question of scale

The concept of scale

To explain the diversity of matter we need the concept of scale. In this context, scale means a typical "size" of detail for a measurement. Consider the three photographs of snow on the right. On a scale of ten meters snow, looks smooth and featureless. But the closer you zoom in, the more detail you see.

10 meters

At a scale of ten centimeters, you can see snow contains individual particles. On a scale of one millimeter, you can see that each particle is a tiny crystal with a characteristic six-fold symmetry. On a scale a million times smaller than a snowflake, we find that ice is made of water molecules in a hexagonal crystal that underlies the six-fold symmetry of the snowflake.

10 centimeters

The description of the macroscopic scale

Observations are on the macroscopic scale when they are large enough for us to see, or measure directly in experiments. In chemistry, the macroscopic scale includes things the size of cells that you need a microscope to see. Light microscopes can enlarge an image around 1,000 times. The size of atoms is far smaller than can be seen with the highest power light microscope.

10 millimeters

The description of the microscopic scale of atoms

In chemistry (and physics) the microscopic scale refers to details that are the size of atoms and smaller. Atoms are so small, a single snowflake contains around 10^{18}, or one billion billion (1,000,000,000,000,000,000) atoms! Matter as small as atoms is also sometimes referred to as the atomic scale. In chemistry we traditionally will refer to the atomic scale as the microscopic scale even though this differs from how biologists would use the term.

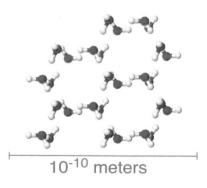
10^{-10} meters

One snowflake =

1,000,000,000,000,000,000 atoms!

The microstructure of matter

Matter is made of particles

All of the matter you experience is made of atoms. Atoms are so small that we experience only the collective average behavior of many trillions of them - completely washing out the effects of individual particles. On the microscopic scale however, every property of matter we experience from the slipperiness of oil to the hardness of steel, ultimately comes from how the individual atoms behave. The following four central ideas define our fundamental understanding of how ordinary matter is made.

The atomic nature of matter

Salt	ice	hydrogen, helium, lithium, beryllium, boron, carbon, nitrogen, oxygen, fluorine, neon ...	water, citric acid, fructose, cellulose, retinoic acid, ascorbic acid
Sodium and chlorine atoms		carbon dioxide / ethyl alcohol	
Matter is composed of very tiny particles (atoms, ions, molecules)	The particles form solids, liquids, and gasses.	92 types of atoms combine to create compounds.	Most matter is a mixture of compounds.

The arrangement of particles in matter

The actual particles of matter might be single atoms but more commonly they are groups of atoms bonded together. For example, a single particle of water (molecule) is two hydrogen atoms bonded with a single oxygen atom. The forms of matter we observe arise from the way the particles of matter are connected to each other.

1. In a solid such as cubic zirconia, oxygen and zirconium atoms are bonded into a crystal which locks the atoms in place.
2. In a liquid, such as water, the particles are still attracted to each other so they stay close together, but individual particles have enough energy that they slide easily around each other and exchange places.
3. In a gas the particles are no longer connected to each other; they move very fast, and individual particles are relatively far away from each other.

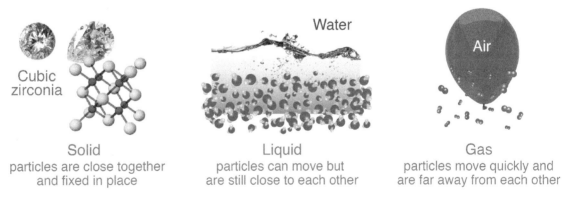

Solid	Liquid	Gas
particles are close together and fixed in place	particles can move but are still close to each other	particles move quickly and are far away from each other

Common matter is a mix

Many common substances are mixtures of solids, liquids, and gases as well as mixtures of compounds and elements. A good example is food. Many foods contain solid, liquid, and gas mixed together. More than a thousand different chemical compounds are in a typical dinner ranging from simple substances such as water to complex substances such as proteins, fats, and carbohydrates.

The elements

Atoms and elements

Atoms are the fundamental building blocks that make up matter. But, not all atoms are the same. We find 92 naturally occurring types of atoms on Earth. Each unique type of atom is an **element**. All atoms of the same element are similar to each other, but differ from atoms of any other element. Oxygen, silicon and calcium are examples of elements. Oxygen atoms have a different structure and properties compared to silicon atoms, calcium atoms, or any other element.

oxygen   silicon   calcium

All atoms of the same element are similar to each other and different from atoms of any other element.

The element symbol and atomic number

Each element has a name and a one or two letter **element symbol**. For example, the element oxygen has the symbol "O" and the element calcium has the symbol "Ca." The element symbol is bold in the table below. Each element also has an **atomic number (Z)** which starts at one for hydrogen, the lightest element, and runs to 92, for uranium, the heaviest naturally occurring element. The elements past atomic number 92 were all created artificially in a laboratory. The atomic number is a whole number and it is unique for each element. The atomic number is the number to the left of the element name in the table below.

Abundance of the elements

The chart below shows the first 100 elements arranged by increasing atomic number along with their abundance in Earth's crust. Notice that the most abundant element in Earth's crust is oxygen! Most of us associate oxygen with the gas that makes up 21% of our atmosphere, but oxygen and silicon make up the bulk of rocks in compounds called silicates such as SiO_2, silicon dioxide, which can form quartz, sand, and glass. After silicon, the most abundant elements in the crust are aluminum, iron, calcium, magnesium, and sodium.

The first 100 elements and their abundance in Earth's crust

atomic # element	symbol	abundance	atomic # element	symbol	abundance	atomic # element	symbol	abundance	atomic # element	symbol	abundance	atomic # element	symbol	abundance	atomic # element	symbol	abundance
1 Hydrogen	H	0.15%	21 Scandium	Sc	<.01%	41 Niobium	Nb	<.01%	61 Promethium	Pm	None	81 Thallium	Tl	<.01%
2 Helium	He	<.01%	22 Titanium	Ti	0.66%	42 Molybdenum	Mo	<.01%	62 Samarium	Sm	<.01%	82 Lead	Pb	<.01%
3 Lithium	Li	<.01%	23 Vanadium	V	0.02%	43 Technetium	Tc	None	63 Europium	Eu	<.01%	83 Bismuth	Bi	<.01%
4 Beryllium	Be	<.01%	24 Chromium	Cr	0.01%	44 Ruthenium	Ru	<.01%	64 Gadolinium	Gd	<.01%	84 Polonium	Po	None
5 Boron	B	<.01%	25 Manganese	Mn	0.11%	45 Rhodium	Rh	<.01%	65 Terbium	Tb	<.01%	85 Astatine	At	None
6 Carbon	C	0.18%	26 Iron	Fe	6.30%	46 Palladium	Pd	<.01%	66 Dysprosium	Dy	<.01%	86 Radon	Rn	None
7 Nitrogen	N	<.01%	27 Cobalt	Co	0.00%	47 Silver	Ag	<.01%	67 Holmium	Ho	<.01%	87 Francium	Fr	None
8 Oxygen	O	46%	28 Nickel	Ni	0.01%	48 Cadmium	Cd	<.01%	68 Erbium	Er	<.01%	88 Radium	Ra	<.01%
9 Fluorine	F	0.05%	29 Copper	Cu	<.01%	49 Indium	In	<.01%	69 Thulium	Tm	<.01%	89 Actinium	Ac	None
10 Neon	Ne	<.01%	30 Zinc	Zn	0.01%	50 Tin	Sn	<.01%	70 Ytterbium	Yb	<.01%	90 Thorium	Th	<.01%
11 Sodium	Na	2.30%	31 Gallium	Ga	<.01%	51 Antimony	Sb	<.01%	71 Lutetium	Lu	<.01%	91 Protactinium	Pa	<.01%
12 Magnesium	Mg	2.90%	32 Germanium	Ge	<.01%	52 Tellurium	Te	<.01%	72 Hafnium	Hf	<.01%	92 Uranium	U	<.01%
13 Aluminum	Al	8.10%	33 Arsenic	As	<.01%	53 Iodine	I	<.01%	73 Tantalum	Ta	<.01%	93 Neptunium	Np	None
14 Silicon	Si	27%	34 Selenium	Se	<.01%	54 Xenon	Xe	<.01%	74 Tungsten	W	<.01%	94 Plutonium	Pu	None
15 Phosphorus	P	0.10%	35 Bromine	Br	<.01%	55 Cesium	Cs	<.01%	75 Rhenium	Re	<.01%	95 Americium	Am	None
16 Sulfur	S	0.04%	36 Krypton	Kr	<.01%	56 Barium	Ba	0.03%	76 Osmium	Os	<.01%	96 Curium	Cm	None
17 Chlorine	Cl	0.02%	37 Rubidium	Rb	0.01%	57 Lanthanum	La	<.01%	77 Iridium	Ir	<.01%	97 Berkelium	Bk	None
18 Argon	Ar	0.00%	38 Strontium	Sr	0.04%	58 Cerium	Ce	0.01%	78 Platinum	Pt	<.01%	98 Californium	Cf	None
19 Potassium	K	1.50%	39 Yttrium	Y	<.01%	59 Praseodymium	Pr	<.01%	79 Gold	Au	<.01%	99 Einsteinium	Es	None
20 Calcium	Ca	5%	40 Zirconium	Zr	0.01%	60 Neodymium	Nd	<.01%	80 Mercury	Hg	<.01%	100 Fermium	Fm	None

The periodic table

The periodic table

The **periodic table** of elements is a chart that lists all the natural and human-made types of elements that are known to exist. The table is organized in a way that lets you predict each element's properties and chemical behaviors. The periodic table is a chemist's most important tool. You can find the full-size version of the periodic table on the inside cover of the printed version of this textbook. There is an interactive periodic table on the toolbar of the electronic version of Essential Chemistry. Early chemists did not have advanced technology to help them distinguish one type of atom from another. They had to rely on their observational skills to recognize patterns. Chemists noticed that oxygen atoms always combined with the metals lithium, sodium, and potassium in a ratio of one oxygen atom to two metal atoms. Lithium, sodium, and potassium were then grouped together. Scientists arranged the periodic table of the elements based on careful observations.

Metals, non-metals, and metalloids

The primary grouping on the periodic table is metals, non-metals, and metalloids. Metals tend to be shiny, dense and solid at room temperature. Some metals like mercury are liquid at or just above room temperature. Metals like copper or zinc are good conductors of heat and electricity. Non-metals include solids like carbon and liquids like bromine. Most nonmetals are gases at room temperature, such as oxygen. Non-metals tend to be good insulators because they are usually poor conductors of heat and electricity. Metalloids like silicon have properties of both metals and nonmetals.

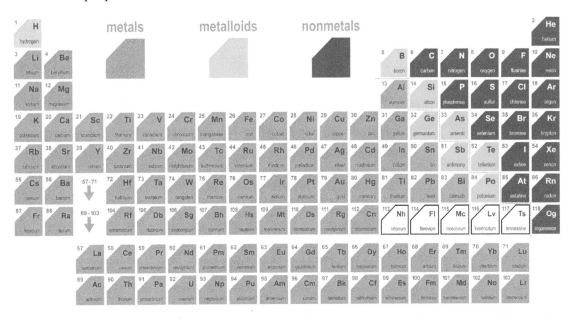

The metal/non-metal line

This periodic table above shows a thin dotted line running though the metalloids. Notice how elements that touch the line are metalloids (except for aluminum, Al). Metals are to the left except for hydrogen, H, and nonmetals are to the right. Remember where this line belongs on the periodic table. Knowing whether an element is a metal, nonmetal, or metalloid is important for a number of concepts in chemistry.

Section 3.1: Atoms and the Periodic Table

Information on the periodic table

Element symbols Each element in the periodic table is represented by a one or two-letter symbol. For example, C is the element symbol for the element carbon and Ca is the symbol for calcium. The first letter in the element symbol is always capitalized and the second letter is always lower-case. In word form, element names do not need to be capitalized. Note that the element symbols do not always correspond to English letters in the element name. Gold has the symbol **Au** which is the first two letters of aurum, the Latin word for gold.

Atomic number Elements are arranged in the periodic table in order of increasing atomic number starting with atomic number 1 for the element hydrogen. The atomic number is the identification number for each element. If the atomic number is 6, the element is carbon. If the atomic number is 20, it is calcium. The atomic number is also the number of protons in the nucleus. Hydrogen has 1 proton, helium has 2 protons, and lithium has 3 protons.

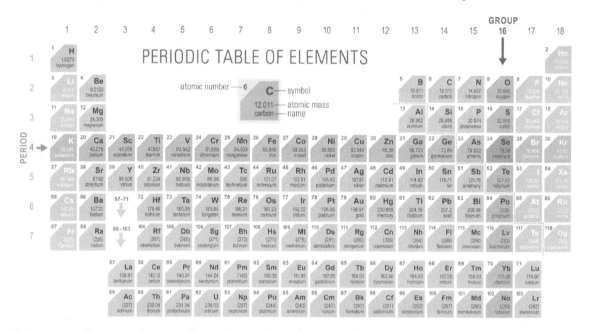

Protons equal electrons In a neutral atom, the number of negative electrons is equal to the number of positive protons. Lithium has three protons and therefore, three electrons. Beryllium has four protons and four electrons. The atomic number of a neutral atom tells you the number of protons and also the number of electrons.

Solved problem How many protons and electrons are in a neutral atom of sulfur, S?

  Relationships According to the periodic table, sulfur's atomic number is 16
  Solve If atomic number is 16, then there are 16 protons. If the atom is neutral, then there are also 16 electrons to cancel the positive charge.
  Answer A neutral atom of sulfur has 16 protons and 16 electrons.

Atomic mass The number beneath each element symbol is the **atomic mass**. Atomic mass gives the average mass of protons and neutrons in the nucleus of an element. Only particles found in the nucleus are included in the atomic mass; the mass of electrons is not included.

Horizontal periods and vertical groups Each horizontal row on the periodic table is called a **period**. Each vertical column is called a **group**. There are 7 periods and 18 groups. The first period has only 2 elements: hydrogen is in group 1, and helium is in group 18. Parts of the 6th and 7th periods are dropped below the main table so they fit on a standard page.

64 Chapter 3: Classifying Matter

Elements and life on Earth

You are made mostly of CHON

Here is an amazing fact: take any living animal or plant on Earth, including any human, and you would find that 96% of the mass of the plant or animal is only four elements! The four elements are hydrogen, carbon, oxygen, and nitrogen. Sometimes abbreviated CHON, these four elements form the majority of all biological molecules from cell membranes to DNA. There is a small, but critically important number of other elements. For example, calcium gives strength to your bones and your red blood cells cannot do their job without iron. In fact, most of the elements on the periodic table exist in your body in a **trace amount**.

Top fifteen elements in the human body

element	% by mass	element	% by mass	element	% by mass
oxygen	61%	phosphorus	1.1%	magnesium	0.027%
carbon	23%	sulfur	0.20%	silicon	0.026%
hydrogen	10%	potassium	0.20%	iron	0.0060%
nitrogen	2.6%	sodium	0.14%	fluorine	0.0037%
calcium	1.4%	chlorine	0.12%	zinc	0.0033%

The human body

The average human body is 60% water - which is made of hydrogen and oxygen. Of the remaining 40%, the largest fraction are proteins. Proteins are very large molecules that make up muscles, hormones, antibodies, skin, and virtually every other part of the body. Proteins are made of simpler building blocks called amino acids. All life on Earth, plant and animal, contains the same 20 amino acids, of which three are shown in the diagram below. Can you see how these molecules are made of mostly carbon (gray), hydrogen (white), oxygen (red), and nitrogen (blue)?

three of 20 amino acids that make up proteins

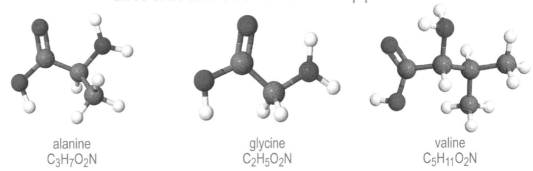

alanine $C_3H_7O_2N$ glycine $C_2H_5O_2N$ valine $C_5H_{11}O_2N$

Why carbon?

The chemistry of living things always involves plenty of carbon because it is the lightest element that can connect to four other atoms. Carbon atoms are also found everywhere and are very stable. Notice that the gray carbon atoms are the backbone of the amino acid molecules in the diagram. Because of carbon's ability to make four connections to other atoms, virtually all the molecules in living organisms, apart from water, are built around carbon. The differences in the estimated 50,000 different proteins in the human body come from the many different ways the carbon atoms are arranged. Can you see the difference in the amino molecules above? Can you see how some parts are the same?

Macronutrients and trace elements

Elements in small amounts

Eleven of the top fifteen elements in the human body are considered **macronutrients**. Potassium, K, and calcium, Ca, are among these eleven. Both elements are used in many body functions, from nerve signaling to bone structure. There are also fifteen **trace elements** which are beneficial in very small amounts but toxic in large quantities. For example, a tiny bit of arsenic, As, is a normal trace element in the body but too much is extremely toxic. You only need about 0.00001 g (1×10^{-5} g) of arsenic each day to be healthy. At higher levels of arsenic exposure, a human may experience short term discomfort similar to food poisoning. Very high levels of exposure to arsenic can cause cancer or death.

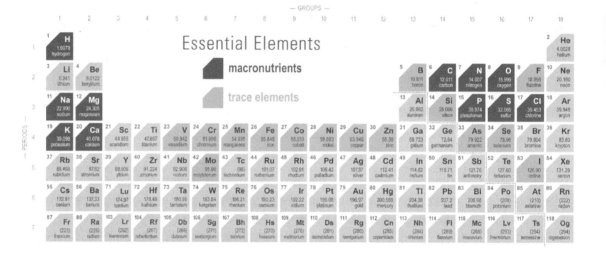

The important role of trace elements

Trace elements often have specialized functions in specific biological molecules. You might be surprised to see iron, Fe, identified as a trace element. Very small amounts of iron are needed in your body to support hundreds of normal body functions such as cell growth, healing, proper functioning of the immune system and transporting oxygen in your blood. Special proteins, called enzymes, depend on trace elements like iron to help them perform the biological functions that keep your body systems working properly. You would not be able to stay alive for very long if your enzymes didn't work. As you probably know, it is important to consume iron-rich foods yet low iron is the most common nutrient deficiency in the United States. How can this be true when popular foods like cereal boast they are "fortified with iron?" Iron deficiency can be avoided by eating a balanced diet that includes fresh, unprocessed foods.

Roles of macro-nutrients

Macronutrients are used to build molecules and structures in your body. For example, you need countless carbon, hydrogen, oxygen, nitrogen, and other atoms to make carbohydrates, fats, DNA, and proteins. You need calcium to build strong teeth and bones. Sodium, magnesium, potassium, and chlorine have important roles such as sending electrical and chemical signals to fire nerves that help you sense your surroundings (shown to the right) and regulating substances that enter and leave your cells.

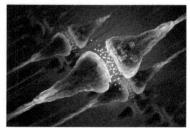

nerves need macronutrients to send and receive signals

Section 1 review

Chapter 3

Atoms exist at the atomic scale meaning that they are too small to be seen with even the most advanced microscopes. Even at the atomic scale atoms are mostly empty space. Neutral atoms have the same number of protons and electrons. The number of protons or atomic number is what distinguishes different elements. The number of protons plus the number of neutrons determines the atomic mass of an element.

The elements are organized into the periodic table, which organizes all the known elements into rows (called periods) and columns (called groups) based on their properties and behaviors. Hydrogen is the most prevalent element in the universe and in our bodies. Living things are primarily made up of a relatively small number of elements along with many trace elements that exist in very small amounts.

Vocabulary words: scale, element, atomic number (Z), element symbol, periodic table, atomic mass, period, group, trace amount, macronutrients, trace elements

Key relationships

Review problems and questions

1. Write the symbol for an element that has 2 protons in its nucleus.

2. Find bromine on the periodic table (atomic number = 35). Is bromine a metal, non-metal, or metalloid?

3. How many protons does a neutral bromine atom have in its nucleus? How many electrons does the atom have?

4. What group is calcium, Ca, in?

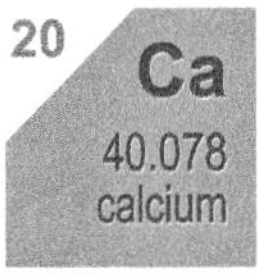

5. An element has 33 protons. What group and period is it in?

Section 3.1: Atoms and the Periodic Table

3.2 - Compounds and Molecules

The element sodium, Na is highly reactive when it is found by itself as a single atom. In fact, it is so chemically reactive that it bursts into flame when it touches water! When a sodium atom is combined with one or more other elements, it becomes much more stable. Table salt is a substance called sodium chloride. Sodium chloride certainly does not burst into flame when it touches water. Elements transform their properties when they combine with other elements to form compounds.

Compounds

Compounds are made of elements

You will rarely come across pure elements in nature. Most matter exists as compounds. A **compound** is a substance that contains atoms of two or more different elements that are chemically bound together. A **chemical bond** is a strong attraction between atoms that hold the elements in a compound together. With a few notable exceptions like helium and platinum, elements quickly form bonds with other elements to make compounds. Almost everything you eat and everything you interact with is a chemical made of a compound or mixture of compounds.

Written chemical formulas

The **chemical formula** of a compound tells you the number of atoms of each element in the compound. The chemical formula for water is H_2O because each particle of water is made of two hydrogen atoms and one oxygen atom. A subscript is a small number that appears at the bottom right of an element symbol. The subscript 2 is what tells you there are two hydrogen atoms in water. When there is no subscript the number 1 is assumed - such as with the oxygen in H_2O. The number 1 nevert written as a subscript in a chemical formula.

Reading a chemical formula

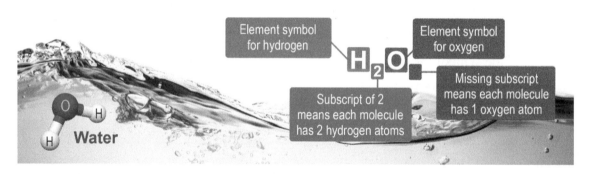

Drawings of formulas

In drawings, atoms are in their elemental or atomic form if they are separated and there are no lines connecting them. If atoms overlap with each other, or they are connected with a solid line, then they are combined as a compound. Overlapping atoms or solid lines represent chemical bonds. As shown below, only the compound H_2O has chemical bonds which are shown as lines that connect atoms and also as overlapping atoms.

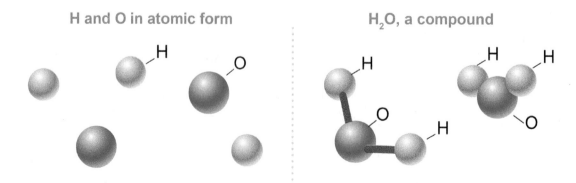

Chemical formulas

All matter is chemical

If you tried to avoid chemicals in the food you eat, it would be impossible because all matter is made from chemicals! Everything that exists has at least one chemical formula listing the elements that make it up, such as the chemical called water, H_2O.

Chemicals are neither "good" nor "bad"

There is no such thing as a good chemical or a bad chemical. Generally, chemicals that provide just enough energy and the right kinds of elements to keep systems functioning are good. But even good chemicals can be unhealthy in large amounts. Unhealthy chemicals are those that provide too much or not enough energy or atoms of a particular kind. Unhealthy chemicals cause harm when they interfere with normal living or nonliving systems.

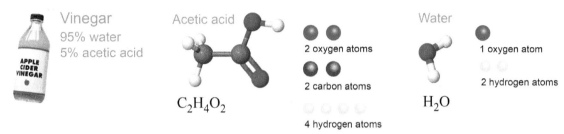

Chemical formulas

Vinegar is a common chemical used to make salad dressing, ketchup, and other foods. The main ingredients of vinegar are 95% water and 5% acetic acid. Both ingredients are compounds. Water has the chemical formula H_2O meaning each particle (molecule) of water has one oxygen (O) atom connected to two hydrogen (H) atoms. Acetic acid has the chemical formula $C_2H_4O_2$ meaning each particle of acetic acid is composed of two carbon (C) atoms, the oxygen (O) atoms and four hydrogen (H) atoms connected in the specific way shown in the diagram above. There are no spaces between the atoms in the formula $C_2H_4O_2$. If you see spaces or extra words, that means there is more than one chemical. For example, H_2O & $C_2H_4O_2$ indicates two chemicals are present: water and acetic acid.

Formulas may hint at structure

CH_3OH is the chemical formula for methanol, or wood alcohol. But why write the formula as CH_3OH instead of writing it as CH_4O? Formulas can be written to reflect how atoms are arranged in three-dimensional space. CH_3OH is written to show that one carbon and three hydrogen atoms are bound together, followed by oxygen and hydrogen.

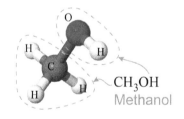

Solved problem

Write a chemical formula for calcium carbonate, commonly called limestone, which has one atom of calcium, one atom of carbon and three atoms of oxygen.

Given 1 calcium atom, Ca; 1 carbon atom, C; 3 oxygen atoms, O
Relationships Add a subscript after the element to tell the number of each element in the compound. The number one does not need to be written.
Answer $CaCO_3$

Solved problem

Count the total number of each type of atom in ethanol, C_2H_5OH.

Given C_2H_5OH
Relationships Subscripts indicate the number of atoms. The number one is implied in the formula.
Answer 2 carbon atoms, C; 6 hydrogen atoms, H; 1 oxygen atom, O

Representing molecules

Structure

The chemical composition (elements) and the structure determine the properties of a compound. In chemistry, "structure" means "how the atoms are connected to each other." In the acetic acid molecule three hydrogen atoms are attached to one carbon atom and the fourth hydrogen atom is attached to an oxygen atom. The simple chemical formula $C_2H_4O_2$ cannot show these connections. Chemists would write the formula for acetic acid as CH_3COOH. This tells you that the atoms are connected in a CH_3 structure attached to a COOH structure. Both are common structures and chemists recognize that the COOH structure means the compound is an organic acid.

Three ways to describe acetic acid

$C_2H_4O_2$ — simple chemical formula

CH_3COOH — functional chemical formula

structural diagram

Limitations of the chemical formula

A chemical formula such as $C_2H_4O_2$ gives information about which atoms are in a molecule but it does not tell you how they are arranged. Writing the formula as CH_3COOH tells you the structure for acetic acid but this strategy does not work for larger molecules.

Structural diagram

A structural diagram is the most simple way to show the bonds between atoms in a molecule. Atoms are identified by their element symbols. Notice the double line between one of the oxygen atoms and carbon atoms in the drawing below. A double line means there are two bonds between those atoms instead of one bond.

Ball and stick model

The ball and stick model helps you see where atoms are placed in a molecule in three dimensional space. The balls represent the centers of the atoms and the sticks represent bonds between atoms. The double bond is represented by two parallel "sticks."

Three dimensional ways to represent the structure of acetic acid

  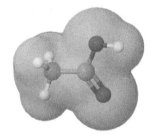

ball and stick model — space filling model — molecular surface model

Space filling model

Atoms are much closer together when they are bonded to each other than shown in a ball and stick model. In fact, atoms overlap when they form bonds. The space filling model gives you an idea of how atoms in a molecule overlap one another. The large surfaces give a better sense of how much space atoms take up compared to the ball and stick model.

Molecular surface model

When you combine the ball and stick model with the space filling model, you get the molecular surface model. The gray shading in the diagram gives you an idea of how the outer surface of the molecule is shaped. Atoms in are often modeled as hard and shiny spheres in molecules, but in reality the outer surface of a molecule is not sharply defined.

Isomerism

Pure elements and compounds

A large fraction of people drink caffeinated beverages such as coffee and tea. Among other biological effects, caffeine stimulates the central nervous system by blocking the action of adenosine, which is linked to feeling drowsy. The caffeine molecule has the chemical formula $C_8H_{10}N_4O_2$. However, the chemical formula really does not tell you what the compound is because there are more than 700 different compounds known that have the exact same chemical formula as caffeine!

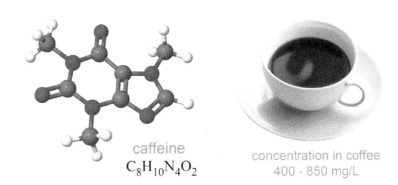

caffeine
$C_8H_{10}N_4O_2$

concentration in coffee
400 - 850 mg/L

concentration in tea
125 - 400 mg/L

Isomers

The biological effects of caffeine come from the way the atoms are arranged in the caffeine molecule. Even if one single oxygen atom trades places with a carbon or hydrogen atom, the result is a new molecule with new properties. The same 24 atoms in caffeine can be rearranged into other compounds. **Isomers** are compounds that have the same chemical formula but different structures. Even though isomers contain the same atoms, they are different compounds and have very different properties.

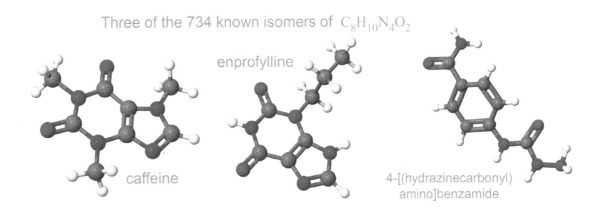

Three of the 734 known isomers of $C_8H_{10}N_4O_2$

caffeine

enprofylline

4-[(hydrazinecarbonyl)amino]benzamide

Isomers are different compounds

Three isomers of $C_8H_{10}N_4O_2$ are shown above. Can you see the structural differences? Enprofylline is a medication that is used to open the air passages in the lungs of asthma patients. Because the chemical structure is different, enprofylline does not interfere with adenosine and therefore does not have the same stimulant effect of caffeine.

Structure determines function

Throughout chemistry we will find that the structure of a molecule determines its function. In this context "function" means both how the compound interacts with other compounds, its chemical properties, and how the compound interacts with the environment, or its physical properties. Different isomers might be solid or liquid at the same temperature. Some might dissolve in water and others may not.

Section 3.2: Compounds and Molecules

Chapter 3
Section 2 review

Pure elements have certain known properties but elements can combine by forming chemical bonds, and the resulting compounds can have very different properties than the elements they are made up of. Compounds have a chemical formula that identifies which atoms constitute the compound. A subscript tells us how many of each type of atom is in the compound or molecule when that number is greater than one. The chemical formula hints at a compound's structure and together they determine that compound's properties. Since the structure of a molecule is important we have developed several ways of representing the arrangement of atoms including the chemical formula, structural models, ball-and-stick models, space filling models, and molecular surface models.

Vocabulary words: compound, chemical bond, chemical formula, isomers

Key relationships

$$C_6H_{12}O_6$$

subscripts indicate the number of each atom present in a formula

Review problems and questions

1. Suppose you read ahead to Chapter 8 and come across a formula that looks like this: $3SO_4^{2-}$. Which number (or numbers) is a subscript?

2. Does methane (CH_4) have chemical bonds? How can you tell?

3. Identify which elements are present in CH_4 and give the number of each type of atom.

4. If you want to know what shape a molecule's surface makes in "real life," which type of model would be best to look at? Explain your choice.

Carbon dioxide (CO_2)

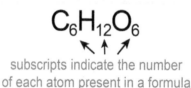

Chemical formula
CO_2

Structural diagram
O—C—O

Ball and stick model

Space-filling model

Molecular surface model

3.3 - Pure Substances and Mixtures

You are surrounded by millions of different kinds of matter. Think about every different element and compound in the air, soil, your body, classroom, and everything else. How do we make sense of such variety?

Categorizing Pure Substances

Substances are elements or compounds

The first step in categorizing types of matter is to search for pure substances. A **pure substance** is made of only one element or compound. Helium gas is a pure substance because it is made of one element, He. Carbon dioxide is a pure substance because it is made of only one compound, CO_2. Vanilla ice cream is not a pure substance because it contains milk, sugar, vanilla flavor, and other substances that can be easily separated. Pure substances cannot be physically separated into different kinds of matter by sorting, filtering, drying, dissolving, heating or cooling. Every particle of a pure substance is identical to its neighbor.

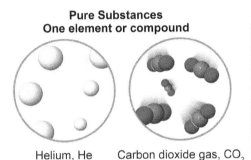

Pure Substances
One element or compound

Helium, He   Carbon dioxide gas, CO_2

Mixtures
Two or more elements or compounds

Mixtures have more than one substance

The word "pure" appears on all kinds of labels such as "pure olive oil". To a chemist, olive oil is not pure because it can be separated into fats, vitamins, pigments and other chemicals. A **mixture** is made of more than one element or compound. Hand sanitizer is a mixture. Many pure substances can be separated from it including water, alcohol, oils, and other chemicals. Chemists always assume pure means made of only one element or compound.

Mixtures can be separated

Mixtures are made of more than one pure substance. They can be physically separated into their components by sorting, filtering, heating or cooling. For example, you could separate chili into its ingredients by using different-sized strainers and filters. Then you could boil the liquid solution left behind to separate water from dissolved solids floating around in the liquid. Separation does not change the properties of each substance. After being separated, you still have the same water, spices, salt, fats, meat and beans. You could easily re-create the chili from the separate substances by mixing them together.

Solved problem

Classify each as either a pure substance or mixture:
a) pudding; b) orange juice; c) mercury; d) water; e) ketchup

Asked Distinguish between pure substances (elements & compounds) and mixtures.

Relationships Pure substances are either elements or compounds. Mixtures are made of two or more kinds of elements or compounds.

Solve Think of what each substance is made of: Pudding is made of milk, flavorings, sugar, etc. Orange juice can be strained and evaporated to dry ingredients. Mercury is found as Hg on the periodic table. Water is H_2O. Ketchup is made of tomato paste, salt, sugar, vinegar, etc.

Answer a, b, and e are mixtures; c and d are pure substances

Homogeneous mixtures

Substances vs. mixtures

Sometimes it isn't easy to tell the difference between a pure substance and a mixture just by looking at it. "Pure" water is almost always chemically a mixture. Small amounts of dissolved minerals are usually present. You need more information about what each sample is made of in order to know whether it is a pure substance or mixture. Think about zooming in to the matter at the molecular scale. How many different kinds of elements or compounds make up the sample of matter? Is there more than one element? Are the elements combined in more than one kind of chemical formula, or are they found in only one compound? Are the components evenly spread out in the sample?

Is this a pure substance or a mixture?

Parts of a mixture retain their identity

A mixture contains two or more elements or compounds that are not chemically bound together. The name gasoline was first used in 1864 to describe a flammable product used to light lanterns. Gasoline is a complex mixture often containing more than 100 different compounds. Refining separates some of the compounds in gasoline by molecular weight and boiling point. Others can be separated by chemical processing. Being a mixture of many compounds, there is no such thing as chemically "pure" gasoline.

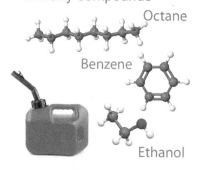
Gasoline is a mixture of many compounds
Octane
Benzene
Ethanol

Solutions are homogeneous mixtures

Gasoline is a **homogeneous mixture** because all of the individual molecules are evenly mixed on the molecular level. For example, there are not "clumps" of benzene floating in a sea of octane. Each different kind of molecule bumps around all the other kinds until all types of molecules are evenly spread out. A homogeneous mixture in which all components are evenly mixed on the molecular level is called a **solution.**

Air is a solution

Solutions can be any phase of matter, including solids and gasses. Air is a gaseous solution of nitrogen, oxygen, argon, water vapor, carbon dioxide, and other trace gasses. Our atmosphere however, is also a mixture because it contains dust particles, pollen, water droplets, and other matter that is "clumped" in aggregates larger than one molecule.

Air is a solution of nitrogen, oxygen, argon, water vapor, and trace gasses.

The atmosphere is a mixture because it contains dust, water droplets, and particles.

14 carat gold is a solid solution of 58.3% gold and 41.7% silver.

Alloys are metal solutions

Gold is a solid solution of silver and gold. If you look at gold at the atomic level, you can see that the atoms are evenly mixed no matter where you look. The "carat" of gold refers to the percentage of the element gold in the solution. 24 carat gold is pure. 14 carat gold is 14 parts out of 24 (58.3%) gold and 10 parts out of 24 (41.7%) silver. Pure, 24 carat gold is too soft a metal even for most jewelry. The addition of silver greatly increases the hardness and durability without detracting from gold's chemical ability to resist tarnishing.

Heterogeneous mixtures

Heterogeneous mixtures have an uneven distribution

A mixture whose components have an uneven distribution is called a heterogeneous mixture. The pure substances that make up heterogeneous mixtures show clumping which may or may not be visible to the naked eye. You can understand why beef stew is a heterogeneous mixture and the different components in pudding stone are equally obvious. But blood is also a heterogeneous mixture even though it looks uniform to the eye because white blood cells, platelets, and other solids are unevenly clumped within liquid plasma. Many solid and liquid heterogeneous mixtures can be easily separated by filtration. For example, you can use a strainer to remove pulp from fresh-squeezed orange juice, and you can use a sifter to remove rocks and pebbles from soil.

Heterogeneous mixtures

Beef stew

Puddingstone

Blood

Flow chart to classify matter

Use the flow chart below to help you classify different types of matter as an element, compound, homogeneous mixture (solution), or heterogeneous mixture.

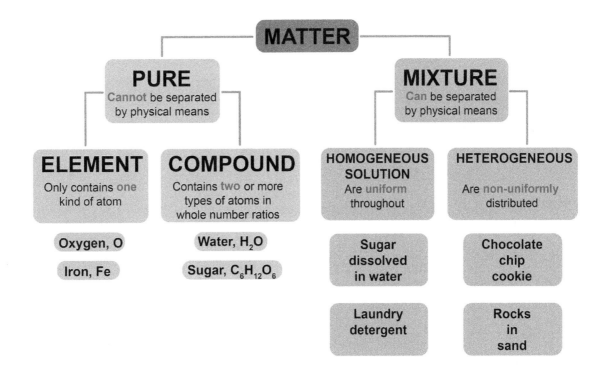

Section 3.3: Pure Substances and Mixtures

Tap water vs. pure water

"Pure" tap water is a solution

Water from a faucet tastes different than bottled water. Why do different sources of water taste different if water is supposed to be a flavorless pure substance with uniform composition? The reason is that "pure" tap water is not pure H_2O in the chemical sense but is actually a solution with many dissolved salts and minerals. Most tap water comes from water supplies that are also treated with chloride to kill bacteria and fluoride to improve dental health.

Drinking water is not pure

You might have wondered why water, a compound that is supposed to be tasteless, has different "flavors" depending on where the water comes from. Minerals that are important to our health such as iron, calcium, and magnesium may be present in "pure" bottled water, spring water and tap water. The ratio of these minerals is what gives drinking water its "flavor." Keep in mind, "pure" bottled water is not "pure" because it contains water with other substances mixed in it.

Distilled water is pure

If you want truly pure water as defined by chemists, you will need distilled water. Distilled water has been boiled in order to separate dissolved impurities from pure water, H_2O. We use distilled or deionized water when we make solutions in the chemistry lab. Dissolved substances like minerals can interfere with the normal behavior of chemicals or systems. For example, when adding antifreeze and tap water to your car, you may introduce chemicals into your car's cooling system that will cause solids to build up and reduce its efficiency.

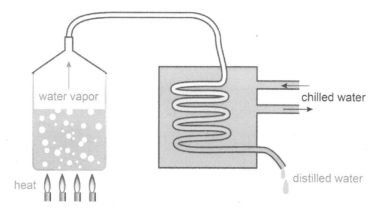

Distillation

Water containing impurities is boiled. The water vapor is pure because the impurities do not boil and stay in the liquid. The vapor passes through a chiller that condenses the vapor back to liquid water again.

Distillation can separate homogeneous mixtures

How can tap water or drinking water be purified? The drawing to the right is a diagram from an old chemistry textbook that shows you how to set up a distillation apparatus for water purification. Modern systems are very similar. To purify water by distillation, water is heated to its boiling point in the flasks. Water as steam rises and moves into the sealed tube while impurities stay behind in the flask. The coiled portion of tubing is surrounded by cool running water. The cool temperature outside the tube causes steam inside the tube to condense into liquid water. The purified water is then collected and stored in a perfectly clean, sealed bottle until it is ready to use.

Section 3 review

Chapter 3

Substances are said to be chemically "pure" when they are made up of either one type of element or one type of compound and cannot be physically separated further. On the other hand, mixtures contain more than one pure substance and can be separated by physical means such as filtering or heating.

A homogeneous mixture or solution is one in which the different atoms and/or molecules are evenly mixed, giving the mixture a uniform distribution. Mixtures with uneven distribution are known as heterogeneous mixtures.

Vocabulary words: pure substance, mixture, homogeneous mixture, solution, heterogeneous mixture, tap water, distilled water, distillation

Key relationships

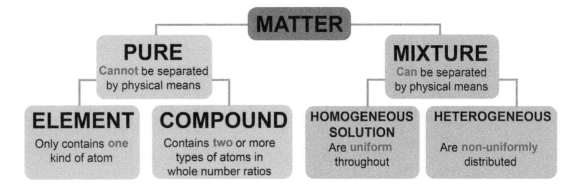

Review problems and questions

1. Classify each of the substances shown below as a pure substance or a mixture. Use the matter classification scheme found in this section for help.

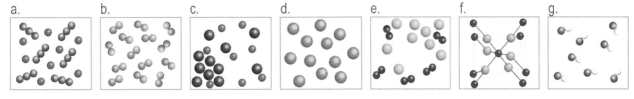

2. Further classify each of the substances above as an element, compound, homogeneous mixture or heterogeneous mixture.

3. Classify each substance below as an element, compound, homogeneous mixture or heterogeneous mixture.
   a. cotton candy
   b. methane, CH_4
   c. rocky road ice cream
   d. carbon
   e. dirt
   f. hand sanitizer
   g. bottled water
   h. salt water (NaCl and H_2O)

3.4 - Physical and chemical changes

Chemistry describes and explains how matter changes. Some changes are reversible; a bent wire can be made straight again. Ice that has been melted can be re-frozen back into ice again. Other changes are not so easy to reverse; iron that is exposed to water rusts and no amount of drying will make the rusty metal revert back to its original shiny state. This section discusses the differences between physical changes such as melting, and chemical changes such as iron turning to rust.

Physical Properties and Changes

Physical properties

A property you can measure with tools or your senses is called a **physical property**. For example, you can see that water is a colorless liquid at room temperature. Color is a physical property. So is temperature. You can measure temperature with a thermometer. The temperature at which steel melts is another physical property. The temperature at which any substance melts, freezes and boils are all physical properties because all can be measured without changing a substance's chemical identity.

Some physical properties: color, volume; temperature

Some physical changes: bending, melting, breaking

Malleability and brittleness

Copper is a relatively soft metal that is **malleable** which means it can be bent into a shape without breaking. The opposite of malleable is **brittle**. Brittle materials like glass break instead of bending. Copper is also **ductile** because it can be stretched into a thin wire without breaking. Malleability, ductility and brittleness are examples of physical properties of solid materials.

A physical change can be reversed

A **physical change** occurs when there is a change in the physical properties of matter. Physical changes can be reversed without changing the chemical identity of a substance. Examples of physical changes include bending, heating, cooling, dissolving, melting, freezing, boiling and breaking. When solid iron melts, it undergoes a physical change from solid iron to liquid iron. This does not change the iron into a new substance because the iron is still iron when it is melted. Physical changes are reversible, which means they can be changed back though physical means (not chemical). For example, changing iron from solid to liquid and back to solid requires only a change in temperature.

Volume is a physical property

Volume is a physical property of matter. When water freezes, the change in volume is a physical change. Water's mass does not change during freezing because you still have the same number of water molecules as before, they are just arranged in a more organized fashion. The change can easily be reversed by melting the water. Some physical changes are easy to observe first-hand. It is easy to see when a phase change from solid to liquid happens. Other physical changes are not as easily seen. You usually cannot directly observe the volume and density of matter change as a result of a temperature change unless there is a phase change.

Thinking about physical changes

Physical changes

How can you tell if a change is a physical change? Fundamentally, if a compound does not turn into different compound, then the change is physical. For example changing the shape of a substance is usually a physical change. Twisting, slicing and deforming are physical changes. It also means crushing or grinding into powder is a physical change. For example, rock salt, table salt and fine grain salt are all made of the same chemical substance. The only difference between them is their different crystal sizes.

Dissolving and mixing do not change identity

Mixing food color into water evenly spreads dye molecules through the water. The dissolved dye molecules have not changed into different molecules. This is evidence that dissolving is a physical change. No matter how thoroughly you mix, creating a mixture is still a physical change. Sometimes mixtures stay dissolved until the water evaporates from them. Other mixtures only last for a short time. For example, you can force a temporary oil-and-water mixture to form with a blender that looks "smooth." Under a microscope however, you still see a heterogeneous mixture of water and oil. The oil droplets become very small, but each droplet still has millions of atoms. Changing the size of the oil droplets is a physical change.

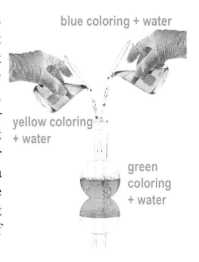

Reversible color changes are physical changes

Color can indicate physical or chemical change. A crayon adds color to paper by shedding a layer of colored wax on the paper's surface. The paper is still paper and the crayon is still crayon. Coloring is a reversible physical change. Removing color from clothes with bleach however, is irreversible so it is not a physical change. Bleaching is a chemical change because it causes a permanent color change. Hair coloring can be either physical or chemical depending on the type of color and the process used.

Bleach is a solution of sodium hypochlorite NaClO

Drying is usually a physical change

Drying is another change that can be physical or chemical. When you dry wet clothes the water is removed from the cloth and this is a physical change. In chemistry, "drying" refers to the purely physical process of removing a particular liquid from a mixture without changing any of the molecules.

When drying is NOT a physical change

Drying paint is different from drying clothing! When paint dries, the paint chemically changes and dry paint does not become liquid again when you add water. This type of drying is not a physical process because it is not easily reversible. For example, some paints combine with oxygen to make new, permanently-dry molecules. This is a chemical change and is not easily reversible.

Chemical changes are not easily reversed

Physical changes

A marshmallow is an edible foam made of a sugar called sucrose and other chemicals. The chemical formula for sucrose is $C_{12}H_{22}O_{11}$. Each sucrose molecule has 12 carbon, 22 hydrogen and 11 oxygen atoms. No matter how hard you squeeze, you cannot change the C:H:O atom ratio in the sucrose molecules. That is why squeezing is a physical change. Squeezing does not rearrange the atoms that make up the substance. Physical changes are easily reversible because the substance does not change its chemical makeup.

Chemical changes

Physical and chemical changes are fundamentally different, but they can occur together. All of the changes that occur when you toast a marshmallow can be classified as either a physical or chemical change. A physical change in a substance does not change what the substance is made of. A chemical reaction occurs during a **chemical change**. Atoms are rearranged to make a new substance. Energy is either given off or absorbed.

before toasting
- white
- solid
- very sweet
- spongy foam texture
- volume: 9.92 cm³
- mass: 8.06 g

after toasting
- brown and white
- gooey
- less sweet
- brittle, crunchy, hard
- volume: 8.71 cm³
- mass: 6.54 g

Chemical and physical changes can occur together

Roasting a marshmallow over fire changes many of its properties. Why? When you heat a marshmallow, a physical change occurs as sugar melts. If the marshmallow begins to toast, something new is created. The flavor, color and smell permanently change on the toasted part because it is chemically different from the non-toasted part. Toasting is a chemical change because it rearranges sugar molecules into different molecules with different C:H:O ratios. When heated enough, atoms in sucrose molecules interact with water molecules in the marshmallow. The result is new sugars with different atom arrangements called fructose and glucose, more water vapor (H_2O), and caramelin ($C_{125}H_{188}O_{80}$). This process is known as caramelization. A chemical change occurs when molecules change their structure. As you can see below, the types of atoms you start with are the same as the atoms you end with. The atoms are just arranged differently.

caramelization in a marshmallow:
sucrose & water molecules form fructose & glucose, which then rearrange into caramelin and water molecules

Matter is conserved in physical and chemical changes

The toasted part of a marshmallow has a different chemical makeup than the non-toasted part. A chemical change starts when heat causes a water molecule to break sucrose apart into fructose and glucose sugars. These sugars then rearrange into a larger molecule called caramelin and water molecules. A toasted marshmallow weighs less than an non-toasted one because some atoms leave the marshmallow when water vapor floats into the air as steam. Matter is conserved as atoms get rearranged into new structures during the chemical change. If you counted the number of C, H and O atoms before and after the chemical change, they would be equal. The picture does not show all of the molecules involved because there is not enough space to show them. Still, physical and chemical changes always obey the law of conservation of matter.

Not easily reversed

The marshmallow undergoes a chemical change. Carbon, hydrogen and oxygen atoms in sucrose molecules combine with water to form caramelin and new water molecules. Caramelin cannot revert to sucrose unless another series of chemical changes occurs.

Chemical properties and chemical change

Chemical properties

A property that can be observed when a substance is transformed into a different substance is called a **chemical property**. When you leave something made of metal outdoors for some time, a chemical reaction causes rust to form on the metal. Metals contain iron atoms. A chemical property of iron is that it combines with oxygen in the air to form rust. Iron reacts with oxygen gas and makes a new substance. Iron, Fe, is a different substance than rust, Fe_2O_3. Rust has different physical and chemical properties than the iron and oxygen that produced it. Matter is conserved because the same number of oxygen and iron atoms are present before and after rusting. Count the number of iron and oxygen atoms on either side of the arrow in the picture to verify matter is conserved during rusting.

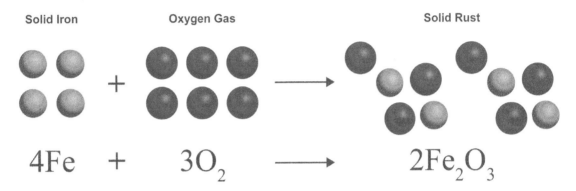

$$4Fe \;+\; 3O_2 \;\longrightarrow\; 2Fe_2O_3$$

Chemical changes

When a substance changes into a different substance, a chemical change has occurred. When iron forms rust, the iron has chemically changed in a way that is not reversible by physical means. For example, rusted iron will not turn shiny again if you take oxygen away. Even if you used sandpaper to remove rust from the nail, you cannot easily turn the rust that is stuck on sandpaper back to solid Fe and O_2 gas. When iron is combined in the new substance, Fe_2O_3, it does not have the same reactivity with oxygen as the pure element Fe does. Fe_2O_3 does not react with oxygen gas. In other words, the chemical properties of elemental Fe are different from the chemical properties of iron when it is in the compound Fe_2O_3. The chemical properties of chemical formula as well as reactivity have both change as a result of a chemical reaction.

We depend on chemical changes

We depend on chemical changes to create useful products or to produce energy. On a cold day, you might burn wood in a fireplace to warm up. Burning is a chemical change that releases energy. Baking relies on chemical change. Dough rises when baking soda contacts an acid and forms carbon dioxide gas bubbles. When meat is cooked, proteins and other substances are changed into new substances that look, taste, and smell completely different. Bleach is a good disinfectant because it changes bacterial cells and kills them. Bleach removes stains because it changes the properties of molecules that produce stains. Plastics that surround you are all artificially created in a factory through a series of chemical changes.

Cooking food and bleaching clothes both produce chemical changes.

Clues for chemical change

Chemical change results in one or more new substances

If you are asked to judge whether a chemical change happened or not, pay close attention to the substances involved in the change. For instance, if you are told water and salt were changed into salt water, no chemical change occurred because the starting substances still have the same chemical identity after the change. Water, H_2O, and salt, $NaCl$, have the same chemical formulas—H_2O and $NaCl$—when they form salt water. When methane, CH_4, combines with oxygen in the air, O_2, carbon dioxide and water are formed. Carbon dioxide is CO_2 and water is H_2O. New substances are formed with a new molecular composition, so a chemical change has occurred.

Identifying chemical change

A chemical change occurs during a **chemical reaction**. A chemical reaction is any process in which one substance changes into a different substance. The ability of a substance to undergo a chemical change and become rearranged into a different substance is called **reactivity**. Reactivity also describes how easily one substance combines with another. But how do you know when one substance changes into another if you can't see molecules, especially when the substances look the same before and after a chemical reaction? The key is to make careful observations during the chemical reaction.

Chemical change is accompanied by evidence

A trained eye can spot clues to distinguish a chemical change from a physical change. Chemical reactions are always accompanied by one or more signs of chemical change. The reaction of clear silver nitrate and solid copper is very visible. The chemical reaction between silver nitrate and copper is written below.

$$2AgNO_3 + Cu \rightarrow Cu(NO_3)_2 + 2Ag$$
$$\text{clear} \quad \text{solid} \quad \text{blue} \quad \text{solid}$$

The liquid changes from colorless to blue. Fuzzy crystals appear in the place where solid copper used to be, and some crystals fall off to the bottom of the flask. The copper does not disappear; it is rearranged. It is now part of the solution, giving it its blue color. The blue solution is copper(II) nitrate and the crystals are solid silver. If you touch the flask, you will notice a temperature change. It is warmer while the reaction is taking place. Once the reaction is finished, the flask returns to room temperature.

Signs of a chemical change

Color change

Color change is a sign of a chemical change when it cannot be easily reversed by ordinary physical means. Permanent hair dye chemically changes your hair, but temporary color washes out because it only coats your hair. A tattoo is a seemingly permanent skin color change, but tattoos only physically change the skin. Tattoo pens work by inserting ink below the surface of the skin. Skin remains skin, and ink remains ink so no chemical change has occurred.

Color changes are evidence of chemical changes

Bubble formation

New bubbles that appear during a reaction that have a new chemical formula signal chemical change. When solid magnesium is added to hydrochloric acid, new hydrogen gas is released and is visible as bubbles:

$$Mg + 2HCl \rightarrow H_2 + MgCl_2$$

Sometimes you see bubbles in a glass of water. These bubbles came from air that dissolved into the water. The air is not newly formed, so this is not a sign of chemical change. Bubbles that you see during boiling are a newly formed gas, however, the gas does not have a different chemical formula. When a substance boils, the "new" gas is actually the same substance in the gas phase. When water boils, the bubbles are made of water vapor, gaseous H_2O. That's why boiling is always a physical change.

Bubbling or foaming can be a sign of chemical change ...

... or just dissolved gas escaping.

Odor change

Change in odor or flavor signals chemical change. Rotten food definitely smells and tastes different than fresh food! Raw food also smells and tastes different than cooked food. Cooking and rotting both produce new substances through chemical changes. When a batch of cookies bakes in an oven, you can easily smell the evidence of chemical change!

Precipitate formation

When clear liquid from the flask contacts the clear liquid in the beaker on the right, the liquid turns yellow. Did you also notice that the yellow substance is cloudy? That's because the cloudy yellow substance is made of tiny solid crystals. If two liquids form a solid on contact, a chemical reaction has taken place. The term **precipitate** refers to a solid that forms from two liquids as a result of a chemical change.

Energy change

A chemical change always involves a change in energy. Some chemical reactions, such as burning release energy, causing temperature to rise. Others absorb energy, such as the reaction in an instant ice pack, causing a temperature drop. Glow sticks and fireworks release energy as light! The color of the light depends on the identity of the chemicals that are reacting.

Energy and change

Energy is needed to change matter

Any change in matter requires an exchange of energy. Think about crushing a can. The more energy you add, the more change you will cause. You can easily put a dent in an aluminum can, but with a little more effort, you can crush it. If you add enough energy to heat the can to 1,221 °F or (660.3 °C), the can will change from solid to liquid. Energy is needed because forces at the molecular level must be overcome to cause change.

It takes energy to crush a can.

Energy is central to chemistry
Interactive simulation

One obvious physical property of matter is its *state* of being solid, liquid, or gas. Every substance has a temperature at which it changes state. For example, when water absorbs energy it turns from solid to liquid (melts) at 0 °C and from liquid to gas (boils) at 100 °C. The interactive simulation allows you to change the temperature and see what happens to the atoms as they transition through the three phases.

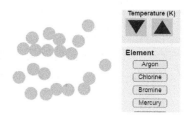

The forces acting between molecules

Chemical and physical changes happen because matter constantly seeks the lowest possible energy arrangement. The forces that hold particles together in solids and liquids are called **intermolecular forces** (IMFs). Substances that are solid at room temperature have slightly stronger IMFs than substances that are liquid. Room temperature gases have very small IMFs. That's why gases can spread out to fit their container. *Physical phase changes involve only intermolecular forces.* When you melt a solid such as chocolate, you are adding enough energy to partially overcome the IMFs and allow the molecules to slide around past each other and become liquid.

Milk chocolate melts between 30°C and 32°C.

The forces acting within molecules

Much stronger **intramolecular forces** called *chemical bonds* hold the atoms together *within* a molecule. Chemical changes that rearrange chemical bonds involve higher energy because intramolecular forces are so much stronger. For example, it takes *more than 100 times* as much energy to separate the hydrogen and oxygen atoms in water molecules than it does to melt ice into liquid.

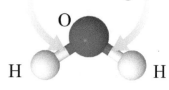
Strong interatomic forces hold molecules together

Intramolecular forces are involved in chemical changes

The graph shows the types of changes that occur in water as you add energy to it. It takes 333 J of energy to melt one gram of solid ice into liquid water at 0 °C. Melting is a physical change. By comparison, it takes 51,000 J of energy to cause a chemical change and break the intramolecular bonds in one gram of water. When you break the chemical bonds in water, you end up with separate hydrogen and oxygen atoms!

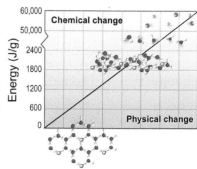

Section 4 review

Chapter 3

Elements and compounds have physical properties that we can observe or measure, such as the phase it is in (solid, liquid or gas), as well as temperature, malleability, ductility, density and more. A physical change is one that changes physical properties but does not change what the substance is. Physical changes include heating or cooling, dissolving, splitting or grinding a substance, and such changes are reversible.

A chemical change is a change in which substances are changed chemically into a different substance. Chemical changes that occur during chemical reactions are not easily reversed. Some signs that a change is chemical rather than physical are a permanent color change, bubble or gas formation, odor production, a precipitate being formed, and energy being absorbed or given off. Physical changes do not affect the bonds that hold molecules together while chemical changes alter those bonds, converting one substance into something completely different.

Vocabulary words: physical property, malleable, brittle, ductile, physical change, chemical change, chemical property, chemical reaction, reactivity, precipitate, intermolecular forces, intramolecular forces

Review problems and questions

1. Classify each property as either physical or chemical.
   a. conducts electricity
   b. reacts
   c. chemical formula
   d. volume
   e. mass
   f. length

2. Classify each change as either a physical or chemical.
   a. rearranged its formula
   b. bent when pushed
   c. broke when pulled
   d. bubbled violently on contact
   e. turned from clear to bright pink
   f. boiled

3. What are the 5 signs of a chemical change you can see or measure in the lab?

4. Each of the following diagrams are marked with at least one red 'x'. For each diagram, determine whether the 'x' is pointing out an intermolecular force or an intramolecular force.

a.
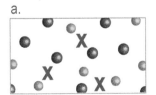

b.
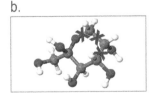

c.
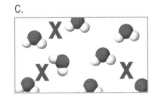

Section 3.4: Physical and Chemical Changes

Magical Chemical Reactions

essential chemistry

Did you ever see a performer turn a magic wand into a bouquet of flowers, or turn a pigeon into a rabbit? Of course this performance is a carefully constructed illusion. But, real transformations of matter happen all the time. These transformations can be explained by science, but they are no less magical!

Would you believe it if I told you that simply by adding some energy to a small sample of matter, you can make its volume expand to over 1000x greater than it started? It sounds like magic, but this is exactly what happens to a sample of water when it is boiled. When you add energy to a cup of water in a pot on the stove, those water molecules now have the ability to escape each other. They can spread out from that cup and fill the entire room, and you don't even have to say abracadabra!

Just like magicians seem to be able to sort and select individual items from a deck of cards, scientists can sort and select individual substances out of a complex mixture. Instead of using sleight of hand, scientists use the physical and chemical properties of the individual components to separate the substance. Distillation is a process by which mixtures are separated by taking advantage of a substance's boiling point. The simplest example of this is distilled water. An impure water-based homogeneous mixture is heated to boiling, and the evaporated water is cooled and condensed into a different container. Voilà. Pure water is pulled from the air, just like an ace is pulled from the middle of a deck of cards!

Molecules can also be separated by size, or by attraction to a particular substance. This is how chromatography works. Have you ever seen paper with black ink that gets wet? The black ink spreads out into a rainbow of colors. This happens because some of the dyes in the ink mixture are very attracted to the water, and other dyes in the ink are very attracted to the paper. The color of the dyes, and attraction of the dyes to water or paper is based on the properties of the molecules that make up the dyes. Chemists can separate those dyes even more by taking advantage of these properties—no sorting hat needed!

Magical Chemical Reactions - 2

What if I told you there was a wand that would attract metal to itself? Does this sound like magic? Nobody put a spell on the wand, it simply has a metal magnet on the end, and that metal is iron. Being strongly attracted to a magnet is one of the physical properties of iron. In fact, the term for being strongly attracted to a magnet is ferromagnetic because ferrum is Latin for iron.

Chemical changes can be even more transformative than physical changes. There is a metal that is so reactive that it creates a fire when placed in water! It sounds like a magic trick, but it can be explained by the chemical properties of the metal and the changes it goes through during a reaction. This metal loses electrons easily and reacts rapidly. If it comes in contact with water, it produces hydrogen, a combustible gas. If there is enough energy in the system, the hydrogen will combust creating a flame around the dancing piece of metal on the surface of the water. This metal is sodium, atomic number 11 on the periodic table. In the laboratory it is stored in a glass jar and covered with oil. The oil coats the surface of the metal so that it cannot react with any moisture in the air.

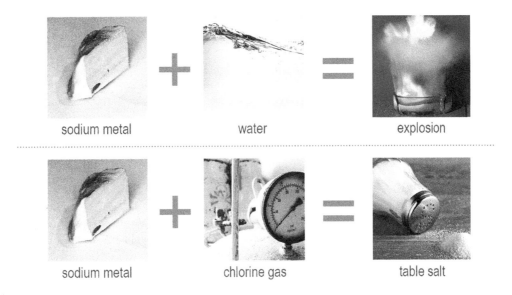

sodium metal + water = explosion

sodium metal + chlorine gas = table salt

Sodium metal can react with other substances besides water. There is a yellow greenish gas that is toxic to ingest and hazardous to both animals and humans. This gas is chlorine, atomic number 17, and it is so dangerous that is has been used as a weapon in war! Do you know what happens if reactive sodium metal comes into contact with this dangerous gas? Table salt, technically called sodium chloride, is "magically" produced! When combined, these two toxic substances react to form something that is not only safe, but is also necessary. The body needs sodium chloride for maintaining the correct concentrations of fluids in your tissues, transmitting nerve signals, and helping muscles contract.

Matter is amazingly diverse, and the physical and chemical changes that matter can undergo appear to be magical, but it can all be explained by science.

Chapter 3 review

Vocabulary
Match each word to the sentence where it best fits.

Section 3.1

atomic mass	atomic number (Z)
element	element symbol
group	macronutrients
period	periodic table
scale	trace amount
trace elements	

1. A unit of measure that shows a particular amount of detail is called a(n) _____.
2. The _____ shows how many protons are in the nucleus.
3. A horizontal row of the periodic table is called a(n) _____.
4. A(n) _____ is a vertical column on the periodic table.
5. K is the _____ for potassium.
6. _____ are elements that are needed by living organisms which are obtained from food.
7. Even though tiny amounts are present in living organisms, too much of many of the _____ can be deadly.
8. A(n) _____ of almost every element on the periodic table can be found in our bodies.
9. A(n) _____ is a substance that cannot be broken down through chemical means into simpler substances.
10. The _____ organizes the elements into rows and columns.
11. The _____ is an average mass of an atom's protons and neutrons.

Section 3.2

| chemical formula | compound |
| isomers | |

12. While a(n) _____ shows the ratio of the different elements in a compound, it does not show how the individual atoms are connected.
13. Two or more different elements bound together chemically form a(n) _____.
14. _____ are compounds that have the same chemical formula but different structures.

Vocabulary
Match each word to the sentence where it best fits.

Section 3.3

distillation	distilled water
heterogeneous mixture	homogeneous mixture
mixture	pure substance
solution	tap water

15. Water that has been boiled into steam and condensed back into liquid again is called _____.
16. _____ is a process used to purify water.
17. Another word for a homogeneous mixture is a(n) _____.
18. A(n) _____ contains an irregular distribution of particles.
19. Since _____ contains dissolved minerals, it is actually a solution.
20. A sample containing only one kind of atom or compound is classified as a(n) _____.
21. A(n) _____ is matter that can be separated by physical means into more than one pure substance.
22. A(n) _____ contains more than one substance but is uniformly mixed at the molecular level.

Section 3.4

brittle	chemical change
chemical property	chemical reaction
ductile	intermolecular forces
intramolecular forces	malleable
physical change	physical property
precipitate	reactivity

23. A(n) _____ is the process by which a chemical change occurs.
24. A(n) _____ can be determined directly from the substance.
25. The molecules of the substance do not change during a(n) _____.
26. Chemical bonds are broken and reformed to create new substances during a(n) _____.
27. The forces between molecules are called _____.
28. A(n) _____ substance can be pulled and stretched without breaking.
29. _____ are the forces between atoms within a molecule.
30. A(n) _____ is an insoluble compound that forms from a reacting solution.
31. _____ describes how easily one substance combines with another.

Chapter 3 review

Section 3.4

brittle	chemical change
chemical property	chemical reaction
ductile	intermolecular forces
intramolecular forces	malleable
physical change	physical property
precipitate	reactivity

33. Glass is _____ because it breaks instead of bending.

34. A(n) _____ substance is one that can be flattened and pressed without breaking.

35. The ability to react with water is an example of a(n) _____.

Conceptual questions

Section 3.1

36. Would you expect phosphorous to be a good conductor of electricity? Why or why not?

37. Name two elements that were not created in any substantial anount in the Big Bang at the beginning of the universe?

38. The universe is mostly composed of which element?

39. What does it mean for an atom to be electrically neutral?

40. ❰ Explain the different characteristics of metals, nonmetals and metalloids on the periodic table.

41. Describe briefly how the elements other than the first three lightest (H, He, Li) were formed.

42. ❰ What is the role of trace elements in your body?

43. Even though multivitamins can help provide you with necessary trace elements, why would it not be good to take too many multivitamins at one time?

44. How would you find out how many electrons are in a particular type of neutral element?

45. The average person size human body contains 4×10^{28} atoms. 4×10^{23} of them are copper - the same as the copper in wires. Why is copper considered a trace element even though there are 400 trillion, billion copper atoms in the average body?

46. Mercury thermometers, once common in science classrooms, have been largely replaced due to safety concerns. The "fire diamond" and the global harmonization system (GHS) are two methods used to label a mercury hazard. Locate and reproduce images of both labels and evaluate their relative strengths and weaknesses.

47. What is the most useful number for identifying an element on the periodic table? Defend your choice.

48. What is the most abundant element in Earth's crust? What type of compounds in the crust is it found in?

49. What neutral atom has 35 electrons?

50. What element has 43 protons?

51. What element has an atomic mass of 44.956 amu?

52. For each of the following elements give a different element that has similar characteristics:
    a. Iodine
    b. Xenon
    c. Copper
    d. Barium

53. Determine the following properties for the element zinc by using a periodic table.
    a. group number
    b. period number
    c. number of protons
    d. Atomic mass

54. Determine the following properties for the element argon by using a periodic table.
    a. group number
    b. atomic mass
    c. period number
    d. number of protons

55. For the neutral element nitrogen, use the periodic table to determine the following quantities:
    a. number of protons
    b. group number
    c. period number
    d. number of electrons

56. Use the periodic table to determine the following quantities for the element silver:
    a. group number
    b. period number
    c. element symbol
    d. number of protons

57. Choose a vitamin to research. Read at least 2 articles, one written by a scientist and one posted on a health-related website.
    a. Write a 1-page summary of your research. Cite your sources according to your teacher's directions.
    b. Does this vitamin improve health? If so, how? Present arguments made by both authors.
    c. Share evidence each author used to back up their statement.
    d. Comment on the reliability of each source.
    e. Do not use direct quotes. Your 1-page paper must be in your own words.

Chapter 3 review

Section 3.1

58. Identify which element that has 22 protons in its nucleus.

59. Use the periodic table to determine how many electrons are in a neutral cesium atom.

Section 3.2

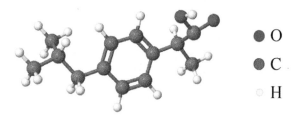

60. For the molecule in the above picture:
    a. Write the chemical formula of the compound.
    b. Which type of model is being used to represent the molecule?

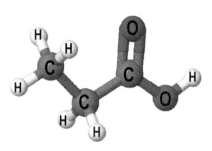

61. What is the chemical formula for the molecule above?

62. For the compound sodium bicarbonate, $NaHCO_3$:
    a. Write the names of each element in the compound.
    b. How many atoms of each element are in the compound?

63. For the compound ammonium acetate, $NH_4CH_3CO_2$:
    a. Write the names of each element in the compound.
    b. How many atoms of each element are in the compound?

64. An unknown compound has a chemical formula of $C_8H_{16}O_3$:
    a. Write the names of each element in the compound.
    b. How many atoms of each element are in the compound?

65. Aluminum oxide has a chemical formula of Al_2O_3:
    a. Write the names of each element in the compound.
    b. How many atoms of each element are in the compound?

66. Which atomic particle is responsible for chemical bonds?

67. What is the chemical formula of sodium sulfate, a compound with two sodium atoms, one sulfur atom, and four oxygen atoms?

68. If you wanted to represent a molecule that showed how atoms are connected to each other, but you do not care about how the atoms are arranged in space, which model is best?

69. What are the two things that result in a compound's chemical and physical properties?

70. How can two compounds with the same chemical formula have different chemical and physical properties?

71. What is the difference between an atom and a compound?

72. Which type of model is the simplest representation of a molecule?

73. Is N_2 an element or a compound? Explain your answer.

74. Lactic acid is found in milk. This compound has 3 carbons, 6 hydrogens, and 3 oxygens. Write the chemical formula for lactic acid.

Section 3.3

75. What is the difference between a heterogeneous mixture and a homogeneous one?

76. Classify each of the following as either a pure substance or mixture. Reference a periodic table if necessary.
    a. granite, a type of rock
    b. neon gas
    c. benzene (C_6H_6)

77. Classify each of the following as either a pure substance or mixture. Reference a periodic table if necessary.
    a. air sample
    b. distilled water
    c. fruit juice

78. Classify each of the following as either a pure substance or mixture. Reference a periodic table if necessary.
    a. tap water
    b. sodium
    c. sugar water

79. Further classify each substance in question 76 as an element, compound, homogeneous mixture, or heterogeneous mixture.

80. Further classify each substance in question 77 as an element, compound, homogeneous mixture, or heterogeneous mixture.

81. Further classify each substance in question 78 as an element, compound, homogeneous mixture, or heterogeneous mixture.

Chapter 3 review

Section 3.3

82. Give two different examples of how a homogeneous mixture can be physically separated in a lab setting, and two different examples of how a heterogeneous mixture can be physically separated in a lab.

83. Describe one difference between solid sugar and sugar dissolved in water in two or more sentences.

What is in your tap water?

84. ❮ Research the tap water near your home or school. See if you can determine the following information.
   a. List the concentrations of any dissolved minerals such as calcium, or sodium.
   b. Does the water contain fluoride and if so, how much?
   c. How often is the water tested?
   d.
   e. What is the source of the water? Well? Reservoir?

Section 3.4

85. Are physical changes reversible? Why or why not?

86. For what reason are chemical changes irreversible?

87. Provide your own example of an irreversible change.

88. Provide your own example of a reversible change.

89. You are trying to determine whether or not a certain compound will react with water. After you mix the compound in water, what signs will look you for to see if a chemical change took place?

90. Decide if the following phenomenon are chemical or physical changes:
   a. fruit growing
   b. condensation
   c. molten iron solidifies

Frying an egg and eating it.

Physical changes?

Chemical changes?

91. Decide if the following breakfast time events are chemical or physical changes:
   a. frying an egg
   b. the egg cools down
   c. chewing the egg
   d. the egg is digested

92. Intramolecular and intermolecular forces are different in what ways? Use a drawing to support your answer.

93. An unbroken glow stick has a mass of 80 g. When you bend it, a glass vial inside the glow stick breaks open and releases a chemical that mixes with the surrounding liquid.
   a. Which sign(s) of a chemical change do you to see as a result of the reaction ?
   b. What is the mass of the glow stick once the reaction is over? Explain your answer.

94. Are each of the following observations chemical or physical properties?
   a. the substance boils at -75 °C
   b. the substance corrodes with acid
   c. the substance rusts with moisture

95. Are each of the following observations chemical or physical properties?
   a. the substance is transparent
   b. the substance does not react with bases
   c. the substance is light blue

96. A compound is observed to have the following properties. Designate each one as either chemical or physical.
   a. brittle
   b. ignites at 225 °C
   c. is at room temperature

chapter 4 Temperature and Heat

When you are outside on a chilly evening, nothing warms you up like a warm bowl of soup. So after you go inside, you pour some soup into a metal pot and turn on the stove. A few minutes after putting the pot on the stove, the soup is steaming and bubbling and ready to warm you up!

Did you ever stop to think about how that soup got hot? Somehow the hot stove transferred energy to the cold pot. That energy heated the pot, which in turn heated the soup. The result is the soup got hot as the energy came from the stove and passed through the pot. If you could measure all the temperatures you would find that the the metal pot gets hotter than the soup.

Why does the metal pot heat up so much faster than the soup that it contains? Why do the plastic handles on the pot stay cool enough to touch, even when the soup is steaming? Some materials heat up faster than others. Some materials also transmit heat better than others. Heat flows faster through metal compared to plastic.

If the soup is too hot, you can add a cube of cold ice to the bowl. After a few seconds, the soup is cooler and the ice cube has disappeared. Did the ice give off some "cold"; or did the soup lose some of its "hot"?

All of this has to do with temperature, energy, and how that energy moves in the form of heat. As we explore these concepts in this chapter, you get a deeper understanding of matter and energy and what it means to be "hot" or "cold"!

Chapter 4 study guide

Chapter Preview

At the molecular level, temperature is a measure of the average kinetic energy of the particles of matter. As temperature decreases, the kinetic energy of atoms or molecules decreases. When kinetic energy reaches the lowest possible level there is a minimum possible temperature known as absolute zero. While temperature is the average kinetic energy of atoms or molecules, thermal energy or heat is a measure of the total energy of all the molecules of the matter present. Heat flows from hot to cold, and therefore into or out of a system based on its temperature relative to its surroundings, stopping when equilibrium is reached. Substances change phase as energy is added or removed from them as seen in the water cycle. The addition of heat energy to a system or its surroundings is known as an endothermic change, and the removal of heat energy is known as an exothermic change. When a system removes energy, its surroundings receive and absorb that energy. When a system absorbs energy, the energy comes from its surroundings so the surroundings lose energy.

Learning objectives

By the end of this chapter you should be able to:
- convert temperatures among the Kelvin, Celsius and Fahrenheit scales;
- explain connection between kinetic energy and temperature;
- distinguish between temperature and heat;
- distinguish between open, closed and isolated systems and how they interact with the surroundings;
- calculate the heat (in joules) associated with a temperature change of a substance;
- interpret a heating curve diagram to explain the changes that a substance undergoes when heat is added or removed from it; and
- determine whether a change is exothermic or endothermic.

Investigations

4A: Temperature and thermal energy
4B: Specific heat
4C: Energy and food
4D: Heat of fusion
4E: Project: Design an Insulator
4F: Research Presentation Enhancement: Insulators in the Home

Pages in this chapter

94 Temperature
95 Temperature scales
96 Temperature measures motion
97 Motion, energy and temperature
98 Temperature is an average
99 Absolute zero
100 Section 1 Review
101 Thermal Energy and Heat
102 Thermodynamics and systems
103 Energy bar diagrams to track changes
104 Heat transfer
105 Thermal equilibrium
106 The first law: Energy conservation
107 Specific heat
108 Solving temperature and heat problems
109 Solving heat and energy conservation problems
110 Energy and food
111 Calculating calories and Calories
112 Section 2 Review
113 Heat and Changes of State
114 Melting and the heat of fusion
115 Boiling and heat of vaporization
116 Evaporation and condensation
117 Phase changes and the water cycle
118 Energy, chemical changes and physical changes
119 Section 3 Review
120 Color thermometers
121 Color thermometers - 2
122 Chapter review

Vocabulary

thermometer	thermistor	thermocouple
temperature	Fahrenheit scale	Celsius scale
Brownian motion	kinetic energy	kinetic molecular theory
absolute zero	Kelvin scale	thermal energy
heat	joule	calorimetry
calorie	second law of thermodynamics	system
surroundings	open system	closed system
isolated system	conduction	thermal conductor
thermal insulator	thermal equilibrium	first law of thermodynamics
specific heat	calorimeter	adiabatic
phase change	endothermic	exothermic
melting point	heat of fusion	boiling point
heat of vaporization	evaporation	condensation
electrolysis		

4.1 - Temperature

Humans experience temperature, but we really only have a vague sense of hot or cold. You can feel when an object is warm or cold, but you cannot feel its exact temperature. If you spent some time outdoors on a cold winter day then entered a 68 °F room, the room feels warm. The same room at the same temperature will feel cool if you come in from outdoors on a warm day in summer. We often sense temperature by comparison. In fact, you and your friends might not even agree if the room you are in is warm or cool. Scientists have developed numerous instruments to accurately record properties that can be measured like temperature.

Experiencing temperature

How glass thermometers work

Temperature is a measure of the average kinetic energy of the individual atoms or molecules. A **thermometer** is a tool that directly measures temperature. A glass thermometer is filled with liquid alcohol that either expands or contracts depending on temperature. The thermometer responds to small temperature changes because the bulb at the base stores a large volume of alcohol. When temperature increases, the alcohol expands to take up more space. This forces the alcohol to rise up the glass tube. When temperature decreases, alcohol contracts and moves down the tube into the bulb. Temperature is measured by the height the alcohol reaches in the glass tube.

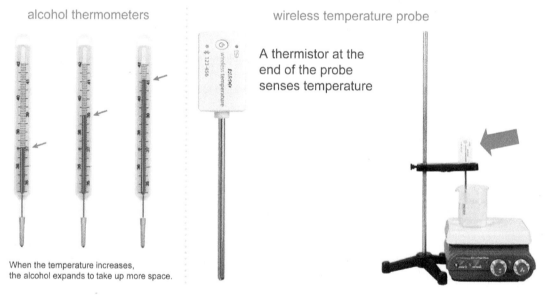

Other types of thermometers

A **thermistor** reports temperature by changing its electrical properties as the temperature changes. A thermistor gets its name from the phrase thermally sensitive resistor, which means the way electricity flows through the thermistor changes depending on the temperature. The temperature sensor you use in your laboratory experiments has a thermistor at the end of its stainless steel tube. A **thermocouple** is another electrical sensor that measures temperature by directly responding to a difference in temperature.

Thermocouples have two or more wires made of different metals that heat up or cool down at different rates. The temperature difference between the two wires generates an electrical signal that can be interpreted in different ways depending on the device. Many appliances in your home rely on thermistors and thermocouples to maintain normal operating temperatures and to keep you comfortable.

Temperature scales

The Fahrenheit scale

Two commonly used temperature scales are the **Fahrenheit scale** and **Celsius scale**. Water freezes at 32 degrees and boils at 212 degrees on the Fahrenheit scale. Water's freezing point and boiling point are 180 Fahrenheit degrees apart. Temperature is commonly measured in Fahrenheit in the United States. A temperature of about 74 °F is a pleasant day.

The Celsius scale

The Celsius scale places the freezing and boiling points of water 100 degrees apart, instead of 180. Water freezes at 0 °C and boils at 100 °C. Most scientists and engineers measure temperature in Celsius. Most countries outside the United States use the Celsius scale for all descriptions of temperature, including daily weather reports. A pleasant 74 °F day is equal to about 23 °C.

Celsius is used outside the U.S.

If you travel outside the United States, you will need to know the difference between the Fahrenheit and Celsius scales. A temperature of 23 °C in Paris, France is a pleasant spring day, but a temperature of 23 °F is a cold winter day in Minneapolis, Minnesota!

how to convert between Celsius and Fahrenheit

$$T_{Fahrenheit} = \frac{9}{5} T_{Celsius} + 32 \qquad T_{Celsius} = \frac{5}{9}(T_{Fahrenheit} - 32)$$

Temperature conversions

The formulas to convert Fahrenheit degrees to Celsius degrees are shown above. Fahrenheit degrees are smaller than Celsius degrees. A temperature change of 9 °F is the same as a temperature difference of 5 °C. Additionally, we must account for the fact that 0 °C equals 32 °F. When converting from °C to °F, first scale the °C by 9/5 and then add 32 degrees.

Solved problem

What is the Fahrenheit equivalent of 10 °C?

Given   10 °C

Relationships   °F = $\frac{9}{5}$ °C + 32

Solve   °F = $\frac{9}{5}$ 10 °C + 32 = 18 + 32 = 50 °F

Answer   10 °C is equal to 50 °F

Temperature measures motion

The relationship between temperature and energy

You have seen temperature scales, and you are familiar with how thermometers are used to take a temperature reading. But do you know what temperature is actually measuring? Whether it looks like it or not, particles that make up matter are always moving. Temperature is a direct measurement of how much energy matter has from the movement of the tiny particles that make it up.

Moving matter

When you look at a glass of milk, it appears uniform. However, if you looked at a drop of milk under a microscope, you would see tiny fat globules suspended in water. These tiny globules of fat would be jiggling around randomly. The motion of the fat globules requires some energy. Where does this energy come from?

Modeling matter

Imagine you are at stadium concert and a very large inflatable ball is in the middle of the audience. If ten people push on the ball from the right, and ten push from the left, the ball will not move. But, if there is an imbalance of forces pushing on the ball, then the ball will change directions. As people from the crowd randomly push on the ball it will appear to jump around in all directions. If you looked down on the stadium from above, you wouldn't be able to distinguish the moving people, but you would see the ball jiggling around.

 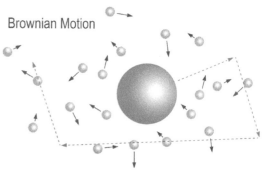

Brownian Motion

Brownian motion

The ball gets its energy from the surrounding people in the audience. Fat globules in milk are totally surrounded by water molecules, and this is where they get their energy. If there is an imbalance of forces based on the motion of individual water molecules, the globule will move. The random movement of the fat globule based on the impact from surrounding water molecules is known as **Brownian motion**. We can't see water molecules; but Brownian motion is evidence that matter exists in tiny discrete particles. It is also evidence that water molecules are constantly moving around. Temperature is the average motion of particles in a substance. Some of the particles are moving faster than others, so therefore some of the particles are contributing to Brownian motion more than others.

Interactive simulation

Watch a simulation of how particles move among one another in the interactive simulation titled Brownian Motion. Trace the movement of a random molecule to observe the interactions of a single particle among its neighbors. Look for evidence of particles moving at different speeds and reflect on the meaning of "average kinetic energy."

Motion, energy and temperature

Why does Brownian motion occur?

Brownian motion occurs for two important reasons:

1. Matter (including water) is made of atoms;
2. Atoms and molecules are in constant, rapid, random motion.

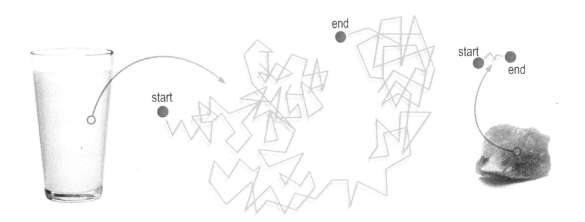

When you look at a still glass of milk without a microscope, you don't see any movement. This is true for all ordinary matter. But, Brownian motion demonstrates that there are constantly jiggling particles in a fluid. This is true for all ordinary matter, even in solids, although particles of a solid do not demonstrate Brownian motion unless they are dissolved in a fluid. Even if you can't see it, energy is causing microscopic motion among particles of matter. The energy of motion among molecules is called **kinetic energy**.

Kinetic energy and motion

If particles are always moving, why doesn't the glass of milk jump around on the table? Even something as tiny as a grain of salt appears to stand still, but the particles that make up the salt are constantly moving. How can this be? These three factors explain the phenomenon:

1. Particles of matter are tiny. The particles may be atoms, ions, or molecules.
2. There are lots of particles and they are typically close together. This is especially true for solids and liquids. A single grain of salt could have trillions of individual atoms. In a gas, the particles are farther apart.
3. A gas particle will rapidly move in one direction until it collides with another atom, then it will change directions. Particles in liquids have less freedom to move, and solid particles have the least freedom of movement.

These tiny particles, that are very close together, are constantly banging into each other. At any given moment there are as many particles bouncing in one direction as there are in the other direction. The average directional motion of the whole group is zero, so to you the matter appears to be standing still. However, the individual atoms are constantly moving. The idea that all matter contains the particles that are constantly moving is the basis of the **kinetic molecular theory**.

Temperature

Temperature is a measure of the average kinetic energy of individual particles such as atoms or molecules. When temperature increases, the kinetic energy of motion increases and particles move more rapidly.

Section 4.1: Temperature

Temperature is an average

Molecules at the same temperature may have different energies

We say temperature is a measure of the average kinetic energy of particles, because not all particles have the same energy. Most particles have energy that is close to the average but some atoms or molecules have more energy than the average, and some have less than the average energy. This is true for solids, liquids and gases but it is easier to visualize a gas. The graph below shows how much kinetic energy individual gas molecules have on the horizontal (x) axis and the number of molecules on the vertical (y) axis. As you can see, individual molecules in a sample at a given temperature have a wide range of energy.

Distribution of kinetic energy among three gas samples at different temperatures

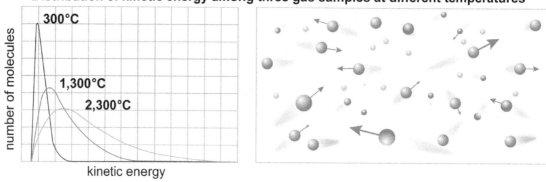

Changes in temperature cause changes in energy

Notice that the graph changes in two ways as the temperature increases.

1. The average energy increases with increasing temperature. The peak of the graph, with the highest number of molecules shifts to the right.

2. The range of energies increases with increasing temperature. The graph gets wider and flatter because there are fewer molecules that are close to the average.

Kinetic energy distribution among a gas compared to a liquid/solid at the same temperature

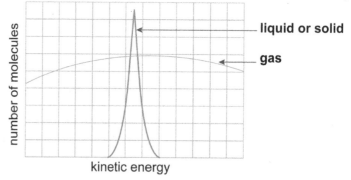

Molecular energy and states of matter

In a gas, the particles are very far apart. The average distance between gas molecules is thousands of times the size of a single molecule. That would be like your closest neighbor being a mile away. Since they are so spread out, gas molecules have few opportunities to bump into each other and exchange energy. So there are always some molecules with low kinetic energy and some molecules with very high kinetic energy, causing the range of energies to be very wide. Molecules are very close together in a liquid or solid, separated by a distance less than their own size. These molecules are constantly bumping into each other and exchanging energy. Fast molecules tend to slow down, and slow molecules tend to speed up. These short-distance interactions cause energy among particles to even out and produce a narrow energy curve.

Absolute zero

Absolute zero

When temperature decreases it tells us the kinetic energy of atoms is also decreased. Kinetic energy cannot get lower than zero therefore it is logical that there is a lower bound to temperature as well. The lowest possible temperature at which molecular kinetic energy is essentially zero is called **absolute zero**. Absolute zero is −273 °C or −459 °F. It is not possible to have a temperature lower than absolute zero, or −273.15 °C. Because of quantum effects, the actual kinetic energy cannot become exactly zero, but for most purposes molecules have no free kinetic energy at absolute zero. Scientists have recorded a temperature within one hundredth of absolute zero, but 0 K has never been reached.

The Kelvin scale

The **Kelvin scale** starts at absolute zero and has the same increments of degrees as the celsius scale. A difference of 1K is the same as a difference of 1°C. There cannot be a negative Kelvin temperature. The Kelvin scale is used because it measures the absolute energy of atoms. A temperature in Celsius or Fahrenheit measures only the relative energy, relative to the freezing point of water. The freezing point of water is an arbitrary reference point. Absolute zero is a much more logical reference point based on a universal property of matter. Note: When using Kelvin it is customary not to use the degree sign, instead express temperature in "Kelvin" as opposed to "degrees Kelvin."

Converting from Celsius to Kelvin

One Kelvin (K) temperature unit is the same size as one Celsius unit. To convert a Celsius temperature to Kelvin temperature, simply add 273 to the Celsius temperature.

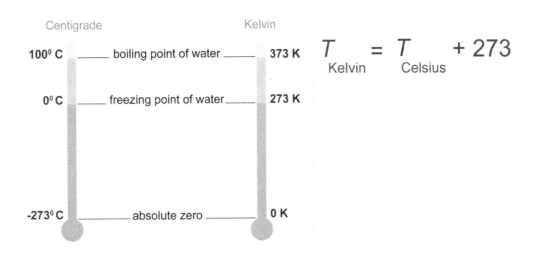

$$T_{Kelvin} = T_{Celsius} + 273$$

Solved problem

It is a sunny 75 °F day. Express this temperature in Kelvin.

Given 75 °F

Relationships $T_C = (T_F - 32) \times \frac{5}{9}$ $T_K = T_C + 273$

Solve Convert °F to °C then convert °C to K.

$$T_C = (T_F - 32) \times \frac{5}{9} = (75 - 32) \frac{5}{9} = 23.8 \text{ °C}$$

$$T_K = T_C + 273 = 23.8 + 273 = 297 \text{ K}$$

Answer 75 °F is equal to 297 K.

Chapter 4

Section 1 review

Human beings can only sense changes in temperature relative to their body temperature. Accurate temperature measurements are made using a device called a thermometer. Temperature can be measured using the Fahrenheit scale or the Celsius scale, which are based on the freezing and boiling points of water. Temperature is a result of the kinetic theory of matter: atoms or molecules that make up matter are in constant motion and thus each have kinetic energy. In liquids and solids the particles are close and energy-exchanging collisions ensure that most particles have kinetic energy close to the average. In gases, the particles are widely spread out with few collisions, and a wide range of kinetic energies exists between particles. When kinetic energy of particles reaches the lowest possible level, the temperature is termed absolute zero as it is the lowest possible temperature. The Kelvin scale uses the same unit of temperature as the Celsius scale but starts at absolute zero.

Vocabulary words	thermometer, thermistor, thermocouple, temperature, Fahrenheit scale, Celsius scale, Brownian motion, kinetic energy, kinetic molecular theory, absolute zero, Kelvin scale

Key relationships	• $T_{Fahrenheit} = \frac{9}{5} T_{Celsius} + 32$ and $T_{Celsius} = \frac{5}{9}(T_{Fahrenheit} - 32)$ • $T_{Kelvin} = T_{Celsius} + 273$

Review problems and questions

1. A thermometer gives you a temperature reading, but what is it actually measuring?

2. What is the coldest possible temperature called? What is its value in °C? What is its value in K?

3. Why is temperature an average? Why don't all particles in a sample of matter have the same temperature?

4. Convert the temperature 15 °C to degrees Fahrenheit and to kelvin.

5. How does the kinetic energy of matter shown in Picture 1 compare with the kinetic energy of matter in Picture 2?

4.2 - Thermal Energy and Heat

Temperature is a measure of the average kinetic energy per molecule. The total energy of all particles in a sample is called the **thermal energy**. Thermal energy is the energy stored in matter. The amount of energy stored is proportional to temperature and the amount of matter present.

Thermal energy

temperature
the average kinetic energy per particle

thermal energy
the total kinetic energy of all particles in a sample

Average vs sum

If you want to warm up a tub of water at 20 °C, would you rather add a drop of water at 100 °C, or add 1 L of water at 50 °C? The 1 L of 50 °C water of course! Even though the 1 L of water is at a lower temperature, it has a higher thermal energy because there are many more molecules in the sample. The size of the orange area around each particle above shows individual particles within a sample have different amounts of energy. Temperature is an average energy per particle, while thermal energy is the total energy of all particles.

Heat not cool

To increase the temperature of matter, thermal energy must be added. Thermal energy that is being added or subtracted from matter is called **heat**. Heat is the flow of thermal energy from higher temperature matter to lower temperature matter. When the temperature of matter decreases, it is not because "cool" is being added! In science, there is no such thing as cool. If the temperature goes down the reason is because heat is being removed. Air conditioners add cool air to a room, but the air is cooled by removing heat from it. Heaters add warm air to a room by transferring heat to the air.

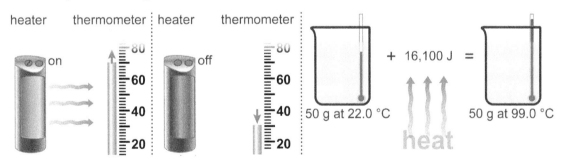

The Joule: a unit for heat and energy

The **joule** (J) is the SI unit for energy and heat. One joule is a small amount of energy. Heating 50.0 g of water from room temperature (22.0 °C) to nearly boiling (99.0 °C) requires about 16,100 joules of heat.

Calories vs calories

The study of heat flow in chemistry is called **calorimetry**. The **calorie** is a historical unit of thermal energy that is still sometimes used in chemistry. One calorie is the amount of thermal energy which will raise the temperature of one gram of water by one degree Celsius. One calorie equals 4.184 joules so a single calorie is more energy than a single joule. Nutritionists also use food Calories to describe the energy content of foods. In fact the food Calories on a nutrition label, with capital "C," are actual kilocalories. One food Calorie = 1 kcal = 1,000 cal = 4,184 J. A 100 Calorie snack really contains 100,000 calories!

Thermodynamics and systems

The second law of thermo-dynamics

Skin is your body's largest organ. It has three major roles: sensation, protection, and body temperature regulation. A swimsuit on a frozen lake would not be a wise choice of clothing because your skin is at a much higher temperature than the air near your skin. Without additional clothing for insulation, heat flows quickly from the higher temperature skin to the lower temperature air due to the **second law of thermodynamics**. A consequence of the second law is that heat (thermal energy) flows from higher temperature to lower temperature.

How to distinguish a system from its surroundings

To track the flow of heat it is necessary to define a **system**. The system is a collection of matter and energy that we choose to keep track of by drawing an imaginary boundary around it. Everything else in the universe outside the system boundary is called the **surroundings**. If you wanted to study the rate at which heat flowed out of an unclothed body, you would define the system to be the body. The boundary would be an imaginary surface just above the skin, and the surroundings would be the air and everything else.

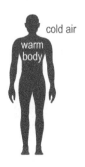 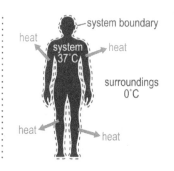

systems and surroundings

We choose the system to include the things we want to keep track of.

The boundary separates the system from the surroundings.

Open, closed and isolated systems

We use systems to focus our attention on how the flow of matter and energy affects the system inside and outside the boundary. In an **isolated system** the boundary is impermeable to matter and energy. Nothing goes in and nothing goes out. True isolated systems are nearly impossible to create but the approximation is useful. The opposite is an **open system** which allows both matter and energy to cross the system boundary. Clothing creates an open system through which a body can exchange heat (energy), sweat, and breath (matter) with the surroundings. In a **closed system**, only energy is exchanged between the system and surroundings. An example of a closed system is a thin plastic film over the skin that prevents breathing but still exchanges heat.

Systems depend on what is being studied

Whenever we talk about the flow of energy, a system is always implied, whether explicitly chosen or not. The body is a good system to study the effectiveness of clothing as an insulator. In a chemical reaction between water and iron, the water and iron are the system. The choice of a system depends on what you are trying to study.

Energy bar diagrams to track changes

Energy bar graphs

Energy bar diagrams are a way to track energy as it is exchanged during a physical or chemical change. Materials before the change are on the left and after the change are on the right. In the center, there is a circle that represents energy flow in and out of the system.

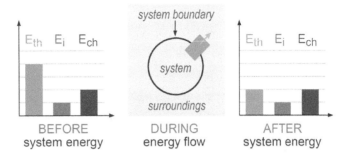

Types of energy

Above the before and after titles, there are three types of energy. E_{th} is thermal energy. This is the kinetic energy all objects have above absolute zero. E_i is interaction energy. This is the potential energy substances have based on their state of matter. E_i changes when there is a change of state. Finally, E_{ch} is the chemical energy. This is the energy associated with chemical change. When a physical change occurs, E_{th} changes but there is no change in E_{ch}. Of course, when there is a chemical change there will be change in E_{ch}. The diagram shows a change in E_{th} with no change in E_{ch}, therefore a physical change occurred. E_{th} is shown as leaving the system during the change, so this is an exothermic physical change. There is no change in E_i. This means a substance is simply losing heat without changing its phase. This occurs when a solid, liquid, or gas gets cooler but does not change its phase. All substances have chemical energy, but since there is no chemical change occurring, E_{ch} could be left blank.

Using the graph

Let's consider a hot cup of tea cooling on a table. We will define the system as the hot tea and a cup. Everything outside of the system is the surroundings. The table is warmed by the cup of tea but we have chosen not to make it part of the system so therefore the table must be surroundings. The parts of the system are written in the center circle. The tea is a liquid so therefore it has some interaction energy. We will say that a solid has one bar of E_i, a liquid two and a gas four. No chemical change is occurring so E_{ch} is left blank. The amount of bars we use is representational and not actual amounts.

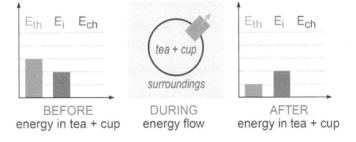

Interpreting the energy graph

E_{th} decreased from 3 bars to 1 because the tea is cooling. The diagram is correct no matter how many bars are used, as long as the number of bars decreases. The 2 "lost" bars of E_{th} are accounted for in the energy flow part of the diagram. The E_{th} is moving across the system boundary from the tea to the surroundings during the change. There is no chemical change, so E_{ch} is not displayed in the diagram. You can see E_i did not change. Even though the hot tea lost energy it did not change from a liquid to a solid. You would only expect to see E_i decrease from 2 bars to 1 if the tea lost enough energy to freeze solid.

Heat transfer

Conduction

Suppose we pour hot and cold water into a divided cup so they can't mix. As an isolated system, heat flows through the divider, but not into the surroundings. The hot water cools down as it loses heat. The cold water gains heat from the other side and gets warmer. Heat transfer through materials by direct contact of the matter is called **conduction**.

Particle collisions

To understand how heat transfers through conduction, let's look back at the kinetic theory of matter. Remember, temperature is a measure of the average kinetic energy of the molecules. The water molecules at 80 °C move around with a higher average kinetic energy than the water molecules at 10 °C. As the high energy water molecules collide with the divider, they transfer some of their energy through the divider to the lower energy water molecules. The result is that the high temperature molecules lose energy, ending at a lower temperature, while the low temperature molecules gain energy.

Thermal conductors and insulators

Some materials are better at allowing heat transfer than others. Glass and metal are both considered a **thermal conductor** because they transfer heat well. Think about touching a metal pot full of boiling water. Metal conducts energy, so heat flows rapidly from the hot water through the metal pot to your skin. A **thermal insulator** is a material that conducts heat poorly, like a polystyrene foam cup. You can comfortably hold a cup of hot chocolate in a foam cup. Heat flows slowly through the foam so the temperature of your hand does not rise quickly. Foam is a good insulator because it contains many spaces of trapped air. Gases are poor conductors because the particles are too far apart to efficiently transfer heat. Double walled metal vacuum bottles or flasks are also surprisingly good insulators. The layer of air between the metal walls has been removed or vacuumed out. Since there is very little air between the walls, there are very few molecules to collide and transfer the heat.

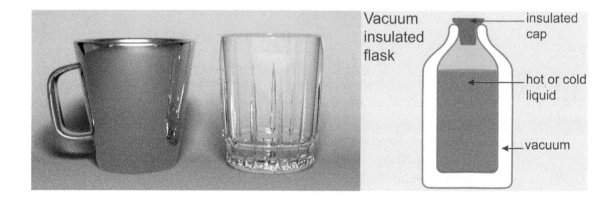

Chapter 4: Temperature and Heat

Thermal equilibrium

Why does heat stop flowing?

When does heat flow stop? When hot and cold water are in an isolated system with a divider, why doesn't the hot water keep losing heat instead of coming to a stable temperature? Obviously, the hot water does not get colder than the cold water.

Heat flow eventually stops

The water molecules in the different halves of the cup are still colliding with the divider, but there is no longer a net transfer of energy. This can only be true if the molecules on both sides have the same average kinetic energy. So they must be at the same temperature! Heat flow eventually stops, but that doesn't mean that collisions aren't still happening.

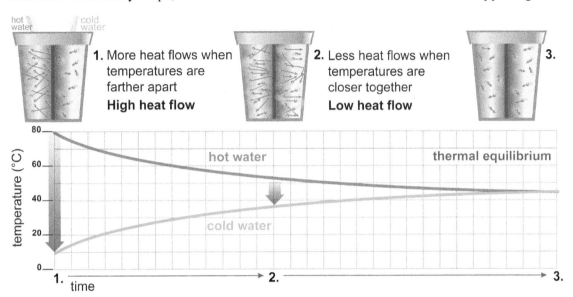

Thermal equilibrium

Two objects have reached **thermal equilibrium** when they have equal temperatures. During thermal equilibrium, no overall heat flows when particles collide because on average, particle temperatures are the same. Heat naturally flows from hot to cold until thermal equilibrium is established. The divider in the cup only slows down the heat flow. The hot and cold water take longer to reach thermal equilibrium, but the end result is the same.

Heat flow rate

When the temperature difference is large, the heat flow rate is fast. Heat flow gradually slows down as the temperatures get closer together. The rate of heat flow gets smaller as the temperature difference gets smaller. As two objects approach thermal equilibrium, the rate of heat flow between them becomes zero when both reach the same temperature.

Heat transfer in humans and animals

Living things rely on heat flow because biological processes absorb and release energy. Your body maintains a constant temperature because heat is distributed by your blood. Your internal body temperature averages 37 °C. Humans are comfortable when air is around 25 °C because at this temperature, body-generated internal heat leaves the body as quickly as it is generated. If the air is 10 °C, you get cold because heat flows out of your body faster than it can be generated. If the air is 40 °C, you feel hot because your body's internal heat cannot flow to the warmer surroundings. Heat flows from a warmer substance to a cooler substance until temperatures are the same. Different parts of your body release heat at different rates by changing the size or diameter of blood vessels. Your blood vessels dilate or open wide when heat needs to be removed from your body. This increases blood flow which helps more heat escape from your body. Extremities like hands and feet will feel warm in this case. If your body needs to retain heat, your blood vessels constrict or get smaller. This reduces blood flow and causes extremities to feel cold.

The first law: Energy conservation

The first law The **first law of thermodynamics** states energy can neither be created nor destroyed. All the energy lost by one system must be gained by the surroundings of another system. Consider an experiment where you mix 50.0 grams of 82.00 °C water and 50.0 grams of 4.00 °C water. When they are poured together, assume no energy or mass is allowed to flow into the surroundings. The total thermal energy after mixing must be the same as it was before mixing. The total thermal energy just gets redistributed over the entire mixture.

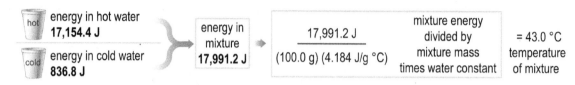

Conservation of thermal energy We need to account for the mass and the temperature of all the initial components. We also need to incorporate how water stores thermal energy. Each gram of water stores 4.184 J of energy for every degree Celsius. We can add up all the energy of the components in the system because thermal energy is a sum. In this case, all the energy stays in the system but will now be distributed over the mass of the entire mixture. The problem can be solved by calculating what temperature the mixture has to be to contain all the energy in the system.

Solved problem How hot is a mixture of 50.0 g of 82.00 °C water and 50.0 g of 4.00 °C water?

Asked What is the temperature of the mixture?

Given 50.0 grams of 82.00 °C water and 50.0 grams of 4.00 °C water

Relationships It takes 4.184 J to raise 1 gram of water by 1 °C. If you multiply this value by the sample mass and its temperature, you can find the amount of thermal energy stored in the sample.

Solve 1. Find the relative thermal energy of the hot and cold water.

$$E_{hot} = (50.0 \text{ g})(82.00 \text{ °C})(4.184 \tfrac{J}{g \cdot °C}) = 17{,}154.4 \text{ J}$$

$$E_{cold} = (50.0 \text{ g})(4.00 \text{ °C})(4.184 \tfrac{J}{g \cdot °C}) = 836.8 \text{ J}$$

2. Sum thermal energies together to find the total energy in the system.
$$17{,}154.4 \text{ J} + 836.8 \text{ J} = 17{,}991.2 \text{ J}$$

3. Divide the sum over the entire mixture.

$$T_{mixture} = \frac{17{,}991.2 \text{ J}}{(100.0 \text{ g})(4.184 \tfrac{J}{g \cdot °C})} = 43.0 °C$$

Answer The total 100.0 g of water has a temperature of 43.0 °C.

Distribution of energy In this case the temperature was exactly between the initial temperatures of the hot and cold samples because they both had the same initial mass, and they were both water so they each had the same ability to store thermal energy per gram. This will not always be the case, but energy will always be conserved and distributed over the entire system based on the component masses, temperatures and ability to store energy.

Specific heat

Heat exchange of different materials

If you take a bite from a cheese pizza that has just come out of a 450 °F oven, which part is least likely to burn you: the crust, the cheese, or the sauce? Although it might be hot, you can probably take a bite of the crust without any harm. But if you try to take a bite of pizza with sauce, watch out - that sauce is much hotter than the crust and you could get burned! How can this be if moments ago the entire pizza had a temperature of 450 °F? It turns out different materials exchange heat at different rates.

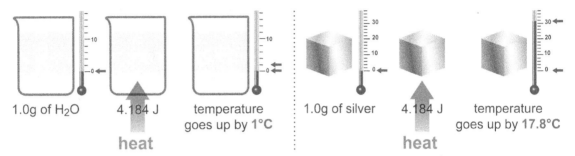

Variation among materials

It takes 4.184 joules of heat energy to raise the temperature of one gram of water by 1 °C. If you add the same amount of heat to one gram of silver metal, the temperature goes up by 17.8 °C! The same amount of heat causes a different change in temperature in different materials. The temperature difference exists because different materials have different abilities to store thermal energy.

specific heat of common substances

materials	specific heat, C_p (J/g·°C)	materials	specific heat, C_p (J/g·°C)
wax	2.100	wood	2.500
water	4.184	concrete	0.880
aluminum	0.900	glass	0.800
gold	0.129	polystyrene foam	0.927
silver	0.235	air at sea level	1.006

Specific heat

The ability of a material to store thermal energy is called **specific heat**. This is defined as the amount of energy it takes to raise the temperature of one gram of a substance by one degree Celsius and has units of J/g·°C. Water's specific heat is 4.184 J/g·°C. You must add 4.184 joules of heat energy for each 1 °C temperature increase per gram of water. Silver's specific heat is 0.235 J/g·°C. It is easier to heat up silver because you only need 0.235 J to heat up a gram of silver by 1 °C. Water's specific heat is 17.8 times higher than silver.

heat equation

$$q = m \times C_p \times (T_2 - T_1)$$

- mass (g)
- temperature change (ΔT, °C)
- specific heat (J/g°C)

The heat equation

The heat equation is used to calculate how much heat energy (q) it takes to make a temperature change (ΔT) in a mass of a material with specific heat (C_p). The temperature change (ΔT) is equal to the final change (T_2) minus the initial change (T_1) as in the equation, $\Delta T = T_2 - T_1$.

Solving temperature and heat problems

Solving specific heat problems

Most specific heat problems include one or both of the following two calculations.
1. Calculate the temperature change from a given heat input.
2. Calculate how much heat is needed to reach a specified temperature.

Solved problem

A welder adds 93,100 Joules of energy to a 100.0 g piece of steel. How much does the temperature of the steel increase?

Asked What is the change in temperature?
Given heat added, $q = 93,100$ J; mass of steel, $m = 100.0$ g; from the table on the previous page, the specific heat of steel, $C_p = 0.470 \frac{J}{g \cdot °C}$
Relationships $q = m \times C_p \times \Delta T$
Solve Rearrange the equation to solve for ΔT: $\Delta T = \frac{q}{m \times C_p}$

$$\Delta T = \frac{93,100 \text{ J}}{100.0 \text{ g} \times 0.470 \frac{J}{g \cdot °C}} = 1,980.85106 \xrightarrow{\text{round}} 1,980 \text{ °C}$$

Answer When 93,100 J of energy are added to a 100.0 g piece of steel, the temperature will change by 1,980 °C.

What is negative heat?

When energy is added to the system the temperature will increase. The positive answer for ΔT verifies energy is absorbed. If energy was released, temperature would decrease and the change in temperature would be negative. Incorporating a negative into the heat equation will result in a negative value for heat. This does not mean that the heat energy is negative! It simply means that the heat was emitted, or given off by that part of the system.

Solved problem

A 50.0 gram piece of aluminum foil comes out of the oven and cools from 180.0 °C to room temperature of 25.0 °C. How much heat is emitted by the foil?

Asked How much heat is emitted by the foil?
Given Mass of foil, $m = 50.0$ g; initial temperature of the foil $T_1 = 180.0$ °C; final temperature of the foil $T_2 = 25.0$ °C; from the table on the previous page, the specific heat of aluminum, $C_p = 0.900$ J/g·°C.
Relationships $q = m \times C_p \times \Delta T$ and $\Delta T = (T_2 - T_1)$
Solve
1. First, determine the change in temperature.
$\Delta T = (T_2 - T_1) = (25.0 \text{ °C} - 180.0 \text{ °C}) = -155.0 \text{°C}$

2. Next, use $q = m \times C_p \times \Delta T$ to calculate heat.

$$50.0 \text{ g} \times 0.90 \frac{J}{g \cdot °C} \times -155 \text{ °C} = -6,975 \xrightarrow{\text{round}} -6,980 \text{ J}$$

The negative value for q verifies that heat is released.

Answer As the 50.0 g piece of aluminum foil cools from 180.0 °C to 25.0 °C, the amount of heat emitted is 6,980 J.

Solving heat and energy conservation problems

Heat released equals heat absorbed

A hot cup of cocoa left on a table cools down. Heat moves from the hot cocoa to the cup, to the table, and to the air. The thermal energy of the hot coffee is decreased and the thermal energy of the nearby air and immediate surroundings are increased. Everything balances; the increase in thermal energy of the nearby air and surroundings is the same as the decrease in thermal energy of the coffee.

Use the first law and the heat equation to solve problems

Energy can neither be created nor destroyed. The first law of thermodynamics helps you think through everyday energy calculations such as heat released by a cup of coffee as it cools down to room temperature. As coffee cools, it heats up nearby air. The heated air rises and moves away from the coffee cup. The heat equation can be used with the first law of thermodynamics. The amount of heat given off by the coffee cup must be equal to the amount of heat absorbed by the surroundings.

Solved problem

What is the temperature change in a kitchen when 450.0 grams of water at 90.0 °C cools down to 23.0 °C? Assume all the heat lost by the hot water is transferred to the air in the kitchen. The kitchen has a volume of 120.0 cubic meters of air, with a mass of 134,400 g.

Asked What is the temperature change in the air?

Given Water values:
- mass of water, m_{water} = 450.0 g;
- initial water temperature, $T_{i\ water}$ = 90.0 °C;
- final water temperature, $T_{f\ water}$ = 23.0 °C;
- specific heat of water = $C_{p\ water}$ = 4.184 J / g·°C (from the table)

Air values:
- mass of air, m_{air} = 134,400 g;
- specific heat of air, $C_{p\ air}$ = 1.006 J/g·°C (from the table)

Relationships $q = m \times C_p \times \Delta T$ and $\Delta T = (T_2 - T_1)$

Solve
1. To find the amount of heat given off by the water, solve q_{water}.

   First, solve ΔT_{water} = (23.0°C - 90.0 °C) = -67.0°C

   Next, solve q_{water} = 450.0 g × 4.184 $\frac{J}{g\cdot °C}$ × -67 °C

   = -126,147.6 round -126,000 J

2. Apply that same amount of heat to the air (but change the sign since it is now being absorbed!)

   q_{water} lost = q_{water} gained, so $q_{air} = m_{air} \times C_{p\ air} \times \Delta T_{air}$ = +126,000 J

3. Calculate the change in air temperature.

   $$\Delta T_{air} = \frac{q_{air}}{m_{air} \times C_{p\ air}} = \frac{126,000\ J}{134,400\ g \times 1.006 \frac{J}{g\cdot °C}}$$

   = 0.93191 round 0.932 °C

Answer The air in the room gets warmer by 0.932 °C

Energy and food

Energy from food

Your body gets the energy it needs from the food you eat. That energy is measured in units called calories. One calorie is the amount of energy it takes to raise the temperature of one gram of water one degree Celsius. From the definition, you may be wondering why the energy you get from food is defined by how much water changes temperature. What is the connection between energy of food and water?

calorie is a unit of energy

We cannot directly measure the energy you get from eating food. The amount of energy your body contains before you eat a food cannot be determined. Therefore, the amount of energy you gain cannot be determined. So scientists use water to catch the energy released when foods are burned in a device called a calorimeter.

Calories vs. calories

There are two types of calories. One thousand lowercase calories are equal to one uppercase Calorie (1,000 cal = 1 Cal = 1 kcal). Food is labeled in uppercase Calories. This notation is confusing, so sometimes we refer to uppercase Calories as kilocalories and write it as kcal.

DIY calorimeter made from styrofoam cups and cardboard

Calorimeter

A **calorimeter** is a well-insulated device in which a reaction takes place, measuring the amount of heat absorbed or released during physical or chemical processes. The reaction is housed in a steel container called a bomb and is started by a hot electric coil. This type of calorimeter is known as a bomb calorimeter. The bomb is surrounded by water which heats up as the reaction progresses. It is assumed that all of the energy from the reaction is used to heat the water. By measuring the temperature increase in the water we can determine the amount of energy that was in the food.

Adiabatic

You can assume no energy is lost from the calorimeter to the surroundings. The bomb, its contents and the surrounding water is the system. If this system were perfectly insulated it would be considered to be **adiabatic**. Adiabatic containers lose no energy from the system to the surroundings. Adiabatic containers are theoretical and are used to make calculations easier. Let's see what happens when we burn a potato chip in a bomb calorimeter.

Energy diagram for the calorimeter

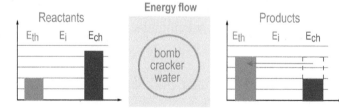

Energy from the chip

In the diagram above, the chip has higher energy in its chemical bonds than the ashes it leaves after burning. It begins with over four bars of chemical energy and ends with two. Since we are considering the system to be adiabatic, no energy is lost to the surroundings. All of the energy from the burning chip is used to heat water, as you can see by the increase in thermal energy. Energy does not move to the surroundings, so nothing is shown to cross the system/surroundings boundary in the center of the diagram.

Calculating calories and Calories

Calculating energy

To find out how much energy is in a potato chip, we will use an equation we have seen before, $q = m \times C_p \times \Delta T$.

Solved problem

A 0.10 g potato chip is burned in a bomb calorimeter that contains 500.00 g of water. The water's temperature changed from 25.0 °C to 50.0 °C. How many Calories are in the chip?

Asked How many Calories are in the chip?

Given 500.0 g of water; initial temperature water is 25.0 °C; final temperature of water is 50.0 °C; chip mass is 0.10 g.

Relationships The specific heat of water is $1.0 \frac{cal}{g \cdot °C}$; there are 1,000 cal in 1 Cal

Solve $q = m \times C_p \times \Delta T = 500.0 \; g \times 1.0 \frac{cal}{g \cdot °C} \times (50.0° \; C - 25.0° \; C) =$

12,500 calories $\xrightarrow{round}$ 13,000 calories or 13 Calories

Answer A single potato chip has 13 Calories. The mass of the chip did not make a difference in this problem because a calorie is defined by the temperature change in a mass of water.

Calorie-burning chart for various activities
approximate Calories burned per hour by a 150-pound woman

exercise	Calories/hour
sleeping	55
sitting	85
standing	100
walking (3 mph)	280+
jogging (5 mph)	500+
step aerobics	550+
power walking	600+
skipping rope	700+
running	700+

one potato chip = 13 Calories

Calories burned during activities

The table above shows the amount of Calories a 150-lb woman will burn per hour doing various activities. Notice that even sleeping uses 55 Calories per hour. Processes like breathing and a beating heart use energy and burn Calories. In the course of using energy, your body is constantly burning Calories. Suppose you wanted to know how long it takes to walk off the 13 Calories in a potato chip.

Solved problem

How long will it take to walk off 13 Calories in a single potato chip?

Given 13 Calories

Relationships Walking requires 280 Cal/hr

Solve

$$\frac{13 \; Cal}{1} \times \frac{1 \; hr}{280 \; Cal} \times \frac{60 \; min}{1 \; hr} = 2.78571 \xrightarrow{round} 2.8 \; mins$$

Answer Walking 2.8 mins will burn off the calories in a 13-Calorie potato chip.

Chapter 4

Section 2 review

While temperature is a measure of the energy per molecule of a substance, heat is a measure of the total energy in a collection of molecules. This means that heat depends not only on the average energy per molecule but also how many molecules there are. Heat flows from higher temperatures to lower temperatures through collisions of moving particles. The rate of the flow is governed by whether the materials in the system are good conductors or good insulators. When thermal equilibrium is reached, the flow of heat ends. Different substances require different amounts of energy per unit of mass to raise their temperature. This is known as specific heat, measured in J/g °C. The heat equation uses specific heat to calculate the energy needed to create a temperature change in a substance.

The food we eat contains the energy our bodies need to survive. The energy in food can be measured using a calorimeter. A calorimeter is an insulated device that measures the amount of heat absorbed from or released to surrounding water. When food is burned in a calorimeter the increase in water temperature is used to calculate the energy of that food.

Vocabulary thermal energy, heat, joule, calorimetry, calorie, second law of thermodynamics, system, surroundings, open system, closed system, isolated system, conduction, thermal conductor, thermal insulator, thermal equilibrium, first law of thermodynamics, specific heat, calorimeter, adiabatic

Key relationships
- 1 cal = 4.184 J   and   1 food Calorie = 1,000 cal = 1 kcal
- $\Delta T = T_{final} - T_{initial}$
- $q = m \times C_p \times \Delta T$ for change in energy during heating or cooling (no phase change)
- Temperature of a water mixture $T_{H_2O\ mixture} = \dfrac{q_{mixture}\,(J)}{m_{mixture} \times C_p}$
- C_p of water = 4.184 $\dfrac{J}{g\,°C}$
- Density of water = 1.00 $\dfrac{g}{mL}$

Review problems and questions

1. When you feel "hot," is energy moving from your body to your surroundings, or is it moving from your surroundings to your body? Explain your answer.

2. A cup of water left uncovered at room temperature will eventually evaporate.
    a. Is the uncovered cup an open system or closed system?
    b. What would happen to the water if the cup was covered? Explain your answer.

3. A 25.2-g piece of metal absorbs 275.01 J of energy when it is heated from room temperature, 23.0 °C, to 35.1 °C. What is the specific heat of the metal?

4. Two cups of water are mixed. The first cup has 50.8 mL of water at 15.3 °C and the second cup has 35.1 mL of water at 52.9 °C. What will the temperature of the mixture be?

4.3 - Heat and Changes of State

The addition or removal of heat causes temperature changes and changes in the state of matter. Adding heat to solid water (ice) at 0 °C causes solid to melt into liquid. Adding heat to water at 100 °C causes some of the water to turn from liquid to gas, boiling into steam. Melting and boiling are a change of state, or **phase change**. In a phase change, a substance changes the microscopic organization of its constituent particles (atoms, ions, or molecules) without changing its chemical makeup. All substances change phase at different temperatures, even metals such as iron. Iron melts at 1,535 °C and boils into a gas at 2,750 °C.

Changes in states of matter

Why do phase changes occur?

States of matter result from competition between attractive forces and temperature. Attractive forces between the particles of matter cause them to stick together and form solid and liquid structures. As temperature rises, the molecules have a higher average kinetic energy. Highly energetic molecules move back and forth so much they cannot stay together in an orderly structure like a solid.

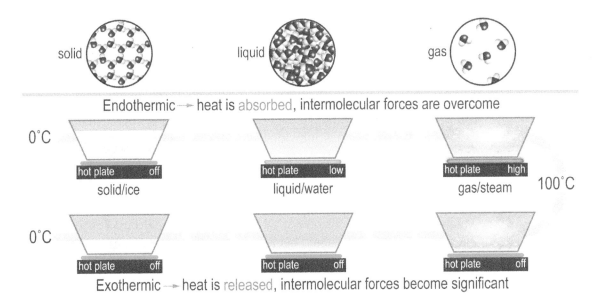

Exothermic and endothermic changes

Phase changes can be classified by whether they absorb or release heat. If you are adding energy to cause a change to happen, then the process is **endothermic**. An example of this is going from a liquid to a gas. Energy must be added for this phase change to occur. According to the law of conservation of energy, the same amount of energy must be released when the process happens in reverse. So when a gas is turning back into a liquid, energy is being given off. Energy release is an **exothermic** process.

Melting and the heat of fusion

Phase change requires energy

It takes energy to rearrange the tighter-bound particles in a solid into the looser-bound particles in a liquid. This is true even when the temperature stays constant! A good way to demonstrate this is to add heat at a constant rate to solid ice. The graph below shows what happens when heat is added to ice that starts at –50 °C. At first, the added heat causes the temperature to increase - as expected. However, once the ice reaches 0 °C, the temperature stops increasing even though energy is still being added! Once the ice starts melting (changing phase) into liquid water, any added heat is absorbed by the process of changing solid into liquid. Energy absorbed by the phase change is NOT available to also raise the temperature.

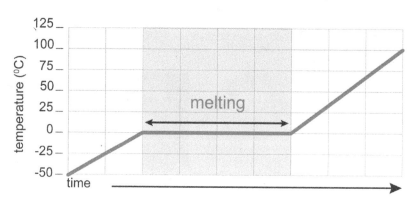

Why temperature stays flat at 0°C

The heating curve stays flat (no temperature rise) until all the solid has been converted into liquid. Once the sample is all liquid the temperature starts to rise again. This can easily be observed with an ordinary thermometer in a well-stirred experiment. The flat section of the graph clearly shows that 0 °C is the temperature at which the phase change occurs. The graph also shows that it takes much more energy to melt the ice into liquid than it does to warm the ice from –50 °C to 0 °C. In general, it takes more energy to change phase than it does to change temperature.

Melting point

The temperature at which a substance changes phase from liquid to solid (or solid to liquid) is called the **melting point**. Melting occurs when the thermal energy of individual particles becomes larger than the attractive forces between particles. Temperature does not change while ice melts because the added energy does not impact particle motion or kinetic energy. Instead, the energy goes into loosening the attractive forces so the particles change their state from solid to liquid.

The heat of fusion

The **heat of fusion** is the amount of energy it takes to change one gram of material, at its melting point, from solid to liquid, or from liquid to solid. The table below gives some values for the heat of fusion (ΔH_{fus}) for common materials. It takes 335 J of energy to melt 1 g of ice into liquid water. This also means that 1 g of liquid water freezing into ice would release 335 J of energy.

heat of fusion for selected substances

substance	heat fusion, ΔH_{fus} (J/g)
water	335
aluminum	321
iron	267
ammonia (NH$_3$)	33
silver	88

Boiling and heat of vaporization

The boiling point

A liquid changes to a gas (boils), or a gas changes to a liquid (condenses) when the temperature is at the **boiling point**. Oxygen boils at -183 °C (-297 °F) and is therefore a gas at room temperature. Water boils at 100 °C (212 °F) and is therefore a liquid at room temperature. Both boiling points are given at a pressure of one atmosphere (1 atm). The boiling point depends strongly on pressure.

Phase change diagram for water

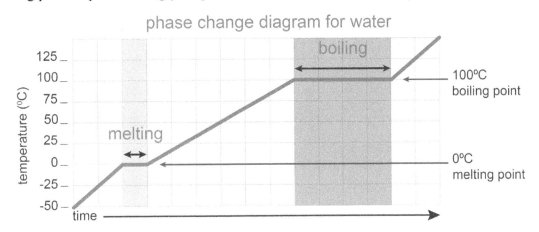

Boiling on a temperature plot

The graph above is called a phase change diagram. It shows the temperatures at which melting and boiling occur. The phase change diagram looks different for different substances. The diagram shows the temperature of water as heat is added at a constant rate. When the temperature reaches 100 °C the added heat is absorbed by the liquid to gas phase change. The temperature stays constant at 100 °C until all the liquid has been boiled into gas (steam). The temperature starts rising again when all the liquid has been converted to gas.

Heat of vaporization

It takes much more energy to completely break intermolecular attractions between particles in boiling compared to overcoming the intermolecular bonds in melting. The **heat of vaporization** (ΔH_{vap}) is the amount of energy it takes to convert one gram of liquid at its boiling point into one gram of gas at the same temperature. The table below shows some values for the heat of vaporization. Note that it takes 335 J to melt a gram of ice into water but 2,256 J to turn 1 g of boiling water into steam! For all substances the heat of vaporization is typically much larger than the heat of fusion.

heat of fusion and vaporization for selected substances

substance	heat of fusion, ΔH_{fus} (J/g)	heat of vaporization, ΔH_{vap} (J/g)
water	335	2,256
aluminum	321	11,400
iron	267	6,265
ammonia (NH$_3$)	33	6,265
silver	88	2,336

Condensation

Water's high heat of vaporization also means that 2,256 J of energy is released when 1 g of steam at 100 °C condenses back into liquid water. This is the reason why burns from steam are often much more severe than burns from hot water. When hot steam condenses on skin the heat of vaporization is released. A similar (but safer) phenomenon occurs when dew condenses on plants at night. As water vapor turns back into liquid it warms both the air and the plants.

Section 4.3: Heat and Changes

Evaporation and condensation

Why do you sweat?

How does sweating cool your body and what does it have to do with phase changes? When your body temperature is too high, your nervous system stimulates sweat glands to release water. The water from your sweat then evaporates. During **evaporation**, molecules change from the liquid phase into gas at a temperature below the boiling point!

EVAPORATION

Surface molecules have no water molecules above to bump them back in.

Deeper molecules are bumped back down by water molecules above.

Molecular explanation of evaporation

Water molecules at the surface of a sample do not have the same fate as water molecules deep below the surface. All molecules are constantly moving and bumping around in all directions. Deeper molecules stay in the water, but molecules on the surface have no molecules above them to bump them back down into the water. When a surface molecule gains enough energy from a collision with a molecule below it, the surface molecule is ejected from the water and evaporates as a gas into the air.

Evaporation carries heat away from liquids

Particles in the gas phase have more energy than in the liquid phase. Evaporation of one gram of liquid to gas transfers the heat of vaporization from the liquid to the gas. Every gram of water that evaporates from your skin carries away $\Delta H_{vap} = 2{,}256$ J of heat. This loss of energy cools the remaining liquid. That is why wet skin after a shower or swim feels cooler than dry skin. Evaporating water, or sweat, transfers heat away from your skin.

Rate of evaporation

The rate of evaporation depends on the temperature of the system and the surroundings, the boiling point of the liquid, and how saturated the air already is. Liquids with lower boiling points evaporate faster. If you have ever put nail polish remover (acetone) on your hand, you know that it feels cold and evaporates quickly. Water has a higher boiling point than acetone and therefore evaporates slower. Evaporation also increases with temperature. The closer the temperature is to the boiling point the faster the rate of evaporation.

Condensation

The opposite process of evaporation is **condensation**. Condensation occurs when a molecule in the gas phase returns to the liquid phase. Morning dew on grass forms from water vapor (gas) in the air that has condensed back into liquid. Condensation is a warming process. Each gram of vapor then condenses and releases the heat of vaporization to the liquid, the air, and the surface on which condensation occurs.

CONDENSATION

Molecules in the gas phase release energy to their surroundings as they return to the liquid phase.

Phase changes and the water cycle

Evaporation, condensation and saturation

Water molecules in the gas phase that randomly hit a liquid water surface sometimes stick and condense into liquid. Molecules in the liquid phase sometimes escape into gas and evaporate if they are near the surface. Evaporation and condensation are the forward and reverse processes in the equilibrium between liquid and gas. Two factors determine whether production of liquid or gas will be favored:

1. Concentration: If more water molecules are in the air, then more water molecules hit a liquid surface and condense.
2. Temperature: The higher the temperature, the more energy the molecules in the liquid have, and the faster they evaporate from the surface.

When the two processes occur at equal rates, the amount of water that evaporates equals the amount that condenses and the air is saturated. Saturated air cannot accept any more evaporating water molecules because an equal amount of water condenses.

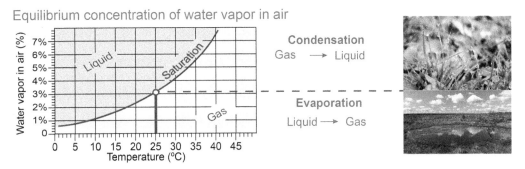

When condensation occurs

The graph above shows the saturation concentration of water in air is about 3.1% at a temperature of 25 °C. If air that already has 3.1% water vapor gets warmer, it is no longer saturated and more water can evaporate. As air temperature increases, so does the amount of water vapor it can hold. However, if air containing 3.1% water vapor gets colder, then some of the water vapor must condense out into liquid.

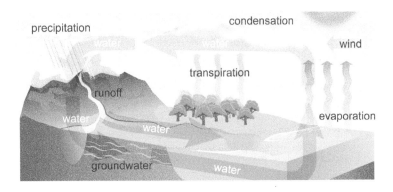

The water cycle

Evaporation and condensation create Earth's water cycle. The sun warms up ocean water and it evaporates. This water vapor rises into the atmosphere. The upper atmosphere is colder so water vapor condenses into tiny droplets of liquid to make clouds. Clouds bring condensed liquid water onto land in the form of rain and snow. Runoff from rain and snow eventually return liquid water to the ocean to repeat the water cycle.

Clouds warm the atmosphere

Clouds are made of liquid water, not vapor. Energy given off during condensation transfers heat from the oceans to the upper atmosphere, warming the atmosphere significantly. Atmospheric water vapor is a significant component of Earth's energy balance.

Energy, chemical changes and physical changes

Energy of physical and chemical change

Evaporation and condensation are physical changes because they do not change water into a different compound. The heat of fusion (ΔH_{fus} = 335 J/g) and the heat of vaporization (ΔH_{vap} = 2,256 J/g) are characteristic of the energies in physical changes. Consider the energy needed to cause the chemical change where water is decomposed into hydrogen and oxygen. The energy needed to cause this reaction is 15,880 J/g!

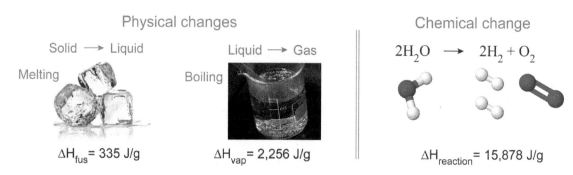

Chemical changes involve much more energy

The chemical change involves 7 - 50 times more energy than the physical change. The energy involved is a major difference between physical and chemical change. Chemical changes involve intramolecular forces (interatomic) which are stronger and take more energy to overcome. Physical changes involve only intermolecular forces, which are weaker and take less energy to overcome.

Exothermic changes

An exothermic change releases heat during the change. Exothermic changes can be either physical or chemical. Freezing is an exothermic process even though you might not associate heat with freezing. Your refrigerator must extract heat from liquid water to freeze ice cubes. Condensation is also an exothermic process. Condensing water vapor into droplets in clouds warms the upper atmosphere. The chemical reaction that combines hydrogen and oxygen to produce water is exothermic. In fact, this reaction is so energetic it is used to fuel rocket engines and has been investigated to fuel cars.

$$\text{Endothermic changes}$$

Melting	Boiling	Dissociation
Solid + 335 J/g $\rightarrow$ Liquid	Liquid + 2256 J/g $\rightarrow$ Gas	$2H_2O + 15{,}878 \text{ J/g} \rightarrow 2H_2 + O_2$

$$\text{Exothermic changes}$$

Freezing	Condensing	Combustion
Liquid $\rightarrow$ Solid + 335 J/g	Gas $\rightarrow$ Liquid + 2256 J/g	$2H_2 + O_2 \rightarrow 2H_2O + 15{,}878 \text{ J/g}$

Endothermic changes

Endothermic changes absorb heat energy during the change. Endothermic changes do not occur without a source of energy. Both melting and boiling are endothermic changes. Both require the input of heat energy to cause the change. The decomposition of water into hydrogen and oxygen gas, known as **electrolysis**, is also an endothermic change. Electrolysis is typically done with electric current supplying the input energy. The instant cold pack is another common example of an endothermic change. A cold pack contains separate compartments of ammonium salt and water. When you squeeze the cold pack, you break the compartments open so the components can mix. Once they mix, they draw in heat as they react, and feel cold. If you touch the cold pack, it will take heat away from you.

Section 3 review

Chapter 4

Adding or removing heat to a substance creates a physical change. That change could be the substance remaining in the same phase (solid, liquid, gas or plasma) and increasing temperature, or at certain temperature points it can mean a change in phase. Two such points are the melting point and the boiling point and are different for different substances. At the melting point or boiling point all heat energy, added or removed, is used to change phase rather than temperature. The amount of energy needed per gram to change phase for a particular substance is known as the heat of fusion for melting or freezing, and the heat of vaporization for boiling or condensing. It can be used to calculate the amount of heat exchanged in a phase change.

Vocabulary words: phase change, endothermic, exothermic, melting point, heat of fusion, boiling point, heat of vaporization, evaporation, condensation, electrolysis

Review problems and questions

1. Is melting endothermic or exothermic? Explain your answer.

2. Make up a 1-paragraph story about how this energy bar diagram was made. In your story, identify the system and its surroundings, explain the direction energy moved through the system and surroundings, and be creative!

3. Why do water molecules at the surface of a body of water evaporate more easily than molecules just below themselves?

5. Why are sections b and d flat in the phase change diagram shown to the right?

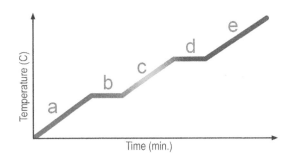

5. Based on where phase changes are happening in the diagram above, which letter best represents how the ΔH_{fus} of this substance compares to its ΔH_{vap}?

    a. The ΔH_{fus} is much smaller than the ΔH_{vap}.
    b. The ΔH_{fus} is much larger than the ΔH_{vap}
    c. The ΔH_{fus} is about the same amount as the ΔH_{vap}.

6. Observe the shape of the heating curve diagram as you move from point (c), to point (b), to point (a).

    a. Describe what happens to the substance as you move through that portion of the diagram.
    b. Is this change endothermic or exothermic? Explain your answer.

Color thermometers

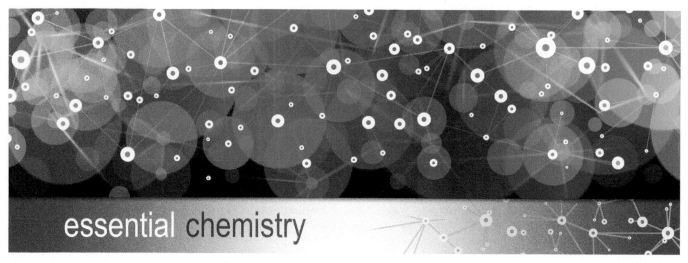
essential chemistry

Have you ever seen a coffee mug where the addition of hot chocolate causes an image to suddenly appear, or used a forehead thermometer that changes color to let you know if you have a fever? There are a lot of items that can change color with a change in temperature. How is this possible, and what does this have to do with chemistry?

That color-changing mug isn't just a fun item; it actually is an application of core chemistry concepts! The image on the mug is printed with a special dye called a leuco dye in the form of microscopic capsules. A leuco dye is a type of material that is thermochromic. As the name implies, this means that it changes color (chromic) with changes in heat (thermo). The dye in the capsule is mixed with a solid compound that is either very acidic or very basic.

Where does the leuco dye get the heat, and why does it change color? If the beverage is at a higher temperature than the mug, then heat will flow from the beverage to the mug. As the heat reaches the microscopic capsules, it provides enough energy to cause a phase change. The solid compound melts! When the compound melts, it causes a chemical reaction near the leuco dye. This causes the leuco dye molecules themselves to subtly change their shape and structure. As the dye molecules change structure, they absorb and reflect light differently causing the image to appear. As the beverage cools, the opposite happens. The compound re-crystalizes and the leuco dye molecules shift back to their original structure and the image fades.

Inexpensive battery testers also use leuco dyes. There are three thin layers in a battery tester. On the bottom is a thin conductive layer. A printed design layer indicating power, is in the middle and a layer containing the leuco dye is on top. By touching the tester in the right spots, electricity will flow through the conductive layer. As the electricity flows, it heats up the conductive layer. Here is where the leuco dye comes in! The heat causes the dye to go from black to clear revealing the graphic. Now you know if your battery is good or if needs to be replaced.

Color thermometers - 2

Leuco dyes are materials that have been used on everything from mugs and battery testers to pencils and clothing. But their temperature sensitivity is rather limited. They can really only tell you if something is "hot" or "not hot". For example, you can find out if your hot chocolate is still hot, but you can't hold a color changing mug to your forehead to check for a fever!

For more precise measurements, you will need to use a different material called a thermochromic liquid crystal. You already know what thermochromic means, but what is a liquid crystal? Liquid crystals are a bit like liquids and a bit like solids and are used for displays in everything from televisions and calculators to cell phones and digital watches. Thermochromic liquid crystals can give you better quantitative information and accuracy than a leuco dye.

If you have used a color-changing thermometer for your forehead, or placed one in an aquarium, then you used something with a thermochromic liquid crystal (TLC). Just like a leuco dye, there has to be some sort of physical or chemical change when heat is added that causes the color to change. In a color changing display, the liquid crystals are in microscopic capsules. Inside the capsules the liquid crystals are arranged in spaced layers, something like matches in a box. When the temperature changes, the liquid crystal's structure twists. This twisting alters the color of light that gets absorbed and reflected by the material. If different wavelengths of light get reflected, then we see a different color. The color change can be finely tuned by changing the type of TLC that is used.

Surprisingly, mood rings also use thermochromic liquid crystals. Modern mood rings, and other mood jewelry, use a thin strip of TLC inside the stone or band. When you buy it, you might think that it will indicate your mood, but really they are just reading your body temperature! Supposedly there is some correlation between your mood and your body temperature. For example, if you are relaxed, then your body temperature should be about normal and the color would be blue-green. If your mood ring turns black, then it might indicate that you are stressed … but we know that it definitely indicates that your finger is cold!

So the next time you see a color changing item, don't just think about the image that will appear, think of how heat, energy, and temperature are related to matter and change.

Section 4.4: Connections

Chapter 4 review

Vocabulary
Match each word to the sentence where it best fits.

Section 4.1

absolute zero	Brownian motion
Celsius scale	Fahrenheit scale
Kelvin scale	kinetic energy
kinetic molecular theory	temperature
thermistor	thermocouple
thermometer	

1. In 1773 it was discovered that extremely small pollen particles danced around even in still water. This was the first observation of _____ caused by the constant collisions of water molecules.
2. _____ is directly proportional to the average kinetic energy of random motion of individual atoms and molecules.
3. The temperature scale where the boiling temperature of water is 212 degrees is the _____.
4. _____ is the energy of motion.
5. Temperature is measured with a device called a(n) _____.
6. _____ is the theoretical temperature at which the kinetic energy of motion of atoms and molecules is so low it cannot get any lower.
7. The _____ sets the boiling point of water at 100 degrees.
8. The idea that matter is composed of small particles in constant motion is the basis for _____.
9. An electronic temperature sensor may use a(n) _____ to sense changes in temperature by the effect of temperature on electrical properties.
10. The _____ is the temperature scale that starts at absolute zero.
11. Temperature can be measured with a thermometer, a thermistor, or a(n) _____.

Section 4.2

adiabatic	calorie
calorimeter	calorimetry
closed system	conduction
first law of thermodynamics	heat
isolated system	joule
open system	second law of thermodynamics
specific heat	surroundings
system	thermal conductor
thermal energy	thermal equilibrium
thermal insulator	

12. One _____ is the amount of heat needed to change the temperature of one gram of water by one degree Celsius.
13. A(n) _____ is a group of interacting objects and effects that are selected for examination.
14. According to the _____, heat always moves from a system of higher temperature to one of lower temperature.
15. Heat does not flow across a(n) _____ very easily.
16. When the temperature of two systems is the same, _____ occurs and heat no longer flows between them.
17. A(n) _____ is made of material that easily permits the flow of heat across it.
18. The study of heat flow in chemistry is called _____.
19. The _____ is sometimes called the law of conservation of energy.
20. The energy required to raise the temperature of 1 gram of a substance by 1 °C is called that substance's _____.
21. The total energy in a collection of molecules is called the _____.
22. The area outside of the system is called the _____.
23. _____ is the thermal energy that is transferred between objects of matter.
24. The SI unit of energy and heat is the _____.
25. In a(n) _____, neither matter nor energy can be exchanged between the system and the surroundings.
26. A(n) _____ allows both matter and energy to cross the system boundary.
27. A(n) _____ is one where only energy can travel between the system and surroundings.
28. The transfer of heat through materials by direct contact of the matter is called _____.
29. A(n) _____ is a device used to measure the energy changes in a chemical process.
30. A process in which no energy is exchanged with the surroundings is called _____.

Section 4.3

boiling point	condensation
electrolysis	endothermic
evaporation	exothermic
heat of fusion	heat of vaporization
melting point	phase change

31. The _____ of a substance is the temperature at which it changes from a liquid to a gas.
32. A chemical reaction is considered _____ if it releases heat.

Chapter 4 review

Section 4.3

boiling point	condensation
electrolysis	endothermic
evaporation	exothermic
heat of fusion	heat of vaporization
melting point	phase change

33. Energy must be absorbed for a substance to melt, therefore melting is an example of a(n) _____ process.

34. The _____ of water is the amount of heat required to change one gram of solid ice into liquid at a constant temperature of 0°C.

35. The _____ is the energy required to completely break the attractions between particles during a phase change.

36. The formation of water droplets on the outside of a cold glass is an example of _____.

37. Through the process of _____, an uncovered liquid in a container will slowly transform into a gas over time.

38. When the organization of molecules in a substance changes, leaving the molecules themselves the same, we say a _____ has occurred.

39. The _____ of a substance is the temperature at which a substance changes from a solid to a liquid.

40. _____ is usually performed with electric current providing the energy needed for the reaction to take place.

Conceptual questions

Section 4.1

41. Explain why a temperature of -280 °C is impossible.

42. Describe how temperature affects the random motion of atoms and molecules.

43. What two properties of matter cause Brownian motion to occur? Explain how these properties result in Brownian motion.

44. How are the Fahrenheit and the Celsius temperature scales different with respect to the boiling and freezing points of water?

45. What units are used to measure temperature? Which units are used most commonly in the lab internationally?

46. What is the relationship between temperature and the kinetic energy of a system?

47. Why does a gas have a larger range of kinetic energies than a solid at the same temperature?

48. « What happens to the range of random kinetic energy for a collection of atoms as the temperature increases?

Section 4.2

49. Warmer than normal temperatures due to human alterations to natural surfaces is a consequence of a phenomenon called the "heat island effect." Research the heat island effect and explain the pros and cons of at least 3 approaches to avoid creating a heat island. Make sure to consider the needs of humans as well as the living and nonliving things in the environment.

50. Which picture represents a higher temperature? How can you tell?

51. Copper has a specific heat of 0.34 J/g °C. What does the value of the specific heat tell you about copper compared to water (specific heat of water = 4.184 J/g °C)?

52. Why can adding heat to a substance cause an increase in temperature?

53. You have a 1000 g sample of water at 45 °C and a 1 g sample of water at 50 °C. Which sample would you estimate to have a higher thermal energy? Defend your choice.

54. Samples of two metals are both 10.0 g. One has a high specific heat and the other has a low specific heat. After heating them for 1 minute with the same amount of heat, which one will have the greatest temperature? Explain.

55. A hot bar of steel that weighs 25 g at 90 °C is placed on a 15 g bar of copper at 25 °C. Explain how energy will flow at the molecular level.

56. Explain why a double walled metal vacuum bottle can be used to keep your water cold even though metals are great thermal conductors.

57. Heat is transferred from an object and undergoes no phase change. What data is needed to calculate the amount of heat that is lost?

58. Explain the process of thermal equilibrium in your own words.

59. Explain how the specific heat of a substance is affected when you increase its mass, for example, from 1 g to 10 g?

60. Suppose two metal cubes at different temperatures come into contact. What is the relationship between the heat lost by one and gained by the other if you assume the cubes are an isolated system? Which law states that this relationship must be true?

61. Bomb calorimeters measure the amount of heat absorbed or released during physical or chemical processes. Are bomb calorimeters adiabatic? Why or why not?

Chapter 4 review

Section 4.2

62. Why does your body burn calories when you are sleeping?
63. How can a calorimeter be used to measure the amount of calories stored in food?
64. Sketch an energy flow diagram for the measurement of the amount of energy contained within a cracker. The reaction takes place inside a bomb calorimeter.

Section 4.3

65. Describe the process of evaporation in your own words.
66. A graph shows the average temperature of a mass of water as heat is added at a constant rate. Temperature is on the y-axis and time is on the x-axis. The water begins as solid ice at -20°C and ends as steam at 150°C. Sketch the graph labeling the temperature of important transitions. Identify the solid, liquid and gas phases.
67. Consider compounds with molecular weight similar to water, such as methane, CH_4. The heat of fusion for methane is 59 J/g and the heat of vaporization is 537 J/g. It only takes 539 J to turn methane form liquid to gas. For this reason methane melts at -182.5°C and boils at -164°C staying liquid over a range of only 20.5°C. The heat of vaporization of water (2,260 J/g) is much larger than the heat of fusion (335 J/g). Discuss the implications of these values, and the difference between them, on living organisms.
68. What do you know about the melting point of a substance if it is a liquid at room temperature?
69. Which is always higher for any given substance: the heat of fusion or the heat of vaporization? Explain why this is the case.
70. Given enough time, a glass of water left out at room temperature will eventually evaporate even though it never reaches its boiling point. How can this evaporation occur?
71. Sketch a cooling curve for molten iron at 2,000 °C cooling to room temperature. The melting point of iron is 1,535 °C.
72. What is indicated by the flat portions of a temperature vs time graph?

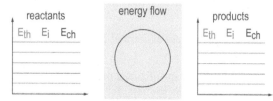

73. Use the energy bar chart above to track the energy when 10 g of room temperature liquid benzene initially freezes.
74. Use the above energy bar chart to track the energy changes when a liquid sample of the organic compound acetone is heated to boiling and evaporates.
75. Use the above energy bar chart to track the energy changes when steam condenses on a glass. Consider that the glass is in the surroundings.
76. Use the above energy bar chart to track the energy changes when some chocolate chips are melted in a microwave.

77. Why does your skin become cooler when you sweat?
78. Is the burning of wood in a barbecue an exothermic or endothermic chemical change? Explain.
79. Suppose there is an insulated flask filled with a mixture of water and ice at equilibrium. What is the temperature of the water? How do you know this?

Quantitative problems

Section 4.1

80. If the average human body temperature in 98.7 °F, what is the temperature in Celsius?
81. The liquid organic compound ethylene glycol is known as antifreeze. It freezes at −12 °C. What is the freezing temperature on the Fahrenheit scale?
82. The flash point of paper is about 451 °F. What is this temperature on the Celsius scale?
83. If liquid nitrogen boils at −196 °C. What is the temperature of boiling nitrogen in Fahrenheit?
84. Convert 243 K to Fahrenheit
85. What is the boiling point of water in Kelvin?
86. Water boils at 100 °C at sea level, but at 93.4 °C at 6,600 feet altitude. What is the temperature difference in kelvin?
87. Zinc is most commonly used as an anti-corrosion agent, and galvanization is the most familiar form. It has a melting point of 787.2 °F and a boiling point of 1,665 °F.
    a. What is the melting point temperature in kelvin?
    b. What is the boiling point temperature in kelvin?
    c. A piece of zinc metal is cooled from 1,408 °F to 822 °F. Calculate the temperature difference in kelvin.

Section 4.2

88. You are trying to determine the specific heat of an unknown liquid. You heat a steel block that has a mass of 108 g to 97.2 °C. The steel block is placed into an insulated flask containing 124 g of the unknown liquid at 23.2 °C. The final solution temperature stabilizes at 35.1 °C. If the specific heat of steel is 0.470 J/g °C, what is the specific heat of the liquid?
89. Adding 8.56 g of $KNO_3(s)$ to 250.0 g of water resulted in a temperature decrease from 25.0 °C to 22.3°C. Calculate the heat change for dissolving KNO_3 in water. Assume the solution has the same specific heat as water (4.184 J/g °C) and that no heat was lost to the atmosphere.
90. How many joules of heat are needed to increase the temperature of 19.3 g of lead from 22 °C to 41 °C? (Specific heat of lead = 0.128 J/g °C)
91. 47.6 g of an unknown metal substance absorbs 440 J of energy and changes temperature from 25 °C to 49 °C. Identify the metal by comparing your answer against the table of physical properties in the appendix.

Chapter 4 review

Section 4.2

92. 32 grams of pure gold are taken out of a jeweler's heating bath at a temperature of 180°C. The gold cools to room temperature of 21°C. How much heat is given up to the surroundings? The specific heat of gold is 129 J/g°C.

93. A 86.5 kg piece of aluminium metal is heated from 23.1 °C to 31.0 °C. Calculate the heat absorbed in kJ by the metal. (Specific heat of Al = 0.900 J/g°C)

94. What is the change in temperature for an iron block that absorbs 918 J of heat if the iron block weighs 67.5 g? (Specific heat of Fe = 0.45 J/g °C)

95. A 92.7 g sample of a metal was heated to 99.7 °C and then plunged into 250 g of water at 23.0 °C. The resulting mixture obtained a temperature of 30.0 °C. (The specific heat of water is 4.184 J/g °C)
    a. How many joules of heat did the water absorb?
    b. How many joules of heat did the metal lose?
    c. What is the specific heat of the metal?

96. An insulated flask contains 137 g of water at 21.4 °C and a second beaker contains 62.9 g of water at 98.7 °C. What is the final temperature of the water after the hot water is poured into the insulated flask? (The specific heat of water is 4.184 J/g °C)

97. A 99.5 °C piece of lead has a mass of 84.1 g. It was added to a beaker containing 65.10 g of water at 27.3 °C. When the water and the metal come to thermal equilibrium, the temperature is 30.7 °C. What is the specific heat of lead? (The specific heat of water is 4.184 J/g °C)

98. 56.4 g of iron (specific heat = 0.450 J/g °C) was heated to 109.3 °C. The iron was dropped into 50.0 g of an unknown solution at 25.3 °C. If the final temperature was 38.1 °C:
    a. How many kilojoules of heat were absorbed by the solution?
    b. What is the specific heat of the solution?

99. An insulated container filled with 100.0 mL of water contains a 0.593 kg cube. The cube and water are in thermal equilibrium at 25.2 °C. After another 100.0 mL of 99.7 °C water are added to the container, the temperature steadies at 49.3 °C. What is the specific heat of the cube? (Specific heat of water = 4.184 J/g °C)

100. The specific heat of liquid ethanol is 2.46 J/g °C. What mass of ethanol releases 357 J of energy to the surroundings when it cools by 37 °C?

101. A 5.77 g pure gold ring is advertised at an amazing price. You decide to test whether or not the ring is authentic by measuring its specific heat. You measure 50.1 g of 25.3 °C water into an insulated flask. After heating the ring to 99.8 °C, you add the ring to the flask. After thermal equilibrium is reached, the temperature is 26.7 °C. If the specific heat of gold is 0.129 J/g °C, is the ring made of pure gold? Support your answer with calculations.

102. A piece of cheesecake is burned in a calorimeter. It heats 10,000.0 g of water from 25.0 °C to 65.0 °C. If one pound of fat is equal to 3,500 Calories, what is the maximum number of pounds gained consuming the cheesecake.

103. A cookie is burned in a calorimeter. It heats 400.0 g of water from 25.0 °C to 65.0 °C. How many calories and Calories are in the cookie?

104. A nut is burned in a calorimeter. It heats 200.0 g of water from 25.0 °C to 45.0 °C. How many calories and Calories are in the nut?

105. A cracker is burned in a calorimeter. It heats 550.0 g of water from 25.0 °C to 75.0 °C.
    a. How many Calories is the cracker?
    b. According to the chart above, how many minutes would you have to swim to burn off the calories in the cracker?

106. A hamburger is burned in a calorimeter. It heats 6,250.0 g of water from 25.0 °C to 85.0 °C. According to the chart above, how long do you have to jog to burn off the calories in the hamburger?
    a. How many Calories is the hamburger?
    b. According to the chart above, how long do you have to jog to burn off the calories in the hamburger?

chapter 5: Chemical Compounds

How can patterns help with naming chemicals?

At first glance, the world can look disorganized and chaotic. But if you look at things a little differently, you will see that nature is full of patterns. Plants have leaves or flowers that are symmetrical—meaning they are the same on both sides. Snowflakes also exhibit a symmetrical pattern. Snowflakes may be different from one another, but if you look close enough, the "arms" on an individual snowflake are symmetrical. Spirals are another common pattern in nature. You can see examples of this on the horns of a ram, a leaf arrangement on plants and even shells of a mollusk.

Patterns are not limited to individual objects. Sometimes patterns only become obvious when you look at something over a long period, or look at a larger collection of similar things. Throughout the year, seasonal weather changes exhibit patterns, as do the phases of the moon. The DNA double helix is itself a pattern and the molecules that make it up always combine in patterns of specific pairs.

Being able to recognize patterns is important in science, mathematics and engineering because it causes us to question the form, function and underlying principles of the patterns. In fact, one of the biggest advances in all of science came as result of the careful organization of elements based on their properties and recognizing the patterns that emerged. Knowing about these patterns helps us understand, classify and name the world around us.

Chapter 5 study guide

Chapter Preview

Countless scientific discoveries have resulted from recognizing patterns in nature. In this chapter you'll learn how Mendeleev created the first periodic table by noticing patterns among the known elements. He organized the elements in rows that demonstrate trends when moving along them and groups or families of elements that all share similar properties. Over time many more elements were added, resulting in the periodic table we use today. Organized by increasing atomic number, the periodic table provides information about how two elements will or won't react. For instance, metals tend to lose electrons to form positively charged ions while non-metals tend to gain electrons to form negatively charged ions. Together they can form neutral ionic compounds. In contrast, two non-metals tend to share electrons, forming covalent bonds. Ionic and covalent compounds have very different properties. Nomenclature is a set of rules for naming these covalent and ionic compounds.

Learning objectives

By the end of this chapter you should be able to:

- identify major groups of elements on the periodic table;
- predict properties of an element based on its location on the periodic table;
- relate the charge of an ion with the number of electrons lost or gained;
- classify a compound as ionic based on its properties;
- name ionic compounds and write ionic formulas;
- name covalent compounds and write covalent formulas;
- identify the properties of covalent compounds; and
- distinguish between ionic and covalent compounds.

Investigations

5A: Patterns and trends

5B: Naming ionic compounds

5C: Store labels and model building

Pages in this chapter

128 The Past and Present Periodic Table
129 Observations leading to the periodic table
130 The modern periodic table
131 Organization of the periodic table
132 Atoms and elements
133 Chemical bonds
134 Section 1 Review
135 Ions
136 Cations and anions
137 Ionic compounds
138 Ionic crystal structures
139 Properties of ionic compounds
140 Polyatomic ions
141 Section 2 Review
142 Naming and Writing Ionic Compounds
143 The crisscross method: Binary compounds
144 Counting atoms and counting ions
145 Ternary and quaternary ionic compounds
146 The crisscross method: Ternary compounds
147 Ionic formulas with Roman numerals
148 Section 3 Review
149 Covalent Compounds
150 Diatomic molecules
151 Molecular compounds
152 Properties of covalent compounds
153 Naming and writing covalent formulas
154 Mixed chemical formula review
155 Section 4 Review
156 Ban DHMO!
157 Ban DHMO! - 2
158 Chapter review

Vocabulary

periodic
cation
formula unit
polyatomic ion
ternary compound
diatomic molecule

family
anion
ionic compound
binary compound
quaternary compound
network covalent

ion
ionic bond
dissociation
nomenclature
covalent bond
nanotechnology

5.1 - The Past and Present Periodic Table

Chemistry evolved from the mysterious art of alchemy into the modern science we know today. By the 1860s, much progress had been made in the science but there was no convenient way to organize information about elements. Scientist and chemistry professor Dimitri Mendeleev was looking for some kind of pattern he could use to organize elements in a way that made sense. Little was known about atoms in Mendeleev's time, except that elements could either be arranged by atomic mass, or by physical and chemical properties. Mendeleev's careful work laid the foundation for today's periodic table. To be a successful problem-solver like Mendeleev, you have to look for clues that give you ideas on how to solve a problem. You also must systematically test your ideas and keep track of the results.

Dimitri Mendeleev

The early periodic table

Looking for patterns

Mendeleev approached the task of looking for patterns like a game. First, he wrote the properties of the 63 known elements on cards. Then he arranged the element cards first by atomic mass, already an accepted way of organizing elements. His big breakthrough was seeing a repeating pattern of physical and chemical properties among the elements. He used these patterns to arrange elements into the groups shown in the table below. None of the group 18 elements had been discovered yet, but they would all fit perfectly in a row beneath F, Cl, Br and I in the table.

Repeating chemical and physical properties

In the table above, find the circled region between zinc, Zn, and arsenic, As. Mendeleev placed two question marks after zinc because he did not think it made sense to place arsenic, the next known element, immediately after zinc. Arsenic's chemical and physical properties most closely matched phosphorus, P, and antimony, Sb. For this reason he placed arsenic between phosphorus and antimony instead of between aluminum, Al and indium, In. Mendeleev proposed there must be undiscovered elements, and he predicted their properties. Many scientists did not accept Mendeleev's work at first, but within 15 years, the first missing element was discovered—and Mendeleev's predicted properties were correct!

Mendeleev predicted unknown elements

Mendeleev made the observation that chemical and physical properties of elements followed a **periodic** pattern. This means patterns in the way elements looked and behaved followed a regular, repeating pattern. This is how the periodic table got its name. Each new row on the periodic table begins so elements with similar properties end up in the same group. If you know the properties of sodium, you can make some educated guesses about the properties of the next element below it, potassium.

Observations leading to the periodic table

Density is periodic

Mendeleev observed repeating patterns in physical properties such as density. When he graphed the density versus the atomic mass of the known elements, Mendeleev found that the density rose and fell in a repeating cycle. From this graph you can see that H, Li, Na, K, Rb, and Cs have low densities. Where are these elements located on the periodic table? These low-density elements all belong in Group 1 on the modern periodic table.

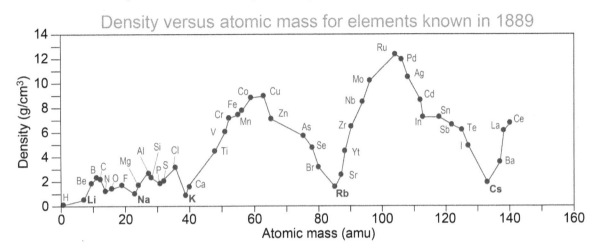

Chemical properties are periodic

In Mendeleev's time, the ratio of either hydrogen or oxygen in a chemical formula was known. When elements are arranged in order of increasing atomic mass and then ordered by the compounds they form with oxygen or hydrogen, another pattern emerged. The repeating pattern in density matches up well with the repeating property of forming an oxide or a hydride. Look for the patterns in the chemical formulas in each column below.

first six rows of oxides and hydrides arranged by Mendeleev

	R_2O	RO	R_2O_3	RH_4 / RO_2	RH_3 / R_2O_5	RH_2 / RO_3	RH / R_2O_7	similar propeties, or RO_4
1	H_2O							*arranged in order stable oxide formed: RO_4, RO_3, RO_2, RO
2	Li_2O	BeO	B_2O_3	CO_2	N_2O_5	OH_2	FH	Fe Co Ni Cu
3	Na_2O	MgO	Al_2O_3	SiO_2	P_2O_5	SO_3	ClH	Ru Rh Pd Ag *
4	K_2O	CaO	?	TiO_2	V_2O_5	CrO_3	MnH	Os Ir Pt Au *
5	Cu_2O	ZnO	?	?	As_2O_5	SeO_3	BrH	
6	Rb_2O	SrO	Yt_2O_3	ZrO_2	Nb_2O_5	MoO_3	?	

Many revisions over time

Mendeleev published a revised periodic table based on oxide and hydride formulas a few years later. It resembled the table above. Thanks to the work of many scientists, it would continue to be revised until it took on its current form in the 1940s.

The noble gases

The important "turnaround" elements for each row, the noble gases, were still unknown during the time Mendeleev was working out the periodic table. While helium was discovered in 1868, the discovery was made observing spectral lines in the Sun! Scientists named the hypothetical new element helium after Helios, the Greek god of the sun. It would be 27 more years before helium was discovered on Earth in 1895. Argon was first observed in 1894, neon in 1898, and xenon also in 1898.

The modern periodic table

In order of atomic number

The accepted atomic model of Mendeleev's time incorrectly assumed the smallest particle of matter was the atom. Subatomic particles like protons were not yet discovered so the logical ordering by atomic number was not obvious. Today's table includes many more elements and is ordered by atomic number, not by atomic mass.

Patterns repeat in rows

Today's periodic table is similar to Mendeleev's system. Elements with similar chemical properties are kept together in vertical columns (groups). The beginning and end of each horizontal row (period) represents a repeated pattern of physical or chemical properties. One example of a property that occurs across a period is metallic character. The most metallic metals are found on the left side of the periodic table. As you move from left to right, atoms become less metallic. More patterns will emerge as we continue to develop the properties of the elements.

Atomic mass

Below each element symbol is the atomic mass in units of atomic mass units (amu). This is the average mass of a single atom. In future chapters we will also interpret the atomic mass as the mass of one mole of atoms in units of grams. Both interpretations are correct.

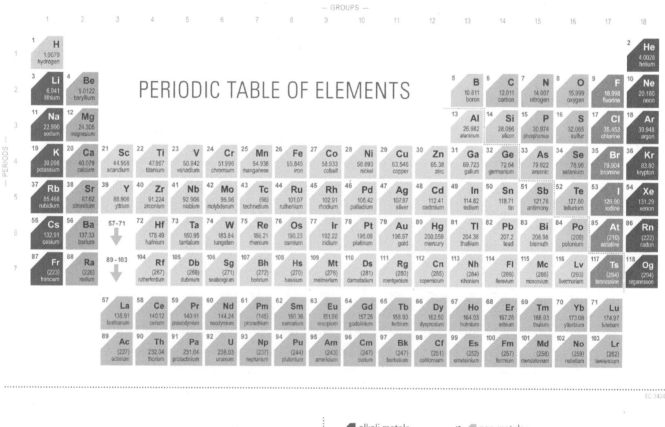

Organization of the periodic table

Groups are also called families

Elements that share physical and chemical characteristics are placed together in groups. Each group is called a *family* because family members have similar physical and behavioral traits. Each family has a name. For example, the most reactive metals are placed together in group one, which is called the alkali metal family.

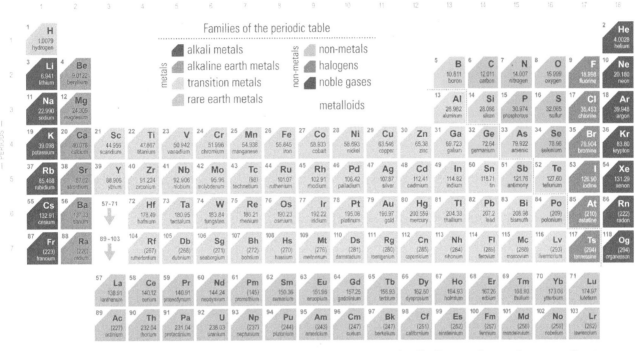

Alkali and alkaline earth metals

Group 1 elements, except hydrogen, are named alkali metals, and group 2 are named alkaline earth metals. Alkali and alkaline metals are soft, shiny, and so reactive that they are rarely found in their elemental form in nature. They instantly react with oxygen in air to form a white film similar to rust on their surface. Some are so reactive, they explode on contact with water!

Halogens and noble gases

Halogens in group 17 are also highly reactive, but they are non-metals. The most reactive element on the periodic table, fluorine, is a halogen. The neighboring noble gases in group 18 are the least reactive elements on the periodic table. The term inert is also used to describe a lack of reactivity. You are able to take in helium gas from a balloon to temporarily change the sound of your voice without worry of chemical harm to your body because the element helium is an inert noble gas.

Lanthanides and actinides

The lanthanide elements span from lanthanum through lutetium. These elements belong in period 6, and all are in group 3 because they have similar chemical behavior. Lanthanides are also known as rare earth metals. Actinides include actinium through lawrencium. Actinides belong in period 7 and only the first four are naturally occurring. Uranium is the last naturally occurring element; everything past uranium is created by scientists in laboratories. All actinides are unstable and radioactive.

Main group and transition

Groups 1-2 and 13-18 are called the main group elements while transition metals span from groups 3-12. The lanthanides and actinides are also known as the inner transition metals. Transition metals have typical metallic properties like malleability and conductivity, but they are less reactive than the alkali and alkaline earth metals.

Atoms and elements

The meaning of the atomic number

Elements are made of atoms and atoms themselves are made of protons, neutrons and electrons. The atomic number is equal to the number of protons in an atom of a specific element. In a neutral atom, the atomic number is also equal to the number of electrons in the atom.

1. All elements are made of atoms.
2. The smallest particle of any element is one atom.
3. Atoms of the same element are alike and are different from atoms of all other elements.
4. The atomic number is the number of protons (or electrons) in atoms of that element.

Protons, neutrons, and electrons

Atoms are made of smaller particles we call electrons, protons, and neutrons. Protons have an electric charge of +1 and attract the electrons which have an electric charge of -1. Positive and negative electric charges attract each other and this attraction is the force that holds atoms together. In a different way the force between electric charges also creates the bonds between atoms in compounds.

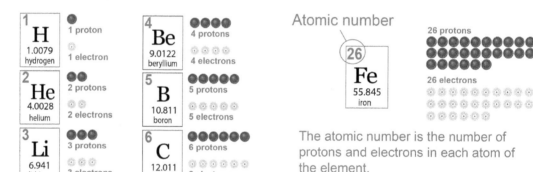

The nucleus

Inside each atom all of the protons and neutrons are together in the very center in a very tiny nucleus. The nucleus holds all the positive charge and virtually all the mass of the atom. Neutrons "glue" the protons together but are otherwise "non-players" in ordinary chemical changes because they stay confined in the nucleus deep inside each atom.

The electron cloud

The atom's volume is occupied by the electrons which form a "cloud" around the tiny nucleus. The number of electrons and protons are equal so the total electric charge of an atom exactly zero. Because the electrons form the outside of the atom, electrons determine all the chemical properties of the elements. Virtually everything in chemistry comes from electrons of one atom interacting with electrons of neighboring atoms.

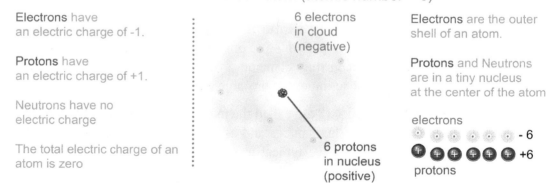

132   Chapter 5: Chemical Compounds

Chemical bonds

Chemical bonds

To form compounds, atoms make chemical bonds with each other. Chemical bonds are formed by electrons in neighboring atoms. Inside the atom, electrons are organized in energy levels by the rules of quantum mechanics (more in a future chapter). When atoms are near each other, their electrons can lower the overall energy by forming a chemical bond. Each element has a different number of electrons and makes different types and numbers of bonds with other elements. There are three main kinds of chemical bonds, though many bonds have varying degrees of electron interaction.

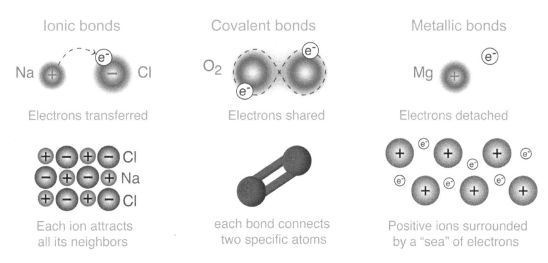

Ionic bonds

Ionic bonds form when one atom "transfers" an electron with another atom. Sodium chloride (NaCl) is an example of a compound formed with ionic bonds. The chlorine atom gains an electron and becomes a negatively charged ion, Cl^-. The sodium atom loses an electron and becomes a positively charged ion, Na^+. The negative chlorine ion attracts all nearby sodium ions and the positive sodium ion attracts all neighboring chlorine ions. An ionic bond is not localized to a single pair of atoms.

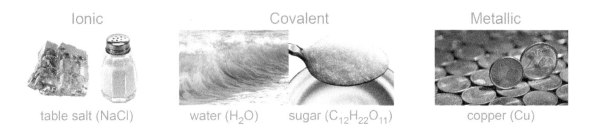

Covalent bonds

Covalent bonds form between atoms that are identical or close to each other on the periodic table, usually in non-metals. The bonds between hydrogen and oxygen atoms in H_2O are covalent bonds. Each covalent bond is a single shared electron pair. Covalent bonds are local - they connect two specific atoms.

Metallic bonds

Metallic bonds form between atoms that have very loosely bound outer electrons. For example, the outer electrons in solid copper form a "sea" around the positive ions. The electrons that participate in bonding are completely disassociated from any single atom. That means the electrons are able to move freely and this is the reason metals such as copper conduct electricity. Metallic bonds are similar to ionic bonds in that the forces act between many atoms and electrons instead of between a single pair of atoms as with covalent bonds.

Chapter 5

Section 1 review

Scientific discoveries often begin with the recognition of patterns. Dimitri Mendeleev created the first periodic table by sorting the elements based on their properties, a pattern which even predicted as yet undiscovered elements. Over time the periodic table was refined to its present form as understanding grew. Repeating or periodic patterns of physical or chemical properties are found when moving across each row or period. The columns are called groups or families because the elements have similar characteristics to one another and can be identified by the group number or the family name. The main group elements are groups 1, 2, and 13 to 18. Groups 3-12 are the transition metals, including the lanthanides and actinides, referred to as the inner transition metals.

Elements are made of atoms and a single atom is the smallest particle of any element. Atoms are made from protons, neutrons, and electrons. The atomic number is the number of protons (and sometimes, electrons) in an atom. Chemical bonds occur through interactions of the electrons in neighboring atoms. There are three basic types of chemical bonds: ionic, covalent, and metallic.

Vocabulary words: periodic, family

Review problems and questions

1. Identify the family names of Groups 1, 2, 17 and 18.

2. What are the elements in the middle of the periodic table called?

3. What property of the noble gases makes them "noble"?

4. What did Mendeleev look for when arranging the periodic table?

5. In the sample of the periodic table above, which of the elements has 32 electrons and how do you know?

6. A chemical bond is an interaction between two atoms formed by which of the following?
   a. protons
   b. electrons
   c. neutrons

134 Chapter 5: Chemical Compounds

5.2 - Ions

If you have ever seen lightning or received a static shock you have experienced electric charge. Everything, including you, is made of atoms that have positively and negatively charged particles. Atoms in carpet fibers transfer a tiny fraction of electrons to your shoes sliding across the carpet on a dry day. If your shoes collect enough extra electrons, your body releases them with a zap when you touch a metal object such as a door knob. Individual atoms can also transfer electrons to other atoms to create ions.

The charge of ions

Ions are atoms with a + or - charge

An **ion** is an atom or molecule that is either positively or negatively charged. An ion is created when a neutral atom accepts or donates one or more electrons. An electron has a charge of 1–. An atom that accepts an electron becomes a negatively charged ion with a charge of 1–. An atom that donates an electron becomes a positively charged ion with a charge of 1+.

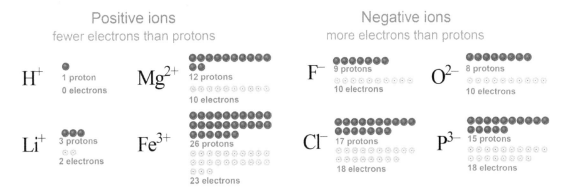

Unequal protons and electrons

Ions have unequal numbers of protons and electrons. Negative ions such as Cl⁻ have more electrons than protons. Positive ions such as Na⁺ ion have fewer electrons than protons. For example, a neutral magnesium atom has 12 protons and 12 electrons. The magnesium ion has a 2+ charge and only 10 electrons.

Writing the symbol for ions

The symbol for the magnesium 2+ ion is Mg^{2+}. Note that the charge of the ion is written as a superscript after the chemical symbol. If the charge is 1+ or 1– the plus and minus symbols are used without a number such as Na^+ or F^-. For charges other than 1+ or 1– the number is written before the plus or minus sign in the superscript, such as Mg^{2+} or O^{2-}.

Solved problem

Calculate the number of electrons each ion has lost or gained: Li^+, S^{2-}, Al^{3+}

Asked Use atomic number and charge to determine how many electrons have moved into or out of each ion.

Relationships Positive charges result from electron loss, negative charges result from electron gain. Use atomic number for number of protons. Compare the number of protons to the charge to calculate the number of electrons lost or gained.

Solve Li^+: Positive charges indicate lost electrons: 1+ = 3 protons vs. 2 electrons
S^{2-}: Negative charges indicate gained electrons: 2– = 16 protons vs. 18 electrons
Al^{3+}: Positive charges indicate lost electrons: 3+ = 13 protons vs. 10 electrons

Answer Li^+ lost 1 electron; S^{2-} gained 2 electrons; Al^{3+} lost 3 electrons.

Cations and anions

Metals form positive cations

Elements on the left side of the periodic table, including all common metals, are likely to lose electrons to become positive ions. A positively charged ion is called a **cation**. Notice how group 1 alkali metals all have a 1+ charge. Lithium (Li^+), sodium (Na^+), and potassium (K^+) exclusively form cations with a charge of 1+. Group 2 alkaline earth metals form cations with a 2+ charge such as Be^{2+}, Mg^{2+}, and Ca^{2+}. The transition metals are more complex and can form cations with different charges depending on which compound they are in. For example, iron can form both 2+ and 3+ cations Fe^{2+} and Fe^{3+}. A complete periodic table with ion charges is available in the navigation bar in the digital textbook. Look for the elements symbol.

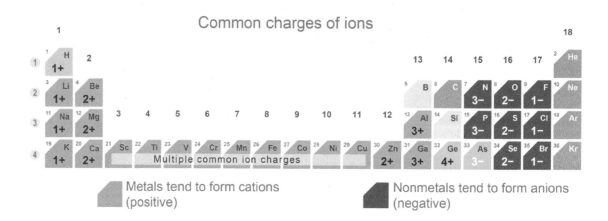

Non-metals form negative anions

Non-metals typically gain electrons to form a negative **anion**. All of the halogens exclusively form anions with a charge of 1– including fluorine (F^-), chlorine (Cl^-) and bromine (Br^-). The oxygen group elements tend to form ions with a charge of 2– including oxygen (O^{2-}), sulfur (S^{2-}) and selenium (Se^{2-}).

Hydrogen usually forms a cation

Hydrogen is in group 1 even though it is a non-metal because it is very reactive, just like the alkali metals. Even though hydrogen is a no-nmetal, hydrogen atoms almost always lose their one electron to make H^+ ions. Under certain conditions hydrogen will also form the hydride ion, H^-, such in the photosphere of the Sun.

Metalloids and noble gases

Metalloids have properties of metals and non-metals. Boron, a metalloid, tends to give up its valence electrons and become a cation. Silicon and carbon can become cations or anions depending on the other elements in the compound. On the far right of the periodic table (group 18) the noble gases do not form ions and typically do not form chemical bonds of any kind.

Solved problem

Use the periodic table to determine the charge of each atom in the list when they become ions. Write each element in its ion symbol form.
a) lithium; b) sulfur; c) strontium; d) zinc; e) iodine; f) aluminum

> Asked — Find the charge of each atom when it is an ion.
> Relationships — Metals tend to form positive ions; non-metals tend to form negative ions.
> Solve — The charge is identified on the periodic table above. Notice how the charge is the same for elements within the same group.
> Answer — a) Li^+; b) S^{2-}; c) Sr^{2+}; d) Zn^{2+}; e) I^-; f) Al^{3+}

Ionic compounds

Ionic compounds

An **ionic compound** is a substance in which attractive forces act between positive and negative ions. Positive and negative ions attract each other just as protons and electrons do. The standard for an ionic compound is table salt, or sodium chloride, (NaCl). Sodium chloride is also known as the mineral halite. Within the sodium chloride compound, each sodium atom gives up one electron to form a sodium ion, Na^+ and each chlorine atom accepts the electron to become a chloride ion, Cl^-. Like all ionic compounds, NaCl is overall electrically neutral because every electron lost by a sodium atom is gained by a chlorine atom.

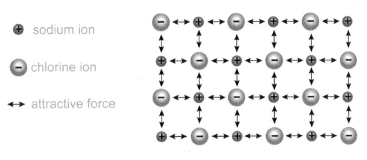

Ionic compounds

The attractive forces in ionic compounds act between every ion and all its neighboring ions.

Sodium chloride (NaCl)

Ionic bonds do not form molecules

The attraction between ions in an ionic compound is called an **ionic bond**. Ionic bonds act between all neighboring ions, and not just two specific ions. In NaCl, each sodium ion is bonded to every neighboring chlorine ion. Each chlorine ion is likewise bonded to every neighboring sodium ion. There are no molecules in ionic compounds such as sodium chloride because a single sodium ion is not uniquely bonded to a single chlorine atom. Ionic bonds are different from the covalent bonds between oxygen and hydrogen atoms in a water molecule. In a water molecule the bonding forces are restricted to one pair of atoms.

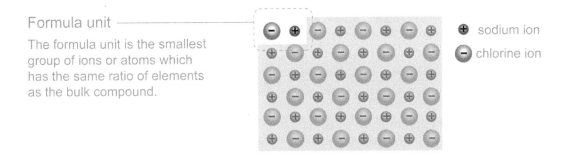

Formula unit

The formula unit is the smallest group of ions or atoms which has the same ratio of elements as the bulk compound.

The formula unit

The chemical formula for an ionic compound is the reduced, simplest whole number ratio of ions in a the compound. Salt (NaCl) has a ratio of one sodium ion per chloride ion because there is always one sodium atom for every chlorine atom. One NaCl "particle" is called a **formula unit** even though the NaCl structure does not have separate NaCl units in the same way that water is made of H_2O molecules. NaCl formula units are the simplest ion ratio, but the compound is many Na^+ ions bonded to many Cl^- ions. For this reason, the term molecule is never used for ionic compounds.

Common ionic compounds

Ionic compounds are found throughout nature and technology. Many minerals such as wurtzite (ZnS, FeS), corundum (Al_2O_3), and olivine (Mg_2SiO_4, Fe_2SiO_4) are ionic. Chalk is an ionic compound called calcium carbonate ($CaCO_3$) which is found in rocks. Calcium carbonate is used in calcium supplements and antacids. Milk of magnesia ($Mg(OH)_2$) is another ionic compound that is used as a antacid.

Ionic crystal structures

Ionic materials and their properties

Ionic substances tend to have very similar properties. They typically:

- are hard yet easily shatter
- are solid at room temperature
- have very high melting points; it takes a lot of energy to melt them
- conduct electricity if heated to a liquid state, but not in a solid state
- conduct electricity when dissolved in water or other liquid
- separate into their component ions to some degree when dissolved in water

Atomic level structures

The behavior of ionic substances are explained by their atomic structure. Ionic compounds are formed by an attraction between a positively charged cation and a negatively charged anion. Oppositely charged ions attract each other in three dimensions, forming multiple ionic bonds with one another. You can see the multiple bonds coming from each ion in the sodium chloride crystal below.

how ionic bonds create a crystal lattice structure for sodium chloride, NaCl

every sodium (Na^+) ion forms an ionic attraction to the six nearest chloride ions

every chloride (Cl^-) ion forms an ionic attraction to the six nearest sodium ions

billions of ionic attractions occur in a small fraction of a single salt crystal

Ion attraction

The ratio of sodium ions to chloride ions in a salt crystal is 1:1. Every positively charged sodium ion is attracted to, and surrounded by, 6 neighboring negatively charged chloride ions and vice versa. This pattern results in large numbers of ionic bonds no matter what geometric shape the ionic crystal forms.

Positive and negative charges balance

For neutral ionic crystals, the total charge of positive ions must be canceled out by an equal total negative charge from negative ions. If the charges did not cancel each other out, the unbalanced charges would cause areas of repulsion and the crystal would break down. This does not mean the ratio of positive ions to negative ions is always one-to-one. Below are some examples of other ionic substances. Notice the ratio of one atom to another in the compound is not always one-to-one.

CaF_2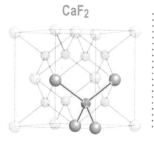

Fe_2O_3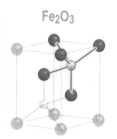

Na_2S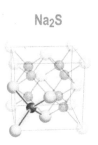

$CaCl_2$

Properties of ionic compounds

Ionic solids form crystals

Ionic bonds are relatively strong. The strong attraction between all neighboring ions tends to force virtually all ionic solids into crystalline structures. A crystal is a regular, repeating three dimensional pattern. The ions in sodium chloride form a cubic crystal with every sodium ion surrounded by six chlorine ions arranged in a cube and vice versa. In cubic zirconia (ZrO_2) zirconium and oxygen form a crystal structure that scatters light so well it is used as a synthetic diamond.

Ionic compounds often form crystals

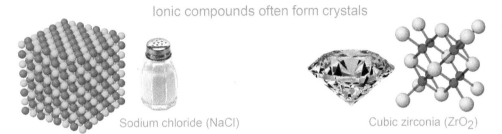

Sodium chloride (NaCl)   Cubic zirconia (ZrO_2)

Hardness, brittleness, and high melting point

The billions of attractions between ions keep particles close together and give ionic crystals high strength. The high bond strength makes ionic substances hard and brittle, and explains their high melting points. Ionic compounds are always solid at room temperature. To change from a solid to a liquid state, ions in a crystal need enough energy to break free of all of the surrounding ionic attractions in order to flow by each other.

Melting point of common ionic compounds

substance	Chemical formula	Melting point (°C)
lithium chloride	LiCl	605°C
sodium chloride	NaCl	801°C
lithium fluoride	LiF	845°C
potassium fluoride	KF	858°C
sodium fluoride	NaF	993°C
calcium carbonate	$CaCO_3$	1,339°C

Dissociation in solution

When ionic substances dissolve they undergo **dissociation** into separate positive and negative ions. In water, NaCl dissociates into separate Na^+ and Cl^- ions. This is not true of molecular substances such as glucose (blood sugar). When glucose dissolves in water the molecules of $C_6H_{12}O_6$ separate from each other but they do not come apart.

Soluble ionic solids dissociate into ions when dissolving

$$NaCl(s) + H_2O(l) \xrightarrow{dissolve} Na^+(aq) + Cl^-(aq) + H_2O(l)$$

Electrical properties

In the solid state, ions are stuck in their crystal structure, unable to move around freely. This means ionic solids are very good electrical insulators. However, when melted or dissolved, charged ions are free to move around. When anions and cations move freely in the liquid or dissolved state, they conduct electricity. This makes molten or dissolved ionic compounds electric conductors. Water with dissolved ions becomes an electrical conductor and the human body makes good use of this fact. The tiny electrical signals of nerve impulses are carried by dissolved potassium and calcium ions.

Polyatomic ions

Polyatomic ions are charged molecules

Sodium acetate, $NaC_2H_3O_2$, is an ionic compound. The cation is Na^+ and the anion is acetate, $C_2H_3O_2^-$. The acetate ion is a **polyatomic ion**. The Greek poly- prefix means many and the acetate ion contains many atoms. When sodium acetate dissolves in water, it dissociates into Na^+ and $C_2H_3O_2^-$ and the acetate ion stays together as a single unit. Even though compounds with polyatomic ions look complicated, they are still only made up of 2 things: cations and anions. Think of a polyatomic ion as a "charged molecule."

Atoms are bound together

Polyatomic ion
a "charged molecule" that tends to stay together through processes such as dissolving

acetate $C_2H_3O_2^-$

carbonate CO_3^{2-}

ammonium NH_4^+

Typical polyatomic ions

A typical polyatomic ion is formed when two or more non-metals bond to one another and form a negatively charged structure. Look at the table below for exceptions, such as when a metal bonds with non-metals. Polyatomic cations are less common. The ammonium cation (NH_4^+) is a critical part of the nitrogen cycle that makes nitrogen available to plants.

common polyatomic ions

NH_4^+	ammonium	OH^-	hydroxide
CH_3COO^- or $C_2H_3O_2^-$	acetate	NO_3^-	nitrate
HCO_3^-	bicarbonate*	NO_2^-	nitrite
CO_3^{2-}	carbonate	$C_2O_4^{2-}$	oxalate
ClO_3^-	chlorate	MnO_4^-	permanganate
CrO_4^{2-}	chromate	PO_4^{3-}	phosphate
CN^-	cyanide	SO_4^{2-}	sulfate
$Cr_2O_7^{2-}$	dichromate	SO_3^{2-}	sulfite

*bicarbonate is the common name for the hydrogen carbonate ion

Crystals with polyatomic ions

Polyatomic ions can form crystal structures. They pack together in a regular pattern with each cation attracting to all of its negative neighbors, and each anion attracting to all of its positive neighbors. Below is an example of calcium carbonate, $CaCO_3$, also known as the minerals calcite and limestome. Calcium carbonate is an important mineral in Earth's carbon cycle and is formed from dissolved calcium and carbonate ions in seawater. Ocean-dwelling organisms such as clams, shrimp, and diatoms incorporate these dissolved ions in their shells and exoskeletons. When these creatures die the calcium carbonate settles to the ocean floor and over millions of years becomes limestone.

Calcium carbonate crystal structure

calcium ion Ca^{2+}

carbonate ion CO_3^{2-}

calcium carbonate $CaCO_3$

Section 2 review

Chapter 5

Neutral atoms have an equal number of protons and electrons. However, outer layer electrons can be transferred in or out of an atom's electron cloud, creating a charged atom called an ion. Since electrons are negatively charged, a neutral atom that gains one or more electrons becomes a negatively charged ion known as an anion. A neutral atom that loses one or more electrons becomes a positively charged ion known as a cation. Metals and hydrogen tend to lose electrons, while non-metals tend to gain electrons. Cations and anions are electrically attracted to one another and can form neutral, ionic compounds. Ionic bonds exist between all the neighboring anions and cations in an ionic compound. These compounds are crystalline in shape and are generally hard but brittle, with high melting points and the ability to conduct electricity when dissolved in water.

Vocabulary words: ion, cation, anion, ionic bond, formula unit, ionic compound, dissociation, polyatomic ion

Key relationships
- A negative ionic charge equals the number of electrons gained (Example: S^{2-} means a sulfur atom gained 2 electrons to become an anion).
- A positive ionic charge equals the number of electrons lost (Example: Mg^{2+} means a magnesium atom lost 2 electrons to become a cation).

Review problems and questions

1. An atom loses 3 electrons when it forms an ion.
    a. What is its charge? Include a positive or negative sign with your answer.
    b. Is it a cation or anion?

2. What kind of ion will a metal form: a cation or an anion? Explain your answer.

3. Explain why it is incorrect to call ionic substances "molecules."

4. List at least four physical properties of ionic compounds.

5. The atom pictured to the right has experienced a change in its number of electrons. Has it become a cation or an anion? How can you tell? What will its charge be?

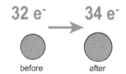

6. The atom pictured to the right has experienced a change in its number of electrons. Has it become a cation or an anion? How can you tell? What will its charge be?

5.3 - Naming and Writing Ionic Compounds

How did chemists decide to call NaCl sodium chloride? What would you call a compound with the chemical formula $NaHCO_3$? This section is about how to decipher and write chemical names for ionic compounds. The rules are not that difficult and once you know them you will be able to recognize and name many useful compounds. The collective set of rules that help you name chemical formulas is known as **nomenclature**.

Compound names

Binary ionic compounds

The simplest ionic compounds are made from two monatomic ions, such as NaCl. Sodium chloride is an example of a **binary compound** which means there are only two kinds of elements: one cation and one anion. Although it appears complex, sodium acetate ($NaC_2H_3O_2$) is composed of only two ions: Na^+ and $C_2H_3O_2^-$. Sodium acetate is not binary because it contains more than two kinds of elements. When naming binary compounds the following rules apply. In the compound name, metal cations are named the same as on the periodic table while anions change their ending to –ide. In NaCl, the Na cation keeps it name – sodium – while the anion changes from chlorine to chloride. Cations are always written first in an ionic formula, followed by the anion.

Naming binary ionic compounds

① the cation (+) goes first

② metal cations use the element name

③ the anion goes second ending with "ide"

NaCl → sodium chloride

$MgCl_2$ → magnesium chloride

Solved problem

Name each of the following binary ionic compounds:
a) CaO; b) $SrCl_2$; c) LiI; d) ZnF_2; e) KBr; f) BeS

Asked: Apply naming rules to the compounds above.

Solve: The first element in each compound is a metal cation. Use its element name. The second element is a non-metal anion. Change its ending to –ide

Answer: a) calcium oxide; b) strontium chloride; c) lithium iodide
d) zinc fluoride; e) potassium bromide; f) beryllium sulfide

Ionic charges cancel in a compound

Ions combine to create a chemical formula that is electrically neutral. All positive charges are canceled by negative charges and vice versa. Beryllium chloride has the formula $BeCl_2$, composed of one Be^{2+} beryllium ion and two Cl^- chloride ions. Two 1− chloride ions are required to cancel out the 2+ charge of a single beryllium ion.

Balancing ionic charge in compounds
The overall compound must have a total charge of zero.

NaF	$BeCl_2$	Rb_2S
one I^- ion balances one Na^+ ion — Na^+ F^- / +1 −1	two Cl^- ions balance one Be^{2+} ion — Be^{2+} Cl^- Cl^- / +2 −2	two K^+ ions balances one S^{2-} ion — Rb^+ Rb^+ S^{2-} / +2 −2
sodium fluoride	beryllium chloride	rubidium sulfide

The crisscross method: Binary compounds

The crisscross method

Ionic compounds must be electrically neutral therefore the total amount of positive and negative charge must be the same. This is the rule that determines how many of each ion species are needed in the chemical formula. The crisscross method uses ion charges to determine how many of each ion must be present in the neutral chemical formula.

The formula for aluminum oxide

Let's use the crisscross method to figure out the chemical formula for the compound called aluminum oxide. The two ions are Al^{3+} and O^{2-}. Follow these steps to find the correct chemical formula.

The crisscross method of balancing ionic charge in compounds

1. Find cation (+) and anion (-) charges and write cation first. $Al^{3+}\quad O^{2-}$

2. Remove the + and - signs. $Al^3 \quad O^2$

3. Exchange the superscripts to make them subscripts for the *other* ion. $Al_2 \quad O_3$

4. Check if the ratios can be reduced by a common factor. 2:3 cannot be reduced Al_2O_3

5. Check that the charges balance

$$\begin{array}{cc} Al^{3+} & O^{2-} \\ Al^{3+} & O^{2-} \\ & O^{2-} \\ \hline +6 & -6 \end{array}$$

Check your work

You can check your work by making sure positive and negative charges cancel in the formula you wrote. The aluminum ion has a 3+ charge while the oxygen ion has a 2- charge. The formula Al_2O_3 contains two Al^{3+} ions and three O^{2-} ions. Based on the calculations shown, the charges cancel. Your answer is correct.

Your answer: Al_2O_3

$$\begin{array}{ll} 2\,(Al^{3+}) & = +6 \\ 3\,(O^{2-}) & = -6 \\ \hline & \text{Total: 0} \end{array}$$

Solved problem

Write the formula for calcium sulfide.

Asked Use the crisscross method to find the chemical formula

Solve Follow the four steps of a crisscross:
Step 1: Calcium: Ca^{2+} Sulfide: S^{2-}

Step 2: Remove + and − signs: $Ca^2 \quad S^2$

Step 3: Crisscross numbers to the opposite element:

$$Ca^2 \!\!\times\!\! S^2 \rightarrow Ca_2S_2$$

Step 4: Ca_2S_2 has a ratio of 2:2, which can be reduced to 1:1. The number 1 is invisible in chemical formulas.

$$Ca_{\cancel{2}}S_{\cancel{2}} \rightarrow CaS$$

Answer The formula for calcium sulfide is CaS.

Section 5.3: Naming and Writing Ionic Compounds

Counting atoms and counting ions

Counting atoms with parentheses

Why do you sometimes see parentheses in a chemical formula? You will use the distributive property to account for all atoms in a formula that has parentheses such as ammonium sulfate, $(NH_4)_2SO_4$. The subscript $_2$ outside the parentheses is distributed only to the atoms inside the parentheses by multiplication as shown below. The sulfate ion, SO_4^{2-} is unaffected by the parentheses and the subscript outside of it. Only the atoms that make up the NH_4^+ ammonium ion inside the parentheses are multiplied by the subscript.

COUNTING ATOMS

$(NH_4)_2SO_4$

$2(N) = 2\ N$ ●●

$2(H_4) = 8\ H$ ●●●●●●●●

$S = 1\ S$ ●

$O_4 = 4\ O$ ●●●●

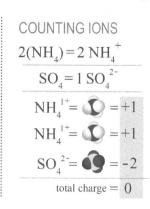

COUNTING IONS

$2(NH_4) = 2\ NH_4^+$

$SO_4 = 1\ SO_4^{2-}$

$NH_4^{1+} =$ ◯ $= +1$

$NH_4^{1+} =$ ◯ $= +1$

$SO_4^{2-} =$ ◯ $= -2$

total charge $= 0$

Counting ions with parentheses

When you count ions in a formula that contains parentheses, you must remember each polyatomic ion is a small molecule that functions as one unit. The ammonium molecule unit is NH_4^+. The ammonium unit in $(NH_4)_2SO_4$ needs parentheses around it to show there are two NH_2^+ units for every one SO_4^{2-} unit. Ions seek to cancel each other out when they form compounds. The ratio of two NH_4^+ ions to one SO_4^{2-} ion is required to create the neutral $(NH_4)_2SO_4$ compound.

If you forget parentheses

Without parentheses, the formula would look like this: $NH_{42}SO_4$. As you can see, when the parentheses are left out, the formula no longer represents the NH_4^+ ion unit. The atom count is also incorrect when parentheses are removed. The formula without parentheses shows only 1 nitrogen atom instead of 2, and 42 hydrogen atoms instead of 8.

Superscripts vs. subscripts

We looked only at charges in the example above. Any time you are looking at ions, remember this important distinction: superscripts tell you the charge, and subscripts tell you the number of atoms.

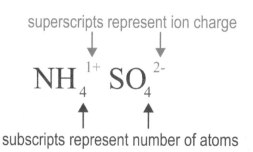

superscripts represent ion charge

$NH_4^{1+}\quad SO_4^{2-}$

subscripts represent number of atoms

Subscripts on polyatomic ions

The rules for writing a chemical formula included moving a superscript from one ion to the opposite ion. This rule will not change. Soon you will work with formulas that contain polyatomic ions that have subscripts. Subscripts that were originally part of a polyatomic ion formula must remain untouched and unmodified when writing a formula. Subscripts that are present before a crisscross do not participate in a crisscross.

144   Chapter 5: Chemical Compounds

Ternary and quaternary ionic compounds

3 or more atoms and 4 or more atoms

Sodium sulfate, Na_2SO_4, is an example of a **ternary compound** because it contains three different elements. Sodium acetate, $NaC_2H_3O_2$, is a **quaternary compound** because it contains 4 different elements. Ternary and quaternary compounds contain at least one polyatomic ion. Naming these compounds is similar to binary compounds because ternary and quaternary compounds consist of a cation and an anion. With the exception of the ammonium ion, NH_4^+, the cation is usually a metal. You will always find the cation written first in the chemical formula, followed by the anion.

Split the formula into two ions

To name a compound with one or more polyatomic ions, start by splitting the compound into its two ions. As shown below, $CaSO_3$ is split right after calcium because calcium is the cation. Unless you see parentheses, you can split a ternary or quaternary compound right after the first element. $(NH_4)_2CO_3$ is split right after $(NH_4)_2$ because NH_4 is a polyatomic cation, as indicated by the parentheses.

Naming ionic compounds with polyatomic ions

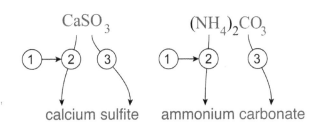

① the cation (+) goes first

② metal cations use the element name or the ion name

③ the name of the anion goes second, use "ide" ending if not polyatomic.

calcium sulfite   ammonium carbonate

A compound with one polyatomic ion

For $CaSO_3$, Ca receives its regular elemental name, calcium. You can look up the exact name of SO_3 on the polyatomic ion chart given earlier in this chapter, and in the Appendix. The chemical formula will not show charges, but subscripts should match. According to the polyatomic ion chart, SO_3^{2-} is called sulfite. Put the two ion names together with the cation first to get the compound name, calcium sulfite.

A compound with two polyatomic ions

Many compounds have two polyatomic ions. In the $(NH_4)_2CO_3$ example, the cation is ammonium (NH_4^+) and the anion is carbonate (CO_3^{2-}). For naming purposes you can ignore the subscript 2 outside of parentheses which tells you there are two ammonium ions. Use what is inside the parentheses when identifying a polyatomic ion. The name of $(NH_4)_2CO_3$ is ammonium carbonate. Finally, if you come across a ternary ionic compound with a monatomic anion, don't forget to change the ending to –ide. For example, $(NH_4)_2S$ is named ammonium sulfide.

Solved problem

Name the following compounds:
a) $Ca(OH)_2$; b) $RbNO_2$; c) NH_4Cl; d) $Al_2(SO_4)_3$

Asked   Apply naming rules to the compounds above.

Relationships   Cations receive their element name except for ammonium, NH_4. Polyatomic anions receive the name listed in the polyatomic ion chart. Monatomic anions change their ending to -ide.

Solve   Split each compound after the cation. Give each half its appropriate name:
a) Ca | $(OH)_2$; b) Rb | NO_2; c) NH_4 | Cl; d) Al_2 | $(SO_4)_3$

Answer   a) calcium hydroxide; b) rubidium nitrite;
c) ammonium chloride; d) aluminum sulfate

Section 5.3: Naming and Writing Ionic Compounds

The crisscross method: Ternary compounds

Formulas containing polyatomic ions

Writing ionic formulas with polyatomic ions is the same as writing them with monatomic ions. Use the crisscross method to get positive and negative charges to cancel out. But be careful, you need to use parentheses when there is more than one polyatomic ion in the resulting crisscrossed formula. An ion like nitrite, NO_2^- is a single, small molecule with a negative charge. We treat nitrite like a single atom. To write a formula that says you need more than one NO_2^- ion you need to put parentheses around the ion.

Crisscross method for barium nitrite

The crisscross method of balancing charge with polyatomic ions

① Find cation (+) and anion (−) charges and write cation first. Ba^{2+} NO_2^-

② Remove the + and − signs. $Ba\ ^2$ $NO_2\ ^1$

③ Exchange the superscripts to make them subscripts for the *other* ion. Use parentheses around polyatomic ions. $Ba\ ^{②}$ $NO_2\ ^{①}$
   Ba_1 $(NO_2)_2$

④ Check if the ratios can be reduced by a common factor. 1:2 cannot be reduced. $Ba(NO_2)_2$

⑤ Check that the charges balance

$$\frac{Ba^{2+}}{+2} \quad \frac{NO_2^-\ NO_2^-}{-2}$$

Check your answer

Check to see that the formula is correct by making sure the total positive and total negative charges are equal. $Ba(NO_2)_2$ has 1 barium ion (Ba^{2+}) and 2 nitrite ions (NO_2^-). As you can see above, the 2+ charge from the barium ion correctly balances the 2− charge from two nitrite ions. The formula is correct.

Hydroxide is often overlooked

The hydroxide ion, OH^-, is polyatomic even though the ion has no subscripts. If it receives a subscript other than the "invisible 1" during a crisscross, you must use parentheses around the ion. The example below shows this for the yellow iron ore, iron(III) hydroxide.

The formula for iron(III) hydroxide

① Find cation (+) and anion (−) charges and write cation first. Fe^{3+} OH^-

② Remove the + and − signs. Use a 1 for single charges. $Fe\ ^3$ $OH\ ^1$

③ Exchange the superscripts to make them subscripts for the other ion. Use parentheses around polyatomic ions. $Fe\ ^{③}$ $OH\ ^{①}$
   Fe_1 $(OH)_3$

④ Check if the ratios can be reduced by a common factor. 1:3 cannot be reduced. $Fe(OH)_3$

⑤ Check that the charges balance

$$\frac{Fe^{3+}}{+3} \quad \frac{OH^-\ OH^-\ OH^-}{-3}$$

Do not modify original subscripts

Suppose a ternary or quaternary compound has a formula that looks like it could be reduced, such as: Na_2SO_4. At first glance it looks like the formula could reduce to NaSO, but this is incorrect. Formulas can only be reduced based on the ion ratio. The formula Na_2SO_4 has 2 ions: Na^+ and SO_4^{2-}. The chemical formula has the ion ratio $2\ Na^+ : 1\ SO_4^{2-}$, which cannot be reduced.

Ionic formulas with Roman numerals

Some metals require Roman numerals

Some metals always form the same ion with the same charge, but many metals can form ions with different charges. Copper, for example, can form an ion with either a 1+ or a 2+ charge. In the table below, there are several names for monatomic cations followed by a Roman numeral. A Roman numeral specifies which charge the ion should have when naming the compound or crisscrossing to write the formula. When you see a cation with Roman numeral 3, III, the ion's charge is 3+. When you see a cation with II, the charge is 2+, and so on.

common monatomic cations

Aluminum	Al^{3+}	Lead (II)	Pb^{2+}
Barium	Ba^{2+}	Lead (IV)	Pb^{4+}
Calcium	Ca^{2+}	Magnesium	Mg^{2+}
Cobalt (II)	Co^{2+}	Mercury (I)	Hg_2^{2+}
Copper (I)	Cu^+	Mercury (II)	Hg^{2+}
Copper (II)	Cu^{2+}	Potassium	K^+
Chromium (II)	Cr^{2+}	Silver	Ag^+
Chromium (III)	Cr^{3+}	Sodium	Na^+
Hydrogen*	H^+	Tin (II)	Sn^{2+}
Iron (II)	Fe^{2+}	Tin (IV)	Sn^{4+}
Iron (III)	Fe^{3+}	Zinc	Zn^{2+}

*Hydrogen is a group 1 nonmetal that can also form an anion with a 1 - charge

a Roman numeral indicates the charge on a positive ion

Cu^{2+}
copper (II)

1 = I	6 = VI
2 = II	7 = VII
3 = III	8 = VIII
4 = IV	9 = IX
5 = V	10 = X

Check for Roman numerals when writing ionic names

The transition metals and group 14 and 15 metals can form more than one charge. You need to check the common cation table above before you name an ionic compound that includes one of those metals. Except for silver and zinc, most transition metals you will work with have Roman numerals. Silver is usually found as a 1+ ion and zinc is commonly found as a 2+ ion. Neither silver nor zinc need a Roman numeral in formula names.

Check if Roman numeral is necessary for the orange colored elements (above).

Determine the ion charge from the formula

Let's name the formula $PbCl_2$. Lead, Pb, is one of the ions that could have more than one charge according to the table above. To determine whether the compound has lead(II) or lead(IV), check the anion's charge. The periodic table shows the chloride ion charge as Cl^-. If the formula requires 2 Cl^- ions, the total negative charge is 2-. That must mean the total positive charge is 2+. Therefore, the lead ion must be Pb^{2+} or lead(II). The compound is named lead(II) chloride.

Chapter 5

Section 3 review

The set of rules for naming chemical compounds is known as nomenclature. For simple ionic compounds the cation is named first and then the anion with its ending changed to –ide. Because ionic compounds are neutral, their charges (shown by superscript) must be balanced. The crisscross method can determine how to balance the charges, and subscripts show the relative ratio of ions in the compound. Polyatomic ions are molecules (atoms chemically bonded to one another) that as a whole carry a charge shared among all the atoms. When balancing charges, polyatomic ions are placed in parentheses with the subscript applied to the whole polyatomic unit. Some metals can form more than one ion with different charges. Roman numerals are used to show which ion is being referred to.

Vocabulary words	binary compound, nomenclature, ternary compound, quaternary compound

Review problems and questions

1. If "chlorine" changes its name to "chloride" in a binary ionic compound, why doesn't "calcium" change its name to "calcide"?

2. Why is $CaCl_2$ a correct formula while Cl_2Ca is incorrect?

3. Why would you use parentheses on $Zn_3(PO_4)_2$ but not on NaOH or $CaCl_2$?

4. Iron(III) oxide is commonly called rust.
   a. What is the purpose of the Roman numeral in the chemical name?
   b. How do you know when to use a Roman numeral in a chemical name?

5. Write formulas for compounds a-c, and write names for compounds d-f:

a.	barium sulfate	d.	$CoSO_3$
b.	tin(II) hydroxide	e.	RbF
c.	aluminum nitrate	f.	$Be_3(PO_4)_2$

5.4 - Covalent Compounds

Ibuprofen is a non-steroidal anti-inflammatory medicine that eases fever, headaches, and muscle swelling from injuries. The atoms in a molecule of ibuprofen, $C_{13}H_{18}O_2$, are not held together with ionic bonds. Ibuprofen is soluble in water but remains intact as a molecule instead of dissociating into ions. This section is about molecular compounds, such as ibuprofen, sugar, octane, paraffin, and many others.

Covalent bonds

Covalent bonds act within a molecule

Each connection between two atoms in the ibuprofen molecule is a **covalent bond**. Electrons in a covalent bond are shared between atoms instead of being transferred. This makes the covalent bond attractive only between the atoms sharing electrons. The strong forces of a chemical bond act between atoms in the ibuprofen molecule but do not exert strong forces on neighboring molecules. This is very different from the forces in ionic substances.

Ibuprofen molecule
Interactive simulation

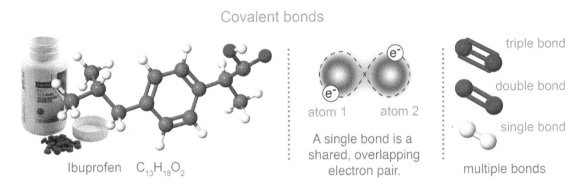

Non-metals tend to form covalent bonds

For reasons that are explained in a future chapter, non-metals tend to form covalent bonds with other non-metals. The ibuprofen molecule contains only carbon, hydrogen, and oxygen and these three elements are all non-metals. A compound containing only non-metals is likely to be covalent. The exceptions are compounds with polyatomic ions, such as nitrate, NO_3^-. The bonds between oxygen and nitrogen within the nitrate ion are covalent bonds. However, compounds that include the nitrate ion may be ionic compounds.

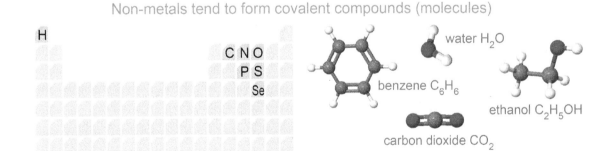

Solved problem

Decide whether each of the following is an ionic or covalent compound:
a) H_2O; b) $ZnCl_2$; c) NH_3; d) CH_3Cl; e) H_2SO_4; f) $MgSO_4$; g) I_2

Asked Classify compounds as ionic or covalent.

Solve Ionic compounds are composed of a metal and non-metal, with the exception of compounds with the ammonium cation; covalent compounds are composed of all non-metals.

Answer a) covalent; b) ionic; c) covalent; d) covalent; e) covalent; f) ionic; g) covalent

Diatomic molecules

Diatomic molecules

A **diatomic molecule** is a compound made of only two atoms with a covalent bond. Carbon monoxide, CO, is a diatomic molecule made of two different atoms. Oxygen gas, O_2, is a diatomic molecule made of two of the same atoms.

7 elements form diatomic molecules

There are seven elements on the periodic table that are found in nature in their pure form bonded to each other as a diatomic molecule. These seven elements are: hydrogen, nitrogen, oxygen, fluorine, chlorine, bromine, and iodine. The seven are all non-metals, they exist as gases at room temperature, and they have the chemical formula X_2 where X represents their element symbol: H_2, N_2, O_2, F_2, Cl_2, Br_2, I_2. When you are told one of these elements is involved in a chemical reaction, you must assume it is in its diatomic form, not its elemental or atom form.

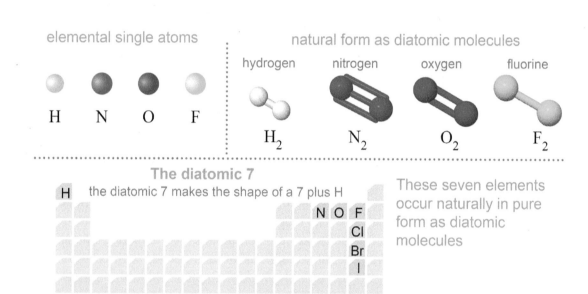

How to remember diatomic elements

Forgetting to write the formula for oxygen as O_2 instead of O could cause much frustration for you in the future. It's well worth taking the time to memorize the seven elements that form diatomic molecules. Make up a saying to help you remember, such as "Have No Fear Of Ice Cream Bars" (H_2, N_2, F_2, O_2, I_2, Cl_2, Br_2), or make up a strange name like "Honcl Brif" (Honcl = H_2, O_2, N_2, Cl_2 and Brif = Br_2, I_2, F_2). Or, remember the "diatomic seven" shown above.

Solved problem

a) Write formulas for gold, bromine, barium, carbon, calcium, and hydrogen.
b) Name F_2, Fe, Xe, N_2, and K.

Asked Distinguish between diatomic molecules and atoms in order to write formulas and name each substance properly.

Solve "Honcl Brif" diatomic elements will receive the 2 subscript.
Diatomic molecules are named the same as their elemental form.

Answer a) Au, Br_2, Ba, C, Ca, H_2
b) fluorine, iron, xenon, nitrogen, potassium

Molecular compounds

Molecules vs. compounds

A molecule is a group of atoms with covalent bond that stays together as a group. Diatomic oxygen, O_2, is a molecule but it is not a compound! A compound must include at least two different elements bonded together. Not all compounds are molecules, and not all molecules are compounds. Water, H_2O is both a molecule and a compound because the atoms have covalent bonds, and there are two kinds of elements in the molecule.

Molecules can be classified by size

A very wide range of molecules exist that range from two atoms (O_2) to more than a hundred thousand atoms in biological molecules such as DNA. Small molecules are made of 2 - 100 atoms and include moderately complex molecules such as ibuprofen. Medium molecules tend to form chains with a good example being paraffin (wax). Large molecules include proteins and network structures which form complex three dimensional shapes.

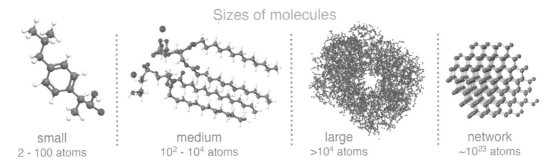

Sizes of molecules

small — 2 - 100 atoms
medium — $10^2 - 10^4$ atoms
large — $>10^4$ atoms
network — $\sim 10^{23}$ atoms

Network covalent structures

A quartz crystal is made of a repeating silicon dioxide (SiO_2) unit in which each silicon atom has a covalent bond to each of its own two oxygen atoms, plus covalent bonds to two neighboring oxygen atoms. The entire crystal can be considered a single molecule! Quartz is an example of a **network covalent** structure. Similar to ionic crystals, but even more tightly bound, network covalent structures tend to be very strong (carbon fiber) and have very high melting and boiling points.

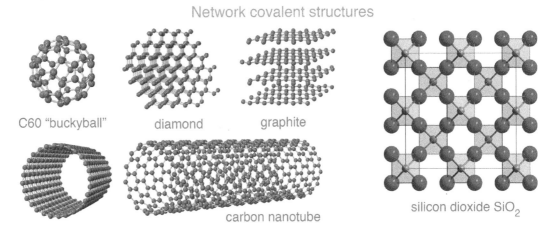

Network covalent structures

C60 "buckyball" diamond graphite carbon nanotube silicon dioxide SiO_2

Carbon networks

A carbon atom can form a great variety of network covalent structures. Carbon compounds are used in the field of **nanotechnology**, or the use of technology to give materials amazing properties made possible by manipulating individual atoms or molecules. Carbon nanotubes are synthesized in labs but some scientists think they could be found in nature. They are stronger than steel even though they are much thinner than a human hair. Carbon nanotubes and diamonds get their strength from their carbon bonding pattern. Diamond and graphite are the most common carbon networks found in nature. Buckyballs are mainly synthesized in a lab but they can be found in nature.

Section 5.4: Covalent Compounds

Properties of covalent compounds

Molecular = covalent

Molecular compounds are held together with covalent bonds, which is why you can use the term molecular and covalent interchangeably. Molecular compounds have a greater variety in properties compared to ionic compounds. Covalent compounds have properties that are largely opposite of ionic compounds. The table below provides a summary of properties displayed by most ionic and most covalent compounds.

Comparing properties of ionic and covalent compounds	
ionic	covalent
Hard and brittle	Hard and brittle to soft and flexible
Solid at room temperature	Solid, liquid, or gas at room temperature
Does not conduct electricity in the solid state	Does not conduct electricity in the solid state
Conducts electricity as a liquid or solution	Does not conduct electricity as a liquid or solution
High melting and boiling points	Boiling and melting points may be high or low

Interactive simulation

In the interactive simulation titled Charged and Neutral Atoms, see how a temperature change affects covalent and ionic particle attractions. The simulation starts out as a molecular substance but when you select "Charge" the substance changes to ionic. Can you tell when a phase change occurs? How does the type of particle attraction (ionic vs. covalent) affect the melting and boiling temperature?

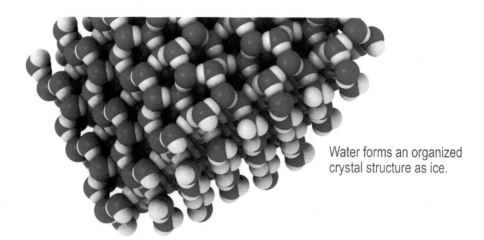

Water forms an organized crystal structure as ice.

Molecular structures can vary

Together, the structure of the individual molecule and the attractions between molecules have a huge influence on the properties of covalent compounds. For example, water molecules in the liquid phase are in constant motion but are so attracted to one another that small amounts of water form beads or spheres. As a solid, water can form an orderly crystal structure that we call ice. As you know, ice is hard and it can shatter which makes it seem like an ionic compound. The strong attractions and high degree of organization between water molecules in ice are responsible for its different properties compared to liquid water. As you saw with carbon networks, a substance that can form a great variety of structures also gives it a great variety of properties.

Naming and writing covalent formulas

Covalent naming is similar to binary ionic naming rules

Let's look at binary covalent compounds. Binary covalent compounds have a simple naming procedure:

1. Use the same rules for naming simple binary ionic compounds.
2. Add a prefix to each name that indicates how many of that atom are in the molecule.
3. If there is only one atom of the first element in the chemical formula, leave out the "mono" prefix. This rule applies only to the first element in the formula.

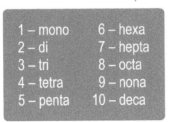

Prefixes for covalent compounds

1 – mono   6 – hexa
2 – di   7 – hepta
3 – tri   8 – octa
4 – tetra 9 – nona
5 – penta 10 – deca

Two hydrogen atoms with one sulfur atom.

H_2S

dihydrogen monosulfide

If there is only one of the first element drop the mono.

PCl_5

phosphorus pentachloride

Solved problem

Write the name for each of the following formulas: N_2O_4, S_2F_{10}, SO_3

Asked Apply naming rules for covalent compounds.

Relationships The name of the formula is constructed from the simple ionic name with prefixes before each element, indicating the number of each atom.

Solve N_2 = dinitrogen and O_4 = tetraoxide; S_2 = disulfur and F_{10} = decafluoride; S = sulfur and O_3 = trioxide

Answer N_2O_4 = dinitrogen tetraoxide, S_2F_{10} = disulfur decafluoride, SO_3 = sulfur trioxide

Writing formulas from names

There are many rules to remember when writing an ionic formula. However, there is no need to worry about charges, crisscrossing, or Roman numerals to get the subscripts correct on a binary covalent formula. The prefix for each element tells you which subscript to use.

Solved problem

Write a formula for each: carbon tetrachloride, tetraphosphorus trifluoride

Asked Write formulas for covalent compounds.

Relationships Use prefixes to determine the number of each element.

Solve Carbon tetrachloride tells you there is one C and four Cl atoms. Tetraphosphorus trifluoride tells you there are four P and three F atoms.

Answer carbon tetrachloride = CCl_4, tetraphosphorus trifluoride = P_4F_3

Section 5.4: Covalent Compounds

Mixed chemical formula review

Identify as ionic or covalent first

You will be asked to figure out chemical formulas or names. First, identify the compound as ionic or covalent. Next, apply the appropriate rules. Use the periodic table, ion chart, and prefix chart to practice naming and writing chemical formulas.

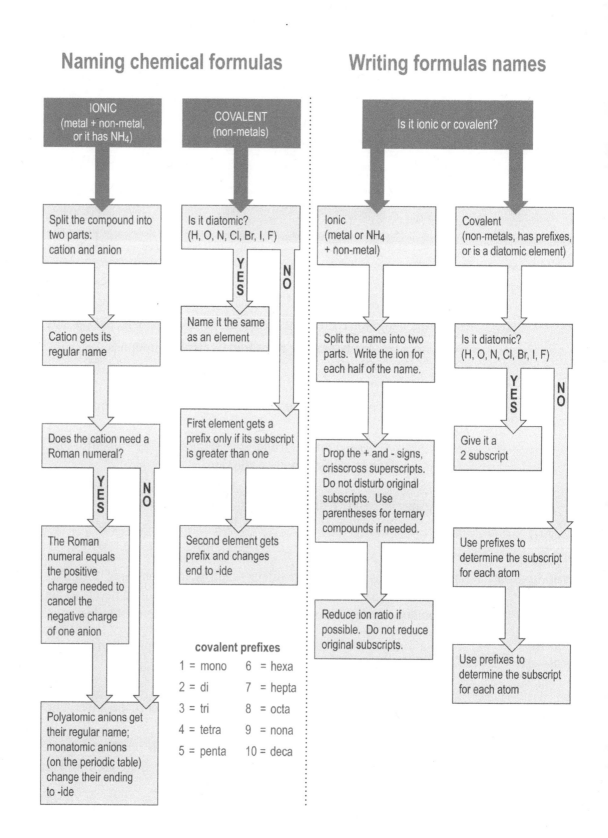

Section 4 review

Chapter 5

Ionic bonds are formed between metals and non-metals that have lost and gained electrons respectively. Covalent chemical bonds are formed as a result of two non-metals sharing electrons to form molecules. Molecules can be as simple as a pair of oxygen atoms forming a molecule of oxygen gas, or as complex as a protein made up of thousands of atoms or highly structured networks like those found in diamonds. Nomenclature for covalent compounds is similar to ionic compounds but with prefixes that state how many of each atom are in the compound. Seven elements exist as diatomic molecules – hydrogen, oxygen, nitrogen, chlorine, bromine, iodine and fluorine ("HONCl BrIF").

Vocabulary words   covalent bond, diatomic molecule, network covalent, nanotechnology

Review problems and questions

1. You have to use different naming rules based on whether you are working with a covalent or ionic compound. How can you tell by looking at a formula whether it is ionic or covalent?

2. Why is O_2 a molecule while neither O nor LiCl is a molecule?

3. When do you use the crisscross method for covalent compounds?

4. When do you use prefixes in ionic compounds?

5. Write names for substances a-d:
    a. I_2
    b. CCl_4
    c. $NaC_2H_3O_2$
    d. $NiCl_2$

6. Write chemical formulas for compounds e-j:
    e. iron(III) chlorate
    f. nitrogen dioxide
    g. sodium bicarbonate
    h. magnesium oxalate
    i. chlorine
    j. sulfur

Ban DHMO!

If you look up the chemical dihydrogen monoxide (DHMO) you would find some interesting facts about the substance, including that it:

- contributes to soil erosion
- is used by athletes to enhance performances
- is the major component of acid rain
- is used as an industrial solvent and coolant
- accelerates corrosion and rusting
- in certain forms, can cause burns and contribute to the greenhouse effect

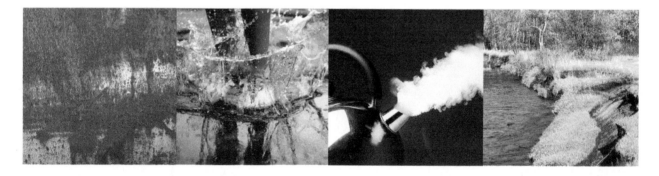

Reading the name and how it is used, a reader might think that dihydrogen monoxide was a serious threat to people and the environment. In fact, there are numerous images online that want to "Ban DHMO!" After this chapter, you know better—and are in on the joke! Dihydrogen monoxide is a molecular compound. When you decode the formula from the name, you get H_2O! Those facts really are about water, but to an uninformed reader they sound dangerous when combined with a chemical name like dihydrogen monoxide. Water is a common name for a very common chemical. There are other common chemicals that have common names. Sometimes these names make sense, but they can also cause confusion because they don't actually tell you about the substance.

Water is a common name for a very common chemical. There are other common chemicals that have common names. Sometimes these names make sense, but they can also cause confusion because they don't actually tell you about the substance.

You might be familiar with some common names for chemicals in your house. Table salt, baking soda, cream of tartar and other chemicals with relatively simple formulas that you could find in your kitchen. You may know that table salt is the common name for sodium chloride, NaCl. Baking soda is another household chemical with a variety of cooking and cleaning uses. Baking soda has the formula $NaHCO_3$. Its chemical name is sodium bicarbonate and it is a compound with sodium ions bonded to polyatomic bicarbonate ion. You can also find cream of tartar in your house. Cream of tartar can be used as a stabilizing agent in whipped cream and egg whites. Cream of tartar seems like an odd name since it is not a cream at all. This powdery white substance is actually potassium bitartrate, and it has the chemical formula of $KHC_4H_5O_6$.

You can find bleach in kitchen cleansers and in your laundry room. Bleach is great when it takes out stains and not-so-great when it accidentally discolors your clothes! It readily reacts with many substances causing them to break down and lose their color. Chlorine is the basis for most bleaches. The chlorine in bleach is not in the form of Cl_2 but rather in the form of polyatomic ion hypochlorite, OCl^-. Common bleach is simply a solution with sodium hypochlorite, NaOCl, as the active ingredient.

There are numerous other compounds with common names. Some of the more familiar ones contain alkali metals (group 1) or alkaline earth metals (group 2). These are very reactive metals not found in nature in their elemental form, and are easily combined with anions of sulfates, carbonates, and hydroxides. You may have heard of Epsom salts that are used sometimes in baths. This is a hydrated or water-containing version of magnesium sulfate with the formula $MgSO_4 \cdot 7H_2O$. Another magnesium containing compound is used as an antacid. Milk of magnesia is a suspension that contains magnesium hydroxide, $Mg(OH)_2$ as the active ingredient. If you are in the garden and you need to add potash to the soil, you are actually adding some potassium as a fertilizer in the form of potassium carbonate, K_2CO_3. If your drain is clogged, you might add some drain cleaner with lye as an active ingredient. Lye is the common name for sodium hydroxide, NaOH.

Common names aren't restricted to ionic compounds. Laughing gas is commonly used in surgery and dentistry. Laughing gas is also known as nitrous, nitro, or NOS and it has the formula N_2O. According to nomenclature rules for molecular compounds, its official name is dinitrogen monoxide.

Some elements also have common names. Quicksilver was the name given to mercury, Hg, because of its color and the fact that it is a liquid at room temperature. Elemental sulfur, S was historically called brimstone. Elemental carbon, C has many common names. Depending on how the carbon atoms are arranged, you could be talking about diamond or graphite. In this case the common name is very useful, because if you asked for some carbon jewelry and received a graphite ring, you would be pretty disappointed!

The key to any nomenclature is to make sure you can effectively communicate with those around you. Common names, like quicksilver and limestone, are colorful and descriptive. Chemical names, like dihydrogen monoxide, can sometimes sound too scientific or even frightening. Now you know that chemical names simply follow a predictable pattern to tell you about the atoms and type of bonding in a compound. These names are easily understood with some basic nomenclature rules.

Chapter 5 review

Vocabulary
Match each word to the sentence where it best fits.

Section 5.1

| family | periodic |

1. A quantity that is _____ has some regularly repeating pattern.
2. A(n) _____ is a group on the periodic table.

Section 5.2

anion	cation
dissociation	formula unit
ion	ionic bond
ionic compound	polyatomic ion

3. If an atom gains or loses an electron and now carries a charge, it is called a(n) _____ .
4. Na^+ and Cl^- make a neutral compound when their opposing charges attract each other. The neutral compound is held together by a(n) _____ .
5. _____ is the separation of a substance into ions.
6. A(n) _____ is an ion with a positive charge.
7. A(n) _____ is an ion with a negative charge.
8. The simplest ion ratio that makes up an ionic compound is the _____ .
9. If a small molecule has an overall positive or negative charge, it is considered a(n) _____ .
10. In a(n) _____ , the positive and negative ions attract each other to keep matter together.

Section 5.3

| binary compound | nomenclature |
| quaternary compound | ternary compound |

11. If a chemical formula contains two atoms or ions, it is a(n) _____ .
12. A(n) _____ is made up of three different elements.
13. A(n) _____ describes a chemical formula that contains four different elements.
14. _____ describes the system used to name chemical compounds.

Section 5.4

| covalent bond | diatomic molecule |
| nanotechnology | network covalent |

15. A structure that contains a large number of atoms covalently bonded together forming a 3D web of connection is called a(n) _____ structure.
16. The manipulation of individual atoms or molecules to create materials with unique properties is called _____ .
17. A(n) _____ is composed of two of the same or two different atoms covalently bonded together.
18. When electrons are shared between atoms, they are in a(n) _____ .

Conceptual questions

Section 5.1

19. Use the periodic table below to identify the letter that represents a(n):

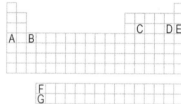

   a. actinide
   b. noble gas
   c. halogen
   d. lanthanide
   e. transition metal
   f. alkali metal

20. Use the periodic table from the previous questions to identify letters that match the chemical or physical property. Each property can have 1 or more letters associated with it:

   a. radioactive
   b. conductive
   c. highly reactive
   d. least reactive

21. The Russian scientist Dmitri Mendeleev is most famous for what contribution to chemistry?
22. How was the organization different in the first periodic table compared to the modern periodic table?
23. Explain how Mendeleev used density patterns to create the first periodic table.
24. What does the term "family" refer to in chemistry?

Chapter 5 review

Section 5.1

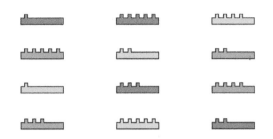

25. Make a table that organizes the shapes above according properties that make them different or similar to each other. Hint: Like Mendeleev, you will predict some missing shape!

26. Sketch the missing shapes that would complete your "periodic table."

Section 5.2

27. An ion has formed because of electron loss. Will the ion have a positive charge, or will it be negative? Explain your reasoning.

28. What happens to the electron lost by one atom during the formation of an ionic bond?

29. Write the ion symbol for each of the following, such as $X^{\#-}$ where X is the element symbol and # is the charge amount. Use – or + as appropriate.
    a. aluminum
    b. iodine
    c. cesium
    d. oxygen

30. What is the particle that moves between atoms during chemical bonding and within the atom, and where is this particle located?

31. Classify each as either cations or anions when they form ions.
    a. aluminum
    b. iodine
    c. cesium
    d. oxygen

32. Give the major reason why the particles in ionic substances are not called "molecules."

33. What is a formula unit in an ionic bond?

34. Is chocolate held together by ionic bonds, based on ionic bond properties? Explain.

35. List two physical processes that will break or change a crystal structure?

36. If ions are flowing freely among each other in a substance, is the substance in the solid or liquid state? Explain.

37. Why is Li^+ considered an ion of lithium?

38. In what form can ionic substances not conduct electricity? Explain.

Section 5.3

39. When you perform a crisscross with polyatomic ions, what happens to the subscripts that are present from the beginning of the crisscross?

40. Why is it inappropriate to reduce the formula $Na_2C_2O_4$ to $NaCO_2$?

41. When is it appropriate to add Roman numerals to a chemical name?

42. When is it appropriate to add parentheses to a chemical formula?

43. If ZnF_2 has three atoms, why is it considered a binary ionic compound?

44. What should the overall charge of a compound be after the ions exchange electrons?

45. Why is the formula ZnF_2 correct rather than ZnF?

46. When you use the crisscross method, what is being crisscrossed?

47. Ionic compounds are never called "molecules", yet it is possible to find molecules in an ionic compound. Explain why.

Section 5.4

48. Do each of the following describe an ionic compound, a covalent compound, or both?
    a. conducts electricity when added to water
    b. particles are barely attracted to one another at room temperature
    c. bends easily
    d. shatters when dropped

49. Is Br_2 a compound? Explain.

50. Do the properties below describe an ionic or covalent compound, or both?
    a. melts at a low temperature
    b. vaporizes at -5 °C
    c. does not conduct electricity as a solid
    d. conducts electricity when melted

51. Explain the difference in electron behavior in covalent bonds versus ionic bonds.

52. Is Br_2 a compound? Explain.

53. Is KOH a molecule? Explain.

54. Is KOH a compound? Explain.

55. Which elements form diatomic molecules?

56. Explain why NaCl is not a diatomic molecule but CO is.

57. What is similar between a network covalent structure and an ionic crystal structure?

Chapter 5 review

Quantitative problems

Section 5.2

58. Calculate the number of electrons lost or gained when the following atoms form ions.

    a. cesium
    b. oxygen
    c. aluminum
    d. iodine

59. Identify whether each of the following atoms will lose or gain electrons when they form ions.

    a. aluminum
    b. iodine
    c. cesium
    d. oxygen

60. Mystery element X is a metal with 3 electrons in its outer layer. Write the ion formed by element X.

61. Mystery element Y forms an ion: Y^{2-}. Has the element given up electrons or gained electrons? How many?

62. Calculate the number of electrons each ion has lost or gained:

    a. Ag^+
    b. Cr^{6+}
    c. Mn^{4+}
    d. Ba^{2+}
    e. I^-
    f. Se^{2-}
    g. N^{3-}

63. Write the most common ion that forms from these elements.

    a. phosphorus
    b. nitrogen
    c. silver
    d. sodium
    e. oxygen

Section 5.3

64. What is the chemical formula for each of the following compounds?

    a. potassium bromide
    b. calcium fluoride
    c. aluminum chloride
    d. boron iodide
    e. aluminum sulfide

65. Write the formula for each of the following names.

    a. tin(IV) carbonate
    b. cobalt(II) chlorate
    c. barium hydroxide
    d. sodium sulfide

66. Write the name for each of the following formulas:

    a. $Al(C_2H_3O_2)_3$
    b. $CuSO_4$
    c. $KHCO_3$
    d. $Sn(NO_3)_2$
    e. $(NH_4)_2SO_4$

67. What are the formulas for the compounds with the following names?

    a. chromium(III) sulfate
    b. copper(II) nitrate
    c. copper(I) chloride
    d. trisodium phosphate
    e. iron(III) sulfate

68. Use your knowledge of the crisscross method to write the formulas for the compound that forms between each pair of elements.

    a. calcium and oxygen
    b. magnesium and sulfur
    c. potassium and chlorine
    d. silver and chlorine

69. Give a chemical name for the following compounds:

    a. $AgNO_3$
    b. K_2CO_3
    c. FeO
    d. Fe_2S_3
    e. $ZnBr_2$

70. Give a chemical name for each of the following compounds?

    a. Cs_2O
    b. MgI_2
    c. $SrCl_2$
    d. $PbCl_4$

Section 5.4

71. Give a chemical name for each of the following compounds.

    a. $SiCl_4$
    b. C_2H_6
    c. BN
    d. P_4O_{10}

72. Give a chemical name for each of the following compounds.

    a. NH_4NO_3
    b. PF_5
    c. C_6H_6
    d. NH_3

73. Give the chemical formula for each of the following compounds.

    a. chlorine monoiodide
    b. carbon tetrafluoride
    c. sulfur dioxide
    d. nitrous oxide

Chapter 5 review

Section 5.4

74. Write the names for each of the following.
    a. Cl_2
    b. H_2
    c. C
    d. Hg
    e. K

75. Give the chemical formula for each of the following compounds?
    a. cobalt(II) sulfate
    b. chromium(III) sulfide
    c. chromium(II) chloride

76. Write the formula for each covalent compound.
    a. phosphorus trifluoride
    b. disulfur tetrafluoride
    c. boron tribromide
    d. carbon tetrabromide

77. Name each compound.
    a. $SrBr_2$
    b. P_2S_5
    c. NH_4I
    d. SO_2

78. Write the formula for each covalent compound.
    a. dichlorine monoxide
    b. dinitrogen tetrachloride

79. Identify if the compound is ionic or covalent.
    a. PCl_4
    b. CCl_4

80. What are the names for the following compounds?
    a. PCl_4
    b. CCl_4

81. Name the following molecule:
    a. O_2
    b. Br_2
    c. F_2

82. Write a formula for the compound that forms between sodium and each polyatomic ion.
    a. hydroxide
    b. acetate
    c. phosphate

83. Write a formula for the compound that forms between aluminum and the following:
    a. fluorine
    b. chlorine
    c. nitrogen

84. Write a formula for the compound that forms between magnesium and each polyatomic ion.
    a. carbonate
    b. bicarbonate
    c. nitride

85. Give the chemical formula for the following compounds.
    a. uranium hexafluoride
    b. oxygen
    c. carbon monoxide
    d. silicon dioxide

86. Identify the following as ionic or covalent.
    a. copper(II) chloride
    b. chlorine monoxide
    c. lead(II) chromate

87. Write the formula for the following.
    a. copper(II) chloride
    b. chlorine monoxide
    c. lead(II) chromate

88. State whether the compound is ionic or covalent.
    a. potassium chromate
    b. cobalt(II) bromide

89. What is the formula for the following compounds?
    a. potassium chromate
    b. cobalt(II) bromide

90. Identify each of the following compounds as ionic or covalent.
    a. KF
    b. CO

91. What are the names of the following compounds?
    a. KF
    b. CO

92. Give the following compounds their appropriate name.
    a. Li_2S
    b. Na_2O
    c. $SnCl_4$
    d. Mg_3N_2

93. There are different ways to memorize the seven diatomic molecules that exist naturally. State your favorite way.

Section 5.6: Chapter Review

chapter 6 Moles

There are different ways to determine how much of something is present. In our daily lives, we usually look at masses and volumes. This is especially true when cooking or baking. A recipe might call for measuring a pound of chicken, a cup of flour and a tablespoon of salt. Assembly instructions are a little different and require counting. In this case, the actual number of items is more important than the mass or the volume. If you have 6 bolts to install a shelf for a bookcase, then you should have 6 nuts to go with them. It would be confusing to say that you needed ½ cup of screws to go with 1/3 cup of nuts. With recipes, it is easy to measure the amount of something with the mass or volume. You would not want to count the little grains of salt or pieces of flour because the particles are so small. With assembly instructions, it is easier to count the number of items.

To understand chemical formulas and reactions, chemists need both ways of counting. Chemists need to be able to count atoms. When forming a compound, atoms go together in specific ratios that we represent as whole numbers in formulas. Just as you need 1 bolt for every 1 nut, a water molecule has 1 oxygen atom for every two hydrogen atoms. However, atoms are so small that it is unrealistic to actually count them. It is more practical to measure the mass or the volume of water. In this chapter, you will learn about another unit of measure that is essential to chemists as they combine these different approaches of measuring and counting.

Chapter 6 study guide

Chapter Preview

To measure particle masses the atomic mass unit (amu) is used, based on a standard carbon atom. The amu is used to express the relative masses of all the elements using an average, and differentiates between various isotopes of elements. Avogadro's number is an important and very large number that is used to group collections of atoms or molecules. One Avogadro's number of something is known as a mole. The molar mass is the mass in grams of one mole of a substance. One mole of any gas has a known volume at a given pressure and temperature. Percent composition is a measure of the percent by mass of each element in a compound and can be extended to the water in hydrates. The molecular formula gives the actual composition of a molecule of a substance while the empirical formula only states the most simplified ratio of the elements of that compound.

Learning objectives

By the end of this chapter you should be able to

- calculate molar mass;
- convert between the number of moles and the number of atoms or molecules in a substance;
- convert between the number of moles and mass of a substance as well as the volume of a gas at STP;
- determine percent composition for compounds, including hydrates; and
- calculate empirical formulas and use them in calculations involving molecular formulas.

Investigations

6A: Counting by weight
6B: Molar mass
6C: Percent concentration of a hydrate
6D: Empirical formula of magnesium oxide

Pages in this chapter

164 The Mole
165 How big is a mole?
166 Calculation with moles
167 Section 1 Review
168 The Mass of a Compound
169 Calculating the molar mass of a compound
170 Using the molar mass
171 Solving problems with moles
172 Calculating atoms or moles within a compound
173 Volume of a gas
174 Section 2 Review
175 Percent Composition
176 Solving percent composition problems
177 Hydrate percent composition
178 The empirical formula
179 Finding the mole ratio in an empirical formula
180 Calculating molecular formulas
181 Putting it all together
182 Section 3 Review
183 Nanotechnology: Molecular machines
184 Nanotechnology - 2
185 Nanotechnology - 3
186 Chapter review

Vocabulary

atomic mass unit (amu)	atomic mass	molar mass	Avogadro's number
mole	formula mass	molecular mass	molecular weight
standard temperature and pressure (STP)	standard molar volume	hydrate	anhydride
empirical formula	molecular formula		

6.1 - The Mole

Atoms and molecules are so small it is all but impossible to do anything tangible with individual particles. When a pharmacist measures out 100 milligrams of ibuprofen they cannot count individual molecules. Even a single tablet contains 4×10^{20} molecules. Instead, chemists have come up with a standard unit called the mole. Instead of working with atoms, we work with moles of atoms. Moles of atoms and molecules typically have masses in grams which are easily measured with laboratory equipment.

Interpreting the atomic mass

The atomic mass unit

The mass of a carbon atom is 1.994×10^{-23} grams. This is such an inconvenient number to work with that scientists define the **atomic mass unit (amu)** as 1.66×10^{-24} grams. Expressed in amu, the mass of a carbon atom is on average 12.011 amu, which is much easier to understand. The **atomic mass** for each element is listed on the periodic table in amu. The average atomic masses of the 92 naturally occurring elements range from hydrogen at 1 amu to uranium at 238 amu. The word "average" means the value accounts for the fact that many elements have several different isotopes or versions of themselves. The most common form of carbon is called carbon-12. It has a mass of exactly 12.000 amu, because there are 6 protons and 6 neutrons in the nucleus. One amu equals 1/12 of the mass of a carbon-12 atom.

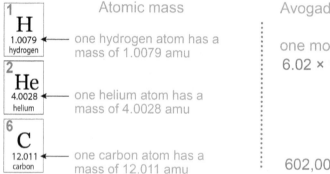
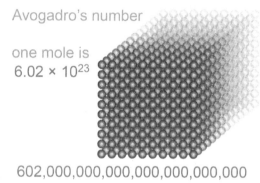

Moles and Avogadro's number

The formula for carbon dioxide, CO_2, tells us that we need two oxygen atoms per carbon atom. Atoms are so small that they cannot be counted individually. Even counting by millions or billions would still be too small. Instead, chemists work in units of one **mole**. One mole is 6.02×10^{23}. This number is called **Avogadro's number** after Italian chemist Amadeo Avogadro who contributed to molecular theory in the 18th century. Just as one dozen represents 12 of something, one mole represents 6.02×10^{23} of something.

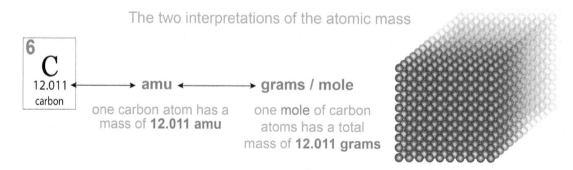

Two ways to interpret atomic mass

The atomic mass represents two things. It is both the average mass of a single atom in amu and the mass of one **mole** of atoms in **grams**. This second interpretation is known as **molar mass**, and it is far more useful and important. Molar mass allows us to use grams, which we can measure, to count atoms. We need to count atoms in order to do chemistry.

How big is a mole?

How big is a mole?

One mole, abbreviated mol, is a very large number. When written out, it looks like this: 602,000,000,000,000,000,000,000. Let's consider one mole of rice grains. One mole of rice has a mass of 1.20×10^{19} kg. When you consider that there are over 7 billion people on Earth, this would give us 1.7×10^9 kg of rice per person. In other words, each person has to eat almost 1.7 billion kg to get rid of one mole of rice! To make things worse, the rice would cover the land area of the Earth close to 50 meters deep. If one mole of rice were made into a cube, it would be 193 km on each side. Using the mole to measure things as large as rice grains is not practical!

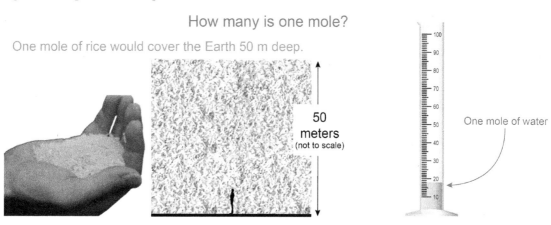

How many is one mole?
One mole of rice would cover the Earth 50 m deep.
50 meters (not to scale)
One mole of water

What do you measure with a mole?

The mole is used to measure things the size of a molecule and smaller. Compared to a molecule, rice grains are very large. One mole of water is just over 18 ml. That isn't even a decent glass of water! Think about how many times smaller a water molecule is than a grain of rice. One mole of rice covers the entire world and one mole of water wouldn't cover your desk. In both cases, they have the same number of particles, 6.02×10^{23}.

Solved problem

How many atoms are in 1.33 moles of silver?

Asked   Convert from moles to atoms
Given   1.33 moles of silver
Relationships 6.02×10^{23} atoms is one mole.
Solve $\dfrac{1.33 \text{ moles silver}}{1} \times \dfrac{6.02 \times 10^{23} \text{ atoms of silver}}{1 \text{ mole silver}} =$

$= 8.0066 \times 10^{23} \xrightarrow{\text{round}} 8.01 \times 10^{23}$ atoms of silver

Answer   1.33 moles of silver contains 8.01×10^{23} atoms of silver.

Solved problem

How many moles are in 5.25×10^{27} atoms of mercury?

Asked   Convert from atoms to moles
Given 5.25×10^{27} atoms of mercury
Relaitonships 6.02×10^{23} atoms is one mole
Solve $\dfrac{5.25 \times 10^{27} \text{ atoms mercury}}{1} \times \dfrac{1 \text{ mole of mercury}}{6.02 \times 10^{23} \text{ atoms of mercury}} =$

$= 8720.93023256 \xrightarrow{\text{round}} 8{,}720$ moles of mercury

Answer 5.25×10^{23} atoms of mercury equals 8,720 moles of mercury.

Section 6.1: Measuring the Very Small

Calculation with moles

Moles are central to chemistry

Moles are a central concept in chemistry. Moles are a means for chemists to relate the macroscopic scale of things that you can see to the microscopic scale of atoms and molecules. Mass is related to moles through molar mass and atoms are related to moles through Avogadro's number. Because a mole represents a particular number of atoms, chemists will often compare things on a per mole basis.

Calculation map between grams, moles, and particles

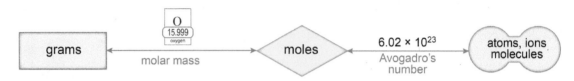

Atomic mass in problems

Converting between moles and grams is very similar to dimensional analysis. The atomic mass on the periodic table is the conversion factor and it is different for each element. According to the periodic table, one mole of calcium atoms has a mass of 40.078 g. The calculations below show how to find moles if you are given grams and how to find grams if you are given moles.

grams to moles

$$\frac{25.0 \text{ g calcium}}{1} \times \frac{1 \text{ mol of calcium}}{40.078 \text{ g calcium}}$$

$$= 0.624 \text{ moles of calcium}$$

moles to grams

$$\frac{0.624 \text{ mol of calcium}}{1} \times \frac{40.078 \text{ g of calcium}}{1 \text{ mol of calcium}}$$

$$= 25.0 \text{ g of calcium}$$

Solved problem

How many grams of calcium do you need to measure if you want to have 1.50 moles of calcium?

- Asked: Convert from moles to grams.
- Given: The element is calcium and there are 1.50 moles.
- Relationships: One mole of calcium has a mass of 40.078 grams.
- Solve: $\frac{1.50 \text{ moles Ca}}{1} \times \frac{40.078 \text{ grams Ca}}{1 \text{ mole Ca}} = 60.117 \xrightarrow{\text{round}} 60.1 \text{ grams Ca}$
- Answer: 1.50 moles of calcium has a mass of 60.1 grams.

Grams to atoms

You can take the mole calculation one step further if you are asked to convert from grams to atoms. Add Avogadro's number to solve for atoms once you have converted grams to moles. You will find in most calculations, the mole is a key conversion unit.

Solved problem

How many atoms are in 210.0 grams of oxygen?

- Asked: Convert from grams to atoms.
- Given: The element is oxygen and there are 210.0 grams of it.
- Relationships: One mole of oxygen atoms has a mass of 15.999 grams.
- Solve: $\frac{210.0 \text{ grams S}}{1} \times \frac{1 \text{ mole O}}{15.999 \text{ grams S}} \times \frac{6.02 \times 10^{23} \text{ atoms of O}}{1 \text{ mole S}} =$

$$= 7.90174 \times 10^{24} \xrightarrow{\text{round}} 7.902 \times 10^{24} \text{ atoms of O}$$

- Answer: 100.0 grams of sulfur contains 7.902×10^{24} atoms of oxygen.

Section 1 review

Chapter 6

To quantify the tiny mass of an atom we use the atomic mass unit (amu), equal to 1/12 the mass of a carbon atom with 6 protons and 6 neutrons. On the periodic table atomic masses of elements are averages because although a given element always has the same number of protons, the number of neutrons can vary. Avogadro's number is a collection of 6.02×10^{23} of something, generally atoms or molecules and is known as a mole. One mole of a substance is the number of particles required to turn the molar mass, in amu, into an equivalent mass in grams.

Vocabulary words: atomic mass unit (amu), atomic mass, molar mass, Avogadro's number, mole

Review problems and questions

1. One mole of rice would cover Earth's surface almost 75 m deep.

    a. One mole of elemental carbon will easily fit in the palm of your hand, but obviously, one mole of rice will not fit in the palm of your hand. Why?

    b. How many carbon atoms are in a mole of carbon atoms?

More than one mole of carbon atoms are found in one handful of coal.

2. Calculate the number of atoms in 2.8 moles of silicon atoms.

3. Calculate the number of moles in 32.1 g iron atoms.

4. How many atoms are in a 0.50 g sample of helium?

5. Calculate the mass of 3.01×10^{23} atoms of carbon.

6.2 - The Mass of a Compound

Most of the truly interesting chemistry occurs with compounds and not pure elements. What we learned with single elements extends more usefully to compounds. The mole and the atomic mass of the elements allows us to find the mass of a mole of a compound. This will be the mass of the molecule in the case of a molecular compound such as water. For ionic compounds it will be the mass of the formula unit.

The mass of a compound

The formula mass

What is the mass of one molecule of CH_3OH (methanol) if we know the mass of oxygen, hydrogen and carbon atoms separately? The mass of one mole of a compound is called the **formula mass**. For molecular compounds, the formula mass is often called the **molecular mass**. The diagram below shows the answer which you may already surmise - to find the formula mass you simply add up the mass of every atom in the formula. One molecule of methanol has a mass of 32 amu.

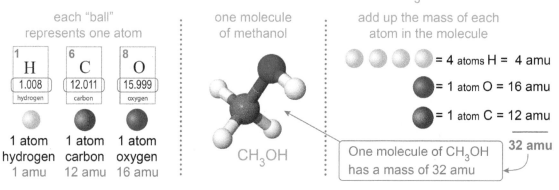

Solved problem

What is the formula mass of aluminum sulfite?

Asked What is the formula mass.
Given The name of the formula.
Relationships The criss-cross method can be used to determine the chemical formula. Then use the periodic table to get the total mass of all atoms in the formula.
Solve Aluminum sulfite is ionic, and the formula is $Al_2(SO_3)_3$.
Al = 2 x 26.982 = 53.964 amu
S = 3 x 32.065 = 96.195 amu
O = 9 x 15.999 = 144.00 amu
formula mass = 53.964 amu + 96.195 amu + 143.991 amu = 294.15 amu
Answer The formula mass of aluminum sulfite is 294.15 amu.

Calculate the formula mass of aluminum sulfite - $Al_2(SO_3)_3$

element	atomic mass		number of atoms		total mass
aluminum	26.982 amu	×	2	=	53.964 amu
sulfur	32.065 amu	×	3	=	96.195 amu
oxygen	15.999 amu	×	9	=	143.991 amu
				total mass	**294.15 amu**

Calculating the molar mass of a compound

Using molar mass

The mass of a molecule in amu is interesting but not very practical. Practical chemistry is done in grams and moles. For a compound such as methanol, CH_3OH, we need the mass of one mole of methanol, known as the molar mass. This is where the correspondence between amu and grams per mole is crucial - the molar mass in grams per mole is the same as the formula mass in amu. One mole of methanol has a molar mass of 32 grams/mole and one molecule of methanol has a mass of 32 amu. This is the reason the formula mass is called the **molecular weight** for molecular compounds. The diagram below shows the calculation of the molar mass for methanol except each "ball" represents one mole instead of one atom.

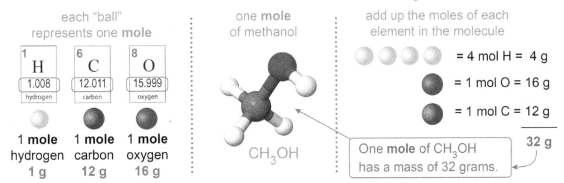

Calculating the **molar** mass of methanol, CH_3OH

Solved problem

What is the mass of 1 mole of ethanol, which has the chemical formula CH_3CH_2OH?

Given: Ethanol, CH_3CH_2OH contains 2 carbon atoms, C, 1 oxygen atom, O, and 5 hydrogen atoms, H.

Relationships: Molar mass is the sum of the molar masses for each atom in the compound.

Solve: Multiply the molar mass by the number of atoms in the formula (CH_3CH_2OH):
Carbon, C: 12.011 g/mol × 2 = 24.022 g/mol
Oxygen, O: 15.999 g/mol × 1 = 15.999 g/mol
Hydrogen, H: 1.0079 g/mol × 5 = 5.0395 g/mol
Now add masses together: 24.022 + 15.999 + 5.04 = 45.0605 g/mol

Answer: One mole of ethanol, CH_3CH_2OH has a mass of 45.0605 grams.

Calculate the molar mass of methane - CH_4

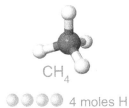

4 moles H
1 mole C

element	atomic mass		number of moles		total mass
carbon	12.011 g/mol	×	1	=	12.011 g/mol
hydrogen	1.008 g/mol	×	4	=	4.032 g/mol
				molar mass	16.043 g/mol

Solved problem

What is the molar mass of hydrogen peroxide, H_2O_2?

Given: Hydrogen peroxide, H_2O_2 contains 2 hydrogen, H and 2 oxygen, O atoms.

Relationships: Add up molar masses for each atom in the compound.

Solve: 2H: (2 × 1.0079) = 2.0158
2O: (2 × 15.999) = 31.998
Total: 34.0138 g/mol

Answer: The molar mass of H_2O_2 is 34.0138 g/mol.

Using the molar mass

Counting by weighing

Chemists use moles and the molar mass to convert ratios of atoms in compounds to ratios in grams that can be measured. For example, the compound silicon carbide has the chemical formula SiC. Silicon carbide is very hard and used as an abrasive, a semiconductor, and even as a synthetic diamond! From the chemical formula we know the ratio of silicon atoms to carbon atoms is 1:1. From the periodic table we know that silicon and carbon have different molar masses. How many grams of silicon and carbon will produce a ratio of one silicon atom per carbon atom?

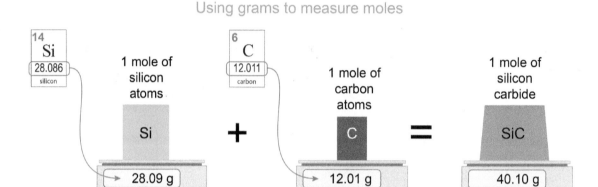

The molar mass converts between moles and grams

From the periodic table we see that one mole of carbon has a mass of 12.011 g and one mole of silicon has a mass of 28.086 g. But the digital balance in a lab may only read to two decimal places. To make silicon carbide we follow the steps below.

1. We need one atom of silicon per one atom of carbon.
2. This is the same ratio as one mole of silicon per one mole of carbon.
3. This is the same as 28.09 grams of silicon per 12.01 grams of carbon!

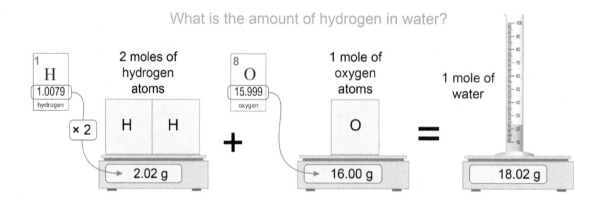

How much hydrogen is in water?

The molar mass is written in units of grams per mole, abbreviated g/mol. A chemical formula tells us the number of moles of each element in a compound. The molar mass tells us how many grams that represents. The same calculation can be done for any compound, such as H_2O. There are two hydrogen atoms per oxygen atom, however, hydrogen has a molar mass of 1.079 g/mol and oxygen has a molar mass of 15.999 g/mol. The formula for H_2O is is the same as 2.02 grams of hydrogen per 16.00 grams of oxygen. Water is 67% (2/3) hydrogen by atoms but only 11% hydrogen by weight!

Solving problems with moles

Why use moles? Whether you are talking about elements or compounds, moles are central in chemistry to relating large quantities to very small things, like atoms, formula units and molecules. As a problem solving strategy, take the given quantity and change it to moles. From there, we can convert moles to just about anything else!

Calculation pathway

Calculation map between grams, moles, and particles

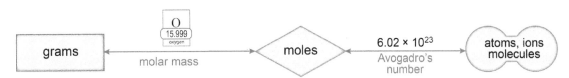

Ionic vs. covalent compounds Recall that it is inappropriate to call an ionic compound a "molecule." You can only call covalent compounds "molecules." When using Avogadro's number to count numbers of ionic compound particles such as lead(II) iodide, PbI_2, you can use the term "formula unit" or "particles" instead.

Solved problem How many formula units are in 325.0 g of lead(II) iodide, PbI_2?

    Asked Convert from grams to formula units.
    Given 325.0 g of PbI_2
    Relationships The molar mass for PbI_2 = 461.00 g/mole.
    Solve Starting on the left side of the calculation pathway, you see that you have to use the molar mass to go from grams to moles. Then use Avogadro's number to go from moles to formula units.

$$\frac{325.0 \text{ g } PbI_2}{1} \times \frac{1 \text{ mol } PbI_2}{461.00 \text{ g } PbI_2} \times \frac{6.02 \times 10^{23} \text{ formula units } PbI_2}{1 \text{ mol } PbI_2} =$$

$$= 4.24403 \times 10^{23} \xrightarrow{\text{round}} 4.244 \times 10^{23} \text{ formula units of } PbI_2$$

    Answer 325.0 g of PbI_2 is equal to 4.244×10^{23} formula units of PbI_2.

Solved problem Propane is used as a fuel for portable devices such as camping stoves and lanterns. What is the mass of 2.30×10^{23} molecules of propane, C_3H_8?

    Asked Convert from molecules of propane to grams of propane.
    Given 2.30×10^{23} molecules of C_3H_8
    Relationships The molar mass for C_3H_8 is 44.0954 g/mol.
    Solve You are starting on the right of the calculation pathway and need to use Avogadro's number to get to moles, and then use the molar mass to go from moles to grams.

$$\frac{2.30 \times 10^{23} \text{ molecules } C_3H_8}{1} \times \frac{1 \text{ mol } C_3H_8}{6.02 \times 10^{23} \text{ molecules } C_3H_8} \times \frac{44.0954 \text{ g } C_3H_8}{1 \text{ mol } C_3H_8} =$$

$$16.84708 \xrightarrow{\text{round}} 16.8 \text{ grams } C_3H_8$$

    Answer 2.30×10^{23} molecules of propane, C_3H_8 have a mass of 16.8 g.

Calculating atoms or moles within a compound

Finding the number of atoms

In many situations in chemistry you are given a quantity of a compound and want to calculate the number of atoms or moles of the constituent elements. For example, how many moles of hydrogen are in 2 moles of ammonia, NH_3? The answer uses the chemical formula. One mole of NH_3 contains three moles of hydrogen so two moles of NH_3 contain six moles of hydrogen.

Nitrogen in fertilizer

Quantities of individual elements in a compound are important because sometimes only one element is needed for a reaction or process. For example, fertilizers contain ammonium (NH_4^+) and nitrate (NO_3^-). These "fixed nitrogen" compounds are readily used by plants but it is the nitrogen that the plants actually use. One of the important measures of a fertilizer is the total nitrogen content - which must be summed for all nitrogen-containing molecules or ions.

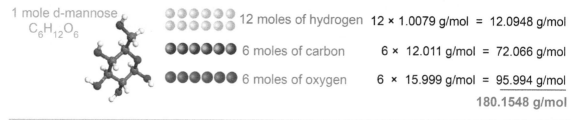

1 mole d-mannose $C_6H_{12}O_6$

12 moles of hydrogen 12×1.0079 g/mol = 12.0948 g/mol
6 moles of carbon 6×12.011 g/mol = 72.066 g/mol
6 moles of oxygen 6×15.999 g/mol = 95.994 g/mol
                                                    180.1548 g/mol

Solved problem

How many oxygen atoms are in 100.0 g of the nutritional supplement d-mannose, $C_6H_{12}O_6$?

Asked Find the number of oxygen atoms in 100.0 g of $C_6H_{12}O_6$.

Given 100.0 grams of $C_6H_{12}O_6$

Relationships The molar mass of d-mannose is:
$6C + 12H + 6O = (6 \times 12.011) + (12 \times 1.0079) + (6 \times 15.999) =$
$= 180.1548$ g/mole

Solve First, we find how many moles are in 100 grams. From the calculation pathway, you are starting on the left side and must first use the formula mass to go to moles.

$$\frac{100.0 \text{ grams } C_6H_{12}O_6}{1} \times \frac{1 \text{ mole } C_6H_{12}O_6}{180.1548 \text{ grams } C_6H_{12}O_6} = 0.55508 \text{ moles } C_6H_{12}O_6$$

Next, we find how many molecules are contained in 0.55508 moles. You are now in the middle of the calculation pathway and can use Avogadro's number to go to molecules.

$$\frac{0.55508 \text{ moles } C_6H_{12}O_6}{1} \times \frac{6.02 \times 10^{23} \text{ molecules } C_6H_{12}O_6}{1 \text{ mole } C_6H_{12}O_6} =$$
$$= 3.34158 \times 10^{23} \text{ molecules } C_6H_{12}O_6$$

Then we find how many oxygen atoms are contained in the amount of d-mannose molecules present.

$$\frac{3.34158 \times 10^{23} \text{ molecules } C_6H_{12}O_6}{1} \times \frac{6 \text{ atoms O}}{1 \text{ molecule } C_6H_{12}O_6} =$$
$$= 2.00495 \times 10^{24} \xrightarrow{\text{round}} 2.005 \times 10^{24} \text{ atoms O}$$

Answer There are 2.005×10^{24} oxygen atoms in 100.0 g of d-mannose, $C_6H_{12}O_6$.

Volume of a gas

Standard temperature and pressure (STP)

It is hard to measure the mass of a gas, but relatively easy to measure the volume. If the temperature and pressure are kept constant, the volume of gas is directly proportional to the mass. This means volume can be used to determine the mass of a quantity of a gas. Because volume changes with temperature and pressure, chemists prefer to compare gases at **standard temperature and pressure (STP)**. The standard temperature is 0.0 °C and the standard pressure is 1 atmosphere = 101.325 kPa.

STP and volume

The **standard molar volume** is one mole of any gas has a volume of 22.4 L at STP. We can express this as a ratio and use it in a manner similar to the way we used 6.02×10^{23}.

The standard molar volume of a gas
One mole of any gas at STP occupies a volume of 22.4 liters

How can all gases have the same volume?

How can all gases have the same volume at STP when some gas molecules are larger than others? Gases have large spaces between particles. In 5 L of a gas, the space between the molecules takes up most of the volume. The size of the molecules themselves adds almost nothing to the volume. This is like our solar system; most of the volume is empty. We could make the planets larger or smaller and it still would not change the overall size.

Calculation map between grams, moles, particles and gas volumes

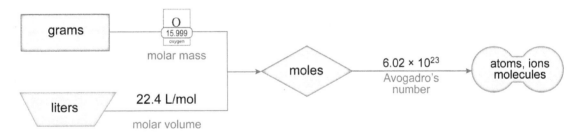

Solved problem

How many liters of oxygen gas are in 456.22 g of oxygen gas at STP?

Given 456.22g of O_2

Relationships Formula mass of O_2 : $(2 \times 15.999) = 31.998$ g/mole

Standard Molar Volume: $\frac{22.4L}{1 \text{ mole}}$

Solve $\frac{456.22 \text{ grams } O_2}{1} \times \frac{1 \text{ mole } O_2}{31.998 \text{ grams } O_2} \times \frac{22.4 \text{ liters of } O_2}{1 \text{ mole } O_2} =$

= 319.374 round 319 liters of O_2 at STP

Answer 456.22g of O_2 at STP occupy a volume of 319 L

O atoms vs. O_2 gas molecules

Some elements like oxygen form diatomic molecules. Is a question asking for atoms of oxygen (O, molar mass = 15.999 g/mol) or molecules of oxygen (O_2, molar mass = 31.998 g/mol)? What about elemental oxygen (O atoms), or oxygen gas (O_2 molecules)? Be sure to choose the correct formula and formula mass for H, O, N, Cl, Br, I and F.

Section 6.2: Masses of Compounds

Chapter 6

Section 2 review

The formula or molecular mass of a compound (in amu) is found by adding up the atomic masses that make up that compound. The molar mass is the mass of one mole of the substance in grams. While the formula mass and molar mass use the same number, they are using different units. That's the value of Avogadro's number: it is the number of particles needed to convert from amu to grams, and allows us to convert between the number of particles and the mass of all of those particles. With gases, mass can be tough to measure directly. But at standard temperature and pressure one mole of gas has a volume of 22.4 L creating a useful conversion factor.

Vocabulary words	formula mass, molecular mass, molecular weight, standard temperature and pressure (STP), standard molar volume
Key relationships	• Formula mass = sum of masses in a single unit of a chemical formula (amu) • Molar mass = the mass of 1 mole of particles of a substance (g/mol) • STP = standard temperature and pressure (T = 0 °C and P = 1 atm) • Molar volume of any gas at STP: $\frac{22.4 \text{ L}}{1 \text{ mol}}$

Review problems and questions

1. Explain why it is incorrect to ask this question: "How many atoms of calcium hydroxide ($Ca(OH)_2$) are in 411.2 g?" and find the molar mass of $Ca(OH)_2$.

2. Calculate the number of moles in 1.99×10^{22} molecules of sugar, $C_6H_{12}O_6$.

3. How many molecules are present in 50.0 g nitrogen gas, N_2?

4. The 50.0 g of N_2 gas mentioned above is at STP. What is the volume of the gas?

5. How many hydrogen atoms are in 0.25 moles of H_2O?

6. What is the formula mass for mercury(I) dichromate, $Hg_2Cr_2O_7$?

7. How many moles are in 6.23×10^3 g of zirconium(III) nitride, ZrN?

8. What volume will 2.55 moles of chlorine gas occupy at STP?

9. How many grams are in 7.820×10^{22} molecules of N_2I_6?

10. How many moles are in 2.4×10^{28} molecules of H_2SO_4?

11. What is the volume of mercury if there are 75.0 moles at STP?

12. What is the volume of xenon if there are 325 moles at STP?

6.3 - Percent Composition

How was it determined that salt is NaCl and not Na_2Cl? One of the many tasks faced by a chemist is to determine the chemical formula of an unknown substance. Modern tools such as a mass spectrometer can tell us the quantities of different elements in a sample. We can determine the formula of an unknown substance and identify it by comparing the ratios of different elements that make it up. The field of forensic chemistry focuses on identifying unknown substances from crime scenes, archeological digs, biology labs, and even meteoroids!

Finding the percent composition

Percent composition

You can use the percentage by mass, or percent composition of each element in a compound to determine its identity. To calculate percent composition, you need to two kinds of information: the molar mass of the compound, and the total mass each individual element contributes to the molar mass.

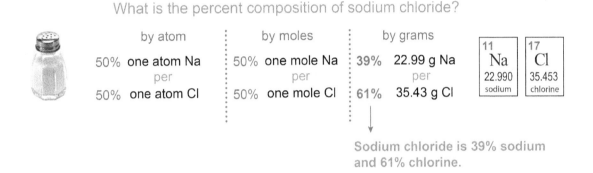

Sodium chloride is 39% sodium and 61% chlorine.

How to work from grams to moles

A common problem is when you know the composition by mass and want to find the composition by moles. The composition by moles gives you the chemical formula. While not always providing the identity of the compound, the chemical formula is an important clue. The diagram below is a calculation map from percent composition by mass to composition by moles for a compound that is 83% potassium and 17% oxygen.

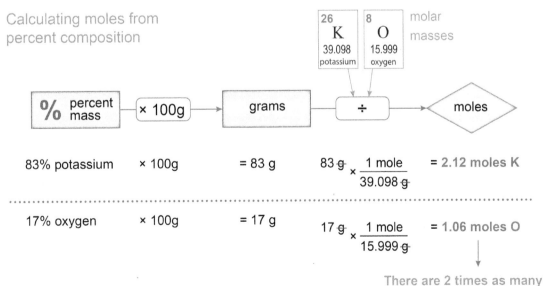

Section 6.3: Percent Composition and Empirical Formula

Solving percent composition problems

Matching known and unknown compositions

A common problem in forensic chemistry is to evaluate whether an unknown compound might match a compound with a known chemical formula. This means comparing the composition of the known compound with the unknown to see how closely they match.

1. Calculate the masses of each element in the known compound.
2. From the masses in step (1), calculate the percent compositions of each element.
3. See if the compositions of known and unknown compounds match within measurement error.

Limitations of percent composition

The percent composition technique can definitively rule out a compound if the percent compositions do not match. However, it cannot identify a larger compound without more information because there are different compounds that have the same elements in the same ratios. For example, C_2H_4 and C_4H_8 are different compounds that have the same percent composition. Aspirin and acetyl benzol peroxide both share the chemical formula $C_9H_8O_4$, which means they will also have the same percent composition.

A 2.5 gram sample of this white substance has the following elements.
1.05 g carbon
0.16 g hydrogen
1.29 g oxygen
Could this substance be sucrose (table sugar)

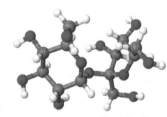

sucrose $C_{12}H_{22}O_{11}$

Solved problem

Based on the information given in the graphic above, could the unknown substance be sucrose?

Asked Is the unknown table sugar?

Given The mass of each element in the powder and the formula for sugar.

Relationships The % of each element should match the percent composition of sugar if the powder is a sample of that substance.

Solve

Find the mass of each element in sucrose;

Total mass of carbon $12 \times 12.011 = 144.132$ g/mole
Total mass of hydrogen $22 \times 1.0079 = 22.1738$ g/mole
Total mass of oxygen $11 \times 15.999 = 175.989$ g/mole
Total mass of sucrose   342.295 g/mole

Find the percent of each element in sucrose:

%C = $\frac{144.132}{342.295} \times 100 = 42.11\%$ C

%H = $\frac{22.1738}{342.295} \times 100 = 6.48\%$ H

%O = $\frac{175.989}{342.295} \times 100 = 51.41\%$ O

Find the percent of each element in the unknown:

%C = $\frac{1.05}{2.5} \times 100 = 42.0\%$ C

%H = $\frac{0.16}{2.5} \times 100 = 6.4\%$ H

%O = $\frac{1.29}{2.5} \times 100 = 51.6\%$ O

Answer The percent composition is similar to the known value. Yes, the sample could be sucrose.

Hydrate percent composition

Hydrates

Have you ever seen the "do not eat" packet in a new pair of shoes or electronics? That packet contains a chemical classified as a desiccant. A desiccant is a substance that attracts water molecules. The water molecules become part of the substance and are called a **hydrate**. They are placed in product packaging to keep the contents from molding or corroding.

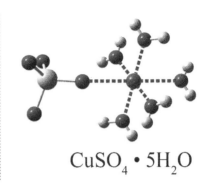

$CuSO_4 \cdot 5H_2O$

Writing hydrates

There is a special way of writing the formula of hydrates. You write the formula of the compound, then a dot and the number of water molecules. For example, $CuSO_4 \cdot 5H_2O$ means that each copper(II) sulfate molecule has five water molecules attached to it. It is called copper(II) sulfate pentahydrate. When heated, the waters come off and the copper(II) sulfate becomes an anhydride. An **anhydride** means without water.

Solved problem

A sample of hydrated cobalt(II) chloride is heated and turns from red to blue. Steam also comes out of the test tube. The hydrated sample had a mass of 21.63g. After heating, it has a mass of 11.80. What is the percent water in the hydrate?

Asked   What is the percent water in the hydrate?
Given   The sample has a mass of 21.63 g and 11.80 g after heating.
Relationships   Find the percentage by dividing water mass by total hydrate compound mass, and multiply by 100.
Solve   First, determine the mass of water that left the test tube.

$$\text{mass before heating} - \text{mass after heating} = \text{water mass}$$

$$21.63 \text{ g} - 11.80 \text{g} = 9.83 \text{ g}$$

Next, divide the water mass by the beginning mass of the hydrate, then multiply by 100 to find the percent water in the hydrate.

$$\frac{9.83}{21.63} \times 100 = 45.45\%$$

Answer   The percent composition of water in the cobalt(II) chloride hydrate, $CoCl_2 \cdot nH_2O$ is 45.45 %.

Note: The "n" in front of H_2O indicates the coefficient is not known, but could be determined mathematically.

Section 6.3: Percent Composition and Empirical Formula

The empirical formula

What is an empirical formula?

An **empirical formula** is the simplest whole-number ratio of elements in a compound. All ionic formulas are already simplified, therefore ionic formulas are always empirical formulas. For example, table salt has a one-to-one ratio of sodium to chloride. Molecular formulas can be empirical as well, but many are not. The molecular compound glucose, $C_6H_{12}O_6$ is clearly not written in its simplest ratio. One of the first steps to determining the chemical formula for an unknown substance is to find the empirical formula from the percent composition.

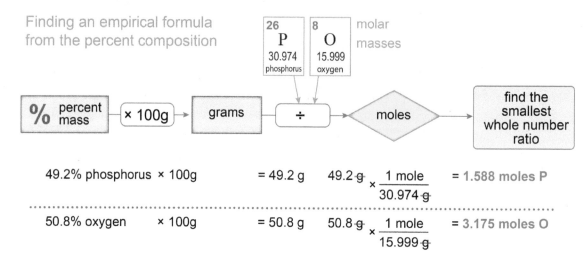

Finding an empirical formula from the percent composition

49.2% phosphorus × 100g = 49.2 g $49.2 \text{ g} \times \dfrac{1 \text{ mole}}{30.974 \text{ g}}$ = 1.588 moles P

50.8% oxygen × 100g = 50.8 g $50.8 \text{ g} \times \dfrac{1 \text{ mole}}{15.999 \text{ g}}$ = 3.175 moles O

Find the smallest whole number ratio of moles

$\dfrac{\text{moles O}}{\text{moles P}} = \dfrac{3.175}{1.588} = 1.999 \sim 2$ therefore the empirical formula is PO_2

The whole number ratio

The trickiest step in the process above is to determine the smallest whole number ratio of moles. This can take a few iterations and is typically not exact. In many cases you will round off 1.85 or 1.9 as approximately equal to 2.

Experimental measurement

The empirical formula is the actual ratio of elements (by atom) in the compound, and is the same as the formula unit for ionic substances. However, for molecular substances the **molecular formula** can be multiples of the empirical formula. For example, the molecular formula for diphosphorus tetraoxide is P_2O_4 but this compound has an empirical formula of PO_2. It takes another scientific step to find the molar mass of the molecule which can then be used to get the molecular formula.

Solved problem

Which of the following is an empirical formula and which is a molecular formula? CH_4, C_6H_6, P_2O_{10}, $CaCl_2$ and Al_2O_3

Asked Identify empirical and molecular formulas.
Relationships The subscripts have to be in the simplest ratio to be an empirical formula.
Solve
- CH_4, $CaCl_2$ and Al_2O_3 are in the simplest ratio
- C_6H_6 and P_2O_{10} are molecular formulas and can be simplified to CH and PO_5.
- Methane, CH_4 is an example of a substance that is the simplest and the actual ratio.

Answer CH_4, $CaCl_2$ and Al_2O_3 are empirical formulas; C_6H_6 and P_2O_{10} are molecular formulas. CH_4 is both an empirical and molecular formula.

Finding the mole ratio in an empirical formula

Calculating the mole ratio of atoms

For a two-element compound divide the larger number of moles by the smaller. For a compound with more than two elements find the element with the smallest number of moles. Divide the moles of each of the other elements by the moles of the element with the smallest number of moles.

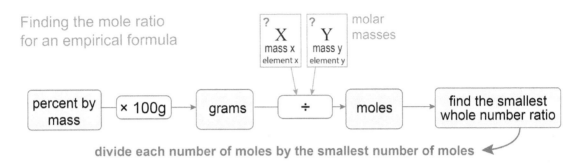

Finding the mole ratio for an empirical formula

divide each number of moles by the smallest number of moles

Solved problem

You find a jar of powder labeled nickel chloride, but as you know nickel can be Ni^{2+} or Ni^{3+}. The Roman numeral is missing from the name. When you analyze 54.50 g of the sample, you find that it is made of 19.40 g of nickel and 35.10 g of chlorine. What is the name and formula of the powdered compound?

Asked Find the correct name and formula for the nickel compound.

Given The sample contains .

Relationships The simplest mole ratio of elements will reveal the empirical formula.

Solve Use given masses to determine moles of each atom. Then divide moles of each atom by the lowest number of moles.

$$\frac{19.4 \text{ g Ni}}{1} \times \frac{1 \text{ mol Ni}}{58.693 \text{ g Ni}} = 0.331 \text{ mol Ni}; \quad \frac{0.331 \text{ mol}}{0.331} = 1 \text{ mol Ni}$$

$$\frac{35.1 \text{ g Cl}}{1} \times \frac{1 \text{ mol Cl}}{35.453 \text{ g Cl}} = 0.990 \text{ mol Cl}; \quad \frac{0.990 \text{ mol}}{0.331} = 3 \text{ mol Cl}$$

Answer The formula is $NiCl_3$ nickel(III) oxide.

Solved problem

A five gram sample of a material has the following composition: 2.003 g of calcium, 0.600 g carbon and 2.397 g oxygen. What is the empirical formula of the compound?

Asked What is the empirical formula?

Given 2.003 g calcium, 0.600 g carbon, 2.397 g oxygen

Relationships The mole ratio of elements can be simplified to give an empirical formula.

Solve

convert grams to moles

$Ca = 2.003 \text{ g} \times \frac{1 \text{ mole}}{40.078 \text{ g}} = 0.0499 \text{ mol}$

$C = 0.600 \text{ g} \times \frac{1 \text{ mole}}{12.011 \text{ g}} = 0.0499 \text{ mol}$

$O = 2.397 \text{ g} \times \frac{1 \text{ mole}}{15.999 \text{ g}} = 0.150 \text{ mol}$

simplify mole ratios

$\frac{\text{moles O}}{\text{moles C}} = \frac{0.150 \text{ mol}}{0.0499 \text{ mol}} = 3$

$\frac{\text{moles Ca}}{\text{moles C}} = \frac{0.0499 \text{ mol}}{0.0499 \text{ mol}} = 1$

formula is $CaCO_3$

Answer The empirical formula is $CaCO_3$.

Calculating molecular formulas

Empirical vs. molecular formula

The empirical formula tells you the ratio of elements in a molecule, but not how many atoms of each there are. A good example of the difference is glucose which has the molecular formula $C_6H_{12}O_6$. However, a mass-analysis of pure glucose would result in an empirical formula of CH_2O because this is the ratio of elements in the molecule.

The difference between molecular and empirical formula

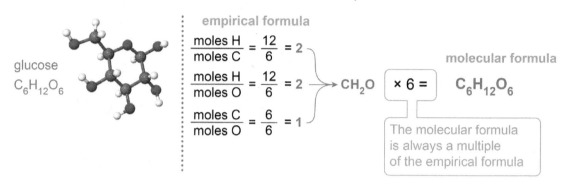

Ionic vs molecular compounds

If you examine the ratios of elements in glucose 6:12:6, you will see that by dividing everything by 6 you can simplify the ratio to 1:2:1, giving the empirical formula CH_2O. An important rule to remember is that the molecular formula is a multiple of, or equal to, the empirical formula. For ionic compounds which dissociate into separate ions in solution, the empirical formula **is** the chemical formula.

Molecular and empirical can be the same

CH_2O is the molecular formula for formaldehyde, not glucose. For formaldehyde, the empirical formula **is** the same as the molecular formula. For any molecular compound the multiplier can be one making the empirical and molecular formula the same.

Molecular formula and empirical formula

To find a molecular formula from an empirical formula you need the molar mass of the compound. The molecular formula will always be a multiple of the empirical formula, therefore the molar mass of the molecule will be a multiple of the molar mass of the empirical formula. The diagram shows a calculation of the molecular formula for a compound with an empirical formula of CH and a molar mass of 78.11 g/mol.

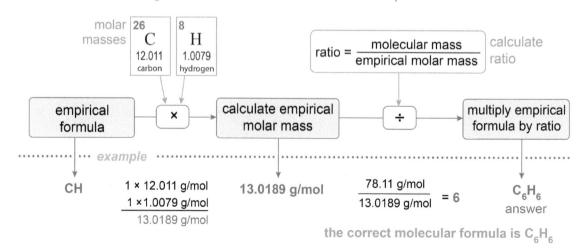

Calculating the molecular formula from the empirical formula

Putting it all together

Finding the actual formula

Let's put it all together now and find the empirical formula and the molecular formula for a compound. There are laboratory techniques that we can use to find the molar mass of an unknown substance. They are discussed in a later chapter.

A 100 gram sample of a molecular substance has the following analysis.

30.4 g nitrogen
69.6 grams oxygen
molar mass = 92.0 g/mol

Determine the molecular chemical formula

Step one: empirical formula

The first step in the process is to find the empirical formula from the percent composition data. This goes in two steps: first convert to moles, then find the smallest whole number ratio.

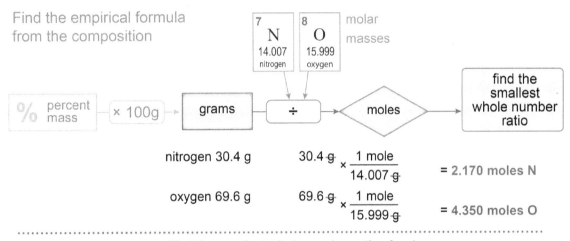

Step two: molecular formula

Once you have the empirical formula, the last step is to determine the molecular formula. This is done by finding the ratio of the molar mass of the empirical formula to the molar mass of the molar mass of the molecule.

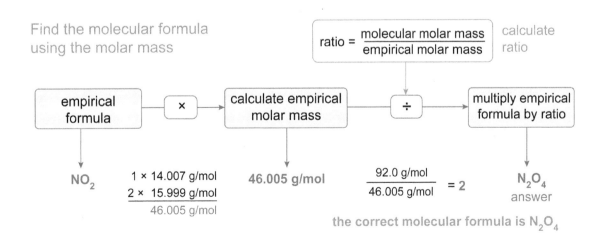

Section 6.3: Percent Composition and Empirical Formula

Chapter 6

Section 3 review

Percent composition states the percent by mass of each of the elements present in a compound and can be used to identify unknown substances. A hydrate is a compound formed by water and another substance. The chemical formula of a hydrate is the formula of the compound and then the number of water molecules attached to each molecule of the compound. The water molecules are typically released when a hydrate is heated to form an anhydride. The mass change from hydrate to anhydride gives the percent composition of water in the hydrate.

Analyzing compounds using percent composition (by mass) yields the empirical formula which is the smallest whole-number ratio of the elements in a compound. For ionic and some molecular compounds the empirical formula is also the molecular formula or formula unit. For other compounds the molecular formula is a multiple of the empirical formula.

Vocabulary words: hydrate, anhydride, empirical formula, molecular formula

Key relationships

- Percent composition, $\% = \frac{\text{total mass of one kind of element}}{\text{formula mass}} \times 100\%$

- Percent composition of a hydrate, $\%_{H_2O} = \frac{\text{water mass}}{(\text{formula mass} + \text{water mass})} \times 100\%$

Review problems and questions

1. What kind of information can you learn about a chemical by its percent composition?

2. A chef has a cold, and can neither taste nor smell anything. She wants to know whether a container of white powder is sugar or starch. The known percent compositions for sugar and starch are:

    Sugar, $C_6H_{12}O_6$: 42.0% C, 6.4% H, 51.6% O
    Starch, $C_6H_{10}O_5$: 61.5% C, 4.3% H, 34.2% O

    The label on the container says the white substance contains 288.264 g C, 20.158 g H, and 159.99 g O. Is the substance sugar or starch?

3. You are working with a hydrate of iron(III) nitrate. The sample mass before heating was 10.0 g. After heating, the mass dropped to 9.25 g. What is the percent water in the hydrate?

4. You are given the molar mass of a molecular compound and the mass of each atom contained in one mole of the compound. What steps do you need to follow to find the molecular formula?

5. Find the molecular formula of a substance that contains 2.04 g of C and 0.355 g of H. Its molar mass is 27.8 g.

Nanotechnology: Molecular machines

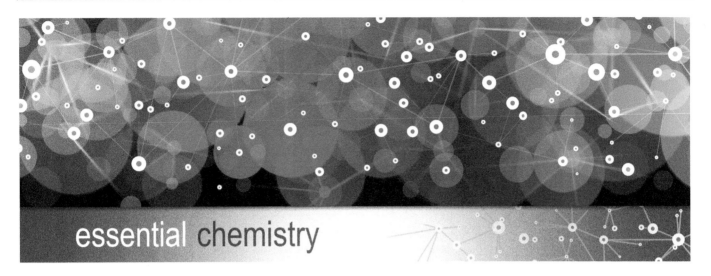

essential chemistry

Over time, scientists have become better at manipulating and visualizing substances at the atomic level and molecular levels. The next step is to build molecular machines that can do work. This is not the realm of science fiction. It is part of an emerging field called nanotechnology. One nanometer (nm) is equal to 10^{-9} meters. In other words, one meter equals one billion nanometers. How small is a nanometer? The diameter of a human hair is about 50,000 nm, and a DNA double helix is about 2 nm in diameter. The term nanotechnology usually refers to the creation, handling, or use of structures having dimensions around 1 nm to 100 nm.

In 2016, the Nobel Prize in chemistry was awarded to a trio of scientists for their work in nanotechnology. Their work advanced the development of tiny molecular machines. Bernard Feringa, Jean-Pierre Sauvage and Sir. J. Fraser Stoddart won the prize for building machines that are a thousand times smaller than the width of a hair on your head!

The first step of tiny machinery is to have components that are "connected" but can still move freely. This was accomplished in 1983 by Sauvage. His group succeeded in linking molecules together to form a chain. It is important to note that the molecules were mechanically linked, but were not chemically bonded together.

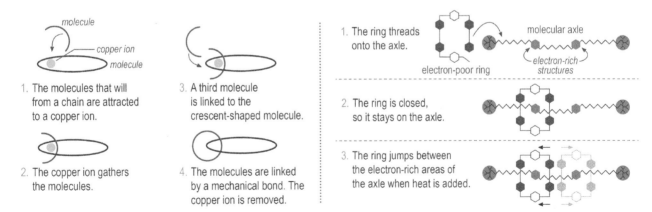

The next step is to make the components move independently. This was accomplished in 1991 when Stoddart made an axle and put a molecular ring on it. The ring stayed on the axle because the axle and the ring had chemical groups that attracted each other. The attraction was loose enough that the ring was still free to move. When heat was added, electrons on different segments of the axle were excited causing the ring to move. The ring could move up and down the axle based on the energy that was added! Compared to large machinery that we see in our everyday life, this may not seem very significant.

Nanotechnology - 2

Many machines require a motor to move. Can a motor be built out of molecules? For a motor to work, the motor must spin around in one direction and not move back and forth randomly.

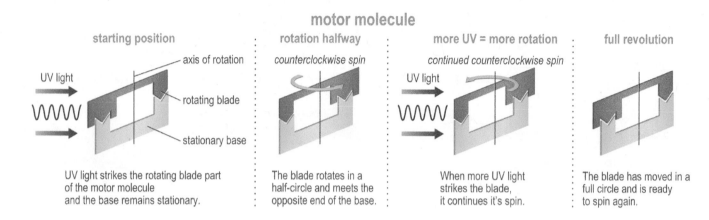

Feringa set about this task and his contribution has moved molecular machinery another step forward. In 1999 he used energy from light and heat to drive a molecular motor consisting of a blade that moves about an axis. Large chemical groups are attached to each other through a bond. When UV light is added, the blades moves 180°. This rotation also causes tension on the bond. To release the tension, the blades snaps over the other blade. This prevents backward rotation. As more UV light is added, the top blade keeps rotating about the axis and snapping into place. Feringa has gotten so good at building a molecular motor that by 2014 he had his motor spinning at 12,000 revolutions per minute. Using these tiny molecular motors, he has rotated a glass cylinder that is more than 10,000 times bigger than the motor itself!

What's the next step after building a motor? Why building a car of course! This is exactly what the Feringa group has done. They have prototyped a nanocar with four of these rotating blades as the wheels and axles attached to another molecule that is acting as the chassis.

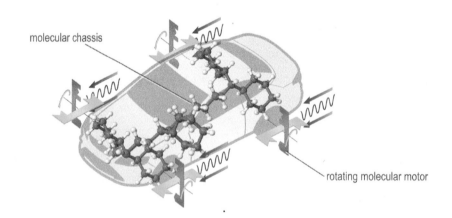

Why build these molecular machines? We aren't going to shrink ourselves down to a billionth of our size and take a spin in the nanocar!

From a biological perspective, molecular machines already exist. Proteins in cells can be considered complex molecular machines. There are motor proteins such as myosin, which is responsible for muscle contraction. Kinesin is another molecular motor protein. It is responsible for moving cargo away from the nucleus by transporting them along microtubules.

Synthetic molecular machines are not as complex as biological "motors." At this point, these man-made molecular machines are at the same stage of development as electric machines were in the 1830s. To many people they look like a novelty, but the scientists themselves keep pushing the boundaries of what is possible. In the 1830s people were creating electrically driven spinning wheels and cranks. Although for the most part these devices were not seen as practical, they did serve to inspire future scientists. Electric machines eventually lead to washing machines, cars, and planes.

Now chemists can visualize, build and even control the movement of molecules inspiring the next generation of scientists. Who know what future materials and devices may come!

Chapter 6 review

Vocabulary
Match each word to the sentence where it best fits.

Section 6.1

atomic mass	atomic mass unit (amu)
Avogadro's number	molar mass
mole	

1. The _____ is a unit of mass just a tiny fraction less than the mass of a single hydrogen atom.
2. The _____ is the mass of one mole of a substance and is written in grams/mole.
3. A common number in chemistry that is named after an Italian scientist and is equal to 6.02×10^{23} particles is referred to as _____ .
4. The _____ listed on the periodic table for every element is primarily due to the average total number of protons and neutrons in atoms of that element.
5. The unit that represents 6.02×10^{23} atoms is called a(n) _____ .

Section 6.2

| formula mass | molecular mass |
| standard molar volume | standard temperature and pressure (STP) |

6. The _____ of 22.4 L/mol is often used in calculations with gasses at standard temperature and pressure (STP).
7. _____ is the mass of one mole of an ionic substance.
8. _____ is formula mass of a molecular compound.
9. When doing calculations with gasses it is usually best to be sure all quantities are expressed at the _____ , which is one atmosphere and 0° C.

Section 6.3

| anhydride | empirical formula |
| hydrate | molecular formula |

10. A compound in which water is part of the chemical structure is known as a _____ .
11. Many shoe boxes contain a dry compound in a packet often marked "Do Not Eat." The compound is a(n) _____ and is included to absorb water from the atmosphere.
12. The _____ is the actual ratio of atoms in a compound.
13. The _____ is the simplest ratio of atoms in a compound.

Conceptual questions

Section 6.1

14. What number of stones are in 1 mole of stones?
15. What evidence on the periodic table supports the fact that there is more than one type of carbon atom?
16. Would all the ants on planet Earth equal to one mole of ants? Explain.
17. Does 200.559 g of mercury have more particles than 10.811 g of boron? Support your answer with calculations.
18. There are two interpretations of atomic mass.
    a. What are they?
    b. Which one is more useful in typical laboratory chemistry?
    c. Why is it more useful?
19. It is unrealistic to measure things as big as rice grains in units of moles.
    a. Explain why.
    b. Give another example of something that would be too big to measure.
20. Counting by weighing is a strategy that is used everywhere.
    a. Explain how this strategy works.
    b. How is this strategy useful in Chemistry?
21. How are grams related to atoms?

Section 6.2

22. How do you find the formula or molecular mass of a compound?
23. Why is it important to be able to find individual element contributions of a sample?
24. ❪ Suppose the temperature increases for a fixed amount of gas, but the pressure stays the same. Does the volume increase, decrease, or stay the same?
25. How is it possible that all gases have the same volume at the same temperature and pressure?
26. Why are gases measured at standard temperature and pressure?
27. How are formula mass and molar mass different?
28. A student is given a chemistry problem. The problem states the number of moles of something and asks for the weight. What can the student use to convert from moles to grams?

Chapter 6 review

Section 6.3

29. How does percent composition help in determining the empirical formula of an unknown substance?
30. Which would increase in mass on a humid day, a hydrate or an anhydride? Explain.
31. How can a hydrate become an anhydride?
32. The "do not eat" packet that comes in shoe boxes and other various objects contains a desiccant. What is the role of the desiccant?
33. What is the difference between an empirical formula and a molecular formula?
34. Give an example of a molecular formula that has the same value as its empirical formula.
35. Give an example of a molecular formula that has a different value than its empirical formula.
36. There is one difference in finding the empirical formula in comparison to finding the molecular formula. What is the one difference?
37. A forensic chemist frequently compares unknown compounds to compounds that have a known chemical formula. In your own words, state the steps the forensic chemist goes through for this process.
38. How can the percent composition technique display the correct ratio of elements but not identify the right compound?

Quantitative problems

Section 6.1

39. A student finds a container of buttons. The student counts 697 buttons. How many moles of buttons did the student count?
40. ❰ Suppose the Moon were made of peas. An average pea has a mass of 0.25 g. The Moon has a mass of 7.35×10^{22} kg. How many moles of peas would equal the mass of the moon?
41. Calculate the mass in grams of each of the following quantities.
    a. 1.0 mole of silver, Ag.
    b. 2.5 moles of aluminum, Al
42. To demonstrate why chemists work in moles, calculate the mass (in grams) of a single atom of each of the following.
    a. 1 atom of silver, Ag
    b. 1 atom of aluminum, Al
43. Calculate the number of atoms in each of the following quantities.
    a. 36.0 grams of sulfur, S
    b. 36.0 g of iron, Fe
    c. 36.0 g of hydrogen, H
    d. 36.0 g of platinum, Pt

44. How many grams are in 4.85 moles of argon, Ar?
45. How many moles of nickel, Ni, are in 987.25 g of nickel?
46. How many moles are in the following quantities?
    a. 36.0 g of sulfur, S
    b. 36.0 g of iron, Fe
    c. 36.0 g of hydrogen, H
    d. 36.0 g of platinum, Pt
47. Convert the following grams to atoms:
    a. 364 g of phosphorus, P
    b. 94.4 g of calcium, Ca
    c. 8.46 g of fluorine, F
    d. 254 g of silver, Ag
48. How many grams are in the following amount of atoms:
    a. 8.79×10^{20} atoms of tin, Sn
    b. 4.49×10^{21} atoms of potassium, K
    c. 4.35×10^{21} atoms of iron, Fe
    d. 9.46×10^{22} atoms of iodine, I
49. How many particles are in the following:
    a. 485 g of sodium
    b. 7.19 g of lithium
    c. 63.4 g of aluminum
    d. 94.2 g of cobalt
50. Calculate how many gold, Au, atoms are in 54.5 grams of gold.
51. Convert the following moles to grams of the following:
    a. 8.89 moles of H
    b. 7.49 moles of N
    c. 1.82 moles of Na
    d. 4.75 moles of C
52. A student wants to conduct an experiment. The student's teacher gives the student 2.50 grams of nickel. In order to conduct the experiment, the student must know the amount of particles and how many moles are in the sample of nickel.
    a. Convert the grams to moles.
    b. Using your answer from above, calculate the amount of particles in the sample.

Section 6.2

53. Calculate the molar mass, in g/mol, of the following compounds:
    a. acetic acid, $C_2H_4O_2$
    b. $NaHCO_3$
    c. $CaCO_3$
    d. H_2SO_4
54. How many moles are in each of the following quantities?
    a. 100. g water, H_2O
    b. 100. g calcium carbonate, $CaCO_3$
    c. 100. g sodium sulfate, Na_2SO_4
    d. 100. g octane, C_8H_{18}

Chapter 6 review

Section 6.2

55. Determine how many molecules are in 6.5 moles of water, H_2O.

56. Convert 4895.87 grams of carbon dioxide to molecules of carbon dioxide.

57. Calculate the mass of 8.76×10^{15} atoms of lead, Pb.

58. Convert the grams to moles in each of the following compounds.
    a. 22.0 g of ammonia, NH_3
    b. 35.0 g of sodium chloride, NaCl
    c. 12.0 g of carbon dioxide, CO_2
    d. 18.0 g of nitrogen dioxide, NO_2

59. How many moles does 1.05 grams of water have?

60. Calculate the formula mass, in amu, for each of the following compounds.
    a. Carbon dioxide, CO_2
    b. methane, CH_4
    c. magnesium nitrate, $Mg(NO_3)_2$
    d. caffeine, $C_8H_{10}N_4O_2$
    e. Iron oxide, Fe_2O_3
    f. octane, C_8H_{18}

61. Calculate how many hydrogen and oxygen atoms are present in 2.02 grams of water?

62. Calculate the mass in grams of each of the following quantities.
    a. 2.50 moles of silicon dioxide, SiO_2
    b. 2.50 moles of propane C_3H_8
    c. 2.50 moles magnesium nitrate $Mg(NO_3)_2$
    d. 2.50 moles of ammonia, NH_3
    e. 2.50 moles of calcium silicate, Ca_2SiO_4

Nutrition Facts
Serving size: 12 fl oz (355 mL)
Servings per container: about 3
Ingredients
Carbonated water, high fructose corn syrup, caramel color, citric acid, natural and artificial flavors, caffeine

Ingredient labels are in order of decreasing percentage by mass.

63. ❰ The two main ingredients in soda pop are carbonated water and high fructose corn syrup. Carbonated water is a weak solution of carbonic acid, H_2CO_3, combining water and carbon dioxide. A student buys 58.7 g of carbonic acid to make a batch of soda pop. How many particles of carbonic acid does the student buy?

64. Milk of magnesia is a combination of magnesium hydroxide and water, which is used as an antacid to treat an upset stomach. A pharmacist gives a patient 10 doses, that contain 2.0 grams of $Mg(OH)_2$ each.
    a. How many moles of magnesium hydroxide does the patient have?
    b. Calculate how many particles of magnesium hydroxide are in each dose.

65. How many molecules are in the following?
    a. 8.2 g of water H_2O
    b. 20.1 g of octane C_8H_{10}

66. The main ingredient of a particular cough medicine, that can be in syrup or tablet form, is an organic compound with the name of dextromethorphan. The formula for this compound is $C_{18}H_{25}NO$. Your mom instructs you to go to the pharmacy to buy cough medicine and to ensure that you get 3.99×10^{21} particles of dextromethorphan. However, at the pharmacy, the label of the package only states the amount of the drug in grams. If there are 12 tablets in the package and 0.05 grams of dextromethorphan in each tablet, how many packages of the cough medicine would you need to buy?

67. At STP, what is the volume of 5475.88 g of methane, CH_4?

68. There is 58.9 L of ammonia gas, NH_3, present at STP. Convert to grams of ammonia gas.

Section 6.3

69. What number should you get if you add up all the % of each element in a compound?

70. Given the empirical formula, CH_2O, state three possible molecular formulas.

71. ❰ Calculate the percent composition by mass for each element of the compound.
    a. Octane, C_8H_{18}
    b. magnesium nitrate $Mg(NO_3)_2$
    c. iron oxide, Fe_2O_3
    d. sodium bicarbonate $NaHCO_3$

72. What is the empirical formula for each of the following compounds?
    a. NO_2
    b. C_6H_6
    c. P_4H_{10}
    d. H_2O_2
    e. $Na_2(SO_4)_2$

73. ❰ Cyclohexane has an empirical formula of CH_2. Its molar mass is 84.18 g/mol. Given this information, what is the molecular formula for cyclohexane?

Chapter 6 review

Section 6.3

74. (What is the empirical formula for a substance for which 100 grams of the substance is analyzed to contain 66 g of iron and 34 g of chlorine?

75. Acetaminophen is a common drug that people, especially children, take to reduce fever and treat minor aches. The formula for this drug is $C_8H_9NO_2$. What is the percent composition of each element in acetaminophen?

Epsom salt

76. Magnesium sulfate heptahydrate is more commonly known as Epsom salt. Epsom salt is used as a home remedy for a number of things, such as soothing sore muscles. A 23.07 g sample of magnesium sulfate heptahydrate is heated to remove the water. The final result weighs 11.27 g. Calculate the percent water in the hydrate.

77. Ethylenediaminetetraacetic acid, more commonly known as EDTA, is used for chelation therapy. It is the treatment for metal poisoning in which EDTA binds to metals in the body and removes them without causing harm. EDTA is composed of only carbon, hydrogen, nitrogen and oxygen.

   a. Given the following percent composition, find the empirical formula for this compound.

   C = 41.099%
   H = 5.5181%
   N = 9.5859%
   O = 43.797%

   b. The molar mass for the compound is 292.2424 g/mol. Using the molar mass and the information above, find the molecular formula of the compound.

78. ((100 g of pure tungsten powder is heated in a pure oxygen atmosphere. A chemical reaction is observed and after the reaction is complete the new substance has a mass of 126 g. What is the empirical formula of the new compound?

79. Given the following % composition, what is the empirical formula of this substance?

   40% carbon
   6.7% hydrogen
   53.3% oxygen

80. (A compound has an empirical formula of CH. Its molar mass is 195 g/mol. What is the molecular formula for this compound?

81. ((A chemist determines that the molar mass of an unknown compound is 60.0 g/mol. A 1.00 gram sample of this compound a sample of this compound contains 0.24 g of carbon, 0.04 g of hydrogen and 0.53 g of oxygen. Determine the molecular formula of the unknown compound.

82. ((A chemist begins an experiment with 10.00 g of an unknown compound that is known to contain chlorine. The compound is reacted with another substance that removes chlorine. At the end of the reaction all that is left is 0.78 g of carbon. What is the empirical formula of the unknown compound?

83. ((A student receives an unknown substance to analyze from a teacher. The unknown weighs 8.5 g. The teacher informs the student that the substance only contains carbon, hydrogen, and oxygen with the following compositions:

   6.41g carbon
   0.38g hydrogen
   1.71g oxygen

   a. Find the percent composition of each element in the compound.
   b. From the percent composition, find the empirical formula of the compound.
   c. What is the molecular formula of the unknown substance, if the molar mass of the compound is 318.3266 g/mol?

chapter 7 | Chemical Reactions

Chemical reactions happen all around us. Cooking and baking involves mixing ingredients and applying heat so chemical reactions can occur to create the flavors and aromas that you want. Plants take energy from the sun, causing a chemical reaction with carbon dioxide and water to produce sugar and oxygen. In fact, our bodies are an amazing chemical reaction factory. Digestion of food is a chemical reaction that extracts essential fuels and nutrients. The oxygen we breathe that is produced by plants, reacts with sugars from the food we eat to produce water, carbon dioxide and energy. Water and energy are used to drive body processes, but excess carbon dioxide and water are exhaled out. Chemical attractions among blood cells, oxygen, carbon dioxide, tissues, and other body components make the chemical changes that keep us alive possible.

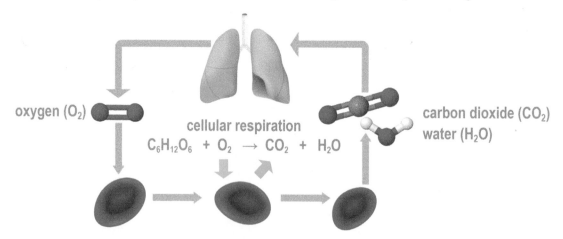

There are certain bits of evidence like color changes, the formation of a gas or solid, and changes in temperature that indicate a chemical change may have occurred. But seeing that evidence doesn't actually tell you what is happening. We know new substances form during a chemical reaction but, where does it come from and how is energy involved? To truly understand chemical reactions, we need to look at them from the perspective of the atoms and compounds that are involved in the process. Knowing how different atoms and compounds behave enables us to see patterns in chemical reactions. Seeing patterns, classifying reactions, and relating them to energy are fundamental parts of understanding chemistry.

Chapter 7 study guide

Chapter Preview

Chemical reactions govern the behavior of the world around us. When two or more molecules interact and change chemically, a reaction occurs through the breaking of bonds and the formation of new bonds. Chemical reactions can generally be classified as synthesis, decomposition, single replacement, double displacement, combustion and polymerization. Following the Law of Conservation of Mass, all chemical reactions must be balanced in terms of the number of atoms involved. They should also relay pertinent information about the reaction, such as the state of matter of the reactants and products. The formation of bonds releases energy, and the breaking of bonds requires the input of energy. Based on the reactants and products, some reactions are exothermic while others are exothermic.

Learning objectives

By the end of this chapter you should be able to:
- interpret symbols used in chemical equations;
- construct and balance a proper chemical equation;
- predict products of a chemical reaction based on the type of reaction;
- predict when a precipitate will form as a result of a chemical reaction;
- distinguish between an endothermic and exothermic chemical equation; and
- calculate the enthalpy change of a reaction using the enthalpies of formation.

Investigations

7A: Balancing chemical equations
7B: Chemical reactions
7C: Solubility rules

Pages in this chapter

- 192 Chemical Equations
- 193 Writing a chemical equation
- 194 State of matter and reaction mechanisms
- 195 Conservation of matter
- 196 Balanced chemical equations
- 197 The rules for balancing chemical equations
- 198 Equation balancing tips
- 199 A structured method - part 1
- 200 A structured method - part 2
- 201 Using the structured balancing method
- 202 Section 1 Review
- 203 Types of Chemical Reactions
- 204 Decomposition reactions
- 205 Single replacement reactions
- 206 Double replacement reactions
- 207 Combustion reactions
- 208 Solving chemical story problems
- 209 Solubility and precipitation
- 210 Softening water using precipitation
- 211 Net ionic equations
- 212 Polymers and polymerization reactions
- 213 Section 2 Review
- 214 Surprising Reactions and Accidental Discoveries
- 215 Surprising Reactions and Accidental Discoveries - 2
- 216 Chapter review

Vocabulary

chemical reaction
product
aqueous (aq)
balanced chemical equation
single replacement
combustion
polymer

chemical equation
insoluble
law of conservation of matter
synthesis
double replacement
hydrocarbon
polymerization

reactant
soluble
unbalanced chemical equation
decomposition
precipitate
spectator ion

7.1 - Chemical Equations

What comes to mind when you hear the word restaurant? From this single word you can envision a business where food is served. When you add one more word, different information is revealed. Seafood restaurant helps you predict what will be on the menu. If you add "seafood restaurant" to a search on a digital map, all kinds of symbols will pop up that tell you cost (in dollar signs), customer ratings (in stars), and driving distance (in miles). A chemical equation also stores a wealth of information in just a few numbers, letters and symbols.

What is a chemical equation?

Chemical reactions describe chemical change

Any time there is a chemical change, a **chemical reaction** must have occurred. A chemical reaction converts one or more substances into new substances. Atoms are neither created nor destroyed during the process of a chemical reaction, but they are rearranged to form new substances. A **chemical equation** is a series of symbols that represents a chemical reaction. The chemical equation describes the chemical change by telling us exactly which atoms are rearranged into which compounds.

The language of chemistry

element symbols	H, He, Li, O, Na, Fe	alphabet of chemistry
chemical formula	H_2O, CO_2, $NaHCO_3$	words of chemistry
chemical equations	$2H_2 + O_2 \rightarrow 2H_2O$	sentences of chemistry

Interpreting chemical equations

A **reactant** is a substance that is changed in the reaction into a different substance. Reactants appear on the left of the chemical equation. A **product** is a substance that is produced by the reaction. Products appear on the right hand side of the chemical equation. The arrow tells the direction of the reaction - which is usually from reactants to products. If we were to read the chemical equation in the above example, it says this: "two moles of hydrogen plus one mole of oxygen react to produce two moles of water."

Coefficients

coefficients tell you the number of moles of each substance

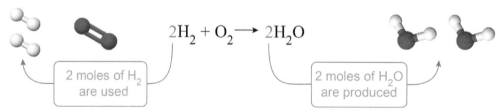

Moles or atoms?

A chemical equation can be interpreted two ways: in terms of atoms or in terms of moles. The atom interpretation is useful when you want to understand how molecules change during the reaction. The moles interpretation is more useful when you want to use the reaction to calculate quantities of products or reactants.

Writing a chemical equation

An example chemical reaction

Let's write the chemical equation for a reaction that takes place while you cook on a gas stove top. Oxygen (O_2) from the air reacts with natural gas (CH_4) to produce carbon dioxide gas (CO_2) and water (H_2O). In this reaction, O_2 and CH_4 are the reactants since they are "used up" by the reaction. Carbon dioxide and water are the products. In words, the reaction can be written as below. For now, you can ignore the coefficients that appear in front of O_2 and H_2O.

Writing a chemical equation

as a sentence .... methane reacts with oxygen to produce carbon dioxide and water.

as words .... methane + oxygen ⟶ carbon dioxide + water

as a chemical equation .... $CH_4 + 2O_2 \longrightarrow CO_2 + 2H_2O$

Writing chemical equations using symbols

The chemical equation is the most accurate and least ambiguous way to describe the reaction. The equation tells exactly what changes to what, and the relationships between the quantities of each which will be used and produced. When natural gas burns completely, one mole of methane (CH_4) combines with two moles of molecular oxygen (O_2) to produce one mole of carbon dioxide (CO_2) and two moles of water (H_2O). You don't see the water because it is vapor at the temperature of this reaction.

Chemical equations for common reactions

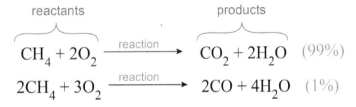

$$\underbrace{CH_4 + 2O_2}_{\text{reactants}} \xrightarrow{\text{reaction}} \underbrace{CO_2 + 2H_2O}_{\text{products}} \quad (99\%)$$

$$2CH_4 + 3O_2 \xrightarrow{\text{reaction}} 2CO + 4H_2O \quad (1\%)$$

Every chemical change can be described by one or more reactions.

Energy in chemical reactions

Does the chemical equation $CH_4 + 2O_2 \rightarrow CO_2 + 2H_2O$ give a complete representation of the burning of natural gas? When we look at a fire, we see that fire gives off light and we feel that fire gives off heat. So, if the chemical equation does accurately represent the phenomenon of burning then where is the heat and where is the light? A more complete accounting of the reaction, including energy, is later in the chapter. Energy can be either a reactant or a product. In the case of methane, the energy is a product because it is given off by the reaction.

Energy in chemical equations

$$CH_4 + 2O_2 \longrightarrow CO_2 + 2H_2O + \text{energy}$$

Almost every chemical change involves either the absorption or the release of energy.

Section 7.1: Chemical Equations

State of matter and reaction mechanisms

State of matter

Symbols that represent the state of matter are often added to chemical equations. These symbols give us more information about how the reaction is carried out. The state of matter is important because gases tend to react faster than solids. The particles in a gas are already moving around each other and can find other reactants very easily. Solid particles do not move very quickly and only react at their surface. This makes solid-solid reactions slow.

State symbols

As shown in the table, the symbol (s) after a chemical formula means the substance is a solid. A gas is symbolized by (g) and liquids by (l). The equation, $2H_2(g) + O_2(g) \rightarrow 2H_2O(l)$ tells us, two moles of hydrogen gas and one mole of oxygen gas react to form two moles of liquid water. Notice how numbers in front of compounds are translated as the number of moles. When a coefficient is not present, assume there is one mole.

(g)	gas
(l)	liquid
(s)	solid
(aq)	aqueous
$\xrightarrow{\Delta}$	heat
$\xrightarrow{cat}$	catalyst

Reaction mechanisms

Chemical equations can indicate the reaction mechanisms by notations over the arrow. The diagram shows notations for catalysts and heat. An element symbol, such as Pt for platinum, indicates that a platinum catalyst is needed. Heat is sometimes indicated by the capital Greek letter Delta, Δ.

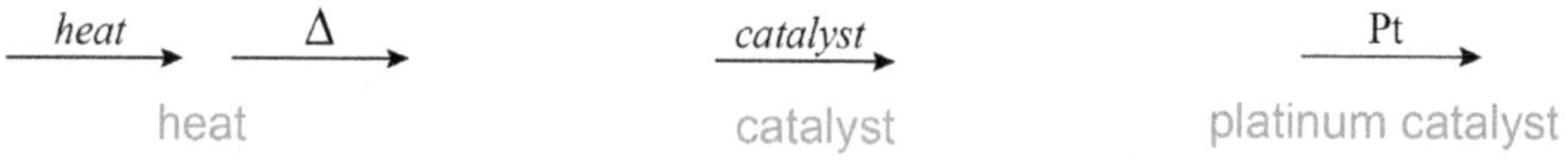

$\xrightarrow{heat}$ heat $\xrightarrow{\Delta}$ $\xrightarrow{catalyst}$ catalyst $\xrightarrow{Pt}$ platinum catalyst

Aqueous verses liquid

The symbol (aq) stands for **aqueous (aq)**. This means dissolved in water. Do no confuse aqueous with liquid. Aqueous iron, Fe^{2+}(aq) is a rusty water mixture but liquid iron, Fe(l) is molten iron at 1,538 °C or 2,800 °F. In this case, the difference could be deadly!

Aqueous solution reactions

Many reactions take place in solution because particles in solution are mobile and can contact each other to react. A good example is the reaction between salt, NaCl, and barium nitrate, $Ba(NO_3)_2$. These are both ionic substances and dissolve into separate ions. NaCl dissociates into sodium ions, Na^+, and chloride ions, Cl^-. Barium nitrate dissociates into Ba^{2+} ions and NO_3^- ions. For the reaction, dissolved NaCl reacts with dissolved barium nitrate, $Ba(NO_3)_2$.

Chemical equations can identify changes of state

$$2NaCl(aq) + Ba(NO_3)_2(aq) \longrightarrow BaCl_2(s) + 2NaNO_3(aq)$$

salt solution + barium nitrate solution → sodium nitrate solution — solid barium chloride

BaCl₂ becomes a solid

The symbol (s) beside barium chloride, $BaCl_2$, means that the $BaCl_2$ is a solid product - not a solution. This means that it is **insoluble** and precipitates, or falls out of the solution. The reaction tells us the other product, sodium nitrate, $NaNO_3$(aq) stays in solution. The sodium and nitrate ions are **soluble** throughout the reaction.

Conservation of matter

Chemical reactions conserve matter

One of the fundamental principles of chemistry is the conservation of matter. In fact, this is so important that scientists call it the **law of conservation of matter**. The law states that the total mass of the reactants is equal to the total mass of the products. In other words, you cannot create or destroy mass during a chemical reaction.

The number of atoms does not change

The term conservation means unchanged. When you hear that "matter is conserved during a chemical reaction", it really means the total number and type of atoms has not changed before and after the reaction. Every single reactant atom must be accounted for in the products. Reactant atoms do not disappear; instead they are rearranged to make products.

Conservation of mass

$$6CO_2 + 6H_2O \longrightarrow C_6H_{12}O_6 + 6O_2$$

Each and every atom in the **reactants** must also appear in the **products**.
Each and every atom in the **products** must also appear in the **reactants**.

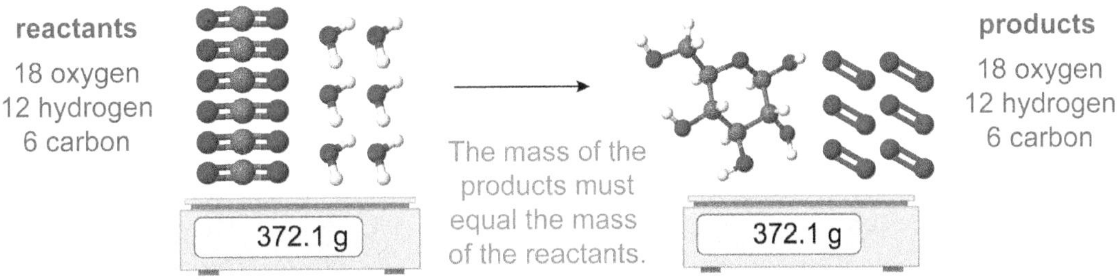

reactants
18 oxygen
12 hydrogen
6 carbon

372.1 g

The mass of the products must equal the mass of the reactants.

products
18 oxygen
12 hydrogen
6 carbon

372.1 g

Photosynthesis

For example, consider the photosynthesis reaction which takes gaseous carbon dioxide from the atmosphere, liquid water from the ground and produces solid glucose and gaseous oxygen. This reaction uses the energy of the sun to reorganize six carbon atoms, twelve hydrogen atoms and eighteen oxygen atoms from one configuration in the reactants to another configuration in the products. The same atoms appear on both sides of the reaction and that is why mass is conserved.

Photosynthesis

reactants $\quad$ products

$$6CO_{2}(g) + 6H_2O(l) \xrightarrow{light} C_6H_{12}O_6(s) + 6O_2(g)$$

264.0 g $\quad$ 108.1 g $\quad$ 270.1 g $\quad$ 192.0 g

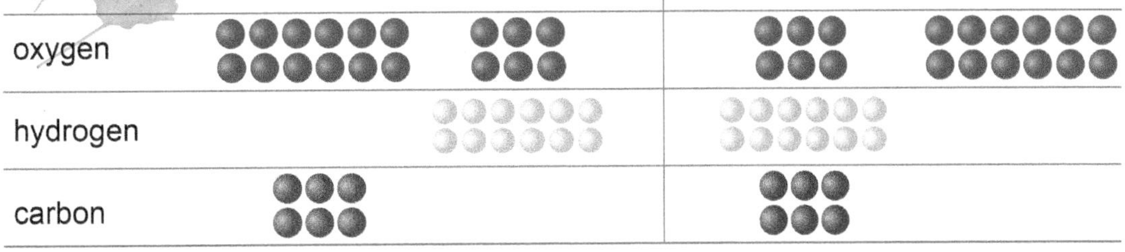

oxygen	●●●●●●●●●●●●	●●●●●●	●●●●●●	●●●●●●●●●●●●
hydrogen		○○○○○○○○○○○○	○○○○○○○○○○○○	
carbon	●●●●●●		●●●●●●	

Why this was not obvious

Conservation of matter seems obvious today but it was not obvious at all when chemistry was being first understood. The problem is that scales do not weigh gases. The masses of each compound in the reaction are shown above. If both the CO_2 and O_2 were undetected it would look like 108.1 grams of water became 270.1 grams of glucose!

Section 7.1: Chemical Equations

Balanced chemical equations

Unbalanced equations are not accurate

Consider a reaction that combines hydrogen gas with oxygen to make water. This reaction occurs in a hydrogen fuel cell and releases a great deal of energy. Many companies are trying to make hydrogen powered cars because the product is water and this would greatly lessen air pollution and carbon dioxide emissions that contribute to climate change.

Why this equation is wrong

The equation $H_2 + O_2 \rightarrow H_2O$ is not quite right. If 1 mole of oxygen reacts with 1 mole of hydrogen, there would be more than just 1 mole of water molecules at the end of the reaction. There would be 0.5 moles of leftover oxygen molecules along with 1 mole of water molecules. Look closely at the reaction summarized in the table below. Notice how there are 2 oxygen atoms in the reactants but there is only 1 oxygen atom in the products.

Unbalanced hydrogen-oxygen reaction

reactants $H_2(g)$ + $O_2(g)$ $\xrightarrow{spark}$ $H_2O(l)$ products
2.0 g   32.0 g   18.0 g

oxygen	2	●●	●	1 (unbalanced)
hydrogen	2	○○	○○	2

Unbalanced chemical equation

The chemical equation $H_2 + O_2 \rightarrow H_2O$ is an **unbalanced chemical equation**. The unbalanced equation identifies reactants and products but it does not correctly account for how much of each are involved. The amount of hydrogen molecules, oxygen molecules, and water molecules must be adjusted until the same type of each atom is equal on both sides of the chemical equation.

The balanced equation

The **balanced chemical equation** adds a coefficient of 2 to the hydrogen molecule on the reactant side and the water molecule on the product side. The coefficient tells you that there are two moles of hydrogen molecules reacting with one mole of oxygen molecules to produce two moles of water molecules. The chart below shows that the number of atoms of each type are now the same on the reactant and product sides of the equation.

The balanced hydrogen-oxygen reaction

reactants $2H_2(g)$ + $O_2(g)$ $\xrightarrow{spark}$ $2H_2O(l)$ products
4.0 g   32.0 g   36.0 g

oxygen	2	●●	●●	2
hydrogen	4	○○○○	○○○○	4

Balanced equations are correct

Chemical equations are balanced by adjusting the coefficients in front of each compound until the total number of each type of atom is the same on both sides of the equation. Only the properly balanced chemical equation tells us the correct relationship between the quantities of reactants and products.

The rules for balancing chemical equations

Balancing an equation

To balance any chemical equation we adjust the coefficients until the same number of atoms appear on the reactant and product side of the equation. This is not always an easy task! Consider the following reaction in which nitric oxide (NO) reacts with hydrogen (H_2) to produce ammonia (NH_3) and water. This reaction is used to synthesize ammonia for fertilizer and other applications.

The unbalanced nitric oxide-hydrogen reaction

$$NO_{(g)} + H_{2(g)} \longrightarrow NH_{3(g)} + H_2O_{(g)}$$

	reactants	products
nitrogen	1	1
oxygen	1	1
hydrogen	2	5

Rule #1 for balancing equations

You cannot change subscripts in chemical formulas because this would change the identities of the substances in the reaction and make it a different reaction. For example, it's impossible to change H_2 to H_5 to get three more hydrogen atoms and balance the equation because H_5 does not exist. Even if H_5 did exist it would be a different compound than H_2.

Rules for balancing chemical equations

You <u>cannot</u> change subscripts.   You <u>can</u> change coefficients in whole numbers.

H_2 to $3H_2$

Rule #2 for balancing equations

You can only change the coefficients in whole-numbers to adjust the number of each molecule in the reaction. For example we can change H_2 into $3H_2$ which increases the number of hydrogen atoms from two to six. We cannot use a fractional coefficient such as 2.5 - that would divide a molecule and create a different compound.

The balanced nitric oxide-hydrogen reaction

$$2NO_{(g)} + 5H_{2(g)} \longrightarrow 2NH_{3(g)} + 2H_2O_{(g)}$$

	reactants	products
nitrogen	2	2
oxygen	2	2
hydrogen	10	10

A balanced equation tells you the amounts of each substance

The balanced equation tells us that when we combine 2 moles of nitric oxide with five moles of diatomic hydrogen we can obtain 2 moles of ammonia and 2 moles of water. This is the recipe for making ammonia from nitric oxide and hydrogen. It correctly represents the amount of each substance and satisfies mass conservation because the total number of each type of atom is equal on both sides of the equation.

Equation balancing tips

Techniques for balancing equations

One way to balance chemical equations is by trial and error. This is where you try using different coefficients until the numbers work out. This approach is also known as the inspection method. The inspection method works fine for many reactions, however, a more structured approach is needed for reactions that involve a number of reactants and products. We'll develop it in the next few pages.

Single molecules

Single molecules in a chemical equation do not show a coefficient of 1. For example, O_2 in a chemical equation means 1 molecule of oxygen.

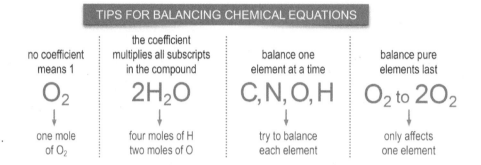

TIPS FOR BALANCING CHEMICAL EQUATIONS

- no coefficient means 1: O_2 → one mole of O_2
- the coefficient multiplies all subscripts in the compound: $2H_2O$ → four moles of H, two moles of O
- balance one element at a time: C, N, O, H → try to balance each element
- balance pure elements last: O_2 to $2O_2$ → only affects one element

Coefficients multiply all subscripts

The coefficient in front of a substance applies to all atoms in the molecule. For example, $2CO_2$ means there are two carbon dioxide molecules. Each carbon dioxide molecule has two oxygen atoms, so the total number of oxygen atoms is four.

One element at a time

Start by balancing only one element at first, then balance the next element, and so on, one element at a time. Every time you change a coefficient, go back and check the all elements you already balanced to make sure you have not unbalanced any of them.

Modify pure elements last

It is always easiest to balance pure elements last. Adding a coefficient to O_2 will change only the number of oxygen atoms. But adding a coefficient to a compound like H_2O will change both the number of hydrogen atoms and the number of oxygen atoms.

Solved problem

Glucose biomass can decompose to ethanol and carbon dioxide:

$$C_6H_{12}O_6 \rightarrow C_2H_6O + CO_2.$$

What is the balanced reaction?

Asked Find the coefficients in order to balance the chemical equation.
Given The unbalanced chemical equation, $C_6H_{12}O_6 \rightarrow C_2H_6O + CO_2$
Relationships The same number of each type of atom must appear on each side.
Solve All atoms involved in the reaction are unbalanced.

unbalanced reaction

	reactants	products
carbon	6	3
oxygen	12	6
hydrogen	6	3

Notice the number of product atoms are all half of the number of reactant atoms.
↓
Add coefficient of 2 to all product compounds.

Answer The balanced equation is $C_6H_{12}O_6 \rightarrow 2C_2H_6O + 2CO_2$

A structured method - part 1

Balancing tougher equations

If you are unable to balance an equation by inspection, use a structured method instead. Let's balance the reaction between blood sugar (glucose, $C_6H_{12}O_6$) and oxygen in a structured way. This reaction, called cellular respiration, provides the internal energy for all living animals, including you, and produces carbon dioxide (CO_2) and water (H_2O).

Cellular respiration - unbalanced reaction

$$C_6H_{12}O_{6(s)} + O_{2(g)} \longrightarrow CO_{2(g)} + H_2O_{(l)}$$

	reactants	products
oxygen	8	3
hydrogen	12	2
carbon	6	1

There are 3 different elements to balance in the equation: C, O and H. All 4 substances involved in the reaction are made up of different combinations of these 3 elements.

How to begin

An effective method to balancing chemical equations is to begin with the elements that you see in the fewest compounds and end with the element that you see in the most substances.

Start with the least common element

The respiration reaction shows both hydrogen and carbon are involved in only one reactant and one product. You could start with either carbon or hydrogen, so let's choose hydrogen. A coefficient of 6 in front of the H_2O product will balance the number of hydrogen atoms. Note that adding this coefficient also changes the total amount of oxygen on the product side from 3 to 8 moles.

$$C_6H_{12}O_{6(s)} + O_{2(g)} \longrightarrow CO_{2(g)} + 6H_2O_{(l)}$$

	reactants	products
oxygen	8	8
hydrogen	12	12
carbon	6	1

Next element

Let's move on to carbon next. You can balance carbon by placing a coefficient of 6 in front of the CO_2 product. Inserting this coefficient changes the amount of oxygen on the product side from 8 moles to 18 moles.

$$C_6H_{12}O_{6(s)} + O_{2(g)} \longrightarrow 6CO_{2(g)} + 6H_2O_{(l)}$$

	reactants	products
oxygen	8	3
hydrogen	12	12
carbon	6	6

Section 7.1: Chemical Equations

A structured method - part 2

Leave balancing pure elements until last

We left oxygen to be balanced last since it is involved in the most number of compounds. Also, oxygen is a pure substance so changing the coefficient specifically on O_2 will not affect the other elements that are already balanced. Let's count the oxygen atoms as they appear in the equation below.

$$C_6H_{12}O_6(s) + 6O_2(g) \longrightarrow 6CO_2(g) + 6H_2O(l)$$

oxygen	18	18
hydrogen	12	12
carbon	6	6

Count atoms of each element on both sides

By adding a coefficient of 6 to O_2 on the reactant side, the reaction is balanced. The 18 moles of oxygen on the reactant side balance the 18 moles of oxygen on the product side. You have now addressed all atoms in the reaction: H, C and O. Before you decide that you are done, check to see that the equation is balanced. Count the number of atoms for each element on either side of the equation. This equation is correctly balanced because the number of atoms for each element is the same on both sides of the equation.

Make sure coefficients cannot reduce

The final step is to make sure the coefficients are reduced to the smallest whole numbers possible. The same balanced reaction could be written with 2 moles of glucose and 12 moles of everything else. This would not be the smallest whole numbers because we can divide them all by two to get a simpler form with smaller coefficients that is also balanced.

$$2C_6H_{12}O_6(s) + 12O_2(g) \longrightarrow 12CO_2(g) + 12H_2O(l)$$

divide all coefficients by 2

$$C_6H_{12}O_6(s) + 6O_2(g) \longrightarrow 6CO_2(g) + 6H_2O(l)$$

Summary of the structured method

Here is a summary of how to use a structured method to balance chemical equations.

1. Write the unbalanced chemical equation.
2. Find the element that occurs in the fewest number of compounds on both sides of the equation and balance it first. If there is a tie between more than one element, balance them in any order.
3. Move on to the rest of the elements. Balance pure elements last.
4. Re-count the number of each element on both sides to make sure they are equal.
5. Make sure the coefficients cannot reduce. Coefficients must be the smallest possible whole numbers.

Using the structured balancing method

Apply the 5 steps

Let's look at an example of how to apply the 5 steps to balance an equation. You might have a portable appliance at home that burns propane. Methane, C_3H_8 reacts with oxygen, O_2 to produce carbon dioxide, CO_2 and water. First, write the equation with no coefficients. Sometimes you get lucky and it is already balanced! This time there are different numbers of elements on both sides so the equation is unbalanced.

Unbalanced reaction for burning propane

$$C_3H_8 + O_2 \longrightarrow CO_2 + H_2O$$

oxygen	2	3
hydrogen	8	2
carbon	3	1

balance elements that appear once each side

Next, identify an element that occurs in only one compound on both sides of the equation and balance it first. This is true for both hydrogen, H and carbon, C but not for oxygen, O in this example. Let's start with carbon. There are 3 C atom in the reactants and 1 C atom in the products. Adding a coefficient of 3 to CO_2 balances carbon.

$$C_3H_8 + O_2 \longrightarrow 3CO_2 + H_2O$$ balance carbon by adding a coefficient of 3 to CO_2

$$C_3H_8 + O_2 \longrightarrow CO_2 + 4H_2O$$ balance hydrogen by adding a coefficient of 4 to H_2O

Hydrogen is the next element that appears only once on each side, so let's balance hydrogen before oxygen. Adding a coefficient of 4 to H_2O balances hydrogen.

Balance remaining elements

All that is left is to balance oxygen. This is easy because oxygen occurs as a pure element on the reactant side. Adding a coefficient of 5 to O_2 on the reactant side balances the equation. The atom count chart shows that the same number each element appear on both sides of the equation.

$$C_3H_8 + 5O_2 \longrightarrow 3CO_2 + 4H_2O$$ balance oxygen by adding a coefficient of 5 to O_2

balanced reaction - all atoms equal on both sides

oxygen	10	10
hydrogen	8	8
carbon	3	3

Check for smallest numbers

We are now done because the coefficients on both sides of the reaction are reduced to the smallest possible whole numbers. The coefficients are: 1:5:3:4. These whole numbers cannot be reduced, so the equation has been correctly balanced.

Chapter 7

Section 1 review

A chemical reaction is a process in which one or more substances are converted into new substances. The initial substances are the reactants, and the substances that result are called the products. As an equation this change is represented with reactants on the left with an arrow leading to products on the right. The phase of each substance is noted as solid, liquid, gas or aqueous, meaning dissolved in water. Special symbols over the arrow indicate when heat is added or a catalyst is used to speed up the reaction. The law of conservation of mass aids in balancing equations and creating a true picture of the reaction. When balancing equations it is useful to balance one element at a time, remembering that coefficients apply to whole molecules. It is also a good strategy to balance pure elements last.

Vocabulary words: chemical reaction, chemical equation, reactant, product, insoluble, soluble, aqueous (aq), law of conservation of matter, unbalanced chemical equation, balanced chemical equation

Key relationships
- Chemical equation: Reactants → Products
- "+" separates different reactants or products
- Law of conservation of mass: mass of reactants = mass of products

Common Symbols	
(g)	gas
(l)	liquid
(s)	solid
(aq)	aqueous (dissolved in H_2O)
→	forms, produces, yields, etc.
$\xrightarrow{\Delta}$	reacts when heat is added
$\xrightarrow{cat}$	reacts with an added catalyst

Review problems and questions

1. Explain why it is impossible to balance the following equation: $C_3H_8 \rightarrow H_2 + CO_2$

2. Translate the following chemical equation to words: $2H_2(g) + O_2(g) \rightarrow 2H_2O(l)$

3. Balance the reaction: $P_4 + O_2 \rightarrow P_2O_5$

4. Balance the reaction: $ClO_2 + H_2O \rightarrow HClO_2 + HClO_3$

5. Balance the reaction: $Al_2(SO_4)_3 + Ca(OH)_2 \rightarrow Al(OH)_3 + CaSO_4$

7.2 - Types of Chemical Reactions

There are many different types of chemical reactions. You can predict the outcome of a reaction based only on the reactants if you understand patterns in written chemical equations.

Synthesis reactions

Patterns in chemical equations

Although many different combinations of reactants are possible, you will see just a few structural patterns emerge. Understanding these patterns will help you predict the products of a reaction. Five types of chemical reactions are:

Types of chemical reactions

A + X → AX	Synthesis
AX → A + X	Decomposition
A + BX → AX + B	Single replacement
AX + BY → BX + AY	Double replacement
X + O_2 → XO	Combustion, or
C_xH_x + O_2 → CO_2 + H_2O	Combustion of a hydrocarbon

Synthesis reactions

A **synthesis** reaction occurs when two or more reactants combine to form one product. The general form of a synthesis reaction is shown below. An example of a synthesis reaction is when hydrogen and oxygen combine to form water. Synthesis reactions combine different substances, so they are also known as combination reactions. The number of products is less than the number of reactants, however, the number of atoms remains the same.

Synthesis reactions
Two reactants combine to form one product.

A + X → AX

$2H_2(g) + O_2(g) → 2H_2O(l)$

Compounds as reactants

The example above shows the combination of two elements, but two or more larger compounds can also combine into one larger compound. One example of this is the reaction of magnesium oxide, MgO, with water to make magnesium hydroxide, $Mg(OH)_2$, also known as milk of magnesia: MgO + H_2O → $Mg(OH)_2$. Milk of magnesia is a common over-the-counter medication used to neutralize excess stomach acid.

Solved problem

Find the product that forms when potassium and chlorine react, then balance the reaction.

Relationships   You can tell this is a synthesis reaction because the question asks you to find "the" product; this means there is only one product.

Solve
1. Write the formula for each reactant. Both reactants are elements, so apply the diatomic molecule rule. Chlorine forms a diatomic molecule, but potassium does not: K + Cl_2 → ?
2. Combine K and Cl to form a compound using ionic formula rules:
   - Write the ion form for each substance: K^+ and Cl^-. Ignore the fact that chlorine is in diatomic form.
   - Criss-cross charges. In this case both charges are "1": KCl
3. Rewrite the reaction with the new compound: K + Cl_2 → KCl
4. Balance the reaction: 2K + Cl_2 → 2KCl

Answer   The balanced synthesis reaction is: 2K + Cl_2 → 2KCl

Decomposition reactions

Decomposition reactions

A **decomposition** reaction breaks one reactant compound into two or more products. Chemically, decomposition is the reverse of synthesis. One example is the decomposition of carbonic acid, H_2CO_3, in carbonated water: $H_2CO_3(aq) \rightarrow H_2O(l) + CO_2(g)$. Under pressure (unopened) the carbonic acid stays in solution because one of the products of the reaction is a gas. When pressure is released the reaction splits some of the dissolved carbonic acid into water and carbon dioxide gas. The released gas makes the "fizz."

Decomposition reactions

One reactant breaks down into two or more products.

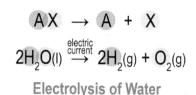

Electrolysis of Water

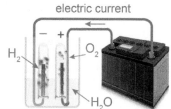

Electrolysis of water is endothermic

One important decomposition reaction is the electrolysis of water. Electric current passing through water causes a decomposition reaction which generates hydrogen and oxygen gas in a 2:1 ratio as seen in the balanced reaction. Hydrogen gas is produced at the negative battery terminal and oxygen gas is produced at the positive terminal. Similar to electrolysis, most decomposition reactions are endothermic. It takes an input of energy to break the chemical bonds in the reactant compound. This is often more than the energy released by forming the products. Energy changes during a reaction can be observed in the form of heat, electrical current, light, or even pressure.

Products of decomposition

At first glance you may think a reactant can decompose into any combination of elements or compounds, but there is a predictable pattern associated with some compounds:

a. Compounds made of only 2 elements will break up into their component elements like water shown above.

b. Carbonates like carbonic acid, H_2CO_3 shown above break up into the oxide of the cation and carbon dioxide.

c. Chlorates like potassium chlorate, $KClO_3$ break down into oxygen gas and a binary salt that contains no oxygen: $2KClO_3(s) \rightarrow 2KCl(s) + 3O_2(g)$.

Solved problem

Predict the products and write the balanced reaction for the breakdown of solid barium chlorate, $Ba(ClO_3)_2(s)$.

Relationships There is only one reactant so this is a decomposition reaction. Chlorates decompose into a binary salt and oxygen gas.

Solve 1. Break the compound up into molecular oxygen and a salt.
- Molecular oxygen is $O_2(g)$
- Barium and chlorine will form the salt. Write the charges for each ion and perform a criss-cross. Write the cation first in the formula.
  - Charged ions: Ba^{2+} and Cl^-
  - Barium chloride: $BaCl_2$
2. Write the reaction with the states: $Ba(ClO_3)_2(s) \rightarrow BaCl_2(s) + O_2(g)$
3. Balance the reaction: $Ba(ClO_3)_2(s) \rightarrow BaCl_2(s) + 3O_2(g)$

Answer The decomposition reaction is balanced as:
$Ba(ClO_3)_2(s) \rightarrow BaCl_2(s) + 3O_2(g)$.

Single replacement reactions

Single replacement

A **single replacement** reaction occurs when an element takes the place of another element in a compound. The example shows the reaction that occurs when solid copper is immersed in a solution of silver nitrate, $AgNO_3$. The copper in the wire replaces silver in the compound to make copper(II) nitrate, $Cu(NO_3)_2$. The silver plates out as silver whiskers on the wire and the solution turns blue because copper(II) nitrate is dissolved in the solution.

Single replacement reactions

One reactant replaces part of another reactant. Reactants and products each include an element and a compound.

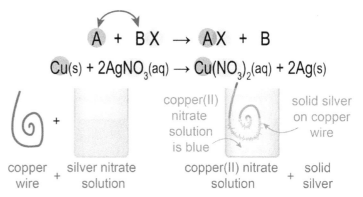

Predicting products

During single replacement reactions, also called single displacement reactions, an element replaces the part of the compound that makes an ion with the same charge. In the reaction above, silver makes an ion with a positive charge, copper(II) makes a positive ion and nitrate makes a negative ion. Silver and copper will trade places because both form a positively-charged ion. Copper forms a new ionic compound with negatively-charged nitrate while silver exists as a single element. When writing compounds, it is important to remember that polyatomic ions stay together. Most polyatomic ions are negative except for the ammonium ion, NH_4^+. In the reaction above, notice how the nitrate ion, NO_3^- stays intact as its cation is exchanged.

Solved problem

Write the balanced reaction that occurs when chlorine gas and solid sodium bromide react.

Relationships An element and compound react in a single replacement reaction.

Solve 1. Translate reactants to formulas.
- Apply the diatomic molecule rule to the element chlorine: Cl_2.
- Write the formula for sodium bromide.
  - Write ions: Sodium = Na^+ and bromide = Br^-
  - Criss-cross charges: Sodium bromide = NaBr
2. Start writing the reaction; include states of matter: $Cl_2(g) + NaBr(s) \rightarrow$
3. Determine the products based on chlorine's charge as an ion.
- Ignore the diatomic nature of chlorine and write it as an ion: Cl^-.
- Chlorine is negative so it will replace the negative part of NaBr.
  - When Cl replaces Br, Br will be on its own. Apply the diatomic molecule rule: Br_2.
  - Cl^- combines with Na^+. Write the positive ion first and criss-cross charges to get the compound: NaCl.
4. Write the complete reaction: $Cl_2(g) + NaBr(s) \rightarrow NaCl(s) + Br_2(g)$.
5. Balance the reaction: $Cl_2(g) + 2NaBr(s) \rightarrow 2NaCl(s) + Br_2(g)$.

Answer The balanced single replacement reaction is:
$Cl_2(g) + 2NaBr(s) \rightarrow 2NaCl(s) + Br_2(g)$.

Double replacement reactions

Cations "trade" partners

In **double replacement** reactions, reactant cations have been interchanged to form products. One example is the reaction between aqueous barium chloride, $BaCl_2$, and aqueous sodium sulfate, Na_2SO_4. The positive ions trade places to form solid barium sulfate, $BaSO_4$, and salt, NaCl. Both reactants and products are compounds. Notice how the polyatomic sulfate ion stays intact while the cations are exchanged.

Double replacement reactions

Two reactants exchange cations. All of the reactants and products are compounds.

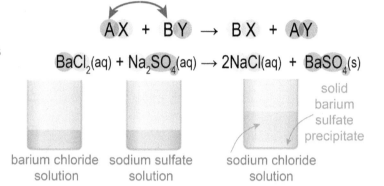

Precipitates

Double replacement, also known as double displacement reactions often form a **precipitate**. A precipitate is a solid that forms when reactant ions in solution produce an insoluble compound, meaning it cannot dissolve. Look for a solid barium sulfate precipitate in the reaction above. The insoluble compound may settle out on the bottom of a beaker, or may remain suspended. A suspended precipitate turns a solution from clear to cloudy.

Predicting products

To predict products of a double replacement reaction, write ions for both reactants. Trade cation places and criss-cross charges to determine the products formed. Cations always come first in an ionic compound, so only the first ion in each compound will trade places.

Solved problem

Write the balanced equation for the reaction of lead(II) nitrate with potassium sulfide.

Relationships: Two compounds are reacting; this must be double replacement. You must exchange cations to determine products.

Solve:
1. Write the ions and criss-cross charges to determine reactant formulas:
   - Ions in lead(II) nitrate: Pb^{2+} and NO_3^-
     - Formula after criss-cross: $Pb(NO_3)_2$
   - Ions in potassium sulfide: K^+ and S^{2-}
     Formula after criss-cross: K_2S
   - Start writing the reaction: $Pb(NO_3)_2 + K_2S \rightarrow ? + ?$
   - Write new formulas for each cation by pairing it with a new anion.
     - For lead (II):
       - Write ions that will combine: Pb^{2+} and S^{2-}
       - Criss-cross charges; in this case both charges cancel: PbS
     - For potassium:
       - Write ions that will combine: K^+ and NO_3^-
       - Criss-cross charges: KNO_3
   - Write the reaction: $Pb(NO_3)_2 + K_2S \rightarrow PbS + KNO_3$
   - Balance the reaction: $Pb(NO_3)_2 + K_2S \rightarrow PbS + 2KNO_3$

Answer: The balanced reaction is: $Pb(NO_3)_2 + K_2S \rightarrow PbS + 2KNO_3$

Combustion reactions

Combustion reactions

Combustion reactions always involve the oxygen molecule, O_2 as a reactant. The general form of a combustion reaction is substance X reacts with oxygen to form a new XO compound. Most often, we think of fire and combustion as being the same thing. However, when iron rusts, solid iron reacts with O_2 in the atmosphere. Technically, rusting iron is a combustion reaction even though fire is not produced.

Combustion reactions

A substance reacts with oxygen.

$$X + O_2 \rightarrow XO$$

$$4Fe_{(s)} + 3O_{2(g)} \rightarrow 2Fe_2O_{3(s)}$$

If the substance is a hydrocarbon, the products are carbon dioxide and water.

$$C_xH_x + O_2 \rightarrow CO_2 + H_2O$$

$$CH_{4(g)} + 2O_{2(g)} \rightarrow CO_{2(g)} + 2H_2O_{(g)}$$

Hydrocarbons

A **hydrocarbon** is a molecule that only contains hydrogen and carbon. Methane (CH_4), propane (C_3H_8), and butane (C_4H_{10}) are examples of hydrocarbons. When hydrocarbons are burned, they always form carbon dioxide and water. Combustion reactions can be very challenging to balance! You often have to go through the balancing process many times.

Solved problem

Write the balanced reaction for the combustion of magnesium metal.

Relationships In a combustion reaction, the given reactant combines with oxygen to form a product.

Solve
1. Write the reactants.
   - Magnesium does not form a diatomic molecule: Mg
   - Oxygen forms a diatomic molecule: O_2
   - The reactants are: $Mg + O_2$
2. Find the charges of the reactants and criss-cross to form a compound.
   - Magnesium = Mg^{2+} and oxygen = O^{2+}
   - Criss-cross charges; in this case the charges cancel: MgO
3. Write the reaction: $Mg + O_2 \rightarrow MgO$
4. Balance the reaction: $2Mg + O_2 \rightarrow 2MgO$

Answer The balanced reaction is: $2Mg + O_2 \rightarrow 2MgO$

Solved problem

Write the balanced reaction for the combustion of pentane, C_5H_{12}.

Relationships This is the combustion of a hydrocarbon. The hydrocarbon will react with oxygen, and the products will be carbon dioxide and water.

Solve
1. When a hydrocarbon undergoes combustion, it reacts with oxygen to produce carbon dioxide and water. No criss-crossing is necessary.
   - $C_5H_{12} + O_2 \rightarrow CO_2 + H_2O$
2. Balance the reaction:
   - Try balancing carbon first, then hydrogen, and balance oxygen last. Keep going back and forth until the reaction is balanced.
   - $C_5H_{12} + 8O_2 \rightarrow 5CO_2 + 6H_2O$

Answer The balanced reaction is: $C_5H_{12} + 8O_2 \rightarrow 5CO_2 + 6H_2O$

Section 7.2: Types of Chemical Reactions

Solving chemical story problems

Story problems

Do you groan when you read "story problems?" You may have to write a balanced equation from the words in a story problem. The next chapter has questions like this: 56.0 g of aluminum reacts with excess copper(II) nitrate. How much copper is produced? This may sound scary, but remember the following tips to help you meet the challenge.

Key words

For the time being let's ignore the 56.0 g and focus on finding the balanced equation. There are helpful key words. It says aluminum reacts with copper(II) nitrate. These substances must be reactants and on the left side of the arrow. The combination of products helps you see what kind of reaction will take place.

aluminum reacts
with copper(II) nitrate

$$Al + Cu(NO_3)_2 \longrightarrow$$

Mass is conserved

The problem goes on to say that copper is produced. Therefore copper is a product and on the right side of the arrow. So far it looks like aluminum and nitrate have disappeared. They are not mentioned as products in the problem. However, from the law of conservation of mass we know they must still be present.

copper is produced ...

$$Al + Cu(NO_3)_2 \longrightarrow Cu + ?$$

Determine reaction type

To find out where the missing aluminum and nitrate belong, we need to determine what type of reaction is taking place.

This looks like a single replacement reaction.
AX + B → BX + A
is the same as
A + BX → B + AX

$$Al + Cu(NO_3)_2 \longrightarrow Cu + ?$$
$$A + BX \longrightarrow B + AX$$

A single element and a compound react to form at least copper, that must mean the other product is a compound. This must be a single replacement reaction. Therefore, the aluminum and nitrate are like the A and X and should be together in a compound on the right. Aluminum has a positive three charge and nitrate has a negative one charge. Therefore, one aluminum needs three nitrates to balance the charge.

Determine what compound is formed by Al^{3+} and NO_3^-

$$Al + Cu(NO_3)_2 \longrightarrow Cu + Al(NO_3)_3$$
$$A + BX \longrightarrow B + AX$$

Balance the equation

Now that we have the same kinds of atoms on both sides of the equation, the last step is to balance the equation.

Balance the equation

$$2Al + 3Cu(NO_3)_2 \longrightarrow 3Cu + 2Al(NO_3)_3$$

We can now deal with the 56.0 g. But why don't we leave that for the next chapter!

Solubility and precipitation

Precipitate formtion

A precipitate forms when a product of an aqueous reaction is insoluble. Precipitates typically form from reactions between ions in solutions of ionic compounds. The solubility rules are a guide that helps you determine which reaction products will form solid precipitates. Any reaction product that is insoluble will precipitate out of solution.

solubility rules for common compounds

soluble compounds	insoluble compounds (except with group 1 metal ions and NH_4^+)
group 1 metal ions Li^+, Na^+, K^+, Rb^+, Cs^+	carbonates, CO_3^{2-}
ammonium, NH_4^+	hydroxides, OH^-, except Ba^{2+}
acetate, $C_2H_3O_2^-$ or CH_3COO^-	chlorides of Cu, Pb, Ag and Hg
nitrates, NO_3^-	bromides of Cu, Pb, Ag and Hg
sulfates, SO_4^{2-}	iodides of Cu, Pb, Ag and Hg
chlorides, except with Cu, Pb, Ag and Hg	sulfides, S^{2-}
bromides, except with Cu, Pb, Ag and Hg	phosphates, PO_4^{3-}
iodides, except with Cu, Pb, Ag and Hg	

The solubility table

Lead(II) nitrate, $Pb(NO_3)_2$, and potassium iodide, KI solutions undergo the reaction: $Pb(NO_3)_2(aq) + 2KI(aq) \rightarrow PbI_2(?) + 2KNO_3(?)$. How can you tell what state the products are in? Use the solubility rules above to see why the reactants are aqueous and to predict the state of products formed. Nitrate compounds are soluble, so $Pb(NO_3)_2$ is dissolved in an aqueous solution (aq). For KI, potassium ions are soluble and so are iodides of potassium; potassium iodide is a soluble compound (aq). Iodides of lead are not soluble so PbI_2 will form a solid precipitate (s). Nitrates of potassium are soluble so KNO_3 will remain aqueous (aq). Any soluble compound dissociates into its ions in solution, and insoluble compounds precipitate out as a solid. Notice the solid PbI_2 at the bottom of the beaker shown below.

Precipitation reactions

One or more product is insoluble and precipitates out of the solution as a solid. Or, the precipitate makes the solution look cloudy.

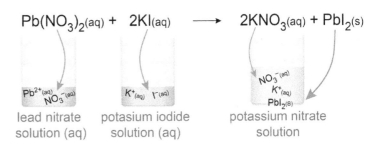

Solved problem

Write the balanced equation for the reaction between magnesium chloride and sodium carbonate solutions. Include states for each compound.

Relationships If two compounds react, a double replacement reaction will occur.

Solve
- Write ions for all reactants: Mg^{2+} Cl^- Na^+ CO_3^{2+}
- Criss-cross to get formulas; switch cations to get products:
  $MgCl_2 + Na_2CO_3 \rightarrow MgCO_3 + NaCl$
- Balance the reaction: $MgCl_2 + Na_2CO_3 \rightarrow MgCO_3 + 2NaCl$
- According to solubility rules, everything except $MgCO_3$ is soluble.
  $MgCl_2(aq) + Na_2CO_3(aq) \rightarrow MgCO_3(s) + 2NaCl(aq)$

Answer $MgCl_2(aq) + Na_2CO_3(aq) \rightarrow MgCO_3(s) + 2NaCl(aq)$

Softening water using precipitation

Dissolved minerals create "hard" water

Hard water is a label commonly used to describe water that has a significant concentration of dissolved metal ions. Hard water can leave a film on your skin after a shower because soaps and oils do not lather and rinse as easily. Magnesium and calcium are two offending species in creating hard water. Iron, Fe^{2+} and manganese, Mn^{2+} also cause problems. Iron can stain clothes, and dissolved minerals build up on water pipes.

Precipitation of magnesium and calcium ions

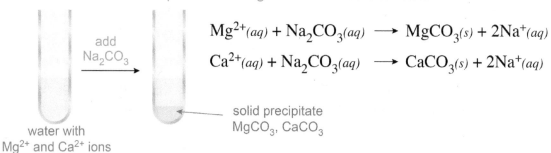

$$Mg^{2+}(aq) + Na_2CO_3(aq) \rightarrow MgCO_3(s) + 2Na^+(aq)$$
$$Ca^{2+}(aq) + Na_2CO_3(aq) \rightarrow CaCO_3(s) + 2Na^+(aq)$$

Removal of dissolved minerals

The process of "water softening" uses substances that precipitate dissolved minerals. Many rural water treatment plants add hydroxide, OH^- and carbonate, CO_3^{2-} to the water supply. These ions react with Mg^{2+} and Ca^{2+} ions to form insoluble precipitates. For example, Mg^{2+} reacts with carbonate, CO_3^{2-} in a water treatment holding tank. The $MgCO_3(s)$ precipitate that forms settles to the bottom of the holding tank where it is easily removed by filtering.

Gas forming reactions

Hard water can affect your plumbing by precipitating out some calcium carbonate. These precipitates, called pipe scaling, can be removed (descaled) with a dilute acid which creates another double replacement reaction that dissolves the scale into soluble compounds, water and gas. For example, when solid magnesium carbonate, $MgCO_3$, and hydrochloric acid, HCl, are mixed, Mg^{2+} and H^+ ions exchange places resulting in soluble magnesium chloride, $MgCl_2$, H_2O, and carbon dioxide gas, CO_2, that just bubbles away.

Descaling pipes with acid to remove mineral deposits

$$MgCO_3(s) + 2HCl(aq) \rightarrow MgCl_2(aq) + H_2O(l) + CO_2(g)$$

Solved problem

What reaction occurs when a solution of copper(II) nitrate ($Cu(NO_3)_2$) is combined with a solution of potassium chloride (KCl)? Write the complete balanced reaction including reactant and product states.

- Asked: Write a reaction between $Cu(NO_3)_2$ and KCl, and identify states.
- Given: Formulas for both reactant solutions.
- Relationships: Because both reactants are solutions, both are aqueous. Both reactants are ionic compounds so a double replacement reaction will occur. The solubility rules state K^+, a group 1 metal ion, and nitrates are soluble but chlorides of copper are insoluble.
- Solve: $Cu(NO_3)_2(aq) + 2KCl(aq) \rightarrow CuCl_2(s) + 2KNO_3(aq)$
- Answer: According to the solubility table, potassium nitrate is soluble and dissociates to $K^+(aq)$ and $NO_3^-(aq)$. $CuCl_2$ forms a precipitate because it does not have a group 1 metal ion or nitrate. All ions are balanced. Copper(II) chloride is insoluble and settles out as a solid.

Net ionic equations

Spectator ions

Some ions appear in solution as both products and reactants as seen in the reaction below. Potassium and nitrate ions remain dissolved in solution throughout the reaction. A **spectator ion** is unchanged in its same dissolved, aqueous form before and after the reaction. The potassium and nitrate ions are considered "spectators" because both are present but neither participate in the reaction. Both ions remain dissolved in solution while the two remaining ions, lead(II) and iodide, combine to form a solid precipitate.

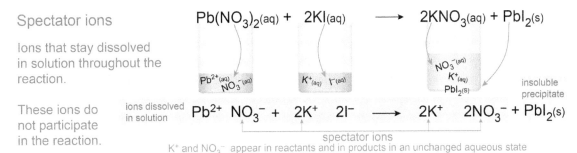

Spectator ions

Ions that stay dissolved in solution throughout the reaction.

These ions do not participate in the reaction.

K^+ and NO_3^- appear in reactants and in products in an unchanged aqueous state

How to write a net ionic equation

When working with solutions it is common to see only actively reacting ions in the chemical equation. This is called a net ionic equation. Double replacement reactions are frequently written as net ionic equations as well as neutralization and redox reactions you will see in future chapters. The net ionic equation ignores spectator ions and shows only the ions that react to form a new product. Let's write the net ionic equation for the reaction above: $Pb(NO_3)_2(aq) + 2KI(aq) \rightarrow 2KNO_3(aq) + PbI_2(s)$.

1. Break apart the aqueous compounds into ions but leave solids intact. Carry states and coefficients to each ion when present. When subscripts indicate the number of ions, bring the number in front of the ion as a coefficient as seen on nitrate below.

   $Pb^{2+}(aq)\ 2NO_3^-(aq)\ +\ 2K^+(aq)\ 2I^-(aq) \rightarrow 2K^+(aq)\ 2NO_3^-(aq)\ +\ PbI_2(s)$
   Note: Make sure the equation is balanced.

2. Cancel all ions that appear in the same state on both sides of the equation.

   $Pb^{2+}(aq)\ \cancel{2NO_3^-(aq)}\ +\ \cancel{2K^+(aq)}\ 2I^-(aq) \rightarrow \cancel{2K^+(aq)}\ \cancel{2NO_3^-(aq)}\ +\ PbI_2(s)$

3. Rewrite the equation with the remaining substances.

   $Pb^{2+}(aq) + 2I^-(aq) \rightarrow PbI_2(s)$

Solved problem

Write the net ionic equation for: $MgCl_2(aq) + Na_2CO_3(aq) \rightarrow MgCO_3(s) + 2NaCl(aq)$.

Relationships The net ionic equation shows only substances that react.

Solve
1. Break apart aqueous substances and leave solids intact. Include states and coefficients.

   $Mg^{2+}(aq)\ 2Cl^-(aq) + 2Na^+(aq)\ CO_3^{2-}(aq) \rightarrow MgCO_3(s) + 2Na^+(aq)\ 2Cl^-(aq)$

2. Cancel ions that appear the same on both sides

   $Mg^{2+}(aq)\ \cancel{2Cl^-(aq)}\ +\ \cancel{2Na^+(aq)}\ CO_3^{2-}(aq) \rightarrow MgCO_3(s) + \cancel{2Na^+(aq)} + \cancel{2Cl^-(aq)}$

3. Rewrite the reaction with the remaining substances.

   $Mg^{2+}(aq) + CO_3^{2-}(aq) \rightarrow MgCO_3(s)$

Answer The net ionic equation is: $Mg^{2+}(aq) + CO_3^{2-}(aq) \rightarrow MgCO_3(s)$

Polymers and polymerization reactions

Polymers are long chains molecules

Polyethylene is the most common plastic in the world with production of over 80 million tons per year. Used in plastic bags, plastic films, toys, and bottles, many kinds of polyethylene are produced with molecular weights from a few thousand to more than one million g/mole. Polyethylene is an example of a **polymer**. A polymer is a large molecule made of many identical parts called monomers. For polyethylene the monomer is ethylene, C_2H_4 and the polymer is a chain of 100,000 or more monomers.

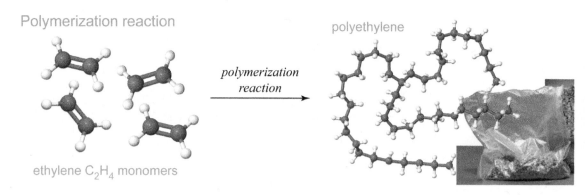

Polymerization

A **polymerization** reaction is a repeated addition of monomers into active sites at the end of chains. Depending on the process the chains can be straight or branched. Notice that the very short polyethylene example above has a branched chain. Modern production of ultra high density polyethylene (UHDPE) can produce straight or branched chains with molecular weights up to 3.5 million g/mol.

Natural polymers

Polymers occur through nature as well as in human technology. Plants produce glucose through photosynthesis then convert it into starch by a polymerization reaction. Starch is a long chain of glucose monomers many thousands of units long. Starch molecules have the CH_3OH side chain on each glucose monomer oriented in the same direction. Woody plants use a variant of the same reaction to makes cellulose. Cellulose is another polymer made from glucose monomers. In cellulose each successive monomer alternates the orientation of the CH_3OH side chain. Starch is also a highly branched molecule while cellulose is a straight chain. The differences make it impossible for most animals to digest cellulose while virtually all can digest starch!

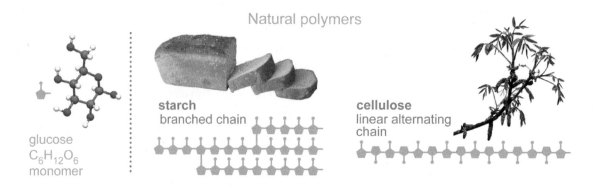

Starch is a vital carbohydrate

Starch is a basic carbohydrate of animal diets, including humans! All grains including bread and pasta are mainly composed of starch. The body reverses the synthesis process and decomposes starch back into glucose during digestion.

Section 2 review

Chapter 7

Chemical reactions are classified into several different types. Synthesis reactions occur when two reactants combine to form one product. Decomposition reactants occur when one reactant is split to form two or more products. In a single displacement reaction one element or polyatomic ion is transferred from one reactant to the other as the reaction proceeds. In a double displacement reaction two reactants exchange parts to form two new products. Combustion reactions involve hydrocarbons with oxygen gas, normally resulting in heat, carbon dioxide and water as products. In aqueous solutions sometimes one of the products is not soluble in water, forming a precipitate. The solubility rules are used to determine when a precipitate will form. Polymerization is the formation of a long chain of molecules built from identical smaller units.

Vocabulary words: synthesis, decomposition, single replacement, double replacement, precipitate, combustion, hydrocarbon, spectator ion, polymer, polymerization

Key relationships
- Synthesis reaction: $A + X \rightarrow AX$
- Decomposition reaction: $AX \rightarrow A + X$
- Single displacement/single replacement reaction: $A + BX \rightarrow B + AX$
- Double displacement/double replacement reaction: $AX + BY \rightarrow BX + AY$
- Combustion reaction: $X + O_2 \rightarrow XO$
- Combustion reaction of a hydrocarbon: $C_xH_x \rightarrow CO_2 + H_2O$
- Polymerization reaction: [monomer + monomer]$_n$ → polymer

Review problems and questions

1. When zinc reacts with hydrochloric acid, zinc chloride and hydrogen gas are produced: $Zn + 2HCl \rightarrow ZnCl_2 + H_2$. What type of reaction is this?

2. Two reactants combine to form a single product. What type of reaction is this?

3. Classify the type of reaction that occurs when silver nitrate reacts with magnesium chloride to form silver chloride and magnesium nitrate.

4. Write a balanced equation for the above reaction. Remember to criss-cross charges for ionic compounds.

5. Use the solubility rules to identify which products, if any, will form a precipitate.

6. In a combustion reaction, if the substance is a hydrocarbon, what are the products?

7. Write the balanced reaction for the combustion of octane, C_8H_{18}.

8. The molecular links in the polymer chain are called _____, and they are formed from one or more molecules called _____.

9. In a combustion reaction, a substance reacts with oxygen. Give an instance where a substance reacting with oxygen is not a combustion reaction.

Section 7.2: Types of Chemical Reactions

Surprising Reactions and Accidental Discoveries

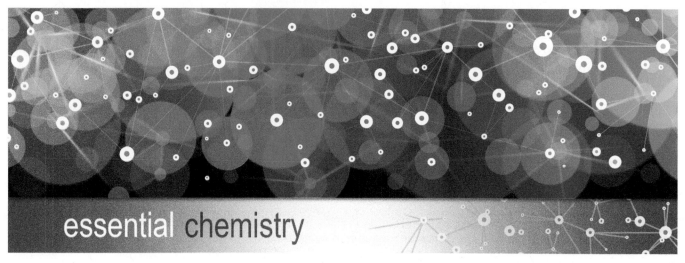

Isaac Asimov said, "The most exciting phrase to hear in science, the one that heralds new discoveries, is not "Eureka!" but "That's funny!" Throughout the history of science, careful observations and detailed characterizations of accidental reactions, unexpected products, or failed experimental waste, led to discoveries that changed the world.

Alexander Fleming generally gets credit for the discovery of penicillin when he famously noticed that mold growth on his petri dishes inhibited the growth of nearby bacteria. But Fleming was never able to extract any penicillin in usable quantities in his lab. Ten years later, a pharmacologist and a team of chemists (yes chemists!) figured out a way to purify penicillin in usable quantities. Finally, in 1944 Margaret Hutchinson Rousseau, a chemical engineer, converted the purification process into a full-scale production.

In fact, the entire modern chemical industry owes itself to one of these accidental discoveries. In the 19th century there was a new kind of waste called coal tar that was left over from turning coal into gaslight. August Wilhelm von Hoffman had an idea to investigate the usefulness of coal tar. He directed an 18-year-old chemist, William Perkin, to try to convert coal tar into quinine. Quinine was in high demand because it could be used to fight malaria, but at the time it was only found in the bark of trees found in the Andes.

Perkin's attempts did not result in the product he was looking for but persistence and careful observations paid off in a surprising way! His first attempts to carry out the reaction produced a red-brown precipitate instead of the hoped for white crystalline product of quinine. After changing one of the reactants, he tried the reaction again this time forming a black residue. Then he tried to isolate the black residue by adding alcohol. When he did, the black residue transformed into a deep purple. He took this purple pigment and applied it to his sister's white silk blouse. In 1868, he patented the violet dye, which came to be known as mauve.

Surprising Reactions and Accidental Discoveries - 2

Perkin's dye created a fashion craze and sparked an industry in its infancy. At the time of Perkin's accidental creation of mauve, colored dyes were hard to come by and were very expensive. In fact, purple dyes were created by using exotic crushed snails! Now Perkin came along and made mauve from coal tar waste. Perkin's odd purple color was favored by fashionistas of the time, including Queen Victoria and Napoleon III's wife Empress Eugenie. It was inexpensive enough that even the middle-class masses could afford to wear it instead of the drab greys, browns, and off-whites that were common at the time. Perkin amassed a fortune and retired at the age of 36. His discovery revolutionized the industry of artificial dyes and made organic chemistry exciting, useful and profitable.

German biologist Walther Flemming used aniline dyes, created from coal tars, to study cells under a microscope. The staining process revealed structures in the cell and he was able to visualize the threadlike structures as cells divided. Since they were dyed, he called these bundles chromosomes, after the Greek "chromos" for color. Much of what we know today about cell division originated from Flemming's observations.

On Perkins' lead, chemical factories began popping up searching for uses of coal tar. This lead to even more profitable accidents. One such happy accident that happened in 1879 was based on careful observation and a bit of carelessness. Constantin Fahlberg was working with compounds and products from coal tar. One evening he stayed at the lab late and rushed off to dinner without stopping to wash his hands. He noticed that the bread he was holding was incredibly sweet. He rinsed with water and grabbed his napkin to dry his mouth. To his surprise, the napkin was even sweeter than the bread. Eventually it dawned on him that his hands were the source of the sweetness. He had created and discovered something that out-sugared sugar. Fahlberg and his lab mates called this super-sweet substance saccharin. Saccharin was the first artificial sweetener. It has had a colorful history since its accidental discovery, but it is still in little pink packs in most restaurants.

Saccharin and Teflon™ were both accidental discoveries.

In the 20th century there were even more accidental reactions and discoveries. In the late 1930's researchers at DuPont were working with refrigerants. One day a new mixture of chemicals reacted to form a powder that made stuff slippery. DuPont used this material, eventually called Teflon, to make nonstick surfaces, and wire cables that are non-conductive. When drawn into thin fibers and woven together it is used to create waterproof clothing.

These historically significant feats would not have been possible without an accidental discovery, and some reaction know-how from chemists and chemical engineers that were observant and persistent enough to recognize that they stumbled upon something special.

Chapter 7 review

Vocabulary
Match each word to the sentence where it best fits.

Section 7.1

balanced chemical equation	chemical equation
chemical reaction	insoluble
law of conservation of matter	product
reactant	soluble
unbalanced chemical equation	

1. When a chemical equation satisfies the law of conservation of mass, it is called a(n) _____ .

2. The _____ is the starting material of a chemical reaction.

3. The expression that describes the changes that happen in a chemical reaction is the _____ .

4. The substance that is created or released in a chemical reaction is called the _____ .

5. The _____ states that the mass of the starting materials must equal the mass of the materials produced by the reaction.

6. A process that creates chemical changes is called a(n) _____ .

7. If a substance is able to dissolve in water, it is considered _____ .

8. A chemical equation that does not satisfy the law of conservation of mass is considered to be a(n) _____ .

9. If a substance is unable to dissolve in water, it is considered _____ .

Section 7.2

combustion	decomposition
double replacement	hydrocarbon
polymer	polymerization
single replacement	synthesis

10. The reaction in which a single substance breaks apart to make two or more new substances is considered a(n) _____ reaction.

11. AB + DX → AX + DB, this reaction is considered a(n) _____ reaction.

12. A reaction in which a single element replaces another element in a compound is called a(n) _____ reaction.

13. A long-chain of molecules that is formed by connecting small repeating units with covalent bonds is called a(n) _____ .

14. A molecule that is only made up of carbon and hydrogen atoms is called a(n) _____ .

15. A(n) _____ reaction assembles a polymer through repeated additions of smaller molecular fragments.

16. A chemical reaction that involves the rapid combination of a fuel with oxygen is called a(n) _____ reaction.

17. _____ reactions include reactions which combine two or more compounds to produce a third compound.

Conceptual questions

Section 7.1

18. What is a chemical reaction?
    a. Describe in your own words what a chemical reaction is.
    b. Give three examples of chemical reactions.

19. A chemical equation is different than a chemical reaction.
    a. Describe what a chemical equation is in your own words.
    b. Give an example of a chemical equation.
    c. Identify the the products and reactants.

20. A student writes the following sentence to describe the chemical reaction below. What is wrong with the student sentence and what would you change to make it correct?

    $$4Fe(s) + 3O_2(g) \rightarrow 2Fe_2O_3(s)$$

    "Four grams of iron react with three grams of oxygen produce 2 grams of iron oxide."

21. From the list below, choose which of the chemical equations conserve mass.
    a. $N_2(g) + H_2(g) \rightarrow NH_3(g)$
    b. $H_2(g) + Cl_2(g) \rightarrow 2HCl(g)$
    c. $2N_2H_4(g) + N_2O_4(g) \rightarrow 2N_2(g) + 4H_2O(g)$

22. Write a sentence that has the same meaning as the following chemical equation.

    $$2CH_3OH + 3O_2 \rightarrow 2CO_2 + 4H_2O$$

Chapter 7 review

Section 7.1

23. A balanced chemical equation must satisfy a fundamental law. What is this law called?

24. ❰ Two of the four chemical equations below are not balanced. Identify and balance the ones which are unbalanced.

    1. $C_2H_6 + O_2 \rightarrow H_2O + CO_2$
    2. $NaHCO_3 + C_2H_4O_2 \rightarrow NaC_2H_3O_2 + CO_2 + H_2O$
    3. $NaOH(g) \rightarrow Na^+(aq) + OH^-(aq)$
    4. $N_2(g) + H_2(g) \rightarrow NH_3(l)$

25. One reason water is essential to living organisms is as a solvent allowing dissolved molecules to move and react. But there are also chemical processes involving water. Describe at least one chemical reaction in living organisms which directly involves water as a reactant or product.

26. Identify the state of matter for each compound or element in the following chemical equation that describes the burning of carbon.

    $C + O_2 \rightarrow CO_2$

27. What is the procedure for balancing chemical equations? Describe each step.

28. What does the statement: "The language of chemistry requires proper grammar to correctly describe chemical processes." mean?

Section 7.2

29. Answer the following questions concerning predicting chemical reactions.

    a. What is the first step to take in order to predict the products of a reaction?
    b. What are the different types of chemical reactions?
    c. How do you predict the products of a Synthesis reaction?
    d. How do you predict the products of a Double Replacement reaction?
    e. How do you predict the products of a Combustion reaction?

30. Categorize each of the following reaction as: combination, decomposition, single replacement, double replacement or combustion.

    a. $CH_4 + 2O_2 \rightarrow 2H_2O + CO_2$
    b. $2Ca(s) + O_2(g) \rightarrow 2\ CaO(s)$
    c. $BaCl_2(aq) + Na_2SO_4(aq) \rightarrow BaSO_4(s) + 2NaCl(aq)$
    d. $NH_4NO_3(aq) \rightarrow N_2O(g) + 2H_2O(g)$
    e. $2Fe(s) + 3Cl_2(g) \rightarrow 2FeCl_3(aq)$

31. Balance the equations and predict the products for the following reactions.

    a. ___ Na + ___ FeBr$_3 \rightarrow$
    b. ___ NaOH + ___ H$_2$SO$_4 \rightarrow$
    c. ___ C$_2$H$_4$O$_2$ + ___ O$_2 \rightarrow$
    d. ___ H$_2$O$_2 \rightarrow$

32. Cellulose makes up the structural part of plants. Cellulose is produced by a polymerization process. Research and describe this process.

33. Based on your knowledge of solubility rules, determine which of the following compounds are soluble and which are insoluble.

    a. $CuCO_3$
    b. Na_3PO_4
    c. $Pb(NO_3)_2$
    d. K_2SO_4
    e. $MgSO_4$

34. In your own words, what are the steps to writing a Net Ionic Equation?

Quantitative problems

Section 7.1

35. Propane (C_3H_8) in a gas grill undergoes a combustion reaction with oxygen and produces carbon dioxide and water vapor.

    a. Write the general chemical equation.
    b. Balance the chemical equation.

36. Aqueous ethyl alcohol (C_2H_5OH) and carbon dioxide are formed when sugar ($C_{12}H_{22}O_{11}$) ferments and reacts with water.

    a. Write the general chemical equation for the reaction above.
    b. Balance the chemical equation.

37. Potassium, K, is a highly reactive alkali metal. When a small piece of solid potassium is placed in water it goes through a vigorous reaction and produces aqueous potassium hydroxide (KOH(aq)) and hydrogen gas (H_2).

    a. Write the general chemical equation for this reaction.
    b. Balance the chemical equation.

Chapter 7 review

Section 7.1

38. Aluminum is abundant in many minerals but aluminum metal was so hard to purify that it was once worth more than gold! Natural aluminum occurs in ores such as bauxite. Bauxite is refined in to alumina, $Al_2O_3(s)$. The alumina is reacted at high temperature with molten carbon compounds to release pure aluminum metal.

   a. Write and balance one possible reaction that starts with alumina, Al_2O_3, and carbon, C, and produces aluminum and carbon dioxide, CO_2.

   b. Write and balance another possible reaction that starts with alumina, Al_2O_3, and carbon, C, and produces aluminum and carbon monoxide, CO.

aspartame

39. Aspartame, ($C_{14}H_{18}N_2O_5(s)$), is a low-calorie, artificial sweetener that many people use as a sugar substitute. When metabolized in the body, aspartame is combined with oxygen, $O_2(g)$, to produce carbon dioxide, nitrogen, and liquid water.

   a. Write a possible chemical equation that represents the reaction of aspartame with oxygen to produce CO_2, N_2, and water. Note that this reaction actually proceeds in many steps in the body.

   b. Balance the equation.

40. Catalytic converters are required on car exhaust systems to reduce pollution. One of the major chemical processes that takes place in a catalytic converter is to reduce the emissions of oxides of nitrogen (NO_x). Balance the following equation for this reaction.

$$NO(g) + CO(g) \rightarrow N_2(g) + CO_2(g)$$

41. Urea is a dry white organic compound and it is highly soluble in water. Due to its solubility, it is very suitable for fertilizers. The reaction to prepare urea is described by the equation below. Balance the equation.

$$NH_3(g) + CO_2(g) \rightarrow CO(NH_2)_2(s) + H_2O(g)$$

42. The following equation describes the combustion of octane. Balance the chemical equation:

$$C_8H_{18} + O_2 \rightarrow CO_2 + H_2O.$$

43. Acid rain is a major source of environmental pollution in many areas of the world. There are several reactions that create acid rain. In the most common reaction, sulfur dioxide from coal-burning reacts with water and oxygen in the atmosphere to form dilute sulfuric acid, $H_2SO_4(aq)$.

   a. Write a chemical equation that describes a possible reaction to produce sulfuric acid from sulfur dioxide, water, and oxygen.

   b. Balance the chemical equation.

44. Balance the following reaction:

$$Al + O_2 \rightarrow Al_2O_3$$

45. Balance the chemical equations below. If they are balanced, write "balanced".

   a. $HCl + NaOH \rightarrow H_2O + NaCl$

   b. $C_4H_{10} + O_2 \rightarrow CO_2 + H_2O$

   c. $Ca + H_2O \rightarrow Ca(OH)_2 + H_2$

   d. $SiCl_4 + H_2O \rightarrow SiO_2 + HCl$

46. When ultraviolet (UV) radiation interacts with ozone (O_3) in the upper atmosphere, it is absorbed by the ozone molecule. This causes ozone to decompose into molecular oxygen, leaving a thinner ozone layer or "hole" in the upper atmosphere. A thinner ozone layer cannot absorb UV radiation so more of it reaches Earth. UV radiation is very damaging to living organisms.

   a. What must be the other product that results from this reaction, if the ozone molecule has a coefficient of one in the balanced equation?

   b. Write the balanced chemical equation.

47. Determine which equations below are unbalanced and balance them.

   a. $NO_2 + H_2O \rightarrow HNO_3 + NO$

   b. $CaCl_2 + Na_2CO_3 \rightarrow CaCO_3 + NaCl$

   c. $N_2O_5 \rightarrow NO_2 + O_2$

   d. $CH_3OH + O_2 \rightarrow CO_2 + H_2O$

48. Balance the following equation for the combustion of methylcyclohexane C_7H_{13}, one of the components of gasoline.

$$C_7H_{13} + O_2 \rightarrow CO_2 + H_2O$$

Chapter 7 review

Section 7.2

49. Balance and classify each of the following chemical reactions as synthesis, decomposition, combustion, single replacement, or double replacement.

   a. $Ca(NO_3)_2 \rightarrow CaO + N_2 + O_2$

   b. $NH_4Cl(s) \rightarrow NH_3(g) + HCl(g)$

   c. $Na_2O(s) + H_2O \rightarrow NaOH(aq)$

   d. $C_6H_{12}O_6(s) + O_2(g) \rightarrow CO_2(g) + H_2O(g)$

50. $Pb(s) + AgNO_3(aq) \rightarrow Pb(NO_3)_2(aq) + Ag(s)$

   a. Balance the above chemical equation.
   b. Classify the reaction by its type.

51. Given the following reactants,

   $NaBr + BaCl_2 \rightarrow$

   a. Predict the products.
   b. Balance the chemical equation.
   c. Classify the equation based on the types you have learned.

52. Given the following reactants,

   $C_6H_{14} + O_2 \rightarrow$

   a. Predict the products.
   b. Balance the chemical equation.
   c. Classify the equation based on the types you have learned.

53. Given the following reactants,

   $Cl_2 + KBr \rightarrow$

   a. Predict the products.
   b. Balance the chemical equation.
   c. Classify the equation based on the types you have learned.

54. Using the following reactants

   $AgNO_3(aq) + HCl(aq) \rightarrow$

   a. Complete and balance the equation carried out in aqueous solution.
   b. Write the total ionic and net ionic equation from the balanced equation.

55. Using the following reactants

   $Cu(OH)_2(s) + HBr(aq) \rightarrow$

   a. Complete and balance the equation carried out in aqueous solution.
   b. Write the total ionic and net ionic equation from the balanced equation.

56. Using the following reactants

   $Pb(NO_3)_2(aq) + H_2S(aq) \rightarrow$

   a. Complete and balance the equation carried out in aqueous solution.
   b. Write the total ionic and net ionic equation from the balanced equation.

57. Using the following reactants

   $BaBr_2(aq) + H_2SO_4(aq) \rightarrow$

   a. Complete and balance the equation carried out in aqueous solution.
   b. Write the total ionic and net ionic equation from the balanced equation.

58. What are the products when $Al_2(SO_4)_3$ and NH_4Cl react in aqueous solution?

59. Iron metal is produced from iron oxide ore. This chemical reaction takes place in two steps. Balance the equation for each step below.

   a. First step: $Fe_2O_3(s) + CO(g) \rightarrow Fe_3O_4(s) + CO_2(g)$
   b. Second step: $Fe_3O_4(s) + CO(g) \rightarrow Fe(s) + CO_2(g)$

60. ❰ Barium bromide ($BaBr_2$) and sodium sulfate (Na_2SO_4) react in solution to form products by a double substitution reaction.

   a. Write the balanced chemical equation and identify any compounds which are precipitates (solid)
   b. Write the net ionic equation.
   c. Identify any spectator ions.

61. ❰ Identify each of the following compounds as soluble or insoluble.

   a. potassium nitrate, KNO_3
   b. calcium carbonate, $CaCO_3$
   c. ammonium carbonate, $(NH_4)_2CO_3$
   d. silver chloride, $AgCl$
   e. lithium chloride, $LiCl$

62. Fill in the missing reactant or product that would complete the balanced reaction.

   a. _____ + $2H_2O \rightarrow Ca(OH)_2 + H_2$
   b. $4NH_3$ + _____ $\rightarrow 2N_2 + 6H_2O$
   c. $Fe_2O_3 + Al \rightarrow$ _____ + Al_2O_3
   d. _____ + $5O_2 \rightarrow 3CO_2 + 4H_2O$
   e. Ni + _____ $\rightarrow NiCl_2 + H_2$

chapter 8 Stoichiometry

If you poured some baking soda into a cup full of vinegar, you would see foaming bubbles. Knowing the chemical names and formulas for baking soda (sodium bicarbonate) and for vinegar (acetic acid), we can write a balanced chemical equation for the reaction.

$$NaHCO_3(s) + HC_2H_3O_2(aq) \rightarrow NaC_2H_3O_2(aq) + CO_2(g) + H_2O(l)$$

The balanced equation symbolizes that baking soda and vinegar produce sodium acetate, carbon dioxide and water. Those foaming bubbles that you see contain gaseous carbon dioxide. If you pour some more baking soda, more bubbles will form. If you keep adding baking soda, something odd happens. Eventually no more bubbles are produced! Does the balanced equation suddenly become wrong?

Of course, the equation is not wrong! The chemical equation above also tells us that the reaction uses the acetic acid in the vinegar. Eventually, the acetic acid is completely reacted into sodium acetate. Once the acetic acid is gone, adding more baking soda does not cause new bubbles because there is nothing left to react with.

Chemical changes are at the heart of chemistry. Being able to describe and predict what will happen when things are reacted is essential to chemists. With models, you can see and balance the substances in a reaction. You should even be able to visualize the atoms rearranging during a chemical change. But there is another important part—chemists also need to know "how much" will happen. This is crucial when creating a new drug, producing a new material, or in studying the human impact on the natural environment.

Chapter 8 study guide

Chapter Preview

Stoichiometry is the calculation of quantities of reactants and products in a chemical reaction. Think of it as the math or accounting of chemical reactions. A balanced equation tells us not only what the reactants and products are in a reaction, but the proportion of each of them or the mole ratios. This is key for being able to convert between the number of moles and the mass in grams or the number of particles of a reactant or product. Percent yield provides a ratio of the actual results of a reaction compared to the calculated or theoretical results. A limiting reactant determines the maximum amount of product for a given reaction. In this chapter you will learn the rules of stoichiometry so that you can calculate masses, moles, and percentages using chemical equations.

Learning objectives

By the end of this chapter you should be able to:
- determine the mole ratio from a balanced equation and use it to calculate moles needed or produced in a reaction;
- make gram to gram conversions of substances in a chemical reaction using formula mass and the mole ratio;
- make gram to particle conversions using Avogadro's number and gram to volume conversions using standard molar volume;
- calculate percent yield using a calculation of theoretical yield; and
- distinguish limiting reactants from excess reactants and perform related calculations.

Investigations

8A: Conservation of mass
8B: Percent yield
8C: Modeling limiting reactants
8D: Determining limiting reactants
8E: Project: Design an Airbag
8F: Research Presentation Enhancement: Airbags and Consumers

Pages in this chapter

- 222 Using the Chemical Equation Builder
- 223 Analyzing a Chemical Reaction
- 224 Interpreting balanced equations
- 225 Mole ratios
- 226 Mole-to-mole problems with mole ratios
- 227 Calculating quantities in grams
- 228 Solving gram-to-gram problems
- 229 Calculating volumetric quantities
- 230 Grams to particles
- 231 Solving problems with mixed units
- 232 Completion problems
- 233 Section 1 Review
- 234 Percent Yield
- 235 Measuring percent yield
- 236 Solving percent yield problems
- 237 Two-step percent yield problems
- 238 Section 2 Review
- 239 Limiting Reactants
- 240 Identifying the limiting reactant
- 241 Determining the limiting reactant
- 242 Solving limiting reactant problems
- 243 Section 3 Review
- 244 Engineering and Green Chemistry
- 245 Engineering and Green Chemistry - 2
- 246 Chapter review

Vocabulary

stoichiometry   mole ratio   percent yield   theoretical yield
actual yield   limiting reactant   excess reactant

Using the Chemical Equation Builder

The equation builder

The icon in the upper right of the navigation bar in the Essential Chemistry electronic book opens the equation builder and stoichiometry calculator.

This button opens the chemistry equation builder

Set up a balanced equation

The first two steps in using the equation builder are to set up a chemical equation and then to balance the equation. You can build or modify reactants and products by clicking on the periodic table to add or select an element, or choose pre-built compounds or ions. The equation balancer will show you a running count of each element on the product and reactant side of the equation as you adjust the coefficients.

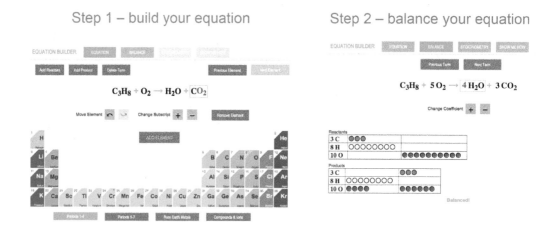

Stoichiometry calculations

Once the balanced equation is set up, the builder automatically calculates the molar masses and displays them under each compound in the equation. Enter a value into any of the other boxes and the equation builder will recalculate all the other quantities based on the value you entered.

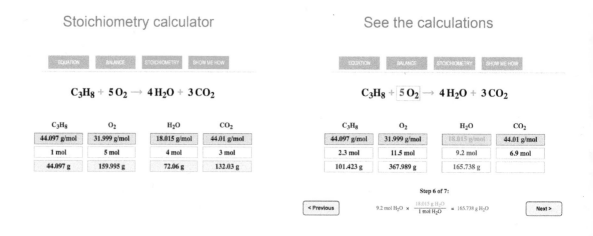

8.1 - Analyzing a Chemical Reaction

A balanced chemical equation tells us how elements and compounds are rearranged from reactants to products. The next step is to use the balanced equation to determine the actual quantities for a reaction, and the real quantities of products that are created. A good example is a recipe for making a cake. The recipe gives the proportions of ingredients for one cake. The actual amount of each ingredient depends on how many cakes you want to make. A chemical equation gives the proportions of each compound. The actual amount of each compound depends on the proportion of each compound and the total amount of reactant or product.

Keys to stoichiometry

What is stoichiometry?

Stoichiometry is the part of chemistry that applies the balanced chemical equation to determine the quantities of reactants and products. For example, methane is widely distributed as natural gas for cooking. We know the balanced chemical equation. How do we use the balanced equation to calculate quantities in moles or grams for any amount of any of the products or reactants? This is the purpose of stoichiometry.

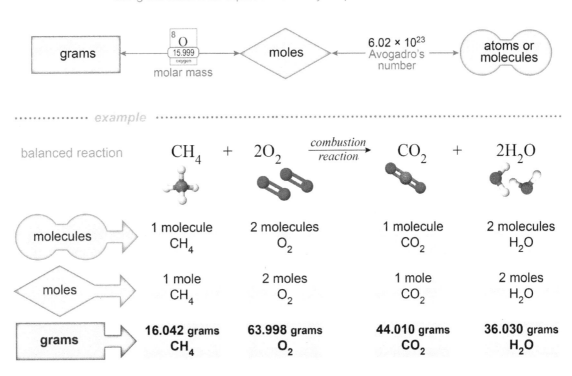

Interpreting the balanced equation

The key to understanding stoichiometry is knowing how to use the mole relationships between reactants and products from the balanced equation. As written they can be interpreted two ways: in terms of molecules and in terms of moles. We will often use stoichiometry to calculate from moles to grams. The gram information is contained in the balanced chemical equation, but there is a step of calculation necessary to uncover it.

Why moles are important

The mole is so important because atoms combine in whole number ratios. Single atoms and molecules are too small to work with. Moles are reasonable sized quantities and ratios of moles are numerically the same as ratios of atoms. Moles are easily converted to grams and an amount in grams can then be weighed out on a balance.

Section 8.1: Analyzing a Chemical Reaction

Interpreting balanced equations

Balanced chemical equations

In virtually all practical applications of chemistry, the balanced chemical equation is interpreted as moles of reactants and products. In the reaction for combustion of methane, one mole of methane combines with two moles of oxygen to produce one mole of carbon dioxide and two moles of water.

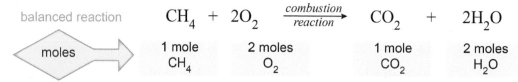

Compare quantities of atoms in moles

Always compare compounds in terms of moles because the ratio of moles is the same as the ratio of atoms. One mole of methane per two moles of oxygen is the same as one molecule of methane per two molecules of oxygen. It is not the same as one gram of methane per two grams of oxygen! Methane (CH_4) and oxygen (O_2) have different molar masses.

Recipes vs. chemical equations.

A kitchen recipe conveys the same information about how much of each ingredient makes up the product. However, kitchen recipes tend to use different units for each ingredient. Don't make the mistake of using grams or any other unit. Chemical equations are always in terms of moles - always!

Cars, tires, and moles

Let's say your job is to make cars. You know that one car needs four tires, not including the spare. It makes no sense to tell your supplier that you need 90 kg of tires to go with 2000 kg of car. Although it is a true statement, it makes building a car far more difficult. You have to convert from kilograms to cars and tires every time you want to make a car. It is far better to talk about the number of parts. Four tires are needed to go with one car. In a similar way the chemical equation is written in the number of particles, not the mass of particles.

four tires are needed to go with one car

Mole ratios

The mole ratio

A **mole ratio** gives the relationship between any two compounds in a balanced chemical equation. The compounds can both be from the reactants, both from the products, or one of each. A mole ratio uses the coefficients from the balanced equation to single out the mole relationship between two compounds that you may wish to do a calculation with.

Mole ratios

A **mole ratio** is the ratio of any two compounds in a balanced chemical equation, *with their coefficients*, from products, reactants, or both.

balanced reaction $\quad C_3H_8 + 5O_2 \longrightarrow 3CO_2 + 4H_2O$

·············· example of 12 mole ratios from the C_3H_8 combustion reaction ··············

propane : **oxygen** $\qquad$ propane : **carbon dioxide** $\qquad$ propane : **water**

① $\dfrac{5\,O_2}{1\,C_3H_8}$ ② $\dfrac{1\,C_3H_8}{5\,O_2}$ ③ $\dfrac{3\,CO_2}{1\,C_3H_8}$ ④ $\dfrac{1\,C_3H_8}{3\,CO_2}$ ⑤ $\dfrac{4\,H_2O}{1\,C_3H_8}$ ⑥ $\dfrac{1\,C_3H_8}{4\,H_2O}$

water : **oxygen** $\qquad$ oxygen : **carbon dioxide** $\qquad$ carbon dioxide : **water**

⑦ $\dfrac{5\,O_2}{4\,H_2O}$ ⑧ $\dfrac{4\,H_2O}{5\,O_2}$ ⑨ $\dfrac{3\,CO_2}{5\,O_2}$ ⑩ $\dfrac{5\,O_2}{3\,CO_2}$ ⑪ $\dfrac{4\,H_2O}{3\,CO_2}$ ⑫ $\dfrac{3\,CO_2}{4\,H_2O}$

Possible mole ratios

Any given chemical equation typically has many mole ratios. The example for the combustion of propane lists twelve mole ratios just from one single equation with four compounds! Each possible pair of compounds in the reaction has two mole ratios - A/B - and its inverse B/A.

Interactive simulation

Explore the interactive simulation titled Getting Ratios Right to verify the statement: No matter what the starting amounts are, reactants will always combine in the same ratio for a given reaction. Select one ratio at a time and wait for the reaction to complete in each case. Observe the number of hydrogen molecules and the number of chlorine molecules that react with each other to determine the ratio. Is the ratio always the same, or does it depend on the number of reactant molecules available? Is it possible to have too much of one reactant?

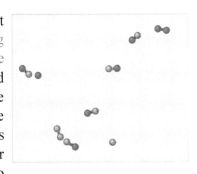

Solved problem

Give two mole ratios that relate aluminum metal, Al, with chlorine gas, Cl_2 in the reaction $2Al(s) + 3Cl_2(g) \rightarrow 2AlCl_3(s)$.

Asked $\quad$ Give two mole ratios involving Al and Cl_2.
Given $\quad 2Al(s) + 3Cl_2(g) \rightarrow 2AlCl_3(s)$
Answer

$$\dfrac{2 \text{ mol Al}}{3 \text{ mol } Cl_2} \text{ and } \dfrac{3 \text{ mol } Cl_2}{2 \text{ mol Al}}$$

Section 8.1: Analyzing a Chemical Reaction

Mole-to-mole problems with mole ratios

Calculating the moles of a product

Mole ratios are used to calculate the amount of product from a given amount of reactant. The mole ratio provides a one-step calculation that directly leads to the answer. The example below applies the mole ratio to find the amount of CO_2 produced from combustion of 20 moles of propane.

How many moles of CO_2 are produced from burning 20 moles of methane?

balanced reaction $\quad CH_4 + 2O_2 \longrightarrow CO_2 + 2H_2O$

$$\frac{20 \text{ mol } CH_4}{1} \times \underbrace{\frac{1 \text{ mol } CO_2}{1 \text{ mol } CH_4}}_{\text{mole ratio}} = 20 \text{ mol } CO_2$$

Burning 20 moles of CH_4 produces 20 moles of CO_2.

Calculating moles of a reactant

A second common use of mole ratios is to calculate the amount of reactants needed to completely react with a given quantity of another reactant. For example it takes 235 moles of oxygen, O_2, to completely react with 47 moles of propane, C_3H_8.

How many moles of oxygen are required to burn 47 moles of methane?

balanced reaction $\quad CH_4 + 2O_2 \longrightarrow CO_2 + 2H_2O$

$$\frac{47 \text{ mol } CH_4}{1} \times \underbrace{\frac{2 \text{ mol } O_2}{1 \text{ mol } CH_4}}_{\text{mole ratio}} = 94 \text{ mol } O_2$$

Burning 47 moles of CH_4 requires 94 moles of oxygen, O_2.

Solved problem

Interactive simulation

Hematite, Fe_2O_3, has the same chemical composition as rust. It is produced from a synthesis reaction:

$$4Fe(s) + 3O_2(g) \rightarrow 2Fe_2O_3(s)$$

Once the iron has rusted, it is exposed to high pressures within the earth and becomes hematite. How many moles of hematite, Fe_2O_3 can be produced from 100.0 moles of oxygen gas?

Asked: Moles of hematite, Fe_2O_3 that can be produced from 100.0 moles of oxygen.

Given: 100.0 moles of O_2

Relationships:
$$\frac{3 \text{ mol Fe}_2}{2 \text{ mol Fe}_2O_3} \quad \text{or} \quad \frac{2 \text{ mol Fe}_2O_3}{3 \text{ mol } O_2}$$

Solve: $4Fe(s) + 3O_2(g) \rightarrow 2Fe_2O_3(s)$

$$\frac{100.0 \text{ moles } O_2}{1} \times \frac{2 \text{ mole Fe}_2O_3}{3 \text{ moles } O_2} = 66.66667 \xrightarrow{\text{round}} 66.67 \text{ moles Fe}_2O_3$$

Answer: 66.67 moles of Fe_2O_3 are formed from 100.0 moles of O_2.

Calculating quantities in grams

Balances measure in grams

If there were a device that directly measures moles, the mole ratios would be sufficient to do practical chemistry. However, no such device exists and therefore chemistry is mostly done in grams. To calculate chemistry problems we use the following three steps:

1. Use the molar masses of compounds to convert quantities from grams to moles.
2. Calculate reaction quantities in moles.
3. Convert calculated reaction quantities in moles back to grams.

Calculation pathway: grams to moles to grams

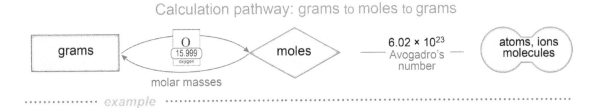

······· *example* ·······

How many grams of sodium sulfate are produced from 25.33 grams of sodium phosphate?

$$2Na_3PO_4 + 3CaSO_4 \longrightarrow Ca_3(PO_4)_2 + 3Na_2SO_4$$

1 calculate molar masses

Na_3PO_4	$CaSO_4$	$Ca_3(PO_4)_2$	Na_2SO_4
163.94 g/mol	136.14 g/mol	310.18 g/mol	142.04 g/mol

2 grams to moles

$$\frac{25.33 \text{ g } Na_3PO_4}{1} \times \frac{1 \text{ mol } Na_3PO_4}{163.94 \text{ g } Na_3PO_4} = 0.15450 \text{ mol } Na_3PO_4$$

(molar mass)

3 moles to moles

$$\frac{0.15450 \text{ mol } Na_3PO_4}{1} \times \frac{3 \text{ mol } Na_2SO_4}{2 \text{ mol } Na_3PO_4} = 0.23175 \text{ mol } Na_2SO_4$$

(mole ratio)

4 moles to grams

$$\frac{0.23176 \text{ mol } Na_2SO_4}{1} \times \frac{142.04 \text{ g } Na_2SO_4}{1 \text{ mol } Na_2SO_4} = \boxed{32.92 \text{ g } Na_2SO_4}$$

(molar mass)

The reaction produces 32.92 g of Na_2SO_4

Why moles are necessary

The diagram above shows that you cannot get directly from grams of one substance to grams of another substance! You must go through the unit of moles because the relationships in the balanced chemical equation are in moles.

Doing the calculation in one line

You can do the grams-to-moles-to-grams calculation in a single line. This is similar to using multiple conversion factors in dimensional analysis. The key is using the units of each ratio to choose which value goes in the numerator and which in the denominator. All units should cancel except for the one you want.

The same calculation in one line

$$\frac{25.33 \text{ g } Na_3PO_4}{1} \times \frac{1 \text{ mol } Na_3PO_4}{163.941 \text{ g } Na_3PO_4} \times \frac{3 \text{ mol } Na_2SO_4}{2 \text{ mol } Na_3PO_4} \times \frac{142.042 \text{ g } Na_2SO_4}{1 \text{ mol } Na_2SO_4} = \boxed{32.92 \text{ g } Na_2SO_4}$$

(molar mass)   (mole ratio)   (molar mass)

Section 8.1: Analyzing a Chemical Reaction

Solving gram-to-gram problems

Another gram-to-gram problem

Many stoichiometric problems follow the pattern of grams-to-moles-to-grams. The mole ratio can involve reactants, products, or a reactant and a product. The problem below is a good example of a grams-to-moles-to-grams problem. You should be able to follow the steps in both the three-line and single line solutions.

How many grams of oxygen are used to burn 100.0 grams of methane?

$$CH_4 + 2O_2 \xrightarrow{\text{combustion reaction}} CO_2 + 2H_2O$$

① calculate molar masses

CH_4	O_2	CO_2	H_2O
16.043 g/mol	31.998 g/mol	44.009 g/mol	18.015 g/mol

··

Three line solution

② grams to moles

$$\frac{100.0 \text{ g CH}_4}{1} \times \frac{1 \text{ mol CH}_4}{16.043 \text{ g CH}_4 \text{ (molar mass)}} = 6.2332 \text{ mol CH}_4$$

③ moles to moles

$$\frac{6.2332 \text{ mol CH}_4}{1} \times \frac{2 \text{ mol O}_2}{1 \text{ mol CH}_4 \text{ (mole ratio)}} = 12.466 \text{ mol O}_2$$

④ moles to grams

$$\frac{12.466 \text{ mol O}_2}{1} \times \frac{31.998 \text{ g O}_2}{1 \text{ mol O}_2 \text{ (molar mass)}} = \boxed{398.9 \text{ g O}_2}$$

Burning 100 g of methane uses 398.9 grams of oxygen!

··

One line solution

$$\frac{100.0 \text{ g CH}_4}{1} \times \underbrace{\frac{1 \text{ mol CH}_4}{16.043 \text{ g CH}_4}}_{\text{molar mass}} \times \underbrace{\frac{2 \text{ mol O}_2}{1 \text{ mol CH}_4}}_{\text{mole ratio}} \times \underbrace{\frac{31.998 \text{ g O}_2}{1 \text{ mol O}_2}}_{\text{molar mass}} = 398.9 \text{ g O}_2$$

Solved problem

When pentane in gasoline is burned the products are carbon dioxide and water. How much of the greenhouse gas, CO_2, is produced when 156.74 g of pentane are burned?

$$C_5H_{12}(l) + 8O_2(g) \rightarrow 6H_2O(g) + 5CO_2(g)$$

Asked How many grams of CO_2 are produced?

Given 156.74 g of pentane

Relationships

$$\frac{5 \text{ mol CO}_2}{1 \text{ mol C}_5H_{12}} \text{ or } \frac{1 \text{ mol C}_5H_{12}}{5 \text{ mol CO}_2}$$

Solve

$$\frac{156.74 \text{ g C}_5H_{12}}{1} \times \frac{1 \text{ mol C}_5H_{12}}{72.149 \text{ g C}_5H_{12}} \times \frac{5 \text{ mol CO}_2}{1 \text{ mol C}_5H_{12}} \times \frac{44.009 \text{ g CO}_2}{1 \text{ mol CO}_2} =$$

$$= 478.03674 \text{ round } 478.04 \text{ g CO}_2$$

Answer 156.74 grams of C_5H_{12} produce 478.04 grams of CO_2.

Calculating volumetric quantities

Calculations with volumes of gas

When a reaction involves gases it is often convenient to work with volumes instead of grams. Volumes of gas are easy to measure while masses are not. Volume can be related to moles at standard temperature and pressure (STP) using the Standard Molar Volume, of 22.4 L per mole. The calculation method is similar to the one we used for grams except it goes from volume to moles and back to volume.

From liters at STP to moles and back to liters

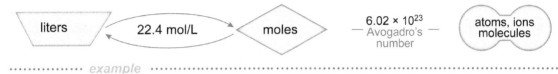

example

How many liters of carbon dioxide are produced from burning 15.3 liters of oxygen at STP?

$$CH_4(g) + 2O_2(g) \xrightarrow{combustion\ reaction} CO_2(g) + 2H_2O(g)$$

Three line solution

② liters to moles $\dfrac{15.3\ L\ O_2}{1} \times \dfrac{1\ mol\ O_2}{22.4\ L\ O_2}$ (molar volume) $= 0.6830\ mol\ O_2$

③ moles to moles $\dfrac{0.6830\ mol\ O_2}{1} \times \dfrac{1\ mol\ CO_2}{2\ mol\ O_2}$ (mole ratio) $= 0.3415\ mol\ CO_2$

④ moles to grams $\dfrac{0.3415\ mol\ CO_2}{1} \times \dfrac{22.4\ L\ CO_2}{1\ mol\ CO_2}$ (molar volume) $= \mathbf{7.65\ L\ CO_2}$

7.65 liters of carbon dioxide.

One line solution

$$\dfrac{15.3\ L\ O_2}{1} \times \underbrace{\dfrac{1\ mol\ O_2}{22.4\ L\ O_2}}_{molar\ volume} \times \underbrace{\dfrac{1\ mol\ CO_2}{2\ mol\ O_2}}_{mole\ ratio} \times \underbrace{\dfrac{22.4\ L\ CO_2}{1\ mol\ CO_2}}_{molar\ volume} = 7.65\ L\ CO_2$$

Converting from grams to liters

A quantity of gas in grams can be converted to liters using the standard molar volume. First convert grams to moles then convert moles to liters using 22.4 L/mol. For example, 100.0 grams of CO_2 are equal to 50.90 liters at STP.

From grams to liters at STP

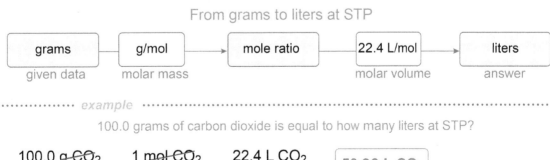

example

100.0 grams of carbon dioxide is equal to how many liters at STP?

$$\dfrac{100.0\ g\ CO_2}{1} \times \underbrace{\dfrac{1\ mol\ CO_2}{44.009\ g\ CO_2}}_{molar\ mass} \times \underbrace{\dfrac{22.4\ L\ CO_2}{1\ mol\ CO_2}}_{molar\ volume} = \boxed{50.90\ L\ CO_2}$$

100.0 g CO_2 = 50.90 L at STP

Section 8.1: Analyzing a Chemical Reaction

Grams to particles

Convert to atoms

Some chemistry problems ask for the number of molecules, ions, or atoms. This calculation is easier than the ones on the last few pages because there is one less molar mass to calculate. Instead of the molar mass, you use Avogadro's number, 6.02×10^{23} to calculate the number of particles from the number of moles.

Converting from grams, moles, or liters to particles

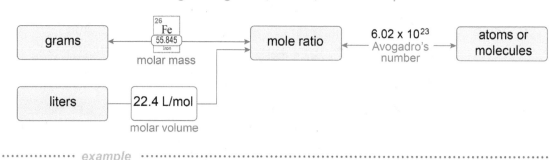

example

Calculate the number of methane molecules in 12.5 liters of methane at STP.

$$\frac{12.5 \text{ L CH}_4}{1} \times \frac{1 \text{ mol CH}_4}{22.4 \text{ L CH}_4} \times \frac{6.02 \times 10^{23} \text{ molecules CH}_4}{1 \text{ mol CH}_4} = 3.3595 \times 10^{22} \text{ molecules CH}_4$$

There are 3.36×10^{22} molecules of methane in 12.5 L at STP.

Dissociation of ions

For some types of aqueous reactions you may need to calculate the number of ions from a quantity in grams or moles. The subscript on the ion in the chemical formula tells you how many ions are produced from one mole of the compound. For example, iron(II) nitrate has the chemical formula $Fe(NO_3)_2$. That means each mole of $Fe(NO_3)_2$ contributes two moles of nitrate ions, NO_3^-.

Solved Problem

Silver is produced when iron reacts with silver nitrate. How many atoms of silver are produced by 275.0 g of Fe(s)?

$$Fe(s) + 2AgNO_3 (aq) \rightarrow Fe(NO_3)_2 (s) + 2Ag(s)$$

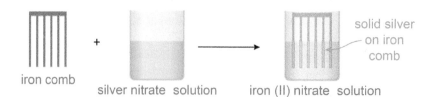

Asked How many atoms of Ag are formed?
Given 275.0 g of Fe
Relationships Fe: 55.845 g/mol, $Fe(s) + 2AgNO_3 (aq) \rightarrow Fe(NO_3)_2 (s) + 2Ag(s)$
Solve $\frac{275.0 \text{ grams Fe}}{1} \times \frac{1 \text{ mole Fe}}{55.845 \text{ grams Fe}} \times \frac{2 \text{ moles Ag}}{1 \text{ mole Fe}} \times \frac{6.02 \times 10^{23} \text{ atoms Ag}}{1 \text{ mole Ag}} =$

$= 5.92891 \times 10^{24}$ round 5.93×10^{24} atoms Ag

Answer 5.93×10^{24} atoms of Ag are produced from 275.0 g of Fe.

Solving problems with mixed units

The chemistry of breathing

One of the difficulties of space travel is getting rid of exhaled carbon dioxide. Normal outdoor air contains 0.04% CO_2 by volume. The indoor limit is 0.5% and at 4% the concentration of CO_2 becomes fatal since it suppresses breathing. Air contains 21% oxygen. When you exhale the O_2 content is reduced to about 16%. The average human exhales about 12 L per minute, of which 5% (0.6 L) of oxygen is used and exhaled as 0.6 liters of CO_2. This adds up to 36 liters of CO_2 exhaled per person every hour.

CO_2 in spacecraft

Spacecraft absorb carbon dioxide with solid lithium hydroxide, LiOH, which reacts with gaseous CO_2 to produce solid lithum carbonate. During the Apollo 13 space mission in 1970 an explosion enroute to the Moon forced the crew to evacuate their main cabin due to high CO_2. They had to quickly improvise a new LiOH absorber or they would not have survived.

How many grams of LiOH does it take to absorb CO_2 from three people over a period of 72 hours? Each person exhales 36 liters of CO_2 per hour.

$$2LiOH_{(s)} + CO_{2(g)} \longrightarrow Li_2CO_{3(s)} + H_2O_{(l)}$$

LiOH canister

Solved problem

Interactive simulation

How many grams of LiOH does it take to absorb CO_2 from 3 people over 72 hours at STP? Each person exhales 36 L of CO_2 per hour.

Asked   How many grams of LiOH are needed
Given   36 liters of CO_2 per hour, per person, 3 people, 72 hours
Relationships $2LiOH(s) + CO_2(g) \rightarrow Li_2CO_3(s) + H_2O(l)$, 22.4 L/mol
Solve   Calculate the total liters of CO_2 produced by the three person crew.

$$\frac{72 \text{ hours}}{1} \times \frac{36 \text{ L } CO_2}{1 \text{ hour}} = 2{,}592 \text{ L } CO_2 \times 3 \text{ people} = 7{,}776 \text{ L } CO_2$$

Calculate the total moles of CO_2 produced.

$$\frac{7{,}776 \text{ L } CO_2}{1} \times \frac{1 \text{ mol}}{22.4 \text{ L}} = 347.1 \text{ mol } CO_2$$

Use the mole ratio to calculate the total moles of LiOH required.

$$\frac{347.1 \text{ mol } CO_2}{1} \times \frac{2 \text{ mol LiOH}}{1 \text{ mol } CO_2} = 694.2 \text{ mol LiOH}$$

Calculate the mass in grams of LiOH from the molar mass.

$$\frac{694.2 \text{ mol LiOH}}{1} \times \frac{23.9479 \text{ g LiOH}}{1 \text{ mol LiOH}} = 16{,}624 \text{ g LiOH}$$

Answer   It takes 16.62 kilograms of LiOH to absorb the CO_2

Section 8.1: Analyzing a Chemical Reaction

Completion problems

Reaction completion problems

Balanced equations are often not given and they have to be derived from context of the problem. The problem below has many clues that will help you balance the equation. Look for key words such as **reacted, produced, makes** and **yields.** These help you identify the reactants and products. Other key phrases such as "decomposes to form" or "undergoes combustion" indicate the type of reaction.

Solved problem

What mass of solid sodium chloride is formed from the synthesis of excess solid sodium and 56.98 L of chlorine gas at STP?

Asked How many grams of sodium chloride are created?
Given 56.98 L of chlorine gas at STP
Solve Start by writing the equation...
$$Na(s) + Cl_2(g) \rightarrow NaCl(s)$$
...and then balance it:
$$2Na(s) + Cl_2(g) \rightarrow 2NaCl(s)$$

Next, use the balanced equation in a stoichiometry problem.

$$\frac{56.98 \text{ L Cl}_2}{1} \times \frac{1 \text{ mol Cl}_2}{22.4 \text{ L Cl}_2} \times \frac{2 \text{ mol NaCl}}{1 \text{ mol Cl}_2} \times \frac{58.443 \text{ g NaCl}}{1 \text{ mol NaCl}} =$$

$$= 297.32876 \text{ round } 297.3 \text{ g NaCl}$$

Answer 56.98 L of chlorine gas produce 297.3 g of sodium chloride.

Missing reactants

Sometimes reactants or products aren't mentioned. The law of conservation of matter states that matter cannot disappear. In a statement like, "calcium oxide is produced from the decomposition of calcium carbonate," carbon is not mentioned on the product side of the reaction but it must be there.

$$CaCO_3 \rightarrow CaO$$

Look for common substances. In this case, one carbon and two oxygens are needed on the product side. Carbon dioxide is a common substance and completes the reaction.

$$CaCO_3 \rightarrow CaO + CO_2$$

Solved problem

How many grams of silver chloride are produced when 135.23 g of silver nitrate react with excess calcium chloride?

Asked How many grams of silver chloride are produced?
Given 135.23 g of silver nitrate
Solve Write the reaction...
$$AgNO_3 + CaCl_2 \rightarrow AgCl + Ca(NO_3)_2$$
...and balance it:
$$2AgNO_3 + CaCl_2 \rightarrow 2AgCl + Ca(NO_3)_2$$

Use the balanced reaction in a stoichiometry problem:

$$\frac{135.23 \text{ g AgNO}_3}{1} \times \frac{1 \text{ mol AgNO}_3}{169.874 \text{ g AgNO}_3} \times \frac{2 \text{ mol AgCl}}{2 \text{ mol AgNO}_3} \times \frac{143.323 \text{ g AgCl}}{1 \text{ mol AgCl}} =$$

$$= 114.093794 \text{ round } 114.09 \text{ g AgCl}$$

Answer 135.23 g of silver nitrate produce 114.09 g of silver chloride.

Section 1 review

Chapter 8

Stoichiometry is the use of a balanced equation to calculate the amounts of reactants and products in a chemical reaction. A balanced equation gives a "recipe" for a reaction, telling us the relative amounts of each substance needed either in molecules or, more usefully, in moles. The coefficients in a balanced equation provide the ratios of moles in a reaction. These mole ratios can be used to determine the number of moles of other substances in a reaction when one is known. Additionally, the formula mass can be used to convert from moles to grams for a given substance. Avogadro's number is used to convert between moles and the number of particles. Finally, the Standard Molar Volume of 22.4 liters per mole can be used to convert between volumes of gases, moles, and grams.

Vocabulary words: stoichiometry, mole ratio

Key relationships

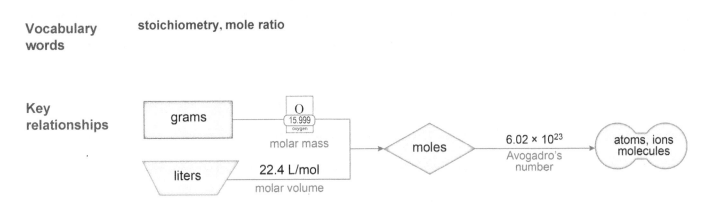

Review problems and questions

1. Calculate the moles of oxygen needed to react with 2.5 moles of iron according to the reaction: $4Fe(s) + 3O_2(g) \rightarrow 2Fe_2O_3(s)$.

2. Use the same reaction from Question #1 to calculate the mass of iron(III) oxide produced when 0.5 moles of iron react with excess oxygen.

3. Use the same reaction from Question #1 to calculate the mass of iron(III) oxide produced when 211.2 g of iron react with excess oxygen.

4. A serving of vanilla ice cream has 15 g of sugar. How many molecules of oxygen does your body need to burn 15 g of sugar ($C_6H_{12}O_6$) during respiration? $C_6H_{12}O_6(s) + 6O_2(g) \rightarrow 6CO_2(g) + 6H_2O(g)$

5. If ammonia (NH_3) decomposes to its component gases at STP, how many liters of hydrogen gas can you expect to form from 3.5 mol NH_3? $2NH_3(g) \rightarrow N_2(g) + 3H_2(g)$

Section 8.1: Analyzing a Chemical Reaction

8.2 - Percent Yield

Nothing is perfect, not even in chemistry! Most reactions do not convert all the reactants into products. The percent yield is a way to compare how much product a reaction actually produces compared to how much was theoretically possible.

Calculating percent yield

Percent yield

Every fall the cranberry farmers in New England flood their fields and float the cranberries off the bushes. The seasonal harvest has become a tourist attraction because the surfaces of the cranberry bogs become bright red with floating berries. Of course not all the berries are ripe! The growers must sort out the unripe and damaged berries. In the example, 160 kg of cranberries are harvested but only 138 kg are ripe. The percent yield is 86.2%.

Percent yield
The percent yield of a process is the actual amount of product divided by the maximum amount of product theoretically possible.

The New England cranberry harvest

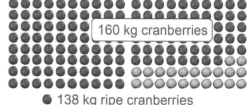
● 138 kg ripe cranberries
○ 22 kg unripe cranberries

$$\text{percent yield} = \frac{\text{number of ripe cranberries}}{\text{total number of cranberries harvested}} = \frac{138 \text{ kg}}{160 \text{ kg}} = 86.2\%$$

Theoretical and actual yields

The **percent yield** is the actual yield of a process divided by the theoretically expected yield. In the cranberry example the theoretically perfect yield is 160. kilograms of ripe cranberries. The actual yield was 138 kilograms making the percent yield 86.2 percent. In a chemical reaction, the **theoretical yield** is the amount of the products formed if the maximum possible amount of the reactants are converted to products. The **actual yield** is the actual amount of products produced by the reaction.

Why percent yield may not be 100%

Very few chemical reactions convert 100% of the reactants to products. Consider the reaction below to decompose calcium carbonate, $CaCO_3$, into carbon dioxide and quicklime, or calcium oxide, CaO. The theoretically perfect reaction would convert all of the $CaCO_3$. A more realistic case is that not all of $CaCO_3$ reacts and some is left unconverted. The percent yield of 94% shows that 6% of the $CaCO_3$ did not react. Note: 94% yield does not mean 94 grams of CaO were produced from 100 grams of $CaCO_3$! The maximum theoretical yield was 56.029 g CaO.

Percent yield

$$CaCO_{3(s)} \longrightarrow CO_{2(g)} + CaO_{(s)}$$
calcium carbonate   carbon dioxide   calcium oxide

theoretical yield	56.029 g
actual yield	52.667 g

$$\text{percent yield} = \frac{\text{actual yield CaO}}{\text{theoretical yield CaO}} = \frac{52.667 \text{ g}}{56.029 \text{ g}} \times 100 = 94.0\%$$

Measuring percent yield

In the lab

A typical measurement of percent yield looks like the diagram below. The reaction starts with 10.000 grams of reactants. After the reaction there are 4.870 grams of CaO.

actual yield

Analyzing the results

The balanced equation is used to calculate the theoretical yield. For this experiment the 10.000 grams of $CaCO_3$ is 0.0999 moles. The mole ratio between $CaCO_3$ and CaO is 1/1 therefore a perfect reaction would produce 0.0999 moles of CaO, which is 5.603 grams.

Analysis of percent yield

$$CaCO_3 (s) \longrightarrow CO_2 (g) + CaO (s)$$
calcium carbonate   carbon dioxide   calcium oxide

step 1: calculate theoretical yield

$$\frac{10.000 \text{ g } CaCO_3}{1} \times \frac{1 \text{ mol } CaCO_3}{100.086 \text{ g } CaCO_3} \times \frac{1 \text{ mol } CaO}{1 \text{ mol } CaCO_3} \times \frac{56.077 \text{ g } CaO}{1 \text{ mol } CaO} = \boxed{5.60283 \text{ g CaO}}$$

molar mass   mole ratio   molar mass   theoretical yield is 5.6028 g CaO

step 2: calculate the percent yield

$$\text{percent yield} = \frac{\text{actual yield CaO}}{\text{theoretical yield CaO}} = \frac{4.870 \text{ g}}{5.6028 \text{ g}} \times 100 = 86.9\%$$

Reactions do not usually yield 100%

Real reactions rarely convert 100% of the reactants to the desired products. Common reasons are impurities in the samples, small surface area, and insufficient time to react. Many chemical reactions also produce side reactions that produce different products than the ones you want. Chemists often "stop" a reaction to produce the most amount of the desired product and the least amount of undesired other products. Stopping a reaction always results in less product than is theoretically possible.

Causes of incomplete yields

To get the highest percent yield, a chemist might repeat the experiment at different temperatures, and/or in a vacuum. The vacuum causes water particles to move from the sample into the dryer air. This helps drive the reaction toward producing more CO_2 and therefore more CaO as well. Some other steps for increasing percent yield might be:

1. heating longer to complete more of the reaction;

2. drying the reactants completely, some reactant is H_2O and not $CaCO_3$;

3. drying the product completely - CaO absorbs water from the atmosphere;

4. care in the procedure, experimental error often reduces yield.

Solving percent yield problems

Applying percent yield

Most percent yield problems have a similar structure. Starting from the balanced chemical equation you first determine the theoretical yield from the amount of reactants you start with. Once you have the theoretical yield, you calculate the percent yield by dividing the actual yield of product produced by the theoretical yield.

Calculating percent yield

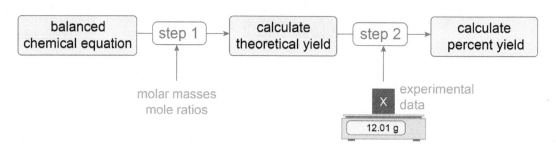

Solved problem

When you burn 2.0 L of propane gas (C_3H_8) at STP during an experiment, you find 5.6 L of carbon dioxide has been produced. What percentage of propane did not burn? Use the reaction:

$$C_3H_8(g) + 5O_2(g) \rightarrow 3CO_2(g) + 4H_2O(g)$$

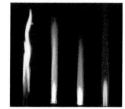

Given: 2.0 liters propane gas and 5.6 liters carbon dioxide gas

Relationships: $C_3H_8(g) + 5O_2(g) \rightarrow 3CO_2(g) + 4H_2O(g)$
and the standard molar volume 22.4 mol/L

Solve: From the standard molar volume the mole ratio of $C_3H_8(g)$ to $CO_2(g)$ is the same as the ratio of gas volumes.

$$\frac{2.0 \text{ L } C_3H_8}{1} \times \frac{1 \text{ mol } C_3H_8}{22.4 \text{ L}} \times \frac{3 \text{ mol } CO_2}{1 \text{ mol } C_3H_8} \times \frac{22.4 \text{ L}}{1 \text{ mol } CO_2} = 6.0 \text{ L } CO_2$$

That means the theoretical yield of $CO_2(g)$ should be 6.0 liters. The actual yield is 5.6 liters. Therefore the percent yield of the reaction is 93.3%.

$$\frac{\text{actual yield}}{\text{theoretical yield}} = \frac{5.6 \text{ L}}{6.0 \text{ L}} \times 100 = 93.3\%$$

Answer: The reaction produced 93% yield of CO_2 therefore 7% of the propane did not burn.

Two reactants never react completely

As the reactants in any two-reactant system get used up, the concentration of reactant molecules goes down. When few molecules of reactants are left, the chance of them randomly bumping into each other to react gets very low. The rate of the reaction slows down to zero as the percent yield approaches 100%. Since the rate of the reaction is going toward zero the reaction never completes to the point where 100% of both reactants are used up. In the next section we will see that an excess of one reactant is required to get complete reaction of the other reactant.

Two-step percent yield problems

Multiplying percent yield

Many reactions proceed in two or more steps. Each step has a percent yield and the overall reaction also has a percent yield. The percent yield for the overall reaction is the product of the percent yields for each step. For example, if the first step of a two-step reaction has a percent yield of 50% and the second step also has a percent yield of 50%, then the overall percent yield is 25% (50% × 50%).

Solved problem

Titanium is a very light, very strong metal used in applications for which the strength to weight advantage is worth the high cost. Titanium ore contains TiO_2 which is refined in two steps to titanium metal.
  Step 1: $TiO_2 + 2Cl_2 + 2C \rightarrow TiCl_4 + 2CO$
  Step 2: $TiCl_4 + 2Mg \rightarrow Ti + 2MgCl_2$
Calculate the percent yield of $TiCl_4$ when 250.0 grams of titanium(IV) oxide, TiO_2, produce 400.0 grams of $TiCl_4$.

titanium replacement knee joint

Asked percent yield of $TiCl_4$
Given 400.0g $TiCl_4$ and 250.0 g TiO_2
Relationships $TiO_2 + 2Cl_2 + 2C \rightarrow TiCl_4 + 2CO$
Solve First calculate the molar masses: TiO_2 has a molar mass of 79.865 g/mol and $TiCl_4$ has a molar mass of 189.679 g/mol.

Now calculate theoretical yield from the mole ratio and the molar masses:

$$\frac{250.0 \text{ g } TiO_2}{1} \times \frac{1 \text{ mol } TiO_2}{79.865 \text{ g } TiO_2} \times \frac{1 \text{ mol } TiCl_4}{1 \text{ mol } TiO_2} \times \frac{189.679 \text{ g } TiCl_4}{1 \text{ mol } TiCl_4} = 593.7 \text{ g } TiCl_4$$

Next, calculate percent yield:

$$\frac{\text{actual yield}}{\text{theoretical yield}} \times 100 = \frac{400.0 \text{ g } TiCl_4}{593.7 \text{ g } TiCl_4} \times 100 = 67.4\%$$

Answer The percent yield is 67%.

Solved problem

In the second step of the reaction $TiCl_4$ vapor passes over molten magnesium at a very high temperature (1,000 °C). What mass of Ti metal is produced if the percent yield of the second step of the reaction is 80%?

Asked grams of Ti metal produced
Given percent yield of second step is 80%
Relationships $TiCl_4(g) + 2Mg(l) \rightarrow Ti(s) + 2MgCl_2(s)$
Solve Use the actual yield of $TiCl_4$ given in the previous problem to calculate the mass of Ti produced in the second step, then apply the percent yield:

$$\frac{400.0 \text{ g } TiCl_4}{1} \times \frac{1 \text{ mol } TiCl_4}{189.679 \text{ g } TiCl_4} \times \frac{1 \text{ mol } Ti}{1 \text{ mol } TiCl_4} \times \frac{47.867 \text{ g } Ti}{1 \text{ mol } Ti} =$$

$$\frac{19{,}146.8 \text{ g Ti}}{189.679} = 100.94317 \text{ g Ti} \times 80\% = 80.75 \text{ g Ti metal}$$

Answer This process produces 80.75 grams of Ti metal starting with 250.0 g of TiO_2.

Section 8.2: Percent Yield

Chapter 8

Section 2 review

The yield of a chemical reaction is the amount of a desired product produced by the reaction. In a perfect reaction, the maximum possible amount of reactants complete the reaction to become the maximum possible amount of products. The theoretical yield is the maximum amount of product that could be produced from a given amount of reactants. The theoretical yield is calculated from the quantities of reactants using the balanced chemical equation.

In a real chemical reaction the actual amount of product produced is usually less than the theoretical yield. There are many reasons for this, impurities in the chemicals, side reactions, and product lost during purifications. The percent yield is the ratio of the actual yield divided by the theoretical yield. Percent yield is typically determined through experiment and can be affected by many factors including temperature, time, concentration, and other factors.

Vocabulary words: percent yield, theoretical yield, actual yield

Key relationships:
- Percent Yield = $\frac{\text{actual yield}}{\text{theoretical yield}} \times 100$

Review problems and questions

15.0 grams pure iron powder

17.6 grams iron oxide, Fe_2O_3

1. A student left 15.0 g of pure iron exposed to the air to form iron(III) oxide by the reaction: $4Fe + 3O_2 \rightarrow 2Fe_2O_3$. She returned several weeks later to see 17.6 g of iron(III) oxide had formed. What is her percent yield?

2. Why didn't the student get a 100% yield? Identify at least two possibilities.

3. How many grams of iron(III) oxide would need to form for the student to measure a 95% yield?

4. Assume the reaction had 100% yield. Explain why the mass of iron(III) oxide formed can be greater than the mass of the pure iron reactant you started with. Is it possible to have a percent yield greater than 100%?

8.3 - Limiting Reactants

the number of cheeseburgers is determined by the limiting reactant

13 buns | 12 patties | 2 slices of cheese (limiting reactant) | 100 pickles | 2 cheeseburgers

Suppose you had the ingredients above and were making cheeseburgers for a party. How many can you make? For each cheeseburger you need the following ingredients.

1 bun + 1 hamburger patty + 1 slice of cheese + 2 pickles = 1 cheeseburger

You have enough hamburger patties to make 12 burgers but you only have two slices of cheese. You have plenty of everything else but one ingredient - the cheese - limits the number of cheese burgers you can make. A similar situation occurs with most chemical reactions.

What is a limiting reactant?

Limiting reactants

When performing reactions in the laboratory it is common to completely use up one reactant to make products while other reactants have some left over. A reactant that is used up completely is called the **limiting reactant**. This reactant is "limiting" because it puts a limit on the amount of product that can be formed. When you completely run out of one reactant the reaction stops. No more products can form no matter how much of the other reactants are left over. Any reactant that is left over after a reaction stops is called the **excess reactant**.

Interactive simulation

The chemical that creates the least amount of product limits a chemical reaction in the same way. In the interactive simulation titled Limiting Reactant, compare a reaction that has the "perfect" amount of ingredients (reactants) with a reaction that has a limiting reactant.

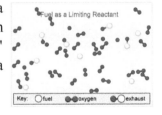

Limiting reactants are common

When a fuel burns in air the molecules of fuel react when they bump into a molecule of oxygen. Suppose there was just enough oxygen to react with every molecule of fuel - a perfect mixture. As both fuel and oxygen get used up, the chance of a fuel molecule bumping into an unreacted oxygen molecule gets slimmer and slimmer. The result is that after any finite time, some fuel is left unburned. The solution is to always have an excess of oxygen. That way every fuel molecule, down to the last, has a good chance to react with an oxygen molecule. Virtually all chemical reactions are designed to have an excess of one reactant to ensure complete reaction of the other reactant.

Identifying the limiting reactant

Chemical example

Gasoline is a mixture of many hydrocarbons with 4 - 12 carbon atoms. Octane, C_8H_{18}, is a good average molecule to represent gasoline even though typical gasoline contains a mixture of fifty or more different molecules. The balanced chemical equation for the combustion of octane is written below, along with the molar mass of each compound in the reaction.

Combustion of octane

$$2C_8H_{18(l)} + 25O_{2(g)} \longrightarrow 16CO_{2(g)} + 18H_2O_{(g)}$$

octane   oxygen   carbon dioxide   water
114.2302 g/mol   31.998 g/mol   44.009 g/mol   18.0148 g/mol

What is limiting?

Consider a reaction between ten grams of octane and 19.7 liters of oxygen at STP. Ten grams of octane are about 7.5 mL. What is the limiting reactant: octane or oxygen?

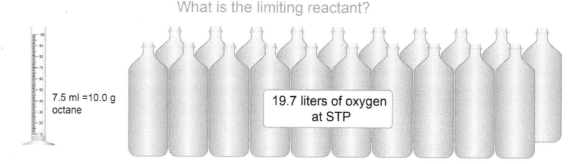

What is the limiting reactant?

7.5 ml = 10.0 g octane

19.7 liters of oxygen at STP

The practical air/fuel mixture

To engineers who design car engines this is a crucial question. To ensure that 100% of the fuel burns, there must be excess oxygen. The engine control computer adjusts the amount of fuel to ensure that fuel (not oxygen) is always the limiting reactant, unlike this problem!

(1) Calculate the available moles of each reactant

$$\frac{10.0 \text{ g } C_8H_{18}}{1} \times \frac{1 \text{ mol } C_8H_{18}}{114.2302 \text{ g } C_8H_{18}} = 0.087543 \text{ mol } C_8H_{18}$$

molar mass

$$\frac{19.7 \text{ L } O_2}{1} \times \frac{1 \text{ mol } O_2}{22.4 \text{ L } O_2} = 0.87946 \text{ mol } O_2$$

molar volume

(2) Choose octane as a reference and calculate the required moles of oxygen

$$\frac{0.087543 \text{ mol } C_8H_{18}}{1} \times \frac{25 \text{ mol } O_2}{2 \text{ mol } C_8H_{18}} = 1.0942875 \text{ mol } O_2$$

mole ratio

(3) Compare the moles required to the moles available to determine the limiting reactant

The balanced reaction requires 1.0942875 moles of oxygen to react with 0.087543 moles of octane. Only 0.87946 moles of oxygen are available therefore, oxygen is the limiting reactant.

Determining the limiting reactant

Finding the limiting reactant

The steps to determine a limiting reactant are always the same.

1. Find out how many moles of each reactant you have.

2. Pick any one reactant and calculate the required moles of the other reactants using the mole ratios from the balanced equation.

3. The limiting reactant is the one with the lowest ratio of moles available to moles required

Note: the reactant with the least mass may not be the limiting reactant! The limiting reactant is determined by moles, not by mass.

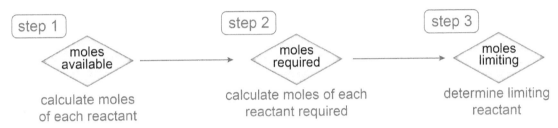

Thermite welding

A clever technique for welding iron rails is to use a chemical reaction called thermite. In the thermite reaction iron(III) oxide reacts with aluminum and gives off a great quantity of heat, enough to melt steel. Two railroad rails can be butted against each other under pressure and when the reaction occurs the heat melts the iron and fuses the rails together.

$$Fe_2O_3(s) + 2Al(s) \rightarrow 2Fe(s) + Al_2O_3(s)$$

Welding tracks with the thermite reaction

Solved problem

When 550.36 g of Fe_2O_3 react with 361.67 g of Al, which will be used up first?

Asked Which reactant is used up first?

Given 550.36 g Fe_2O_3 and 361.67 g Al

Relationships $Fe_2O_3(s) + 2Al(s) \rightarrow 2Fe(s) + Al_2O_3(s)$

Solve Calculate the available moles of each reactant.

$$\frac{550.36 \text{ g } Fe_2O_3}{1} \times \frac{1 \text{ mol } Fe_2O_3}{159.687 \text{ g } Fe_2O_3} = 3.44649 \text{ mol } Fe_2O_3$$

$$\frac{361.67 \text{ g Al}}{1} \times \frac{1 \text{ mol Al}}{26.982 \text{ g Al}} = 13.40412 \text{ mol Al}$$

$$\frac{3.44649 \text{ mol } Fe_2O_3}{1} \times \frac{2 \text{ mol Al}}{1 \text{ mol } Fe_2O_3} = 6.89298 \text{ round} \rightarrow 6.8930 \text{ mol Al}$$

Answer The Fe_2O_3 gets used up first because 6.8930 moles of Al is needed to react with 550.36 g of Fe_2O_3 and there are 13.40412 moles of Al available.

Solving limiting reactant problems

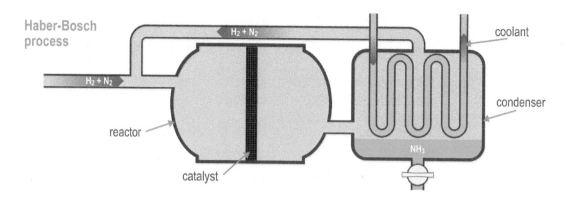

Creating ammonia, NH₃

Crop fertilizers based on ammonia are responsible for approximately one third of the Earth's agricultural output per person. Without synthetic fertilizers the current amount of farmland could not support the words population because nitrogen-fixing soil bacteria are not fast enough to keep pace with agriculture. Developed in 1913, the Haber-Bosch process is an industrial technique for producing ammonia from readily available nitrogen from the air and hydrogen gas.

$$N_2(g) + 3H_2(g) \rightarrow 2NH_3(g)$$

Safety and efficiency

The Haber-Bosch process operates at high pressure and high temperature. Nitrogen gas is plentiful, cheap, and safe. Hydrogen gas is expensive and explosive. To ensure that all the hydrogen is converted into ammonia, the process operates with excess nitrogen. With enough nitrogen, virtually 100% of the hydrogen is converted to ammonia.

Solved problem

A Haber-Bosch ammonia reactor takes in 1 kg of H₂ per second. Calculate the quantity of nitrogen, N₂, gas required to be twice the required amount.

Asked: Quantity of N₂ to be 200%
Given: 1 kg H₂ per second
Relationships: $N_2(g) + 3H_2(g) \rightarrow 2NH_3(g)$
Solve: First, calulate the moles of hydrogen.

$$\frac{1000.0 \text{ g H}_2}{1} \times \frac{1 \text{ mol H}_2}{2.0158 \text{ g H}_2} = 496.08 \text{ mol H}_2$$

Next, calculate the moles required and multiply by 2.

$$\frac{496.08 \text{ mol H}_2}{1} \times \frac{1 \text{ mol N}_2}{3 \text{ mol H}_2} \times 2 = 330.7 \text{ mol N}_2$$

Finally, convert moles of N₂ to grams of N₂.

$$\frac{330.7 \text{ mol N}_2}{1} \times \frac{28.014 \text{ g N}_2}{1 \text{ mol N}_2} = 9{,}265 \text{ g N}_2$$

Answer: The reactor takes in 9.265 kg of N₂ per 1 kg of H₂ to ensure 100% excess N₂.

Section 3 review

Chapter 8

A chemical reaction will stop if there is not enough of a reactant present. Real reactions typically include more than one reactant. There are usually different quantities of each reactant. One reactant may get "used up" by the reaction before the other reactants. In this situation the amount of product that is formed by a reaction is "limited" by the reactant that gets used up first.

The reactant that is totally consumed first is known as the limiting reactant. Any "unused" reactant that remains is known as the excess reactant. Limiting reactants are important because they are used to determine the theoretical yield of a reaction. In practical chemistry, there is always an excess of one or more reactants, usually the least expensive or most abundant ones! Having an excess of inexpensive or common reactants ensures that all of the more expensive or scarce reactants are converted into products.

Finding the limiting reactant involves the previously explored concepts of converting grams to moles using the formula mass and the balanced equation. The limiting reactant is identified from the mole ratios from the balanced equation compared to the moles of actual reactants that are given.

Vocabulary words   limiting reactant, excess reactant

Review problems and questions

1. Why does the limiting reactant determine how much product can form from a reaction?

2. Consider the reaction, $Mg(s) + 2HCl(aq) \rightarrow MgCl_2(s) + H_2(g)$. If you have 5.00 g each of Mg and HCl, which substance is the limiting reactant?

3. Which substance is the excess reactant?

4. If the reaction proceeded with the amounts given in Question 2 and only 0.189 mol hydrogen gas was produced, what is the percent yield?

5. The information below was derived from an Interactive Equation. Based on the information, which reactant would be left over after the reaction if you started with equal amounts of moles of both reactants? Explain your answer.

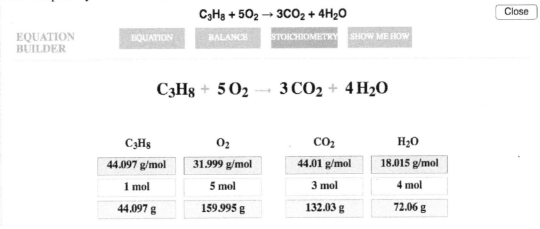

Section 8.3: Limiting Reactants

Engineering and Green Chemistry

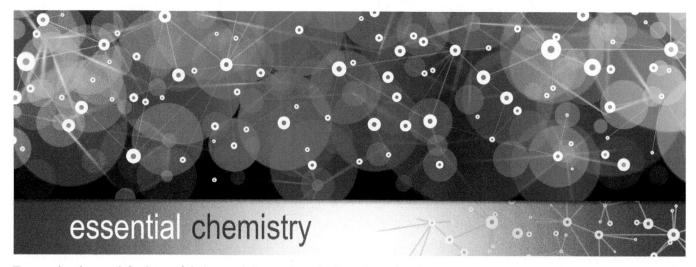

essential chemistry

From plastics and fuels, to fabrics and dyes, chemical engineering has a role to play in the production of many products that you use in your daily life. Chemical engineering takes reactions that happen at a small scale and makes them useful at a much larger scale so the products can reach many, many more people. The goal of chemical engineering and industrial production is typically to get the most amount of product, using the most efficient processes.

Chemical engineers have a lot of factors to consider when they design a process. They need to think about the types of reactants that are involved, other products that are part of the reaction, the most efficient reaction conditions, and the yield. When thinking about the reactants and other products it is important to consider not only how much they cost, and the quantity of "waste" generated, but also if the substances are hazardous.

You have probably seen or heard the term "green" when you are looking at a product or reading about a business. Green is meant to imply that something is environmentally friendly. But, you might be surprised to hear that there is something called "green chemistry." Green Chemistry is a U.S. Environmental Protection Agency (EPA) program that focuses on the design of chemical products and processes that reduce or eliminate the use or generation of hazardous substances.

Instead of just looking at what is produced, green chemistry applies across the entire life of a product. This includes the reactants, the product design, the manufacturing process, the use and even the disposal of the product. Green chemistry is also known as sustainable chemistry. Green chemistry is different than just cleaning up pollution. Cleaning up pollution, also called remediation, is taking care of the problem after it happens. It includes separating hazardous chemicals from other materials, then treating them so they are no longer hazardous or concentrating them so they can be collected for safe disposal.

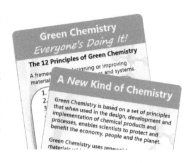

Green chemistry tries to take care of a problem before it even happens. This means generating less pollution by reducing or eliminating the amounts and hazards of the materials that go into making the product, as well as the hazards of the products themselves.

Green chemistry is a philosophy that can be important for all areas of chemistry. There are twelve principles that guide green chemistry. The overall goal of the principles is to apply innovative scientific solutions to real-world problems.

Engineering and Green Chemistry - 2

You have learned about percent yield, which is one way to track efficiency of a chemical process. Percent yield relates the experimental amount of a product to the amount of the product that you should have from the chemical reaction. Percent yield is a great measure of your effectiveness in producing one thing. If there is only one product to a process and a high percent yield, then all of the mass and atoms of the reactants are going to that one product.

What if there is more than one product? How much reactant is "wasted" going to products that you don't want or need? This is where the atom economy comes in. The atom economy, principle #2 of green chemistry, is a way of tracking the efficiency of a reaction and preventing pollution by focusing on the reactants. The idea of the atom economy is to answer the question "what amount of the reactants are incorporated into the final desired product(s), and what atoms are wasted." You can calculate the % atom economy of a reaction as follows.

$$\% \text{ Atom Economy} = \frac{\text{Formula mass of atoms in desired products}}{\text{Formula masses of all the reactants}} \times 100$$

For example, let's look at two reactions that could be used to produce hydrogen gas.

Reaction 1: $Zn(s) + 2HCl(aq) \rightarrow ZnCl_2(s) + H_2(g)$
Molar Masses: $65.380 + 2(36.461) \quad 136.286 + 2.016$

Reaction 2: $CH_4(g) + 2H_2O(l) \rightarrow CO_2(g) + 4H_2(g)$
Molar Masses: $16.042 + 2(18.015) \quad 44.010 + 4(2.016)$

Both reactions produce hydrogen gas. If done carefully and under the right conditions, you could get close to 100% yield. But, that 100% yield simply means that you collected all the hydrogen gas that you expected based on the limiting reactant. What about the other products? What about the atoms from the reactants that did not make the hydrogen gas?

Looking closer at the first reaction, you can get 2 grams of hydrogen for every 138 grams of reactants that are added. Even with a 100% yield, this is only a 1.45% atom economy! The second reaction yields 8 grams of hydrogen for every 52 grams of reactant. At 100% yield this is a better atom economy of 15%. Atom economy is just one part of green chemistry. You could have a really high atom economy by using reactants that are hazardous. You don't have to be a chemical engineer to realize that this would not be an ideal solution.

Chemical engineers apply principles of chemistry, biology, physics and math to solve problems that involve the production of many of the materials that you use daily. By considering principles of green chemistry, engineers and scientist can protect and benefit the economy, people and the planet.

Chapter 8 review

Vocabulary
Match each word to the sentence where it best fits.

Section 8.1

mole ratio	stoichiometry

1. _____ is a content area of chemistry that concerns calculating or analyzing the amounts of reactants and products.

2. A(n) _____ compares the numbers of atoms of two compounds involved in a chemical reaction.

Section 8.2

actual yield	percent yield
theoretical yield	

3. We can calculate our _____ based on the assumption that the reaction yields 100% of the product.

4. In actual experiments in a laboratory, the _____ is obtained and is generally less than the expected amount.

5. The _____ describes the amount of actual product obtained from an experiment compared to the maximum that could have been obtained.

Section 8.3

excess reactant	limiting reactant

6. The _____ is the substance that restricts the amount of product that can be formed during a chemical reaction.

7. The reactant that is remaining after the chemical reaction has formed all products is called the _____.

Conceptual questions

Section 8.1

8. The mole is often referred to as a "chemist's dozen". What is the meaning of this comparison?

9. Explain the meaning of the term $13O_2(g)$ in the following equation for the combustion of butane. Butane is commonly used in fireplace and candle lighters

$$2C_4H_{10}(l) + 13O_2(g) \rightarrow 8CO_2(g) + 10H_2O(g)$$

10. (In the above reaction for combustion of butane, 532.2 g of butane and oxygen produce 532.2 g of water and carbon dioxide. However, 15 moles of butane and oxygen produce 18 moles of water and carbon dioxide. Why is mass conserved but moles are not?

11. (Write a chemical equation that is described in the following sentence.

"In the production of concrete, solid calcium carbonate decomposes under heat to form solid calcium oxide and gaseous carbon dioxide."

12. Write the following sentence as a chemical equation.

"one mole of methane combines with two moles of oxygen to produce two moles of water and one mole of carbon dioxide."

13. Write a sentence that represents the following chemical equation using the word "moles".

$$2C_4H_{10} + 13O_2 \rightarrow 8CO_2(g) + 10H_2O$$

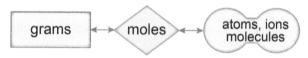

14. Which statement best describes the meaning of this flowchart?

   a. The number of grams multiplied by the number of moles equals the number of atoms, ions or molecules
   b. The number of molecules (atoms, ions) is greater than the number of moles which is greater than the number of grams.
   c. To calculate the number of molecules, ions, or atoms from an amount in grams you must first calculate the number of moles.
   d. In order of importance, grams come first, followed by moles, followed by the number of atoms, ions or molecules.

15. Give one example of a situation in which a chemist or chemical engineer would want to calculate the yield of a chemical reaction? (amount of product produced)

16. Write a sentence that represents the following chemical equation using the word "moles".

$$2C_4H_{10} + 13O_2 \rightarrow 8CO_2(g) + 10H_2O$$

17. The rocket fuel, hydrazine, (N_2H_4) is synthesized by reacting ammonia (NH_3) with hydrogen peroxide, (H_2O_2)

$$2NH_3 + H_2O_2 \rightarrow N_2H_4 + 2H_2O$$

A student proposes that two grams of ammonia and one gram of hydrogen peroxide will produce one gram of hydrazine and two grams of water. Explain why the student's answer is incorrect.

Chapter 8 review

Section 8.2

18. Mechanical harvesting equipment picks fruit without being able to tell if the fruit is ripe or not. How does the concept of percent yield apply to the output of a mechanically harvested crop such as fruit?

19. Compare the actual yield of a chemical reaction with the theoretical yield. Which of the two must be smaller or equal to the other?

20. (When sodium bicarbonate (baking soda) $NaHCO_3(s)$ is heated, it decomposes into three products:
    - sodium carbonate: $Na_2CO_3(s)$
    - water: $H_2O(g)$
    - carbon dioxide: $CO_2(g)$

    In an experiment the products cool to a solid and a lab technician measures the mass of products to be greater than the theoretical yield of sodium carbonate! Can you solve the riddle of where the extra mass came from?

Section 8.3

21. Consider the following reaction for burning propane:

    $$C_3H_8(g) + 5O_2(g) \rightarrow 3CO_2(g) + 4H_2O(g)$$

    Ten moles of oxygen are combined with one mole of propane and the reaction is started. Which of the following statements are TRUE. (There may be more than one)

    a. Three moles of carbon dioxide will be produced.
    b. Six moles of carbon dioxide will be produced.
    c. Four moles of water will be produced.
    d. Two moles of propane will be reacted.

22. Explain one real life example of a limiting reactant not presented in this book.

23. "Hard" water usually refers to the presence of dissolved metal ions such as Mg^{2+} and Ca^{2+}. Water "softeners" are chemical compounds which remove the ions. One strategy is to add a chemical that combines with the metal ions to form an insoluble precipitate. For example, sodium bicarbonate added to water will precipitate out the calcium ions according to the reaction:

    $$NaHCO_3(aq) + Ca^{2+}(aq)$$
    $$\rightarrow 2Na^+(aq) + H_2O(l) + Ca(CO_3)_2(s)$$

    Which strategy would be best for removing the most dissolved Ca^{2+} ions. Also answer why you think so.

    a. Use a minimal amount of sodium bicarbonate so it gett completely reacted.
    b. Use an amount of sodium bicarbonate just enough to react with the estimated amount of Ca^{2+} ions.
    c. Use an excess amount of sodium bicarbonate to be sure that all Ca^{2+} ions react.

24. Is the excess reactant always the reactant present in a larger amount? Why or why not?

25. Lithium nitride (Li_3N) is a reddish-purple solid used in prototype hydrogen-powered vehicles. LIthium nitride can be prodiced by the following the following chemical reaction:

    $$6Li(s) + N_2(g) \rightarrow 2Li_3N(s)$$

    If 9 moles of Li(s) are combined with 2 moles of N_2, what is the limiting reactant?

Quantitative problems

Section 8.1

26. Calculate the number of moles contained in the given quantities of each of the following compounds.

    a. 50.0 g of silicon dioxide, SiO_2
    b. 50.0 g of tungsten, W
    c. 50.0 g of nitrogen, N_2
    d. 50.0 g of nitric acid HNO_3
    e. 50.00 g of calcium hydroxide $Ca(OH)_2$
    f. 50.0 g of hydrogen, H_2

27. Elemental sulfur reacts with sulfuric acid to produce sulfur dioxide and water according to the (unbalanced) reaction below.

    $$S(s) + H_2SO_4(aq) \rightarrow SO_2(g) + H_2O(l)$$

    a. Balance the equation.
    b. How many moles of $SO_2(g)$ are formed from 50.0 g of sulfur, S(s), reacting completely with excess sulfuric acid, $H_2SO_4(aq)$?
    c. How many grams of SO_2 form given the amount in part b above?
    d. A reaction is observed to yield 32.0 g of sulfur dioxide, SO_2. Assume 100% yield. How much sulfuric acid was consumed in the reaction (in grams)?
    e. In a commercial chemical engineering process the less expensive reactants are always supplied in excess, to ensure complete reaction of the reactant that is the most expensive. Per gram of sulfur, how much sulfuric acid would be required so that the sulfuric acid is 200% of the minimum required amount for the reaction?

28. Combustion needs oxygen, but it doesn't have to come from O_2. Magnesium can burn in pure carbon dioxide once the reaction is started.

    a. Write and balance a displacement reaction for magnesium buring in carbon dioxide (CO_2)
    b. 10.0 grams of magnesium is burned in excess carbon dioxide. How much carbon is produced?
    c. How much carbon dioxide is consumed to burn the 10.0 g of magnesium?
    d. What is the minimum volume of carbon dioxide gas at STP required to completely burn 10.0 g of Mg.

Chapter 8 review

Section 8.1

29. (A fascinating chemical reaction demo called "Elephant toothpaste" relies on a reaction between hydrogen peroxide, (H_2O_2) and a solution of dissolved potassium iodide, (KI). The reaction produces water and oxygen gas in two steps.

$$H_2O_2(l) + I^-(aq) \rightarrow H_2O(l) + IO^-(aq)$$

$$H_2O_2(l) + IO^-(aq) \rightarrow H_2O(l) + O_2(g) + I^-(aq)$$

   a. Where does the iodide ion come from?
   b. What happens to the iodide ion in the second reaction?
   c. Write a single balanced equation that represents the decomposition of the hydrogen peroxide into water and oxygen. The I^- ion is an example of a catalyst that helps the reaction take place but is not used up in the process.
   d. How much oxygen is generated from 10 grams of hydrogen peroxide?

30. The reaction for burning propane gas is given below.

$$C_3H_8(g) + O_2(g) \rightarrow CO_2(g) + H_2O(g)$$

   a. Balance the equation
   b. A propane camping stove burns 1.00 gram of propane per 30 seconds. What volume of oxygen at STP is required to completely react with the gram of propane.
   c. By volume, air at STP is 21% oxygen and 79% nitrogen. How many liters of air must pass through the stove in 30 seconds?
   d. A small camping tent has a volume of 3 m³ which is 3,000 L. How long would the stove take to consume 50% of the oxygen in the tent.

31. Silver bromide and magnesium nitrate are produced from the reaction of silver nitrate and magnesium bromide.

   a. Write a balanced chemical equation for the reaction.
   b. If 55 grams of silver bromide was produced from the above reaction, what mass of magnesium bromide reacted?
   c. Calculate the mass of magnesium nitrate produced from 80.0 g of silver nitrate.
   d. If 46.6 g of silver bromide formed, what mass of magnesium nitrate was formed?
   e. Using information from 4 above, what mass of magnesium bromide reacted?
   f. Using information from 4 above, what mass of silver nitrate reacted?

32. (A sequence of two reactions is used to produce hydrogen gas from methane and steam. The hydrogen is used for the synthesis of ammonia for fertilizer.

   1. $CH_4(g) + H_2O(g) \rightarrow CO(g) + 3H_2(g)$
   2. $CO(g) + H_2O(g) \rightarrow CO_2(g) + H_2(g)$

   Consider the entire two-step process in answering the following.

   a. Rewrite the overall two-step reaction as a single balanced reaction.
   b. How many moles of methane does it take to produce 1 mole of hydrogen gas?
   c. What is the mole ratio between methane and hydrogen

33. $4Fe(s) + 3O_2(g) \rightarrow 2Fe_2O_3(s)$
   Rust is another name for iron(III) oxide, which occurs when iron or is exposed to oxygen for a long period of time. How many liters of oxygen will be used to completely rust 2250.0 g of iron at STP?

34. Two moles of iron reacts with three moles of tin(IV) chloride to give two moles of iron(III) chloride and three moles of tin(II) chloride.

   a. Write the balanced chemical equation.
   b. If 45.0 grams of tin(IV) chloride react, how many moles of tin(IV) chloride are there?
   c. How many grams of tin(II) chloride forms if 88.9 grams of tin(IV) chloride reacts?
   d. When reacting with iron, how many moles of iron(III) chloride will form from 18.0 moles of tin(IV) chloride?

35. Nitrogen trifluoride is formed from the reaction of nitrogen gas and fluorine gas.

   a. Write a balanced chemical equation for this reaction.
   b. If 6.0 moles of nitrogen gas reacts, how many grams of nitrogen gas are there?
   c. If 22.60 moles of fluorine gas reacts, how many grams of nitrogen trifluoride is formed?
   d. When reacting 40.0 grams of nitrogen gas in the presence of excess fluorine, how many moles of nitrogen trifluoride will form?

36. In a very violent reaction called a thermite reaction, aluminum metal reacts with iron(III) oxide to form iron metal and aluminum oxide.

   a. Write a balanced chemical equation for the thermite reaction.
   b. What mass of aluminum is needed to react with 171.6 g of iron(III) oxide?
   c. How many moles of aluminum oxide can be made from the reaction if 99.0 g of aluminum is consumed?
   d. If 1.004 moles of iron is produced in the reaction, what mass of iron(III) oxide reacted?

Chapter 8 review

Section 8.1

37. When magnesium burns, some of it reacts with the N_2 in the air to create Mg_3N_2.
    a. Write the balanced equation for the reaction.
    b. If 24.7 grams of Mg_3N_2 is formed, how many liters of N_2 was required at STP?

38. Copper nitrate ($Cu(NO_3)_2$) in solution reacts with dissolved sodium carbonate (Na_2CO_3) to create an insoluble precipitate of copper carbonate ($CuCO_3$).

    $$Cu(NO_3)_2(aq) + Na_2CO_3(aq) \rightarrow CuCO_3(s) + 2NaNO_3(aq)$$

    How much cupper carbonate is produced if 50 g of copper nitrate reacts completely with sodium carbonate?

39. Calcium and water can react to produce calcium hydroxide ($Ca(OH)_2$) and hydrogen gas. A calcium hydroxide is often called limewater.

    $$Ca(s) + 2H_2O(l) \rightarrow Ca(OH)_2(s) + H_2(g)$$

    For the above equation, if 9 moles of Ca(s) react with excess water, how many liters of hydrogen gas, $H_2(g)$, are produced at STP?

40. Silver nitrate ($AgNO_3$) reacts with salt (NaCl) to produce insoluble silver chloride (AgCl) and another product.
    a. Write the balanced equation for the reaction.
    b. If 10.0 grams of silver nitrate react, how many moles are reacting?
    c. How many grams of AgCl will form?

Section 8.2

41. A student performs a reaction which yields 5.77 grams of potassium chloride, KCl. The pre-lab calculations, predict a theoretical yield of 6.00 grams of potassium chloride. What is the percent yield?

42. The world produces about 350,000 tons of phosphorus trichloride (PCl_3) per year for industrial chemistry. The reaction for producing PCl_3 is given below.

    $$P_4(s) + 6Cl_2(g) \rightarrow 4PCl_3(l)$$

    a. What quantity in grams of PCl_3 is produced when 3.00 grams of P_4 reacts with with excess chlorine. Assume the reaction has an 85% yield.
    b. How much P_4 is left unreacted?

43. In the production of concrete, naturally occuring calcium carbonate in limestone ($CaCO_3$) is heated and decomposed into calcium oxide (CaO) "quicklime" and carbon dioxide.
    a. Write a balanced equation for the reaction.
    b. A sample of 34.5 grams of $CaCO_3$ is heated, it decomposes and the released CO_2 is measured at 12.67 grams. What is the percent yield?
    c. How much quicklime is produced?

44. In the refining of titanium, the first step in the reaction is to turn titanium dioxide (TiO_2) in ore into titanium chloride ($TiCl_4$) using the reaction below.

    $$TiO_2 + Cl_2 + C \rightarrow TiCl_4 + CO$$

    a. Balance the reaction
    b. Calculate the amount (in grams) of TiO_2 required to produce 100 grams of $TiCl_4$ if the reaction has a 75% yield.
    c. How many grams of carbon monoxide, a toxic gas, are produced?

45. In the combustion of methane gas, oxygen is consumed. To ensure complete burning, natural gas powered vehicles must mix an excess of oxygen with the methane. The reaction is:

    $$CH_4(g) + 2O_2(g) \rightarrow CO_2(g) + 2H_2O(g)$$

    Calculate the liters of oxygen required at STP to have 100% excess oxygen, or 2 × the minimum needed amount.

46. Arsenic, a naturally occurring element, is found throughout the environment; for most people, food is the major source of exposure. In a commercial production of this element, arsenic(III) oxide is heated with carbon to form carbon dioxide and elemental arsenic.
    a. Write a balanced chemical equation for the reaction.
    b. If 9.02 grams of arsenic(III) oxide is used in the reaction and 7.10 grams of elemental arsenic is produced, what is the percent yield?
    c. What mass of arsenic will result from the reaction of 6.0 moles of arsenic(III) oxide if there is 64.1 percent yield in the reaction?

47. In a laboratory, solid lead was reacted with potassium nitrate to produce lead(II) oxide and potassium nitrite.
    a. Write a balanced chemical equation for the reaction.
    b. In the laboratory, when 40.0 g of potassium nitrate was reacted with excess lead, 25.2 g of potassium nitrite was obtained. What was the percentage yield of potassium nitrite?

Chapter 8 review

Section 8.2

48. The Mannheim process is a chemical reaction that uses natural salt (NaCl) and sulfuric acid (H$_2$SO$_4$) to produce hydrochloric acid (HCl) and sodium sulfate Na$_2$SO$_4$ which is used in the production of soaps.

   a. Write the balanced chemical equation.
   b. 1000 grams of salt reacts to with 1000 grams of sulfuric acid to produce 1000 grams of sodium sulfate. What is the percent yield of the reaction?

49. Carbon tetrachloride is a solvent that was once used in large quantities in dry cleaning. Because it is a dense liquid that does not burn, it was also used in fire extinguishers. Unfortunately, its use was discontinued because it was found to be a carcinogen.

   a. Write a balanced chemical reaction for carbon disulfide reacting with chlorine gas to form carbon tetrachloride and disulfur dichloride.
   b. What is the percent yield of carbon tetrachloride if 722 kg is produced from the reaction of 420. kg of carbon disulfide?
   c. If 47.5 g of chlorine gas is used in the reaction and 29.5 g of disulfur dichloride is produced, what is the percent yield?
   d. If the percent yield of the industrial process is 86.6%, how many kilograms of carbon disulfide should be reacted to obtain 5.00 x 10^4 kg of carbon tetrachloride?
   e. Using information from 4 above, how many kilograms of disulfur dichloride will be produced, assuming the same yield for the product?

Section 8.3

50. A small company that builds bicycles checks their stock room and finds that they have only 155 handlebars, 121 wheels, 126 pedals, and 95 seats left. They have plenty of the other parts that make up a bicycle.

   a. How many total bicycles can the company build?
   b. What parts are left after all the bicycles that can be made are assembled?
   c. What additional parts are required to use up all of the parts and make them into bicycles?

51. Silicon Carbide SiC is frequently used as an abrasive on sandpaper. It is created from silicon dioxide and carbon using the following reaction.
    SiO$_2$ + 3C → SiC + 2CO

   a. If you have 10.0 g of SiO$_2$ and 9.00 g of C, which reactant will be used up first?
   b. Explain your answer above.
   c. How many grams of silicon carbide can be produced?
   d. Calculate the amount of excess reactant remaining following the reaction.

52. Titanium nitride is an extremely hard ceramic material, often used as a coating on cutting tools. The balanced reaction for creating titanium nitride is shown below.
    2Ti(s) + N$_2$(g) → 2TiN(s)

   a. If you have 25.0 g of Ti and 20.0 g of N$_2$, which reactant will be used up first?
   b. Explain your answer above.
   c. How many grams of TiN can be produced?
   d. Calculate the amount of excess reactant remaining following the experiment.

53. 25 grams of potassium hydroxide (KOH) react with 25 grams of ammonium sulfate ((NH$_4$)$_2$SO$_4$) to produce water (H$_2$O), ammonia (NH$_3$) and potassium sulfate (K$_2$SO$_4$).

   a. Write the balanced reaction.
   b. What is the limiting reactant?
   c. How many grams of potassium sulfate are produced if the reaction has 100% yield?
   d. How much ammonium sulfate is left unreacted?

54. Sodium fluoride (NaF) can be produced from diatomic fluorine gas (F$_2$) and sodium iodide (NaI) in a single replacement reaction:

   a. Write the balanced chamical equation.
   b. What is the limiting reactant if 24.63 grams of fluorine gas are reacted with 58.52 grams of sodium iodide?
   c. How much of the excess reactant will be left.

Alkaline battery

55. An ordinary alkaline battery makes electricity by a reaction in which zinc metal (Zn) reacts with manganese dioxide (MgO$_2$) to produce zinc oxide (ZnO) and manganic oxide Mn$_2$O$_3$

   a. Write the balanced equation for the reaction.
   b. Assume an unused AA battery contains 15.0 g of zinc and manganese oxide combined. Calculate the amounts of Zn and MnO$_2$ so the reaction is "stoichiometric" meaning all reactants would be completely used up.
   c. Suppose the battery design engineer needs excess MnO$_2$ because the zinc is solid and concentrated in the core of the battery while the manganese dioxide is dispersed around the outside. Calculate the quantities of both compounds needed for a 20% excess of MnO$_2$ **by moles.**

Chapter 8 review

Section 8.3

56. A sample of zinc (Zn) metal with a mass of 36 grams reacts with 20 grams of hydrochloric acid (HCl) dissolved in water. Zinc chloride ($ZnCl_2$) and hydrogen gas are produced.

    a. Write the balanced chemical reaction.
    b. Which is the limiting reactant?
    c. How much of the zinc is used and how much remains?
    d. What volume of hydrogen gas is produced at STP?

57. (10.0 grams of iron powder is allowed to react with 100. liters of air at STP to form rust (Fe_2O_3) Assume the air is 21% oxygen by volume.

    a. Write the balanced chemical equation.
    b. What is the limiting reactant?
    c. How much rust is created if the reaction has 100% yield?
    d. What is the percentage of oxygen left in the air after the rusting reaction has occurred?

58. 13.0 g of potassium reacts with 13.0 g of iodine gas. Calculate the following.

    a. Write the balanced chemical equation for the reaction.
    b. Which is the limiting reactant?
    c. How much product is made?

59. In the production of metallic iron, iron(III) oxide is reacted with carbon monoxide gas at high temperature. To discover which compound can produce the most iron for a given mass, answer the following questions.

    a. Write the balanced equation for the reaction of iron(III) oxide and carbon monoxide to produce iron and carbon dioxide.
    b. Assuming the reaction begins with 20.0 g of each reactant, which is the limiting reactant?
    c. How much iron will be produced from the above reaction?
    d. Calculate the grams of excess reactant remaining after the reaction.

60. In a single replacement reaction, potassium bromide reacts with chlorine to give potassium chloride and bromine.

    a. Write a balanced chemical equation for the reaction.
    b. Assuming 25.32 grams of chlorine reacts with 47.23 grams of potassium bromide, what is the limiting reactant?
    c. How much of the excess reactant will be left over?

61. Nitrous oxide (N_2O), also known as laughing gas, can be made in the laboratory by reacting sodium nitrate ($NaNO_3$) with ammonium sulfate (($NH_4)_2SO_4$ to produce nitrous oxide, sodium sulfate (Na_2SO_4) and water.

    a. Write the balanced chemical reaction.
    b. 100 grams of sodium nitrate are reacted with 100 grams of ammonium sulfate. Which is the limiting reactant?
    c. How much of the ammonium sulfate is used up abnd how much remains?
    d. What volume is nitrous oxide gas is produced at STP?

62. A hiker uses a camp stove that burns compressed propane according to the reaction below.

$$C_3H_8(g) + 5O_2(g) \rightarrow 3CO_2(g) + 4H_2O(l)$$

The fuel storage tank contains 33.0 liters of propane gas at STP compressed to a much higher pressure to fit in a 0.5 liter tank.

   a. If 1.00 liters per minute (at STP) of propane are released, how much oxygen is needed?
   b. Suppose the movement of air drops the oxygen flowing into the stove burner to 5.0 grams per minute. What fraction of the propane is unburned? Give your answer in percent by moles (volume).

63. In a laboratory, a student attempts the reaction between solid magnesium and oxygen gas.

   a. Write a balanced chemical equation for the reaction.
   b. If the reaction mixture contains 68.3 grams of magnesium and 53.1 grams of oxygen; what is the limiting reactant?
   c. Using information from 2 above, calculate the theoretical yield.

64. By selective oxidation, ethanol, C_2H_5OH in the presence of oxygen produces acetic acid and water.

   a. Write a balanced chemical equation for the given reaction.
   b. What is the limiting reactant when 10.0 grams of ethanol and 4.0 grams of oxygen are consumed in the reaction?
   c. Calculate the grams of acetic acid that will form.
   d. Calculate the grams of excess reactant remaining.

Section 8.5: Chapter Review

chapter 9 | Atomic Structure

When you picture an atom, what do you see? In your head, do you see a small solid marble of matter? Or maybe you are picturing the nucleus with even tinier electrons whirling around, like planets around a sun. Like all models, each of those models are partially correct. A model car isn't a real car, but it can be illustrative in showing how the steering, wheels, or doors work. Models of atoms are useful in explaining or predicting physical and chemical properties in some circumstances, but not in others.

3D model of a car

When you are using a model, it is important to remember that all models have limitations. Envisioning an atom as a marble is great when you are picturing the structure of a solid or building a molecule with a kit, but it doesn't explain why certain chemicals react. Picturing the atom as small solar system can help explain some of the behaviors of electrons, but it gets very complicated when you are trying to explain properties for larger elements.

atomic models

 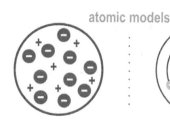 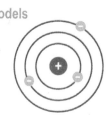 

It is also important to remember that models can and do change. As our knowledge about the structure of the atom has gotten deeper, our model changed to reflect the latest understanding. Our description of an atom has evolved from a solid marble, to something mushy resembling rice pudding with raisins, to a solar system model with orbiting electrons, to the current nuclear atom with an "electron cloud."

So, the question remains... how do we know what we know about atoms, and why is that understanding essential to chemistry?

Chapter 9 study guide

Chapter Preview

While the concept of atoms has existed for thousands of years, our understanding of atoms has taken huge leaps within the last few hundred years. In this chapter you will learn about atoms, focusing on their structure and properties. Atoms are made up of a tiny but very dense nucleus of protons and neutrons surrounded by an electron cloud. The number of protons determines the element but the number of neutrons may vary. Spectral lines provided the clue to quantum numbers and electron orbitals. The periodic table maps electron configurations of the elements. The chemical behavior of atoms depends on electrons in the highest energy level. Depending on the number of valence electrons they may form ions or covalent bonds to create a full outer shell. The periodic table illustrates trends of ionization energy and atomic radii.

Learning objectives

By the end of this chapter you should be able to:
- compare properties of isotopes of the same element;
- describe how atomic theory evolved over time;
- explain the balance of nuclear forces within the nucleus;
- explain how spectral lines provide evidence for the quantum model of the atom;
- write the electron configuration for a given atom; and
- use trends on the periodic table to predict chemical and physical properties of atoms.

Investigations

9A: Isotopic composition
9B: What is a wave?
9C: Light energy
9D: Flame test

Pages in this chapter

254 Atoms Have Structure
255 Electrons, protons, and neutrons
256 Electric charge and the electromagnetic force
257 The nucleus and the strong nuclear force
258 The atomic number and mass number
259 Isotopes and atomic mass
260 Atoms, ions, and the electron cloud
261 Section 1 review
262 The Quantum Atom
263 The early evidence
264 The gold foil experiment
265 What is a wave
266 Light and atoms
267 The puzzle of spectral lines
268 Planck's constant and the uncertainty principle
269 The wave / particle model
270 Orbitals, electron waves and quantum states
271 Energy levels
272 The exclusion principle, Hund's rule, and hybridization
273 Emission and absorption of light
274 Spectroscopy
275 Section 2 Review
276 Electron Configurations
277 The noble gases
278 Group I and group II metals
279 The halogens
280 Carbon, nitrogen, and oxygen
281 Valence electrons
282 The transition metals and noble gas core notation
283 Using electron configurations
284 Electron configurations and the periodic table
285 Atomic radius
286 Ionic radius
287 Ionization energy
288 Section 3 Review
289 Fireworks & Color
290 Fireworks & Color - 2
291 Fireworks & Color - 3
292 Chapter review

Vocabulary

strong nuclear force
isotopes
spectral line
photon
Aufbau principle
spin
ground state
spectroscopy
halogens
atomic radius
ionization energy

atomic number (Z)
atomic mass
Planck's constant (h)
quantum state
energy level
Hund's rule
excited state
spectrometer
valence electrons
ionic radius

mass number (A)
ion
Heisenberg's uncertainty principle
orbital
Pauli exclusion principle
hybridization
spectrum
electron configuration
transition metals
ionization

9.1 - Atoms Have Structure

One oxygen atom bonds with two hydrogen atoms to make water (H_2O). One carbon atom bonds with four hydrogen atoms to make methane (CH_4). What is different about oxygen, carbon, and hydrogen that causes these behaviors? The question extends to all the elements. What makes atoms of one element different from atoms of another element? Are atoms the most basic structures of matter, or are there smaller things inside atoms? It was once thought that atoms were the smallest possible particles of matter. Near the turn of the 20th century we discovered that atoms themselves were made of even smaller particles! The discovery that atoms had internal structure informs our modern understanding of chemistry.

The birth of the atomic theory

A thought experiment	Imagine cutting a piece of aluminum foil into smaller and smaller pieces. When you get to microscopic-size pieces, imagine cutting those into even smaller pieces. How small could you get? Could you get infinitely small?
The first idea of atoms	Democritus (460-370 BC), a Greek philosopher, logically argued that you could not divide matter up into infinitely smaller pieces. Eventually the process of "cutting" must reach a single particle of matter that could no longer be cut again. Democritus named this particle atomos, or atom. Democritus' atoms could not be cut any smaller—they were the smallest possible particles of matter. To get a sense of what I mean, think about a brick house that looks uniform from far away. Close-up, you can see the house is made from individual bricks. Now imagine each brick is indestructible. The smallest possible particle of the house would be one brick, just as Democritus proposed that one atom is the smallest possible particle of matter.
The first modern atomic theory	For the next twenty centuries the idea of atoms was one of many philosophical ideas, untestable, and mostly forgotten. However, as the modern science of chemistry developed, the idea of atoms resurfaced. In 1808 an English school teacher named John Dalton published the first modern hypothesis of the atom. Dalton's four postulates on the Theory of Atoms remain mostly true today, even though our understanding of atoms has vastly increased.
Dalton's postulates	Dalton's four postulates on the Theory of Atoms are:

1. All elements are made of atoms.

2. All atoms of the same element are alike and are different from atoms of any other element.

3. Chemical reactions rearrange atoms but do not create, destroy, or change atoms from one element to another.

4. Compounds are composed of atoms combined in whole-number ratios.

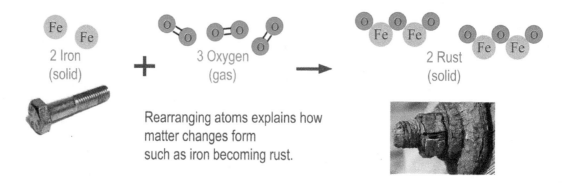

Rearranging atoms explains how matter changes form such as iron becoming rust.

Electrons, protons, and neutrons

Atoms are made of three particles

Atoms are not the simple "smallest particles of matter" as Democritus thought. Atoms themselves are made from even smaller particles called electrons, protons, and neutrons. These three particles have many properties, but for our purpose we will discuss mainly two: mass and charge. The masses of the particles in the atom are very small. The chart below shows that a single electron has a mass of 9.109×10^{-31} kilograms. The proton has positive electric charge and 1,836 times more mass than the electron.

ELEMENTARY PARTICLES IN THE ATOM		Mass	Electric charge
	Proton	1.673×10^{-27} kg	$+1.602 \times 10^{-19}$ C
	Neutron	1.675×10^{-27} kg	0
	Electron	9.11×10^{-31} kg	-1.602×10^{-19} C

The structure of the atom

The incredible diversity of matter comes from the complex way protons, neutrons, and electrons are arranged inside each atom. This internal arrangement is what we call the structure of the atom. The next few chapters explain how the structure of the atom is the key to understanding the richness of chemistry. For example, atomic structure explains why one oxygen atom bonds with exactly two hydrogen atoms to make a water molecule.

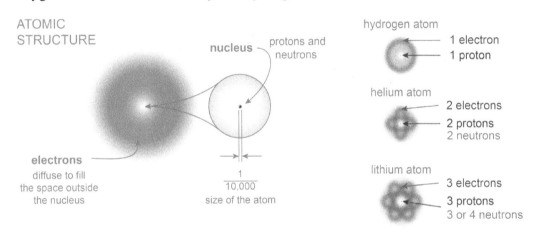

The big ideas

1. Atoms are made of smaller particles called neutrons, protons and electrons.

2. The number of protons determines the element. All atoms of hydrogen have one proton, all atoms of helium have 2, lithium has three, and so on.

3. In a neutral atom, the number of protons and electrons are equal. For example, a hydrogen atom has one proton and one electron. A helium atom has two protons and two electrons.

4. Protons and neutrons are confined to a tiny nucleus at the center of the atom. The nucleus is 10,000 times smaller than the atom but contains almost all the mass.

5. The electrons form a "cloud" around the nucleus. The "size" of an atom is really the size of the electron cloud.

6. Because the electrons define the "outside" of an atom, the chemical properties of elements are determined by electrons. Atoms interact with each other via their electrons.

Electric charge and the electromagnetic force

Electric charge

Electric charge (like mass) is a fundamental property of matter. Unlike mass however, there are two kinds of electric charge, called positive and negative. Positive and negative charges attract each other like north and south poles of a magnet. Two positive charges repel each other and two negative charges also repel each other.

Forces between electric charges

Opposite charges attract   Like charges repel

Protons and electrons balance

Protons and electrons have very different masses but they have the exact same amount of electric charge! The difference is that the charge on the electron is negative and the charge on the proton is positive. The equality and the opposite signs are the reason electric charge usually remains hidden inside atoms. The positive charge on each proton exactly cancels with the negative charge on each electron leaving an atom with zero total charge. Only in special cases does electric charge become obvious, such as static electricity and lightning.

What holds the atom together

The positive protons and the negative electrons attract each other through the electromagnetic force. At the scale of atoms, the electromagnetic force is extremely strong, around 10^{39} times stronger than gravity! The electromagnetic force causes every chemical interaction and almost all the observable properties of matter, such as the phases of solid, liquid or gas, and why atoms form chemical bonds.

ELECTRIC FORCES IN THE ATOM

negative electrons are attracted to the positive nucleus and repelled by other electrons

repulsive forces

Electron cloud

Attractive forces

positive nucleus

CHARGES BALANCE

electrons

$\begin{array}{r} -6 \\ +6 \\ \hline 0 \end{array}$

protons

An atom has zero net charge because equal positive and negative charges cancel each other.

Why atoms are neutral

Atoms are generally uncharged; that is, they have the same number of electrons as they do protons. Any atom missing an electron would have a net positive charge, which creates a powerful electromagnetic attraction and immediately attracts another electron to restore neutrality (zero total charge). A hydrogen atom has one electron to match its one proton. A helium atom has two electrons to match its two protons. Lithium atoms have three electrons to match the three protons and so on.

The nucleus and the strong nuclear force

The strong nuclear force

A helium nucleus contains two protons that repel each other with a very strong force. Why doesn't the repulsive force immediately cause the nucleus to break up? The fact that a helium nucleus does not immediately explode means that there must be an attractive force stronger than the electromagnetic force. This force is (unimaginatively) called the **strong nuclear force** and it attracts protons and neutrons together.

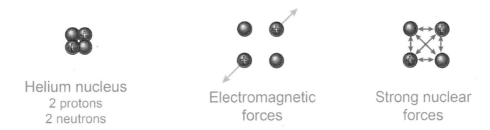

Neutrons help to stabilize the nucleus

Neutrons have zero electric charge and are not affected by the electromagnetic force. If there are enough neutrons, the attractive strong nuclear force overcomes the repulsive electromagnetic force and "glues" the nucleus together. This is the reason every nucleus with more than one proton has at least one neutron. As the number of protons increases, the number of neutrons increases faster to provide enough attractive force to keep the nucleus together.

Why the nucleus is so small

The strong nuclear force has a very short range - only a few times the size of a proton. Imagine two tennis balls moving toward each other. Unless they pass close enough to actually touch, the two balls fly past each other without interacting at all. The strong nuclear force is similar, only attractive instead of repulsive. Protons and neutrons must be almost touching for the strong nuclear force to act. The size of protons and neutrons are the reason the nucleus is about 10,000 times smaller than the atom. If one atom were the size of your classroom, the nucleus would be smaller than a grain of sand in the center. Because of their mass, protons and neutrons make up 99.97% of the mass of an atom.

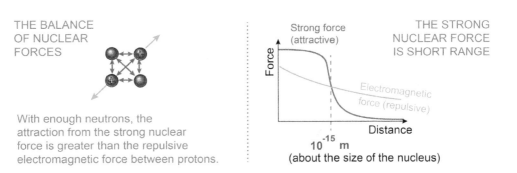

The four forces

There are four primary forces in the Universe: gravity, electromagnetic, weak nuclear force, and strong nuclear force. Although gravity holds our Universe together it is the weakest but the longest reaching force. The electromagnetic force is the next strongest force. Similar to gravity, the electromagnetic force is relatively long-range. The third force is called the weak force and is responsible for some kinds of radioactivity. The fourth force is the strong nuclear force which holds the nucleus together.

Section 9.1: Atoms Have Structure

The atomic number and mass number

The particles in the atom

Incredibly, the entire periodic table - every known element - is made of atoms containing different combinations of just the three particles: electrons, protons, and neutrons. Comparing this to cooking, how many recipes could you make with only three ingredients?

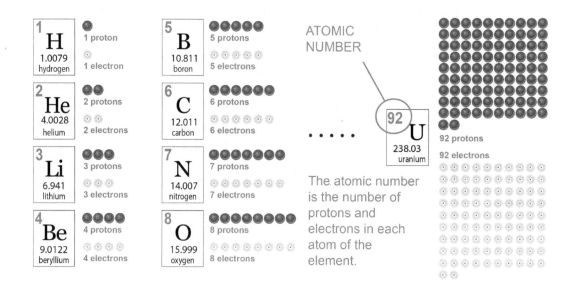

The atomic number

The number of protons is the **atomic number (Z)**. All atoms of the same element have the same number of protons. Every atom of hydrogen has only one proton and one electron, helium has two protons and two electrons. Lithium atoms have three protons and three electrons, and so on. The heaviest naturally occurring element, uranium, has 92 protons and 92 electrons.

Isotopes and mass number

All atoms of the same element have the same number of protons; however, the number of neutrons may differ. For example, there are three isotopes of carbon found naturally. **Isotopes** are variations of an element with different numbers of neutrons. The three most common isotopes of carbon are: $^{12}_{6}C$, $^{13}_{6}C$ and $^{14}_{6}C$. The subscript 6 indicates the atomic number, and the superscripts 12, 13 and 14 indicate the mass number. The **mass number (A)** is the number of neutrons plus protons in the nucleus. We could also write C-12, C-13, and C-14 to indicate the isotope and mass number.

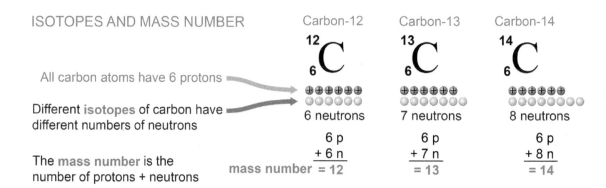

258   Chapter 9: Atomic Structure

Isotopes and atomic mass

Natural elements are a mix of isotopes

Isotopes occur because the stability of a nucleus is a balance between attraction from the strong nuclear force and repulsion from the electromagnetic force. There are also strong quantum effects which we discuss in a later chapter. The result is that some combinations of protons and neutrons are more stable than others; however, many naturally occurring elements have more than one common isotope. Consider the element boron, atomic number five. Boron is used in high-strength fibers and laundry detergents. Two isotopes of boron occur naturally: boron-10 and boron-11.

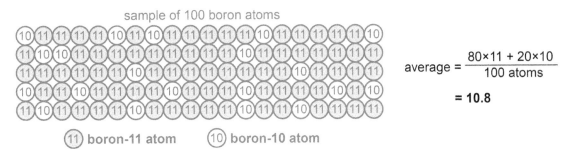

$$\text{average} = \frac{80 \times 11 + 20 \times 10}{100 \text{ atoms}}$$

$$= 10.8$$

The average atomic mass

In a natural sample of boron we find that about 20 atoms in every 100 atoms are boron-10 and 80 atoms are boron-11. The **atomic mass** on the periodic table is an average value of 10.81 g/mol because 80% of boron atoms are boron-11 and 20% are boron-10. In ordinary circumstances every quantity of boron will have the same distribution of isotopes, and quantities of boron are calculated accurately by using the average atomic mass. Note: no single boron atom has a mass of 10.81 amu! The average atomic mass on the periodic table is really meant for calculations with macroscopic quantities - grams or milligrams. When doing calculations with single atoms you will need the atomic mass of a specific isotope.

Stable isotopes

The chart below shows all the isotopes that are stable - meaning not radioactive. Stable isotopes can remain the same forever. Carbon-12 is a good example with six protons and six neutrons. Carbon-13 is also stable with six protons and seven neutrons. The isotope carbon-14 is not stable and eventually decays into nitrogen-14. You will learn more about radioactivity and decay in a later chapter.

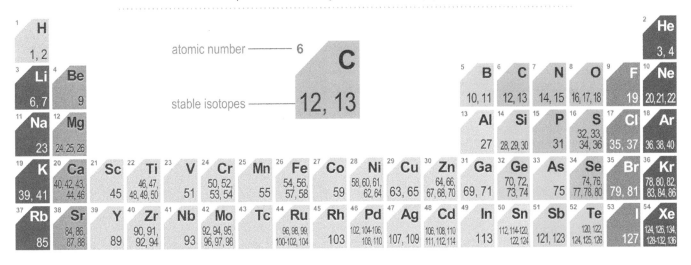

Stable isotopes for the first 54 elements
stable isotopes occur naturally and do not decay (not radioactive)

Section 9.1: Atoms Have Structure

Atoms, ions, and the electron cloud

Why electrons get far from the nucleus

The "size" of an atom is defined by the size of the electron cloud. Because electrons repel each other, the electron cloud of one atom does not normally overlap the electron cloud of another. The electron cloud is large because electrons are 1,836 times lighter than protons (or neutrons). The energy in atoms means electrons have a very high "speed." They are not affected by the strong nuclear force, so the electrons spread far from the nucleus before ultimately being held back by the nucleus' positive charge.

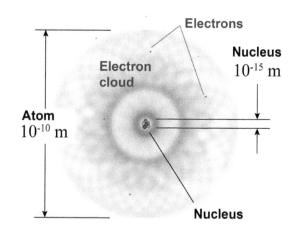

Thinking about the electron cloud

To imagine the electron cloud, think about vigorously shaking a large beach ball full of ping-pong balls. The skin of the beach ball is like the electromagnetic force that keeps the electrons from getting too far from the nucleus. The small, light ping-pong balls represent electrons bouncing wildly around inside the beach ball. The frantically bouncing ping-pong balls are like the electron cloud that fills every atom - except the "surface" is "fuzzy" and not well defined as the plastic skin of a beach ball.

> Except for mass and nuclear behavior, most atomic properties are determined by electrons, including size and chemical bonding.

Electrons determine properties other than mass

Because the outer volume of an atom is made of electrons, electrons dictate how one atom interacts with other atoms. This is why most of chemistry is concerned with electrons and their organization within atoms. The nucleus is located deep inside an atom and it contributes mass, but not much else except through the electrons that are bound to it.

ATOMS BECOME IONS BY ADDING OR LOSING ELECTRONS

Sodium atom

Na (11, 22.990, sodium)
11 electrons
11 protons

A neutral atom has the same number of electrons as protons.

Magnesium ion

Mg^{2+}
10 electrons
12 protons

A magnesium ion loses two of its 12 electrons to have a charge of 2+.

Oxygen ion

O^{2-}
10 electrons
8 protons

An oxygen ion adds two extra electrons making a charge of 2-.

Atoms and ions

A neutral atom starts with the same number of electrons as protons. To become an ion, the atom loses or gains electrons. A monatomic **ion** is a "charged atom" in which the number of electrons and protons is unequal.

Section 1 review

Chapter 9

Electric charge is a fundamental property of matter, similar to mass except there are two kinds of charge: positive and negative. Atoms contain negatively charged electrons, positively charged protons, and neutrons which have no electric charge. The electrons and protons have equal but opposite electric charge. Atoms are held together by the attractive electromagnetic force between electrons and protons.

Inside the atom, the protons and neutrons form a tiny nucleus surrounded by the electrons. The nucleus is 10,000 times smaller than the atom but contains 99.9% of an atom's mass. In the nucleus, protons and neutrons are bound by the strong nuclear force which does not affect electrons. The electrons outside the nucleus form an "electron cloud" that defines the outside of the atom.

The atomic number is the number of protons. Each element has a different number of protons. Isotopes have the same number of protons - and are therefore the same element - but have different numbers of neutrons. Most elements have several naturally occurring isotopes.

Vocabulary words: strong nuclear force, atomic number (Z), mass number (A), isotopes, atomic mass, ion

Key Relationships
- Protons and neutrons make up 99.97% of an atom's mass
- In a neutral atom, the number of protons is equal to the number of electrons
- Positively-charged ions have more protons than electrons
- Negatively-charged ions have more electrons than protons
- Isotope notation for element X: $^A_Z X$ or X-A where A = mass number and Z = atomic number

Review problems and questions

1. It is often said (incorrectly) that an atom is 99.9% empty space. What does this statement mean in terms of mass and volume? Your answer should discuss the contribution of electrons, protons, and neutrons.

2. How many neutrons are in the nucleus of a magnesium-26 atom?

3. If negatively-charged electrons make up most of the size of the atom, why aren't all atoms negatively charged?

4. If like charges repel, how is it possible for the nucleus to stay together when so many positively-charged particles are so close together?

9.2 - The Quantum Atom

The structure of the atom helps to explain its function. Atoms are so small that we have only recently been able to observe them individually. Why do electrons, protons, and neutrons arrange themselves in the precise patterns that create the periodic table? This section presents a few high points of a story that began 2,500 years ago. The theory of quantum mechanics helps to explain atomic structure for chemistry and directly links chemistry and physics. A quantum of anything is the smallest possible unit of that thing.

Quantum theory and chemistry

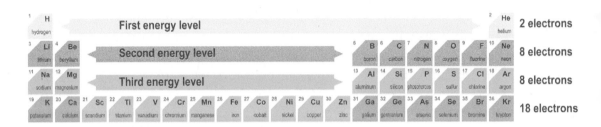

The periodic table and electrons

Why does the first row of the periodic table have only two elements: hydrogen and helium? Why do the second and third rows have eight elements? In 1869, Dmitri Mendeleev developed what would become the modern periodic table from the repeating patterns of chemical properties. However, although various explanations for how atoms connected to one another had been proposed (and many had aspects of the understanding we have today), it would be another 50 years before Lewis proposed his electron pair theory of bonding. Mendeleev could tell his colleagues that elements had repeating patterns, but he could not adequately explain why.

The chemical bond

Lewis's electron explanation for bonding was a breakthrough. 12 years later, Linus Pauling at the California Institute of Technology expanded the explanation using the brand new theory of quantum mechanics. Pauling's 1931 paper "The Nature of the Chemical Bond" explained how the chemical behavior of the elements comes directly from the quantum behavior of electrons in atoms.

Quantum theory and chemistry

At the most fundamental level, Pauling showed that the explanation of chemistry lay in understanding the quantum patterns of electrons in each element. We won't go into the details of quantum theory but instead, will focus on the specific behaviors that "cause" chemistry.

1. Atoms interact with other atoms through electrons. Interactions such as forming chemical bonds occur through electrons.

2. When electrons are confined in an atom, their wave properties force them into specific patterns called quantum states that minimize the total energy of the atom.

3. Each quantum state can hold only one single electron.

4. The electrons in an atom fill up the quantum states in order of lowest energy to highest energy.

5. The electrons in the highest energy levels determine the chemical properties of each element, including bonding, melting, boiling, reactivity and atomic size.

The early evidence

Discovery of cathode rays

In 1870, English scientist William Crookes created a sealed glass tube and pumped out virtually all of the air inside. When he applied a negative voltage to one end of the tube and a positive voltage to the other, an invisible "energy ray" was detected moving from the negative end to the positive end. Crooke called his discovery cathode rays. The race was on to determine what cathode rays were.

CROOKE'S TUBE

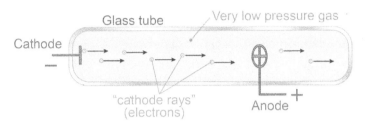

The cathode rays left a sharp shadow of the anode on the end of the glass tube.

Discovery of the electron

It took 27 years before another Englishman, J.J. Thomson, devised a series of experiments that showed cathode rays were deflected by magnets and were also deflected toward a positively charged plate, and away from a negative plate. The rays were always the same whether the electrodes were gold, lead, brass, or any other material. Based on his evidence Thompson concluded in 1897 that cathode rays were a stream of negative particles approximately 2,000 times lighter than the lightest known atom. He called the new particles electrons and proposed that electrons must be inside all atoms.

Interactive simulation

Re-create Thomson's experiment in the interactive simulation titled Thomson's Discovery. Experiment with different atoms and plate charges. Observe the effects on the cathode ray and see how Thomson came to the conclusion that a cathode ray is made of negatively charged particles.

The atom must have structure

If atoms contained electrons then there must be a building block smaller than an atom. Furthermore, Thomson found that electrons were negative even though atoms were neutral, so there had to be something positive inside atoms to cancel the negative charge of the electrons. Scientists around the world searched for answers.

Radioactivity

In 1898, French scientists Marie and Pierre Curie discovered that uranium spontaneously emitted some form of unknown, invisible energy that could expose photographic film. The Curies named their discovery radioactivity. In 1899 Ernest Rutherford's team identified two forms of energy emitted by uranium, which he named alpha and beta. Rutherford determined that alpha particles had a positive charge and had about 8,000 times more mass than electrons. The beta particles were later shown to be electrons.

RADIOACTIVITY
A radioactive nucleus spontaneously changes - and may emit particles and/or energy

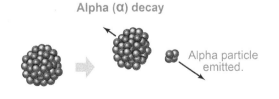
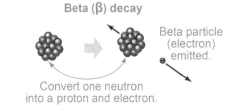

The gold foil experiment

Discovery of radioactivity in 1898

The discovery of alpha particles gave Rutherford a crucial idea. Since alpha particles were smaller than atoms, they might be able to pass through an atom and be detected on the other side. Along with Hans Geiger and Ernest Marsden (a student) the team conducted many experiments to observe alpha particles after having passed through matter. The crucial experiments that built much of our modern explanation of atomic structure were devised and carried out at the Cavendish Laboratory between 1907 and 1910. The experiments showed that all the positive charge and virtually all the mass of an atom were concentrated in a tiny nucleus at the center. Atoms are mostly empty space!

The gold foil experiment

In the most famous experiment, the Cavendish scientists aimed a stream of high speed alpha particles at a thin gold foil. They observed how many alpha particles were detected at different angles after passing through the foil. Rutherford believed (as did everyone at the time) that atoms were essentially solid positive "stuff" with the tiny electrons spread inside like raisins in a loaf of raisin bread. He expected most of the alpha particles to be deflected a little as they crashed through the "bread-like" volume of the target atoms.

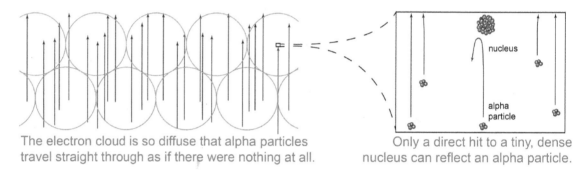

The electron cloud is so diffuse that alpha particles travel straight through as if there were nothing at all.

Only a direct hit to a tiny, dense nucleus can reflect an alpha particle.

What Rutherford's team observed

The results were so unexpected that Rutherford's team repeated the experiments many times over a period of a year. They always got the same results.

1. Virtually all the alpha particles passed through completely unaffected, as if they had somehow completely missed every atom.
2. A very few alpha particles bounced off at large angles and about 1 in 20,000 even bounced backward!

The discovery of the nucleus

From the first observation, the team deduced that most of the space inside an atom must be empty. The second observation could only be explained if nearly all of an atom's mass, and all of the positive charge, were concentrated in a tiny volume, which they called the nucleus. Only a massive, positively-charged nucleus could bounce a fast alpha particle straight backward. From the ratio of particles that "missed" to ones that "hit," Rutherford estimated the size of the nucleus to be 10,000 times smaller than the diameter of an atom!

Interactive simulation

The interactive simulation repeats Rutherford's experiments. Pressing [Run] allows alpha particles to hit the foil. The three detectors (A, B, C) count the number of particles reaching them. Press [Stop] to stop the simulation and see how many particles were strongly deflected (A) compared to how many passed right through (C).

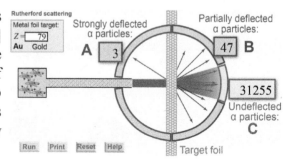

What is a wave

What are waves?

Things that are very small tend to act more like waves than particles. Which brings us to the question what are waves? Waves are regular repeating oscillation. The highest point of a wave is called the crest and the lowest is called the trough. The number of crests to pass a point in one second is called the frequency. Waves with high frequency have high energy. The distance from one crest to another is called the wavelength. The wavelength of a light indicates its color. Waves with short wavelengths are on the blue end of the spectrum and have high frequencies and therefore high energy. The amplitude of a wave is measured as half the distance from the crest to the trough. The amplitude of a wave is the intensity of that wave. For light, it is the brightness. For a sound wave, it is the loudness.

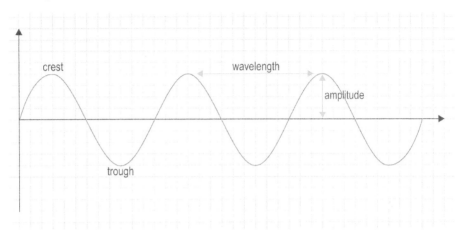

Wave equations

All colors of light move at the same speed, 3.0×10^8 m/s. The speed of light is symbolized as a lower case letter c. The relationship between wavelength and frequency is illustrated by the equation:

$$c = \lambda \nu$$

Wavelength is the Greek letter lambda, λ and frequency is the letter nu, ν. The energy of a wave (E) is calculated by multiplying the frequency of the wave by a constant called Plank's constant (h). The value of Planck's constant equals 6.626×10^{-34} J·s.

$$E = h\nu$$

Solved Problem

What is the energy of an orange light that has a wavelength of 640. nm?

Asked Find the energy (E) of a wave.

Given $\lambda = 640.$ nm or $640. \times 10^{-9}$ m

Relationships $c = \lambda\nu$ and $E = h\nu$

Solve First use the equation $c = \lambda\nu$ to find frequency

$c = \lambda\nu \quad 2.998 \times 10^8 \text{m/s} = 640. \times 10^{-9} \text{m} \times \nu$

$$\frac{2.998 \times 10^8 \text{ m/s}}{640. \times 10^{-9} \text{ m}} = 4.6844 \times 10^{14} \frac{1}{\text{s}} = \nu$$

Then use the equation $E = h\nu$ to find energy.

Planck's constant $h = 6.626 \times 10^{-34}$ J·s.

$E = 6.626 \times 10^{-34}$ J·s $\times 4.6844 \times 10^{14} \frac{1}{\text{s}} = 3.10 \times 10^{-19}$ J

Answer An orange light has 3.10×10^{-19} J of energy.

Light and atoms

How does the electro-magnetic force spread out?

After the discovery of the nucleus, the next crucial mystery was how the electrons "fit" into the atom. The important clue was a long-standing puzzle regarding how light is emitted by atoms. Consider two electrons a distance apart. The first electron exerts a repulsive electromagnetic force on the second. The force does not instantly reach from the first electron to the second, but travels between the two electrons at the incredible speed of 300 million m/s. Next, consider the force from an oscillating electron, such as inside an atom. The oscillating electron creates an oscillating electromagnetic force that carries the oscillation to the second electron.

OSCILLATING ELECTRONS MAKE LIGHT

Two electrons are repelled by their mutual electromagnetic force.

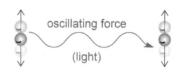
An oscillating electron (inside an atom) creates an oscillating electromagnetic force that travels away and exerts oscillating forces on another electron.
The oscillation of electromagnetic force is a light wave.

Light waves are electro-magnetic

In 1888, Scottish physicist James Clerk Maxwell proved that the oscillating electromagnetic force between the two electrons was a light wave! An electron oscillating up and down 460 trillion times per second creates a ripple of electromagnetic force that spreads outward. If the ripple reaches another electron in the retina of your eye, that electron responds by oscillating 460 trillion times per second. You perceive red light!

Electrons in atoms create light

Virtually all visible light comes from the oscillations of electrons in atoms. What we see as color is actually the different frequencies of light caused by electrons having different amounts of energy. The frequencies of light are so high we use units of terahertz: one THz equals one trillion oscillations per second (1×10^{12} Hz). Blue-violet light has the highest frequency we can see at 780 THz. Deep red light has the lowest frequency at 400 THz.

Wavelength

Because the oscillation travels through space it also has a wavelength. The faster the oscillation, the shorter the wavelength. Red light with a frequency of 4THz has a wavelength of 650 nanometers (650 nm = 650×10^{-9} m). Blue-violet light with a frequency of 780 THz has a shorter wavelength of 385 nm.

Interactive simulation

WAVELENGTH, COLOR, AND ENERGY OF LIGHT

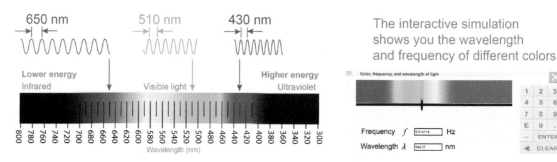

The interactive simulation shows you the wavelength and frequency of different colors

The puzzle of spectral lines

The spectrum

When electricity is passed through hydrogen gas, the gas glows and gives off light. The light, however, is not like the light given off by a light bulb. Instead of a smooth rainbow of colors, hydrogen gas gives off a few very specific colors and nothing in between. The pattern of colors is unique for each element - each **spectral line** is like a fingerprint in color. Every hydrogen atom emits the same spectral lines. Every helium atom emits spectral lines that are different from hydrogen. In 1885, a Swiss school teacher named Johann Balmer discovered that the colors of light in the hydrogen spectral lines obeyed a precise mathematical relationship, named Balmer's formula in his honor.

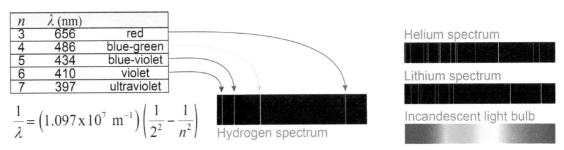

BALMER'S FORMULA

n	λ (nm)	
3	656	red
4	486	blue-green
5	434	blue-violet
6	410	violet
7	397	ultraviolet

$$\frac{1}{\lambda} = \left(1.097 \times 10^7 \text{ m}^{-1}\right)\left(\frac{1}{2^2} - \frac{1}{n^2}\right)$$

Balmer's formula

In Balmer's formula, n is an integer, such as 3, 4, or 5. When n = 3, the formula predicts a wavelength of 656 nm, which matches exactly the red line in the hydrogen spectrum. While Balmer's formula matched the observed pattern of colors, neither Balmer nor anyone else could explain why. The formula implied that something inside hydrogen atoms acted like a series of switches. The atomic switches could be set to any integer, such as 3 or 4, but not any number in between, such as 3.5.

The first Bohr model of the atom

Danish physicist Niels Bohr deduced a brilliant explanation for Balmer's formula in 1913. Bohr proposed that the electron makes circular orbits around the nucleus. Just as the Sun's gravity bends the paths of the planets into orbits, the attraction from the positive nucleus does the same thing with negative electrons. The farther an electron is away from the nucleus, the more energy it has. An electron falling from an outer orbit to an inner one gives up the energy difference as light. The larger the energy difference, the greater the energy of the light. The diagram below shows four transitions between electron orbits. Each gives off a different amount of energy which corresponds to a different color of light. The beginning of the quantum theory was Bohr's proposed explanation for why electrons could only orbit at certain special distances.

BOHR'S FIRST MODEL OF ELECTRON ORBITS

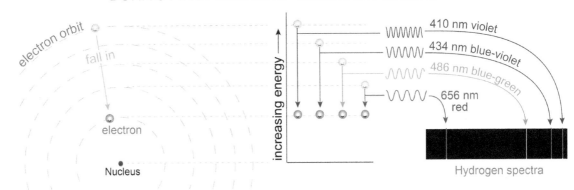

Section 9.2: The Quantum Atom

Planck's constant and the uncertainty principle

When quantum effects are important

Quantum effects become important when the size/mass/energy of a system becomes comparable to a particle wavelength or photon energy given by equations earlier in this section. The common reference for the quantum scale is **Planck's constant (h)**, which has the symbol, h, and a value of 6.626 x 10^{-34} joule-seconds (J·s).

$$\text{Planck's constant: } h = 6.626 \times 10^{-34} \text{ J·s}$$

The world of the very small

In the very small world of quantum mechanics things that seem strange to us are very common. Small particles have characteristics of both waves and particles. A wave will change properties when it bounces off something. This is like throwing a tennis ball against a wall and watching it turn into a golf ball as it bounces away. Waves can also be divided or diffracted as they move though slits. Waves can move through an object like a comb and become many waves on the other side. A ball will not become many different balls after moving through a comb. All objects have both wave like and particle like properties when the object is very large the wave properties are incredible small. Once the object gets small around the Planck length, the wave-like properties become dominant.

Why we don't normally see quantum effects

Consider the red spectral line of hydrogen with a wavelength of 656 nm (10^{-9}m) and energy of 3 x 10^{-19} joules. An electron with this energy has a wavelength of 9 x 10^{-10} m. The diameter of a hydrogen atom is 1 x 10^{-10} m. Since the atom is smaller than the quantum wavelength of an electron, we expect quantum effects to be very strong for electrons in an atom. By comparison, a microscopic dust speck has a diameter of one micron (10^{-6}m) and a mass of 10^{-16} kg. A dust speck drifting through the air at 1 m/s has a quantum wavelength of 1 x 10^{-16} m. The quantum wavelength is a billion times smaller than the dust speck so quantum effects are completely negligible.

The uncertainty principle

Heisenberg's uncertainty principle says that there is a limit to how accurately you can know where an electron is, or its direction and velocity. This limit is a direct consequence of the probabilistic nature of matter on the atomic scale. If we limit the location of an electron to within an atom, the probability of the electron having a certain velocity spreads out. The electron could be zipping around the atom with any velocity in any direction at any moment of time. In fact, the whole concept of the electron particle orbiting the nucleus is wrong. The electron's probability wave spreads out over the whole atom; there is a cloud of probability that describes the chance of the electron being in one place or another.

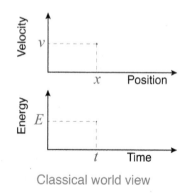

Classical world view

Heisenberg's uncertainty principle

In the quantum world there is a limit to how precisely we can know position, velocity, energy, and time independently of each other.

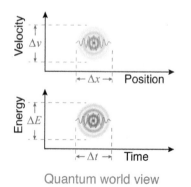

Quantum world view

The wave / particle model

Photons

The word "quanta" means a discrete unit that cannot be subdivided. In 1905, Albert Einstein proposed a quantum theory of light stating that light was a stream of tiny discrete energy bundles. Each energy bundle he defined as one **photon**. A single photon is like a "particle" of light that also has wave-like characteristics of frequency and wavelength. Photons cannot be divided and are the smallest possible quantity of light.

Wave-particle duality

Photons are an example of wave-particle duality. Duality refers to the fact that on the atomic scale, light has particle-like properties (individual photons) and wave-like properties (frequency and wavelength). Ordinary room light contains billions of photons per second so we don't notice the presence of individual photons. But a single atom can absorb or emit energy only in units of whole photons! The particle nature of light dominates the interaction between light and atoms.

Classical electron
Similar to a tiny ball

Classical light wave
Bright / Dim
Can have any quantity of energy

Quantum electron
Properties of wavelength and frequency

Photons
Bright / Dim
Bundles of discrete amounts of energy

The quantum theory of light

The quantum theory of light is very different from the wave theory. Wave theory says that you can reduce the energy of any color light wave as much as you want. The quantum theory says that for any given color (frequency) of light, one photon is the smallest amount of energy you can have.

The quantum electron

From far away, a particle is a tiny speck of matter that has a definite size, mass, and position, like a tiny ball. This is called the classical world view and is based on our intuition about objects in the macroscopic world. On the scale of atoms however, an electron is not like a tiny ball at all, but is "smudged out" into a wave packet with frequency and wavelength. This is the quantum world view, in which a "particle" is not like a tiny ball. If this seems strange and confusing - it is! Our intuition about matter and energy comes from the macroscopic scale - at which quantum effects are "averaged out" over trillions of atoms.

Bohr's hypothesis

In his 1913 paper, Bohr showed that his hypothesis of orbits, together with the photon theory, explained Balmer's formula. The energy of the 652 nm red photon in the hydrogen spectrum was precisely the energy difference between the second and third "orbits" of the electron around a hydrogen nucleus. But Bohr did not know why the electron could occupy only those orbits. Eleven more years of research produced the answer in 1924. That is the year French physicist Louis de Broglie proposed that electrons have a wavelength related to their energy, similar to photons of light. The wavelength of an electron is so small that electrons behave like particles until they are confined to a space comparable to their wavelength, such as the inside of an atom.

Orbitals, electron waves and quantum states

Electrons are trapped in atoms

A light wave trapped between two parallel reflecting mirrors builds up a standing wave of energy. An electron wave confined to an atom behaves in a similar way, only the boundaries of an electron are not parallel mirrors. Geometrically, the inner boundary is the center of the atom. The outer boundary is the farthest distance the electron's energy can raise the electron against the attractive force from the nucleus. These boundaries confine an atomic electron to a spherical shell centered on the nucleus.

The discovery of electron waves in the atom

One year after de Broglie proposed that electrons were a quantum wave, Erwin Schrödinger wrote down an equation for the conservation of energy of de Broglie's new quantum wave. When Schrödinger's equation was applied to an electron bound to a proton, an amazing discovery was made.

1. The electron wave forms a sequence of three-dimensional "standing wave" patterns.
2. Each pattern corresponded to a specific energy for an electron.
3. The differences in energy between one pattern and another were exactly equal to the photon energies of the spectral lines in hydrogen!

Orbitals and quantum states

Each unique standing wave pattern is called a **quantum state**. Because of quantum effects, the probability of finding an electron in a quantum state is virtually 100%. Electrons in atoms are either in a quantum state or moving from one state to another. The existence of quantum states solved Bohr's paradox. Electrons do not have orbits but they do have quantum states with the same energies as Bohr's orbits. For this reason each quantum state inside an atom is called an **orbital**.

Electron orbitals (quantum states)

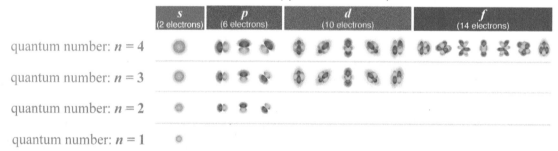

The s, p, d, and f orbitals

The orbitals are named by their quantum number (1, 2, 3, 4, ...) and a letter that describes the type of orbital: s, p, d, and f. The "s" orbital is spherically symmetric around the nucleus and can hold 2 electrons. The "p" orbital has three sub-orbitals aligned with the three coordinate axes. Each suborbital can hold two electrons for a total of six. The "d" orbital has a more complex shape and can hold a maximum of 10 electrons. The "f" orbitals can hold 14 electrons. Higher orbitals exist mathematically but all 92 electrons in the heaviest naturally occurring element fit into available s, p, d, and f orbitals.

Orbitals for the first five quantum numbers

Quantum number	s orbitals	p orbitals	d orbitals	f orbitals
n = 1	**1s** *(2 electrons)*			
n = 2	**2s** *(2 electrons)*	**2p** *(6 electrons)*		
n = 3	**3s** *(2 electrons)*	**3p** *(6 electrons)*	**3d** *(10 electrons)*	
n = 4	**4s** *(2 electrons)*	**4p** *(6 electrons)*	**4d** *(10 electrons)*	**4f** *(14 electrons)*
n = 5	**5s** *(2 electrons)*	**5p** *(6 electrons)*	**5d** *(10 electrons)*	**5f** *(14 electrons)*

Energy levels

Orbitals form energy levels

The quantum numbers and orbitals describe the shape of the "cloud" for each individual electron. Because the average distance from the nucleus is different, the electron energy is also different. The orbital closest to the nucleus (1s) has the lowest energy. Orbitals farther from the nucleus have higher energies. The diagram below shows how the energies of the orbitals compare to each other. The orbitals naturally group into energy levels. An **energy level** is a group of orbitals that have similar energy. Notice that the quantum numbers and energy levels are not the same. The 3d orbital has approximately the same energy as the 4s and 4p orbitals, and therefore is in the fourth energy level (not the third).

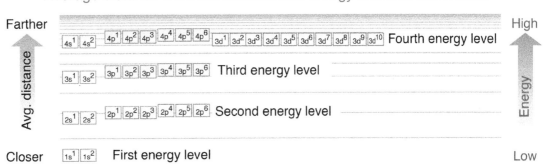

The Aufbau principle

On the scale of atoms the force binding electrons to the nucleus is extremely strong - far, far stronger than gravity. Moving an electron from the 2s to the 3s orbital represents an energy difference equivalent to rolling a ball up a hill four million times the height of Mt. Everest. Because of the energy differences, the **Aufbau principle** states that electrons fill the lowest energy orbitals before higher energy orbitals. For example, the first two electrons in every atom are in the 1st energy level. The next electron goes into the 2s orbital. The Aufbau principle is nicknamed the building up principle since it determines the order in which atoms build up their electron shells.

The order of the energy levels

The number of electrons in each energy level exactly reproduces the row structure of the periodic table. The first energy level (H, He) is the first row of the periodic table. The first row has only two elements because there are only two electrons in the 1s orbital. The second energy level (Li - Ne) is the second row of the periodic table with both s and p orbitals. As the diagram below shows, the periodic table is a map of the electron structure in the atom.

The periodic table is a map of the electron structure in the atom

Section 9.2: The Quantum Atom

The exclusion principle, Hund's rule, and hybridization

The exclusion principle and electron spin

Quantum mechanics forbids any two electrons in the same atom from being in the exact same quantum state at the same time. This rule is known as the **Pauli exclusion principle**, after Physicist Wolfgang Pauli who discovered it in 1925. So why are there two electrons in the 1s orbital? The answer is that electrons have spin. **Spin** is a quantum property that has only two values: +1/2 and -1/2. Each orbital (or sub-orbital) holds two electrons because they do not have the same quantum state: one electron is spin +1/2 and the other is spin -1/2. For example, the 1s orbital holds two electrons and each of the 2p suborbitals holds two electrons.

Hund's rule

Consider the 2p orbitals. There are three suborbitals (p_x, p_y, p_z) aligned along the x, y, and z axes. All three have the same energy because they have the same average distance from the nucleus. However, because electrons repel each other, there are small energy differences that cause the electrons to separately fill the p_x, p_y, and p_z suborbitals before they "double-up." This behavior was discovered by Fredric Hund in 1925 and is named **Hund's rule**. The diagram below shows how the rule applies to the four lightest elements with electrons in p-orbitals.

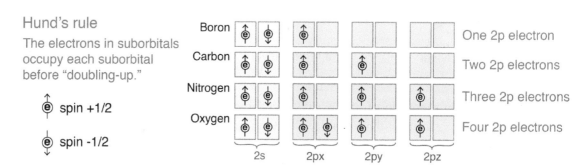

Hybridization and bonding

The 2s and the three 2p orbitals are very close to each other in energy. For this reason, the spin-pairing of electrons makes it favorable for electrons in these orbitals to mix together and form four "sp-hybrid" orbitals. This behavior is known as **hybridization**. Hybridization will be very important as we show how chemical bonds form. The reason is because paired electrons are not available to form chemical bonds with other atoms! The diagram below shows how the 2s and one of the 2p orbitals behave as one hybrid orbital. Both boron and nitrogen have three unpaired electrons that can make chemical bonds. Carbon has four unpaired electrons and oxygen has only two.

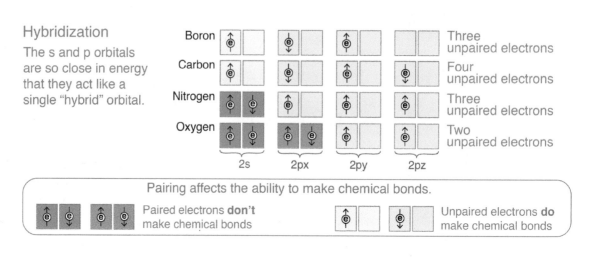

Emission and absorption of light

The ground state

When every electron in an atom occupies the lowest energy orbital (subject to the exclusion principle) the atom is said to be in the **ground state**. The ground state is the lowest energy configuration of an atom. For example, lithium has three electrons. A lithium atom in the ground has two electrons in the 1s orbital and one electron in the 2s orbital.

Atoms absorb energy through electrons

Electrons do not always stay in the ground state! Atoms change their energy by rearranging electrons between energy levels. This may occur through absorbing or emitting light, through electricity, or through collisions with other atoms. For example, one of lithium's 1s electrons might move to a 3s orbital. An atom is in an **excited state** when one or more electrons occupies an orbital higher than the ground state. A lithium atom with an electron in a 3s orbital is in an excited state.

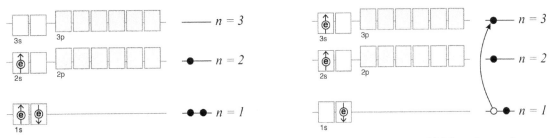

Ground states and excited states

Ground state of lithium (3 electrons)   An excited state of lithium (one of many)

Emission

Any atom in an excited state may lower its energy by emitting a photon of light. The energy of the photon and the color of the light depend on the energy difference (ΔE) between the starting and final energy levels. The frequency is given by Planck's relation $\Delta E = h\nu$. An atom can only lose an amount of energy equal to the difference between two energy levels. For example, a lithium atom that drops an electron from the 4s orbital level to the 2s orbital gives off blue-green light with a wavelength of 497 nanometers.

Absorption

Any atom may go from the ground state to an excited state by absorbing exactly the right amount of energy. An atom may only absorb a photon of light if the photon matches the energy difference between two energy levels and the higher energy level is unoccupied. For example, a lithium atom in the ground state may absorb a photon of blue-green light with a wavelength of precisely 497 nm and promote one electron from the 2s orbital to a 4s orbital.

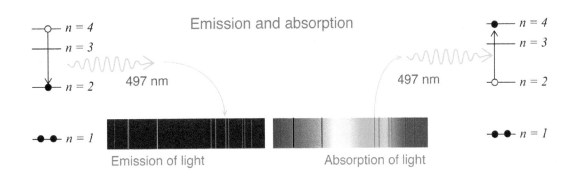

Section 9.2: The Quantum Atom

Spectroscopy

Orbital energy varies by element

While all atoms share the s, p, d, f structure of orbitals, the actual energy of each orbital varies with each element. For example, the 1s orbital of lithium has a different energy than the 1s orbital of hydrogen or carbon. One reason for the difference is that increasing the positive charge in the nucleus increases the attractive force on electrons. Another reason is that inner electrons partially "shield" the nuclear charge from outer electrons.

Spectra and spectral lines

The light emitted by atoms consists of discrete colors. A **spectrum** shows the individual colors that correspond to the energy transitions between energy levels. Each color in a spectrum is usually called a spectral line. When atoms are emitting light, bright lines appear against a dark background in an emission spectrum. When atoms are absorbing light, dark lines appear against a continuous rainbow of color. These black lines are called absorption lines and the resulting spectrum is an absorption spectrum.

Emission and absorption spectra of lithium

Spectroscopy

Every element has a unique signature in its line spectrum. Because the spectra of the elements are well known, it is often possible to deduce the identity of unknown elements from their spectra. For example we know that the composition of a certain distant star might be 79% hydrogen, 18% helium, and traces of other gases from the spectra of light observed through a telescope. **Spectroscopy** is the study of spectra for research and technology.

Emission spectra of different elements

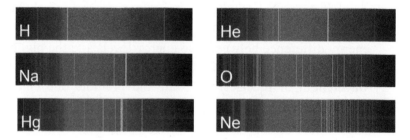

Spectrometers

An instrument that disperses light into its constituent wavelengths is called a **spectrometer**. The simplest kind of spectrometer is a simple prism. Light of any kind, not just white light, can enter the prism through one side and is dispersed into a spectrum as it exits out the other side. A research spectrometer has much greater ability to resolve the fine detail in the spectrum. For example, notice the double yellow lines in the mercury spectrum (Hg) above. A good spectrometer can separate the two into distinct lines. A simple prism would blur them together.

Section 2 review

Chapter 9

The differences in the chemical properties of the elements arise from the quantum behavior of electrons in atoms. Historically the electron was the first sub-atomic particle discovered, followed by the Curie's discovery of radioactivity in 1898. Rutherford's gold foil experiment demonstrated the existence of the nucleus in 1910. The critical clue to deducing electron structure was the observation of spectral lines. Bohr's original "electron orbits" model of the atom (1913) provided a mechanism for creating spectral lines but not an explanation. The theory of quantum mechanics states that particles such as electrons have wave properties, and waves such as light have particle properties. Planck's constant sets the scale at which quantum effects dominate over classical behavior.

An electron confined to an atom can only be stable when occupying a quantum state. The quantum states fall into four families of orbitals: s, p, d, and f. Each orbital can hold up to two electrons and no more. The orbitals have a different average distance from the nucleus and therefore, different energy. The energies of the orbitals group electrons into energy levels. The lowest energy levels fill first (Aufbau principle) to make the ground state of an atom, with each orbital in an energy level getting a single electron before any orbital gets two (Hund's rule). An atom is in an excited state when one or more electrons moves up an energy level from the ground state. An atom can emit photons (light) when an excited electron drops from a higher orbital to a lower orbital. The photon energy equals the energy difference between the electron's initial and final orbital. This is the explanation for spectral lines. Each element has a unique "fingerprint" of spectral lines called a spectrum.

Vocabulary words: spectral line, Planck's constant (h), Heisenberg's uncertainty principle, photon, quantum state, orbital, Aufbau principle, energy level, Pauli exclusion principle, spin, Hund's rule, hybridization, ground state, excited state, spectrum, spectroscopy, spectrometer

Review problems and questions

1. Consider Rutherford's gold foil experiment. Suppose 99% of the alpha particles went straight through but one alpha particle out of every hundred bounced backwards. What conclusion could you draw about the size of the nucleus? Explain the reasoning behind your answer.

2. Identify the total number of electrons each orbital type can hold.

    a. s orbitals have _____ electrons
    b. p orbitals have _____ electrons
    c. d orbitals have _____ electrons
    d. f orbitals have _____ electrons

3. What element has seven electrons in the second energy level, including s and p orbitals?

4. How many electrons does phosphorus have in the third energy level?

5. Write a summary of how the atomic model evolved over time. Include Dalton's model, Thomson's model, Rutherford's model, Bohr's model and the modern quantum model.

6. How does a series of different-colored lines in a line emission spectrum from an element provide evidence for the quantum model?

9.3 - Electron Configurations

In the next chapter we will see that, excepting for mass, every chemical and physical property of the elements depends 100% on how the electrons are organized in the outer "shell" of the atom. The elements in each group of the periodic table have common chemical properties because the electron configuration of the outermost electrons is similar. This section shows you how to find the electron configuration of any element and use the electron configuration to predict the element's chemical properties.

Writing an electron configuration

Reading an electron configuration

An **electron configuration** is a compact way of describing how the electrons are distributed among the s, p, d, and f orbitals. Consider hydrogen with an electron configuration of $1s^1$. The first number (1) is the principal quantum number, n. The letter (s) identifies the orbital type and the superscript (1) tells us there is one electron in the 1s orbital.

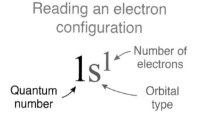

Electron configurations of hydrogen, lithium, and oxygen

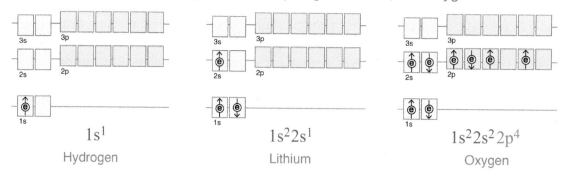

Finding an electron configuration

Lithium has an electron configuration of $1s^2 2s^1$. This tells us that there are two electrons in the 1s orbital and a third electron in the 2s orbital. The electron configuration of oxygen is $1s^2 2s^2 2p^4$. This tells us the 1s and 2s orbitals are both full and there are four electrons in the 2p orbital. There are two rules to determining the electron configuration.

Rules for finding electron configurations

(1) The order of filling orbitals follows this list from left to right.

$1s^2$	$2s^2 2p^6$	$3s^2 3p^6$	$4s^2 3d^{10} 4p^6$	$5s^2 4d^{10} 5p^6$	$6s^2 4f^{14} 5d^{10} 6p^6$	$7s^2 5f^{14} 6d^{10} 7p^6$
Energy level 1	**2**	**3**	**4**	**5**	**6**	**7**

Example: Mg magnesium $1s^2\ 2s^2 2p^6\ 3s^2$

(2) The total of all superscripts adds up the number of electrons.

$1s^2\ 2s^2 2p^6\ 3s^2$ $2 + 2 + 6 + 2 = 12$ Magnesium has 12 electrons

Energy levels and quantum numbers

Quantum numbers help us describe an electron's behavior. For s and p orbitals, the energy level (principal quantum number) corresponds to the row of the periodic table. However, d and f orbitals have more energy than some of the orbitals of higher energy levels. For example, the 3d orbital has a higher energy than the 4s orbital - even though 3d orbitals belong to the third energy level, 4s orbitals fill with electrons before the 3d orbitals.

The noble gases

Group 18 elements are noble gases

At the far right of the periodic table are the noble gases (group 18). The four lightest elements in this group include helium (He), neon (Ne), argon (Ar), and krypton (Kr). Historically these were labeled "noble" because they do not bond with any of the other elements (with a few rare exceptions).

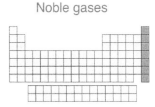
Noble gases

Why the noble gases are important

If you understand why the noble gases don't form chemical bonds, it becomes easier to understand why (and how) other elements do form bonds. Look at the electron configurations of helium, neon, argon, and krypton. What is common to all four?

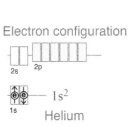
Electron configuration

Helium

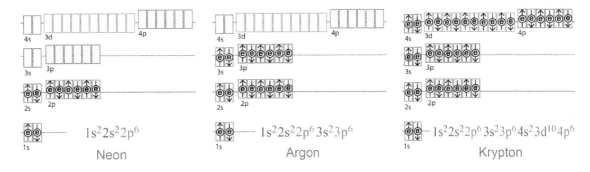
Electron configurations of the noble gases

Highest energy level

Take a close look at the outermost s and p orbitals, that is, those with the highest principal quantum number. Helium only has an s-orbital, but its s-orbital is full. Neon, argon and krypton have full outermost s and p orbitals which is seen as s^2p^6 below.

noble gases have full outermost s and p orbitals

$1s^2$ $1s^22s^22p^6$ $1s^22s^22p^63s^23p^6$ $1s^22s^22p^63s^23p^64s^23d^{10}4p^6$
Helium   Neon   Argon   Krypton

To bond or not to bond?

Elements that have energy levels completely filled with electrons do not make chemical bonds because a full energy level is very stable. As you might have guessed, elements that do make chemical bonds have unfilled energy levels that are able to take in or eliminate electrons. Take a moment to commit these important ideas to your memory.

1. **Electrons in an unfilled energy level are available to form chemical bonds.**

2. **Electrons in completely full energy levels do not form chemical bonds.**

The importance of the outermost electrons

Earlier we stated that electrons determine all the chemical properties of an element. We can now refine this statement significantly. Only the electrons in the highest occupied energy level interact with other atoms. These outermost electrons are the ones that determine chemical behavior.

Group I and group II metals

Electron configuration

The alkali metals (group 1) and alkaline earth metals (group 2) all lose electrons easily. Sodium becomes a positive ion (Na^+) when it gives up an electron. Lithium and potassium also give up an electron to become positive ions (Li^+, K^+). Can you see how this characteristic comes directly from their electron structure?

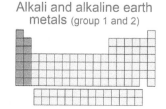
Alkali and alkaline earth metals (group 1 and 2)

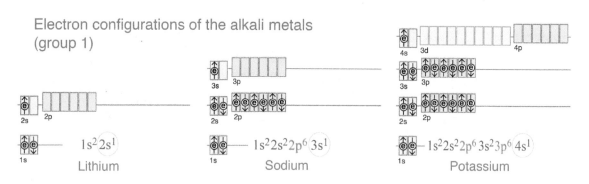

Electron configurations of the alkali metals (group 1)

Lithium $1s^2 2s^1$

Sodium $1s^2 2s^2 2p^6 3s^1$

Potassium $1s^2 2s^2 2p^6 3s^2 3p^6 4s^1$

Why alkali metals form +1 positive ions

All the alkali metals have a single electron in the highest populated energy level. Lithium's third electron ($2s^1$) is alone on the second energy level. Sodium's eleventh electron ($3s^1$) is alone on the third energy level. Potassium's nineteenth electron ($4s^1$) is alone on the fourth energy level. Each of these elements can get to a noble gas electron configuration by losing one electron. Lithium, sodium, and potassium form chemical bonds with other elements by transferring one electron to achieve a full energy level.

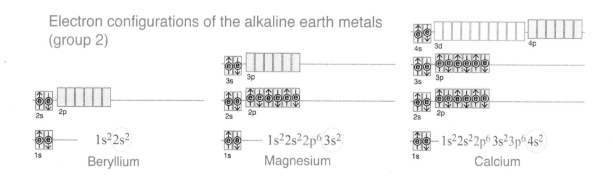

Electron configurations of the alkaline earth metals (group 2)

Beryllium $1s^2 2s^2$

Magnesium $1s^2 2s^2 2p^6 3s^2$

Calcium $1s^2 2s^2 2p^6 3s^2 3p^6 4s^2$

Group two elements form +2 ions

Beryllium (Be), magnesium (Mg) and calcium (Ca) are the most common of the alkaline earth metals (group 2). These elements also tend to lose electrons easily but, they tend to lose 2 electrons instead of 1, making their positive ions Be^{2+}, Mg^{2+} and Ca^{2+}. Can you see how this can be predicted from the electron configuration?

Electron configuration of group two elements

All group 2 elements have two electrons in the highest populated energy level. Beryllium ($2s^2$) has only two electrons in the second energy level, leaving six unfilled orbitals. Magnesium ($3s^2$) and potassium ($4s^2$) both have two electrons in the second energy level, leaving six unfilled orbitals.

How group two elements bond

Beryllium, magnesium, and calcium form chemical bonds with other elements by transferring two electrons to achieve a full energy level; for example, beryllium oxide (BeO), magnesium oxide (MgO), and calcium oxide (CaO).

The halogens

Explaining halide behaviors

The behavior of the noble gases and alkali metals leads to the following statements, which explain much of how and why chemical bonds occur.

> Atoms with completely filled energy levels do not form bonds because they already have the lowest energy they can have.

> Atoms with *unfilled* energy levels can achieve lower energy by changing their electron configuration to match one of the noble gasses.

The electron configuration rules

The **halogens** (group 17) include fluorine (F), chlorine (Cl) and bromine (Br), and are just before the noble gases on the periodic table. Unlike metals, the halogens tend to take one electron instead of giving up any. Fluorine forms the negative ion F⁻, chlorine forms the negative ion Cl⁻ and bromine forms the negative ion Br⁻. The diagram below shows the electron configuration for these three elements. Can you see why these elements tend to gain an electron?

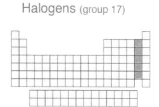
Halogens (group 17)

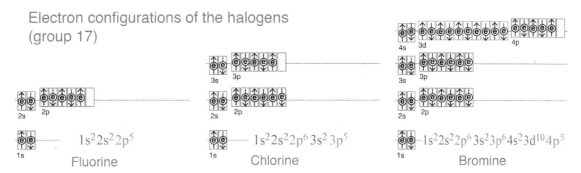

Electron configurations of the halogens (group 17)

$1s^2 2s^2 2p^5$ — Fluorine

$1s^2 2s^2 2p^6 3s^2 3p^5$ — Chlorine

$1s^2 2s^2 2p^6 3s^2 3p^6 4s^2 3d^{10} 4p^5$ — Bromine

Halogens are very reactive

By acquiring one electron, each of the halogens achieves the low-energy state of having completely filled energy levels. The "energy benefit" of attracting the needed electron is so great that the halogens are extremely reactive and do not occur naturally as pure elements. In their laboratory-purified form, these elements form diatomic molecules such as Cl_2 and F_2. Halogens fill their outer s and p orbitals when they gain electrons to form ions.

Electron configurations of halogens as ions

$1s^2 2s^2 2p^{(5+1)}$ $1s^2 2s^2 2p^6 3s^2 3p^{(5+1)}$ $1s^2 2s^2 2p^6 3s^2 3p^6 4s^2 3d^{10} 4p^{(5+1)}$
Fluoride ion, F⁻   Chloride ion, Cl⁻   Bromide ion, Br⁻

Chlorine is a necessary nutrient

All the pure halogens are highly toxic to many organisms. As an example, chlorine is used to sterilize drinking water and swimming pools because free chlorine kills bacteria and other harmful microorganisms. However, when combined, sodium and chlorine become NaCl, or table salt. Salt is a necessary nutrient for life and maintains electrical conductivity in neurons among other important functions. Combined with hydrogen, chlorine forms hydrochloric acid (HCl) which is produced in your stomach to digest food.

Section 9.3: Electron Configurations

Carbon, nitrogen, and oxygen

About carbon, oxygen, and nitrogen

Carbon, nitrogen, and oxygen are among the most important elements to life. Carbon (C) forms the backbone of biological molecules, such as proteins, fats, sugars, and nucleic acids. Nitrogen gas (N_2) makes up 78% of Earth's atmosphere and oxygen gas (O_2) makes up 21% of the atmosphere. Nearly 40% of all the atoms in your body are carbon, nitrogen and oxygen atoms. An extraordinary 86.7% of your body mass comes from carbon, oxygen and nitrogen atoms.

Electron structure makes these elements versatile

The electron structures of carbon, oxygen, and nitrogen allow many different bonding patterns. This versatility is why these elements are so important to life. Oxygen and nitrogen both tend to accept electrons, rather than donate them. Carbon can go either way, sometimes accepting electrons and sometimes donating them.

Carbon, Nitrogen, Oxygen

Electron configurations of carbon, nitrogen, and oxygen

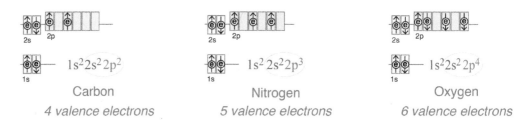

Carbon — $1s^2 2s^2 2p^2$ — 4 valence electrons

Nitrogen — $1s^2 2s^2 2p^3$ — 5 valence electrons

Oxygen — $1s^2 2s^2 2p^4$ — 6 valence electrons

Valence electrons

Electrons that are in unfilled energy levels are called **valence electrons**. Valence electrons participate in chemical bonds. Electrons in "closed shells" do not participate in making chemical bonds. The concept of valence is critical to understanding chemistry. For example, carbon has six electrons. Two are in the full 1s orbital, making a closed shell since all the quantum states are occupied. The second energy level has only four of the eight quantum states occupied. Carbon therefore has four valence electrons, the four in the unfilled second energy level. Each of those four valence electrons is available to form a bond by pairing with an unpaired electron from another atom. In methane (CH_4), for example, each of the four hydrogens contributes its unpaired lone electron to a bonded pair with one of carbon's unpaired valence electrons.

Why carbon is important

Carbon is the lightest, and most abundant element in the universe that can make four bonds. This is the primary reason carbon is so important to life. As we will see in later chapters, life requires complex molecules, such as DNA and proteins. The versatility of carbon's bonding abilities is why all biological molecules are built around structures of carbon.

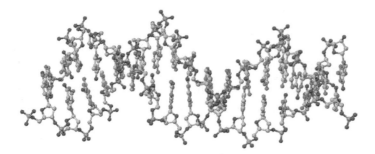

A tiny fragment of a DNA molecule showing the core of carbon (grey) atoms. A full DNA molecule has more than 10 billion atoms.

The hydrogen atoms are not shown.

Valence electrons

Valence electrons

Valence electrons occupy the highest energy level in an atom. Being in the highest energy level makes them the outermost electrons in an atom. Valence electrons are shared in chemical bonds. Valence electrons are also the electrons which are lost or gained when ions are formed. For example, carbon ($1s^2 2s^2 2p^2$) has four valence electrons. The highest energy level is 2 and there are two electrons in the 2s-orbital and two more in the 2p-orbitals. The rules below will help you find the number of valence electrons.

Rules for determining the number of valence electrons

(1) Start with the electron configuration of the ground state.

magnesium (Mg) $1s^2 2s^2 2p^6 3s^2$

(2) Starting from the 1s orbital, neglect all electrons in full energy levels.

$1s^2 2s^2 2p^6 3s^2$

(3) The electrons remaining in the highest populated energy level are the valence electrons.

$1s^2 2s^2 2p^6 \,\boxed{3s^2}$ Magnesium has two valence electrons

The importance of valence electrons

The purpose of this chapter is to understand the chemical properties of the elements. We started with the nucleus and the electron cloud, then explored how electrons occupy quantum states (orbitals) grouped into energy levels. Now we focus on the highest populated energy level, the electrons that form the outer shell of an atom. These outer electrons determine the chemical properties of the element. Valence electrons also explain why chlorine forms ions with a charge of 1− and magnesium forms ions with a charge of 2+.

Solved problem

Determine the number of valence electrons for boron.

- Asked: The number of valence electrons
- Given: The element is boron, atomic number 5
- Relationships: Valence electrons are electrons in the highest populated energy level.
- Solve: Boron has five electrons, therefore its electron configuration is:
  $1s^2 2s^2 2p^1$
  The 1s shell is full so does not count towards valence. There are three electrons in the second energy level, which is the highest energy level in boron's configuration.
- Answer: Boron has three valence electrons.

Solved problem

Determine the number of valence electrons for phosphorus (P).

- Asked: The number of valence electrons
- Given: The element is phosphorus, atomic number 15
- Relationships: Valence electrons are electrons in the highest populated energy level.
- Solve: Phosphorus has 15 electrons, therefore its electron configuration is:
  $1s^2 2s^2 2p^6 3s^2 3p^3$
  The 1s, 2s, and 2p orbitals are full so the first and second energy levels do not count towards valence. There are 5 electrons in the third level.
- Answer: Phosphorus has five valence electrons.

The transition metals and noble gas core notation

What are the transition metals?

The **transition metals** are the elements in the middle of the periodic table. The transition metals include familiar "metals" like iron (Fe), nickel (Ni) and copper (Cu) as well as exotic metals, such as osmium (Os) and "noble" metals, such as gold (Au) and silver (Ag). All the transition metals are solid at room temperature except mercury (Hg) and all are excellent conductors of electricity.

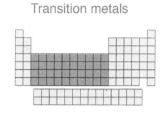
Transition metals

Transition metals have electrons in d orbitals

The transition metals all have electrons in full or partly filled d orbitals. Since there are no d orbitals in the first, second and third rows of the table, the transition metals first appear on the fourth row. The transition metals illustrate the fact that the 3d orbitals have higher energy than the 4s orbitals, even though the principal quantum number (3) of 3d is lower than the principal quantum number (4) of 4s.

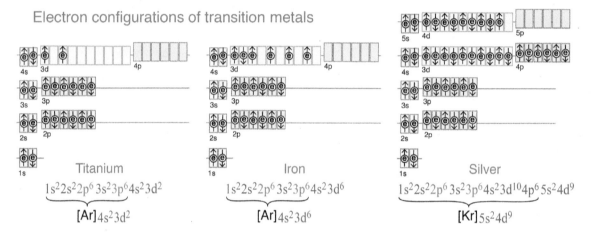

Electron configurations of transition metals

Titanium: $1s^2 2s^2 2p^6 3s^2 3p^6 4s^2 3d^2$ — $[Ar]4s^2 3d^2$

Iron: $1s^2 2s^2 2p^6 3s^2 3p^6 4s^2 3d^6$ — $[Ar]4s^2 3d^6$

Silver: $1s^2 2s^2 2p^6 3s^2 3p^6 4s^2 3d^{10} 4p^6 5s^2 4d^9$ — $[Kr]5s^2 4d^9$

Using the noble gas core notation

For elements past the second row of the periodic it is tedious to write the entire electron configuration, especially when the real importance is only the valence electrons. Chemists have a short-hand way to write complex electron configurations using the noble gas core. For example, the electron configuration of argon is written $[Ar] = 1s^2 2s^2 2p^6 3s^2 3p^6$. As the diagram above shows, the electron configuration of titanium is the same as argon plus $4s^2 3d^2$. The following two ways to write the electron configuration are equivalent.

$$[Ar]4s^2 3d^2 \;=\; 1s^2 2s^2 2p^6 3s^2 3p^6 4s^2 3d^2$$

Noble gas core notation for electron configurations

$[He] = 1s^2$

$[Ne] = 1s^2 2s^2 2p^6$

$[Ar] = 1s^2 2s^2 2p^6 3s^2 3p^6$

$[Kr] = 1s^2 2s^2 2p^6 3s^2 3p^6 4s^2 3d^{10} 4p^6$

$[Xe] = 1s^2 2s^2 2p^6 3s^2 3p^6 4s^2 3d^{10} 4p^6 5s^2 4d^{10} 5p^6$

$[Rn] = 1s^2 2s^2 2p^6 3s^2 3p^6 4s^2 3d^{10} 4p^6 5s^2 4d^{10} 5p^6 6s^2 4f^{14} 5d^{10} 6p^6$

Transition metals and valence electrons

The bonding patterns of the transition metals are complex because they have so many valence electrons. Virtually all these elements have multiple ways they share electrons. For example, iron (Fe) forms FeO (iron(II) oxide), Fe_2O_3 (iron(III) oxide - rust) and Fe_3O_4 (iron(II,III) oxide, magnetite). The oxygen atoms behave the same way but the iron atoms share electrons differently in each of these three compounds.

Using electron configurations

Electron configuration problems

The electron configuration is the basis for understanding chemical bonds, which are the subject of the next chapter.

1. Writing down the electron configuration when given the element or identifying an element from its electron configuration.
2. Predicting or explaining a chemical behavior using the electron configuration.

Problems with electron configuration tend to fall into several types. However, in all cases the same two rules stated earlier still apply.

Rules for finding electron configurations

(1) The order of filling orbitals follows this list from left to right.

$1s^2\ \ 2s^2 2p^6\ \ 3s^2 3p^6\ \ 4s^2 3d^{10} 4p^6\ \ 5s^2 4d^{10} 5p^6\ \ 6s^2 4f^{14} 5d^{10} 6p^6\ \ 7s^2 5f^{14} 6d^{10} 7p^6$

Energy level	1	2	3	4	5	6	7

Example: Mg magnesium $1s^2\ \ 2s^2 2p^6\ \ 3s^2$

(2) The total of all superscripts adds up the number of electrons.

$1s^2\ \ 2s^2 2p^6\ \ 3s^2$ $2 + 2 + 6 + 2 = 12$ Magnesium has 12 electrons

---

Solved problem

Name the first three elements on the periodic table that have a p^2 valence electron configuration. What do all 3 elements have in common?

Asked To identify the first three elements with two valence electrons in a p orbital, and find out what they share in common.

Given The problem specifies valence electrons - so these are the highest populated energy level.

Relationships The order of filling orbitals is:
$1s^2 2s^2 2p^6 3s^2 3p^6\ 4s^2 3d^{10} 4p^6\ 5s^2 4d^{10} 5p^6\ 6s^2 4f^{14} 5d^{10} 6p^6$

Solve The first element with $2p^2$ valence electrons has electron configuration:
$1s^2 2s^2 2p^2$ or $[He]2s^2 2p^2$
This is carbon, because it has six electrons and an atomic number of six.

The next element with $3p^2$ valence electrons has electron configuration:
$1s^2 2s^2 2p^6 3s^2 3p^2$ or $[Ne]3s^2 3p^2$
This element has 14 electrons and therefore is silicon.

The next element with $4p^2$ valence electrons has electron configuration:
$1s^2 2s^2 2p^6 3s^2 3p^6 4s^2 3d^{10} 4p^2$ or $[Ar]4s^2 3d^{10} 4p^2$
This element has 32 electrons and therefore is germanium.

Answer The three elements are carbon, silicon, and germanium. All are in group 14 of the periodic table.

Electron configurations and the periodic table

Explaining Mendeleev's discovery

For the first periodic table, Dmitri Mendeleev grouped the elements in order of how they combine. Elements with 3 electrons (Li), 9 electrons (Na), 19 electrons (K), and 37 electrons (Rb) combine with oxygen in 2:1 ratios to make Li_2O, Na_2O, K_2O, and Rb_2O. Elements with 2, 10, 18, and 36 electrons are all noble gases (He, Ne, Ar, and Xe) that make no chemical bonds with other elements. Elements with 4 electrons (Be), 12 electrons (Mg), and 20 electrons (Ca) combine with oxygen in a 1:1 ratio to make BeO, MgO, and CaO. The repeating pattern of electron configurations explain the repeating pattern of similar chemical behavior.

Valence electron configurations for the first 54 elements

#	Symbol	Config
1	H	$1s^1$
2	He	$1s^2$
3	Li	$2s^1$
4	Be	$2s^2$
5	B	$2p^1$
6	C	$2p^2$
7	N	$2p^3$
8	O	$2p^4$
9	F	$2p^5$
10	Ne	$2p^6$
11	Na	$3s^1$
12	Mg	$3s^2$
13	Al	$3p^1$
14	Si	$3p^2$
15	P	$3p^3$
16	S	$3p^4$
17	Cl	$3p^5$
18	Ar	$3p^6$
19	K	$4s^1$
20	Ca	$4s^2$
21	Sc	$3d^1$
22	Ti	$3d^2$
23	V	$3d^3$
24	Cr	$3d^4$
25	Mn	$3d^5$
26	Fe	$3d^6$
27	Co	$3d^7$
28	Ni	$3d^8$
29	Cu	$3d^9$
30	Zn	$3d^{10}$
31	Ga	$4p^1$
32	Ge	$4p^2$
33	As	$4p^3$
34	Se	$4p^4$
35	Br	$4p^5$
36	Kr	$4p^6$
37	Rb	$5s^1$
38	Sr	$5s^2$
39	Y	$4d^1$
40	Zr	$4d^2$
41	Nb	$4d^3$
42	Mo	$4d^4$
43	Tc	$4d^5$
44	Ru	$4d^6$
45	Rh	$4d^7$
46	Pd	$4d^8$
47	Ag	$4d^9$
48	Cd	$4d^{10}$
49	In	$5p^1$
50	Sn	$5p^2$
51	Sb	$5p^3$
52	Te	$5p^4$
53	I	$5p^5$
54	Xe	$5p^6$

Practice problem

Name two elements that will have similar chemical behavior to calcium (Ca).

Asked To identify two elements with similar chemistry to calcium

Given Calcium has atomic number 20 and electron configuration: $1s^2 2s^2 2p^6 3s^2 3p^6 4s^2$

Relationships Elements with similar valence electrons have similar chemical properties.

Solve Beryllium has a $2s^2$ valence electron configuration and magnesium has $3s^2$ valence electron configuration.

Since both of these elements have the same valence electron configuration they will have similar chemical properties.

For example, each combines with oxygen in a 1:1 ratio to make BeO, MgO, and CaO.

Answer Beryllium and magnesium also have a s^2 valence electron configuration and will have similar chemical properties to calcium.

Atomic radius

The meaning of the atomic radius

The atomic radius is a measure of how close atoms get to each other when connected by chemical bonds. The fuzzy electron cloud really does not have a definite boundary so the radius of a single isolated atom is not well defined. For example, in a molecule the nuclei of the oxygen and carbon atoms are separated by the sum of their atomic radii. The diagram below shows how the atomic radius varies across the periodic table.

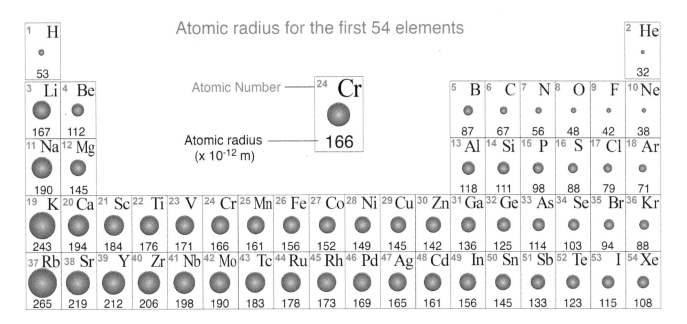

Atoms get smaller across the periodic table

Notice that the atomic radius gets smaller as one moves from left to right across the periodic table. For example, oxygen has a smaller atomic radius than lithium even though oxygen has more electrons! The explanation is that oxygen, with six protons, also has a stronger charge in the nucleus compared to lithium with only three protons. The larger charge of the nucleus attracts electrons more strongly and as a result the atom gets smaller until the energy level is filled. Once an energy level is filled another effect takes over.

atomic radius trend

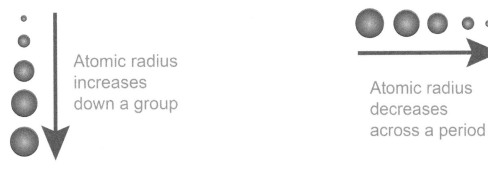

Atomic radius increases down a group

Atomic radius decreases across a period

Atoms get larger with energy level

Atoms get larger as energy level increases. The higher the energy level, the larger the atom. Potassium with four energy levels is larger than lithium with only two energy levels. This is because the higher energy levels are farther from the nucleus and their electrons are also partly shielded by inner electrons. The "jump point" in atomic radius occurs at the closing of each energy level. For example, the atomic radius of argon is only 71×10^{-12} m while the next element, potassium, jumps in atomic radius to 243×10^{-12} m.

Section 9.3: Electron Configurations

Ionic radius

Ionic radius

An *ionic radius* is the approximate size of an atom's ion. Ionic radii are a measure of how close ions get to each other when fixed in a crystal lattice, but this measurement is somewhat uncertain. Ions can change their size depending on other ions that surround them. The ionic size trend is similar to atomic size as you might expect. Noble gases are not included since they do not tend to form ions. If you move across period 2 you will see ionic radius decrease just like atomic size. However, you will notice a drastic jump in ionic radius between carbon and nitrogen.

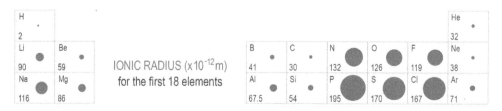

Ions and valence electrons

The jump in atomic size you saw on the previous page was due to the shielding effect of higher energy levels. But there is no shielding effect across a period because the energy level stays the same. The highest energy level across all of period 2 is the second energy level. The main reason why an ionic radius gets bigger depends on whether valence electrons are given up or gained by an ion. Moving across period 2, lithium, beryllium and boron will make positively charged ions because they will give up valence electrons to achieve a noble gas electron configuration. Starting at nitrogen, atoms will gain electrons to achieve a noble gas configuration. When an atom loses valence electrons its electron cloud shrinks. When valence electrons join an atom its electron cloud gets larger.

Trend across a period

Consider the ionic radius trend in terms of the type of ion formed. As you move left to right across a period ionic size decreases for metals. Since nonmetals gain electrons when they form ions, ionic radius will jump higher for the first nonmetal in a period. From there ionic radius will decrease across the remainder of the nonmetals in a period. Noble gases do not tend to form ions, but given enough ionization energy, they could be made to form ions.

Trend down a group

The trend in ionic radius generally increases down a group as you would expect. Ions with fewer energy levels are smaller than ions with more energy levels.

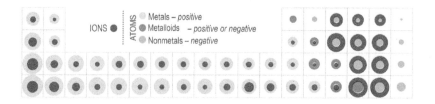

Through the d-block

The trend in ionic radius is less predictable through the d-block transition elements. Transition metals commonly form a variety of ions due to their partially filled d-orbitals.

Ionization energy

Definition of ionization energy

Ionization is the removal of one or more electrons from an atom leaving a positive ion. The amount of energy it takes to remove an electron is closely related to the electron structure. The **ionization energy** is the amount of energy it takes to remove the outermost electron from an atom. The element with the highest ionization energy is helium. All noble gases have high ionization energies because all have a completely full outer electron shell.

Units of ionization energy

Ionization energy is a measure in units of kilojoules (1,000 J) per mole (kJ/mol). One kilojoule per mole is a large amount of energy. As an example, one mole of hydrogen gas has a mass of one gram. To store one kilojoule of energy the gram of hydrogen must be at a temperature of 12,000 degrees! As another consideration it takes only 13.8 kJ/mole to melt iron into liquid. The ionization energy of iron (762 kJ/mol) is enough to boil the liquid iron into gas and then heat the gas to thousands of degrees.

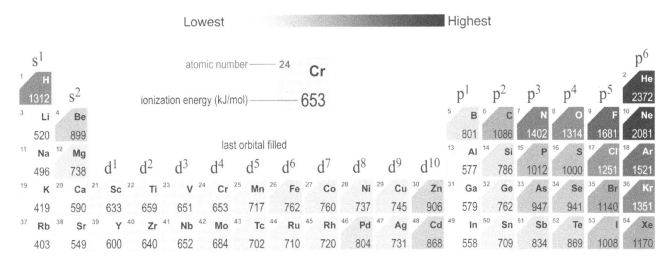

Ionization energies for the first 54 elements

Noble gases have high ionization energy

The noble gases have the highest ionization energies because they have full energy levels. Electrons in full energy levels are tightly bound to the atom. Within the noble gases, increasing atomic number makes the ionization energy lower. For example, the ionization energy of neon is 2,081 kJ/mol and the ionization energy for xenon is only 1,170 kJ/mol. This is because the higher energy levels are farther from the nucleus and the attractive force is partially shielded by electrons in lower energy levels.

Alkali metals low ionization energy

Across the periodic table, the ionization energy increases the closer the element gets to a fully populated orbital. The ionization energy is lowest when there is only a single electron in an orbital. The lowest ionization energies belong to the alkali metals which have a single s^1 electron. Sodium has an ionization energy of only 496 kJ/mol, about one fifth that of helium. Within the alkali metals we see the same trend toward decreasing ionization energy moving down the periodic table from lithium (Li) to rubidium (Rb).

Transition metals

Notice how the ionization energy increases through the transition metals, then decreases dramatically immediately after. For example, zinc is the last transition metal in the fourth period. The ionization energy for zinc is 906 kJ/mol, but the element immediately to the right, gallium, has a much lower ionization energy of 579 kJ/mol. This is because zinc has no partially-filled orbitals, but gallium does.

Chapter 9
Section 3 review

Electron configurations describe how electrons are distributed among orbitals in an atom. Chemical behavior depends on the outermost valence electrons in the highest filled energy level. Noble gases have completely filled energy levels so they make no chemical bonds. Alkali metals easily lose their single valence to form 1+ ions. Halogens are one electron short of a full outer energy level and form 1- ions by acquiring an electron. Carbon, nitrogen, and oxygen have a valence electron structure that allows a rich variety of chemical bonds. Transition metals have complex bonding patterns due to their many valence electrons.

Ionization energy and ionic radius are periodic properties explained by the pattern in repeating electron configurations. Alkali metals have the lowest ionization energy and noble gases have the highest. Because electrons form the "outer shell" of atoms, many element properties are directly related to their electron configuration. In many respects the periodic table is a map of the electron configuration of the elements.

Vocabulary words: electron configuration, halogens, valence electrons, transition metals, atomic radius, ionic radius, ionization, ionization energy

Review problems and questions

1. Name three elements that have a p^4 valence electron configuration.

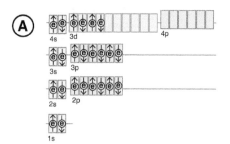 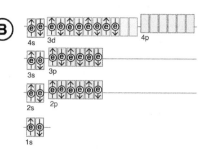

2. For the electron structures (A and B) shown in the diagram above,
    a. identify the element;
    b. write the electron configurations with all energy levels and orbitals;
    c. write the electron configurations using the noble gas core notation.

3. Which of the following pairs of elements have the same valence electron structure, and therefore have similar chemical properties? (There may be more than one.)
    a. potassium and beryllium
    b. chromium and molybdenum
    c. oxygen and sulfur
    d. iron and aluminum

4. Use the electron configuration to predict which of the following elements is chemically unlike the others: aluminum, boron, magnesium, gallium.

5. According to the trend in ionization energies, which element will have a higher ionization energy: beryllium, or fluorine? How is electron configuration related to the higher ionization energy?

6. According to the trend in ionic radius, which element becomes larger when it forms an ion, beryllium or fluorine? How is electron configuration related to the higher ionization energy?

Fireworks & Color

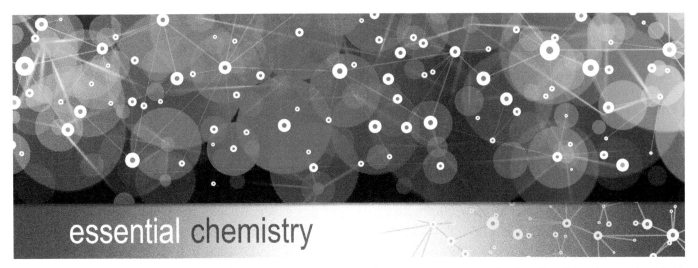
essential chemistry

Imagine a warm summer evening filled with sudden loud noises and bursts of color that illuminate the night sky. Don't panic, it's just a fireworks show! Did you know that there is a surprising amount of chemistry in those spectacular displays of light, sound and color?

Quite a lot has to happen to launch a firework up in the sky. When the fuse is lit, it quickly reaches the lifting charge shown on the diagram. The lifting charge is a mixture of potassium nitrate, charcoal and sulfur – commonly called black powder or gunpowder. When heated with the fuse, this mixture creates an exothermic reaction that drives the firework skyward.

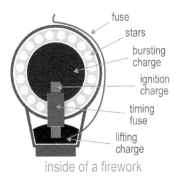
inside of a firework

A second fuse is also part of the firework; look for it in the diagram. This is a timing fuse, and it is designed to go off once the firework reaches the appropriate altitude. When this timing fuse goes off, then the burst charge is ignited. The burst charge is what sets off the spectacular show of sight and sound. The explosion of the burst charge sends out "stars" that were carefully embedded in the firework. The spatial arrangement of the stars in the charge determines the design displayed in the sky. This is how you can get multi-colored patterns and shapes like smiley faces in the sky. Look at the figure and imagine a three-dimensional sphere of these stars that covers the entire surface of a firework. What kinds of color patterns and designs would you want to create?

If it were just about shooting "stars" into the sky, fireworks would not be particularly impressive. But there is so much more to the display – and so much chemistry involved!

Fireworks & Color - 2

The stars contain chemicals that give off a characteristic color. The chemicals in the stars are metal powders or metal salts that impart a known color when "excited." Heat from the explosion provides energy that excites the electrons of the metal atom (or ion). In many fireworks an oxidizer is added to the gunpowder mixture. The oxidizer, usually a nitrate, chlorate, or perchlorate provides oxygen to produce a hot exothermic reaction. Electrons in the metals use this energy to transition from their ground states to their excited states.

oxygen sources	
ClO_4^-	perchlorate
ClO_3^-	chlorate
NO_3^-	nitrate

The excited electrons quickly return to their original energy level and give off the excess energy in the form of light. For example, strontium ions from strontium nitrate absorb energy from the oxidation reaction, then release photons of light in the 650-700 nm range. We see this as a red burst in the sky. The specific electron configuration of the metal atom or ion is the key to the color, but the selection of the compound is important too. Some compounds are hygroscopic, meaning they attract and hold moisture, and these would not make an effective firework even though they contain the correct metal.

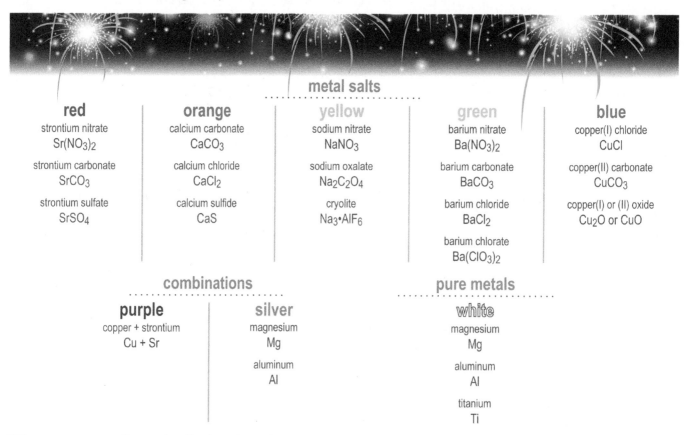

metal salts

red	orange	yellow	green	blue
strontium nitrate $Sr(NO_3)_2$	calcium carbonate $CaCO_3$	sodium nitrate $NaNO_3$	barium nitrate $Ba(NO_3)_2$	copper(I) chloride $CuCl$
strontium carbonate $SrCO_3$	calcium chloride $CaCl_2$	sodium oxalate $Na_2C_2O_4$	barium carbonate $BaCO_3$	copper(II) carbonate $CuCO_3$
strontium sulfate $SrSO_4$	calcium sulfide CaS	cryolite $Na_3 \cdot AlF_6$	barium chloride $BaCl_2$	copper(I) or (II) oxide Cu_2O or CuO
			barium chlorate $Ba(ClO_3)_2$	

combinations

purple	silver
copper + strontium Cu + Sr	magnesium Mg
	aluminum Al

pure metals

white
magnesium Mg
aluminum Al
titanium Ti

When you go to a fireworks display, the colors get a lot of attention, but there are also sounds associated with the explosion. Those bangs, crackles, and whistles are also the result of some interesting chemistry.

Bangs are the result of the explosive mixture of an oxidizer (potassium chlorate or potassium perchlorate), sulfur, and a metal (usually aluminum). When ignited this mixture produces a flash of light and a loud bang. This is used so frequently that it is often referred to as a "flash and bang" mixture.

The crackling of fireworks is the result of the combustion of another compound. Originally it was made with a lead compound mixed with an alloy of aluminum and magnesium called magnalium. An alloy is a solid metal-metal mixture. This combination is then divided up into small pieces. When these small granules are combusted rapidly, a crackling sound is produced.

Since lead compounds are toxic, bismuth compounds like Bi_2O_3 and $(BiO)_2CO_3$ are now used with magnalium to create the crackling effect.

| potassium benzoate | sodium salicylate | picric acid | gallic acid |

The compounds that create whistling are even more interesting. Like many of the explosive mixtures before, an oxidizing agent like potassium perchlorate is present. In this case the oxidizing agent is mixed with gallic acid, sodium salicylate, potassium benzoate or picric acid and tightly packed in a tube. If you look at the pictures of those compounds, you'll notice that they all contain a ring of 6 carbons. You'll learn more about these types of compounds in later chapters. For now, it is important to know that small explosions of these compounds in the tube creates oscillations in the gases created by the combustion. The whistling sound is a result of the standing wave produced from the oscillations.

You might not have realized it before, but now you can see that there is a lot of applied chemistry that goes into making fireworks. There is also a lot of carefully choreographed engineering involved. The lift charge needs to run out just as the second fuse ignites the burst charge. The colors from the stars are timed to explode and ignite with the appropriate whistles, bangs and crackles. If things detonate too early or late, then the firework will be too close to the ground making it less spectacular and more dangerous. These and many other reasons make fireworks dangerous to handle unless you have been highly trained. So, the next time you see a fireworks display, think about safety and essential chemistry!

Chapter 9 review

Vocabulary
Match each word to the sentence where it best fits.

Section 9.1

atomic mass	atomic number (Z)
ion	isotopes
mass number (A)	strong nuclear force

1. The number of neutrons plus the number of protons is called the ____.
2. A(n) ____ is an atom or group of bonded atoms that have an imbalance of protons and electrons.
3. The ____ of an atom is a weighted average of all of the isotopes found in nature.
4. ____ is the number of protons in the nucleus of an atom.
5. The ____ holds the nucleus together.
6. Two or more atoms with the same number of protons and different number of neutrons are called ____.

Section 9.2

Aufbau principle	energy level
excited state	ground state
Heisenberg's uncertainty principle	Hund's rule
hybridization	orbital
Pauli exclusion principle	photon
Planck's constant (h)	quantum state
spectral line	spectrometer
spectroscopy	spectrum
spin	

7. The ____ is the lowest energy state available to an electron in an atom.
8. The ____ states that electrons enter the lowest available energy level before higher energy levels.
9. A dark or bright line in an otherwise uniform and continuous spectrum is called a(n) ____.
10. ____ states that each sub-orbital will fill with one electron before any sub-orbital accepts two electrons.
11. A(n) ____ is a state for an electron in an atom described by a set of quantum numbers.
12. An instrument that measures a spectrum by dispersing light into its constituent wavelengths is called a(n) ____.
13. ____ is a quantum property of electrons that has a value of ± 1/2.
14. An electron is in a(n) ____ if there is a lower energy state that it could be occupying but it is instead occupying a higher one.
15. A(n) ____ is the smallest possible quantity of light.
16. Each quantum state inside an atom is called a(n) ____.
17. The ____ states that no more than two electrons can occupy any given orbital.
18. The ____ states that one cannot know the velocity and the position of a particle simultaneously.
19. Quantum states that have similar energy are said to be in the same ____.
20. The individual colors that correspond to the energy transitions between energy levels are shown in a(n) ____.
21. ____ is commonly used as a reference to see if quantum effects are relevant.
22. Orbitals can undergo ____ where they combine to form new form.
23. The use and study of spectra is called ____.

Section 9.3

atomic radius	electron configuration
halogens	ionization
ionization energy	transition metals
valence electrons	

24. The process of removing an electron from an atom is called ____.
25. ____ are a group of diatomic elements that tend to take one electron instead of giving any up.
26. The amount of energy required to remove one electron from the valence shell of an atom is called the ____.
27. A(n) ____ is a compact way of describing how the electrons are distributed among the orbitals.
28. The ____ is a measure of the distance from the center of the nucleus to the boundary of the adjacent cloud of electrons.
29. Iron, nickel, and copper are examples of ____.
30. ____ are the outermost electrons in an atom.

Conceptual questions

Section 9.1

31. How is Dalton's view of the atom similar to that of Democritus?
32. Dalton thought that chemical reactions could rearrange atoms, but not create or destroy them. What is another name for this idea?
33. What is different in the ^{14}N and ^{15}N isotopes?
34. A neutron and a proton have the same mass of 1 amu. Carbon has six protons and six neutrons. Why is the mass on the chart 12.011 amu instead of 12 amu?
35. What part of the atom gives the atom most of its properties? Explain your reasoning.
36. What is the difference between a molecular and a monatomic ion?

Chapter 9 review

Section 9.1

37. What part of the atom has the most mass?
38. What part of the atom takes up the most room?
39. What is equal in a neutral atom?
40. What changes when the Z number goes from Z = 6 to Z = 7?
41. What parts of the nucleus are attracted to each other and what parts are repulsed by each other?
42. Write the four primary forces in order of strongest to weakest.
43. What is the relationship between the strength of the four primary forces and the distance the forces reach?
44. What is the Z number for tungsten and what does it tell you about the tungsten nucleus?
45. In a negatively charged ion, which are more abundant, protons or electrons?
46. What determines the size of an atom?

Section 9.2

47. According to Rutherford's gold foil experiment:

    a. What takes up 99.99% of the space in an atom?

    b. Explain how he discovered the nucleus.

    c. How did he know the nucleus was very small?

48. What makes a hybrid orbital different from a standard orbital?
49. With reference to ground state and excited state:

    a. What is the difference between the two?

    b. Which is further from the nucleus?

    c. Which is the highest in energy?

50. What is a quantum?
51. What is wave-particle duality?
52. Compare how an electron cloud is described by the classical world view with how an electron cloud is described by the quantum world view.
53. How does Heisenberg's uncertainty principle apply to finding electrons in an atom?
54. Why do electrons only occupy certain orbits?
55. Why can't you know the velocity and position of a particle at the same time?
56. Describe the shapes of "s," "p," and "d" orbitals.
57. What is an energy level?
58. List the types of orbitals from lowest energy to highest energy.
59. State the Aufbau principle.
60. Under what circumstances can an atom absorb a photon?
61. What is the relationship between wavelength and energy? For example, as wavelength increases, what happens to energy?
62. What is spectroscopy?
63. What does a spectrometer do?
64. According to Plank's relation, what is the limit to the amount of energy an atom can emit?
65. Why don't all 1s orbitals have the same energy?
66. What is the difference between emission spectra and absorption spectra?
67. What is the relationship between frequency and energy? For example, as frequency increases, what happens to energy?
68. State Hund's rule.
69. What is the relationship between frequency and wavelength? For example, as frequency increases, what happens to wavelength?
70. What is the difference between the light emitted from an incandescent light bulb and the light emitted from the hydrogen atom?
71. What document did Linus Pauling say was a map for the arrangement of electrons in an atom?
72. How did Thompson know that protons must exist?
73. How did Thompson discover that cathodes rays were electrons?
74. Why does hydrogen only emit four spectral lines?
75. What quality of a photon is directly proportional to its energy?
76. Which travels faster gamma rays ($v \approx 10^{24}$ Hz) or infrared light ($v \approx 10^{12}$ Hz)? Explain your answer.
77. Create a written report with a presentation on the historical theories and experiments leading to our present model of the atom.

    a. Include both text and visual elements to enhance understanding and add interest.

    b. Use digital media, such as a slide presentation.

78. State the Pauli exclusion principle.
79. What is the relationship between the number of electrons in an energy level and the periodic chart?

Section 9.3

80. Why does sodium form a +1 ion?
81. Why does chlorine form a -1 ion?
82. Do alkali metals have high or low ionization energy? Explain your answer.
83. Do halogens have high or low ionization energy? Explain your answer.

Chapter 9 review

Section 9.3

84. What do all transition metals have in common?

85. What is the trend for ionization energy as you go from left to right on the periodic table?

86. What is the trend for ionization energy as you go down a column on the periodic table?

87. What is the trend for atomic radius as you go down a column on the periodic table?

88. What is the trend for atomic radius as you go from left to right on the periodic table?

89. Why is fluorine smaller than beryllium, when fluorine has more electrons than beryllium?

90. Why does the radius increase so rapidly in the alkali metal family?

91. Why do elements in the same family have similar characteristics?

92. What do the electron configurations of alkaline earth metals have in common?

93. What do the electron configurations of iron, ruthenium, and osmium have in common?

94. What is an electron configuration?

95. What do members of the halogen family do to reach a stable electron configuration?

96. What do members of the alkali metal family do to reach a stable electron configuration?

97. Name four transition metals

98. What makes an electron unavailable for bonding?

99. What are the outermost electrons called?

100. When writing electron configurations,
    a. What does the periodic table have to do with the order in which you fill orbitals with electrons?
    b. What do the superscripts in a configuration refer to, for example, the number 3 in $2p^3$?
    c. If you add all superscripts together in an electron configuration, what should the total number represent?

101. What do the electron configurations of the noble gases have in common?

102. What makes an electron available for bonding?

103. Where are the electrons that interact with other atoms located?

104. Why do atoms lose electrons?

105. What electron configuration characteristics do nonreactive atoms share?

106. What do all transition metals have in common?

107. Why do atoms gain electrons?

108. What is ionization?

109. Which family has the lowest ionization energy? Explain.

110. What is the ionic radius?

111. Why does the ionic radii decrease across a period?

112. Arrange the ions N^{3-}, O^{2-}, Mg^{2+}, Na^+, and F^- in order of increasing ionic radius.

Quantitative problems

Section 9.1

113. How many protons and neutrons are in each of the following isotopes?
    a. ^{14}N
    b. ^{15}N
    c. ^{13}N
    d. ^{16}N

114. How many protons and neutrons are in the following isotopes?
    a. ^{244}Pu
    b. ^{242}Pu
    c. ^{239}Pu

115. What is the mass number for an element that has 66 electrons, 66 protons and 96 neutrons?

116. What element has a Z number of 31?

117. What element has a Z number 77?

118. What is the mass number for an element that has 15 electrons, 15 protons and 15 neutrons?

Section 9.2

119. What is the wavelength of the photons with the following energy?
    a. 9.01×10^{-20} J
    b. 3.07×10^{-24} J
    c. 7.48×10^{-15} J
    d. 5.25×10^{-22} J

120. How much energy do the photons with the following wavelengths have?
    a. 659 nm
    b. 8.18×10^{-12} m
    c. 0.0237 m
    d. 5.61×10^{-6} m

121. Visible light has a frequency between 390 nm to 700 nm. A red photon has a frequency of 4.61×10^{14} Hz and a wavelength of 651 nm while a green photon has a wavelength of 511 nm and frequency of 5.87×10^{14} Hz.
    a. Calculate the energy of the red photon.
    b. Calculate the energy of the green photon.

Chapter 9 review

Section 9.2

122. A photon of ultraviolet radiation has 9.94×10^{-19} J of energy. What is its frequency and wavelength?

123. A mole of photons has an energy of 546 kJ. Calculate the wavelength and frequency of the photons.

124. Hydrogen's emission spectrum contains four wavelengths that are visible to humans. For each wavelength, calculate the energy difference between the two energy levels that are responsible for the spectral lines.

    a. 656 nm
    b. 486 nm
    c. 434 nm
    d. 410 nm

Section 9.3

125. How many valence electrons do each of the following atoms have?

    a. Cl
    b. B
    c. S
    d. P
    e. H
    f. He

126. Use the short-hand noble gas core method to write electron configurations for the following elements.

    a. Kr
    b. Hg
    c. Po
    d. Fr
    e. Sb
    f. I

127. Write electron configurations for the following.

    a. H
    b. He
    c. N
    d. O
    e. Na
    f. Ne
    g. P

128. Write electron configurations for the following ions.

    a. P^{3-}
    b. S^{2-}
    c. Cl^-
    d. Al^{3+}
    e. B^{3+}

chapter 10 — Bonding and Valence

Carbon is found everywhere from vitamins, plastics, seashells, and the biomolecules that make up your body. But in nature, you never find single atoms of carbon. Carbon atoms are always found attached to other atoms. For example, in Vitamin C, carbon is bound in a molecule of $C_6H_8O_6$, while seashells are made of calcium carbonate, $CaCO_3$. Even in the elemental forms of carbon, like diamond and graphite, carbon atoms are bound to other atoms. In these cases, the "other atoms" just happen to be more carbon atoms!

vitamin C

calcium carbonate

$C_6H_8O_6$

$CaCO_3$

This is not unique to carbon. Most elements, with the exception of a few of the "noble" ones like helium and neon, are found bonded either to themselves like H_2 and O_2, or to other elements as in $C_6H_{12}O_6$ and $NaSO_4$. The chemical symbol for a metal is written simply as the symbol of the element, even though metal atoms are attracted to other metal atoms. A pure sample of copper, Cu, actually has millions and millions of copper atoms that are attracted to each other.

Atoms stick together. We see examples of this in the diversity of matter. But what causes this to happen? When atoms stick together, what kinds of properties do we observe? Ultimately, it is the electrons in atoms and forces of attraction. The ability to relate atomic properties to attractive forces is essential in our understanding of bonding.

Chapter 10 study guide

Chapter Preview

Atoms combine in predictable ways to form molecules. But what are the rules that govern when bonds will form and when they won't? In this chapter you will learn about the different types of chemical bonds between atoms in a molecule and why they form. Chemical bonds arise from electrons being shared between nuclei to achieve a noble gas electron configuration. Lewis dot diagrams help to visualize the valence electrons in bonding. Energy is released when bonds form, and absorbed when bonds are broken. Electronegativity is a measure of how strongly atoms attract electrons and determines the polarity of molecules. In ionic compounds electrons are donated and accepted rather than shared. In metals, a "sea" of electrons surrounds positive ions. Chemical formulas and VSEPR theory determine the shape of molecules. Intermolecular forces between molecules of a substance help determine physical properties.

Learning objectives

By the end of this chapter you should be able to:

- draw Lewis dot diagrams for given molecules;
- apply the octet rule to determine the types of chemical bonds that will form;
- predict the type of bond that will form between two elements;
- determine the molecular geometry of a given molecule using VSEPR theory;
- diagram isomers, resonance structures, ring structures, and polymers; and
- compare the strength of attraction among different types of intermolecular forces.

Investigations

10A: Types of bonding
10B: Lewis structures and VSEPR
10C: Surface tension

Pages in this chapter

298 Chemical Bonds
299 How chemical bonds form
300 Bonding and electron configurations
301 The octet rule
302 Lewis dot diagrams and electron pairs
303 Lewis dots and molecular structure
304 Double and triple bonds
305 Section 1 Review
306 Bond Types
307 Bonding and energy
308 Electronegativity
309 Non-polar covalent and polar covalent bonds
310 Polar and non-polar molecules
311 Ionic bonds
312 Electron configuration and ions
313 Polyatomic ions
314 Simple ionic compounds
315 Metals and metallic bonds
316 The bond triangle
317 Section 2 Review
318 Molecular Geometry
319 Valence Shell Electron Pair Repulsion (VSEPR)
320 Linear and trigonal shapes
321 Tetrahedral shapes
322 Bond angles and polarity
323 Bent shapes and larger molecules
324 Hybridization, sigma and pi bonds
325 Isomers
326 Resonance structures
327 Rings
328 Polymers
329 Section 3 review
330 Intermolecular Forces
331 Water and hydrogen bonding
332 Polar substances
333 London dispersion and Van der Waals forces
334 Ionic substances
335 Diamond and network substances
336 Metals and alloys
337 Section 4 review
338 Stronger than Steel?
339 Stronger than Steel? - 2
340 Chapter review

Vocabulary

chemical bond
ion
Lewis dot diagram
triple bond
non-polar covalent bond
metal
non-metals
electron density
resonance
monomer
dipole
surface tension
lattice
alloy

polarization
isoelectronic
Hund's rule
single bond
polar covalent bond
metallic bond
semiconductors
hybridization
aromatic hydrocarbon
polymerization
dipole–dipole attraction
London dispersion attraction
network covalent

diatomic molecule
octet rule
double bond
electronegativity
crystal
metalloids
VSEPR (Valence Shell Electron Pair Repulsion)
isomers
polymer
intermolecular forces
hydrogen bond
Van der Waals attractions
ductile

10.1 - Chemical Bonds

The incredible variety of matter we experience, from amethyst to zirconia, occurs because elements make chemical bonds with other elements to form compounds. For example, oxygen and carbon form the compound carbon dioxide, CO_2. The reason why carbon dioxide is CO_2 and not CO_3 is the subject of this chapter. Elements combine with other elements in predictable ways. The periodic table arranges the elements by their chemical behavior and in this chapter we will see that this is a direct result of the electron structure of atoms.

What are chemical bonds?

Bonds are formed by electrons

A **chemical bond** is a relatively strong attraction between electrons of one atom and the nucleus of another atom. If conditions are favorable - such as between carbon and hydrogen atoms - the interaction causes atoms to "stick" together through the electromagnetic force. A molecule building kit represents a chemical bond as a "rod" joining atoms together. Chemical structure diagrams show bonds as lines. The diagrams below shows four different ways to represent a methane molecule (CH_4).

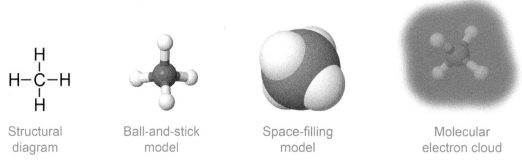

Structural diagram   Ball-and-stick model   Space-filling model   Molecular electron cloud

Different models give different information

What really occurs between atoms does not look like rods or lines! We use rods and lines to show the connections and geometry of how atoms are bonded. Each of the four diagrams are correct ways to represent the methane molecule. Each diagram gives different information about the molecule, but none of the diagrams gives complete picture of the molecule's properties at the atomic scale. The "electron cloud" diagram on the far right above gives the most complete picture of what a molecule is like on the atomic level. Each of the four bonds are an overlapping, shared electron cloud. The molecular cloud shows that the four electron pairs that make the bonds spread out over the whole methane molecule.

Questions about chemical bonds

In the previous chapter we developed the electron structure of the atom, which is the foundation for understanding chemical bonds. In this chapter we will apply what we learned to real chemistry. The following questions will guide our understanding.

1. Why do chemical bonds form between some elements and not others?

2. Why does carbon make four bonds and oxygen make two?

3. Are all chemical bonds similar or are there different kinds of bonds?

4. How can we predict when a chemical bond will form?

How chemical bonds form

Attraction between nuclei

Pure hydrogen naturally exists as a **diatomic molecule** - H_2. The H_2 molecule makes a good starting point for understanding bonding, because each hydrogen only has one proton and one electron. Consider two hydrogen atoms as they approach each other. The electron in the top atom is attracted to its own nucleus, but is also attracted to the nucleus of the bottom atom! The attraction between an electron in one atom and the nucleus of another atom is the fundamental reason chemical bonds form.

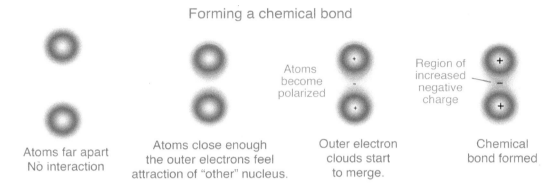

Forming a chemical bond

Polarization

As two atoms get close to each other, the attraction between one atom's valence electrons and the nucleus of the other atom creates a slight **polarization** of each electron cloud. An atom is polarized when the charge becomes unevenly distributed. The attractive forces from the two nuclei pull a slightly higher concentration of negative charge in-between the two atoms. This region of enhanced negative charge attracts both nuclei and creates the chemical bond. The result of the polarization is an attractive force between the atoms.

Forces between atoms

When two atoms get too close to each other, the force between them becomes repulsive due to nucleus-nucleus interactions. If electron clouds overlap too much the cloud of one atom strongly repels the cloud of the other atom. The more the clouds overlap, the higher the repulsive force between the two atoms. Competing attractions and repulsions result in an equilibrium distance at which the force is zero. A chemical bond forms at this equilibrium distance as seen in the diagram.

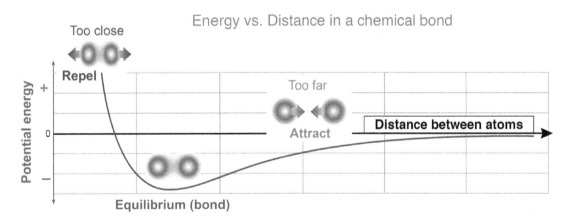

Interactive Simulation

Experiment with the balance between attractive and repulsive forces and note potential energy patterns in the interactive simulation titled Energy of Bond Formation.

Section 10.1: Chemical Bonds

Bonding and electron configurations

Electron orbitals and bonding

In a previous chapter, you learned the role of valence electron configurations in chemistry. Consider the elements sodium and sulfur. Do these elements combine in a stable compound? If so, what will be the chemical formula of the compound? A sodium atom has one valence electron so it makes only one chemical bond. A sulfur atom has six valence electrons. A sulfur atom needs two more electrons to reach full energy levels so sulfur atoms generally make two chemical bonds. The diagram below shows how both elements can reach the lowest-energy condition of having full energy levels by sharing two pairs of electrons. Each shared pair represents a chemical bond.

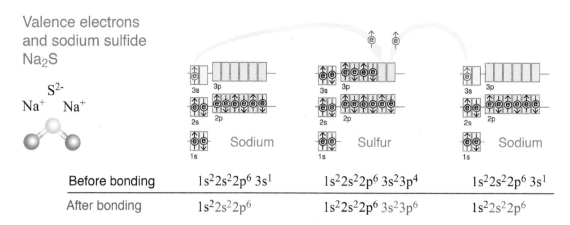

Valence electrons and sodium sulfide Na_2S

	Sodium	Sulfur	Sodium
Before bonding	$1s^2 2s^2 2p^6\ 3s^1$	$1s^2 2s^2 2p^6\ 3s^2 3p^4$	$1s^2 2s^2 2p^6\ 3s^1$
After bonding	$1s^2 2s^2 2p^6$	$1s^2 2s^2 2p^6\ 3s^2 3p^6$	$1s^2 2s^2 2p^6$

Bonding creates a noble gas electron configuration

Before bonding the electron configurations of all three atoms have partially filled energy levels. After bonding, each of the three atoms has a full energy level! Each sodium atom has an electron configuration like that of neon: $1s^2 2s^2 2p^6$. The central sulfur atom has an electron configuration like that of argon: $1s^2 2s^2 2p^6 3s^2 3p^6$. The formula for sodium sulfide is Na_2S because sodium and sulfur must combine in a 2:1 ratio to allow all three atoms to reach the noble gas electron configuration.

Isoelectronic particles

A sodium atom loses an electron to reach the stable $1s^2 2s^2 2p^6$ configuration. The sulfur atom gains two electrons to reach $1s^2 2s^2 2p^6 3s^2 3p^6$. Each sodium atom loses an electron and becomes a sodium **ion** with a charge of 1+. The sulfur atom gains two electrons and becomes a sulfide ion with a charge of 2−. You can predict the charge of an ion by the number of electrons an atom must gain or lose to reach its noble gas configuration. The electron configuration of an ion is the same as a noble gas! Different elements with the same electron configuration are **isoelectronic**. Na^+ and neon, Ne are isoelectronic elements.

Electron configuration of ions (isoelectricity)

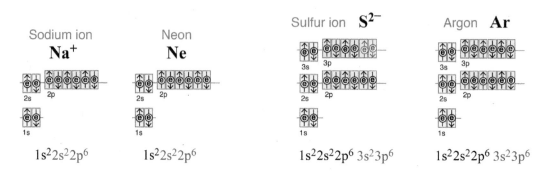

The octet rule

Oxygen tends to form two bonds

Oxygen has six valence electrons in the second energy level. You might think this means oxygen can form six chemical bonds. In fact, oxygen usually forms only two bonds. For example, in water, H_2O, the central oxygen only forms one bond with each of the two hydrogen atoms.

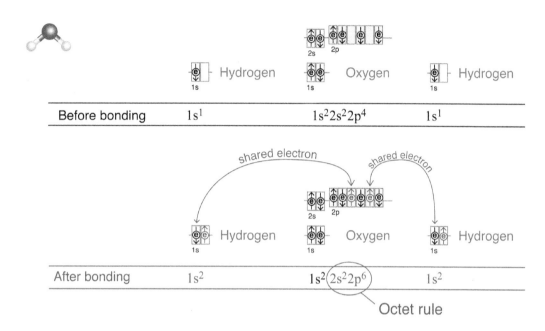

Valence electrons and water, H_2O

Electrons that form bonds

Only single valence electrons form bonds. Four of the six valence electrons in oxygen are paired. These are know as lone pairs of electrons. The other two valence electrons are single. These are the electrons that can form bonds. A single electron in oxygen pairs with a single electron in hydrogen forming a chemical bond.

8 valence electrons are the octet rule

Eight valence electrons completely fill the s and p orbitals. Full s and p orbitals also means the atom has a full energy level. Atoms with full energy levels are energetically very stable. Atoms will lose, gain or share electrons so that they have full energy levels. The rule that compounds form so as to give every atom eight valence electrons is called the **octet rule**.

Light elements

Hydrogen, helium, lithium, and beryllium lose or gain electrons to fill the first energy level. It has s orbitals and no p orbitals. These elements tend to form compounds in which these atoms have a valence configuration of two electrons instead of eight.

Solved problem

Use the octet rule to determine how many atoms of fluorine and carbon will bond together.

Relationships Atoms combine so each has an octet of valence electrons.

Solve
- Fluorine has an electron configuration of $1s^2 2s^2 2p^5$ and therefore has seven valence electrons. It only needs one more electron to reach octet.
- Carbon has four valence electrons. Since carbon has the greater number of single electrons, it will add as many fluorine atoms as possible to reach octet. It needs four more electrons to reach octet.

Answer Four F atoms combine with 1 C atom

Lewis dot diagrams and electron pairs

Lewis dot diagrams

A **Lewis dot diagram** represents one to eight valence electrons arranged around an element. Each electron is represented by a dot. Up to eight electron dots are added around the four sides of an element symbol in a pattern as shown. The numbers represent the order in which valence electron dots are added around the symbol. The Lewis dot diagram for sulfur shows six valence electrons: two pairs of two dots and two unpaired dots. Electrons repel each other so they are added to each side of the element symbol first as single electrons, then as pairs. Unpaired electrons are less stable than paired electrons.

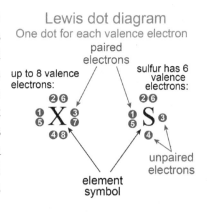

Dot diagrams for elements 1-18

The first dot does not always need to be placed to the left of the symbol, and you can add dots either in the clockwise or counterclockwise direction. Lewis dot diagrams for the first 18 elements are shown below. Hydrogen and helium are small and therefore there are no more than two electrons in the first energy level. Elements 3-18 can have up to eight dots representing the eight valence electrons in the s and p orbitals. The pattern repeats for the second and third rows because elements 3-10 have the same electron configuration as elements 11-18.

Lewis dot diagrams

H							He
Li	•Be	•B•	•C•	•N•	:O•	:F•	:Ne:
Na	•Mg	•Al•	•Si•	•P•	:S•	:Cl•	:Ar:
1	2	3	4	5	6	7	8

Valence electrons

Electron pairing

Electron pairs in the diagrams reflect **Hund's rule**. Compare the diagrams for carbon, nitrogen, and oxygen. Carbon's 4 unpaired dots represent 4 potential bonds. Nitrogen's five electrons make one pair and three unpaired valence electrons. The unpaired electrons are available for making chemical bonds. The diagram for oxygen shows two paired electrons and two unpaired electrons. The two unpaired electrons can each form a bond.

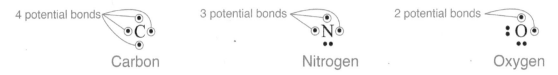

Understanding pairing

Here is a good analogy for bonding and electron pairs. Imagine nitrogen's five valence electrons are five strangers filling up bench seats on an eight-passenger bus. Four people can each have their own seat but the fifth person has to sit beside someone. That leaves three half-filled seats available for "bonding."

Transition metals

Lewis dots are useful for understanding bonding that involves s and p orbitals. The transition metals have more complex bonding patterns because there are more electrons in the d and f orbitals. Chemists do not usually use Lewis dot diagrams for the transition metals, because the octet rule becomes more of a "rule of 18" which is not well represented by a simple dot pattern.

Lewis dots and molecular structure

Lewis dot diagram of CCl_4

How does carbon combine with chlorine? Each chlorine atom contributes one electron to pair with one of carbon's four valence electrons. The Lewis dot structure for carbon tetrachloride (CCl_4) shows that the four chemical bonds in CCl_4 are each between one chlorine atom and the central carbon atom. Carbon tetrachloride is an industrial solvent once used for dry cleaning clothes, as a refrigerant, and in commercial fire extinguishing systems.

Notice that each atom in the molecule is surrounded by eight dots.

Figuring out structure with Lewis dots

Lewis dot diagrams are a useful way to figure out the structure of a molecule. Consider a molecule of ethanol (C_2H_5OH). Notice that each carbon atom is surrounded by eight dots (4 bonds). Each oxygen atom is also surrounded by eight dots (two bonds). Each hydrogen atom in the molecule contributes its single 1s valence electron and forms one bond (two dots). A correct molecular structure is one in which each element in the molecule has either eight dots or two dots corresponding to its valence electron structure.

Notice that each carbon and oxygen atom is surrounded by eight dots and each hydrogen atom has two dots.

Structural diagrams

Lewis dot diagrams describe the electron structure that explains the structural diagram you are already familiar with. Each single bond in a structural diagram represents a pair of electrons.

Section 10.1: Chemical Bonds

Double and triple bonds

Double bonds and triple bonds

Both oxygen and nitrogen exist in the atmosphere as diatomic molecules. To meet the octet rule each atom in O_2 must share two pairs of electrons with its partner in a **double bond**. Each nitrogen atom in N_2 must share three pairs of electrons in a **triple bond**. The picture below shows Lewis dots and structural diagrams for double and triple bonds.

Double bond

Ö::Ö   :Ö=Ö:

Diatomic oxygen O_2

Triple bond

:N:::N: :N≡N:

Diatomic nitrogen N_2

Molecules may mix bonds

Many molecules may contain a **single bond** together with double or triple bonds. Vinegar and glycine are good examples of molecules with single and double bonds. Vinegar is also known as acetic acid and glycine is an important amino acid in proteins. Note that every atom in the vinegar molecule is properly surrounded by either two or eight dots satisfying the octet and duet rules. With the double bonds you must be careful to count the four dots toward the octet rule for both atoms in the bond.

Molecules may contain single, double and triple bonds together

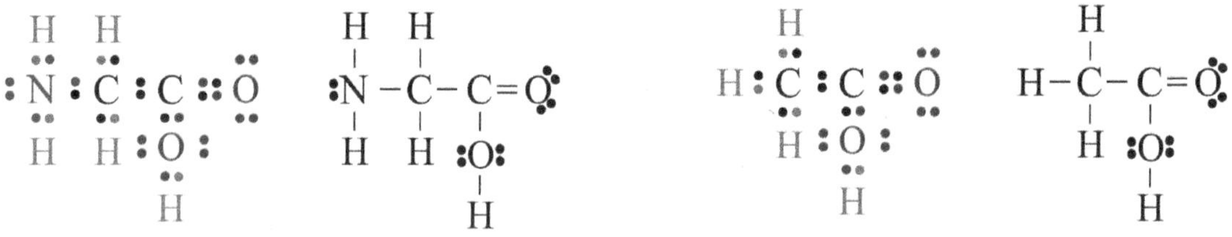

Glycine $C_2H_5NO_2$

Acetic acid $C_2H_4O_2$

Sample problem

Draw the Lewis dot diagram for one possible structure for propene C_3H_6.

Given The chemical formula C_3H_6

Solve Draw Lewis dot diagrams for one C atom and one H atom:

·Ċ·   ·H

Carbon has more single electrons so bond atoms to it first. Connect all 3 C atoms.

·Ċ:Ċ:Ċ·

Now add the 6 H atoms.

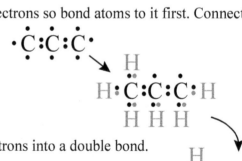

Move leftover single electrons into a double bond.

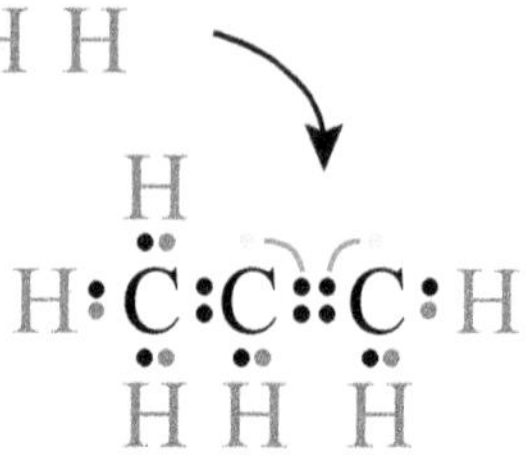

Answer

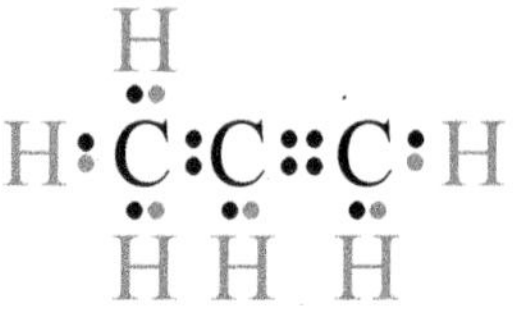

        H
        ··
H:C:C::C:H
  ·· ··
  H H H

Chapter 10: Bonding and Valence

Section 1 review

Chapter 10

A chemical bond is a relatively strong attraction between atoms caused by the sharing or exchange of valence electrons. Atoms form chemical bonds to attain the low-energy state of having completely filled energy levels. For many elements, the number of electrons traded or shared depends on the number of valence electrons to be gained or lost to achieve an octet (8) of valence electrons. The lighter atoms: hydrogen and lithium, form bonds to reach configurations with two valence electrons (instead of 8). A Lewis dot diagram shows how paired and unpaired electrons create bonding patterns. For example, nitrogen has five valence electrons of which the first two form a pair and do not make bonds. Lewis dot diagrams are a very useful way to analyze potential structures for compounds. A structural diagram is actually derived from the Lewis dot diagram by trading each pair of electrons for a single line. A single bond represents one pair of electrons. Double bonds involve two pairs of electrons and triple bonds involve three pairs of electrons.

Vocabulary words: chemical bond, polarization, diatomic molecule, ion, isoelectronic, octet rule, Lewis dot diagram, Hund's rule, double bond, triple bond, single bond

Review problems and questions

1. Which of the following best describes a chemical bond?
    a. a bridge of matter that connects two atoms
    b. a pair of electrons shared between two atoms
    c. an attractive exchange of positive charge between the nucleus of neighboring atoms
    d. when atoms get close enough that gravity pulls them together into a bound pair

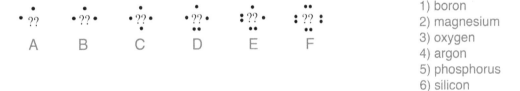

1) boron
2) magnesium
3) oxygen
4) argon
5) phosphorus
6) silicon

2. Use the image and list of elements above to match the correct element with its Lewis dot diagram.

3. Only one of the lewis dot diagrams is correct for a neutral compound of nitrogen and hydrogen. Pick the correct diagram and explain why each of the others is wrong.

4. Draw a lewis dot diagram for a compound with two oxygen atoms, two carbon atoms and three hydrogen atoms ($C_2H_3O_2$). There is more than one correct solution.

10.2 - Bond Types

Earlier in the chapter we introduced covalent and ionic bonds. Carbon dioxide is a common example of a covalent compound and sodium chloride is a common example of an ionic compound. There is also a third type of bond - the metallic bond - which occurs between atoms in a metal. The atoms in aluminum are held together by metallic bonds. Each of the three types of bond has characteristic properties. For example, metallic compounds are electrical conductors while covalent compounds are electrical insulators. In this section, we will learn how the type of bond is determined by two factors. One factor is how strongly each atom in a molecule "attracts" electrons from neighboring atoms (electronegativity). The second factor is whether the bond is localized to a single pair of atoms, delocalized over several atoms (covalent) or acts between an extremely large number of atoms (metallic and ionic).

Overview of the three bond types

Covalent bonds — Covalent bonds form between atoms that are identical or close to each other on the periodic table, usually in the non-metal area. The bonds between oxygen atoms in O_2 are covalent bonds. Each covalent bond is a single shared electron pair. Covalent bonds are localized - they connect two specific atoms by "sharing" electrons between atoms. The bonding force between atoms in one O_2 molecule does not strongly affect nearby O_2 molecules.

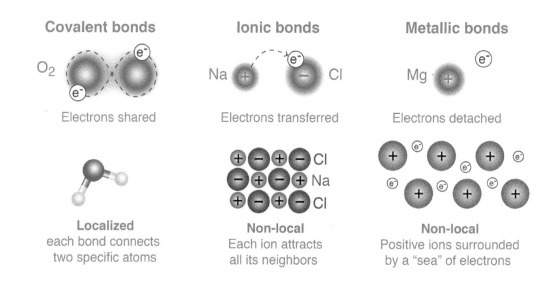

Ionic bonds — Ionic bonds form when one atom "transfers" an electron with another atom. In sodium chloride (NaCl) the chlorine atom gains an electron and becomes a negatively charged ion, Cl^-. The sodium atom loses an electron and becomes a positively charged ion, Na^+. Because of its negative charge, the chlorine atom attracts all nearby sodium atoms! Similarly, the positive sodium atom attracts all neighboring chlorine atoms. An ionic bond is not localized to a single pair of atoms. Sodium chloride forms a tight crystal (table salt) because of the attraction between every sodium ion and chloride ion.

Metallic bonds — Metallic bonds form between atoms that have very loosely bound outer electrons. For example, the outer electrons in solid copper form a "sea" around the positive ions. The electrons that participate in bonding are completely disassociated from any single atom. Metallic bonds are similar to ionic bonds in that the forces act between many atoms and electrons - not a single pair of atoms as with covalent bonds.

Bonding and energy

The octet is a low energy configuration

Chemical bonds occur because systems tend to rearrange themselves into the configuration with the lowest energy. Consider the system made up of a ball rolling up or down a hill. Halfway up the side, the ball feels a force pulling it lower. The system of ball-and-hill has the lowest energy when the ball is resting at the bottom. An atom with electrons is also a system. Quantum calculations are beyond the scope of this book but the energy of a system of many atoms is lowest when there are eight valence electrons around each atom - filling the 2s and 2p orbitals. For the light elements the energy is lowest when there are two electrons - filling the 1s level.

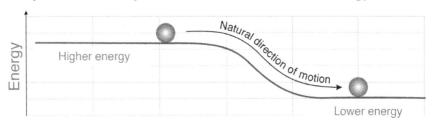

Systems naturally tend to move toward lower energy

Bonds and energy

A system of two neon atoms has the same energy whether or not the atoms are close to each other. This is because each neon atom already has eight valence electrons and so the atoms cannot lower their total energy by forming a bond. A system of two hydrogen atoms is very different. The unfilled electron "vacancy" in the first energy level means a system of two hydrogen atoms can achieve lower energy by forming a bond. In the compound H_2, each atom shares both electrons so that each atom can get to the lowest energy state with full energy levels.

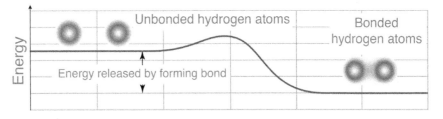

Chemical bonds form because there is an energy advantage

Forming bonds releases energy

By forming chemical bonds, a system of atoms lowers its total energy. Where does the energy go? The energy is released as the bond is formed - often as heat. When a chemical bond is formed, energy is released. Hydrogen gas is explosive in the presence of oxygen because of the energy released when hydrogen atoms and oxygen atoms form bonds to become H_2O.

Energy is needed to break bonds

If you put enough energy into a molecule the bonds can break and the atoms can separate. That is why hydrogen gas does not spontaneously explode! You have to supply a bit of energy - a spark - to break the bonds in the first few hydrogen (H_2) and oxygen (O_2) molecules. The energy released when these recombine into H_2O is enough to break all the bonds in the surrounding H_2 and O_2 molecules. The explosion is a rapid chain-reaction of breaking the H_2 and O_2 bonds and then reforming stronger chemical bonds to make H_2O - releasing the energy difference in the process.

Electronegativity

What is electro-negativity ?

Electronegativity describes how strongly one atom attracts electrons from its companion atom within a chemical bond. An atom with a high electronegativity will be highly attracted to the valence electrons that belong to the atom it is bonded to. The periodic table shows a trend in electronegativity based on valence electrons. In general, small atoms with many valence electrons have high electronegativity while large atoms with few valence electrons have low electronegativity.

Halogens have high electro-negativity

Elements on the right of the periodic table tend to have high electronegativity. These elements have nearly-filled energy levels, making it favorable to capture electrons from neighboring atoms. For example, chlorine strongly attracts electrons from other atoms. Chlorine ($3s^2 3p^5$) has seven valence electrons and needs one more to get to the octet of eight. Chlorine has a high electronegativity because gaining one electron is much more energetically favorable than losing seven. All the halogens (including chlorine) have a high electronegativity. The element with the highest electronegativity (3.98) is fluorine.

Alkali metals have low electro-negativity

On the left side of the periodic table are the alkali metals which have very low electronegativity. The alkali metals, such as sodium, have only one valence electron. The single $3s^1$ valence electron takes little energy to remove and therefore sodium tends to form bonds in which each sodium atom donates one electron. The tendency to lose an electron is the reason for sodium's low electronegativity value of 0.93.

Electronegativity for the first 54 elements

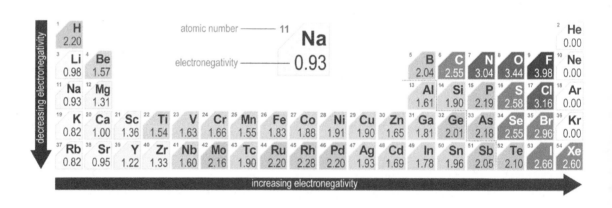

Units of electro-negativity

Electronegativity values have no units. The values are based on a scale developed by Linus Pauling to rank the energy benefit to an atom from gaining one electron. The Pauling scale assigns hydrogen an electronegativity value of 2.20. All other atoms are assigned an electronegativity value based on whether more or less energy is gained by adding an electron - compared to hydrogen.

Electro-negativity difference

The difference in electronegativity between two bonded atoms predicts whether the bond will be ionic, covalent or somewhere in-between. To get the electronegativity difference, subtract the lower electronegativity value from the higher value.

Non-polar covalent and polar covalent bonds

Equal electron sharing

Bonds between different elements are generally classified as covalent when the electronegativity difference is about 1.7 or less. There are two kinds of covalent bonds: polar and non-polar. A **non-polar covalent bond** occurs when electrons are equally shared between bonded atoms. This is always true between atoms of the same element - such as between oxygen atoms in O_2. The double bond between oxygen atoms in O_2 is a non-polar covalent bond because both atoms have the same electronegativity value of 3.44. If you calculate the electronegativity difference between the atoms, you get: 3.44 - 3.44 = 0.

Non-polar covalent bonds have an electronegativity difference less than 0.5

Electronegativity Difference   O O   C H
                                3.44 - 3.44 = 0.00   2.55 - 2.20 = 0.35

Electronegativity difference up to 0.5

Electronegativity difference is not a clear-cut way to determine bond type. Most chemical bonds are neither purely covalent or ionic, but instead are somewhere in-between. Bonds between atoms with an electronegativity difference of 0.5 or less tend to have very small dipole moments. Unless the electronegativity difference is exactly 0, there is an unequal sharing of electrons in the bond. Electrons in a bond spend more time near the element with higher electronegativity than the element with lower electronegativity. For example, carbon and hydrogen have an electronegativity difference of 2.55 - 2.20 = 0.35, making C-H a non-polar covalent bond. Electrons spend more time near carbon than hydrogen.

Polar covalent bonds

Unequal sharing of electrons causes an unequal distribution of electric charge. One atom becomes more negative and the other more positive. The amount of charge separation in a chemical bond is called bond polarity. Polarity has important consequences! Ice is lighter than liquid water because the H-O bonds are polar - making water a polar molecule.

Polar covalent bonds have an electronegativity difference of 0.5 - 2.1

Electronegativity Difference   O H   C O
                                3.44 - 2.20 = 1.24   2.55 - 3.44 = -0.89* = 0.89
                                                      *If you get a negative number, make it positive

Electronegativity difference 0.5 - 2.1

Elements with an electronegativity difference between 0.5 and 2.1 form a **polar covalent bond**. Electrons in a polar covalent bond spend more time near the atom with higher electronegativity. Carbon monoxide has C-O bonds. Carbon has an electronegativity of 2.55 compared to 3.44 for oxygen. The electronegativity difference is 0.89 therefore the bond is polar. An O-H bond is also polar because the electronegativity difference is 1.24.

Interactive simulation

See how electronegativity difference affects charge distribution between two bonded atoms in the interactive simulation titled Electronegativity. What happens to charge distribution when there is a large electronegativity difference compared to a small difference? How does polarity change with the amount of electronegativity difference?

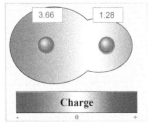

Polar and non-polar molecules

The meaning of polar and non-polar

A polar molecule has a charge separation for the whole molecule - not just an individual bond. Most molecules have many bonds which may be a mixture of polar and non-polar bonds. When chemists describe a substance as polar or non-polar, they are describing the charge across the whole molecule, not just the individual bonds.

Non-polar bonds

Whether a molecule is polar or non-polar depends on the location and polarity of the individual bonds. This simplest case is a molecule with all non-polar bonds. If there are only non-polar bonds then the molecule is usually non-polar. The electronegativity difference between carbon (2.55) and hydrogen (2.20) is only 0.35. Therefore, simple hydrocarbons such as methane are non-polar.

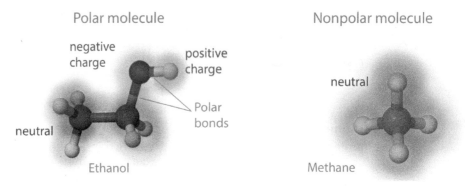

Polar bonds can make polar molecules

The electronegativity difference between carbon (2.55) and oxygen (3.44) is 0.89 making C-O a polar bond. Alcohols such as ethanol are polar because they have a polar bond at one end of a molecule. The polar bond means that one end of the ethanol molecule has a charge compared to the other end. Water is another example of a polar molecule. The oxygen-hydrogen bond is polar and the shape of the water molecule makes one "side" more positive and the opposite side more negative.

Polar bonds can also make non-polar molecules

The geometry of the molecule matters. If a molecule is symmetric - such as carbon dioxide - the molecule can be non-polar even though the bonds are polar. In the case of carbon dioxide the two polar bonds are opposite each other in a linear molecule. Neither end of the molecule has a different charge than the other end so there is no favored orientation and the overall molecule is non-polar.

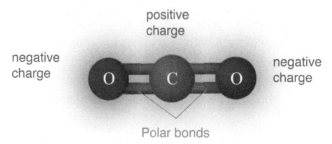

Polarity is important to chemical activity

Because electromagnetic forces are strong (they cause all chemical bonds) the polarity of a molecule has a great effect on both chemical activity and physical properties. Oil and water do not mix because water is a polar molecule and oil is non-polar. Water and alcohol do mix because both are polar molecules.

Ionic bonds

When ionic bonds form

An ionic bond forms when the difference in electronegativity is larger than 1.7. An example is the compound sodium chloride. Sodium has an electronegativity of 0.93 compared to 3.16 for chlorine. The difference in electronegativity is 2.23 therefore this is an ionic bond. The single valence electron from the sodium atom is essentially transferred to the chlorine atom forming two ions: Na^+ and Cl^-. Calcium fluoride (CaF_2), commonly found as the mineral feldspar, is another example of an ionic compound.

Ionic bonds have an electronegativity difference of greater than 2.1

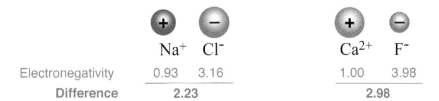

Representing ionic bonds

Because they connect multiple atoms, ionic bonds are **not** represented with lines on structural diagrams. Instead, the + and − signs indicate the charge of each ion. It is understood that the attraction between positive and negative charge is the ionic bond even if the bond is not shown as a line. Notice that an ionic bond can be part of a compound that also contains covalent bonds! A good example is sodium bicarbonate (baking soda) in which the sodium and hydrogen forms an ionic bond with the carbonate ion. The carbonate ion (CO_3^{2-}) contains covalent bonds.

Representing an ionic bond

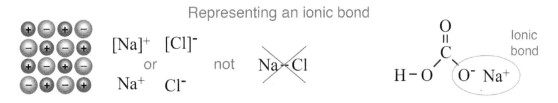

Sodium chloride (NaCl)   Sodium Bicarbonate ($NaHCO_3$)

Elements that form ionic bonds

Most elements can form ions in order to form ionic compounds. Look for the ionic charge pattern across the main-group elements in the table below. Note that metal elements on the left of the periodic table form positive ions and non-metal elements on the right tend to form negative ions. Elements located between metal and non-metal elements form both positive and negative ions.

Common charges of ions

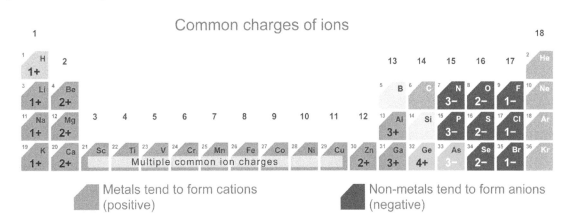

Section 10.2: Bond Types

Electron configuration and ions

Predicting the charge of ions

Like all the alkali metals, sodium will lose its valence electron to become Na⁺. Losing one electron allows sodium to have a noble gas configuration with completely filled energy levels. Why won't sodium instead gain 7 electrons to become Na^{7-}? Losing one electron is much more likely to occur than gaining 7 electrons. We have never observed Na^{7-} ions in nature. The electron configuration of Na^+, Mg^{2+}, O^{2-}, and Ne are shown below. Note that all three ions have the same electron configuration as neon. The charge of an ion depends on how many electrons it must gain or lose to get to a noble gas configuration.

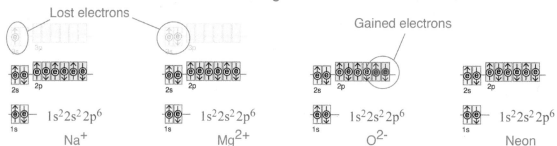
Electron configuration of ions

Polyatomic ions

In many compounds the "ion" is actually a "charged molecule" consisting of several atoms covalently bonded together. The bicarbonate ion HCO_3^- is a good example. From the diagram below you can see that the carbon atom and two of the oxygen atoms have a full octet. The third oxygen atom makes only one bond and therefore needs an extra electron to complete its octet. The extra electron is what gives the bicarbonate ion its charge of -1.

The bicarbonate ion (HCO_3^-)

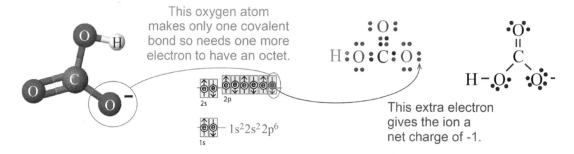

Charges instead of bonds

Polyatomic ions occur when covalently bonded atoms pick up or lose electrons instead of making more bonds with other atoms. In the bicarbonate ion, one oxygen atom gains an extra electron instead of forming a second bond. The diagram below shows that carbonic acid, CH_2O_3, is a neutral molecule in which the "third oxygen" bonds with an additional hydrogen atom to satisfy the octet rule. In water, carbonic acid dissociates to form a bicarbonate ion (HCO_3^-) and an H^+ ion.

Carbonic acid and the bicarbonate ion

Carbonic acid (H_2CO_3)   Bicarbonate ion (HCO_3^-)

Polyatomic ions

What is a polyatomic ion?

Many compounds include ions in which multiple atoms are covalently bonded together and stay together through a chemical reaction. A polyatomic ion is a molecule with a charge and can be negative or positive. The chart below lists some common polyatomic ions.

common polyatomic ions

NH_4^+	ammonium	OH^-	hydroxide
CH_3COO^- or $C_2H_3O_2^-$	acetate	NO_3^-	nitrate
HCO_3^-	bicarbonate*	NO_2^-	nitrite
CO_3^{2-}	carbonate	$C_2O_4^{2-}$	oxalate
ClO_3^-	chlorate	MnO_4^-	permanganate
CrO_4^{2-}	chromate	PO_4^{3-}	phosphate
CN^-	cyanide	SO_4^{2-}	sulfate
$Cr_2O_7^{2-}$	dichromate	SO_3^{2-}	sulfite

*bicarbonate is the common name for the hydrogen carbonate ion

Solving a polyatomic ion structure

The nitrate (NO_3^-) ion is a critical nutrient for plants and an important component of soil. How do we find the structure of the nitrate ion? The diagram below shows the steps. Notice how a negative charge requires you to add electrons to the total.

Step 1: Count valence electrons.

NO_3^-

N O₃
5 + 3(6) + 1 = 24

Step 2: Fill in the electrons starting with most electronegative (O).

After placing 24 electrons, notice that nitrogen has only 6, not 8.

Step 3: Make a double bond to satisfy the octet rule for nitrogen.

The brackets and charge show this is an ion

For nitrate there are three solutions

#1, #2, #3

Partial charges

Notice the brackets and the minus sign (-) on the Lewis dot diagram that indicate that the structure is an ion with a -1 charge. The chemistry of polyatomic ions often depends on which of the atoms carries the charge of the ion. In the case of nitrate, we assign a partial charge to the pair of single-bonded oxygen atoms. Partial charges are symbolized by the lower Greek letter delta, δ. Since oxygen is more electronegative than nitrogen, each single-bonded oxygen atom gets a -1 partial charge. To maintain the overall charge of the ion, the nitrogen atom gets a +1 partial charge. The double-bonded oxygen atom has a zero charge because the double bond makes a complete octet.

Showing partial charges on structural diagrams

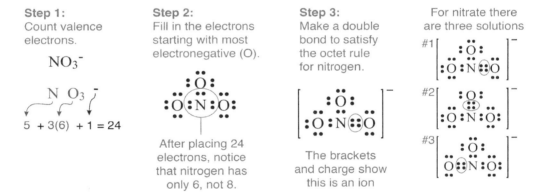

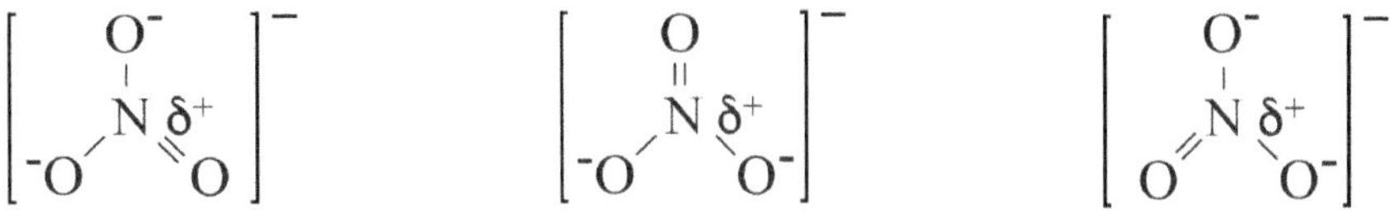

The oxygen is more electronegative so gets the negative charge.

Section 10.2: Bond Types

Simple ionic compounds

The sodium ion

Ionic substances tend to form a **crystal** structure where oppositely charged ions are arranged in a regular pattern. The sodium chloride (NaCl) crystal forms a cube where eight sodium ions form a cube around every chlorine atom and vice versa. Ionic substances are held together through the attraction between countless oppositely-charged ions, but the overall crystal charge is neutral. A neutral crystal has exactly equal numbers of positively- and negatively-charged ions no matter how large the crystal is.

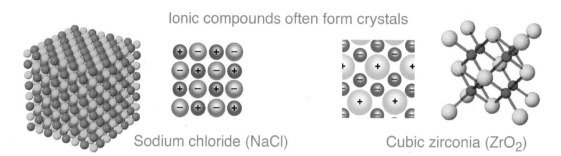

Ionic compounds often form crystals

Sodium chloride (NaCl)   Cubic zirconia (ZrO$_2$)

Non-metal ions

How are subscripts in an ionic chemical formula determined? It comes down to positive and negative charges canceling one another out to form a neutral compound. Consider magnesium chloride. Magnesium makes an Mg^{2+} ion. Chlorine makes a Cl$^-$ ion. To balance the charges two (-1) chlorine ions are needed for every one (+2) magnesium ion. This makes the ratio of magnesium to chlorine 1:2, and the formula is therefore MgCl$_2$.

Solved problem

What is the chemical formula for magnesium fluoride?

Asked Chemical formula

Given magnesium and fluorine

Relationships Magnesium has 2 valence electrons and electronegativity of 1.31.
Fluorine is a halogen and has 7 valence electrons and electronegativity of 3.98.

Solve
1. First we look at the difference in electronegativity to determine the type of bond. In this case the difference is 3.98 − 1.31 = 2.67. This is an ionic bond because the electronegativity difference is greater than 1.7.

2. Next we determine the ion charges. Magnesium is [Ne]2s^2 and loses its two valence electrons so it will have a charge of +2. Fluorine has seven valance electrons and needs one more to form an octet so its ionic charge with be -1.

3. Finally we know the charges must add to zero, therefore it takes two fluorine ions (-1 each) to balance the +2 of magnesium (-2 + 2 = 0).

Answer The formula is MgF$_2$.

Metals and metallic bonds

Definition of a metal

A **metal** is a substance that is a good conductor of electricity (and heat). Conducting electricity means that some electrons are free to move independently - completely detached from any specific atom. To explain why, we need to refer back to the concept of ionization energy. Ionization energy is the amount of energy it takes to remove an electron from an atom. When the ionization energy is low enough, the electrons are not tightly bound to a single atom. Virtually all the metals have ionization energies below about 800 kJ/mol. The heavier metals have electrons in loosely bound 3d, 4d, and higher energy levels far from the nucleus.

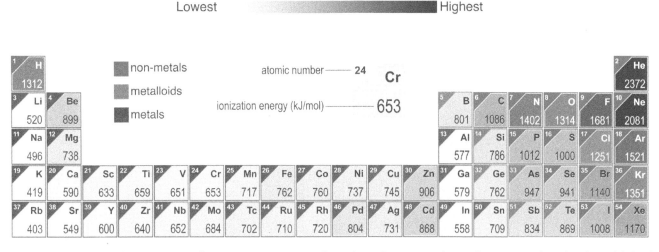

Metallic bonds

When atoms of copper come together they form one large "super-molecule" in which the valence electrons don't just circulate between two copper atoms. Instead, copper's valence electrons circulate among all the copper atoms - bonding them all together. A good analogy is that positive copper ions form a crystal in which the ions are surrounded by a sea of electrons. You might think of a metal as an enormous pool of shared electrons holding together a fixed structure of positive ions. In a **metallic bond**, the valence electrons are shared among all atoms in the entire material. The mobile (detached) electrons are the reason metals are good conductors of electricity.

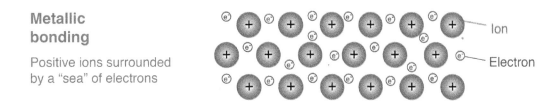

Metallic bonding

Positive ions surrounded by a "sea" of electrons

Metalloids and non-metals

The elements to the right of the periodic table are classified as **non-metals**. The non-metals do not conduct electricity and do not form metallic bonds because they have high ionization energies. Boron, silicon, and germanium, are **metalloids**. Their electronic properties are between that of the metals and non-metals. These elements - especially silicon and germanium - are also known as **semiconductors**. Semiconductors have some free electrons and are the basis for virtually all electronic devices.

Section 10.2: Bond Types

The bond triangle

Electron mobility

The ease with which electrons can move between atoms is called electron mobility. In a pure metallic bond the valence electrons are highly mobile, moving freely between atoms. Another way to say this is that electrons in a metal are delocalized, they move easily between atoms. The opposite occurs when electrons are trapped to a single atom - as with an ionic or covalent bond. In both ionic and covalent bonds the electron is highly localized, confined to one atom or a few atoms in the case of a molecule. As you might guess, delocalized electrons are present in an electrically conductive metal. Localized electrons are found in electrical insulators.

The bond triangle

The bond triangle shows the relationships between ionic, metallic and covalent bonds. Each corner of the triangle represents one of the three pure types: metallic, covalent, and ionic. The bottom of the triangle represents the average electronegativity with metals (low) to the left and halogens (high) to the right. The right side represents the electronegativity difference between the two atoms with zero at the bottom representing a pure covalent bond. The lower left corner represents delocalized electrons.

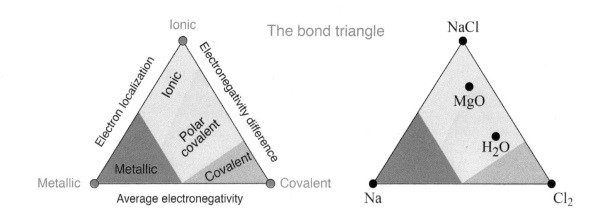

Using the bond triangle

All chemical bonds fall somewhere on the triangle. A pure covalent compound such as chlorine gas (Cl_2) falls in the lower right. The electronegativity difference is zero, the average electronegativity is high, and the bond is highly localized to two atoms. Sodium chloride (NaCl) is a pure ionic bond with a high electronegativity difference, a moderate average electronegativity, and highly localized (trapped) electrons. Pure sodium metal is in the lower left with low electronegativity, zero electronegativity difference and highly mobile electrons (delocalized).

Interpreting the bond triangle

The bond triangle is useful for understanding real chemical bonds which have varying characteristics of all three bond types. For example, water is polar covalent with the hydrogen oxygen bond having an average electronegativity of 2.8 and an electronegativity difference of 1.4. From the position of the water molecule you can predict that the electrons are localized to the water molecule, but not to specific atoms in the molecule. The bond in magnesium oxide (MgO) is on the border between ionic and polar covalent. We expect MgO to be an electrical insulator because the position in the triangle is outside the metal zone.

Section 2 review

Chapter 10

Chemical bonds are classified as covalent (shared electrons), ionic (transferred electrons) or metallic (detached electrons). Bonds form when the overall energy of a system of atoms is lower than the atoms in their previous energy state. Electronegativity describes the ability of an atom to attract electrons. Halogens have the highest electronegativity and alkali metals have the lowest. Electronegativity difference predicts how evenly valence electrons are distributed in a chemical bond between two atoms. A non-polar covalent bond, such as the bond in O_2, has an electronegativity difference of zero. Bonds are classed as covalent when the electronegativity difference is less than 0.5. A polar covalent bond has an electronegativity difference between 0.5 and 1.7. The bonds within a polar molecule create an asymmetric charge separation for the molecule. A molecule can have polar bonds and not be polar if the molecule is symmetric.

Ionic bonds have an electronegativity difference greater than 1.7. Ions are charged atoms in which electrons have been gained or lost. The charge equals the number of electrons gained or lost in order for there to be an octet or duet of valence electrons. A polyatomic ion is a charged molecule where a covalently bonded group of atoms has an excess or deficit of electrons. Positive and negative charges are balanced in an ionic formula.

Metallic bonds occur between elements with low ionization energy. Metallic bonds are characterized by mobile (non-localized) electrons. Mobile electrons allow metals to conduct heat and electricity. The bonding triangle is a way to visualize the degree to which a chemical bond is ionic, covalent, or metallic.

Vocabulary words: electronegativity, non-polar covalent bond, polar covalent bond, crystal, metal, metallic bond, metalloids, non-metals, semiconductors

Review problems and questions

1. Characterize the bonds in silicon dioxide, SiO_2, as covalent, polar covalent, or ionic. Silicon dioxide is found naturally in quartz and silicate compounds make up 10% of Earth's crust. Silicon dioxide is also the primary constituent of ordinary glass used in dishes and windows.

2. The diagram above shows the compound isopropanol, commonly sold in the drug store as rubbing alcohol. Is isopropanol polar or non-polar? Explain your answer.

3. Predict the charge of each ion and the chemical formula for a compound of potassium and nitrogen.

4. A mystery element is determined to have an ionization energy of 620 kJ/mol. Predict whether this element will be an electrical conductor or an electrical insulator.

10.3 - Molecular Geometry

Molecules are three-dimensional and the exact shape of a molecule often determines how it behaves. You may have heard about "left-handed sugar." The same atoms in normal sugar can be arranged in a slightly different shape that still tastes sweet but cannot be digested by the body, and therefore has no calories. The reason the "left-handed sugar" cannot be digested is that the special molecules (enzymes) that digest sugar act on the very specific shape of the sucrose molecule. The sweetness sensation comes from a nerve receptor that acts on a different part of the sucrose molecule. The left-handed sugar is shaped the same as sucrose in the area that triggers the sweet taste and a different shape in the area used by the digestive enzyme sucrase. This section will introduce the geometry of molecules and the basic rules that determine molecular geometry. In later chapters you will explore further how the geometry - and folding pattern - of molecules affects food, medicines and even the chemistry of life itself.

The meaning of molecular geometry

Levels of molecular description

Molecular geometry tells you how the atoms in a compound or molecule are arranged in three dimensions. The methanol (wood alcohol) molecule is a good example. The first level of description, a chemical formula, CH_3OH does not tell you how the atoms are connected. The two-dimensional representations shows each bond in a flat plane. The three-dimensional representations are closest to reality if we could "see" an individual molecule.

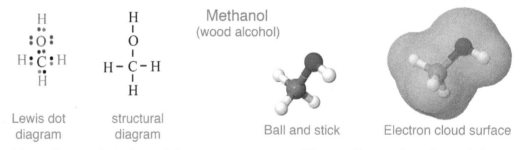

The tetrahedral shape

Notice that around the central carbon atom, the three hydrogen atoms and the oxygen atom form a tetrahedron. The tetrahedron, a four-sided pyramid, is one of the fundamental shapes in which molecules form. The electron structure inside a carbon atom (remember the s and p orbitals) directly causes the methanol molecule to have this shape. A three-dimensional model of a methanol's tetrahedral shape is shown in the interactive simulation.

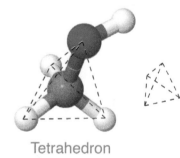

Tetrahedron

Molecules don't stand still

Ball and stick models are good for visualizing molecules but they hide a key fact. Atoms are never still! At ordinary temperatures the atoms are in constant movement and the molecule bends and flexes at every bond. The higher the temperature, the more agitated the atoms are until bonds finally break when the temperature gets high enough. The precise angles in models and diagrams really show the average position of the atoms. The quantum reality is much more active.

Valence Shell Electron Pair Repulsion (VSEPR)

Electron density

The three-dimensional shapes of molecules determine many properties of matter. For example, many medicines work on a "lock-and-key" principle. The specific shape of a drug molecule fits into a particular location on a protein to enhance or block its function. The general principle by which we understand molecular geometry is:

electrons repel each other.

Chemical bonds form regions of **electron density** and these regions repel each other because of their negative charge. The diagram below shows that two areas of electron density repel each other into a linear shape. Three regions repel each other into a trigonal shape and four regions repel each other into a tetrahedral shape.

Regions of electron density repel each other

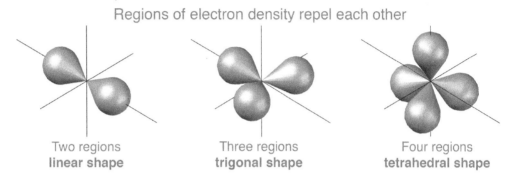

Two regions — linear shape   Three regions — trigonal shape   Four regions — tetrahedral shape

VSEPR

The theory that areas of high electron density repel each other is known by the acronym **VSEPR (Valence Shell Electron Pair Repulsion)**. The VSEPR theory has the following two main points.

1. Chemical bonds tend to arrange themselves as far apart as possible.

2. Unbonded electron pairs are also regions of electron density and repel each other and chemical bonds.

Tetrahedral shapes

Consider the methane molecule. In methane there are four chemical bonds and each bond represents a region of higher electron density. What shape allows each negative electron density region to be farthest away from all the others? The answer is a tetrahedron. The four chemical bonds in a methane molecule arrange themselves at the four corners of a tetrahedron. The four corners of a tetrahedron are as far apart as the bonded electron pairs can get from each other and still be the same distance from the central carbon atom. Molecules with four chemical bonds form tetrahedral shapes.

Regions of electron density repel each other

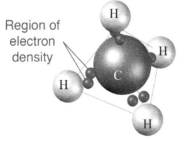

Region of electron density

The four regions of electron density repel each other to create the tetrahedral shape of the methane molecule.

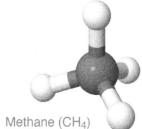

Methane (CH$_4$)

Section 10.3: Geometry of Molecules

Linear and trigonal shapes

Trigonal pyramidal shapes

The VSEPR theory states that the effect of lone pairs of electrons on molecular geometry is significant. Electron pairs that are not part of chemical bonds are still regions of higher electron density. Repulsion from unbonded electron pairs affects the geometry of a molecule. For example, the ammonia molecule (NH_3) has three chemical bonds and one unbonded electron pair. The shape of the ammonia molecule is trigonal pyramidal. The three bonds are repelled away from the unshared electron pair. Molecules with three bonds and one unbonded pair tend to form pyramidal shapes.

Unbonded pairs repel bonds and other pairs

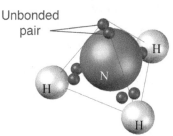

The ammonia molecule has a **trigonal pyrimidal** shape because there are four regions of electron density - one of which is an unbonded pair

Ammonia (NH_3)

Trigonal shapes

In some molecules there are three regions of electron density around a central atom. The formaldehyde molecule, CH_2O is a good example. The double bond groups two electron pairs in a single region. The two bonds to the hydrogen atoms make a second and third region. The three regions of electron density get farthest apart when the atoms surrounding the central carbon are at the vertices of a triangle. A triangle in a plane (flat) is called trigonal planar. Formaldehyde is a trigonal planar molecule.

Three regions of electron density

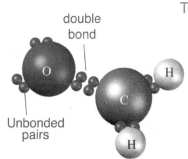

The formaldehyde molecule has a **trigonal planar** shape because there are three regions of electron density - one of which is a double bond.

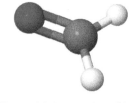

Formaldehyde (CH_2O)

Linear shapes

The carbon dioxide molecule has two double bonds between the central carbon atom and each oxygen atom. The Lewis dot diagram shows the double bonds in the diagram below. There are two regions of electron density around the carbon atom. As you might expect, the double bonds repel each other to the opposite sides of the carbon atom and there are no lone pairs on the center atom to affect the geometry. Therefore the CO_2 molecule is linear. Molecules with two regions of electron density form linear shapes.

Two regions of electron density

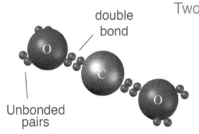

The carbon dioxide molecule has a **linear** shape because there are two regions of electron density (two double bonds).

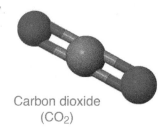

Carbon dioxide (CO_2)

Tetrahedral shapes

Bond angles in drawings

Molecular shape helps you predict properties such as electron distribution, polarity, boiling point, color and chemical activity. The first methane drawing shown below is correct as far as number of atoms and bonds. However, the drawing gives a false impression that hydrogen atoms are bonded 90° apart from each other. The true bond angle between every hydrogen atom in methane is 109.5°, the maximum distance four bonded atoms can be positioned apart from one another in a tetrahedral molecule.

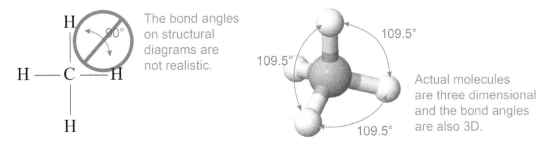

Wedges and dotted lines indicate 3D shape

Use your model kit to build a 3D model of methane. Lay the molecule on the edge of your desk so carbon and 2 hydrogen atoms lie flat in a single plane or flat surface. Notice how one hydrogen atom points above or in front of the plane and one hydrogen atom points below or behind the plane. The wedge in the drawing above represents the atom that points in front of the plane and the dotted line represents the atom that points behind the plane. Bonds that lie flat in the plane are shown as solid lines. Rotate the molecule a few times. You will find that no matter how you rotate the molecule, it will have 2 bonds that lie flat in a plane, with 1 bond sticking out in front of the plane and 1 bond behind the plane.

Geometries with 4 Regions of Electron Density	methane / CH_4	ammonia / NH_3	water / H_2O	hydrochloric acid / HCl
molecular shape	tetrahedral	trigonal pyramidal	bent	linear
# of atoms bonded to central atom	4	3	2	1
# of lone pairs on central atom	0	1	2	3

Tetrahedral arrangement

When it comes to molecular shape we are only concerned about the lone pairs on the central atom. The right combination of bonded atoms and lone pairs around a central atom can cause a molecule to behave like a tetrahedron in three dimensional space even when an individual molecule's geometry is not tetrahedral. Look for a pattern as you move left to right across the molecules shown above. Did you notice how a bonded atom is replaced with a lone pair each time you move to the right? Did you also see how the molecular shape changes but still forms a tetrahedron in the 3D space around itself? The picture below shows how one individual water molecule with a bent geometry actually forms a tetrahedron in 3D space in the solid phase thanks to its 2 lone pairs.

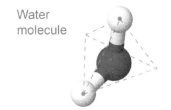

Water molecule

A water molecule has a "tetrahedral" shape when you consider the two lone pairs of the central oxygen atom.

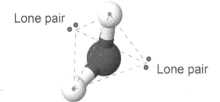

Section 10.3: Geometry of Molecules

Bond angles and polarity

Trigonal planar

A flat triangle results when there are 3 regions of electron density around a central atom. Each region maximizes its bond angles in a flat plane. Bond angles are only equal if electron regions around the central atom are identical. As shown in boron trifluoride (BF_3), all three regions around boron are bonded atoms so bond angles equal 120°.

Geometries with 3 Regions of Electron Density	formaldehyde H_2CO 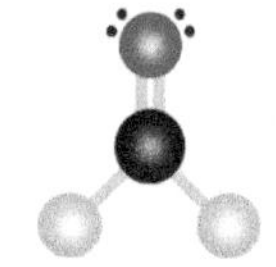	ozone O_3 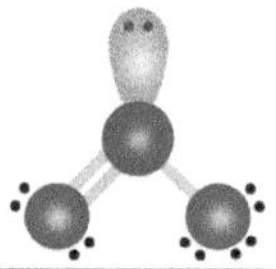
molecular shape	trigonal planar	bent
# of atoms bonded to central atom	3	2
# of lone pairs on central atom	0	1

Varied Bond Angles:
- boron trifluoride BF_3: 120°
- formaldehyde H_2CO: 121.9°, 116.2°
- ozone O_3: 116.8°

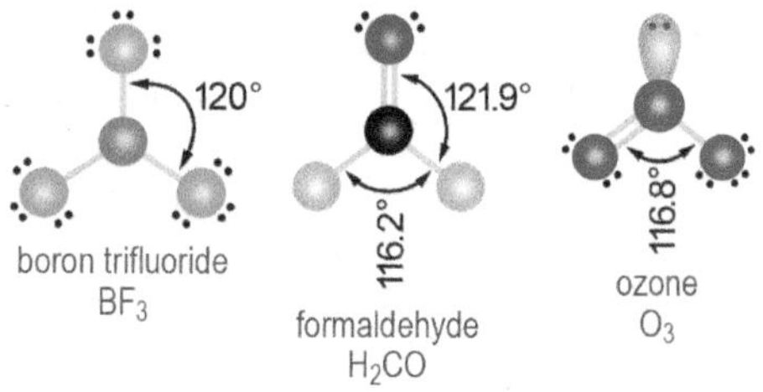

Lone pairs repel more effectively

Bond angles around the double bond in formaldehyde (H_2CO) are larger than 120°. A double bond repels more strongly than a single bond because twice as many electrons are present. You might be surprised to see that the lone pair in ozone (O_3) repels more strongly than a single bond even though both contain only 2 electrons. This is because lone pairs of electrons sit closer to the nucleus than bonding electrons, creating a stronger region of negative charge. Molecular shape is influenced by many things such as attractions and repulsions of bonded atoms, bond length, electronegativity and van der Waals interactions.

Molecular shape and polarity

Simply-shaped bent and trigonal pyramidal molecules are polar. These shapes have lone pairs on the central atom that create an unbalanced distribution of electrons. Ozone, in the table above, and water are polar for this reason. When a central atom has no lone pairs, a molecule is non-polar if electrons are distributed evenly over bonded atoms. If electrons are unevenly distributed, the molecule is polar. Formaldehyde shown above is polar. Even though the central atom has no lone pairs, electrons are not evenly distributed over the bonded atoms. The oxygen atom has 2 lone pairs around it but the hydrogen atoms do not. The oxygen end has a negative charge and the hydrogen ends have a positive charge.

VSEPR summary table

# of electron-dense regions	shape class	shape example	3D drawing	shape name	number of atoms bonded to central atom	number of lone pairs on central atom
2	LINEAR			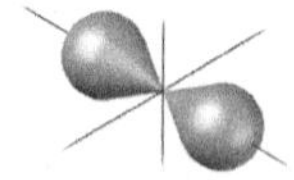 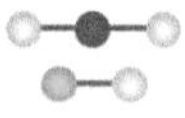 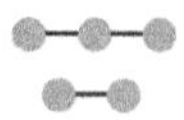 linear	1 or 2	0
3	TRIGONAL			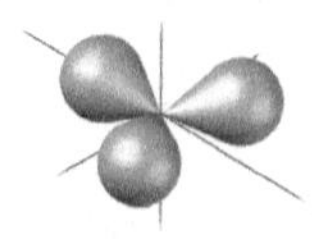  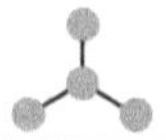 trigonal planar	3	0
3	TRIGONAL			 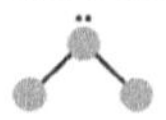 bent or angular	2	1
4	TETRAHEDRAL			   linear	1	3
4	TETRAHEDRAL			 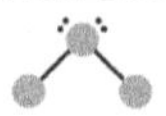 bent or angular	2	2
4	TETRAHEDRAL			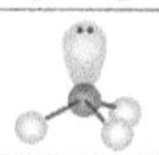 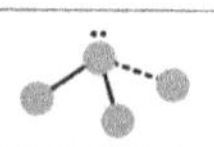 trigonal pyramidal	3	1
4	TETRAHEDRAL			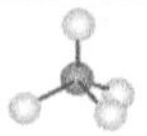 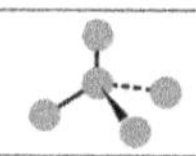 tetrahedral	4	0

Bent shapes and larger molecules

Bent shapes

A water molecule is "bent" because there are two chemical bonds and two unbonded electron pairs. Repulsion from the unbonded pairs forces the bonds with the two hydrogen atoms into a plane with the central oxygen atom. Molecules with two unshared pairs and two chemical bonds tend to form this bent shape.

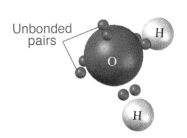

"Bent" molecules

The water molecule has a **bent** shape because the unbonded pairs repel the two O-H bonds.

Water (H_2O)

Parts of molecules

The VSEPR principle applies to every combination of bonds in a molecule. More complex molecules typically include many different shapes. For example, the acetic acid molecule $C_2H_4O_2$ has a tetrahedral shape at one end where the carbon bonds with three hydrogen atoms. The bonds around the other carbon atom are trigonal because the double bond makes three regions of electron density. The C-O-H group forms a bent shape.

Using Lewis dots

For molecules containing less than 10 atoms the Lewis dot diagram helps predict the shape.

1. Each group of dots repesents one region of electron density.
2. Count the number of dot groups surrounding each atom. Each group is a region of electron density.
3. Once you know the number of regions of electron density you can predict the molecule shape.

For example, the acetic acid molecule $C_2H_4O_2$ has a tetrahedral shape at one end where the carbon makes four bonds. There are four groups of dots and therefore four regions of electron density. The bonds around the other carbon atom are trigonal because the double bond makes three groups of dots which means three regions of electron density. The C-O-H group forms a bent shape because the oxygen atom has four regions of electron density but two are bonds and two are unbonded pairs.

Molecules have multiple shapes

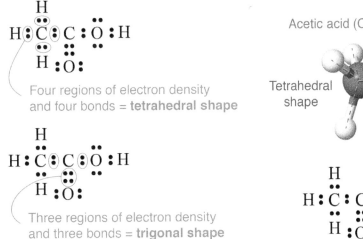

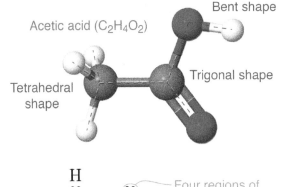

Section 10.3: Geometry of Molecules

Hybridization, sigma and pi bonds

Hybridization The tetrahedral shape of the methane molecule occurs because the s and p orbitals combine - a process called **hybridization**. Normally the s-orbital is spherical and the three p-orbitals are at right angles to each other. The hybridized sp^3 orbital has four regions of electron density aligned with the corners of a tetrahedron. The hybridized orbitals have lower energy than the original s and p orbitals after the atom has formed chemical bonds. Electrons in atoms that have not formed chemical bonds still occupy s and p orbitals.

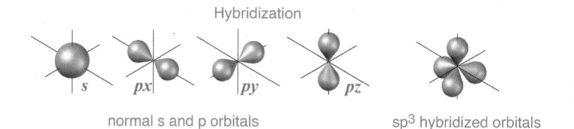

The hybridized orbital Hybridized orbitals spread out as far from each other as they can get. When there are four sp^3 orbitals, they are 109.5° away from each other. Three sp^2 orbitals spread out to a 120° angle and form a trigonal planar molecule.

Sigma and pi bonds For sigma bonds, the orbitals overlap and form a single bond. The overlapping orbitals can either be hybridized or unhybridized. A sigma bond is the strongest of the covalent bonds. Double and triple bonds come from overlaps of sigma and pi bonds. A double bond is a combination of one sigma bond (stronger) and one pi bond (weaker). A triple bond has one sigma bond and two pi bonds.

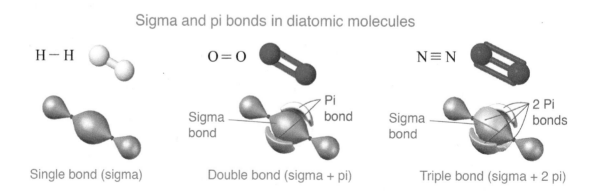

Chapter 10: Bonding and Valence

Isomers

Levels of molecular description

Different molecules that have the same chemical formula are called **isomers**. The compound, C_2H_6O is a good example of a molecule with two isomers. One isomer is ethanol which is distilled from grain and is found in drinks like beer and wine. The diagram below shows the structure of the ethanol molecule with the two carbons bonded together. The second isomer is dimethyl ether, a compound often used in spray cans as a propellant. Note that the oxygen atom is in the center of the dimethyl ether molecule.

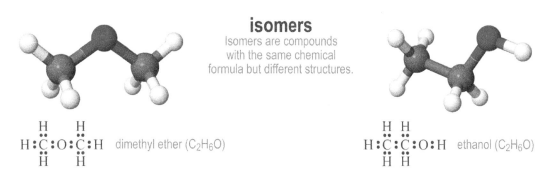

isomers
Isomers are compounds with the same chemical formula but different structures.

Lewis dot diagrams for isomers

Note from the Lewis dot diagrams that both isomers satisfy the octet rule. Any configuration of atoms that did not satisfy the octet rule would quickly rearrange itself into an isomer that does. Ethanol and dimethyl ether are two isomers of C_2H_6O, however other molecules may have many isomers.

Isomers have different properties

Isomers usually have different physical and chemical properties. For example the boiling point of dimethyl ether is -24 °C while the boiling point of ethanol is 78 °C. At room temperature, dimethyl ether is a gas and ethanol is a liquid.

Solved problem

Use Lewis dot diagrams to find three isomers of a compound with the chemical formula CH_5NO. The names of the compounds are for reference - you will not be asked to name compounds of this type. A typical problem is to construct a possible isomer that obeys the rules for bonding.

- Asked: Three different isomers
- Given: chemical formula: CH_5NO
- Relationships: Atoms combine so each has an octet of valence electrons.
- Solve: Connect the nitrogen and oxygen atoms to the carbon atom in as many different combinations as possible. Add hydrogen atoms last. Display non-bonding electrons as lone pairs.
- Answer: The three isomers are shown below with Lewis dot diagrams and structural formulas.

methanolamine   methylhydroxylamine   methoxyamine

Section 10.3: Geometry of Molecules

Resonance structures

Resonance structures

The Lewis dot diagram for the thiocyanate ion (SCN⁻) can be drawn two different ways and both are correct. One way uses two double bonds and the second way uses a triple and a single bond. When a single molecule or ion can have two or more valid Lewis dot diagrams, the molecule is said to have **resonance** structures. A resonance structure has different arrangements of adjacent bonds which can change form. The thiocyanate ion resonates between the two forms, so its structure is an average of the two diagrams. The ion has a minus one charge and therefore one electron has to be added to the structure.

Resonance structures
Thiocyanate ion (SCN⁻)

$$[S - C \equiv N]^- \rightleftharpoons [S = C = N]^-$$

Structure #1   Structure #2

A molecule with more than one valid Lewis Dot structure has *resonance*.

Solving the nitrate ion structure

The nitrate (NO_3^-) ion is important to life as it is a critical nutrient for plants. Fertile soil contains nitrogen in the form of nitrate. Nitrogen has five valence electrons and oxygen has six. Nitrogen has a lower electronegativity (3.04) compared to oxygen (3.44) so symmetry puts the nitrogen atom in the center. Notice that the nitrate ion has three resonant forms that alternate the position of the double bond.

Step 1: Count valence electrons.

NO_3^-

$5 + 3(6) + 1 = 24$

Step 2: Fill in the electrons starting with most electronegative (O).

After placing 24 electrons, notice that nitrogen has only 6, not 8.

Step 3: Make a double bond to satisfy the octet rule for nitrogen.

Step 4: See if there are resonant forms.

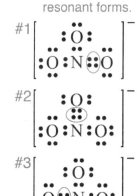

What resonance structures mean

In reality, the nitrate ion assumes none of the three drawn forms but is actually a hybrid of all three. The real nitrate ion shares the double bond equally among the three oxygen atoms so each oxygen essentially gets 4/3 of an electron. That makes the N-O bond shorter than a single bond which is why the nitrate ion stays together in many chemical reactions.

Resonant forms in structural diagrams

Rings

Benzene forms a ring

In the early 1800s it was known that the compound C_6H_6 was important but its structure was not known. The brilliant German chemist August Kekule proposed the correct structure in 1865. Benzene is a ring of six carbon atoms with alternating double bonds. The double bonds do not really alternate like the diagram but instead the six electrons circulate around the ring. The benzene ring is an example of resonance structures. The benzene resonance is particularly stable and benzene rings are found in many compounds.

Molecules with ring structures

Benzene (C_6H_6)

Glucose ($C_6H_{12}O_6$)

Sugars have a ring structure

Glucose, or blood sugar, is a six-sided ring with one oxygen replacing a carbon atom in the ring. Fructose, known as fruit sugar, is a five-sided ring structure with an oxygen atom in the ring. Table sugar (sucrose) is a combination of a glucose and a fructose bonded together in a double ring structure.

Sucrose (table sugar) has two joined rings

Glucose ring

Fructose ring

Aromatic hydrocarbons

Many compounds that create strong smells (and tastes) contain benzene rings. This compound is known as an **aromatic hydrocarbon**. A good example is vanillin, $C_8H_8O_3$. Vanillin is the compound that produces the characteristic smell and taste of vanilla. Natural vanillin is extracted from the vanilla bean, which was first cultivated by the Aztecs in central America. Vanilla was imported to Europe by the Spanish conquistadors around 1520. Synthetic vanillin was first produced in the 1930's as a byproduct of paper pulp processing and today, about 60% of the vanillin in desserts and foods is synthetic.

Many aromatic hydrocarbons have ring structures.

Vanillin

Section 10.3: Geometry of Molecules

Polymers

Linear shapes

Many important molecules are large, containing one hundred or more atoms. Protein molecules can include more than 10,000 atoms. We find that virtually all large molecules are built up of repeating smaller units. An example is the common plastic polyethylene. The plastic baggie that holds your sandwich is made of polyethylene. Polyethylene is a **polymer**. "Poly" means many and a polymer is a chain with many repeating units. The chains in common plastics may be 1,000 to 100,000 monomers long!

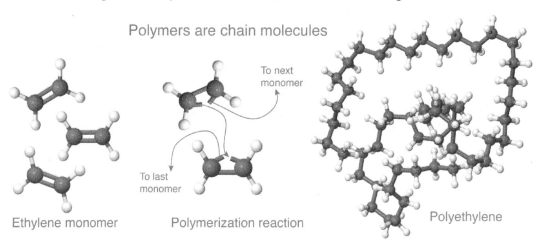

Chains of monomers

The repeating unit of a polymer is called a **monomer**. Polyethylene is built up of ethylene monomers. The double bond in ethylene is broken by heat and processing. The ethylene monomers are connected in a **polymerization** reaction. Every plastic is created through polymerization reactions using different monomers. Common examples of polymers are Teflon™, nylon, polystyrene, and rubber.

Starch and cellulose

Polymers are found in nature as well as human technology. Plants produce glucose from photosynthesis and the glucose is assembled into long chains called starches. Starches are high-energy foods because the body can easily break them down into glucose. Plants also produce the non-digestible polymer cellulose - which is the principle component in wood. Cellulose is also made of chains of glucose but the CH_2OH group alternates. Some insects such as termites can digest cellulose, but humans cannot.

Cis- and trans-fats

You have likely heard that trans-fats are bad for you, but you have heard of cis-fats? Eating too much trans-fat has been linked to low density lipoproteins, or LDLs, which cause cholesterol to deposit in arteries leading to health problems. Natural fat molecules contain chains of hydrocarbons with occasional double bonds that put two hydrogen atoms on the same side of the chain as seen in the cis- configuration above. This causes the chain to "kink" because of the bent bond. Synthetic fat molecules in the trans- configuration have the hydrogen atoms on opposite sides of the double bond as shown. This creates a trigonal shape that leaves the chain straight. The difference between the bent and straight chains has strong biological effects.

Section 3 review

Chapter 10

Molecular geometry describes the three dimensional shape of molecules. Ball and stick models and 3D pictures are good for visualizing shapes but do not show that real molecules are constantly in motion. The VESPR theory states that areas of electron density (chemical bonds and unbonded electron pairs) repel other areas of electron density. This repulsion creates the three-dimensional shapes of molecules. Four areas of electron density create a tetrahedral shape. If one of the four is an unbonded electron pair and three are bonds then a trigonal pyramidal shape results. If two of the four are unbonded electron pairs and two are bonds then a bent shape is created. Three areas of electron density create a trigonal planar shape. Two areas of electron density create a linear shape.

Isomers are different molecules that have the same chemical formula but different configurations of atoms. Isomers have different chemical and physical properties. Resonance structures are the same compound with different arrangements of adjacent single, double, or triple bonds. Common molecular structures include rings such as benzene and sugar. Polymers are long chain molecules built up from simpler units called monomers.

Vocabulary words

VSEPR (Valence Shell Electron Pair Repulsion), electron density, hybridization, isomers, resonance, aromatic hydrocarbon, polymer, monomer, polymerization

Key relationships

# of electron-dense regions	shape class	shape example	3D drawing	shape name	number of atoms bonded to central atom	number of lone pairs on central atom
2	LINEAR			linear	1 or 2	0
3	TRIGONAL			trigonal planar	3	0
				bent or angular	2	1
4	TETRAHEDRAL			linear	1	3
				bent or angular	2	2
				trigonal pyramidal	3	1
				tetrahedral	4	0

Review problems and questions

1. The nitrous oxide (laughing gas) molecule has the chemical formula N_2O.
    a. Propose a Lewis dot structure for nitrous oxide.
    b. How many areas of electron density are there in the nitrous oxide molecule?
    c. Describe the shape of the molecule.

2. A molecule with four areas of electron density, two of which are chemical bonds, will have what shape?
    a. linear   b. bent   c. trigonal planar   d. linear

3. Find three isomers of a compound with the chemical formula CH_3NO.

Section 10.3: Geometry of Molecules

10.4 - Intermolecular Forces

Molecules and polyatomic ions stay together through chemical bonding. Chemical bonds are intramolecular forces, meaning they act between the atoms within a molecule. There are other forces that act outside of or between individual particles. These intermolecular forces hold molecules and ions together in solids and liquids. In general, intermolecular forces are much weaker than chemical bonds. However, intermolecular forces result in many of the most important physical properties of matter - such as freezing, boiling, phase changes, and solubility.

Intermolecular forces

How does atomic behavior result in physical properties?

Consider water, iron, and methane. At room temperature, water is liquid, iron is solid, and methane is a gas. Diamond is the hardest substance known yet the same pure material (carbon) also makes slippery graphite! Why? How do properties on the atomic level explain the different observable properties of these substances? Macroscopic properties, such as boiling point, melting point, and hardness, have their explanation in the microscopic behaviors of atoms, ions, and molecules.

Physical properties that depend on intermolecular forces

	Methane	Olive oil	Water	Wax	Salt	Iron
Room temperature	Gas	Liquid	Liquid	Solid	Solid	Solid
Melting point	-182°C	-6°C	0°C	37°C	801°C	1538°C
Boiling point	-161°C	700°C	100°C	380°C	1413°C	2862°C

What are intermolecular forces?

The forces that hold atoms and molecules together to form solids and liquids are called **intermolecular forces**. In molecular compounds such as wax or oil, the forces are truly intermolecular. In metals, ionic solids (salt), and diamond, the forces are between atoms and ions. However, the term intermolecular forces is still used to describe the forces that hold individual particles together to form bulk matter such as solids and liquids.

Types of intermolecular forces

	Strength	Examples	Bonding type
Non-polar attractions	Weakest	oils, waxes	London dispersion
Polar attractions		water, alcohol	polar, hydrogen bondng
Ionic		salt, minerals	ionic charges
Metallic		iron, metals	metallic bonds
Covalent	Strongest	diamond	network covalent bonds

Relationship with physical properties

The temperature at which a substance melts or boils is a good measure of the strength of intermolecular forces. Weaker intermolecular forces mean lower attraction between neighboring molecules, and lower melting and boiling points. Substances with low intermolecular forces are often gases or liquids at room temperature. Stronger intermolecular forces mean it takes a higher temperature - more energy - to separate neighboring ions, atoms, or molecules. Ionic, metallic, and covalent bonded solids have high melting points due to the strength of intermolecular forces.

Water and hydrogen bonding

Water is a dipole

Water's unique properties come from the shape and bonds in a water molecule. Water is a bent polar molecule because the oxygen atom in the center has a higher electronegativity than the two bonded hydrogen atoms. The charge separation in a water molecule is called a **dipole** because there are two different charged ends - one more positive and one more negative. The symbol δ^- indicates the partially negative end and δ^+ indicates the partially positive end. When water molecules are together in a liquid or solid the oppositely charged parts of neighboring molecules attract each other. The negatively charged oxygen atom in one molecule attracts all the positively charged hydrogen atoms in neighboring molecules.

Water is a polar molecule

Electronegativity of H = 2.20

Electronegativity of O = 3.44

δ^+ = partial positive charge
δ^- = partial negative charge

A water molecule has a dipole at each O-H bond. A large electronegativity difference between O and H causes a charge separation.

Hydrogen bonds are attractions between molecules. Partially negative oxygen atoms are attracted to partially positive hydrogen atoms.

Dipole-dipole forces and hydrogen bonds

The attraction between the charged regions of neighboring polar molecules is called a **dipole–dipole attraction**. Dipole-dipole attractions are relatively strong compared to other types of forces between molecules, but still weaker than ionic or metallic bonds. A **hydrogen bond** is a special type of dipole-dipole attraction in which hydrogen atoms on one molecule are attracted to a highly electronegative atom, such as O, N or F on a neighboring atom.

Effects of Hydrogen Bonding

Water, H$_2$O

- bent shape
- O-H electronegativity difference: 1.24
- forms hydrogen bonds
- melts at 0 °C
- boils at 100 °C
- liquid at room temp.

Hydrogen bonds

No hydrogen bonds

Hydrogen sulfide, H$_2$S

- bent shape
- S-H electronegativity difference: 0.38
- no hydrogen bonding
- melts at -82 °C
- boils at -60 °C
- gas at room temp.

Hydrogen bonding and boiling point

Compare water, H$_2$O, with hydrogen sulfide, H$_2$S. Both molecules have a bent shape, and both are composed of a Group 16 central atom with two bonded hydrogen atoms. Yet the boiling point for H$_2$S is -60 °C while the boiling point for water is 160 degrees higher! Why such a large difference? The difference is that polar water molecules share hydrogen bonds which hold them together much more tightly compared to non-polar H$_2$S molecules.

Surface tension

Surface tension is another property of water that is caused by the strength of hydrogen bonding between water molecules. Surface tension is a force that pulls all exposed surfaces of water together, causing drops to contract into spheres when they leave a container. Water has a relatively high surface tension compared to substances with no hydrogen bonding.

Polar substances

Like dissolves like

The rule "like dissolves like" is an observation of the properties of polar and non-polar molecules. Water is a polar molecule and other polar molecules dissolve easily in water. for example methanol and ethanol both dissolve in water because all alcohols are polar molecules. Benzene does not readily dissolve in water because benzene is a non-polar molecule. If benzene is mixed with water the benzene will remain floating on top.

Like dissolves like - polar substances dissolve other polar substances

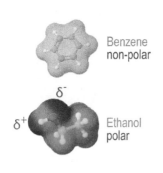
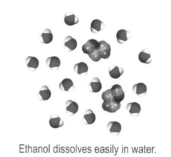
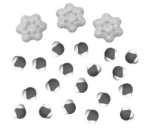

Ethanol dissolves easily in water.   Benzene does not dissolve in water.

Polar and non-polar substances

Polar substances dissolve either polar or ionic substances. Both salt and sugar dissolve easily in water. This is because there is a strong attraction between the water molecules and the dissolving ions or sugar molecules. Non-polar substances such as oil, fat and wax do not dissolve in polar substances like water because the oil lacks attractive forces. Non-polar substances cannot dissolve ionic substances either. For example, salt does not dissolve in oil, but a non-polar substance like wax or fat will dissolve in oil.

how ionic substances dissolve in water

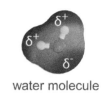

water molecule

Polarity and boiling point

Molecules with large dipoles form strong attractions. The molecules below are arranged by an increasing ability to attract neighboring molecules due to polarity. Hydrogen gas, H_2 is non-polar and particle attractions between molecules are minimal, so it has a low boiling point. Water has a high boiling point because each molecule has two polar bonds that attract many neighboring molecules to itself. Other properties are also affected by polarity - such as the biological activity of medicines and proteins.

boiling point and attractive forces

red < orange < yellow < green < blue
most negative   most positive

H_2 / -252.87 °C NH_3 / -33.3 °C   HF / 19.5 °C H_2O / 100 °C

London dispersion and Van der Waals forces

The London-dispersion attraction

Non-polar substances such as oil form liquids and solids therefore attractive forces between molecules must exist. When two molecules are close together, there is a temporary attractive force between the electron cloud of one molecule and the nuclei in the other molecule. This creates a temporary charge separation - or polarization - that is weaker than a polar bond. This effect is called the **London dispersion attraction**. The London dispersion attraction is responsible for the way oils flow thickly and butter solidifies when cold. All atoms and molecules attract each other through London dispersion, however the dipole attraction is generally stronger for polar molecules.

Boiling point increases with size

The more surface area a molecule has, the stronger the London dispersion attraction. This is easy to see in the boiling point increase of small, straight-chain hydrocarbons. Butane is a common fuel for small lighters. Butane is a gas at room temperature but is relatively easy to liquefy by chilling below -1 °C. Octane is a primary component of gasoline and has a high enough boiling point that it stays liquid over a wide range of environmental temperatures. In the gas phase, octane is explosive - hence it is used as a fuel.

Size of non-polar molecules and boiling point

Butane
C_4H_{10}
boiling point = **−1 °C**

Hexane
C_6H_{14}
boiling point = **68 °C**

Octane
C_8H_{18}
boiling point = **125 °C**

Van der Waals

The London dispersion and dipole-dipole forces between atoms and molecules are collectively know as **Van der Waals attractions** after Dutch scientist Johannes Diderik van der Waals. Van der Waals attractions are largely responsible for the phase changes in molecular compounds such as water, wax, oil, and air. Ionic and metallic compounds have different and stronger attractive forces between neighboring particles.

Attractions vs. bonds

van der Waals attraction and intermolecular forces

London dispersion   Weakest

dipole-dipole (polar) attraction

hydrogen bonding

van der Waals attractions

Chemical bonds

ionic bonds
metallic bonds
network covalent bonds   Strongest

As shown above the "line" between a chemical bond and an attractive force does not actually exist. The difference between a chemical bond and attractive force is largely the strength of attraction between adjacent particles. Stronger attractions between adjacent particles are more likely to result in a chemical bond. The weakest attractive forces are not likely to result in a chemical bond.

Ionic substances

Characteristics of ionic substances

The forces between ions in ionic compounds are comparable to chemical bonds, and much stronger than dipole or London dispersion forces. For this reason, ionic substances tend to have similar physical properties.

- Ionic substances are usually brittle and hard.
- Ionic substances form crystalline solids.
- Ionic substances are solid at room temperature and have very high melting points.
- Ionic substances become conductive when melted or dissolved in water or other polar solvents.

Sodium chloride

A simple ionic compound is sodium chloride, NaCl. Sodium chloride forms a hard, brittle cubic crystal in which every sodium ion (Na^+) is surrounded by six chlorine ions (Cl^-) and vice versa. Sodium chloride melts at 800 °C and makes a conductive salt solution when dissolved in water.

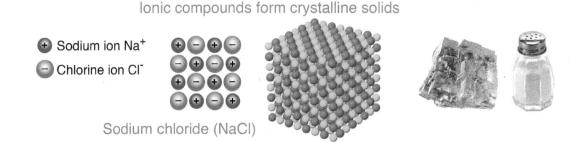
Ionic compounds form crystalline solids
Sodium chloride (NaCl)

Ionic minerals

Many minerals are ionic substances and a good example is calcite, $CaCO_3$. Calcite occurs naturally in many forms and clear calcite crystals were once used in optics. Easily dissolved in acids, calcite is a primary mineral in stalactites and other cave formations formed by dripping water. The structure of calcite alternates calcium ions with carbonate ions and is called a **lattice**.

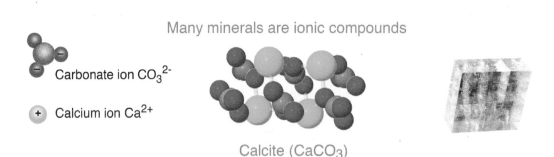
Many minerals are ionic compounds
Calcite ($CaCO_3$)

Ionic solutions

Many ionic substances dissolve in water and acids. When dissolved, the positive and negative ions are surrounded by polar water molecules and prevented from returning to the solid state. Once dissolved, the mobile ions can move and carry electric current making ionic solutions electrical conductors. Batteries rely on conductive solutions as do the nerve impulses in your body! Nerve signals travel through conductive ions including calcium (Ca^{2+}), potassium (K^+), and sodium (Na^+).

Diamond and network substances

Diamond is a network covalent structure

Most molecules are too small to see even with the best microscope, but by some definitions a diamond is a single molecule! The carbon atoms in diamond are bonded together in a **network covalent** structure. In a network covalent structure every atom is bonded to neighboring atoms in an interconnected mesh. Network covalent structures are very strong. To scratch diamond you have to break chemical bonds, not just intermolecular forces.

Network covalent structures of carbon

C_{60} "buckyball"   Diamond   Graphite

Carbon nanotube

Carbon network structures

Carbon and silicon form the most common network covalent substances. In addition to natural diamond, many network covalent structures have been discovered for carbon. Discovered in 1985 the spherical molecule C_{60} nicknamed "buckyball" is one of the fullerenes. Another carbon structure, graphite, is used as a dry lubricant because the atoms are organized in sheets that slide over each other. Carbon nanotubes are the strongest known fiber - many times stronger than steel. Ongoing research efforts seek to make practical lengths of nanotube, the only material strong enough to build a space elevator.

Quartz and glass

Quartz and glass are also network covalent compounds. Both are composed of silicon dioxide (SiO_2). Quartz and glass share similar characteristics of hardness, brittleness, and transparency. The quality of transparency comes from strongly bound electrons. Light traveling through a substance is constantly absorbed and re-emitted by atoms. If the electrons are bound tightly the electron responsible for the absorption is constrained to re-emit the light with the same color and in the same direction. All transparent materials are also electrical insulators!

Quartz and glass are network covalent compounds of silicon dioxide

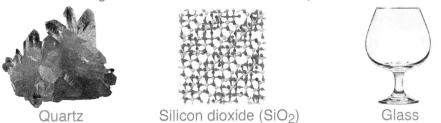

Quartz   Silicon dioxide (SiO_2)   Glass

Metals and alloys

Ductility and metallic bonds

Metals have the unusual properties of being strong but also **ductile**. A material is ductile if it bends or deforms without breaking. The opposite of ductile is brittle. Glass is brittle and breaks before bending much while aluminum can bend quite far before breaking. The ductility of metals comes from the metallic bond's combination of strength with non-localized electrons. Remember that metallic bonds occur through a sea of non-localized electrons.

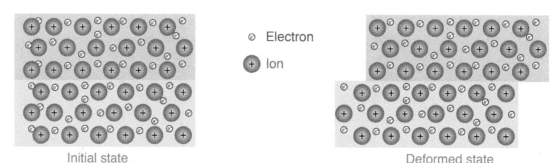

Ductility and metallic bonding

How metals deform

The bonds between metal ions are made through mobile electrons instead of through electrons attached to any specific atom. If the ions become too separated, a void is filled with negative charge which attracts the ions back. This is the explanation for the strength of metals - metals are hard to break. However, when a metal bends, neighboring atoms slide relative to each other maintaining about the same separation between ions. The mobile electrons allow the bonds to move and reform between the atoms easily during the deformation.

Alloys

Metals are crystalline and when a metal deforms the deformation tends to follow the weakest direction - between the atoms along a crystal plane. Anything that disrupts the crystal order will make a metal harder by disrupting the movement of atoms along any plane. Metals are actually polycrystalline which means the atoms are arranged in many tiny crystalline domains. The atoms in a domain are aligned with each other but neighboring domains have different alignments. The polycrystalline structure makes metals harder because it takes more energy to move atoms that are not all aligned along a common crystal plane. Metals are heat treated to change the size of their crystal grains. An **alloy** is a mixture of two or more metals. Alloys are stronger than their pure component metals because the different size atoms (like grain boundaries) disrupt the crystal structure and make it harder for deformations to spread.

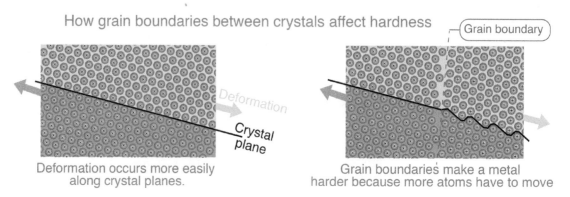

How grain boundaries between crystals affect hardness

Section 4 review

Chapter 10

Intermolecular forces are attractions between a molecule and its neighboring molecules that determine many physical properties of a substance. The strongest intermolecular forces are dipole-dipole interactions. These occur when the part of a molecule that is partially negatively charged interacts with the part of a neighboring molecule that is partially positively charged. Dipole-dipole interactions involving hydrogen atoms of polar molecules are especially strong and called hydrogen bonds. London dispersion forces stem from temporary dipoles and are the weakest of the intermolecular forces.

Ionic forces are much stronger than dipole or London dispersion forces and lead to the common properties of ionic substances. A covalent network is a compound where atoms are in a continuous network of covalent bonds, such as in diamonds. Metallic solids have a network of metal nuclei with a mobile "sea" of electrons moving between them that gives rise to the properties of metals.

Vocabulary words: intermolecular forces, dipole, dipole–dipole attraction, hydrogen bond, surface tension, London dispersion attraction, Van der Waals attractions, lattice, network covalent, ductile, alloy

Review problems and questions

1. Water has a molecular weight of 18 g/mol and methane has a molecular weight of 16 g/mol. Although they are so close in weight, their melting points vary largely. Why does this occur?

2. What is the relationship between the polarity of a molecule and its boiling point?

3. What is the relationship between the length of straight chain hydrocarbon molecules and boiling point?

4. Select ALL of the species below that can form a hydrogen bond. Explain why the other species cannot hydrogen bond.

    a. C_2H_6
    b. HCN
    c. CH_3NH_2
    d. HCl
    e. $CH_3CH_2CH_2OH$
    f. CH_3OCH_3
    g. H_2S
    h. HF

5. Rank the following compounds from weakest intermolecular forces to strongest. Explain your answers.

    I_2, H_2S, H_2O, N_2

6. List ALL types of intermolecular forces that would occur in each of the following:

    a. CCl_4
    b. KCl
    c. PCl_5
    d. CH_3OH
    e. HBr

Stronger than Steel?

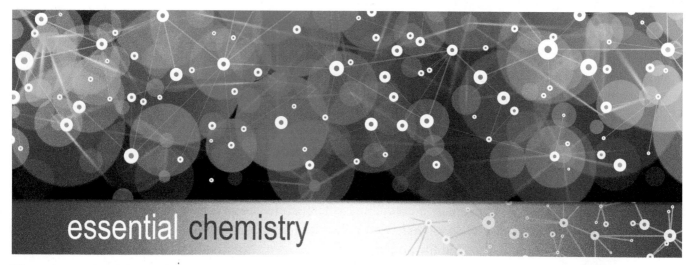

There is a material that is used to make safety clothing, combat helmets and bulletproof vests. You might think that it needs to be a hard, rigid material like steel or even diamond to be strong enough to stop a bullet. But this material can be spun into fibers and incorporated into ropes, racing shoes, socks and gloves. How can one material have such broad applications? The answer, of course, is based on its chemical bonding and structure.

bullet proof vest and helmet

The material capable of doing all this is poly-paraphenylene terephthalamide. That name is a mouthful, so you might know it from its more common name, Kevlar. It was invented in 1965 by Stephanie Kwolek while working at DuPont, a chemical company. In the early 1960s people anticipated an upcoming gasoline shortage. DuPont issued a challenge to its researchers – they were to design a new high performance fiber to replace steel wires in tires. By replacing the heavy steel wires in the tires, cars could use less fuel. Kwolek was up to the task of looking for a super-strong but lightweight fiber, since she was already working with long chain polymers called polyamides.

Stephanie Kwolek

As the name suggests, a polyamide is a polymer made of repeating amide monomers. An amide is a group of atoms that have nitrogen atoms connected to a carbon atom and are double bonded to an oxygen atom. This monomer is made from the reaction of two different chemicals, one containing the nitrogen atoms, and one containing the carbons double bonded to oxygen atoms. As they react, a hydrogen atom comes off one of the reactants and chlorine comes off the other.

The rest of the polymer structure is made from benzene (C_6H_6) ring derivatives. You will learn more about the chemistry of carbon atoms in the organic chemistry and biochemistry chapters.

Kwolek found a solvent for a polymer that created a very peculiar solution. A typical polymer solution is sort of syrupy, but this solution was thin, watery and could appear shimmery with glints of color like an opal. The polyamide that she was working with made a liquid crystal. In a liquid crystal, the long rod-like polymer units arranged themselves parallel to each other. So it was a "liquid" because the units were free to move, but it was a "crystal" because the arrangement followed a consistent stacking pattern.

Polymer solutions are run through a spinner to turn them into fibers. The spinner forces the solution through tiny holes (about 0.001 inches in diameter) to make thin strands. The strands are then spun together to make fibers. The person in charge of the spinner at DuPont initially refused to spin Kwolek's unusual solution fearing that it would plug the holes in his spinner. The mechanical spinning and pulling oriented the molecules correctly and a strong beautiful fiber was pulled. Following this, many other fibers were spun from liquid crystal polymer solutions, including the yellow fiber that would eventually be known as Kevlar.

Kevlar gets its strength not just from the bonds within the polymer, but from the attractive forces between the large polymer molecules. The polymers are generally flat because the carbon atoms are sp^2 hybridized and are trigonal planar. You can see from the image that the 6-membered carbon rings of different polymer molecules can stack on top of one another. This maximizes the London dispersion forces that can interact between the non-polar regions of the polymers. Even stronger than the London dispersion forces, hydrogen bonding exists between the –NH group of one polymer and the C=O group of an adjacent polymer. The polymer units are arranged parallel to each other to maximize attractive forces.

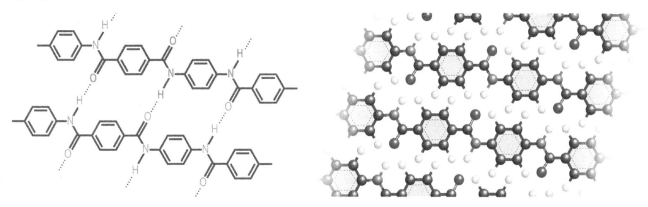

The properties of Kevlar are influenced by the strength of the attractive forces between the molecules. It has a much higher tensile (or pulling) strength than steel even though it has a much lower density. In fact, the tensile strength to weight ratio is five times stronger than steel! These strong attractive forces make it a very good "antiballistic" material. A tightly woven fabric of strands of these parallel polymer molecules are very hard to push or pull apart. Even a bullet has a hard time making it through a Kevlar vest.

Kevlar has gone on to find uses in many applications, from light body armor to super strong ropes for bridges. It is also used to make protective clothing for athletes and scientists. Outside of person-protection, it has been incorporated into the shells of canoes, marching snare drums, and as a protective sheath for underwater cable. It has been used in about 200 different applications!

Kwolek was recognized for her contribution to the science and technology behind Kevlar. In 1994, she was inducted into the National Inventors Hall of fame – at the time only the fourth woman to do so. She died in 2014 at the age of ninety, but not before leaving behind a legacy that curiosity, persistence, and an essential understanding of chemistry can change the world.

Chapter 10 review

Vocabulary
Match each word to the sentence where it best fits.

Section 10.1

chemical bond	diatomic molecule
double bond	Hund's rule
ion	isoelectronic
Lewis dot diagram	octet rule
polarization	single bond
triple bond	

1. A(n) _____ is a bond formed when two pairs of electrons are shared between two atoms.
2. Two atoms that are bonded together naturally are called a(n) _____.
3. The _____ states that elements will have eight electrons in their valence shell when they are stable.
4. A(n) _____ is a bond formed when three pairs of electrons are shared between two atoms.
5. A(n) _____ is a bond formed when one pair of electrons is shared between two atoms.
6. A(n) _____ is a charged particle.
7. Atoms that have the same electron configuration are called _____.
8. The idea that each orbital must contain a single electron before any orbital accepts two electrons, is expressed in _____.
9. A relatively strong attraction called a _____ occurs between a pair of electrons in neighboring atoms.
10. As two atoms approach one another, their electron clouds undergo _____.
11. A(n) _____ contains an element symbol surrounded by dots representing its valence electrons.

Section 10.2

crystal	electronegativity
metal	metallic bond
metalloids	non-metals
non-polar covalent bond	polar covalent bond
semiconductors	

12. _____ are a class of compounds that have properties between conductors and insulators.
13. Diatomic oxygen forms a(n) _____ because it has an electronegativity difference of zero.
14. A(n) _____ is a bond with a slightly unequal sharing of electrons.
15. Fluorine has the highest value of _____ on the periodic table.
16. A(n) _____ arranges oppositely charged ions in a regular pattern.
17. A substance that is crystalline when solid and conducts heat and electricity is called a(n) _____.
18. A bond that is composed of a loosely connected "sea of electrons" is called _____.
19. Boron, silicon, and germanium are examples of _____, which have properties between those of metals and non-metals.
20. _____ are a class of compounds that are insulators.

Section 10.3

aromatic hydrocarbon	electron density
hybridization	isomers
monomer	polymer
polymerization	resonance
VSEPR (Valence Shell Electron Pair Repulsion)	

21. Molecules have _____ when they have two or more valid Lewis dot diagrams.
22. _____ is a reaction that combines smaller molecular fragments into a polymer.
23. The s and p orbitals combine to form sp orbitals in a process called _____.
24. The _____ theory states that the shapes of molecules are influenced by lone pairs of electrons.
25. In order to determine a molecule's geometry, you must count the regions of _____.
26. A(n) _____ is a long chain of molecules built up from smaller molecular units.
27. A chemical formula is said to have _____ if it can be drawn more than one way.
28. A hydrocarbon with a distinct smell and a resonant ring structure is a(n) _____.
29. The building blocks of polymers are called _____(s).

Section 10.4

alloy	dipole
dipole–dipole attraction	ductile
hydrogen bond	intermolecular forces
lattice	London dispersion attraction
network covalent	surface tension
Van der Waals attractions	

30. A(n) _____ is stronger than its pure component metal.
31. _____ is an attractive force in liquids caused by intermolecular forces.

Chapter 10 review

Section 10.4

alloy	dipole
dipole–dipole attraction	ductile
hydrogen bond	intermolecular forces
lattice	London dispersion attraction
network covalent	surface tension
Van der Waals attractions	

32. An intermolecular attraction between hydrogen and a highly electronegative atom in a different molecule are called _____(s).

33. The attraction between the charged regions of neighboring polar molecules is called a(n) _____.

34. The forces present in molecular compounds such as wax or oil are called _____.

35. A(n) _____ contains a separation of positive and negative charges.

36. A(n) _____ describes the pattern of ions in a crystal.

37. A collection of intermolecular forces caused by dipole interactions is called _____.

38. Large molecules, where each atom is covalently bonded to multiple neighboring atoms, form _____ structures.

39. _____ is a physical property of materials that describes the ease to which it stretches.

40. A(n) _____ is a weak attraction between atoms caused by the interaction of electrons.

Conceptual questions

Section 10.1

41. What part of the atom forms bonds?

42. How is a sodium ion different from a sodium atom?

43. What three ions are isoelectronic with Neon?

44. Why do atoms form ions that are isoelectronic with noble gases?

45. Not all valence electrons in nitrogen form bonds. Why?

46. Describe the octet rule.

47. When does the octet rule apply?

48. What do the electron configurations of atoms obeying the octet rule have in common?

49. Describe an equilibrium distance with regard to bonding.

50. Why does hydrogen form a diatomic molecule?

51. Use the octet rule to explain how phosphorus combines with chlorine to form a neutral molecule.

52. Do hydrogen atoms have full energy levels when they form diatomic molecules? Explain your answer.

53. What happens when atoms get too close to each other?

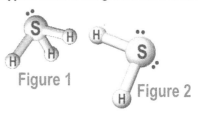

Figure 1   Figure 2

54. Only one of the structures is correct for a neutral compound of sulfur and hydrogen.
    a. Which diagram is correct?
    b. Why is the other diagram incorrect? Explain your answer.

Space-filling   Molecular electron cloud

55. Compare the above methane molecules represented with two different model styles: the space-filling molecular surface model, and the molecular electron cloud model.
    a. Which model has more information about the molecule? What additional information is provided?
    b. Which model most accurately represents the surface of a molecule? Explain your answer.

56. What happens to the electron cloud when an atom becomes polarized?

57. Consider a molecule represented with the ball and stick model.
    a. What does a ball represent?
    b. What does a stick represent?
    c. What kind of information does this model accurately represent?
    d. What kind of information does this model inaccurately represent?

58. What does a Lewis dot diagram represent?

59. How does Hund's rule apply to Lewis dot diagrams?

60. How does the Pauli exclusion principle apply to Lewis dot diagrams?

61. How does the octet rule apply to Lewis dot diagrams?

62. How is the electron cloud around a molecule different from an electron cloud around a single atom?

Chapter 10 review

Section 10.2

63. How is an atom's electronegativity related to its tendency to form either a positive or negative ion?
64. Why does hydrogen form diatomic molecules when neon does not?
65. How do you modify a Lewis dot diagram for an ionic substance?
66. What type of bonds create crystals?
67. Describe a metallic bond.
68. How is a metallic bond different from an ionic bond?
69. How is a metallic bond different from a covalent bond?
70. What does a bond triangle show?
71. What is electronegativity?
72. Where are elements with the highest electronegativity found on the periodic table? Where are elements with the lowest electronegativity found?
73. Hydrogen has, H_2 and oxygen gas, O_2 are both highly explosive gases when given a spark. Why don't H_2 and O_2 instantly react when they contact one another?
74. Two atoms form a bond.
    a. Which has lower energy, the unbonded atoms before the reaction or the bonded atoms after the reaction? Explain your answer.
    b. Based on your first answer, does that mean energy is released, or energy is absorbed when bonds form? Explain.
    c. In order to break the bond, will energy need to be released, or absorbed?
75. What happens to electrons when ionic bonds are formed?
76. What happens to electrons when covalent bonds are formed?
77. Describe a crystal's structure.
78. Why are metals good conductors of electricity?

Section 10.3

79. Why do electrons repel each other?
80. What are the main points of the VSEPR Theory?
81. What orbitals are responsible for a tetrahedral shape?
82. Describe hybridization.
83. Is a sigma bond or a pi bond stronger?
84. What type of bonds are formed from sigma interaction?
85. What is the relationship between isomers and their properties? Give an example.

86. What is a limitation when using a ball and stick model of the atom?
87. In regards to electron density, why does methane form a tetrahedral shape?
88. In terms of sigma and pi bonds, what is a triple bond composed of?
89. Explain how isomers are different from one another.
90. What does it mean when a molecule is resonant?
91. What is the primary component of an aromatic hydrocarbon?
92. If a molecule has 4 areas of electron density, what is the molecular geometry if the central atoms has:

    a. 2 un-bonded electron pairs and 2 chemical bonds

    b. 1 un-bonded electron pair and 3 chemical bonds

    c. 0 un-bonded electron pairs and 4 chemical bonds

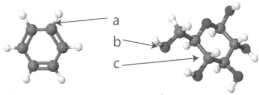

93. The two molecules shown above are benzene (left) and glucose (right).

    a. How many regions of electron density are present for atoms a, b, and c? Hint: Don't forget to account for lone pairs!

    b. According to VSEPR theory, what 3D molecular shape is formed by a, b, and c?

Section 10.4

94. What is the relationship between the strength of the intermolecular forces of a substance and its melting point?
95. List the Van der Waals forces from weakest to strongest.
96. How are intermolecular forces different than bonds?
97. What is the difference between an ionic lattice and a network covalent structure?
98. Ionic substances tend to have similar physical properties. Provide two examples describing the common physical properties of ionic compounds.
99. Why are alloys stronger than their pure component metals?
100. Turn to the page where London dispersion and Van der Waals attractions are introduced in Section 10.4. Find the diagram at the bottom of the page with the margin labeled "Attractions vs. bonds." Analyze how the image distinguishes an intermolecular force of attraction from a chemical bond. Take on the role of a chemistry textbook author. Write a 1-paragraph explanation of how this image helps you distinguish an "attraction" from a true chemical bond.

Chapter 10 review

Section 10.4

101. List the following types of attractions from weakest to strongest.

    a. dipole-dipole attraction
    b. ionic bonds
    c. London dispersion
    d. network covalent bonds
    e. metallic bonds
    f. hydrogen bonds

Quantitative problems

Section 10.1

102. Draw the Lewis dot diagrams for the following molecules.

    a. Cl_2
    b. H_2S
    c. SiH_4
    d. CH_4
    e. C_2H_6
    f. C_4H_{10}

103. Draw a Lewis dot diagram for the following compounds.

    a. C_2H_3OH
    b. C_2HOH
    c. CH_2ClBr
    d. BN
    e. C_2H_2FCl
    f. C_6H_8

104. For the following molecules, draw a Lewis dot diagram that satisfies the octet rule.

    a. C_2H_4
    b. C_2H_2
    c. CO_2
    d. PN
    e. C_3H_6
    f. C_3H_4

105. Use the following two-way frequency table to determine the structure of the unknown molecule.

Unknown Molecule

	carbon-carbon	carbon-oxygen	carbon-hydrogen
Single Bonds	1	2	6
Double Bonds	1	1	0
Total	2	3	6

Section 10.2

106. Draw the Lewis dot diagrams for the following ions.

    a. CN^-
    b. NH_4^+
    c. NO_2^-
    d. SO_3^{2-}
    e. CH_3COO^-

107. Predict the charge of each ion and the chemical formula for a compound of lithium and sulfur.

108. Classify the following bonds as covalent, polar covalent, or ionic based on the difference in their electronegativity.

    a. potassium and iodine
    b. carbon and oxygen
    c. fluorine and bromine
    d. magnesium and oxygen
    e. boron and nitrogen
    f. rubidium and fluorine

Section 10.3

109. Draw two isomers of $C_3H_8O_2$.

110. Draw two isomers for C_4H_{10}.

111. Fill in the blanks on the following table.

Molecular Geometry Table

Regions of electron density	Number of atoms bonded to the center atom	Number of lone pairs of electrons on the center atom	Molecular Geometry
2	2		Linear
3		0	Trigonal Planar
	4	0	Tetrahedral
4			Trigonal Pyramidal
	2	2	

112. Fill out the following two-way frequency table.

Molecular Geometry Table

Geometry	Example	Number of atoms bonded to the center atom	Number of lone pairs of electrons on the center atom
Tetrahedral	CH_4		
Trigonal Pyramidal	NH_3		
Bent	H_2O		
Linear	CN^-		

113. Draw the Lewis structures and determine the geometry for the following molecules.

    a. CBr_4
    b. PN
    c. PF_3
    d. BF_3
    e. SiO_2
    f. H_2S

chapter 11 | Energy and Change

When you are outside on a hike and you see a waterfall you notice that the water always falls down hill. In fact, it would be very surprising if water were to spontaneously fly up to the top of the waterfall! The water falls down because energy is released in the process. We even build dams to use that release of energy and generate electricity. By using pumps or buckets, we could make the water go back to the top of the hill, but we would need to provide the energy to make it happen.

Thermodynamics looks at the interplay between heat and work to see if something will happen. Just like water falling downhill, if a physical or chemical change can result in the release of energy, it will happen spontaneously. Spontaneous can be a misunderstood word. If something is spontaneous, it does not mean it will happen quickly or explosively. It simply means that it will happen. Iron reacts with oxygen to rust, but it is not a particularly fast or dramatic process. Phase changes also occur spontaneously under the right conditions. Ice melts spontaneously if left outside on a warm day. But if you took the water and put it back inthe freezer, it would become solid again. This is part of the interplay between heat and work because the "direction" of a spontaneous process can be changed with temperature.

Scientists often ask the question, "will it happen?" To answer this question, we need to look at the relationshipbetween energy and change, and have an essential understanding of… thermodynamics.

Chapter 11 study guide

Chapter Preview

Thermochemistry is the branch of chemistry that deals with heat changes during chemical reactions. In this chapter you will learn to interpret phase change diagrams and make calculations of the heat energy released or absorbed when a reaction takes place. A phase diagram shows graphically how temperature and pressure combine to determine the phase of a given substance. The First Law of Thermodynamics states that energy cannot be created or destroyed. The 2nd Law of Thermodynamics states that entropy of an isolated system always increases. Chemical reactions are accompanied by a change in enthalpy, giving energy to the surroundings when exothermic and absorbing energy from the surroundings when endothermic. The net energy of a reaction can be found by comparing the bond energies of the products and reactants. The Gibbs-Helmholtz equation determines whether a reaction is spontaneous. Animals gain energy from the food they eat and that energy can be determined using calorimetry. Fats are used to store energy while carbohydrates are available for immediate use.

Learning objectives

By the end of this chapter you should be able to:
- describe the motion of atoms or molecules in different states of matter;
- interpret a phase diagram and explain the concepts of sublimation, deposition, the triple point and the critical point;
- calculate energy requirements for phase changes and temperature changes;
- describe and interpret energy bar diagrams and energy profiles;
- explain the concept of activation energy and its role in reactions;
- calculate reaction energy of a reaction;
- use Gibbs free energy to determine if a reaction is spontaneous; and
- apply Hess's law to determine enthalpy changes for chemical reactions.

Investigations

11A: Evaporative cooling
11B: State changes
11C: Hess's law

Pages in this chapter

- 346 Energy and Changes of State
- 347 The effect of polarity on phase transitions
- 348 Pressure and phase equilibrium
- 349 Reading a phase diagram
- 350 The triple point and sublimation
- 351 The change of state graph
- 352 Change of state problems
- 353 Section 1 Review
- 354 Energy and Chemical Changes
- 355 Endothermic and exothermic reactions
- 356 Bond energy
- 357 Calculating enthalpy change in a reaction from bond enthalpies
- 358 The enthalpy of formation
- 359 Hess's law and reaction enthalpy
- 360 Solving reaction enthalpy problems
- 361 Hess's law and thermochemical equations
- 362 Energy bar diagrams
- 363 Photosynthesis
- 364 Section 2 Review
- 365 Entropy and Spontaneity
- 366 The meaning of entropy
- 367 Entropy, energy, and enthalpy
- 368 Entropy change in chemical reactions
- 369 Gibbs free energy
- 370 Section 3 review
- 371 Following the Water to Mars
- 372 Following the Water to Mars - 2
- 373 Following the Water to Mars - 3
- 374 Chapter review

Vocabulary

condensed matter	phase diagram	critical point
supercritical fluid	triple point	sublimation
deposition	first law of thermodynamics	enthalpy
endothermic	exothermic	enthalpy of formation
Hess's law	state function	thermochemical equation
activation energy	entropy	second law of thermodynamics
spontaneous	Gibbs free energy	Gibbs–Helmholtz equation

11.1 - Energy and Changes of State

Changes in energy provide the underlying "force" that drives all chemical changes. The difference between physical change and chemical change is really a difference in the amount of energy involved. For example, in Chapter 4 we learned that the physical change from solid ice to liquid water requires an energy change of 335 J/g. By comparison, the chemical change that separates water into hydrogen and oxygen involves 16,197 J/g, almost fifty times larger. In this section, we will take a deeper look at energy and change. We will start with revisiting physical changes of state then proceed in the next section to chemical changes.

Changing state

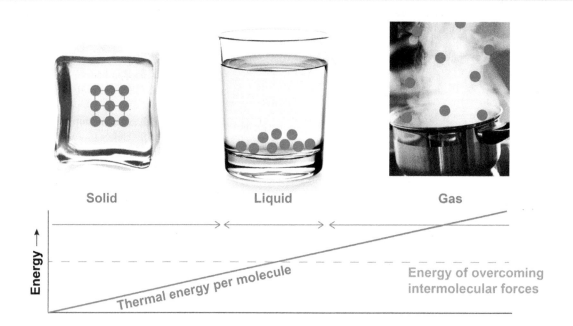

Energy and phase

In general, whether matter is solid, liquid, or gas depends on the ratio of thermal energy per molecule (or atom) compared to the energy needed to break intermolecular bonds between molecules (or atoms). Matter is solid when the energy per molecule is less than the energy stored in intermolecular bonds. Matter becomes liquid when the energy per molecule is roughly equal to the energy of intermolecular bonds. Matter becomes a gas when the energy per molecule is greater than the energy of intermolecular bonds. The relative strength of the intermolecular forces determines when phase transitions occur. The stronger the intermolecular forces, the higher the phase transition temperatures.

The liquid phase and energy

Notice that the liquid state only occurs over a narrow range of temperature. This is because the liquid state corresponds to a narrow band of energy over which the thermal energy is approximately equal to the intermolecular bond energy. Methane has about the same molecular weight as water but methane melts at -182.5°C and boils at -164°C. The range of temperature over which methane is liquid is only 18.5°C! Nitrogen is liquid over a range of 14°C and ammonia remains liquid over a range of 44°C. Water is the exception, and it is one of the many reasons water is essential to life. Water is a polar molecule with particularly strong intermolecular bonds. The strong intermolecular bonds are the reason water remains liquid over a very wide range of 100°C.

The effect of polarity on phase transitions

Polar vs. non-polar

The difference between polar and non-polar liquids is immediately obvious if you walk into a room with an open container of gasoline compared to a room with an open container of water. You would smell the gas right away. This is not just because your sense of smell detects gasoline more readily than water (although that is true). The "air" above an open container of gasoline at room temperature is more than 30% hydrocarbon molecules! The hydrocarbon molecules in gasoline are non-polar with only weak intermolecular forces holding them together. It takes little energy to get light hydrocarbon molecules to go from liquid to gas. By comparison, room temperature air over water only contains only 2% water vapor. Water molecules are polar and have large intermolecular forces. It takes considerably more energy to make polar molecules go from the liquid to the gas state.

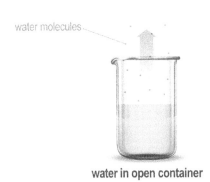
water in open container

gasoline in open container

Water vs methane

Polar molecules attract other polar molecules like magnets attract one another. Their attractions hold the molecules together and it takes more energy to change their state compared to non-polar molecules. The comparison between water and methane is a good example. Both have similar molecular weight but water is polar and methane in non-polar. Water has a higher melting point and boiling point when compared to methane.

physical properties of methane and water

Property	Methane, CH_4	Water, H_2O
Molar Mass	16.042 g/mol	18.015 g/mol
Melting Point	−182 °C	0 °C
Boiling Point	−160 °C	100 °C

Interactive simulation

The interactive simulation (button left) allows you to explore the relative difference between polar and non-polar substances. Change the temperature using the sliding control. You easily will be able to see the transition from liquid to gas. Repeat the simulation for both polar and non-polar substances. Notice what happens to the boiling point when you compare polar and nonpolar molecules.

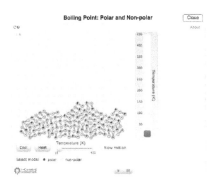

Section 11.1: Changes of State

Pressure and phase equilibrium

Boiling and altitude

At standard pressure, water boils at 100°C. The boiling point of water drops by approximately 1 °C every 294 meters above sea level. For example, water boils at 95°C in the city of Denver, Colorado, at an altitude of 1609 meters! Many recipes need to cook longer in Denver compared to other cities nearer to sea level. At 5,100 meters, the small mining town of La Rinconada in Peru is the highest permanent settlement in the world. At this altitude water boils at 83°C (181°F) and cooking many foods cannot be done without a pressure cooker. A pressure cooker raises the pressure inside so water boils at 121°C instead of 100°C. Foods cook much faster in a pressure cooker.

Water boils at 83°C (181°F)

La Rinconada, Peru

Water boils at 121°C (250°F)

Pressure cooker

Energy favors condensed matter at high pressure

Solids and liquids are two kinds of condensed matter. **Condensed matter** is made of particles such as molecules, atoms or ions that are arranged closely together. At high pressures, there is an energy advantage for matter to remain closely packed - as in liquids and solids. Conversely, at low pressure, the energy advantage shifts to matter being more spread out - favoring the gas phase. This is the reason the boiling point changes with pressure. Later in the chapter we will see this is associated with entropy, a new thermodynamic variable related to energy.

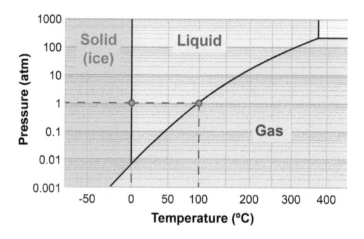

Phase equilibrium diagram for water

The phase equilibrium diagram

A **phase diagram** is a graph of temperature and pressure for a specific substance. The phase diagram for water is shown above. If you trace a line horizontally at 1 atmosphere of pressure, the phase transition from solid to liquid (melting point) occurs at 0°C. The fact that the boundary between solid and liquid is vertical means that pressure has little effect on the melting point. The transition from liquid to gas occurs at 100°C. The liquid-gas transition curve has a strong dependence on pressure. Below one atmosphere of presssure, water boils at a lower temperature than 100°C.

Reading a phase diagram

Reading a log scale

The pressure axis on a phase diagram often has a logarithmic (log) scale. The log scale allows a wide range of pressure to be represented while still being able to read small differences. For example, the diagram represents a range of 0.001 to 100 atmospheres yet can show the difference between 0.2 and 3 atmospheres. This difference would not be noticeable on a linear scale from 0 - 100. The diagram on the left shows you how to read a log scale. The tic marks are not spaced evenly as with a linear scale.

Using the phase diagram

The phase equilibrium diagram is usually used to solve two kinds of problems.

1. Finding the temperature at which water boils at a given pressure.
2. Finding the pressure at which water boils at a given temperature.

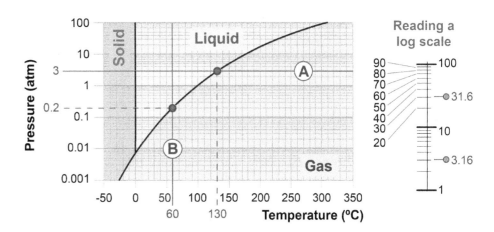

Finding the temperature

A horizontal line (A) is used to find the temperature at which water boils at a given pressure. In the example, the line is drawn at a pressure of 3 atmospheres. The horizontal line (A) intersects the curve separating the liquid and solid phases at a temperature of 130°C. This tells us that water boils at 130°C at a pressure of 3 atmospheres. At three atmospheres water will be a gas above 130°C and a liquid below 130°C.

Finding the pressure

A vertical line (B) is used to find the pressure at which water boils for a given temperature. In the example, the line is drawn at a temperature of 60°C. The vertical line (B) intersects the curve at a pressure of 0.2 atmospheres. This tells us that water at 60°C will boil into gas at pressures below 0.2 atmospheres and will be liquid at pressures above 0.2 atmospheres.

Solved problem

A deep sea submarine has an air pressure of 4 atmospheres. At what temperature will water boil?

Given Pressure = 4 atm.
Relationships Use the liquid/gas curve on the phase equilibrium diagram.
Solve A horizontal line on the phase equilibrium graph intersects the liquid/gas curve at about 145°C.
Answer At 4 atm of pressure, water boils at 145°C.

The triple point and sublimation

The triple point

There is a single point on the phase equilibrium diagram where all three phases converge and water can exist in all three phases at the same time. This is called the **triple point**. The triple point of water is at 0.006 atmospheres and 0.01°C. At this combination of temperature and pressure water can be a solid, liquid, and gas at the same time. The triple point is unique for any substance. A substance can have many boiling and melting points but there may be only one triple point.

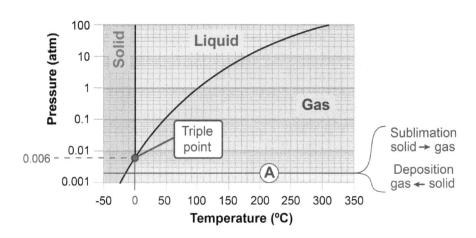

The triple point of water

Sublimation and deposition

At pressures below the triple point there is no stable liquid phase. Consider what happens to a sample of ice at -50°C at 0.002 atm pressure - represented by the horizontal line (A) on the phase equilibrium graph. As the temperature increases, the ice turns directly into gas at -17°C without ever becoming liquid! This is an example of sublimation. **Sublimation** is the phase transition from solid to gas without ever becoming liquid. **Deposition** is the opposite process, where a substance goes directly from the gas to solid.

Dry ice

Dry ice is a nickname for solid carbon dioxide. The triple point of carbon dioxide is -57°C and 5.2 atm. Solid CO_2 is "dry" because it goes directly from solid to gas without ever melting into a liquid. You can see from the phase diagram on the right that one atmosphere is below the triple point. Dry ice sublimates directly to gas and cooled carbon dioxide becomes solid through deposition.

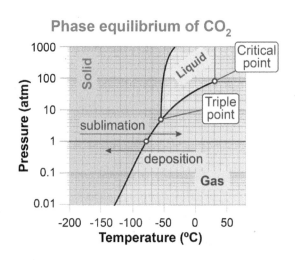

The critical point

The diagram for carbon dioxide shows has a critical point at 31°C and 7.4 atmospheres. Above the critical point, there is no significant difference in energy between the gas and liquid phases. Both phases co-exist in a "mushy snow" that is neither liquid nor solid. Much of the interior of the gas giant planets is believed to be in this region of the phase equilibrium diagram.

The change of state graph

Change of state diagrams

A horizontal line on the phase diagram represents a process during which heat is continuously added or subtracted at constant pressure. A change of state graph plots the temperature of a substance during the process of adding or subtracting energy.

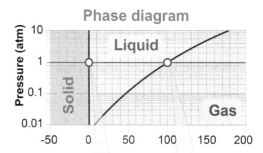
Phase diagram

Reading the change of state graph

Important things to notice about the change of state graph.

1. The temperature only changes when the substance is completely in one phase.

2. Temperature stays constant during phase changes because the energy change is being absorbed by the phase change and there is no energy available to change temperature.

3. The phase change graph is the same whether heating (solid→liquid→gas) or cooling (gas→liquid→solid).

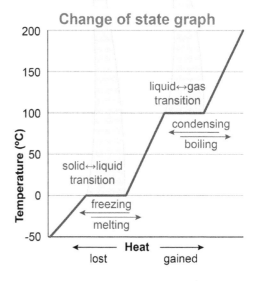
Change of state graph

Interpreting the flat parts of the graph

The flat parts on the phase change graph occur at the melting or boiling points. For water, the melting/freezing point is at 0°C and there is a flat region of the graph that represents the energy absorbed or released by the phase change. The boiling/condensing point is at 100°C and another flat region of the graph represents the energy absorbed or released by the phase change. Melting and freezing are opposite processes that occur at the melting point just as boiling and condensing are opposite processes that occur at the boiling point.

Where the energy goes

The temperature stays constant during a phase change, even though energy is changing, because the energy stored in matter has three parts. One part is thermal energy, E_{th} that changes with temperature. Another part is internal energy, E_i which is associated with the phase and is not dependent on temperature. A gas has the highest internal energy per gram and a solid has the lowest internal energy. The third form is chemical energy E_{ch} from the bonds between atoms in a compound. The difference between heating a solid, and melting a solid is where the energy goes. In heating, the energy goes into thermal energy. In melting, the energy (heat of fusion) becomes internal energy.

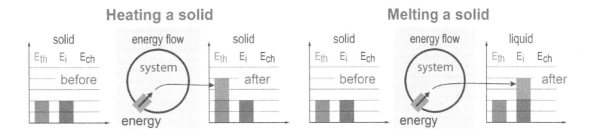

Section 11.1: Changes of State

Change of state problems

Solving temperature and phase change problems

In problems involving temperature and phase changes you must always think about where the energy goes: does it change temperature or does it change phase?

- Energy absorbed or released by changes in temperature is calculated with:

$$q = m \times C_p \times \Delta T$$

- Energy absorbed or released by phase change is calculated with:

$$q = m \times \Delta H_{fus} \quad \text{or} \quad q = m \times \Delta H_{vap}$$

Energy is conserved

Realize that energy is always conserved. Energy that is used for a phase change is not available to change temperature, and vice versa. For example, if 50 J of energy are added to a system and 35 J are used to raise the temperature of the system, then only 15 J remain available to change the phase of any matter in the system.

Solved problem

What amount of heat is required to completely boil 25.0 g of liquid water at 100.0 °C? Water's heat of vaporization = ΔH_{vap} = 2,256 J/g.

- Given: Mass of water, m_{water} = 25.0 g, heat of vaporization = ΔH_{vap} = 2,265 J/g
- Relationships: $q = m \times \Delta H_{vap}$
- Solve: There is no change in temperature since water is already at its boiling point. All of the heat will go toward vaporizing the sample
$q = m \times \Delta H_{vap}$

$$q = \frac{25.0 \text{ g } H_2O}{1} \times \frac{2{,}256 \text{ J}}{\text{g}} = 56{,}400 \text{ J or } 56.4 \text{ kJ for } H_2O$$

- Answer: It would take 56,400 J or 56.4 kJ of energy to boil the water

Solved problem

50.0 g of ice cubes at -19.2 °C are added to a glass of soda. Which absorbs more heat: warming up the ice to 0 °C or melting the ice? The specific heat of ice is 2.0 J/g·°C and the heat of fusion for water is 334 J/g.

- Asked: Compare the energy used to warm ice by 19.2 °C with the energy used to melt ice.
- Given: Initial temperature of the ice = T_1 = -19.2 °C
Specific heat of ice = C_p = 2.0 $\frac{J}{g \cdot °C}$
Heat of fusion of ice = ΔH_{fus} = 334 J/g
- Relationships: $q = m \times \Delta H_{fus}$ and $q = m \times C_p \times \Delta T$
- Solve: Solve for the energy required to warm up 50.0 grams of ice to the energy required to change 50.0 grams of ice to liquid.

$$q_{ice} = m \times C_p \times \Delta T = 50.0 \text{ g} \times 2.00 \frac{J}{g \cdot °C} \times 19.2 °C = 1{,}920 \text{ J}$$

$$q_{melting} = m \times \Delta H_{fus} = 50.0 \text{ g} \times 334 \frac{J}{g} = 16{,}700 \text{ J}$$

- Answer: Melting requires 16,700 J of energy compared to 1,920 J to warm up the ice.

Section 1 review

Chapter 11

Atoms, molecules, or ions (particles) in the solid state do not have enough internal energy to readily exchange places with their neighbors. Particles in a liquid state are still close together but they have enough internal energy to easily change places. Particles in the gas state are far from one another and have relatively high internal energy. The phase diagram shows how the phase of matter varies with pressure and temperature. Under certain conditions, a substance can go directly from solid to gas (sublimation) or from gas to solid (deposition). The triple point is a value of temperature and pressure at which all three phases of matter exist simultaneously. Change of phase is a physical change. No bonds are broken. Energy is used to increase kinetic energy or to overcome intermolecular forces.

Vocabulary words: condensed matter, phase diagram, critical point, supercritical fluid, triple point, sublimation, deposition

Review problems and questions

1. Find letters A through H on the phase diagram. Each letter represents changing conditions for the substance, starting at one pressure and temperature and ending at a different temperature or pressure. For each letter,

    a. identify which variable changed: pressure, or temperature;
    b. describe whether the variable increased or decreased; and
    c. identify the phase change that occured, if any.

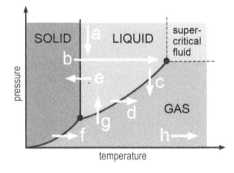

2. Ethanol (C_2H_5OH) is a polar molecule with a molar mass of 46.0 g/mol. Propane (C_3H_8) is a nonpolar molecule with a molar mass of 44.1 g/mol. Which compound will have the higher boiling point and why?

3. Use the graph to match letters to statements. Statements can have more than one correct letter.

    a. Where is a phase change occurring?
    b. Where is there a mix of liquids and gases?
    c. Where would condensation occur?
    d. Where is a solid getting warmer?
    e. Where are there liquids only?
    f. Where would a liquid cool down?
    g. Where are intermolecular forces being overcome?

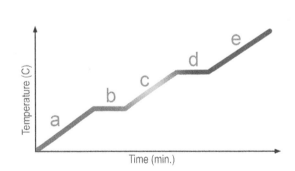

4. The table shows heats of fusion and vaporization for several substances. Find the energy required to boil 75.00 g of alcohol.

heat of fusion and vaporization for common substances

substance	heat of fusion. ΔH_{fus}(J/g)	heat of vaporization. ΔH_{vap}(J/g)
water	335	2,256
alcohol	104	854
liquid nitrogen	25.5	201
iron	267	6,265
silver	88	2,336

11.2 - Energy and Chemical Changes

Long before humans understood chemistry we understood that transforming wood into ashes released heat and light. Wood and ashes are clearly different substances and today we understand that the transformation is a chemical change and all chemical changes involve an exchange of energy. In this section, we will expand on the nature of chemical energy which we introduced with the concept of chemical reactions.

Chemical bonds store energy

The first law of thermodynamics

The first law of thermodynamics The **first law of thermodynamics** states that energy cannot be created or destroyed but it can change from one form to another. Sound familiar? It is also known as the law of conservation of energy. We can change kinetic energy into potential energy and the other way around, but we cannot create energy from nothing. Although batteries seem to create energy from nowhere, inside the battery a chemical reaction converts the potential energy in bonds into moving electrons.

No free lunch Sometimes the first law is referred to as the "no such thing as a free lunch" law. It means that there is a cost for everything. If we drop a ball, it releases kinetic energy. But first we have to give it potential energy by raising it up. Whenever we burn gasoline to get energy, we break the chemical bonds that plants created millions of years ago using the sun's light energy. Every time something happens in the Universe, it took energy to do it. Energy is the cost for every process in the Universe.

Enthalpy, H
Energy associated with temperature and the interactions between the particles of matter such as chemical bonds, phase, pressure, and volume.

Enthalpy does **not** include
other forms of energy such as light energy, electrical energy, nuclear energy, gravitational energy, elastic energy

Change in enthalpy, ΔH
Many processes result in changes in enthalpy including chemical reactions, phase changes, temperature changes, and the expansion or compression of a gas.

Enthalpy Energy has fundamental units of joules (J). In chemistry, we will mostly be concerned with a particular form of energy called **enthalpy**. Enthalpy is specifically the energy associated with temperature and the state of matter including its pressure and volume. Energy in the forms of light, or motion, or gravitational attraction are not included in enthalpy, nor is electrical energy or nuclear energy. To separate it from other forms of energy we represent enthalpy by the letter H (h). A change in enthalpy is represented by ΔH. Many types of enthalpies exist. The energy it takes to boil one mole of a liquid is called the molar heat of vaporization and has the symbol, ΔH_{vap}. The energy that is removed when one mole of a substance freezes is called the molar heat of fusion and has the symbol, ΔH_{fus}. Virtually every important process in chemistry has an enthalpy associated with it.

Endothermic and exothermic reactions

Endothermic and exothermic

Chemical reactions always involve the exchange of energy through enthalpy. An important distinction is whether the reaction absorbs enthalpy from the surroundings, or gives off enthalpy that is transferred to the surroundings. A reaction that absorbs enthalpy is classed as **endothermic**. Endothermic reactions must have an external source of energy to occur. A reaction that releases enthalpy is classed as **exothermic**. Exothermic reactions give off energy, often in the form of heat. Burning is a classic example of an exothermic reaction.

hot pack

$NaC_2H_3O_2(aq) \rightarrow NaC_2H_3O_2(s) + 18 kJ/mol$
heat given off = exothermic

cold pack

$NH_4NO_3(s) \rightarrow NH_4NO_3(aq) - 25.7 kJ/mol$
heat absorbed = endothermic

Absorbing and releasing energy

Cold and hot packs are a good example of endothermic and exothermic changes. In one type of cold pack, ammonium nitrate is released to dissolve in water. For ammonium nitrate, the process of dissolving is endothermic and absorbs 321 J/g from the solution which makes the pack cold. Endothermic changes absorb energy. Heat packs contain a supersaturated solution of sodium acetate. Sodium acetate has a freezing point of 54 °C (130°F). As soon as the solution is disturbed (squeezed) the sodium acetate starts to "freeze" into a solid and gives off its heat of fusion as it warms-up to its frozen state! This process is exothermic because it releases energy. Both hot and cold packs use the energy of phase changes to absorb or release heat.

The change in enthalpy, ΔH

When we evaluate the energy flow in a reaction we define the system to be the products and reactants. The surroundings includes everything else, including water that the reactants or products might be dissolved in. The change in enthalpy, ΔH applies to the system of reactants and products. When ΔH is positive, the system gains energy and the reaction is endothermic. When ΔH is negative, the system loses energy and the reaction is exothermic.

Representing energy in exothermic reactions

$CH_4(g) + 2O_2(g) \rightarrow CO_2(g) + 2H_2O(g) \quad + 882.0 \text{ kJ/mol}$

$CH_4(g) + 2O_2(g) \rightarrow CO_2(g) + 2H_2O(g) \quad \Delta H = -882.0 \text{ kJ/mol}$

Representing energy in endothermic reactions

$NH_4NO_3(s) \rightarrow NH_4^+(aq) + NO_3^-(aq) \quad - 25.7.0 \text{ kJ/mol}$

$NH_4NO_3(s) \rightarrow NH_4^+(aq) + NO_3^-(aq) \quad \Delta H = +25.7 \text{ kJ/mol}$

The sign of ΔH

The sign of ΔH can be confusing. In the top example for burning methane (CH_4) the reaction gives off +882 kJ of energy per mole of methane burned. This is the energy absorbed by the surroundings so it is positive. The system of products and reactants must lose this same amount of energy - because energy is conserved. The thermochemical equation shows ΔH as −882 kJ/mol.

Bond energy

Bond energy

A system of two carbon atoms has lower energy when the atoms are bonded compared to when they are apart. The difference is what we call the energy of the chemical bond. Because of energy conservation, the bond energy is released when the bonds form. The same bond energy must be added to break the bonds and separate the atoms again. The interactive simulation allows you compare the energy required to separate a pair of molecules compared to the energy it takes to separate the atoms within a molecule. Watch the potential energy as you move the atoms or molecules.

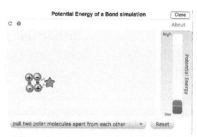

The simulation calculates the potential energy in the bond as the atoms are separated.

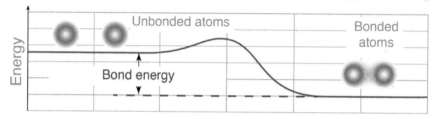

Average bond energy

The average energy of different chemical bonds are measured in units of kJ/mol. For example, in the table below, one mole of single carbon-carbon (C-C) bonds has an energy of 348 kJ, or 348,000J. As you might expect, bonds between different atoms have amounts of energy. Carbon-hydrogen bonds are 413 kJ/mol while carbon-oxygen bonds are 358 kJ/mol. There is a more extensive table of bond energies in the Appendix.

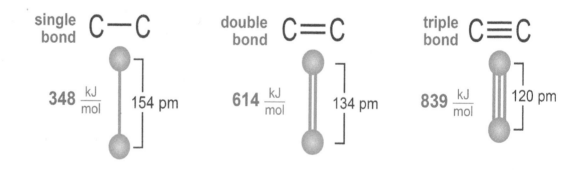

Some average bond energies (kJ/mol)

C-C, 348	C=C, 614	C-H, 413	C-O, 358	C=O, 799	C-N, 293	H-H, 436
O-H, 463	O-O, 146	O=O, 495	N-N, 163	N≡N, 941	N-O, 201	N-H, 391

Double and triple bonds

Single bonds are generally longer and weaker than double or triple bonds. Double and triple bonds bring the atoms closer together and involve more energy. A carbon-carbon double bond has a bond energy of 614 kJ/mol and a triple bond has a bond energy of 839 kJ/mol. The diagram shows that the average bond length is smaller for double and triple bonds compared to single bonds. One picometer is 1×10^{-12} meters.

Calculating enthalpy change in a reaction from bond enthalpies

Reactant bonds break during a reaction

Chemical reactions break bonds in the reactants and reform bonds in the products. To determine the enthalpy change we add up the enthalpies of all the bonds in the reactants and products separately. The difference is the enthalpy change of the reaction. Consider a simple reaction to make water from hydrogen and oxygen. The diagram shows the reaction and the numbers of each type of bond.

Analyzing the reaction

$$2H_2 + O_2 \rightarrow 2H_2O$$

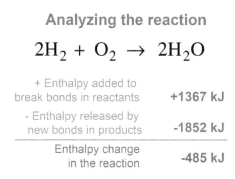
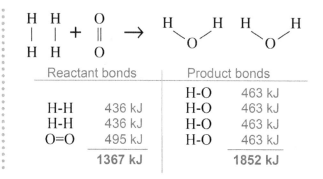

+ Enthalpy added to break bonds in reactants	+1367 kJ
− Enthalpy released by new bonds in products	−1852 kJ
Enthalpy change in the reaction	−485 kJ

Reactant bonds	
H-H	436 kJ
H-H	436 kJ
O=O	495 kJ
	1367 kJ

Product bonds	
H-O	463 kJ
H-O	463 kJ
H-O	463 kJ
H-O	463 kJ
	1852 kJ

Calculating the reaction energy

The reaction releases 485 kJ of enthalpy to produce 2 moles of H_2O. Here is how the enthalpies add up using the balanced chemical equation.

1. There are two moles of H-H bonds at 436 kJ/mole and one mole of O=O bonds at 495 kJ/mol. This enthalpy must be added to the system to break the bonds.
2. The products contain four moles of O-H bonds at 463 kJ/mol. This is the enthalpy given off by the reaction when the new bonds form.
3. The overall enthalpy change for the reaction is ΔH = −485 kJ. The negative sign means the reaction gives off energy and is therefore exothermic.

Solved problem

Calculate the enthalpy change for the combustion of ethane (C_2H_6).

$$2C_2H_6(g) + 7O_2(g) \rightarrow 4CO_2(g) + 6H_2O(g)$$

Bond enthalpy	
C-H	413 kJ
C-C	348 kJ
O=O	495 kJ
C=O	799 kJ
H-O	463 kJ

Relationships $\Delta H_{reaction} = H_{reactant\ bonds} - H_{product\ bonds}$

Solve

Reactants		Products	
12 C-H	12 × 413 = 4956 kJ	8 C=O	8 × 799 = 6392 kJ
2 C-C	2 × 348 = 696 kJ	12 O-H	12 × 463 = 5556 kJ
7 O=O	7 × 495 = 3465 kJ		
reactant bonds	9117 kJ	product bonds	11948 kJ

Answer $\Delta H = H_{reactant\ bonds} - H_{product\ bonds}$
= 9117 kJ − 11948 kJ = −2831 kJ. The reaction releases energy (exothermic).

The enthalpy of formation

The enthalpy of formation

Counting the chemical bonds broken and reformed is useful for understanding the flow of energy in reactions. But it is tedious work and chemists pre-calculated the results for many compounds in the **enthalpy of formation**. The standard enthalpy of formation, ΔH_f, for a number of compounds are listed in the table below. Most of the values are negative because energy is released when the compounds are formed from their pure constituent elements. We are typically concerned with changes in enthalpy therefore, pure elements such as the diatomic gases, O_2, N_2, H_2, are assigned enthalpies of formation of zero.

Standard enthalpy of formation ΔH_f for common substances

substance	Chemical formula	ΔH_f kJ/mol	substance	Chemical formula	ΔH_f kJ/mol
oxygen (g)	O_2	0.0	methanol (l)	CH_3OH	-238.4
hydrogen (g)	H_2	0.0	ethanol (l)	C_2H_5OH	-277.0
nitrogen (g)	N_2	0.0	ferric oxide (s)	Fe_2O_3	-824.2
carbon dioxide (g)	CO_2	-393.5	copper sulfate (s)	$CuSO_4$	-771.4
water (l)	H_2O	-285.5	glucose (s)	$C_6H_{12}O_6$	-1,271
water (g)	H_2O	-241.8	calcium chloride (s)	$CaCl_2$	-795.8
methane (g)	CH_4	-74.87	calcium chloride (aq)	$CaCl_2$	-877.3
propane (g)	C_3H_8	-104.7	ozone (g)	O_3	143
salt (s)	NaCl	-407.3	nitrous oxide (g)	N_2O	82.1
salt (aq)	NaCl	-411.1	acetylene (g)	C_2H_2	227.4

The meaning of ΔH_f

The enthalpy of formation of water is -285.5 kJ/mol. This is the energy released when 1 mole of hydrogen, H_2, combines with one half mole of diatomic oxygen, O_2, to make one mole of water. The value is negative because this is an exothermic reaction and energy leaves the system. Note that the balanced reaction releases 571 kJ but produces two moles of water. The enthalpy of formation per mole is therefore -285.5 kJ/mol. This is the same result as we obtained from counting bond energy.

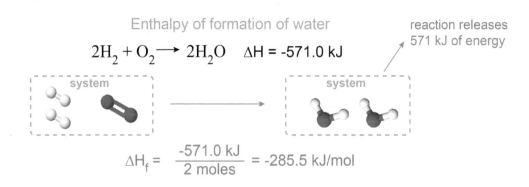

Enthalpy of formation of water

$$2H_2 + O_2 \rightarrow 2H_2O \quad \Delta H = -571.0 \text{ kJ}$$

reaction releases 571 kJ of energy

$$\Delta H_f = \frac{-571.0 \text{ kJ}}{2 \text{ moles}} = -285.5 \text{ kJ/mol}$$

(s), (l), (g), and (aq)

The standard enthalpies of formation identify the state of matter. Standard means under the conditions of 25°C and one atmosphere, not to be confused with standard temperature and pressure (STP), a set of conditions associated with the study of gases. The standard enthalpy of formation is -285.5 kJ/mol for liquid water (l). The enthalpy of formation for water vapor (g) is -241.8 kJ/mol. The difference of 43.7 kJ/mol includes 40 kJ/mole for the heat of vaporization and 3.7 kJ/mol to get from 25°C to 100°C at the boiling point.

Hess's law and reaction enthalpy

Hess's law

Hess's law states the total enthalpy change for any reaction is the sum of the enthalpy changes for any sequence of reactions that produce the same end products. Hess's law is the key to using enthalpies of formation to analyze the energy in a chemical reaction.

State functions

Hess's law works because enthalpy is a **state function**. A state function is a property whose value does not depend on the path taken to reach that specific state. This means that the enthalpy it takes to form one mole of NaCl is the same whether the reaction occurs in one step or one million steps.

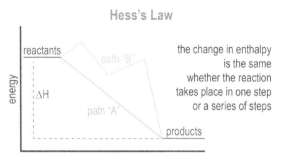

Hess's Law — the change in enthalpy is the same whether the reaction takes place in one step or a series of steps

Application of Hess's law

$$\Delta H_{reaction} = \Delta H_f (\text{products}) - \Delta H_f (\text{reactants})$$

Using Hess's law

The most useful application of Hess's law is that the enthalpy change for any chemical reaction can be calculated as the enthalpy of formation of the products minus the enthalpy of formation of the reactants. The reactions that form the individual reactants or products may be different from the reaction being analyzed, but Hess's law says the ending enthalpies are the same. Be careful of the signs! We are subtracting ΔH_f of the reactants because the chemical bonds in the reactants must be broken.

Calculating ΔH for a reaction

$$CH_4 + 2O_2 \xrightarrow{\text{combustion reaction}} CO_2 + 2H_2O$$

reactants	products
CH_4 1 mol × −74.87 $\frac{kJ}{mol}$ = −74.87 kJ	CO_2 1 mol × −393.5 $\frac{kJ}{mol}$ = −393.5 kJ
$2O_2$ 2 mol × 0.0 $\frac{kJ}{mol}$ = 0.00 kJ	$2H_2O$ 2 mol × −285.5 $\frac{kJ}{mol}$ = −571.0 kJ
Total reactants: −74.87 kJ	Total products: −964.5 kJ

$$\Delta H_{reaction} = \Delta H_f (\text{products}) - \Delta H_f (\text{reactants})$$

$$-964.5 \text{ kJ} - (-74.87 \text{ kJ}) = -889.63 \text{ kJ}$$

ΔH for the reaction
− 889.63 kJ

Use the balanced equation

The calculation shows that burning one mole of methane releases about 890 kJ of enthalpy. The coefficients in the balanced reaction must be used when calculating the reaction enthalpy for each product or reactant. The enthalpy of formation is in units of kJ per mole. Multiplying by the number of moles puts the enthalpy in energy units.

Solving reaction enthalpy problems

Summary of Hess's law

Remember these key points when using Hess's law to solve reaction enthalpy problems.

1. The enthalpy of formation of pure elements is zero, even if they are diatomic.

2. If a reaction is reversed, the algebraic sign of ΔH is also reversed.

$$2H_2(g) + O_2(g) \rightarrow 2H_2O(g) \quad \Delta H = -483.6 \text{ kJ}$$
$$2H_2O(g) \rightarrow 2H_2(g) + O_2(g) \quad \Delta H = +483.6 \text{ kJ}.$$

3. Enthalpies of formation are in kJ per mole and complete reaction enthalpies must consider the moles of each reactant or product from the balanced reaction.

4. Enthalpies of formation are different for different phases of matter. For example, liquid water, $H_2O(l)$, has an enthalpy of formation of -285.5 kJ/mol but water in the gas phase, $H_2O(g)$, has an enthalpy of formation of −241.8 kJ/mol.

5. A reaction is **exothermic** when ΔH is **negative** because the reactants lose (give off) energy in becoming the products.

6. A reaction is **endothermic** when ΔH is **positive** because the reactants absorb energy from the surroundings to become the products.

Solved problem

Magnesium hydroxide reacts with hydrochloric acid to form magnesium chloride and water. Determine whether the reaction is endothermic or exothermic and calculate the reaction enthalpy per mole of magnesium hydroxide that reacts.

$$Mg(OH)_2(s) + 2HCl(aq) \rightarrow MgCl_2(s) + 2H_2O(l)$$

$Mg(OH)_2(s)$	$HCl(aq)$	$MgCl_2(s)$	$H_2O(l)$
$\Delta H_f = -924.7$ kJ/mol	$\Delta H_f = -166.2$ kJ/mol	$\Delta H_f = -641.8$ kJ/mol	$\Delta H_f = -285.8$ kJ/mol

Given: Balanced reaction and ΔH_f for all compounds.
Relationships: $\Delta H_{reaction} = H_f(\text{products}) - H_f(\text{reactants})$
Solve:

	Reactants			Products
$Mg(OH)_2$	1 mol × −924.7 kJ/mol = −924.7 kJ		$MgCl_2$	1 mol × −641.8 kJ/mol = −641.8 kJ
HCl	2 mol × −166.2 kJ/mol = −332.4 kJ		H_2O	2 mol × −285.8 kJ/mol = −571.6 kJ
reactant total		−1257.1 kJ	product total	−1213.4 kJ

Answer: $\Delta H_{reaction} = \Delta H_f (\text{products}) - \Delta H_f (\text{reactants})$
= −1213.4 kJ − (−1257.1 kJ) = +43.7 kJ/mol $Mg(OH)_2$
The reaction is endothermic.

Hess's law and thermochemical equations

Example thermochemical equations

A thermochemical equation is a chemical equation plus the enthalpy change. For example, the equations below are thermochemical equations for synthesizing carbon dioxide. Note that thermochemical equations can be written with the enthalpy change per mole of reactant or product (kJ/mol), or for the enthalpy change of the reaction as written (kJ).

Thermochemical equations

$$C + O_2 \rightarrow CO_2 \quad \Delta H = -393 \text{ kJ/mol}$$ *Enthalpy per mole of reactant or product*

$$2CO + O \rightarrow 2CO_2 \quad \Delta H = -566 \text{ kJ}$$ *Enthalpy change for the reaction as written*

Multistep reactions

Hess's law gives us a way to manipulate thermochemical equations using algebra. This can be useful to find the enthalpy change for an unknown reaction from reactions for which ΔH is known. According to Hess's law, the enthalpy changes for each step in a multistep reaction can be added to give the enthalpy change for the overall reaction. As an example, the reaction for synthesis of carbon monoxide (CO) can be done in two steps using the reactions above for carbon dioxide.

Net reaction

$$2C + O_2 \rightarrow 2CO \quad \Delta H = ?$$

Step one	$2C + 2O_2 \rightarrow 2CO_2$		$\Delta H = -786 \text{ kJ}$
Step two (reverse reaction)	$2CO_2 \rightarrow 2CO + O_2$		$\Delta H = +566 \text{ kJ}$
Subtract O_2 from both sides	$2C + 2O_2$	$\rightarrow \quad 2CO + O_2$	
Net reaction	$2C + O_2$	$\rightarrow \quad 2CO$	$\Delta H = -220 \text{ kJ}$

The enthalpy of formation

If we need the enthalpy per mole we can divide the total enthalpy change by 2 since two moles of CO are produced. The result is the enthalpy of formation for CO of $\Delta H_f = -110$ kJ/mol.

Express in kJ/mol

$$C + \tfrac{1}{2}O_2 \rightarrow CO \quad \Delta H_f = -110 \tfrac{\text{kJ}}{\text{mol}}$$

Solving multi-step reactions

The following steps are useful to solve a multi-step thermochemical problem.

1. Componds that appear on both sides of a chemical equation can be subtracted from both sides.
2. Multiply the reaction enthalpy for each step by the number of moles as needed.
3. Reverse the sign of ΔH for any reverse reaction.
4. The enthalpy for the net reaction is the sum of the enthalpies for each step.

Energy bar diagrams

Exothermic reactions

If you get confused about the sign of ΔH always remember that ΔH describes the energy content of the system, which is the products and reactants. An exothermic reaction means the products and reactants lose energy overall during the reaction. ΔH is negative because the system is losing energy.

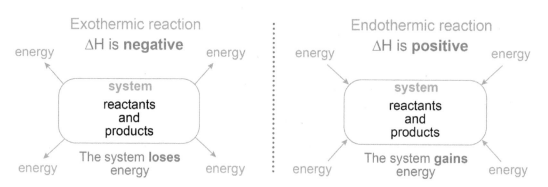

Endothermic reactions

The opposite is true of endothermic reactions. Energy is absorbed by the system so the net change in enthalpy during the reaction is positive and ΔH has a positive sign. It helps to imagine yourself as the products and reactants. Are you gaining or losing energy during the reaction? Energy must be conserved!

Energy diagrams

The energy change can be represented by an energy bar diagram where the circle in the center represents the system of products and reactants. The formation of water from oxygen and hydrogen is a good example. Burning hydrogen is exothermic and the reaction gives off energy. Electrolysis is the opposite. Electrical energy is added to break water into hydrogen and oxygen again. Electrolysis is endothermic.

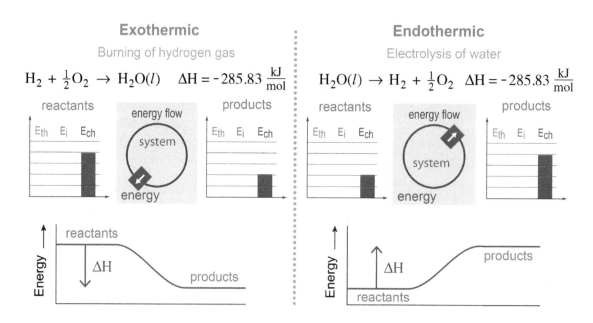

Energy profile

The graphs below the energy bar diagrams are another way to look at reactions. Potential energy is plotted on the Y-axis and time or reaction progress is on the X-axis. Exothermic reactions have higher reactant energy and lower product energy. The direction of ΔH is negative. The opposite is true for endothermic reactions and the direction of ΔH is positive.

Photosynthesis

The energy in glucose

How much solar energy does it take for photosynthesis to create one mole of glucose from the pure elements carbon, oxygen, and nitrogen compared to the way plants do it from carbon dioxide and water? The difference is very important to the fundamental chemistry of most forms of life on Earth!

The formation reaction for glucose

$$6C_{(s)} + 6H_{2(g)} + 3O_{2(g)} \longrightarrow C_6H_{12}O_{6(s)} \qquad \Delta H_f = -1{,}271 \text{ kJ/mol}$$

from the table of standard enthalpies of formation

Photosynthesis

According to the tabel of enthalpies of formation, forming glucose from its constituent elements is an exothermic reaction! That means no external energy is required. However, we know that plants need energy from sunlight to produce the same reaction. Why? The difference is that elemental carbon and hydrogen are not available in the environment. The diagram below looks at the actual photosynthesis reaction involving CO_2 and H_2O which are available in a plant's environment.

The photosynthesis reaction

$$6H_2O_{(l)} + 6CO_{2(g)} \xrightarrow{light} C_6H_{12}O_{6(s)} + 6O_{2(g)}$$

$\Delta H_f = -285.5$ kJ/mol (H_2O)
$\Delta H_f = -393.5$ kJ/mol (CO_2)

$\Delta H_f = -1{,}271$ kJ/mol ($C_6H_{12}O_6$)
$\Delta H_f = 0.00$ kJ/mol (O_2)

reactants

-1×6 mol $\times$ (-285.5 kJ/mol)
-1×6 mol $\times$ (-393.5 kJ/mol)
_____
4,074 kJ

products

1 mol $\times$ ($-1{,}271$ kJ/mol)
6 mol $\times$ (0 kJ/mol)
_____
$-1{,}271$ kJ

net enthalpy change for the reaction

$$\Delta H = 4{,}074 \text{ kJ} - 1{,}271 \text{ kJ} = 2{,}803 \text{ kJ}$$

reaction is endothermic!

Photosynthesis is endothermic

The actual photosynthesis reaction is endothermic once we conside the energy it requires to break apart the carbon dioxide and water in the reactants. A plant must absorb 2,803 kilojoules of energy from light to produce one mole of glucose.

Solved problem

A tree has a leaf area of 50 m². The intensity of sunlight during the day is about 300 J/m² each second and over one full day the tree absorbs 540 million joules of energy. How much glucose could the tree produce if all the energy went into glucose production?

Given 540×10^6 J of energy
Relationships $\Delta H = 2803$ kJ/mol for glucose.
Solve

$$\frac{540 \times 10^6 \text{ J}}{} \times \frac{\text{mol } C_6H_{12}O_6}{2{,}803{,}000 \text{ J}} \times \frac{180.158 \text{ g } C_6H_{12}O_6}{\text{mol } C_6H_{12}O_6} = 34{,}700 \text{ g}$$

Answer The tree can produce 34.7 kg of glucose per day.

Section 11.2: Energy and Chemical Changes

Chapter 11

Section 2 review

Thermodynamics is the study of how energy (often as heat) changes during chemical processes, including reactions. The heat content of a system is known as enthalpy. Nearly every chemical reaction has an associated enthalpy change. Exothermic reactions have a negative enthalpy change, while endothermic reactions have a positive enthalpy change. The change in enthalpy during a reaction comes from breaking and reforming chemical bonds. Different bonds release different amounts of energy when they form. Double and triple bonds release more energy than single bonds. The enthalpy change of a reaction can be calculated by subtracting the sum of the bond energies of the products from the sum of the bond energies of the reactants.

The enthalpy of formation, ΔH_f, is the enthalpy released per mole when a compound is formed from its constituent elements. Hess's law says the total enthalpy change for a reaction is the sum of the enthalpy changes for any sequence of reactions that start with the same reactants and end with the same products, independent of the nature of the intermediate reactions. This allows us to calculate the enthalpy change in a reaction as the enthalpy of formation of the products minus the enthalpy of formation of the reactants. A thermochemical equation shows the chemical reaction and the net change in enthalpy, usually in kJ or kJ/mol.

Vocabulary words: first law of thermodynamics, enthalpy, endothermic, exothermic, enthalpy of formation, Hess's law, state function, thermochemical equation, activation energy

Key relationships

$$\Delta H_{reaction} = \Delta H_f (\text{products}) - \Delta H_f (\text{reactants})$$

Review problems and questions

1. If a reaction has a positive change in enthalpy, $+\Delta H$, is it endothermic or exothermic? Explain.

2. Answer questions a and b based on the energy diagram:
   a. Is the energy bar diagram showing a physical or chemical change? How can you tell?
   b. Is this change endothermic or exothermic? How can you tell?

3. Calculate the reaction energy for: $CH_4 + 2O_2 \rightarrow CO_2 + 2H_2O$ and classify it as exothermic or endothermic.

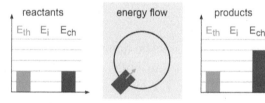

bond type	ΔH kJ/mol
H-C	410
C-C	350
O=O	494
C=O	799
H-O	460

4. Use the enthalpy of formation to calculate the enthalpy change in the oxidation of ethanol.

$$C_2H_5OH + 3O_2 \rightarrow 2CO_2 + 3H_2O$$

C_2H_5OH	O_2	CO_2	H_2O
$\Delta H_f = -269.3$ kJ/mol	$\Delta H_f = 0$ kJ/mol	$\Delta H_f = -393.5$ kJ/mol	$\Delta H_f = -285.8$ kJ/mol

11.3 - Entropy and Spontaneity

We observe heat flows from higher temperature to lower temperature and never the other way around. If a cup of coffee at 80°C is left in a 20°C room, heat always flows from the coffee to the room. We never see heat flow from the cooler air into the warmer coffee. Why? On the microscopic level, molecules can go backward or forward. But heat flows in one direction, from higher to lower temperature. The reason is that the macroscopic world contains trillions upon trillions of particles. The chance of one particle going in reverse may be fifty percent. The chance of all trillions and trillions reversing the same way is so small that it will never happen. The tendency of the universe to increase the information needed to describe the matter in a system is described by a new property called entropy.

The second law of thermodynamics

The second law of thermodynamics
Heat only flows spontaneously from higher temperature to lower temperature and not the other way.

Reversibility

The ability to go equally forward or backward is called reversibility. A reversible process may proceed identically in the other direction when the inputs are reversed. If cooling were reversible, it would be possible for a 20°C coffee to warm back up to 80°C by exchanging heat with room air. The room air certainly has sufficient thermal energy, so energy conservation is not the reason. We observe, however, that any process in which heat flows is irreversible! Why? Why is heat flow irreversible?

Entropy is $S = Q/T$

Suppose the coffee cup loses $Q = 1,000$ J of heat. If we divide the heat by the temperature in Kelvins we get a new property, $S = Q/T$, called entropy. Entropy, S, has units of energy per degree. The coffee cup loses 2.8 J/K of entropy. The 1000 J of heat lost by the coffee is gained by the room air at a lower temperature. The room air gains 3.4 J/K of entropy. If we look at the whole system of coffee and room air together, energy is conserved but entropy increases! The total entropy of the system increases by +0.6 J/K. The second law of thermodynamics is really about entropy. Entropy determines whether a process is reversible. If the cooling were to reverse, then the net change in entropy would be −0.6 J/K and that does not happen. **Processes in the universe move in the direction of increasing entropy.**

Coffee cup
$$\Delta S = \frac{Q}{T} = \frac{1,000 \text{ J}}{353 \text{ K}} = -2.83 \text{ J/K}$$

Room air
$$\Delta S = \frac{Q}{T} = \frac{1,000 \text{ J}}{293 \text{ K}} = +3.41 \text{ J/K}$$

Net change +0.6 J/K
Total entropy increases

The meaning of entropy

Entropy is energy distribution

Entropy, S, describes the way energy and matter are distributed in a system or substance. The more different ways there are to arrange the system, the higher the entropy. In general entropy increases from solid to liquid to gas. In a solid, the particles can only vibrate in place as energy is added. Liquids can use absorbed energy to move particles from one place to another, or rotate and vibrate the particles. Gases have all of the motion a liquid; however, the particles can also move away from each other at great speeds. Since gases have the most ways to distribute energy, they have the highest entropy. Solids have the lowest. In fact, a pure solid at absolute zero has zero entropy.

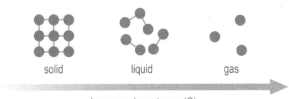

The second law

The **second law of thermodynamics** states that all spontaneous processes have an increase in entropy. This means energy is spreading out in the Universe. It may start out in a compact form like chemical bonds, which are broken to form new bonds. However, not all of the energy goes into forming new bonds. Some of the energy is distributed as heat, light or sound. A system that is 100% efficient and loses no energy to the surroundings, converts all of the available energy into work. There is no such thing as a 100% efficient system so some energy is always being lost to the surroundings.

Spontaneity

The word **spontaneous** has a specific meaning in science. It means that the process will occur under the specified conditions of temperature and pressure. However, it does not mean that it will happen fast. Carbon spontaneously turns into diamonds at very high temperature and pressure. But it takes thousands of years to complete the process.

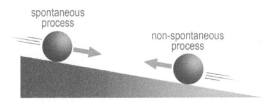

System and surroundings

You might argue that carbon with its random molecular pattern has higher entropy than the highly structured diamond, and you would be right! The Universe of thermodynamics has only two parts: the system and surroundings. Here we have defined the system as carbon turning into diamonds. Everything outside of the system is the surroundings, including the most distant stars. Even though there was a decrease in entropy of the system, there must have been a larger increase in the entropy of the surroundings. Heat energy is transferred to the surroundings, contributing to the ever-increasing entropy of the Universe.

Entropy, energy, and enthalpy

Calculating entropy

The entropy change in a reaction can be calculated as the change in enthalpy, ΔH, divided by the absolute temperature, T, in Kelvins. The units of entropy are J/mol·K in this equation. Remember that we typically express enthalpy in molar quantities (molar enthalpy, in J/mol) - molar entropy is similar. Instead of the entropy of a standard mass or a single particle, we measure the entropy of a mole of a substance, which is a much more usable quantity.

Entropy change

$$\Delta S = \frac{\Delta H}{T}$$

ΔS	Entropy change in J/K
ΔH	Enthalpy change in J
T	Absolute temperature in K

Changes in entropy

The natural tendency toward increasing entropy drives many processes. A good example is the melting of ice into water at the melting point of 0°C. The standard entropy is given the symbol, S° and has units of J/mol·K. When ice melts into water, the entropy increases from 41 J/mol·K to 70 J/mol·K, a change of ΔS = 29 J/mol·K.

Thermochemical equations with entropy

Standard entropy of water

H_2O (s)	$S° = 41$ J/mol·K
H_2O (l)	$S° = 70$ J/mol·K
H_2O (g)	$S° = 189$ J/mol·K

$$H_2O(s) \rightarrow H_2O(l) \quad \Delta S = 29 \frac{J}{mol \cdot K}$$

Entropy, enthalpy and ΔH_fus

If we rearrange the entropy formula to solve for the heat energy, the quantity TΔS has units of J/mol and is a change in enthalpy per mole. Heating and changing phase both affect entropy. For the melting of water at 273K (0°C), TΔS = 7917 J/mol. This is exactly the molar heat of fusion plus the additional heat needed to raise the water from 0°C to 25°C. The temperature of 25°C is the standard for molar entropies given in scientific tables. The big idea here is that the heat associated with phase changes is due to the change in entropy.

Entropy and enthalpy

The quantity $T\Delta S$ is a change in enthalpy ΔH

The heat of fusion, ΔH_{fus} for water

$$T\Delta S = 273 \text{ K} \times 29 \tfrac{J}{mol \, K} = 7917 \tfrac{J}{mol}$$

$$= 6030 \tfrac{J}{mol} + 1887 \tfrac{J}{mol}$$

Molar heat of fusion   Heat to raise H_2O from 0°C to 25°C

Standard conditions for entropy

The entropy listed in most tables is at "standard ambient temperature and pressure", which is different from "standard temperature and pressure" or STP. Entropies in J/molK are typically given at one atmosphere of pressure and 25°C temperature. By contrast, many other thermodynamic quantities are often given at STP, which is one atmosphere of pressure and 0°C temperature.

Section 11.3: Entropy and Spontaneity

Entropy change in chemical reactions

Entropy in reactions

A thermochemical equation for a reaction can be written to show the change in entropy as well as enthalpy. The change in entropy, ΔS for the reaction is the entropy of the products minus the entropy of the reactants.

Entropy change for a reaction

$$\Delta S_{reaction} = S_{products} - S_{reactants}$$

Solved problem

Methanol burns in oxygen to produce water and carbon dioxide. What is the entropy change at 25.0 °C per mole of methanol in the following reaction?

$$2CH_3OH(l) + 3O_2(g) \rightarrow 4H_2O(g) + 2CO_2(g)$$

Asked What is the entropy of reaction?
Given $2CH_3OH(l) + 3O_2(g) \rightarrow 4H_2O(g) + 2CO_2(g)$
Relationships

$$\Delta S_{reaction} = S°_{products} - S°_{reaction}$$

Selected Thermodynamic Data

Compound	$CH_3OH(l)$	$O_2(g)$	$H_2O(g)$	$CO_2(g)$
Standard molar entropy	ΔS° = 127 J/mol·K	ΔS° = 205 J/mol·K	ΔS° = 189 J/mol·K	ΔS° = 214 J/mol·K

Solve First, add up the entropies of the products and reactants.

	Reactants			Products	
$CH_3OH(l)$	2 mol × 127 J/mol·K = 254 J/K		$H_2O(g)$	4 mol × 189 J/mol·K = 756 J/K	
$O_2(g)$	3 mol × 205 J/mol·K = 615 J/K		$CO_2(g)$	2 mol × 214 J/mol·K = 428 J/K	
reactant total		869 J/K	product total		1184 J/K

$\Delta S_{reaction} = S°_{products} - S°_{reaction}$
$\Delta S_{reaction} = 1184$ J/K $- 869$ J/K $= 315$ J/K

This is for two moles of methanol therefore the entropy per mole for the reaction is 167.5 J/mol·K.

Answer The entropy change per mole of methanol is 167.5 J/mol·K.

Gibbs free energy

Gibbs free energy

In 1878, Joshua Willard Gibbs proposed an enthalpy-entropy function called the **Gibbs free energy** that includes both energy from chemical changes and energy from phase changes and temperature. The free energy, ΔG, describes the energy available to do work, like pushing a piston or moving electrons through a circuit. We already know how to account for the chemical bond-energy change through ΔH. By including the term TΔS the Gibbs free energy takes account of energy that goes into temperature and phase changes.

Free energy
Gibbs-Hemholtz equation

$$\Delta G = \Delta H - T\Delta S$$

Free energy = enthalpy of chemical changes − enthalpy absorbed by phase and temperature changes

The sign of ΔG

Some reactions happen as soon as the reactants are combined, and other reactions don't. Some processes, such as evaporation, occur spontaneously even though it takes energy to convert liquid water to gas. Much like enthalpy, the algebraic sign of Gibbs free energy also has meaning. When ΔG is negative, the process is spontaneous under the specified conditions of temperature and pressure. When ΔG is positive, the process is nonspontaneous and will not take place under the given conditions.

symbol	algebraic sign	meaning
ΔH	−	exothermic
ΔH	+	endothermic
ΔG	−	spontaneous
ΔG	+	non-spontaneous

Gibbs-Helmholtz equation

Entropy, enthalpy and Gibbs free energy work together to determine if a process will take place under the given conditions. The **Gibbs–Helmholtz equation** equation tells us that the enthalpy change (ΔH) **and** the way the energy is distributed (TΔS) determine whether a process will take place (±ΔG).

ΔH	ΔS	-TΔS	ΔG	spontaneity
+	−	+	+	non-spontaneous
−	+	−	−	spontaneous
−	−	+	+ or −	low temp: spontaneous / high temp: non-spontaneous
+	+	−	+ or −	low temp: non-spontaneous / high temp: spontaneous

Using the chart

There are numerous thermodynamic charts that list the entropy and enthalpy for many substances. These can be found online or in resource books. The chart above helps you use the algebraic signs for entropy and enthalpy to determine the spontaneity of a process.

First line in the chart

The first line of the chart shows an endothermic process that lowers entropy. If the energy is low enough the reaction could take place. However, the decrease in energy violates the second law of thermodynamics so actually the reaction never takes place.

Last 2 lines in the chart

The spontaneity of the last two rows of the table is impossible to determine from either entropy or enthalpy. These situations are determined by temperature.

Chapter 11
Section 3 review

Heat naturally flows from higher temperature to lower temperature as a result of the second law of thermodynamics. The second law is really about increasing entropy. Processes such as heating or chemical reactions naturally proceed in the direction of increasing entropy. Molar entropy has units of energy per mole per degree, usually J/mol·K. Physically, entropy describes the tendency of energy to spread out and become evenly distributed rather than be concentrated.

A change in entropy, ΔS, can be included in a thermochemical equation in a similar way as a change in enthalpy, ΔH. The quantity $T\Delta S$ has units of enthalpy, or enthalpy per mole. During a phase change, the heat of fusion increases or decreases the entropy of a substance.

The Gibbs free energy is the net energy available from a process that can be released to the surroundings, and do external work. The free energy is the enthalpy change of the reaction, ΔH, minus $T\Delta S$ to account for energy remaining in the products due to phase change and/or temperature change. A process is spontaneous when the sign of the Gibbs free energy, ΔG is negative.

Vocabulary   entropy, second law of thermodynamics, spontaneous, Gibbs free energy, Gibbs–Helmholtz equation

Relationships $\Delta S = \dfrac{Q}{T}$ $\Delta G = \Delta H - T\Delta S$ $\Delta S_{reaction} = S°_{products} - S°_{reactants}$

Review problems and questions

1. The image to the right shows a phase change.
   a. What is the phase change?
   b. Is entropy among the particles increasing or decreasing as a result of this phase change? Explain your answer.

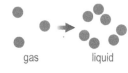

2. A block of hot iron at 90 °C is dropped into a bucket of water at 20 °C. Calculate the net entropy change for the system of the iron and water if the iron loses 500J and the water gains 500 J. You may assume the temperature stays constant during the transfer of heat. This is not actually true as the iron cools and the water warms. But it makes the math much simpler!

3. Calculate the entropy change, ΔS, for the formation of carbon dioxide at 25 °C and one atmosphere.

$$C(s) + O_2(g) \rightarrow CO_2(g)$$

Standard entropy of formation at 25 °C

C(s)	$O_2(g)$	$CO_2(g)$
S° = 6 J/mol·K	S° = 205 J/mol·K	S° = 214 J/mol·K

4. The Gibbs free energy of a reaction is +1.65 kJ/mol. Will the process occur spontaneously?

Following the Water to Mars

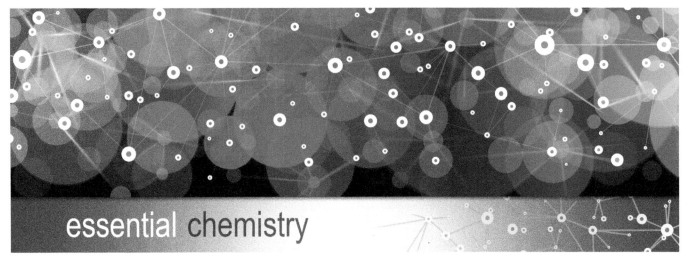

essential chemistry

Water is a very important natural resource. Water is everywhere from the oceans to the polar ice caps, and in the atmosphere. The amazing thing about water is that it can change state and move around the world. Water in the atmosphere precipitates as either rain or snow, and finds its way into rivers, streams, and oceans. Liquid water evaporates back into the atmosphere starting the whole process over again.

The cycling of water moves matter (water) and energy around the planet. Solar energy drives the process by evaporating water from lakes, streams, oceans, and even soil. Water vapor is then transported over large distances and then condenses to release its latent energy into the atmosphere as it forms clouds and ultimately precipitation.

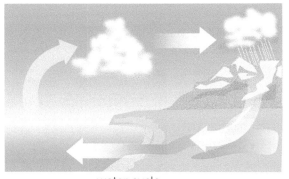

water cycle

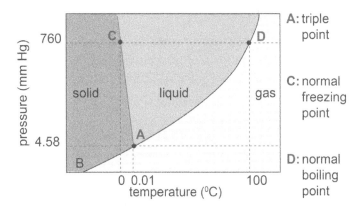

A: triple point
C: normal freezing point
D: normal boiling point

The availability of water and its properties allowed life to develop on Earth. Water is called a universal solvent because its molecular structure enables it to dissolve many substances. As a result, water provides a means of transporting substances to and from cells in all living organisms. To be a good solvent, the water molecules must be close to one another, but free to move around. In other words, it must be in liquid form. From the phase diagram, you can see that water is a liquid under the normal conditions of living things.

The phase diagram of water reveals something else that is interesting about water - liquid water is denser than solid water. How does the phase diagram show this? Imagine that you have a mass of solid water at 0.9 atm and 0.0 °C, which is just to the left of the solid-liquid equilibrium line. If you were to take that mass of water and increase the pressure to 1.1 atm at a constant temperature, you would compress it into a smaller volume, and in the process actually make the water melt! This means that the liquid is more dense than the solid. For most substances, the solid is more dense than the liquid. But water is not like most substances, and that's a good thing. Imagine what would happen if ice did not float. Lakes would freeze from the bottom up, and eventually lead to an entire body of water that is frozen. Aquatic life depends on the fact that ice floats!

Following the Water to Mars - 2

Water is so important in sustaining life that when NASA looks for extraterrestrial life, their motto is "follow the water." There are several possible "ocean worlds" in our solar system. A dwarf planet called Ceres in the asteroid belt between Mars and Jupiter is one of these possible ocean worlds. Scientists estimate that Ceres consists of about 25% water in the form of ice. NASA's Dawn mission arrived at the dwarf planet in 2015 and discovered what is potentially an ice volcano, as well as evidence of carbon-containing compounds that are building blocks of life. A little further out there are multiple moons of Jupiter that scientists believe have an icy layer. The most likely moon to be able to support life is Europa. They believe that a sub-surface ocean lies beneath Europa's icy crust. Beyond Jupiter, moons of Saturn, Neptune, and even the dwarf planet Pluto show some evidence of being an "ocean world."

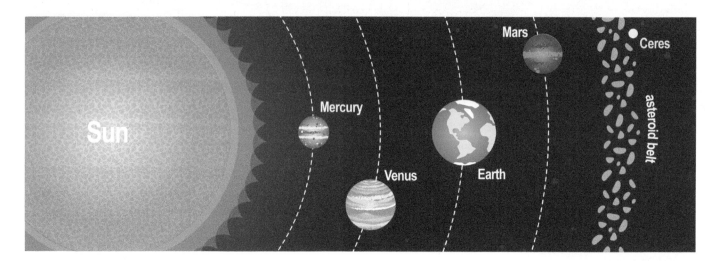

Earth's neighbor Mars is a planet that inspires scientific exploration (and some popular science fiction). Observations and measurements show that water on Mars exists mainly in polar ice caps. Ice can also be found in deposits just under the surface. The image to the right is a color-enhanced photograph of a cliff on Mars. This photo was taken from above the cliff by NASA's Mars Reconnaissance Orbiter (MRO). A thick, multi-layered sheet of ice has been colored blue so you can see it better. The atmosphere of Mars is quite different than Earth's atmosphere. Mars is mostly carbon dioxide and about 0.05 % water, while Earth's atmosphere is only about 0.4 % CO_2 and 1 % water. The temperature can get low enough to freeze Mars' atmosphere during winter, depositing a blanket of dry ice snow, the solid form of CO_2, at the poles.

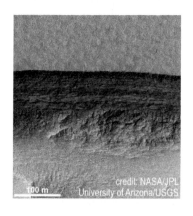

The Martian atmosphere is also much thinner than of Earth. It is so thin that scientists believed there were only two phases of water possible on Mars, solid and gas. If you look at the phase diagram for water you can see how the combination of low temperature and low pressure would support this claim. The pressure is low enough that if you could bring a cup of liquid water to Mars, it would either immediately freeze or immediately boil, based on a couple of degrees in temperature difference.

Following the Water to Mars - 3

Recent findings from the MRO provide evidence that liquid water occasionally flows on the red planet. An instrument on the orbiter detected signals of hydrated materials in the slopes of craters that ebb and flow over time. How can this be? One possible explanation lies in the phase diagram! The global average atmospheric pressure on Mars is 0.006 atmospheres, which is just slightly less than the triple point of water. But there are regions that have higher pressures, making liquid water possible. Another factor is that dissolved salts in water can lower the freezing point making conditions more favorable for the liquid.

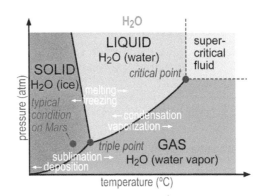

It took several years and several passes of orbiting spacecraft to discover that the Martian surface could hold liquid water. These discoveries show that water is not just Earth's natural resource, but could be a natural resource that supports life elsewhere in the solar system!

Chapter 11 review

Vocabulary
Match each word to the sentence where it best fits.

Section 11.1

condensed matter	critical point
deposition	heat of fusion
phase diagram	plasma
polar	states of matter
sublimation	supercritical fluid
triple point	

1. _____ is matter in which the particles are in contact with each other.
2. Matter exists as a combination of a solid, liquid, and a gas at the _____.
3. A(n) _____ is a diagram in which pressure is plotted against temperature.
4. Dry ice gets its name because it undergoes _____, changing from a solid state directly to a gas.
5. A liquid substance with _____ molecules has a higher boiling point than other substances of similar molar mass that are not _____.
6. _____ occurs when a substance goes from the gas state directly to the solid state.
7. The _____ is the energy absorbed or released in a phase change between liquid and solid.
8. At pressure and temperature above the _____ on a phase diagram, the solid and liquid phases become indistinguishable.

Section 11.2

activation energy	endothermic reaction
enthalpy	enthalpy of formation
exothermic	first law of thermodynamics
Hess's law	photosynthesis

9. The _____ describes the energy released when a compound is formed from its constituent elements.
10. The energy associated with chemical bonds, pressure, and temperature is known as _____.
11. An _____ requires external energy input to occur.
12. The _____ states that the total energy in an isolated system remains constant.
13. A reaction for which ΔH is negative is called _____.
14. _____ states that the enthalpy change for a reaction is independent of what path the reaction proceeds on between the reactants and products.

Section 11.3

entropy	Gibbs free energy
Gibbs–Helmholtz equation	second law of thermodynamics
spontaneous	

15. A _____ reaction occurs on its own without any external energy input.
16. The _____ accounts for the changes in both entropy and enthalpy.
17. The _____ states that entropy increases in all spontaneous occurrences.
18. _____ has units of J/K and describes the way energy is arranged in a system.
19. The energy "left over" after the completion of a reaction and all phase changes have occurred is called _____.

Conceptual questions

Section 11.1

20. How does matter change from one state to another?

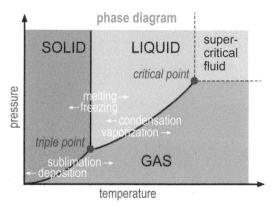

21. According to the phase diagram above, can you boil a liquid without changing temperature? Explain your answer.
22. According to the phase diagram above, what are the conditions of temperature and pressure that will make a substance a solid?
23. According to the phase diagram above, what are the conditions of temperature and pressure that will make a substance a gas?
24. What changes of state occur when:
    a. A cold drink gets wet on the outside of the glass?
    b. Frost forms on the window on a cold day?
    c. Dry ice is used to create fog?
25. On a phase diagram what happens at:
    a. The curve between the liquid and gas?
    b. The curve between liquid and solid?
    c. The curve between solid and gas?
    d. The triple point?

374   Chapter 11: Energy and Change

Chapter 11 review

Section 11.1

26. Condensed matter comprises liquids and solids. What about the arrangement of atoms or molecules and their behavior makes solids and liquids condensed states?

27. Of the four phases, or states, of matter covered in this chapter, describe some differences in the particle arrangements of a gas, liquid, solid.

28. There is a point where three phases of matter converge called the triple point. What happens to a substance at the triple point?

29. Deposition is where a substance changes state from a gas to a solid. Is this likely to occur at a high temperature? Use the phase diagram to support your answer.

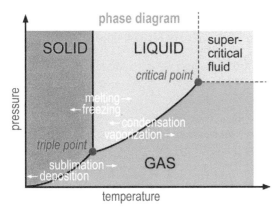

30. Snow disappears from the top of mountains even though the temperature is well below freezing. What process is occurring? Use a phase diagram to support your answer.

31. Temperature is the average kinetic energy of particles. Can a substance change temperature as it changes state? Use a change of state diagram to support your answer.

32. Under what conditions can substances change temperature as heat is added?

33. Can you melt ice without changing its temperature? Use a phase diagram to support your answer.

Section 11.2

34. If the reactants of a chemical reaction have more energy than the products, do the surroundings gain or lose energy?

35. Not all bonds have the same energy. Which is most easily broken in nature, a single, double or triple bond? Explain your answer.

36. Which are favored at low temperatures, exothermic or endothermic reactions? Explain your answer. What happens to the temperature of the system?

37. Endothermic processes occur all around us. Name at least three common endothermic processes. Why are these processes endothermic?

38. Exothermic processes occur all around us. Name at least three common exothermic processes. Why are these processes exothermic?

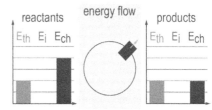

39. Using the above Energy Bar Diagram, answer the following questions.

   a. Is the diagram above illustrating an exothermic or an endothermic reaction? Explain your answer.
   b. Explain how energy is conserved for this reaction.

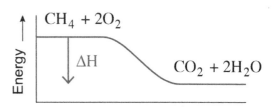

40. For the combustion of methane,

$$CH_4 + 2O_2 \rightarrow CO_2 + 2\,H_2O + \text{energy}$$

Answer the following questions using the energy profile graph above..

   a. Is this an endothermic reaction or an exothermic reaction?
   b. For this reaction, which has more potential energy: the bonds in the reactants or the bonds in the products?
   c. What happens to the energy in this system?
   d. Draw an Energy Bar Diagram illustrating the combustion of methane.

41. What does the change in enthalpy tell you about a reaction? What do the values of ΔH (positive and negative) mean for the reaction?

Chapter 11 review

Section 11.3

42. ❨ When a solid dissolves in water, the entropy (ΔS) is positive. When a gas dissolves in water, the entropy is negative. Why is this?

43. When determining the spontaneity of a reaction, which is most affected by temperature: entropy or enthalpy?

44. Entropy can be seen as the way energy is distributed in a substance. Does plasma have more entropy than gas?

45. What thermodynamic factors work together to determine if a process will take place under given conditions? What equation illustrates this process?

46. The combined enthalpy-entropy function, ΔG, determines the spontaneity of reactions. What does the value of Gibbs Free Energy tell you about a reaction?

47. ❨ For each of the following reactions, will the change in entropy be positive or negative? If there is a change of state, identify the states before and after the reaction.
    a. Dry Ice sublimating.
    b. Ice melting.
    c. Carbohydrates being convert to carbon dioxide and water in a cell.
    d. Two ionic substances forming a precipitate.
    e. Helium escaping from a balloon.

48. What are the trends that natural processes tend to follow for entropy?

Quantitative problems

Section 11.1

49. ❨ 36.5 g of diethyl ether ($C_4H_{10}O$) loses 2,751 J of energy. This lowers the temperature by 16 °C to the boiling point of diethyl ether (-116 °C) and causes some of it to solidify. If the heat of fusion of diethyl ether is 97 J/g and the specific heat is 2.22 J/g·°C, what is the mass of diethyl ether that freezes?

50. How much heat in kJ is released when 1.37 L of water at 0 °C solidifies to ice? (The density of water is 1 g/mL and the heat of fusion of water is 0.335 kJ/g)

51. The enthalpy of fusion, ΔH_{fus}, for water is 335 J/g. If gy one ice cube has a mass of 65.3 g, how much heat ener is absorbed from a drink when a dozen ice cubes melt into liquid water at 0 °C?

52. A solvent reaction contains 100 g of liquid ethanol which must be boiled off. What is the minimum quantity of heat needed if the ethanol is already at the boiling point of 78.4°C. The heat of vaporization is 854 J/g

53. Which requires more heat, melting a 0.50 kg block of ice at 0 °C or boiling 50 g sample of water at 100 °C?

54. Use the following properties of water to answer the questions: ΔH_{fus}= 335 J/g ; ΔH_{vap}= 2,256 J/g MP = 0 °C ; BP = 100 °C C_{ice}= 2.06 J/g·°C ; C_{water}= 4.184 J/g·°C ; C_{gas}= 2.02 J/g·°C
    a. How much heat, in joules, would be required to raise the temperature of a 73.0 g sample of ice from -25.0 °C to the melting point of water?
    b. How much energy is needed to melt 73.0 g of ice?
    c. What quantity of heat is needed to warm 73.0 g of water from its freezing point to its boiling point?
    d. How much heat is required to evaporate 73.0 g of water?
    e. How much energy will raise the temperature of steam from 100 °C to 112 °C?
    f. What is the total amount of energy needed to turn 73.0 g of ice at -25 °C to water vapor at 112 °C?

55. How much heat, in joules, is given off when 7.5 g of water vapor condenses and then cools to 25 °C? (the ΔH_{vap} of water = 2,256 J/g)

Section 11.2

bond type	kJ/mol
H-C	410
C-C	350
O=O	494
C=O	799
H-O	460

56. Use the bond enthalpies to calculate the molar enthalpy change for the following combustion reactions.
    a. $CH_4(g) + 2O_2(g) \rightarrow CO_2(g) + 2H_2O(g)$
    b. $2C_2H_6(g) + 7O_2(g) \rightarrow 4CO_2(g) + 6H_2O(g)$
    c. $C_3H_8(g) + 5O_2(g) \rightarrow 3CO_2(g) + 4H_2O(g)$

Chapter 11 review

Section 11.2

57. Based on the equations in the previous question, calculate how much energy is evolved when:
   a. 50.0g of CH_4 reacts with excess O_2.
   b. 50.0g of C_2H_6 reacts with excess O_2.
   c. 50.0g of C_3H_8 reacts with excess O_2.

58. Using the bond enthalpy table, determine the amount of energy in a carbon-carbon triple bond for the balanced equation below.
   $$2C_2H_2(g) + 5O_2(g) \rightarrow 4CO_2(g) + 2H_2O(g)$$
   ΔH_{comb} = -2,444 kJ/mol.

59. Natural gas (mainly methane) can be used as a reducing agent for Copper(II) Oxide, but the reaction is fairly slow. Use the following reaction to answer the questions below.
   $$4CuO(s) + CH_4(g) \rightarrow 4Cu(s) + 2H_2O(l) + CO_2(g)$$
   a. What is the heat of reaction? Is it exothermic or endothermic?
   b. What is the entropy of reaction? What does this change represent in terms of order?
   c. What is the entropy change at 25.0 °C in the following reaction? Is the reaction spontaneous?

bond type	kJ/mol
F-F	159
O=O	494

60. Calculate the enthalpy required to break one mole of O-F bonds given the following reaction:
   $$2F_2(g) + O_2(g) \rightarrow 2OF_2(g) \quad \Delta H = 56 \text{ kJ}$$
   (Hint: oxygen is the central atom of OF_2)

61. In an experiment to determine the enthalpy change, ΔH, for the neutralization reaction:
   $$NaOH(aq) + HCl(aq) \rightarrow NaCl(aq) + H_2O(l)$$
   a. Calculate the heat of reaction in kJ/mol of water formed for the reaction.
   b. What does the ΔH value tell you about the reaction?

62. Iron ores such as haematite contain iron(III) oxide, Fe_2O_3, in which iron can be extracted in a blast furnace. During a mid step in this process, calcium oxide reacts with silica impurities in the haematite to produce slag, calcium silicate.
   $$CaO(s) + SiO2(s) \rightarrow CaSiO3(l)$$
   Calculate the energy of this reaction.

Section 11.3

63. Are the following reactions increasing or decreasing in terms of entropy?
   a. $2KBr(s) + H_2SO_4(aq) \rightarrow K_2SO_4(aq) + 2HBr(g)$
   b. $2NO(g) + O_2(g) \rightarrow 2NO_2(g)$
   c. $Mg_3N_2(s) + 6H_2O(l) \rightarrow 3Mg(OH)_2(s) + 2NH_3(g)$
   d. $NH_4NO_3(aq) \rightarrow N_2O(g) + 2H_2O(l)$

64. Indicate whether the following reactions are spontaneous or non-spontaneous.
   a. $Fe_2O_3(s) + 3CO(g) \rightarrow 2Fe(s) + 3CO_2(g)$
      $\Delta H°_{rxn}$ = -23.9 kJ/mol
   b. $2N_2O_5(g) \rightarrow 4NO_2(g) + O_2(g)$
      $\Delta H°_{rxn}$ = -110.0 kJ/mol
   c. $Na_2O(s) + SiO_2(s) \rightarrow Na_2SiO_3(s)$
      $\Delta H°_{rxn}$ = -192.0 kJ/mol
   d. $Mn(s) + 2HCl(aq) \rightarrow MnCl_2(aq) + H_2(g)$
      $\Delta H°_{rxn}$ = -221.0 kJ/mol $\Delta S°_{rxn}$= -79.7 J/mol·K

65. By method of observation, determine the changes in entropy for each of the following reactions. State whether the ΔS value will be negative or positive.
   a. $C_6H_6(l) + 3H_2(g) \rightarrow C_6H_{12}(l)$
   b. $4NH_3(g) + 5O_2(g) \rightarrow 4NO(g) + 6H_2O(g)$
   c. $N_2(g) + 3H_2(g) \rightarrow 2NH_3(g)$
   d. $2H_2(g) + O_2(g) \rightarrow 2H_2O(g)$
   e. $PCl_5(g) \rightarrow PCl_3(g) + Cl_2(g)$
   f. $C_8H_{18}(l) + \frac{25}{2}O_2(g) \rightarrow 8CO_2(g) + 9H_2O(g)$

66. Use the following reaction to answer the questions.
   $$CO(g) + 2H_2(g) \rightarrow CH_3OH(g)$$
   a. Calculate the standard entropy change for the reaction of carbon monoxide gas with hydrogen gas.
   b. What does this entropy change tell you about the reaction?

67. Nickel reacts with hydrochloric acid to produce Nickel(II) chloride and hydrogen gas. Use the chemical reaction to answer the following.
   $$Ni(s) + 2HCl(aq) \rightarrow NiCl_2(aq) + H_2(g)$$
   a. What is the entropy change at 25.0 °C?
   b. Is this reaction spontaneous?

chapter 12 Gases

Whether you realize it or not, we live in a sea of gases. When you are walking down a hall, or taking a stroll in the park, you are pushing your way through trillions and trillions of gas particles. And it's a good thing too! The gas particles in the air around you make life on earth possible. Greenhouse gases are necessary to keep the planet warm and life sustaining (although too much greenhouse gas will make the planet too warm). Carbon dioxide and oxygen make photosynthesis and respiration possible. Not only are we moving through this sea of gases, we are at the bottom of it! The atmosphere that keeps us alive extends to about 100 km (60 miles) over your head getting thinner and thinner as it stretches upward.

The study of gases is what propelled chemistry out of the realm of alchemy and into a true science. Early "natural philosophers" began experimenting with the "empty space" around them. As they did, they made measurements, looked for relationships, and discovered empirical laws that govern the properties of gases.

There was a true chemical revolution in the late 18th century. Joseph Priestly (1733-1804) isolated and characterized eight gases including what is now called oxygen. This was before the atomic theory as we know it and he called oxygen "dephlogisticated air" based on the theory of phlogiston. The theory held that some materials contained a fire-like element called phlogiston that made them combustible. Priestly's results were interpreted differently by another chemist, Antoine-Laurent Lavoisier (1743-1794). Lavoisier is considered by many to be the "father of modern chemistry." He studied chemical reactions with the part of the air that he renamed "oxygen," established the law of conservation of mass, and proved that water is made from the gases of oxygen and hydrogen. In addition to sparking the chemical revolution with his work on gases, Lavoisier participated in the French Revolution. During this time, he came under attack and was guillotined.

Antoine-Laurent Lavoisier

Gases are an important and life-sustaining property of matter, and the study of gases has been integral to our essential understanding of the natural world.

Chapter 12 study guide

Chapter Preview

While most of the time we don't see them, gases constantly surround us. They exert pressure on our bodies and make it possible for Earth to sustain life. In this chapter you will explore the basic properties of gases and the chemical laws that govern their behavior. Gases are composed of fast moving particles with large amounts of space between them. Gases expand or contract to fill their container. They exert pressure on surfaces due to collisions of particles with that surface. The pressure exerted by a gas is dependent on the amount of gas, the size of the container and the temperature. Partial pressures can be determined using Dalton's law. The partial pressures of the gases that make up Earth's atmosphere create atmospheric pressure. Boyle's law, Charles' law and Avogadro's law each help explain the behavior of gases in relation to pressure, volume, temperature and the amount of gas present. Together their results can be combined into the ideal gas law.

Learning objectives

By the end of this chapter you should be able to:

- explain how a gas produces pressure;
- describe how gas pressure is affected by volume, heat and the number of particles in the gas;
- use Dalton's law to calculate partial pressures of gases;
- use Boyle's law, Charles' law, the combined gas law and Avogadro's law to predict the behavior of gases;
- describe the kinetic molecular theory; and
- use the Ideal Gas Law under fixed and varying conditions.

Investigations

12A: Volume of a gas

12B: Boyles' law

12C: Charles' law

Pages in this chapter

- 380 Pressure and Gas Behavior
- 381 Pressure
- 382 Changing pressure
- 383 Measuring pressure
- 384 Dalton's law
- 385 Atmospheric pressure
- 386 Section 1 Review
- 387 Gas Laws
- 388 Using Boyle's law
- 389 Pressure in daily life
- 390 Creating a new temperature scale
- 391 Charles' law
- 392 Combining Boyle's law and Charles' law
- 393 Avogadro's law
- 394 Molar volume
- 395 Section 2 Review
- 396 Kinetic Molecular Theory
- 397 The ideal gas law
- 398 The ideal gas law with fixed conditions
- 399 The ideal gas law with varying conditions
- 400 Section 3 Review
- 401 Deep Dives and High Altitudes
- 402 Deep Dives and High Altitudes - 2
- 403 Chapter review

Vocabulary

pressure
vacuum
atmospheric pressure
Charles' law
molar volume
universal gas constant

pascal (Pa)
Dalton's law
Boyle's law
combined gas law
ideal gas

barometer
water displacement
absolute zero
Avogadro's law
kinetic molecular theory

12.1 - Pressure and Gas Behavior

Scientists were curious about the behavior of gases long before the age of modern chemistry. Gases change volume drastically as pressure and temperature change, which makes them easy to measure. A gas doubles its size when the Kelvin temperature doubles. Solids and liquids change very little. Imagine if a solid like your home were to noticeably expand when the temperature goes from 40.0 °F to 80.0 °F. Your house does expand, but only about 0.0000028 inches per degree. Such a small change in expansion could not be detected by researchers 300 years ago, when gases were first studied.

Properties of gases

Properties of Gases

Compared to solids and liquids, gases have a unique set of properties. These properties form the basis of the kinetic molecular theory:

1. Gases have relatively large spaces between particles.
2. Gases have very low density compared to liquids or solids.
3. Gases are highly compressible compared to liquids and solids.
4. Gases can expand or contract to fill their container.

Behavior of gases

There is a large amount of space between gas particles because gases are made of tiny atoms or molecules that are not attracted to one another. This explains why gases have low density and why they are transparent. Particles are too thinly distributed in a gas to be visible or dense.

gas particles are constantly in motion

Other properties of gases

The empty space also explains why gases can either compress or expand to fill the volume of their container. To compress is to reduce the space between particles and to expand is to increase the space between particles. Gas particles are constantly moving at random, causing them to hit each other and the walls of their container. By now you have probably made many observations that support the idea that a gas is made of particles that are far apart and in constant motion. For example, if you have ever taken a balloon outdoors at a time of year when there is a large temperature difference, you would have noticed the balloon expands when temperature increases and it contracts when temperature decreases. The balloon is a sealed container so you know the matter inside is able to change its volume.

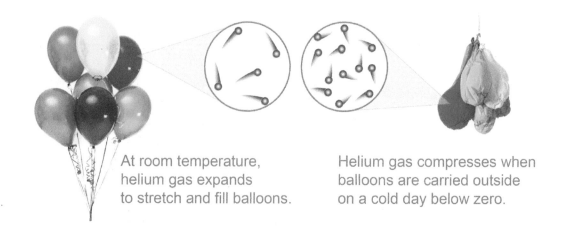

At room temperature, helium gas expands to stretch and fill balloons.

Helium gas compresses when balloons are carried outside on a cold day below zero.

Pressure

Pressure

Consider a brick that has dimensions of 2 x 4 x 8 inches that exerts a force of 3 lbs. The force is created by gravity pulling on the brick. The force on the brick will not change whether it is placed on its small side or its large side. However, the pressure does change. **Pressure** is force per unit area. It is the way force is distributed on a surface. In position a, the 3-lb force of the brick is spread out over the 4 x 8 in. surface. The pressure is 3 lb/32 in.^{2} = 0.094 lb/in.^{2}. In position b, the force is spread out over the 2 x 4 in. surface. The pressure is 3 lb/8in^2 = 0.38 lb/in^2. Position b exerts more pressure even though the force is constant.

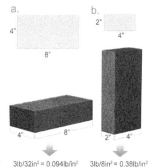

Force

When you dive down to the bottom of a swimming pool you may feel pain in your ears. This is caused by the pressure of water. In this case the volume of the pool is not changing; but as you swim deeper there is more water over your head which creates more force. The pain in your ears is caused by the force of all of the water above you.

Pressure is force per unit area

Gas pressure is the force per unit area that the gas exerts on any surface in contact with the gas. Gas particles have an incredible amount of motion. Some gas particles are moving over 300 miles per hour. That motion is in all directions. The collisions caused by this motion is gas pressure. A balloon holds its shape because gas particles inside the balloon are colliding with the walls of the balloon. When heat is applied the particles move faster and collide with more force. This increased activity makes the balloon bigger.

internal pressure in a balloon

Measuring pressure

Pressure is equal to force divided over area. The SI unit of force is the newton, N. A pressure of 1 N/m^2 means there is a force of one newton distributed over each square meter. The unit N/m$_2$ is also known as the pascal, Pa. One **pascal (Pa)** equals a pressure of one newton of force per square meter of area. A common pressure unit in the United States is pounds per square inch, psi. One psi is equal to one pound of force per square inch of area. One psi is a much larger unit of pressure than one pascal (1 psi = 6,895 Pa).

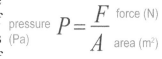

$$1\,\text{Pa} = \frac{1\,\text{N}}{\text{m}^2}$$

Interactive simulation

Did you ever notice how it gets more and more difficult to blow air into a balloon as you approach its maximum size? When this happens, you are experiencing pressure from gas particle collisions! Explore a similar phenomenon in the interactive simulation titled Gas Pressure in a Syringe. Think about how the gas inside and outside the syringe is like the gas inside and outside a balloon. Recognize the syringe is a rigid container while a balloon is not as rigid. Try pushing and pulling the syringe plunger when the cap is on and off. Notice how the syringe and gas particles behave when you let go of the plunger.

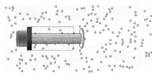

Changing pressure

Particle collisions

Pressure exerted by a gas is due to the collisions of many particles among themselves and with the walls of their container. Anything that affects either the frequency of particle collisions or how hard they collide will change the pressure. Specific changes to the environment of a gas can have even more impact.

how the number of collisions and impact strength affect gas pressure

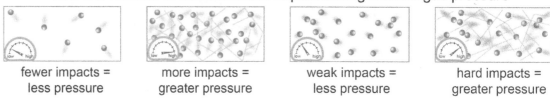

| fewer impacts = less pressure | more impacts = greater pressure | weak impacts = less pressure | hard impacts = greater pressure |

Ways to change pressure

Add particles to a container with a fixed size. When you pump air into a tire that is already full, the size of the tire will not change very much, but the increased number of particles in the tire causes more molecular collisions. Gas pressure increases. If you want to decrease pressure, remove particles from the container.

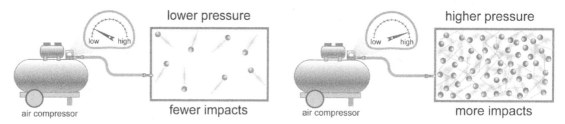

Reduce the container size while keeping the amount of particles the same. When each particle has less space to move around in, there is an increase in molecular collisions and gas pressure increases. Every time you exhale, your lungs decrease their volume to increase the inside pressure. The higher pressure inside forces air out of your lungs. When you inhale, your lungs expand their volume. This lowers the pressure inside your lungs, allowing air to rush in.

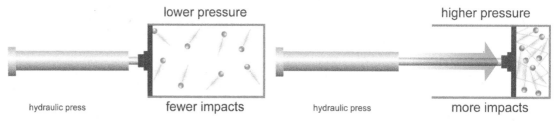

Heat the gas. An increased temperature causes increased kinetic energy among the particles. Particles hit the container walls more often and they hit harder. Both events increase pressure. This is why most aerosol spray cans have a label that warns consumers to keep the cans away from heat and fire. A high temperature can increase pressure inside the can enough to make it explode. If you want to decrease pressure, cool the gas down.

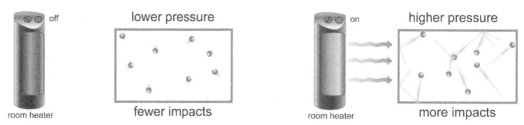

Measuring pressure

Pressure and elevation

The air pressure at the top of Mt. Everest is only about one-third the pressure at sea level. The higher you go up the mountain the less air there is above you, so there are fewer particle collisions. This causes the air pressure around you to decrease. Lower air density also makes it harder to breathe.

Air pressure

What happens if you place your finger over a straw in a glass of water and then lift the straw out of the glass? Water stays in the straw because downward pressure exerted by the water in the straw is less than the upward air pressure. Air below the straw pushes on the water and keeps it in the straw as long as one end is covered. This works until you use a straw that is over 32 feet tall. At this point, the water's weight would exert greater pressure than the air below the straw. Some water would come out until the water's pressure and the air pressure are equal.

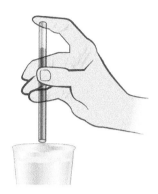

The barometer

The **barometer** was one of the first tools used to measure air pressure. It was invented by Italian scientist Evangelista Torricelli in the 1600s. The torr is a unit of pressure named in his honor. Torricelli used mercury instead of water in his barometer because its density is 14 times greater than water. The column of mercury needed to balance air pressure is much less than 32 feet. Standard air pressure can push a column of mercury to a maximum height of 760 mm. This is where the pressure unit mmHg comes from. As air pressure changes, so does the mercury column height. Decreased air pressure pushes less on the mercury, allowing some mercury to move from the column into the column base. Mercury is able to move up and down the tube because a vacuum exists inside the clear part of the tube.

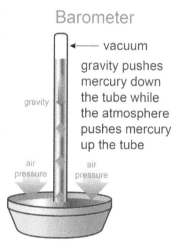

Barometer — vacuum; gravity pushes mercury down the tube while the atmosphere pushes mercury up the tube

What is a vacuum?

Space begins where our atmosphere ends. You may have heard that space is a **vacuum** because "there's no air or matter in space." Space is an imperfect vacuum because a few hydrogen gas molecules can be found in an average cubic meter. Matter is not evenly distributed in space but it is found everywhere. The parts of space that have very little matter come close to being a perfect vacuum. Pressure is very low where there is little matter because only a few particles are available to collide and create pressure. Even a vacuum created in a lab is imperfect so it is more practical to think of a vacuum as a space with less matter and less pressure than its surroundings.

Weather and atmospheric pressure

Most weather maps show areas of high and low atmospheric pressure. Pressure changes are caused by many factors, including air temperature changes that result in air density changes. Large areas of slightly varying pressure have a significant effect on weather patterns. Warm, less dense air rises and releases water vapor as clouds and precipitation. This is why low pressure is associated with cloudy or rainy weather. High pressure is associated with sunny weather because sinking, dense air tends to be dry.

Section 12.1: Pressure and Gas Behavior

Dalton's law

Dalton's Law of Partial Pressure

When you have a mixture of gases, each component of the gas contributes some pressure to the total. These are called partial pressures. **Dalton's law** states that in a mixture of non-reacting gases, the total pressure exerted is equal to the sum of the partial pressures of the individual gases.

$$P_{tot} = P_1 + P_2 + P_3 + \text{etc.}$$

Water displacement

Dalton's law is most often used when collecting gas over water. It is easier to direct the movement of a gas if it is bubbled through water. A jar is filled with water and inverted in a shallow pan. A tube connected to a reaction chamber is inserted under the inverted jar. As the gas is created, water is forced out of the jar. This technique of collecting gas is known as **water displacement**. The gas is collected over the water in the jar.

However, the collected gas is not pure. Water vapor above the liquid mixes with the collected gas, so you must account for the gas pressure caused by water vapor. The pressure of water vapor mixed with the collected gas depends on the temperature. To find the pressure of dry gas we use Dalton's law and a table of known water vapor pressures at given temperatures.

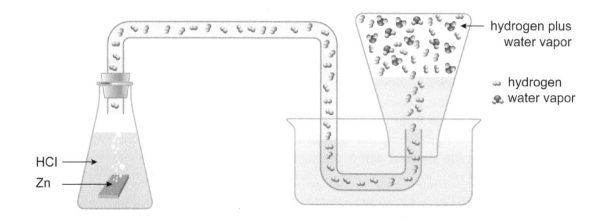

Solved problem

The total pressure in the jar is 101.325 kPa and the temperature is 25.0 °C. What is the pressure of hydrogen gas that was collected over water?

Asked What is the pressure of hydrogen gas?

Given P_{tot} = 101.325 kPa and from the table, P_{H_2O} = 3.1690 kPa

Relationships $P_{tot} = P_1 + P_2$ or $P_{tot} = P_{H_2O} + P_{H_2}$

Solve $P_{tot} = P_{H_2O} + P_{H_2}$
101.325 kPa = 3.1690 kPa + P_{H_2}
P_{H_2} = 101.325 kPa − 3.1690 kPa = 98.156 kPa

Answer The partial pressure of hydrogen gas is 98.156 kPa.

water vapor pressure	
T, °C	P, kPa
10	1.2281
15	1.7056
20	2.3388
25	3.1690
30	4.2455
35	5.6267

Atmospheric pressure

The composition of air

Together the partial pressures of atmospheric gases add up to the atmospheric pressure at sea level as shown. The largest components of our atmosphere are 78.09% nitrogen and 20.95% oxygen. You can see that they also have the largest partial pressures.

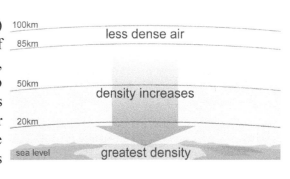

78.1 kPa	$P_{nitrogen}$
20.9 kPa	P_{oxygen}
1.28 kPa	P_{water}
0.97 kPa	P_{argon}
0.05 kPa	$P_{carbon\ dioxide}$
101.3 kPa	P_{tot} at sea level

Air pressure

The air above you is about 100 km (60 miles) high and is being pulled down by gravity. If you could fly straight up into the atmosphere, there will be less air on top of you as you go higher. Air gets less dense at high altitudes because air particles are not packed together as closely. The force per area we experience as air pushes on everything around us is called **atmospheric pressure**.

Pressure at sea level is 14.69 lb/in²

Normal air pressure is higher than you may think. It is 14.69 psi (lb/in²) at sea level. Let's calculate the amount of force air produces constantly on the palm of your hand. The area of your palm is approximately 6" x 3", with a surface area of 18 in². If each square inch has a pressure of 14.69 lb/in² from the air, the total force on your hand is about 264 lbs!

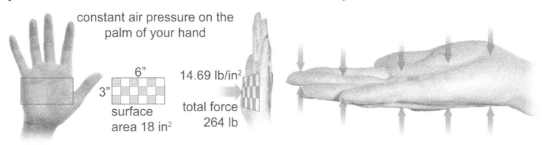

Why don't you feel air pressure?

You might wonder if you had all that force pressing on you, why you don't feel it? Gas particles move very fast and bounce off everything in the room. For example, air particles are pushing down on your hand and pushing up on your hand at the same time. The amount of force pushing up is very close to the amount of force pushing down. In effect, the up and down forces cancel each other out and the net force is close to zero.

Standard pressure equivalencies

Standard pressure is the average pressure at sea level. The SI units for pressure is kiloPascals, kPa. The standard SI value is 101.325 kPa. Whenever a scientist says Standard Pressure, it is ok to use any of the numbers below. You can also use these equivalencies to convert between pressure units. The standard pressure equivalences are:

14.69 lb/in² = 14.69 psi = 1.000 atm = 760.0 mmHg = 760.0 Torr = 101.325 kPa

Solved problem

Convert a pressure of 1.250 atm to kPa.

Given 1.250 atm

Relationships Standard pressure equivalencies (above) are 1.000 atm and 101.325 kPa

Solve Use the standard pressure equivalencies to create a conversion factor:

$$\frac{1.250\ atm}{1} \times \frac{101.325\ kPa}{1.000\ atm} = 126.65625\ kPa \xrightarrow{round} 126.7\ kPa$$

Answer The pressure is equal to 126.7 kPa

Chapter 12

Section 1 review

With vast distances between them, particles of gases move very fast. As a result, gases are of low density and are highly compressible. Pressure is defined as force per unit area and increases for a gas when the gas is heated, compressed or has additional gas particles added to the system. The combined weight of the gases that make up Earth's atmosphere is atmospheric pressure, and is measured using a barometer. Dalton's Law is often used to determine the pressure of a gas when collected over water. According to the law, the total pressure in a mixture of non-reacting gases is equal to the sum of the partial pressures of the individual gases.

Vocabulary words	pressure, pascal (Pa), barometer, vacuum, Dalton's law, water displacement, atmospheric pressure

Key relationships

- Pressure = $\frac{force}{area}$
- 1 mmHg = 1 Torr
- Standard pressure: 14.69 psi = 1.000 atm = 760.0 mmHg = 760.0 Torr = = 101.325 kPa
- Dalton's Law: $P_{total} = P_1 + P_2 + P_3...$

Review problems and questions

1. A friend has given you a helium-filled balloon to celebrate your birthday. When you walk outside into the cold 38 °F weather, your balloon shrinks and sinks to the ground. Explain the balloon's behavior in cold air from the perspective of temperature, gas particle collisions and pressure.

2. The air pressure on Mt. Everest is lower than the air pressure at any beach. Why?

3. You are asked to solve a Dalton's law problem for a system that has 3 gases in it. The partial pressure of Gas 1 is given as 0.98 atm. Gas 2 is 1.02 atm. Gas 3 is 75.923 kPa. What do you have to do with the given pressures before you can use Dalton's equation?

4. A system has 3 gases where the partial pressure of Gas 1 is given as 0.98 atm, Gas 2 is 1.02 atm and Gas 3 is 75.923 kPa. What is the total pressure of the system?

5. Hydrogen peroxide, H_2O_2 has decomposed to water and oxygen gas. The oxygen gas is collected over water in a closed vessel at 20 °C. The pressure in the vessel equals 12.713 kPa. What is the partial pressure of the oxygen gas?

water vapor pressure

T, °C	P, kPa
10	1.2281
15	1.7056
20	2.3388
25	3.1690
30	4.2455
35	5.6267

12.2 - Gas Laws

If you used hairspray this morning, you relied on a gas to force liquid hairspray out of a bottle. Hairspray bottles work because a gas predictably moves from higher pressure to lower pressure. A number of gas behaviors can be predicted using one or more equations called the gas laws. Gas laws are used to predict gas behavior but do not explain that behavior.

Volume and pressure of a gas

Boyle's hypothesis

In the mid 17th century Robert Boyle studied the relationship between the pressure of a gas and its volume. He used Torricelli's ideas to design a device that would allow him to measure the pressure and volume of an enclosed gas. Boyle constructed a J-shaped tube called a manometer that was open on one end and sealed on the other. He poured mercury into the open end to trap a bubble of air inside the closed end of the tube. Boyle started with enough mercury to equal atmospheric pressure, and then changed the mercury level to see the effect on the trapped air volume. By adding more mercury, he increased pressure on the trapped air. Boyle used the change in air volume to determine the change in air pressure. The number of gas particles was kept constant because no additional gas could get in or out. Boyle also kept the temperature constant while taking measurements. The data and graph below show results you would expect to get from such an experiment.

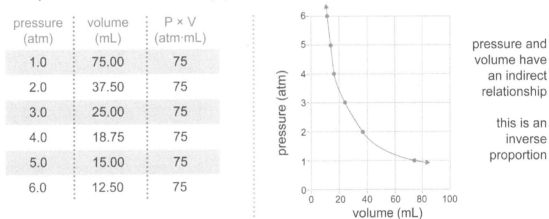

pressure-volume relationship (when temperature and moles of gas are constant)

pressure (atm)	volume (mL)	P × V (atm·mL)
1.0	75.00	75
2.0	37.50	75
3.0	25.00	75
4.0	18.75	75
5.0	15.00	75
6.0	12.50	75

pressure and volume have an indirect relationship

this is an inverse proportion

Boyle's law

Boyle determined that doubling the pressure caused the trapped air volume to be cut in half, and cutting pressure in half caused the trapped air volume to double. When the volume goes down the gas particles are compressed and there are more collisions of particles per area. This causes the pressure to increase. **Boyle's law** states that pressure is inversely proportional to volume and is expressed as P × V = a constant. The constant could change from one experiment to another if the temperature or amount of gas used is varied, but as long as the temperature and moles of gas are unchanged during an experiment, the P × V = a constant relationship is true. The subscript 1 refers to initial values and 2 refer to new values.

Boyle's law
P = pressure
V = volume
k = a constant value*

$$P \times V = k \quad \text{or} \quad P_1 V_1 = P_2 V_2$$

gas pressure and volume are inversely proportional only when the amount of gas (moles) and temperature are constant

(*the value of k depends on the amount of gas and temperature)

Using Boyle's law

Volume affects pressure

Think about what happens to gas particles when their container size increases. In the example below notice that an increase in container volume results in a lower pressure. When the gas expands to fit its container volume, there will be more space between particles. Particles that are spread farther apart will cause fewer impacts on the container walls per second, thus lowering the pressure.

$$P_1V_1 = P_2V_2$$
1.00 atm × 1.2 L = 0.372 atm × 3.20 L
1.2 atm·L = 1.2 atm·L

$P_1 = 1.00$ atm
$V_1 = 1.20$ L

when container size increases, the increased volume results in reduced pressure because gas particles impact the container walls less frequently per unit time

$P_2 = 0.375$ atm
$V_2 = 3.20$ L

Pressure affects volume

Have you ever heard a bottle of water suddenly make a loud noise on an airplane or on a drive to the mountains or beach? Next time you take a trip where there is a large change in elevation, such as up and down a mountain, bring a sealed bag of chips with you. If you begin the drive at a lower elevation such as 500 ft. and reach an elevation of 8,210 ft, air pressure drops by nearly one-fourth. The volume of air inside the sealed chip bag will increase and it will appear to inflate itself.

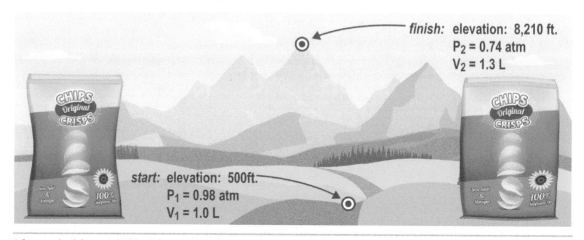

finish: elevation: 8,210 ft.
$P_2 = 0.74$ atm
$V_2 = 1.3$ L

start: elevation: 500ft.
$P_1 = 0.98$ atm
$V_1 = 1.0$ L

Solved problem

If a sealed bag of chips increases its volume from 725 mL to 815 mL on a drive from the beach to a mountain, what is the new pressure inside the bag?

Asked   What is the bag's pressure at the new destination (P_2)?
Given $P_1 = 1.00$ atm; $V_1 = 725$ mL; $V_2 = 815$ mL
Relationships $P_1V_1 = P_2V_2$
Solve $P_2 = \dfrac{1.00 \text{ atm} \times 725 \text{ mL}}{815 \text{ mL}} = 0.88957$ mL $\xrightarrow{\text{round}}$ 0.890 mL

Answer   The bag pressure will decrease to 0.890 atm.

Interactive simulation

Observe how pressure responds when you change the volume of a gas in the interactive simulation titled The Volume-Pressure Relationship. Notice how the number of particle-particle and particle-container collisions responds to changes in volume. How does the number of collisions affect pressure?

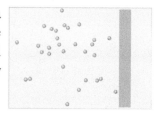

Pressure in daily life

Suction

It seems strange to say that there is no such thing as suction. It feels like you are sucking air into your lungs when you breathe but that is not the case. Air is being pushed, not pulled into your lungs. "Suction" can almost always be explained by unbalanced gas pressure. Gas exerts a pressure by having its particles push on something. There is no way for a gas to pull on anything.

Pressure balance

We don't see things constantly crushed by air pressure because air pressure is balanced nearly everywhere. When you blow up a balloon and tie it closed, a huge amount of gas particle activity occurs. Air particles collide with the balloon's surface both outside and inside the balloon. It doesn't get any bigger or smaller because the pressure inside and outside are equal. If you changed the pressure by adding more or less air to the balloon, the balloon will change its size in order to regain pressure equality with its outside environment.

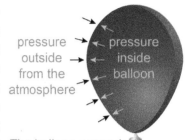

The balloon expands to keep pressure equal inside and outside the balloon.

Your lungs and pressure

Breathing is another application of Boyle's law. When you inhale, lung volume increases. Air particles inside your lungs produce lower pressure because they spread out and reduce their impacts. Air particles outside your lungs push and collide with greater frequency. Outside air particles push their way to the low pressure area inside your lungs until pressure is equal inside and outside your lungs. The pull feeling during inhalation comes from your diaphragm muscle. When you inhale, the muscle contracts and moves downward, pulling the lungs open. When you exhale the muscle relaxes and returns to its upward position. This decreases lung volume and increases air pressure inside your lungs. Air pushes out until pressure equalizes, then it's time to inhale again.

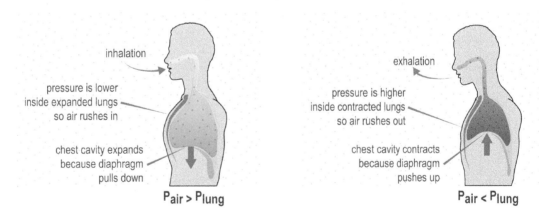

Tools that use pressure

Straws and vacuum cleaners work by creating a region of low pressure for air particles to push themselves into. When you use a straw you expand the volume inside your mouth, lowering the pressure. Atmospheric pressure pushes the liquid up the straw into your mouth. Vacuum cleaners use a pump to create a region of low pressure. Atmospheric pressure outside the vacuum pushes air and dirt carried by the air into the machine. Vacuum cleaners do not create a vacuum anywhere close to the vacuum in space. On average, vacuum cleaners create regions with only 20% less pressure than the surrounding air.

Creating a new temperature scale

Heating a gas and expansion

In the early 1700s, French scientist Guillaume Amontons discovered that a change in temperature caused a change in gas volume. He realized that a gas takes up more space when it is heated; that is, an increase in temperature caused an increase in volume. No formal temperature scale existed at the time, so he was not able to determine the exact relationship between gas temperature and its volume.

A volume and temperature experiment

The Celsius temperature scale was established in 1742. Decades later, Jacques Charles developed an experimentally determined gas temperature-volume relationship. Charles used a sealed container to keep the number of moles of gas particles constant. The container could change size easily, thus keeping pressure constant. Only the temperature and volume were allowed to change. The data and graph below show results you would expect to get from such an experiment. Notice how as temperature is doubled, volume also doubles. Volume and temperature are directly proportional.

volume-temperature relationship (when pressure and moles of gas are constant)

volume (L)	temp. (°C)	V/(T+273) (L/°C)
2.73	0	0.010
3.50	77	0.010
4.75	202	0.010
5.78	305	0.010
6.94	421	0.010

*Note: use kelvin or add 273 to °C for a directly proportional relationship between temp. and vol.

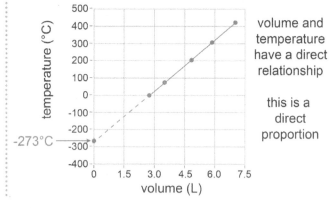

volume and temperature have a direct relationship

this is a direct proportion

The lowest temperature

The graph above shows as temperature decreases, so does volume. Lord Kelvin, a scientist, realized that if you extend the line on the graph to zero mL, you would also reach the lowest possible temperature. Zero volume occurs at −273 °C on the graph. This temperature came to be known as **absolute zero**. Kelvin created a new absolute temperature scale based on absolute zero where no negative temperatures were possible. It was identical to the Celsius scale except temperature starts at 0 instead of −273. The kelvin scale is "absolute" because it begins at zero and only increases from there; the scale has no negative values.

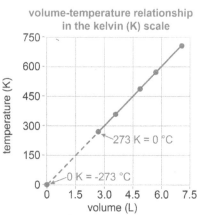

°C to K conversion

To convert a Celsius temperature to the kelvin unit, add 273 to the Celsius temperature.

$$K = °C + 273$$

Matter at absolute zero

When matter is cooled close to absolute zero, energy is at a minimum and we see very strange quantum properties. For example, matter may become a superconductor, which is a substance that allows electricity to flow very easily and efficiently through itself. Other substances take on a phase transition of what appears to be melting. Quantum-level melting is not driven by thermal energy like we are used to seeing on the everyday scale.

Charles' law

Volume and temperature

Charles' law describes the relationship between temperature and volume at a constant pressure. When temperature increase, gas particles move faster and have more collisions. The law states that volume is directly proportional to temperature, and can be expressed as V/T = some constant. If the initial pressure or moles of gas used are changed, the constant could change from one experiment to another. When gas pressure and moles remain the same, the V/T = constant relationship will be true if temperature is reported in Kelvins. The Kelvin temperature scale is used in pressure calculations.

Charles' Law: when pressure & moles of gas are constant, when temperature increases, volume increases in response to increased collision frequency and impact force

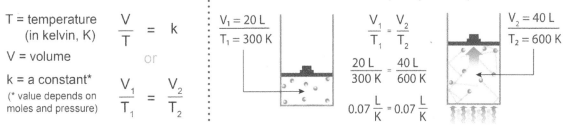

T = temperature (in kelvin, K)
V = volume
k = a constant* (* value depends on moles and pressure)

$$\frac{V}{T} = k$$

or

$$\frac{V_1}{T_1} = \frac{V_2}{T_2}$$

Temperature and gas volume

Let's examine the temperature-volume relationship at the particle level in a container that can change its volume. If gas temperature is increased, there will be more frequent and harder impacts on the container walls. This creates a pressure imbalance between the container and its surroundings. The gas will expand until the particles are spread far enough apart so the pressure inside and outside of the container are equal again.

Solved problem

How small will a 5.0 L balloon become when you take it from your classroom, where the temperature is 22.1 °C, to the cafeteria freezer where it is -18.0 °C?

Asked What is the new balloon volume (V_2) when the temperature drops to -18.0 °C?

Given $V_1 = 5.0$ L, $T_1 = 22.1$ °C, $T_2 = -18.0$ °C

Relationships
- Kelvin temperature conversion, K = °C + 273
- Charles' law, $\frac{V_1}{T_1} = \frac{V_2}{T_2}$ is easier to work with when rearranged as: $V_1 \times T_2 = V_2 \times T_1$:

Solve First, Convert temperatures from °C to K:

- T_1 in Kelvin, 22.1 °C + 273 = 295.1 K
- T_2 in Kelvin -18.0 °C + 273 = 255 K

Next, enter quantities into the equation:

5.0 L × 255 K = V_2 × 295.1 K

Rearrange the equation to solve for V_2:

$$V_2 = \frac{5.0 \text{ L} \times 255 \text{ K}}{295.1 \text{ K}}; \quad V_2 = 4.32057 \xrightarrow{\text{round}} 4.3 \text{ L}$$

Answer The balloon's new volume will be 4.3 L.

Combining Boyle's law and Charles' law

The combined gas law

Recall that Boyle's law predicts how gas pressure varies with volume, but only when temperature and number of particles are constant. Charles' law predicts how gas volume varies with temperature when pressure and moles are kept constant. A law we have not mentioned called Gay-Lussac's law predicts how pressure changes with temperature when moles are kept constant. However, in most situations all of those variables may be changing. Moles of gas can be held constant just by placing the gas in a sealed container so nothing can escape or enter. We combine these laws to give a new formula that allows for pressure, volume, and temperature to vary. See how the formula for the **combined gas law** is derived below.

Combined Gas Law

P = pressure
T = temperature (Kelvin)
V = volume
k = a constant*
(*value depends on moles of gas)

$$\frac{PV}{T} = k \quad \text{or} \quad \frac{P_1 V_1}{T_1} = \frac{P_2 V_2}{T_2}$$

P-V-T relationship when moles of gas are constant

Boyles' Law: $P_1 V_1 = P_2 V_2$ if T and moles are constant

Charles' Law: $\frac{V_1}{T_1} = \frac{V_2}{T_2}$ if P and moles are constant

combined gas law: $\frac{P_1 V_1}{T_1} = \frac{P_2 V_2}{T_2}$ if moles are constant

Solved problem

Meteorologists use weather balloons to measure atmospheric conditions up to 100,000 ft above sea level where pressure equals 1.120 kPa and temperature is about -46.0 °C. A weather balloon with 12.5 L of helium gas is at sea level where the pressure is 101.325 kPa and the temperature is 19.1 °C. How big will the balloon be when it reaches 100,000 feet?

Asked What is the balloon's new volume (V_2) when it reaches 100,000 feet?
Given P_1=101.325 kPa; V_1=12.5 L; T_1=19.1 °C; P_2=1.120 kPa; T_2=-46.0 °C
Relationships Kelvin conversion, K = °C + 273, and the combined gas law, $\frac{P_1 V_1}{T_1} = \frac{P_2 V_2}{T_2}$, which is easier to work with when rearranged as:
$P_1 \times V_1 \times T_2 = P_2 \times V_2 \times T_1$
Solve When converted to Kelvin, T_1 = 292.1 K and T_2 = 227 K
101.325 kPa × 12.5 L × 227 K = 1.120 kPa × V_2 × 292.1 K

$$V_2 = \frac{101.325 \text{ kPa} \times 12.5 \text{ L} \times 227 \text{ K}}{1.120 \text{ kPa} \times 292.1 \text{ K}}$$

V_2 = 878.82601 round 879 L

Answer The balloon's new volume at 100,000 feet is 879 L.

Deriving Boyle's and Charles' law

When temperature is held constant, it drops out of the equation. The combined gas law becomes Boyle's law.

$$\frac{P_1 V_1}{\cancel{T_1}} = \frac{P_2 V_2}{\cancel{T_2}} \text{ becomes } P_1 V_1 = P_2 V_2$$

When pressure is held constant, the combined gas law becomes Charles' law.

$$\frac{\cancel{P_1} V_1}{T_1} = \frac{\cancel{P_2} V_2}{T_2} \text{ becomes } \frac{V_1}{T_1} = \frac{V_2}{T_2}$$

Avogadro's law

Gas volume and the number of particles

In the early 1800s, an Italian scientist named Amedeo Avogadro studied the chemical and physical behavior of gases. Avogadro found a relationship between the amount of a gas present (moles) and the volume it takes up. No matter what gas he studied, Avogadro found equal volumes of gases contained an equal numbers of particles. This would only work if the gases were at standard temperature and pressure (STP: 0 °C and 1 atm).

Volume and moles

The data and graph below show that doubling the number of moles doubles the volume. When the number of moles doubles, the number of collisions between particles also increases. This causes the volume to increase. This means that volume and moles are directly proportional. **Avogadro's law** is expressed as V/n = some constant. If the initial temperature or pressure of the gas was changed, the "constant" could change from one experiment to another. As long as the temperature and pressure were kept the same, the V/n = constant relationship would be true.

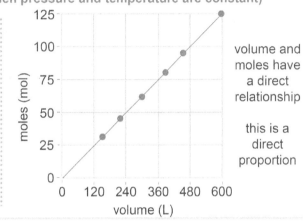

volume-moles relationship (when pressure and temperature are constant)

volume (L)	moles (mol)	V/n (L/mol)
100	20	5
200	40	5
300	60	5
400	80	5
500	100	5
600	120	5

volume and moles have a direct relationship

this is a direct proportion

Avogadro's Law
V = volume
n = moles
k = a constant*
(*value depends on pressure and temp.)

$$\frac{V}{n} = k \quad \text{or} \quad \frac{V_1}{n_1} = \frac{V_2}{n_2}$$

V-n relationship when temperature and pressure are constant

Solved problem

How many moles of gas were added to a balloon to cause it to expand from 5.0 L to 35.2 L at constant temperature and pressure? Initially there were 0.30 moles in the balloon.

Asked   Find n_2

Given V_1 = 5.0 L; V_2 = 35.2 L; n_1 = 0.30 moles

Relationships $\frac{V_1}{n_1} = \frac{V_2}{n_2}$

Solve $\frac{5.0 \text{ L}}{0.30 \text{ moles}} = \frac{35.2 \text{ L}}{n_2}$

$n_2 = \frac{35.2 \text{ L} \times 0.30 \text{ moles}}{5.0 \text{ L}} = 2.112 \text{ round } 2.1 \text{ moles}$

Answer   There were 2.1 moles of gas added to the balloon.

Molar volume

Avogadro's law

Avogadro's law states that doubling the number of gas particles (moles) results in doubling the amount of space they take up (volume). Equal volumes of different gases will have the same number of gas particles no matter what gas is present, as long as the temperature and pressure of both gases are the same. This applies only to gases, not liquids or solids. But why doesn't the identity of the gas matter?

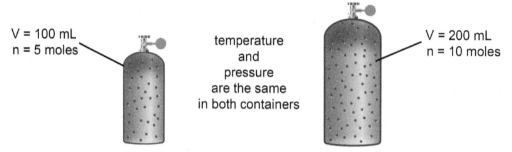

Gas particles and space

Gases are about 1,000 times less dense than solids and liquids. There is so much empty space between particles in a gas that the size of an individual particle has little to do with the amount of space the gas will take up. The size of the container has the biggest influence on how much volume a gas will take up. Since the particles in liquids and solids touch each other, there is almost no empty space between particles. This is why liquids and solids are not easily compressible. The interactions and the distance between particles in solids and liquids are important when considering volume.

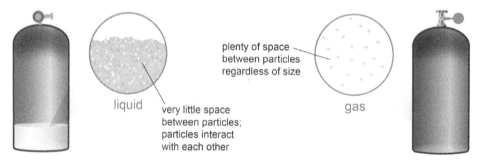

Particle size

Temperature is directly related to the kinetic energy of particles. If two gases are at the same temperature, they must have particles with the same average kinetic energy. If one gas has heavier particles, its particles will move more slowly than the gas with lighter particles. The two different gas particles will exert the same force when they hit the wall of their container. This means both gases will exert the same pressure for the same number of particles and occupy the same volume. This is why at standard temperature and pressure (STP, 0 °C and 1.00 atm) one mole of any gas occupies 22.4 L. This is known as the **molar volume** of a gas.

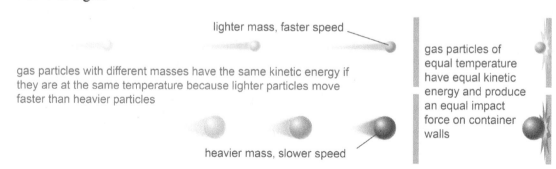

Section 2 review

Boyle's Law states that for a given gas, pressure is inversely proportional to volume. Pressure tends to equalize which helps explain the function of lungs, straws, vacuum cleaners and other instances where an increase in volume creates a region of low pressure. Charles' law states that for a gas at constant pressure, volume is directly proportional to temperature. This relationship allowed Lord Kelvin to mathematically determine absolute zero. Boyle's law and Charles' law together form the combined gas law relating pressure, volume and temperature for a given gas. Avogadro's law states that if temperature and pressure remain constant, the volume of a gas is directly proportional to the number of moles of gas present. This holds true regardless of the size of the individual gas particles.

Vocabulary words   Boyle's law, absolute zero, Charles' law, combined gas law, Avogadro's law, molar volume

Review problems and questions

1. What temperature will a gas reach when it is transferred from a 15-L bottle to a 9.50-L bottle? The initial gas temperature is 37.0 °C.

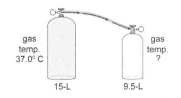

2. A student was solving the following problem, but they did not get the correct answer. "A gas at STP is warmed to 15 °C. Its new volume and pressure were measured as 8.31 L and 0.78 atm. What was the original gas volume?"
Here is the student's work:

$$V_1 = \frac{P_2 V_2 T_1}{P_1 T_2} = \frac{0.78 \text{ atm} \times 8.31 \text{L} \times 0\,°\!C}{1 \text{ atm} \times 15°\!C} = \text{the wrong answer!}$$

What did the student do wrong?

3. More moles of air are added to a container with an adjustable volume. Temperature is constant.
   a. What will happen to the container volume?
   b. Which gas law best describes the changes in this system?

4. A container with an adjustable volume holds a fixed number of moles of gas at a constant temperature. When the container is made smaller,
   a. What will happen to the gas pressure?
   b. Which gas law best describes the changes in this system?

5. Which gas law should you use to solve a problem where pressure, volume and temperature are all changing at the same time?

12.3 - Kinetic Molecular Theory

An **ideal gas** is like the ideal song or ideal food. It does not exist; however, most gases come very close to being ideal. Ideal gases are described by the kinetic molecular theory.

Kinetic molecular theory

The kinetic molecular theory

The **kinetic molecular theory** assumes gases are ideal. The following assumptions must be made about a gas according to the kinetic molecular theory:

1. Particle motion. Gas particles act like spheres that are constantly moving in random directions.
2. Particle collisions. A gas particle travels in a straight line until it either strikes another particle or it hits the walls of its container.
3. Particle size. Gas particles are tiny. There is always plenty of space between a gas particle and a neighboring particle or container wall.
4. No attractive or repulsive forces. The motion of one gas particle is not influenced by either attractive or repulsive forces of a neighboring gas particle. Gas particles move independently of one another.
5. Elastic collisions. A gas particle eventually collides with another gas particle or with the walls of its container. These collisions are elastic, meaning there is no net loss of energy from the system.
6. Kinetic energy. Temperature is defined as the average kinetic energy of a sample. Therefore the average kinetic energy of any gas sample depends directly on the temperature of the gas.

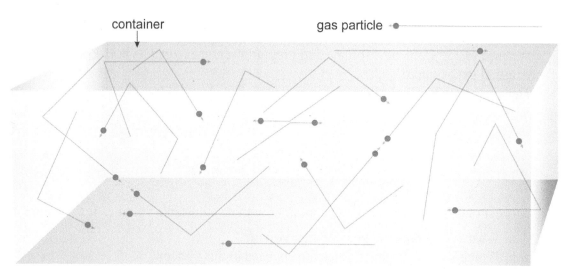

Why we use the theory

So far we have considered all gases to be ideal. However, this theory is a description that does not describe real gases. It is an approach to mathematically model the way gases probably behave on average. Making the assumptions of the kinetic molecular theory makes calculating gas temperature, volume and pressure far easier without much compromise in the final result. For example, the volume of a gas particle is insignificant compared to the space between the particles. So if we do the math required to account for the size of gas particles, it would not make much difference in the final outcome. It would be a lot of work for a low-benefit outcome.

The ideal gas law

The ideal gas law

Boyle's law works only if the moles and temperature of a gas are held constant. Charles' law works only if the moles and pressure of a gas are held constant. Avogadro's law works only if the temperature and pressure of a gas are held constant. These conditions are not difficult to produce in a controlled laboratory setting, but outside of a lab it is more realistic to expect pressure, temperature, volume, and sometimes moles of gas to change all at the same time. We eliminated some constant conditions by combining Boyle's law with Charles' law. We can have a gas law that works for all gases under all temperature and pressure conditions if we also include Avogadro's law. This law is called the ideal gas law, and the equation is: PV = nRT.

Ideal Gas Law

P = pressure
V = volume
n = moles
T = temperature (Kelvin)
R = universal gas constant

$$\frac{PV}{nT} = R \quad \text{or} \quad \frac{P_1V_1}{n_1T_1} = \frac{P_2V_2}{n_2T_2}$$

usually written as: PV = nRT

---

combined gas Law

constant moles: $\dfrac{P_1V_1}{T_1} = \dfrac{P_2V_2}{T_2}$

constant P and moles: $\dfrac{V_1}{n_1} = \dfrac{V_2}{n_2}$

Avogadro's Law

all variables change

$$\frac{P_1V_1}{n_1T_1} = \frac{P_2V_2}{n_2T_2} = R$$

or

PV = nRT

ideal gas law

The Universal Gas Constant: R.

The universal gas constant, R, is the same value under all conditions.

The universal gas constant, R

Like the other gas laws, the ideal gas law requires a calculated constant. However, this constant has the same value for all gases regardless of the initial conditions. That is why the constant is called the **universal gas constant**, represented by the letter "R." Like pi (π), R is a fundamental constant of nature. Only the units used to calculate "R" affects its value. The table below presents two different values for "R" based on different pressure units: atm vs. kPa. A good approach is to choose the version of "R" that matches given units of pressure. If you discover one or more given values do not match the units given in "R" when solving a problem, perform the necessary unit conversions before completing the calculation.

the universal gas constant with different pressure units

V = L	P = atm	V = L	P = Pa
P = 1.00 atm	n = 1.00 mol	P = 101.325 kPa	n = 1.00 mol
V = 22.4 L	T = 273 K	V = 22.4 L	T = 273 K

$$R = \frac{PV}{nT} = \frac{1.00 \text{ atm} \times 22.4 \text{ L}}{1.00 \text{ mol} \times 273 \text{ K}}$$

$$R = 0.0821 \; \frac{\text{atm} \times \text{L}}{\text{mol} \times \text{K}}$$

$$R = \frac{PV}{nT} = \frac{101.325 \text{ kPa} \times 22.4 \text{ L}}{1.00 \text{ mol} \times 273 \text{ K}}$$

$$R = 8.31 \; \frac{\text{kPa} \times \text{L}}{\text{mol} \times \text{K}}$$

The ideal gas law with fixed conditions

Ideal gases

The ideal gas law is only perfectly accurate for an ideal gas, one in which the particles take up no space and have no interactions with each other. This is important when particles are close together, such as when gases are under high pressure and/or low temperature.

Real gases

An ideal gas experiences no attractive forces, but a real gas cooled to very low temperatures may have particles that experience Van der Waals attractions to some extent. An ideal gas would not condense into a liquid at the condensation point as real gases do because there would no sufficient particle attractions. But, the ideal gas law can be quite accurate for real gases of sufficiently high temperature and/or low pressure.

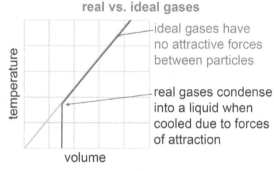

real vs. ideal gases
- ideal gases have no attractive forces between particles
- real gases condense into a liquid when cooled due to forces of attraction

PV = nRT

If you are studying a gas where the conditions are not changing, you can use the $PV = nRT$ version of the ideal gas law. When you do not have comparative final and initial temperature, pressure, etc. values, conditions are not changing. Review the solved problem below to help you understand how to use the universal gas constant with the ideal gas law, $PV = nRT$.

Solved problem

What is the pressure in a 300.0 mL container filled with 150.1 g of argon gas at 23.0 °C?

Asked: What is the tank pressure under these unchanging conditions?

Given: V = 300.0 mL; mass = 150.1 g Ar; T = 23.0 °C

Relationships:
- $PV = nRT$
- $R = 0.0821 \frac{atm \times L}{mol \times K}$
- Molar mass of Ar = 39.948 g/mol

Solve: Convert the given quantities to match the units of R: Convert volume to L; convert temperature to K and mass to moles.

V = 0.3 L, T = 296 K,

$$n = \frac{150.1 \text{ g Ar}}{1} \times \frac{1 \text{ mole Ar}}{39.948 \text{ g He}} = 3.757 \text{ mol Ar}$$

Rearrange $PV = nRT$ as $P = \frac{nRT}{V}$

Plug quantities into the equation:

$$P = \frac{3.757 \text{ mol Ar} \times 0.0821 \frac{atm \times L}{mol \times K} \times 296 \text{ K}}{0.3 \text{ L}}$$

Answer: The pressure of the tank is 304 atm.

Interactive simulation

Investigate the behavior of an ideal gas when moles or temperature of a gas change in the kinetic molecular theory interactive simulation. Change the number of gas molecules or the temperature of gases on either side of a piston. Watch the piston and particles carefully to find evidence of the kinetic molecular theory.

The ideal gas law with varying conditions

Changing conditions

The ideal gas law contains all the gas laws except for Dalton's law. You can use the ideal gas law to solve any problems that involve physical changes in gases other than total pressure problems. You just need to choose the correct version of the ideal gas law. When you solve gas problems where conditions are changing, the universal gas constant "R" is eliminated from the equation. You can use any pressure and volume units you want as long as the initial and final conditions use the same units. However, you still must convert temperatures to the Kelvin temperature scale. Any conditions not mentioned in the problem are assumed to be constant, so the ideal gas equation is able to be simplified into several of the basic gas laws. We can sum up the problem-solving procedure for gases with changing conditions into two steps:

steps for using $\dfrac{P_1 V_1}{n_1 T_1} = \dfrac{P_2 V_2}{n_2 T_2}$

Step 1: Simplify
You can simplify the equation by cancelling out anything that stays constant. For example, if volume is not changing, then V_1 and V_2 both cancel out.

Step 2: Solve
Plug in known quantities for remaining variables in the simplified equation and solve for the unknown variable.

Solved problem

How high does the pressure of a sealed bag of pretzels at 100.291 kPa get when you move it from inside a 20.2 °C grocery store to your car, which is 41.2 °C inside?

Asked What is the new pressure of the bag under changing temperature conditions?

Given $P_1 = 100.291$ kPa; $T_1 = 20.2$ °C; $T_2 = 41.2$ °C

Relationships
- 20.2 °C = 293.2 K; 41.2 °C = 314.2 K
- Ideal gas law under changing conditions:

$$\dfrac{P_1 V_1}{n_1 T_1} = \dfrac{P_2 V_2}{n_2 T_2}$$

Solve Neither moles nor volume are changing, so you can cancel both from the equation:

$$\dfrac{P_1 \cancel{V_1}}{\cancel{n_1} T_1} = \dfrac{P_2 \cancel{V_2}}{\cancel{n_2} T_2}$$

Substitute values: $\dfrac{100.291 \text{ kPa}}{293.2 \text{ K}} = \dfrac{P_2}{314.2 \text{ K}}$

Rearrange and solve for P_2:

$$P_2 = \dfrac{100.291 \text{ kPa} \times 314.2 \cancel{\text{ K}}}{293.2 \cancel{\text{ K}}}$$

$P_2 = 107.47419$ round → 107 kPa

Answer The pressure in the bag will go up to 107 kPa.

Chapter 12

Section 3 review

In chemistry we treat gases as if they are ideal, meaning that they follow specific rules governed by kinetic molecular theory. This allows us to mathematically model gases and make reasonable calculations about their behaviors. Combining the results from Boyle's law, Charles' law and Avogadro's law creates a formula that works regardless of which of the variables of a gas change. This is known as the ideal gas law ($PV = nRT$ where R is the universal gas constant). This law is most accurate for gases at relatively high temperatures and/or low pressures. Under those conditions, real gases adhere to the rules for ideal gases more closely.

Vocabulary words   ideal gas, kinetic molecular theory, universal gas constant

Key relationships
- Ideal gas law: $PV = nRT$ or $\dfrac{P_1 V_1}{n_1 T_1} = \dfrac{P_2 V_2}{n_2 T_2}$ or $P_1 V_1 n_2 T_2 = P_2 V_2 n_1 T_1$
- Universal gas constant, $R = 0.821 \, \dfrac{\text{atm} \cdot \text{L}}{\text{mol} \cdot \text{K}}$ or $R = 8.31 \, \dfrac{\text{kPa} \cdot \text{L}}{\text{mol} \cdot \text{K}}$

Review problems and questions

1. You need to use the ideal gas law to solve for the temperature of a gas where pressure, volume, and amount of gas are given. The given amounts are as follows: Volume = 631.4 mL; Amount = 8.21 g neon gas; Pressure = 1.31 atm. Which version of the universal gas constant should you use (select from the Key Relationships section above)?

2. If you were to solve the problem presented in Question #1, which of the givens needs to be converted, and what units should they be converted to in order to cancel with the units in "R"?

3. What mass, in grams, of hydrogen gas are present in a 510 mL container at 38.7 °C and 101.752 kPa?

4. What conditions would have to be held constant in order to derive Avogadro's law from the ideal gas law? Support your answer by showing how to derive Avogadro's law from the ideal gas law equation when conditions change.

5. The flasks below contain the exact amount of gas molecules. In which flask is the pressure highest?

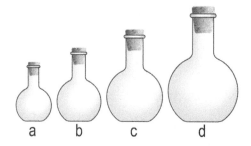

6. What happens to the pressure exerted by a gas if the volume of the gas increases at constant temperature?

Deep Dives and High Altitudes

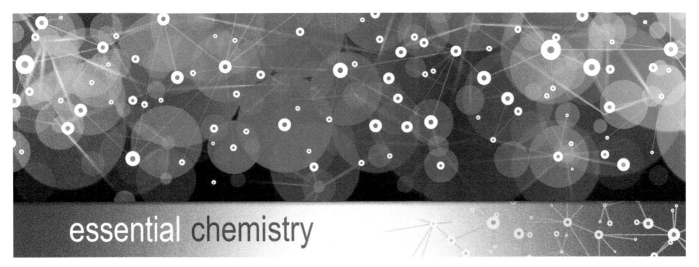

essential chemistry

Gas laws play an important role in the exploration of our environment. At sea level, atmospheric pressure is relatively high and this helps us breathe. The higher pressure enables air, and specifically oxygen from the air, to easily enter our lungs. In environments with higher or lower amounts of pressure, humans have adapted ways to help supply the appropriate amount of oxygen to our bodies.

As oceans cover most of the earth's surface, we have developed specific equipment to support underwater explorations. At sea level the atmospheric pressure is 1 atmosphere (atm) or ~14.7 pounds per square inch (psi). Just diving a few feet underwater results in an increase in pressure that you can feel in your eardrums. The deeper you go the greater the pressure pushing down on you. For every 10 meters of depth, there is an increase by 1 atm (14.7 psi).

For sustained exploration of more than a few feet of depth, divers use SCUBA gear. SCUBA is an acronym for Self Contained Underwater Breathing Apparatus. A standard SCUBA tank holds air that is compressed to 200 atm (3000 psi). It is essentially a mixture of 79% nitrogen and 21% oxygen to match the atmospheric composition that our lungs are used to. A regulator is used to make sure the air that divers breathe from the tank is at the same pressure as the surrounding water.

decreasing pressure **increases volume**

Safe practices of underwater exploration can be explained by the properties of gases. The air leaving the SCUBA tank needs to be at the same pressure as the surrounding water. When the pressures are equal, air moves from the tank to your lungs without causing damage. As divers descend, they breathe increasingly dense air because it is at a higher and higher pressure—there are literally more molecules of air in the same lung volume. Boyle's law explains why divers should exhale when they ascend. If divers held their breath during ascent, the constant amount of air in their lungs would increase in volume as the pressure decreased—this would not be good for your already inflated lungs!

Deep Dives and High Altitudes - 2

Surfacing too fast after a dive can result in decompression sickness, also known as "the bends." As a diver descends, the inert nitrogen from the SCUBA tank dissolves into blood and body tissues. At higher pressures, more nitrogen dissolves. This relationship between pressure and gas solubility is called Henry's law. You can easily see an example of Henry's law with carbonated drinks. When the drink is sealed there is dissolved carbon dioxide in the drink as well as a small volume of pressurized carbon dioxide gas in the container. Once the container is opened, the pressure is released, and whoosh the dissolved carbon dioxide gas comes out of solution and you get lots of bubbles. A diver ascending very quickly is similar to opening the bottle—but with dangerous effects. As the pressure drops during ascent, dissolved nitrogen is released as gaseous nitrogen into the body. Too much of this gaseous nitrogen in the body could cause blood flow blockages, and damage to blood vessels and nerves.

Going from the sea to the mountains, we see that functioning at high altitudes is another challenge related to gas properties. Overall atmospheric pressure decreases with increasing altitude. At 5,000 feet above sea level, the atmospheric pressure is about 0.83 atm, at 10,000 feet above sea level it is about 0.69 atm and on Mount Everest at 29,000 feet it is only about 0.33 atm. So, when they say the air is "thinner" at higher altitudes, they mean there are literally less molecules of air.

Decreasing overall pressure is a problem because it means that there is less available oxygen. The overall composition of the air remains the same regardless of the altitude: about 78% nitrogen, 21% oxygen and an assortment of other gases combining to add up to 100%. At sea level, the partial pressure of 21% oxygen is about 0.21 atm. But at an altitude of 29,000 feet, with a total pressure of 0.33 atm, the partial pressure of oxygen is only about 0.07 atm! High altitude climbers will often bring canisters of bottled oxygen because of this. The first successful summit of Mount Everest by Tenzing Norgay and Sir Edmund Hillary was aided by bottled oxygen.

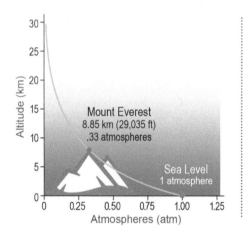

When training at higher altitude your body makes more oxygen-carrying red blood cells (RBCs).

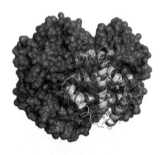

Each RBC has more than 250 million hemoglobin molecules.

Some runners, cyclist, and hikers train at high altitudes to improve their performance. When athletes first go to train at high altitudes, they must allow time (generally 1-3 days) for their bodies to adjust to the lower atmospheric pressure because their bodies need to function with decreased amounts of oxygen. The body adapts to ensure that enough oxygen gets to its tissues by making some changes, such as an increase in breathing rate and an increase in the number of oxygen molecules entering the lungs. The body also makes more hemoglobin that contains red blood cells, which in turn bind more of the available oxygen molecules and deliver them to the tissues. When athletes return to lower atmospheres, the increased ability to circulate more oxygen is thought to be a competitive advantage.

Understanding gas properties has allowed us to explore our world from the depths of the ocean to the tops of the mountains.

Chapter 12 review

Vocabulary
Match each word to the sentence where it best fits.

Section 12.1

atmospheric pressure	barometer
Dalton's law	pascal (Pa)
water displacement	

1. A(n) _____ is an instrument that measures atmospheric pressure.

2. _____ is a method of measuring the volume of the space gas occupies as it pushes liquid out of a container.

3. _____ is caused by the mass of air exerting a force on you as well as all around you.

4. Pressure can be measured in units of _____, the equivalent to one newton per square meter.

5. _____ states that in a mixture of non-reacting gases, the total pressure exerted is equal to the sum of the partial pressures of the individual gases.

Section 12.2

| Boyle's law | Charles' law |
| combined gas law | molar volume |

6. You will find 6.02×10^{23} molecules in a(n) _____ at standard temperature and pressure.

7. _____ states that the pressure and volume of a gas have an inverse relationship at a constant temperature.

8. The law that states that the volume of a gas is directly proportional to the temperature at constant pressure is _____.

9. The _____ is the law that allows you to solve for changes in a gas when pressure, volume, and temperature vary.

Section 12.3

| ideal gas | kinetic molecular theory |

10. The gas laws are only 100% accurate for a(n) _____.

11. Our basic understanding of substances at the molecular level is described by the _____.

Conceptual questions

Section 12.1

12. What is pressure of an object as it relates to force, assuming the force of gravity on the object does not change? What happens to pressure if the orientation of the object changes?

13. A gas exerts force across the walls of its container. What is the cause of this gas pressure?

14. Compare the particles of a gas, liquid, and solid on the molecular level. Base your comparison on:
    a. Particle density.
    b. Ability to compress.
    c. Ability to expand.

15. How is a gas able to exert pressure even though there are large distances between gas particles? What factors can increase or decrease the pressure that a gas is able to exert?

16. You have accidentally stepped in quicksand. If you were not able to immediately relocate, would it better to stay standing or to lie down? Explain.

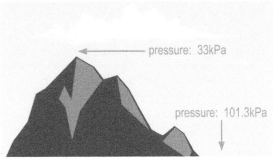

17. Why does atmospheric pressure decrease as you climb in elevation up a mountain?

18. List the properties of gases.

Section 12.2

19. As a sealed syringe is squeezed, what happens to pressure? Explain this on a molecular level.

20. Using what you have learned about gas particles and pressure, describe inhalation as it refers to suction.

21. Two gases have the same temperature and number of particles. Using your knowledge of Avogadro's Law, what else must be equal?

22. At standard temperature and pressure, all gases take up 22.4 L of volume. Why doesn't the size of gas molecules make a difference in gas volume under these conditions?

23. The kelvin is the primary unit of temperature measurement for the physical sciences. Explain why it is useful to use this temperature scale for calculations instead of Fahrenheit or Celsius.

24. A gas is sealed within a container that allows the gas to expand and contract. How can this be used to measure temperature?

25. A gas law states that pressure and temperature are directly proportional to each other as volume and moles remain constant. Using this relationship, predict and what will happen to the temperature of a gas as it increases in pressure. Explain your answer.

26. You should never throw a spray can into a fire or leave it in direct sunlight, even if the can is empty. Why is this?

Section 12.5: Chapter Review 403

Chapter 12 review

Section 12.2

27. Kinetic energy is the energy an object has due to its motion, provided by the random spontaneous movement of its molecules. Explain the relationship between kinetic energy and temperature.

28. Two gases have the same pressure and volume. Using your knowledge of Avogadro's Law, what else must be equal?

29. Use Avogadro's Law to explain why it gets harder to push air into a bicycle tire with a hand pump as the tire gets filled up.

30. Explain why it is impossible for a helium-filled balloon to rise to the top of the atmosphere without popping.

31. Why do you see more tire debris and blowouts on the road in warmer weather compared to cooler weather? Support your answer with an explanation of gas behavior inside a tire on the molecular level.

Section 12.3

32. Ideal gases are described by the kinetic molecular theory. List at least three different aspects of the kinetic molecular theory.

33. The kinetic molecular theory explains why gases have their properties. Describe how the theory explains this on a molecular level.

34. All of the gas laws consider gases to be ideal. Identify the properties of an ideal gas that are not necessarily true for real gases.

35. Water is dissociated into hydrogen gas and oxygen gas in the following reaction when an electric current is run through it.

$$2H_2O(l) \rightarrow 2H_2(g) + O_2(g)$$

If the volume of the products is measured, their ratio is found to be $2H_2 : 1O_2$, the same as the mole ratio in the balanced reaction. If the masses of the hydrogen and oxygen are measured, they are found to be in a mass ratio of $1H_2 : 8O_2$. Why is this?

36. Compare what happens to a real gas as it continuously cools down with what happens to an ideal gas.

37. Why use the Ideal Gas Law if an ideal gas doesn't exist?

Quantitative problems

Section 12.1

38. Use pressure equivalencies below to convert from one pressure to another: 14.69 psi = 1.000 atm = 760.0 mmHg = 760.0 Torr = 101.325 kPa.

   a. 6.57 atm to mmHg
   b. 33.5 psi to atm
   c. 7.00 psi to Pa
   d. 0.89 atm to kPa
   e. 100,435 Pa to atm

39. A tank has a total pressure of 300.0 kPa. It contains only three gases: 73.56 kPa of neon, 154.78 kPa of helium, and an unknown amount of xenon. What is the pressure of xenon in the tank?

40. A tank contains 2.4 atm of nitrogen and 7.1 atm of argon. What is the total pressure in the tank?

41. Indicate which represents the higher pressure in each of the following pairs:

   a. 487 mmHg or 64,745 Pa.
   b. 13.11 psi or 98.723 kPa.
   c. 1.67 atm or 1,120 Torr.
   d. 71.9 kPa or 487 mmHg.
   e. 154 lbs/in^2 or 9.98 atm.

42. When collecting oxygen gas and calculating its partial pressure by displacing water from an inverted bottle, the presence of water vapor in the collecting bottle must be accounted for. The total pressure of the jar is measured to be 745 mmHg and the temperature is 25.0 °C. If the pressure of the water vapor is 3.17 kPa, what is the pressure of the oxygen gas in mmHg?

43. A vehicle has its tires inflated to 32 psi. If it is touching the ground over an area of 125 in^2, how much does the car weigh?

44. Your 150-lb friend wants to go for a walk in the snow.

   a. If she wears stylish shoes that contact the ground over a total of 9.32 in^2, how much pressure does she exert on the ground?
   b. If she wears snowshoes instead, that pressure changes to 1 psi. What must be the total surface area of the snowshoes?

Section 12.2

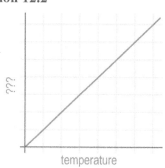

45. Does the graph above show two variables that are directly proportional to each other, or inversely proportional?

   a. Use your observation of the plotted line to support your answer.
   b. Assuming this data was measured from inside a self-adjusting container within which pressure does not change, what measurement must be on the y-axis? Explain your answer.
   c. Which gas law relates the relationship shown in the graph?

46. A balloon was inflated to a volume of 5.00 L at a temperature of 7.00 °C. If it is blasted by heat until it is 47.0 °C, what is its new volume?

Chapter 12 review

Section 12.2

47. A 0.50 ml CO_2 bubble gets heated from 26.4 °C to 150.2 °C. What is the new volume of the bubble?

48. A swim team member inflates a balloon to 2.00 liters at the surface and takes it on a dive to the bottom of the pool. The pressure is 1.21 atm at the bottom of the pool. Assume the pool is at sea level.
    a. What will happen to the balloon volume?
    b. What is the new volume of the balloon?
    c. Why is the pressure so high at the bottom of the pool?

49. A balloon is inflated to a volume of 10.0 liters at a pressure of 0.90 atm. It is sealed and released in the air. It floats up to an altitude where its volume becomes 50.0 liters.
    a. What happened to the pressure outside the balloon?
    b. What is the new pressure inside the balloon?

50. The passenger compartment of a spacecraft has a volume of 1.50×10^4 L at 770.0 mmHg. The spacecraft docks with a larger ship that has a volume of 2.50×10^5 L and no air in it. Assume the temperatures of both vessels are equal.
    a. What it the total volume of the connected spacecraft when the connecting door is opened between the two?
    b. What is the air pressure of the two spacecraft combined togeteher?
    c. Human lungs cannot function properly at the calculated pressure so breathable air would need to be generated inside the combined spacecraft. At what pressure would it be safe for astronauts to remove space suits?

51. What is the volume of oxygen gas if a 38.0 L sample at 27.5 °C is cooled to -90.5 °C?

52. What is the pressure of the gas contained in a 200.0 mL spray can if the pressure at 20.0 °C is 2.8 atm, and it is left in an environment that is 49.0 °C?

53. An aerosol can contains 0.50 L of compressed gas at 8.2 atm pressure. When the gas is injected into a sealed plastic bag, the bag inflates to a volume of 1.98 liters. What is the pressure inside the bag?

54. The first thermometers were made by trapping gas in a container that would allow the gas to expand and contract. The temperature could be measured by looking at the volume of the gas. For one particular thermometer, the known volume of gas is 135.3 mL at a known temperature of 11.0 °C.
    a. What variables would need to be constant in order for the gas thermometer to be most accurate?
    b. If you saw that the thermometer gas volume was 141.2 mL, what would the Celsius temperature be?

55. Draw a graph of two physical measurements of gases that are inversely proportional to each other. Which gas law would this graph illustrate? Explain your answer.

56. You are going to SCUBA dive to a depth where the pressure is 3,016 mmHg and the temperature is -4.00 °C. If you were to take a beach ball with you, what would its new volume be if it contained 2.00 liters of air under a pressure of 748.0 mmHg and a temperature of 26.7 °C at the surface?

57. You blow up a balloon to 100.0 mL in a room that is 27.0 °C. If you take the balloon outside where it reaches equilibrium with an air temperature of -3.0°C, what is its new volume?

58. You fill a balloon with helium to a size of 6.0 liters at standard temperature and pressure. After you tie the balloon shut, you accidentally let go and it floats higher into the atmosphere until the pressure is 0.37 atm and the temperature drops to -14.7 °C. What is the balloon's volume at its new altitude?

59. A 1.30 cm^3 bubble escapes from the bottom of a large fish tank where the pressure is 1.12 atm and temperature is 26 °C. What is the bubble's volume just before it escapes the surface of the fish tank water at 22 °C and standard pressure?

60. You place a sandwich in a zip-sealed bag at standard pressure, and it traps 50.0 mL of air. When you drive up to the mountains, the pressure drops to 700.0 mmHg. Assume the sandwich bag is air-tight and the sandwich stays in the bag inside your temperature-controlled car.
    a. What happens to the size of your sandwich bag?
    b. Support your answer with a calculation to quantify the new volume taken up by the air.
    c. You take the same drive again with another sandwich, except you squeeze all but 12.0 mL of air out of the sandwich bag. The car windows are down on this trip, so temperature changes from 25.2 °C to 19.3 °C. How much does the volume of air in the bag change this time?

61. Suppose you invert an empty beaker and fully submerge it in 26.5 °C water. You can see there is about 250.0 mL of air in the beaker. If the water is heated to 84.3 °C, how much water will get displaced to accommodate the expanding air? Explain your answer.

62. A container hidden on another planet holds 13.1 liters of air during the night. During the day it holds 15.0 liters of air at 31 °C.
    a. What is the night temperature on the planet, in Kelvin?
    b. in °C?

Section 12.3

63. The air pressure of a helium-filled hot air balloon over 3,000 feet in the sky is 89.06 kPa. There are 2.70×10^5 L of hot helium in the balloon at a temperature of 74.0 °C. How many moles of helium are in the balloon?

Chapter 12 review

Section 12.2

64. How many moles of air do you need to add to a balloon to change its volume from 1.5 L to 2.2 L? The balloon already has 0.06 mol of air.

65. Changes in pressure due to changing altitude or an incoming weather system can affect your body. There is on average 0.75 cm³ of air inside your middle ear at any time.
    a. If a massive storm is headed your way, air pressure can drop from 1 atm to 0.89 atm very quickly. What new amount of space will the air in your middle ear take up?
    b. Do you think you would notice this change in volume? Why or why not?

66. How much volume will 3.50 moles of helium gas take up, if it is at STP? How does this compare to the volume an equal number of moles hydrogen gas will take up, under the same conditions?

67. What is the volume of 7.89 g of O_2 gas at 37.3 °C and a pressure of 0.882 atm?

68. A tire has 0.5 L of air and 0.022 mol of nitrogen gas, N_2. What will the new volume be if you add 0.022 more moles? Support your answer with a calculation.

Section 12.3

69. Use the ideal gas law to calculate the volume of 1.00 mole of any gas at standard temperature and pressure.

70. You are trying to fill a 100.0 liter tank to a pressure of 250.0 atm at 27.0 °C. What mass of CO_2 is needed to fill the tank?

71. How many grams of oxygen gas would you have if you had a volume of 76.38 L under a pressure of 432.89 mmHg at standard temperature?

72. A 227.6 L sample of helium gas at 36.0 °C is cooled at a constant pressure to -84.0 °C. What is the new gas volume?

73. At what temperature does 4.00 g of H_2 occupy a volume of 40.0 L at a pressure of 1.20 atm?

74. At what temperature would 3.10 moles of CO_2 gas have a pressure of 1.50 atm in a 30.0 L tank?

75. A weather balloon can inflate to hold 6,000.0 L of helium gas. Knowing that the gas will expand greatly as it rises and the pressure decreases, you put 15.0 moles of He gas into the balloon. What is the pressure when the balloon has completely filled with gas after it rises to a point where the temperature is -5.0 °C

76. What is the pressure of a 248 liter tank containing 36.8 kg of argon gas at 27.0 °C?

77. You have 16.32 g of nitrogen gas stored in a 20.0 liter tank. What is the temperature if the pressure is 0.67 atm?

78. A canister of acetylene, C_2H_2 has a volume of 6.50×10^4 milliliters. The temperature and pressure of the gas are 420 Kelvin and 1.10 atmosphere respectively. How many moles of acetylene are in the canister?

79. A cartridge contains 0.931 moles of CO_2 gas. Calculate the volume of the gas if it has a pressure of 62.2 atm and a temperature of 305 K.

80. A gas contains 22.0 moles at a temperature of 450 K, and a volume of 100.0 liters, what is the pressure of the gas?

81. Chlorine gas is used in water treatment to disinfect drinking water, swimming pools, ornamental ponds and aquariums; sewage and wastewater, and other types of water reservoirs. How many moles of chlorine are in a tank that contains 7,500 L of chlorine gas, with a pressure is 241 atm and a temperature of 31.4 °C.

82. A SCUBA tank has a pressure of 195 atm at a temperature of 286 K. The volume of the tank is 390 L. How many moles of air are in the tank?

83. A tank of oxygen has a volume of 1,950 L. The temperature of the gas is 315 K. What is the pressure the gas if there are 9,750 moles of oxygen in the tank?

84. A tank of argon gas is held at a pressure of 9,901 kPa and a temperature of 902 K. If there are 28.0 moles of gas in the tank, what is the volume of the gas?

85. A 50-cubic feet neon cylinder contains 5.20×10^3 grams of neon gas. It has a pressure of 67 atm at a temperature of 54 °C. What is the volume of the tank?

86. « Many industrially important compounds, such as ammonia, nitric acid, organic nitrates (propellants and explosives), and cyanides, contain nitrogen. Apart from its use in fertilizers and energy-stores, nitrogen is a constituent of organic compounds as diverse as Kevlar used in high-strength fabric and cyanoacrylate used in superglue.
    a. If 16.8 grams of a 22.5-liter can of nitrogen gas are compressed to 748 torr. What is the temperature of the gas in the can.
    b. If a sample of nitrogen gas is collected in a 10-gallon container at a pressure of 2,280 mm Hg and a temperature of 765 °C. How many grams of nitrogen gas are present in this sample?
    c. What is the pressure in atm exerted by 2.84 moles of nitrogen gas in a 250.0 mL container at 58 °C?

87. What is the temperature, in °C, of a 7.53 L container that contains 7.06 moles of nitrogen gas and 8.49 moles of oxygen gas at 726 psi?

88. What is the pressure inside a 56.0 L tank containing 1,576.7 grams of helium at 20.6 °C?

89. A carbonated drink in a bottle contains 6.0 g of carbon dioxide. If carbon dioxide was compressed into the container at 1,216 mm Hg and 15.0 °C, what is the volume of carbon dioxide in the container?

Chapter 12 review

Section 12.3

90. On average, the 12-ounce soda cans sold in the US tend to have a pressure of roughly 120 kPa of carbon dioxide when canned at 4 °C, and 250 kPa when stored at 20 °C.

    a. What is the mass, in grams of carbon dioxide gas in the soda can at 4 °C?

    b. What is the mass, in grams of carbon dioxide gas in the soda can at 20 °C?

91. Use the table below as needed to answer the following questions.

Conversions
1.00 atm = 101,300 Pa
1.00 atm = 101.3 kPa
1.00 atm = 760 mmHg
1.00 atm = 760 torr
1.00 atm = 14.7 psi

    a. Calculate the pressure in pounds per square inch (psi) of a 6,655 L tank containing 1195 moles of gas at 300 °C.

    b. A tank of oxygen has a pressure of 25,000 kPa. The temperature of the gas inside is 35 °C. If there are 4.22×10^5 grams of oxygen in the tank, what is the volume of the gas?

    c. A 110 L ball is inflated to a pressure of 34.0 pounds per square inch (psi). If the ball contains 9.95 moles of air. What is the temperature in Celsius?

    d. Air pressure at the bottom of Death Valley, at 282 feet below sea level, is 776 mmHg. What is the pressure in atmospheres?

    e. If the gas in a container can support 1,640 liters and 330,238 pascals at 962 kelvin. What is the number of moles of gas in the container?

92. Soda manufacturers often inject cold liquid with pressurized carbon dioxide, then bottle the drink under high pressure. Each beverage is designed with a specific carbonation level for optimal flavor, and anti-microbial effects.

    a. A can of a typical cola has 3.7 L of carbon dioxide dissolved in the drink at a temperature of 298 K. The can has an internal pressure of about 55.0 psi. How many grams of carbon dioxide are contained in the can?

    b. A refrigerated can of lemon-lime soda has an internal pressure of about 30 pounds per square inch. Let's say the soda was carbonated to 2.0 L of CO_2 at a temperature of 5 °C. How many moles of CO_2 are present in the can?

chapter 13 Solutions

Water and air, two of our precious natural resources, are both examples of solutions – or uniformly mixed substances that are mixed on an individual particle level. Air is a solution of gases that consists mostly of nitrogen and oxygen. Natural water sources are a mixture of water and dissolved substances like minerals, salts, and gases.

What impact does human activity have on these natural resources? Human activity can result in unwanted, unhealthy, or even climate altering particles being dispersed into the atmosphere. A visible example of atmospheric pollution is smog. The term "smog" comes from a combination of two words, smoke and fog. Now it refers to the haze, a more complex mixture of various air pollutants that interact with sunlight to form ground-level ozone. This haze of chemicals hangs in the air over many cities in industrialized countries. Our waterways can also be impacted by human activity. The most visible example of this is an oil-spill. We clearly see this because the oil and water are not miscible. Since they do not mix, the less dense oil will form a thin layer on top of the water. Other pollutants that do mix with water, like lead ions or fertilizer runoff, can also get into waterways. These are not immediately visible like a layer of oil, but they can still be a significant health or environmental hazard.

The number of particles in the solution plays just as big a role as the types of particles that are mixed. If you had two separate cups of tea and added one small crystal of sugar to one cup and added two teaspoons of sugar to the other, both cups would be a homogeneous mixture of water, tea components and sugar, but they would taste very different! The same is true for pollution. It might not be possible to remove all of a contaminant from a resource, but we need to ensure that the amount is low enough to be safe. We are constantly looking for ways to mediate the impacts of human activities on our natural resources. It turns out that part of the solution to the problem of pollution is an essential understanding of ... solutions.

Chapter 13 study guide

Chapter Preview

A bit of sugar stirred into water will dissolve, while vegetable oil added to water will separate and float on top. Why do some substances mix together to form solutions while others do not? In this chapter you will explore the basic properties of solutions. A solution is one or more solutes dissolved in a solvent. Dissolved ionic compounds are separated into ions while dissolved molecules remain covalently bonded. Whether compounds will dissolve in a given solvent depends on the polarity of the two substances. Increasing temperature, pressure, surface area and adding kinetic energy can all increase the rate at which a solute dissolves. Molarity and molality are two useful measures of the concentration of a solution. Beer's law explains the relationship between light absorbance and solution concentration. The presence of solute generally increases density of a solution while lowering the freezing point and raising the boiling point.

Learning objectives

By the end of this chapter you will be able to:
- distinguish between ionic and covalent dissolving processes at the particle level;
- describe physical means that can be used to separate a solute and solvent;
- explain why "like dissolves like" and why certain substances are immiscible;
- use the solubility rules to determine when a precipitate will be formed in a chemical reaction;
- describe the factors that affect the rate at which a solute dissolves and what is meant by a saturated or supersaturated solution;
- calculate concentration using molarity as well as ppm or ppb; and
- calculate molality and use it to solve for changes in freezing point or boiling point.

Investigations

13A: Electrolytes

13B: Solution concentration

13C: Colored solutions

13D: Project: Design a Purification System

13E: Writing and Discussion Enhancement: Water Purification

Pages in this chapter

- 410 What is a Solution?
- 411 Solution separation techniques
- 412 The dissolving process
- 413 Like dissolves like
- 414 Solids and gases in solutions
- 415 Electrolyte solutions
- 416 Section 1 Review
- 417 Solubility and Solution Formation
- 418 Saturated solutions
- 419 Solubility curves
- 420 Factors that affect solution formation rate
- 421 Effect of temperature and pressure on solubility
- 422 Solubility in reactions
- 423 Section 2 Review
- 424 Solution Concentration
- 425 Molarity of a solution
- 426 Calculating the molarity of a solution
- 427 More solution calculations
- 428 Working with very dilute solutions
- 429 Preparing a solution from a solid solute
- 430 Dilution: Preparing a solution from an existing solution
- 431 Using color to measure concentration
- 432 Section 3 Review
- 433 Properties of Solutions
- 434 Characteristics of solutions
- 435 Colligative properties
- 436 Molality
- 437 Calculating colligative effects
- 438 Section 4 Review
- 439 Don't Drink the Water
- 440 Don't Drink the Water - 2
- 441 Chapter review

Vocabulary

solvent	solute	dissolved
osmosis	reverse osmosis	desalination
chromatography	dissolution	dissociation
solvation	hydration	miscible
immiscible	diffusion	electrolyte
conductivity	concentration	dilute
concentrated	solubility	insoluble
soluble	saturated	aqueous equilibrium
supersaturated	Henry's law	molarity (M)
colorimeter	Beer's law	heat of solution
colligative property	molality (m)	

13.1 - What is a Solution?

A solution is a homogeneous mixture made of two or more elements or compounds. There are no components bigger than a single particle. Solutions can be in any phase as long as the solution particles are evenly distributed throughout the sample. Steel is a solid solution made of iron and other metals. Dry air is a solution of many gases, mainly nitrogen and oxygen. Many solutions in our bodies are made of water and ions. These solutions keep the body functioning properly even under harsh conditions.

Solutes and solvents

Copper(II) sulfate

Solid copper(II) sulfate ($CuSO_4$) forms a bright blue solution when it is dissolved in water. The solution contains liquid water molecules and the ions that make up copper(II) sulfate.

solid copper(II) sulfate
$CuSO_4$ (s)

copper(II) sulfate
aqueous solution

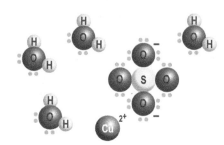
copper ions dissociate from
sulfate ions in an aqueous solution

Parts of a solution

A solution contains a solvent and one (or more than one) solute. The **solvent** is the substance that makes up the majority of the solution. As shown above, the copper(II) sulfate solution uses water as the solvent. The **solute**, copper(II) sulfate, is the dissolved substance in the solution. Although solutions can have more than one solute, there is only one solvent in any solution, and it dissolves the solute.

Distribution of particles

If all the solute particles are able to spread evenly throughout the solvent without clumping or settling, the solute has fully **dissolved**. A copper(II) sulfate solution is transparent with a blue tint. The solution will remain transparent in spite of the blue tint as long as all of the solute particles can dissolve and evenly distribute. Anywhere particles are not evenly distributed, or if too much solute is added, the solution will be less transparent.

Transparency of liquids

Liquid solutions are usually transparent because the solutes are small particles. Milk is not transparent and not a solution. If you observed a sample of milk with a microscope, you would see different-sized fat clumps. Each clump contains thousands of fat molecules that are not evenly spread out. Milk is opaque because the fat clumps are so large they scatter light instead of letting it continue along its path. Milk is a heterogeneous mixture, not a solution.

Everyday "solvents"

You probably know the word solvent from everyday language. It usually means a liquid such as turpentine, alcohol, mineral spirits, acetone (nail polish remover), and similar chemicals. In chemistry, the term "solvent" means anything that dissolves something else. Water is an important solvent in chemistry, and it is the most important solvent to living things.

Solution separation techniques

Filtration, distillation, and evaporation

Solutions can be separated by a number of different physical means. Either distillation or evaporation can be used to separate a solution. Both methods rely on either the solute(s) or solvent to change to a gas and leave the solution. Dissolved particles in a solution are too small to get stuck on filter paper, so you can't use filtration to separate solutions. However, a process called reverse osmosis can be used. **Osmosis** is the tendency of a solution to evenly distribute solutes within the solvent, where the solvent will move to areas of high solute concentration. A semipermeable membrane allows only water molecules to pass through it. In the osmosis part of the diagram, water moves through a semipermeable membrane to the salt water side where solute concentration is higher.

Reverse osmosis

Reverse osmosis is a method used to purify water by using pressure to force water through a semipermeable membrane towards lower solute concentration. The Reverse Osmosis part of the diagram shows when pressure is increased on salt water, water molecules are forced to pass through the membrane to the fresh water side. Salt cannot go through the membrane, so the salt water increases its concentration as water moves to the fresh water side.

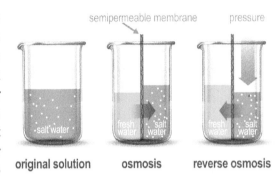

Desalination

Desalination is a process of using reverse osmosis to convert sea water into fresh water. This is an expensive way to purify water on a large scale because it takes a lot of energy to keep large amounts of water at high pressure. Also, the excess salt left behind can threaten ocean ecosystem health. Nevertheless, the threat of long-term drought has made desalination an increasingly popular water purification alternative.

Chromatography

Most pens and markers contain ink that is a solution of many different colors. An ink solution can be separated with a process called **chromatography**. Solutes become separated when they travel through a medium at different rates. A solute's particle size and degree of attraction to the solvent medium determine how far it will travel up the chromatography strip. The diagram shows separation of a water-based ink solution using paper and water. Water is called the mobile phase because it travels from the bottom of the paper strip towards the ink spot. The paper strip is the stationary phase because it does not move. If the ink was non-polar water would not work as the solvent. A non-polar solvent such as acetone would need to be used instead.

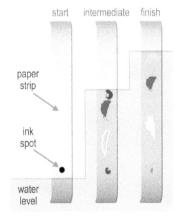

Separation by attraction

Ink molecules that are least attracted to water settle towards the bottom of the strip. Molecules with the greatest attraction to the mobile phase are the most soluble, so they move highest up the paper. In the diagram, the black ink is shown to be a solution of yellow, pink, and blue ink solute molecules. The blue ink traveled the farthest with the mobile phase because it had the highest attraction to the water.

The dissolving process

Water is the universal solvent

Water dissolves so many molecular and ionic substances it is known as the universal solvent. Water is a good solvent because H_2O molecules are small and polar. Polar water molecules are attracted to charged ionic particles in salts and polar regions of molecular substances like sugar. The solution formation process where a solute is dissolved by a solvent is also called **dissolution**. Either ionic or covalent compounds can undergo dissolution. Let's investigate the dissolution of a salt solute in water (the solvent).

Water dissociates ionic solutes

Table salt (NaCl) dissolves in water because polar water molecules are attracted to charged Na^+ and Cl^- ions in a salt crystal. The negative side of a water molecule is attracted to a positive Na^+ ion and the positive end of a water molecule is attracted to a negative Cl^- ion. When a water molecule contacts an ion, the particles transfer energy which causes the ions in the salt crystal to separate in a process called **dissociation**. Once water's attractive forces have pulled an ion away from the crystal, the ion is immediately surrounded by the oppositely charged sides of more water molecules. Because of their strong bonds, covalent molecules do not undergo dissociation. The solvent surrounds the molecule but the molecule stays intact.

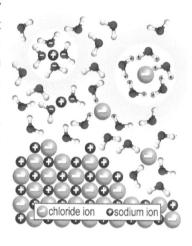

Hydration is a type of solvation

Solvation is the process where a solvent surrounds solute particles. Solvation of a salt in water is called **hydration** because water is the solvent. Hydrated sodium and chlorine ions stay in solution because their accompanying polar water molecules are strongly attracted to the ions. All of the salt will dissolve if there are enough water molecules around.

Interactive simulation

Explore how water molecules orient themselves around cations and anions during dissociation in the interactive simulation titled Dissolving Salt in Water. Add water to dissolve the ionic solute and then begin the simulation. Investigate particle interactions with charges on and off.

Solvation of molecules

A type of covalent molecule called glucose dissolves in water. Glucose molecules have regions of positive and negative charge. Solvation occurs when glucose is mixed with water because the charged areas attract many water molecules. The water molecules orient themselves in a way that cancels the charged parts of glucose making it neutral. Unlike ionic substances, glucose molecules are not separated or dissociated into ions. Molecules stay together during solvation.

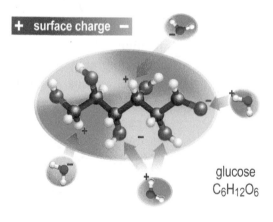

Dissolving is a kinetic process

Dissolving is a kinetic process because solvent molecules transfer energy when they collide with solute molecules. Any kinetic process is affected by changes in kinetic energy—specifically, temperature. Hotter solvent molecules have more energy and are more effective at bringing solute molecules into the solution. You may have noticed that a powder dissolves much more easily in hot liquids than in cold liquids.

Like dissolves like

Solvents other than water

Water is a very important solvent due to its ability to dissolve a variety of substances, but it cannot dissolve everything because not all solvents are polar. Non-polar solvents such as hexane (C_6H_{14}) are found in glue, de-greasers and spot removers that you use to treat stains on fabrics. Hexane is used heavily in industry for a number of processes. It is a non-polar molecule derived from oil, so it is good at dissolving non-polar solutes. Non-polar solutes that do not dissolve in water may dissolve in non-polar hexane instead.

Polar and non-polar do not mix

Oil molecules are non-polar and do not mix with water because water is polar. However, oil is **miscible** in non-polar mineral spirits because it dissolves easily to form a mixture. The phrase "like dissolves like" means solutes and solvents must have "like" or similar polarity to form a solution. Polar solvents dissolve polar solutes, and non-polar solvents dissolve non-polar solutes. Polar and non-polar do not mix.

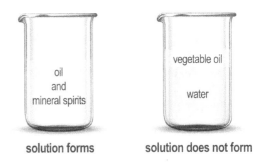

General rules for solvents

Solubility depends on the polarity of the solvent and solute(s). Polar solvents like water can usually dissolve ionic solutes because ions are charged and are able to attract water. Polar and non-polar substances are said to be **immiscible** because they are unable to mix and form a solution. Some solvents such as lightweight alcohols can dissolve either polar or non-polar solutes because their molecules have polar as well as non-polar regions. The table below lists common solvents other than water that you may recognize.

common solvents

name	formula	uses	properties
acetone	CH_3COCH_3	nail polish remover	flammable, toxic
methanol	CH_3OH	solvent, fuel	flammable, toxic
mineral spirits	C_6H_{14}	paint thinner	flammable, toxic
methylbenzene	C_7H_8	glue solvent	flammable, toxic

Interactive simulation

In the interactive simulation titled Like Dissolves Like, discover which substances are polar and which are non-polar based on the type of substance they dissolve in. If a particle in the simulation dissolves in oil, is it polar or non-polar? Which particles could be alcohols?

Section 13.1: What is a Solution?

Solids and gases in solutions

Solutions can be any phase

When you hear the word solution in a science class, you probably imagine a substance in the liquid phase. Solutions can form between solutes and solvents of different phases, such as a gas solute and liquid solvent, as long as the solute and solvent are compatible and are evenly distributed throughout the entire sample.

Solid solutions

Particles cannot easily move around in solids because they are locked in fixed structures. How do solid solutions form if a solvent must come in close contact with a solute to form a solution? Solids do form solutions, but the process is slow unless both solvent and solutes are melted to a liquid state. Once particles are freed from their structure they mix randomly by colliding with one another. When the solution cools back to a solid, particles self-arrange into a crystal structure according to attractions and repulsions. Alloys, like steel, are metal solutions in which iron is the solvent and elements such as carbon, nickel, and chromium are solutes.

Gas solutions

Molecules move around randomly and quickly in a gas, and for the most part we can ignore the attractive forces between gas particles. Gas solute-gas solvent solutions form by **diffusion** where gas particles collide with one another and mix randomly until all gas particles are evenly distributed in their container. In addition to temperature, pressure conditions have a significant impact on the speed of gaseous solution formation, especially when the solvent is a liquid and the solute is a gas.

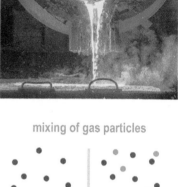

mixing of gas particles

before diffusion   after diffusion

Particle movement in liquids

Look around and you can see that most chemicals are either solid or gas at room temperature. Chances are most of the liquids around you are water-based solutions. This is even true inside your body. Sugar, salt, most proteins, and carbohydrates in your body are either solids or solids dissolved in liquid water. The particle arrangement in a liquid gives it an advantage over solids and gases when forming solutions. Liquids contain many molecules close together, so particles can easily move around and interact with each other. For this reason, most substances used by chemists are made into solutions before proceeding with a reaction.

Importance of water

Like all living things, thousands of complex chemicals interact in your body to keep you alive. Aqueous solutions make it possible for these chemicals to interact. A large number of these chemicals are either ionic or polar—that's why these reactions happen most efficiently in aqueous solutions where water is the solvent. Plasma is an aqueous solution that helps reactants circulate and reach each other throughout the human body. Plasma is the liquid portion of blood that is a solution of water (solvent) and dissolved salts and proteins (solutes).

Electrolyte solutions

Electrical conductivity

Ionic compounds dissociate or split apart into their constituent charged particles called ions when they dissolve to form a solution. Charged particles such as ions are capable of conducting electricity. Aqueous solutions are able to conduct electricity when they contain dissolved ions. An **electrolyte** is a solute that can conduct electricity when it is dissolved in an aqueous solution. You may have noticed a tag on a hair dryer that warns you of an electric shock if you use it near water. Tap water is an electrolyte solution because it contains dissolved ions like chloride, Cl⁻, that can conduct electricity from the electrical socket, to the hair dryer, through the water and anything touching the water including your body! This is why it is unwise to use any electrical device in a bathtub.

Two-ion dissociation

Some salts dissociate into exactly two individual ions: one cation and one anion. The ratio of ions produced in a solution relates to the ratio of ions in the chemical formula. Sodium chloride (NaCl) is one such salt. One mole of NaCl produces two moles of ions: Na^+ and Cl^-.

$$NaCl(aq) \longrightarrow Na^+(aq) + Cl^-(aq)$$
$$1.0 \text{ mol} \longrightarrow 1.0 \text{ mol} + 1.0 \text{ mol}$$

Three-ion dissociation

Magnesium chloride ($MgCl_2$) is an example of a salt that dissociates into three ions. One mole of $MgCl_2$ forms three moles of ions: one mole of Mg^{2+} ions and two moles of Cl^- ions. The ion ratio in the chemical formula is 1:2 ions, totaling three ions per formula unit. However, not all solutes dissociate into the full number of ions that make up a formula unit. The extent of formula unit dissociation depends on many factors including degree of attraction between the solute and itself versus the attraction between the solute and the solvent.

$$MgCl_2(aq) \longrightarrow Mg^{2+}(aq) + 2Cl^-(aq)$$
$$1.0 \text{ mol} \longrightarrow 1.0 \text{ mol} + 2.0 \text{ mol}$$

More ions conduct more electricity

As you can see, equal moles of NaCl and $MgCl_2$ will not produce equal moles of ions in a solution. If 1 mole of NaCl produces 2 moles of ions in a solution and 1 mole of $MgCl_2$ produces 3 moles of ions, the $MgCl_2$ solution can conduct more electricity than NaCl. A greater number of freely-moving ions results in a greater ability to conduct electricity.

Molecular substances do not dissociate

Can molecular substances such as sugar ($C_6H_{12}O_6$) conduct electricity? Molecular substances do not dissociate because their covalent bonds hold the molecule together as a complete unit when dissolved in solution. If there is no dissociation then no ions are formed. Although there can be charged areas on covalent compounds, molecular solutions lack freely-moving individual charged particles.

$$C_6H_{12}O_6(aq) \longrightarrow C_6H_{12}O_6(aq)$$
$$1.0 \text{ mol} \longrightarrow 1.0 \text{ mol}$$

Conductivity

The lack of charged particles means molecular solutions cannot conduct electricity. **Conductivity** in a solution is a measure of the amount of dissolved ions present in solution. Dissolved ions and molecules are too small to be seen, but a conductivity sensor can detect the presence of ions.

How do conductivity sensors work?

A conductivity sensor measures the flow of electricity between two points at the end of the probe. The flow of electricity is affected by the amount (and type) of ions present in the solution and the distance between the points. The sensor reports conductivity as microsiemens per centimeter (μS/cm).

Chapter 13

Section 1 review

A homogeneous mixture of two or more substances is called a solution. A solution is made up of one or more solutes dissolved in a solvent. Solutions can be separated through several means including: evaporation or distillation, reverse osmosis and chromatography. Water molecules are small and polar, making water a particularly good solvent for many things including ionic compounds. Polar solutes dissolve in polar solvents and non-polar solutes dissolve in non-polar solvents. Polar and non-polar substances do not mix to form solutions. In addition to liquids, solids and gases can also form solutions. When dissolved, ionic compounds separate into ions while covalent molecules remain intact. A conductivity sensor can measure the ability of a solution to conduct electricity.

Vocabulary words	solvent, solute, dissolved, osmosis, reverse osmosis, desalination, chromatography, dissolution, dissociation, solvation, hydration, miscible, immiscible, diffusion, electrolyte, conductivity

Review problems and questions

1. Brass is a solution of copper and zinc or other metals. Different types of brass are made by mixing different ratios of metals. Manganese brass is used to make golden dollar coins in the U.S. This type of brass is about 70% copper (Cu), 29% zinc (Zn), and 1% manganese (Mn). Classify Cu, Zn, and Mn as either a solute or solvent. Explain your reasoning.

2. On Planet Janet, widgets and gizmos are polar while doodads and whatchamacallits are non-polar. Will gizmos dissolve in doodads to form a solution? Why or why not?

3. How many ions will each of the following substances produce when dissolved in water?
    a. KCl
    b. CH_3OH
    c. $Mg(NO_3)_2$

4. Which of the compounds above will probably be the best conductor of electricity? Explain your answer.

5. In one complete sentence, summarize the difference between the way an ionic substance dissolves in water compared to the way a molecular substance dissolves in water.

6. If you are asked to separate sugar from a sugar water solution and you had no money to spend on materials, which physical means of separation would you choose: filtration, distillation, evaporation, reverse osmosis, desalination, or chromatography? Explain your choice.

13.2 - Solubility and Solution Formation

A **concentration** tells you how much solute is dissolved within an amount of solution. A **dilute** solution is made mostly of solvent with a small amount of solute. A **concentrated** solution has a lot of dissolved solute compared to the total amount of solution. **Solubility** is defined as the amount of solute that will dissolve in a solvent.

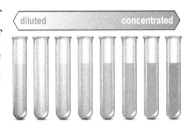

What is solubility?

Solutions are mostly solvent

You have probably heard of concentrated dish soap or fruit juice. The concentrated version of a product has more units solute per unit solvent (soap or fruit juice per water molecule) than the non-concentrated product. Keep in mind, a concentrated solution still has less solute than solvent because a solvent always makes up the majority of a solution.

Solubility

In the table below, barium sulfate has almost zero solubility in room temperature water because it does not dissolve to any great extent. In other words, barium sulfate is **insoluble** in that particular solvent at 20 °C. A **soluble** substance can be dissolved by a solvent. The temperature is always specified along with solubility because temperature strongly affects solubility. The solubility in the table below is given as a percentage of grams solute per 100 grams solution because this is the most accurate way to report solubility.

solubility of common substances in water at 20 °C

substance	solubility (g/100mL)	substance	solubility (g/100mL)
calcium hydroxide $Ca(OH)_2$	0.17	table salt (NaCl)	36.0
barium sulfate ($BaSO_4$)	0.00025	carbon dioxide (CO_2)	0.17
epsom salts ($MgSO_4$)	32.0	glucose ($C_6H_{12}O_6$)	81.0
baking soda ($NaHCO_3$)	9.6	table sugar ($C_{12}H_{22}O_{11}$)	230.9

Most solutions are more dense than solvents alone

The density of a solution varies with concentration. For most solutions, adding more solute raises the solution density because a greater number of particles will fit in the same amount of space. When solute molecules are attracted to solvent molecules, density increases because there is less space between particles. Another reason a solution becomes more dense than its solvent is because an added solute may have a higher density than the solvent in the first place.

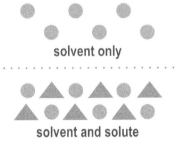

Volume is not additive

If you added 10 grams of solute to a solution, the mass of the solution will increase by 10 grams regardless of whether the solute dissolves or not. But adding 10 mL of a solid solute does NOT increase the volume by 10 mL because solute particles position themselves between solvent particles. The solution volume usually increases, but not as directly as the mass increases. This is the primary reason for changes in density when a solution forms. Liquid solutes are not always additive either. Sometimes the total solution volume ends up being lower than the individual volumes, sometimes higher, and sometimes equal depending on how the solute particles position themselves among the solvent particles.

Saturated solutions

What is a saturated solution?

If you add 220 grams of sugar to 100 mL of water at 20 °C, will all of the sugar crystals dissolve? You can use the table to see that only 201.9 grams will dissolve. The remaining 18.1 g will remain as a pile of undissolved solid sugar crystals that sink to the bottom of your solution. When a solution has dissolved the maximum amount of solute possible, it is **saturated**. According to the table, no more than 201.9 grams of sugar per 100 mL water will dissolve at 20 °C. Any additional solute added will not dissolve. For salt, only 35.9 grams will dissolve at the same temperature. You may have also noticed that a lot more sugar will dissolve with every 20 °C increase in temperature, but not as much for salt.

solubility of table sugar vs. salt in water

solution temperature (°C)	sugar solubility (g/100mL)	salt solubility (g/100mL)
0	181.9	35.7
20	201.9	35.9
40	235.6	36.4
60	288.8	37.0
80	365.1	37.9
100	476.0	37.9

Balance of dissolved solutes

If you create a saturated sugar solution at a raised temperature and hold the temperature constant, the concentration of dissolved sugar will stay the same. However, individual sugar molecules that are dissolved will not always remain dissolved in solution. Even though the saturated solution concentration is not changing, some solute particles re-form tiny solid solute crystals (recrystallize) with each other. The crystals return to solution when the solvent re-dissolves them. A saturated solution has reached **aqueous equilibrium** when the solute recrystallization rate equals the solute dissolution rate. If the solution is not saturated, dissolution happens faster than recrystallization.

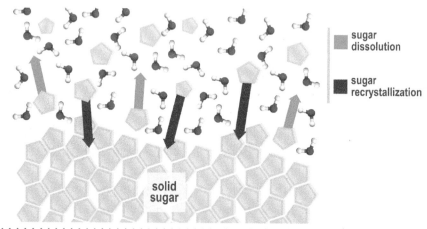

aqueous equilibrium: rate of dissolving = rate of crystal formation

Rates of dissolution and recrystallization

Let's look at a saturated sugar solution. When a molecule of dissolved sugar collides with some undissolved sugar, the dissolved molecule will either bump away and stay dissolved, or it will stick to the undissolved sugar molecules. If it sticks to the undissolved sugar molecules, it will precipitate out of the solution as a recrystallized solid. Recrystallization is not likely to happen frequently if a solution is unsaturated because there is a smaller chance of colliding with and sticking to undissolved molecules. When a solution forms, the amount of undissolved solid decreases and the amount of dissolved solute increases. The more solute there is in solution, the higher the chance that a solute molecule will come out of solution and become solid again.

Solubility curves

Saturated solutions

Consider what happens when you make a saturated sugar solution in water heated to 80 °C. The graph shows that you can dissolve about 365 grams of sugar per 100 mL of water. This is a lot more solute than solvent by mass, but because sugar is the substance that dissolves, it is the solute. The graph also helps you predict what will happen when the solution cools down to room temperature just over 20 °C. The graph shows only around 205 g of sugar are soluble at that temperature. What happens to the other 160 g of sugar?

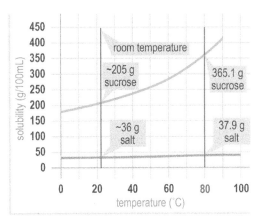

solubility of sugar and salt in water

Supersaturated solution

Only 205 g of sugar can stay dissolved in solution at 20 °C. The remaining solute particles are forced to come out of solution and recrystallize. As it cools, the solution becomes **supersaturated** because the solution contains more dissolved solute than it can hold. A solid sugar crystal known as a seed crystal is put in the supersaturated solution to help solute particles recrystallize on its surface. The seed crystal grows larger as the solution cools and the excess sugar solidifies. Rock candy is made by seeding a supersaturated solution.

Solubility curves

The graph below shows several solubility curves that indicate how much solute dissolves at a specific temperature. As you can see, temperature has varying impact on different solutes. The solubility of potassium nitrate (KNO_3) changes a lot, but not as much for other solutes such as salt (NaCl). Notice ammonia (NH_3) shows a sharp decline in solubility with increasing temperature. You may recall that ammonia is a gas. You will learn more about the solubility of gases in liquids in the next section.

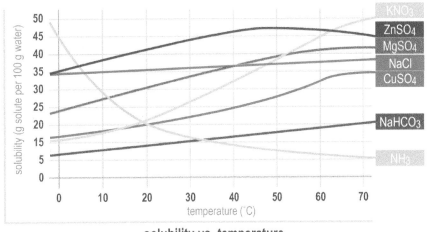

solubility vs. temperature

Section 13.2: Solubility and Solution Formation

Factors that affect solution formation rate

Forming solutions

Once you know a solution will form, you can consider ways to speed up the solution formation process. Dissolving is a physical process. Anything that increases the chances of effective contact between solute and solvent particles will speed up solution formation.

Temperature and pressure

For most solids and liquids, increased temperature speeds up solution formation rate. Greater kinetic energy increases the chances of particle collisions except for gases in a gas-liquid solution. Gases tend to leave a solution when they gain energy. Increased pressure will help liquid-gas solutions form, but pressure has little effect on solids and liquids.

Dissolving

Which dissolves faster: a teaspoon of loose sugar, or a teaspoon of sugar in a sugar cube? Loose sugar dissolves faster because the process of dissolving can only happen at the surface between solvent and solute particles. If solute particles are small, they can spread through the solvent more easily. Increased surface area helps solutes dissolve faster because more solute particles can contact the solvent. You might have noticed that drink mixes that need to be dissolved are crushed into small crystals or ground into a powder so they have a greater surface area. The image below shows what happens to surface area when a given volume is left unaltered, is cubed, and is powdered.

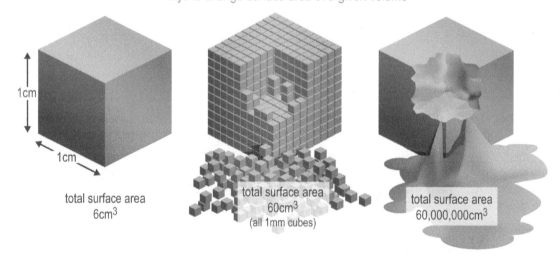

ways to change surface area of a given volume

1cm / 1cm
total surface area 6cm^3

total surface area 60cm^3 (all 1mm cubes)

total surface area 60,000,000cm^3

Increasing the rate of dissolving

Grinding sugar or salt into a powder does not change the amount of either compound that will dissolve in a given volume of water at a given temperature. In other words, increased surface area does not change solubility. If you add more powdered sugar to water than it can dissolve, the solution will become saturated and the extra solute will sink to the bottom of the container as a solid. Changing surface area makes the solute dissolve faster, but it doesn't change how much will dissolve.

Stirring a solution

You have known for a long time that stirring causes a solution to form faster, but now you can think about stirring in terms of kinetic energy. When you stir a solvent, you add kinetic energy which increases the chances of solute-solvent contact. Stirring affects the rate of solution formation, but it does not change the amount of solute that will dissolve.

Effect of temperature and pressure on solubility

Solubility and temperature

Higher temperatures increase the solubility of most liquid and solid solutes in liquid solvents. Warmer solvent particles have more energy. It is easier to overcome attractive forces that hold the solid together when the solvent has more energy. A warm solid is also easier to dissolve because the particles in the solid have more energy. Solids with more energy more easily separate from neighboring particles. Gas solubility generally decreases as temperature increases in an open system at standard pressure.

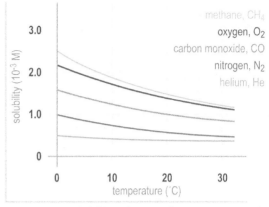

solubility of gases in water

Gases are less soluble in warm solvents

You may have seen tiny bubbles form in a pot of water being heated before it has reached boiling. Those bubbles are made of escaping trapped air. Gases already have minimal attractive forces, so the increased kinetic energy in a warm solvent gives gases enough energy to leave the solution. Increased ocean temperature is a concern for this reason. If gases are less soluble in warmer solutions, less dissolved oxygen will be available to support marine life.

Henry's law

As you know, soda goes flat after it is opened. That is because temperature and pressure both have an impact on the solubility of a gas in a liquid-gas solution. **Henry's law** states that at a given temperature, the solubility of a gas in a liquid depends on the concentration and partial pressure of that gas directly above the liquid. In other words, gases become more soluble in liquids with increased pressure. When soda is carbonated, CO_2 is forced into the liquid by increasing the pressure of CO_2 above the liquid.

A gas-liquid solution

In an unopened soda bottle the main gas directly above the liquid is a small volume of CO_2 that has dissolved out of the soda. You don't see any CO_2 bubbles because the contents of the bottle are under pressure and the CO_2 remains dissolved in the soda. When opened, CO_2 gas rushes out as you can tell by the sound it makes. Henry's law predicts CO_2 will become less soluble in the soda because the pressure goes down, CO_2 will leave the soda solution to spread out into the air. The evidence for this is the appearance of bubbles as CO_2 leaves solution

Bubbles = CO_2 gas leaving the soda solution

Solubility in reactions

Solubility rules

Intermolecular forces of attraction between solute and solvent particles have the largest impact on whether a solute is soluble or not in a particular solvent. For molecular compounds, like dissolves like. Let's revisit the solubility rules from a previous chapter. The solubility rules are for aqueous solutions, so the solvent is always water. Not all ionic compounds are water-soluble. If water's attraction to ionic solute particles can overcome the forces holding the solute together, the solute will dissolve. This explains why some aqueous reactants form insoluble precipitates as products.

solubility rules for common compounds

soluble compounds	insoluble compounds (except with group 1 metal ions and NH_4^+)
group 1 metal ions Li^+, Na^+, K^+, Rb^+, Cs^+	carbonates, CO_3^{2-}
ammonium, NH_4^+	hydroxides, OH^-, except Ba^{2+}
acetate, $C_2H_3O_2^-$ or CH_3COO^-	chlorides of Cu, Pb, Ag and Hg
nitrates, NO_3^-	bromides of Cu, Pb, Ag and Hg
sulfates, except with Ba, Pb, and Sr	iodides of Cu, Pb, Ag and Hg
chlorides, except with Cu, Pb, Ag and Hg	sulfides, S^{2-}
bromides, except with Cu, Pb, Ag and Hg	phosphates, PO_4^{3-}
iodides, except with Cu, Pb, Ag and Hg	

Solved problem

Aqueous silver nitrate, $AgNO_3$ will react with aqueous magnesium bromide, $MgBr_2$. Determine whether each product will form an insoluble precipitate or not, then write the balanced chemical reaction with the appropriate states.

Asked Will the products be soluble? What are the states of the reactants and products?

Given Reactants are aqueous.

Relationships Two ionic compounds will undergo a double replacement reaction

Solve Before you begin, you must predict the products of the reaction.
- First, complete the reaction. Swap either reactant anions or reactant cations, and criss-cross charges to get products:

$$AgNO_3(aq) + MgBr_2(aq) \rightarrow AgBr(?) + Mg(NO_3)_2(?).$$

- Balance the reaction:

$$2AgNO_3(aq) + MgBr_2(aq) \rightarrow 2AgBr(?) + Mg(NO_3)_2(?)$$

Next, use the solubility rules to see if either product is insoluble in water.
- $AgBr$ is insoluble, and $Mg(NO_3)_2$ is soluble.
- The insoluble product will form a precipitate and the soluble product will be aqueous.

Answer $2AgNO_3(aq) + MgBr_2(aq) \rightarrow 2AgBr(s) + Mg(NO_3)_2(aq)$

Section 2 review

Chapter 13

A solution can be described as dilute or concentrated depending on how much solute is dissolved in the solvent. Solubility is a measure of how well a particular substance dissolves in a given solvent. A substance with zero solubility is said to be insoluble in that solvent. A saturated solution is one in which the maximum amount of solute is dissolved in a particular amount of solvent. Such a solution is in equilibrium with solute dissolving and coming out of solution at equal rates. Solubility generally increases with temperature. A supersaturated solution contains more solute than the solvent could dissolve under normal circumstances. Increasing temperature of the solvent, the surface area of the solute, and stirring can all increase dissolving rates. For gases, pressure makes them more soluble in liquid as described by Henry's law. The solubility rules are used to determine if a precipitate will form when a reaction occurs.

Vocabulary words: concentration, dilute, concentrated, solubility, insoluble, soluble, saturated, aqueous equilibrium, supersaturated, Henry's law

Review problems and questions

1. What can you do to test whether a room-temperature solution is saturated or not?

2. Use the solubility curve to see how much sugar you can dissolve in 100 mL of water if you heat it to 40 °C.

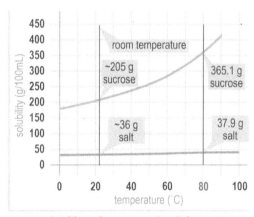

solubility of sugar and salt in water

3. Draw a diagram that shows why crushing a solute increases the speed at which a solution will form. Include an explanation in your drawing.

4. Use solubility rules to identify which of the following will dissolve in water:
   a. silver bromide
   b. lead(II) acetate
   c. barium sulfate

13.3 - Solution Concentration

There are several different ways to express the concentration or ratio of solute to solvent in a solution. Knowing how to interpret different units of concentration helps chemists understand how much of a solute is in a given solution. This is important for determining everything from knowing how much of a solution to use in a reaction to deciding whether a particular solute concentration is harmful to your health.

Concentration

Proper blood potassium concentration

Your blood is a solution that contains many dissolved solutes, including potassium. If you do not have enough potassium dissolved in your blood, you can experience muscle weakness or cramps and an irregular heart beat. Too much potassium can also cause an irregular heart beat. Between 100 and 200 mg of potassium are lost when sweating during vigorous exercise. That's why it is a good idea to stay hydrated with a sports drink that contains potassium.

potassium is lost through sweat

Even if you do not exercise, you need to take in 4 - 5 grams of potassium from the foods you eat each day to stay healthy and maintain a blood concentration of potassium between 140-200 mg potassium per liter of blood (mg/L).

Concentration calculations

Common ways to calculate concentration are: grams per liter, percent and molarity. Grams per liter is a ratio of grams solute per volume of solution in liters (g/L). Percent concentration is a ratio of solute quantity to solution quantity (either by mass or by volume, but units must cancel). Molarity is a concentration expressed with moles of solute per liter of solution (M).

four ways to calculate concentration

$$\text{concentration}_{g/L} = \frac{\text{mass(g) of solute}}{\text{liters of solution}}$$

$$\text{concentration}_\% = \frac{\text{mass of solute}}{\text{mass of solution}} \times 100\%$$

$$\text{concentration}_\% = \frac{\text{volume of solute}}{\text{volume of solution}} \times 100\%$$

$$\text{concentration}_{molarity} = \frac{\text{moles of solute}}{\text{liters of solution}}$$

Notice many of the formulas shown require a quantity of solution. Sometimes you will need to add the amount of solute and solvent together to get a quantity for a solution. Think of a solution as an equation.

$$\text{solution} = \text{solute} + \text{solvent}$$

Solved problem

If you dissolve 5.0 g of salt in 30.0 g of water, what is the mass percent concentration of salt in the solution?

Asked Find the percent of solute in the solution.

Given 5.0 g solute and 30.0 g solvent

Relationships Solution = solute + solvent; % concentration = (solute/solution) x 100

Solve The % mass equation needs the mass of solution. A solution is made of a solvent and its solutes:

$$\text{Solution} = \text{Solvent} + \text{Solute(s)}$$

This solution totals 35.0 g. Now solve for percent solute mass:

$$\% \text{ mass} = \frac{5.0 \text{ g}}{35.0 \text{ g}} \times 100\% = 0.14286 \times 100\% = 14\%$$

Answer The salt solution concentration is 14%.

Molarity of a solution

Molarity

The **molarity (M)** of a solution is the number of moles of solute per liter of solution. Molarity helps chemists know the ratios and numbers of particles in solutions.

$$\text{Molarity of solution (M)} = \frac{\text{moles of solute}}{\text{L of solution}}$$

Using molarity

Molarity helps chemists obtain the correct amount of reactants for a reaction. If you know the molarity of a given volume of reactant solution, you can calculate the moles of solute in the solution. If you need to add a given number of moles to a reaction, you can calculate the exact volume of solution needed. Brackets are an abbreviation for concentration in moles per liter. For example, [1.5] means "a concentration of 1.5 M" or "the concentration is a ratio of 1.5 moles solute per liter solution."

M = mol/L

Molarity expresses the amount of moles solute dissolved per liter of solution. Acetic acid, commonly known as vinegar, is produced by bacteria in rotting fruit. It is used as a household cleaner and often needs to be diluted because even a small concentration of vinegar can give off an odor. If you add 5.50 g acetic acid to 100.0 mL of water, how do you determine the concentration of the solution?

acetic acid

Calculating molarity

This solution formation scenario seems complex, but can be solved using what you know about molarity. If molarity equals moles of solute per liter of solution, first you will need to convert the given mass of solute to moles of solute using formula mass. Then you can determine molarity by dividing the moles of solute by the volume of solution in liters.

Solved problem

5.50 g of acetic acid (HCH_3COO) are added to 100.0 mL of water. What is the molarity of the solution?

Asked Find the molarity of a solution.

Given Amount of acetic acid solute and the volume of solution.

Relationships Molarity (M) = $\frac{\text{mol solute}}{\text{L solution}}$

Solve First, convert the acetic acid mass to moles. The formula mass of the HCH_3COO solute is 60.05 g/mol.

$$\frac{5.50 \text{ g } HCH_3COO}{1} \times \frac{1 \text{ mol } HCH_3COO}{60.05 \text{ g } HCH_3COO} = 0.09159 \text{ mol } HCH_3COO$$

Next, convert the volume of water to liters: 100.0 mL = 0.1000 L.

Now use moles of the solute to calculate solution molarity.

$$M = \frac{0.09159 \text{ mol } HCH_3COO}{0.1000 \text{ L}} = 0.9159 \frac{\text{mol } HCH_3COO}{\text{L}}$$

$$\xrightarrow{\text{round}} 0.916 \frac{\text{mol}}{\text{L}} = 0.916 \text{ M } HCH_3COO$$

Answer The solution's concentration is 0.916 M HCH_3COO.

Calculating the molarity of a solution

Applied solution chemistry

Molarity is useful in reactions that involve aqueous solutions because this concentration tells you the number of molecules (in moles) there are per volume solution so you can calculate the exact volume needed for a given reaction. Having appropriate solute concentrations in solutions is critical to every living system on Earth, from the solutions inside your body to the solutions in the ocean.

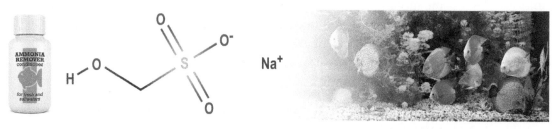

Solution chemistry in a fish tank

You may have experienced a solution chemistry disaster in a fish tank at some time. If toxic substances build up in a fish tank, the entire system could crash. For example, ammonia (NH_3) is a waste product produced by fish that can build up to toxic levels. Ammonia levels must be monitored closely, especially in tanks with a lot of fish. If you notice NH_3 approaching a toxic level, you could either change the water or you could add a solution that reacts with the NH_3 to remove it. A chemical called sodium hydroxymethanesulfonate ($NaHOCH_2SO_3$) is the active ingredient in ammonia remover solution. It works by reacting with NH_3 and forming a salt which can be filtered out and removed from the tank. Adding too much $NaHOCH_2SO_3$ can create a toxic environment for fish, so you need to use solution chemistry to determine the exact amount to add to the tank.

Solved problem

How much 0.5 M ammonia remover solution ($NaHOCH_2SO_3$) should you add to a fish tank in order to completely react with 0.01 mole of excess ammonia (NH_3)?
Balanced reaction: $NH_3 + NaHOCH_2SO_3 \rightarrow NaH_2NCH_2SO_3 + H_2O$

Given Moles NH_3 (0.01 mol), $NaHOCH_2SO_3$ solution concentration (0.5M), and the balanced reaction

Relationships Molarity (M) = $\frac{\text{mol solute}}{\text{L solution}}$

Solve Use stoichiometry to determine moles of $NaHOCH_2SO_3$ needed based on the amount of ammonia present. The mole ratio from the balanced reaction allows you to convert moles of ammonia to moles of $NaHOCH_2SO_3$.

$$\frac{0.01 \text{ mol } NH_3}{1} \times \frac{1 \text{ mol } NaHOCH_2SO_3}{1 \text{ mol } NH_3} = 0.01 \text{ mol } NaHOCH_2SO_3$$

Rearrange the concentration formula to find the required volume of 0.5 M $NaHOCH_2SO_3$ solution.

$$\text{L solution} = \frac{0.01 \text{ mol } NaHOCH_2SO_3}{0.5 \text{ M}} = \frac{0.01 \text{ mol } NaHOCH_2SO_3}{1} \times \frac{1L}{0.5 \text{ mol}} =$$

$$= 0.02 \text{ L} = 20 \text{ mL } NaHOCH_2SO_3$$

Answer You would need to add 20 mL of the ammonia remover solution to react with 0.01 moles of excess ammonia (NH_3) in the fish tank.

More solution calculations

Grams per liter

You must measure ingredients carefully when you are baking so you minimize waste and maximize the quality of your product. The same is true when measuring quantities to create solutions for laboratory use. Suppose you want to conduct an experiment that requires 10.0 g NaCl but you only have a salt solution with a concentration of 50.0 g/L. How much solution do you need to use to get 10.0 g of salt? Use a little algebra to rearrange the solution equation to solve for volume of solute.

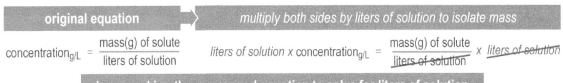

| original equation | multiply both sides by liters of solution to isolate mass |

$$\text{concentration}_{g/L} = \frac{\text{mass(g) of solute}}{\text{liters of solution}} \qquad \text{liters of solution} \times \text{concentration}_{g/L} = \frac{\text{mass(g) of solute}}{\cancel{\text{liters of solution}}} \times \cancel{\text{liters of solution}}$$

| keep working the rearranged equation to solve for liters of solution |

$$\text{liters of solution} \times \text{concentration}_{g/L} = \text{mass (g) of solute}$$

| divide both sides by concentration $_{g/L}$ to isolate liters of solution | rearranged equation |

$$\frac{\text{liters of solution} \times \cancel{\text{concentration}_{g/L}}}{\cancel{\text{concentration}_{g/L}}} = \frac{\text{mass(g) of solute}}{\text{concentration}_{g/L}} \qquad \text{liters of solution} = \frac{\text{mass(g) of solute}}{\text{concentration}_{g/L}}$$

Solved problem

If the concentration of a salt solution is 50.0 g/L, how much solution do you need if you want 10.0 g of salt?

Given 10.0 grams of solute and a solution concentration of 50.0 g/L
Relationships Use the concentration and mass (above) to find volume needed.
Solve

$$L = \frac{10.0 \text{ g salt}}{50.0 \frac{g}{L}} = \frac{10.0 \text{ g salt}}{1} \times \frac{1 \text{ L}}{50.0 \text{ g salt}} = 0.200 \text{ L solution}$$

Answer You would need 0.200 L of salt solution to get 10.0 g of salt.

Rearrange the formula

Frequently you will need to rearrange each of the concentration formulas to solve for the desired quantity. The example below shows how to rearrange a percentage formula.

| calculating solute mass from solution mass and % concentration |

$$\text{concentration}_{\%} = \frac{\text{mass of solute}}{\text{mass of solution}} \times 100\% \qquad \text{mass of solute} = \text{mass of solution} \times \frac{\text{concentration}_{\%}}{100}$$

Solved problem

The fluid in an intravenous (IV) bag contains a 0.90% NaCl solution. If you need to supply 5.0 kg of solution for a hospital, what mass of NaCl do you need?

Given 5.0 kg of IV solution, solute concentration of 0.90%
Relationships Rearrange the % concentration by mass formula as shown above.
Solve

$$\text{mass of solute} = 5.0 \text{ kg} \times \frac{0.9}{100} = 0.045 \text{ kg} = 45 \text{ g NaCl solute}$$

Answer You would need 45 g NaCl to make 5.0 kg of IV fluid.

Section 13.3: Solution Concentration

Working with very dilute solutions

Small quantities

Parts per million (ppm) and parts per billion (ppb) are used to describe small concentrations of solutes. Percent, ppm and ppb are all solute-to-solution concentration ratios. One drop of ink dissolved in a large gasoline tanker truck full of water is a ratio of 1 drop of ink to about 1 billion drops of solution, or 1 ppb. Some solutes are found in such small amounts, a concentration in molarity or percent is hard to grasp. How concentrated is a 0.0024% salt solution? It's easier to think about this small quantity as 24 ppm.

Water quality

Distilled, deionized water is a pure substance because it contains only the compound H_2O. Tap water is a solution because it has many dissolved solutes in addition to water. Clean drinking water normally contains dissolved minerals, but too much of a particular solute can be a concern. Water quality scientists monitor substances that impact the safety of drinking water. They create reports about the concentration of toxic substances like arsenic measured in ppb, and less toxic minerals like calcium measured in ppm.

ppm and ppb

The ratio of 1 part solute to 1 million parts solution equals 1 ppm. In units of time, 1 ppm equals 1 second in 12 days. The ratio 1 part per billion is 1 second in 32 years. The equations to express concentration as ppm or ppb are shown below. For equations that say "units," any mass or volume unit can be used as long as the units cancel.

equations for solutions with small concentrations

$$ppm = \frac{mg\ solute}{Kg\ solution} = \frac{mg\ solute}{L\ solution} = \frac{units\ solute}{units\ solution} \times 10^6$$

$$ppb = \frac{\mu g\ solute}{Kg\ solution} = \frac{\mu g\ solute}{L\ solution} = \frac{units\ solute}{units\ solution} \times 10^9$$

note: μg = microgram; $1\mu g$ = 0.001 mg or 1×10^{-6} g

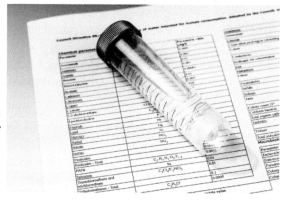

pph — 1 penny in $1.00
ppm — 1 penny in $10,000.00
ppb — 1 penny in $10,000,000.00

Solved problem

Scientists have determined that a lead concentration in drinking water at or above 15 ppb is unsafe. Analysis of a drinking water sample shows there are 2.10 µg lead dissolved per 1.50 liters water. Is the water safe to drink?

Asked Is the drinking water below the limit of 15 ppb?

Given 2.10 µg lead solute, 1.50 L solution

Relationships The best equation to use is: $ppb = \frac{\mu g\ solute}{L\ solution}$

Solve

$$ppb = \frac{\mu g\ solute}{L\ solution} = \frac{2.10\ \mu g\ lead}{1.50\ L\ solution} = 1.40\ ppb$$

Answer Yes, the water is safe to drink because 1.40 ppb is well below the limit of 15 ppb.

Preparing a solution from a solid solute

Stock solutions

The most common unit of concentration in laboratory experiments is molarity (M). A solution of a known molarity helps you accurately measure a quantity of ions or molecules. Chemists often need to make stock solutions of solutions with higher molarity using solid solutes and distilled water. These concentrated solutions can later be diluted to the desired molarity.

volumetric flasks

Preparing a solution

Steps for preparing a solution of known molarity (M) from a solid solute

1. Find the formula mass of the solute to be dissolved.
2. The desired molarity indicates the number of moles needed per liter solution. Use the formula mass to calculate the grams of solute needed. Multiply the required moles by the formula mass to get grams of solute needed.
3. Measure the required grams of solute.
4. Add the solute to a volumetric flask like the ones shown above. Chemists use a volumetric flask to improve accuracy.
5. Add distilled water (solvent): Fill the volumetric flask about two-thirds of the way up to the line with distilled water.
6. Cap and invert the flask several times until all of the solid dissolves.
7. Fill the volumetric flask with distilled water up to the correct volume mark. You may need a pipette to get the meniscus right on the line.
8. Cap and invert the flask several times to ensure complete mixing.

Solved problem

Your experiment calls for 250.0 mL of a 1.5 M $NaHCO_3$ solution. How much solute and solvent do you need to mix together to form the solution?

Given 250.0 mL of solution needed, 1.5 M solution concentration

Solve First, use molarity and volume to determine the number of moles solute. Then convert moles to grams.

- Find the moles of $NaHCO_3$ needed:

$$1.5 \text{ M} = \frac{? \text{ mol } NaHCO_3}{0.250 \text{ L}} = 0.375 \text{ mol } NaHCO_3$$

- Convert to grams:

$$\frac{0.375 \text{ mol } NaHCO_3}{1} \times \frac{84.007 \text{ g } NaHCO_3}{1 \text{ mol } NaHCO_3} = 31.5026 \text{ g } NaHCO_3$$

Round the answer: 31.5026 g = 32 g

Answer You will need to dissolve 32 g $NaHCO_3$ in enough distilled water to form 250.0 mL solution.

Section 13.3: Solution Concentration

Dilution: Preparing a solution from an existing solution

Preparing solutions from other solutions

You can control the concentration of a flavored drink by adding a little (or a lot) of a flavored solution to a glass of water. Water is the solvent, but is the concentrate in the bottle a solute? Yes, and no. The liquid in the container is already a solution of water and other dissolved substances. You are only changing the concentration of that solution by adding more solvent. The original drink mix is concentrated, and your watered-down drink is dilute. You performed a dilution as you added concentrated solution (drink mix) to solvent (water). In the chemistry lab, you may need to dilute a concentrated solution into a more dilute solution. The most concentrated solution is called a stock solution.

dilution: $M_1 \times V_1 = M_2 \times V_2$

$V_{water} = V_2 - V_1$

Dilution

Steps for preparing a solution of known molarity (M) by diluting stock solution:

1. Identify the molarity of the concentrated stock solution; this is the initial molarity, M_1.
2. Identify the molarity you need for the new solution; this is final molarity, M_2.
3. Decide how much of the new, dilute solution you will need. This is final volume, V_2.
4. Use the dilution formula: $M_1 \times V_1 = M_2 \times V_2$ to determine how much original, concentrated solution you need. Solve for V_1, initial volume of concentrated solution.
5. The difference between the final volume (V_2) and the initial volume (V_1) equals the amount of distilled water you'll need to add to the final solution: $V_{water} = V_2 - V_1$

Solved problem

Your experiment calls for 50.0 mL of a 0.10 M NiCl$_2$ solution. How much 1.00 M NiCl$_2$ stock solution and how much distilled water do you need to mix to make the solution?

Given M_1, M_2 and V_2

Solve First, find the volume of stock solution needed.

$$V_1 = \frac{M_2 \times V_2}{M_1} = \frac{0.10 \text{ M} \times 50.0 \text{ mL}}{1.00 \text{ M}} = 5.0 \text{ mL}$$

Next, solve for water needed: $V_{water} = V_2 - V_1$

$V_{water} = 50.\text{ mL} - 5.00 \text{ mL} = 45 \text{ mL}$

Answer You need to mix 5.0 mL NiCl$_2$ stock solution with 45 mL distilled water.

Properly label solutions

When you finish preparing a solution you plan to store, label it immediately. Choose a label with a durable adhesive. Use a permanent marker or ball-point pen. Include your name, date the solution was prepared, the chemical name, concentration, chemical formula, and the appropriate chemical storage code. If the solution is hazardous, make a note if it.

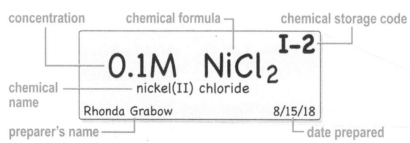

Using color to measure concentration

Solution color

Solutions sometimes have color but are usually transparent. That means light can shine through a solution. The amount of light a solution absorbs relates to the concentration of solute particles in it. A device called a **colorimeter** uses the intensity of colored light shined through a colored solution to measure its concentration. The color or wavelength of light shined through the solution must be equal to the color that is best absorbed by the solution. A colored solution best absorbs its complimentary color, which is found as its opposing color on a color wheel. A pink solution best absorbs green light. You would set up a colorimeter to shine green light through the pink solution to measure its solute concentration.

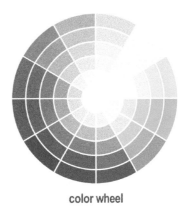

color wheel

Colorimeter

As green light travels through the pink solution, some light is absorbed by solute particles. This is called absorbance. Light that makes it through the solution is called transmittance. Transmitted light strikes a detector. The energy from this light is directly measured by the colorimeter and is reported as transmittance.

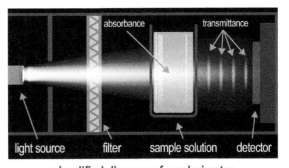

simplified diagram of a colorimeter

Beer's law

Beer's law states absorbance and concentration are directly proportional. The higher the concentration, the more solute particles are present to absorb light. As concentration goes up, absorbance also increases. Transmittance is inversely proportional to concentration. The higher the concentration, the less light makes it to the detector because there are more solute particles to absorb light on its way through the solution.

Determining an unknown concentration

Using Beer's law you can use a colorimeter to determine the unknown concentration of a solute in a solution. If the solution is colored you can directly measure its transmittance as described above. If the solution is not colored, you can add a reactant that will make the solution colored when it reacts with the solute. The more solute particles there are, the more intense the color will become. This allows scientists to measure levels of colorless, potentially toxic substances in the field such as the concentration of nitrates in a river.

Chapter 13

Section 3 review

Concentration can be measured in g/L, percent mass, or using molar concentration or molarity. Molarity is the number of moles of solute per liter of solution. With very dilute solutions, parts per million and parts per billion are used to measure concentration. Preparing solutions requires knowing the formula mass of the solute as well as accurately measuring the mass of solute and volume of solvent. Dilute solutions can be prepared by adjusting the final volume if the initial and final molarities are known. A colorimeter is a tool for measuring concentration that uses light passed through a solution. Beer's law states that light absorbance and concentration are directly proportional to one another.

Vocabulary words: molarity (M), colorimeter, Beer's law

Review problems and questions

1. A student dissolves 2.5 g NaCl in enough distilled water to make 50.0 mL solution. What is the molarity of this solution?

2. If a juice drink advertises it has 150% of the recommended amount of vitamin C per 500 mg serving, how much vitamin C is in the drink?

3. How much dilute solution will be produced when you dilute 50.0 mL of a 1.0 M acetic acid solution to a concentration of 0.20 M?

4. If you were performing the above dilution, how much water would you need to add to the concentrated acetic acid to get the dilute solution?

5. According to Beer's law, which solution of red dye would have a higher absorbance value: 0.1 M or 0.2 M? Explain your answer.

6. A water quality report from your local tap water source says your tap water contains 0.20 mg fluoride per 100.0 mL of tap water. What is the concentration of fluoride in the tap water, in ppm? Hint: pay attention to the volume units!

7. 0.79 milligrams of sucrose is dissolved in 225 grams of water. What is the concentration of sucrose in parts per billion (ppb)? Note: 1 μg = 0.001 mg

8. If 36 grams of potassium chloride is dissolved in 128 grams of methanol. What is the percent of solute in the solution?

9. The solubility of lithium chloride is 0.0822 grams in 5510 grams of water. What is this concentration in ppm?

10. A 0.45 M solution was made using 6.45×10^4 mg of magnesium hydroxide. Calculate the volume of the solution.

13.4 - Properties of Solutions

When an ionic solute dissolves, the solvent attracts solute particles and carries them off into the solution. In aqueous solutions, water molecules surround water-soluble ionic solutes. Ionic bonds between solute particles are broken because they are overcome by solute-solvent particle attractions. During the dissolving process energy is absorbed when bonds are broken and released when bonds are formed. The energy involved the dissolving process is called the **heat of solution**.

Heat of solution

Endothermic reactions and solutions

When you spend more money than you have, you have "negative" dollars. A solute that loses energy when it dissolves has a negative heat of solution. The negative sign tells you the direction energy has moved from the solute's point of view. A negative heat of solution means energy is released from the solute into the solution. The released energy would make a thermometer placed in it rise in temperature. Overall, each mole of calcium chloride ($CaCl_2$) releases 82.8 kJ of energy as it dissolves.

$$CaCl_2(s) \xrightarrow{H_2O\ (l)} Ca^{2+}(aq) + 2Cl^-(aq) + 82.8\ kJ$$

Exothermic reactions and solutions

When the energy appears as a product, heat of solution is negative. This means dissolving $CaCl_2$ is an exothermic process and -82.8 kJ of energy are released per mole dissolved. Instant heat packs use solutes like $CaCl_2$ with a negative heat of solution. Because energy is conserved, the heat released by calcium chloride as it dissolves equals the heat gained by the solution. Heat packs take advantage of this heat flow. A heat pack contains water and $CaCl_2$ in a thin tube. Breaking the tube allows the $CaCl_2$ to dissolve in the water and heat up the solution.

When a solute gains energy from the solution

When energy appears as a reactant, heat of solution is positive. Ammonium nitrate (NH_4NO_3) has a positive heat of solution of +25.7 kJ/mol identified as a reactant below. That means one mole of NH_4NO_3 solute dissolving is an endothermic process that absorbs 25.7 kJ of energy. A thermometer would show a decrease in temperature as this solution forms. The solute absorbs energy from the solution. The solution becomes colder because energy moves from the solution to the solute.

$$NH_4NO_3\ (s) + 25.7 kJ \xrightarrow{H_2O\ (l)} NH_4^+(aq) + NO_3^-(aq)$$

Endothermic reactions

A cold pack contains a tube of NH_4NO_3 in water. Breaking the tube allows the NH_4NO_3 to dissolve in water. So much energy is absorbed by the solute that the solution may get cold enough to freeze water vapor from the surrounding air directly on the outside of the cold pack, forming a visible layer of frost.

Characteristics of solutions

Volume is not additive

A 20 g sample of salt has a volume of about 18 mL. An 80 g sample of water has a volume of 80 mL. When you add 20 g of salt to 80 mL of water you get a solution with a mass of 100 g (20 g + 80 g = 100 g). However, when you add 18 mL of salt to 80 mL of water, you do not get 98 mL of solution. That is because dissolved sodium ions (Na^+) and chloride ions (Cl^-) fit in the spaces in between water molecules. Mass is always additive when solutions are formed because mass is conserved, but volume is not. Solution volume at a given temperature is influenced by particle attractions between solute and solvent and solute particle size.

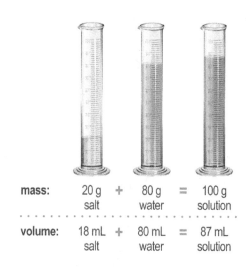

mass: 20 g salt + 80 g water = 100 g solution

volume: 18 mL salt + 80 mL water = 87 mL solution

Density vs. concentration

When the salt and water are mixed, the total volume appears to decrease while mass does not. A solution's density is usually greater than the solvent density alone. Salt water is more dense than fresh water. The graph shows density increases as salinity increases in the ocean. Most solid solutes increase the density of a solution compared to the solvent alone. The graph shows a density of 1 g/mL with 0% salinity, accounting for the density of water alone when no salts are dissolved.

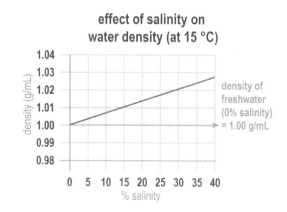

effect of salinity on water density (at 15 °C)

density of freshwater (0% salinity) = 1.00 g/mL

Gases and liquids as solutes

A solute can either increase or decrease the solution density compared to the solvent density alone. Alcohol is less dense than water and is highly water-soluble, so an alcohol-water solution is less dense than pure water. The carbon dioxide gas dissolved in carbonated water is also less dense than pure water, so carbonated water is less dense than plain water.

Freezing points

Solutes expand the range of temperatures where a liquid can exist. Solute particles lower the freezing point of a solution compared to the pure solvent. The freezing point of water decreases by 1.86 °C per mole of solute particles dissolved. The average concentration of dissolved salts in seawater is 3.5%. Sodium chloride alone makes up 3.0% of the total 3.5% salinity. Each formula unit of NaCl dissociates into Na^+ and Cl^- ions, therefore each mole of dissolved NaCl salt adds 2 moles of particles. Dissolved salts cause seawater to freeze at -2.2 °C instead of 0 °C.

Boiling points

A solution has a higher boiling point compared to the pure solvent alone. Greater increases in the boiling point can be achieved by adding more and more solute, to a point. For water, each mole of solute particles increases the boiling point by about 0.5°C. Seawater boils at 100.6 °C instead of 100 °C.

Colligative properties

Physical properties of solutions

If you live where it snows, you are familiar with "salting" roads and sidewalks. Think about how salt interacts with ice at the particle level. Ice freezes at 0 °C (or 32 °F) but when you add salt to ice, it melts on contact even if the salt is colder than the ice! This phenomenon describes a **colligative property** of a solution. Colligative properties are physical properties such as freezing and boiling points of a solvent that are changed by adding solute particles. The new freezing and boiling points depend only on the number of solute particles added to the solution, not on the nature or type of particle.

salt is sprinkled on icy roads and sidewalks

Freezing point depression

When you add solute particles to a solvent, the resulting solution has a lower freezing point. This is known as freezing point depression. It has to get colder for the solution to freeze. Solutions have a direct relationship between the number of solute particles and freezing point depression. For example pure water freezes at 0 °C. When water is frozen solid, molecules become organized and are held tightly in place. Freezing is always associated with a decrease in energy among particles (entropy). Solid water has limited molecule arrangements compared to liquid water. Particles can vibrate but not slide past one another in the solid phase. When solute particles are added to ice, their energy spreads out into the solution. If enough solute particles are added, the increased entropy results in a phase change from solid to liquid, otherwise known as melting.

Entropy of solutions

Hydrogen bonds hold water molecules in place when they form an organized, hexagonal crystal pattern during freezing. Dissolved solute particles increase solution entropy and interfere with hydrogen bonds. This makes it harder for a crystal to form until temperature and energy are low enough for the hydrogen bonds to hold water molecules in place.

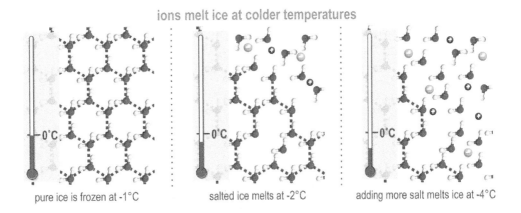

ions melt ice at colder temperatures

pure ice is frozen at -1°C   salted ice melts at -2°C   adding more salt melts ice at -4°C

Amount matters, not identity

Freezing can still happen when solutes are added, but now it happens at a lower temperature. As more particles are added, the freezing point goes lower and lower. The amount of particles has the greatest impact on how low the freezing point will drop, not the identity of the particles.

The amount of particles determines how low freezing point will go, not the identity of the particles.

Section 13.4: Properties of Solutions

Molality

Boiling point elevation

Boiling point elevation is another colligative property. When you add solute to a solvent, the resulting solution has a higher boiling point than the pure solvent. The added solute particles increase a solvent's energy requirement to transition from the liquid to vapor phase. The more solute particles you add, the higher the boiling point will go.

Salt is for flavor

Some people think adding salt to water for boiling pasta will make the pasta cook faster. This might sound correct because adding a solute increases the boiling point of water. However, you would need to add far more salt than a pinch to raise the boiling temperature of water by just 1°C. The practical purpose of adding salt to pasta water is to add flavor.

More particles cause greater colligative effects

Any ionic solution such as 1.0 M NaCl or 1.0 M $MgCl_2$ can be imagined as particles. For every liter of 1.0 M NaCl solution, there is 1 mole of Na^+ particles and 1 mole of Cl^- particles. Think about the number of ions that are formed when NaCl dissociates instead of thinking of NaCl as a single solute. For a 1.0 M $MgCl_2$ solution, 1 mole of Mg^{2+} particles and two moles of Cl^- particles form in every liter of solution. NaCl dissociates into 2 ions, and $MgCl_2$ dissociates into 3 ions. $MgCl_2$ forms more particles per mole, so it will have a greater effect on freezing and boiling point.

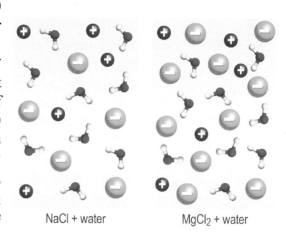

NaCl + water $MgCl_2$ + water

Molality, m

To investigate the change in the freezing point when a solution forms, use the unit of concentration called **molality** (m). Molality is the number of solute moles per kilogram of solvent. You cannot use molarity (M) because when solutions change temperature, density also changes as the volume expands or contracts. If density changes, molarity will be incorrect as it is a mole-volume ratio. Using molality corrects for this because it is a mole-mass ratio. The equation to calculate the molality (m) of a solution is shown below.

$$\text{molality} \implies m = \frac{\text{moles of solute}}{\text{kg of solvent}}$$

Solved problem

If 5.0 g NaCl are dissolved in 0.3247 kg water, what is the solution molality?

Asked Find the molality of the solution.
Given 5.0 g solute and 0.3247 kg solvent
Solve First, convert the mass of solute to moles of solute.

$$\frac{5.0 \text{ g NaCl}}{1} \times \frac{1 \text{ mol NaCl}}{58.443 \text{ g NaCl}} = 0.085553 \text{ mol NaCl}$$

Now solve for m: $m = \frac{0.085553 \text{ mol}}{0.3247 \text{ kg}} = 0.26348 \text{ m} \xrightarrow{\text{round}} 0.26 \text{ m}$

Answer The solution has a molality of 0.26 m.

Calculating colligative effects

Applied colligative properties: Antifreeze

Using antifreeze in a car's radiator is an everyday application of colligative properties. Radiator fluid must remain in the liquid state so it can circulate around an engine no matter how cold it is outside. Antifreeze is a solution of water and a molecular solute called ethylene glycol ($C_2H_6O_2$). Water in the car's radiator resists freezing at cold temperatures when antifreeze is added. The antifreeze solution remains liquid in temperatures below freezing so a car can run properly in very cold weather. It also raises the boiling point of water to prevent an engine from overheating. Antifreeze is also known as coolant.

How to find the change in freezing point

You can purchase antifreeze at the proper concentration for your car at an auto parts store, or you can buy a bottle of concentrated antifreeze to make your own water-based dilution. Use the formula below to determine how much a solution's freezing (or boiling) point will change for a given molal (m) concentration. The change in freezing point (ΔT_f) and molality (m) apply to the solution, and the freezing point depression constant (K_f) applies to the solvent. The number of dissolved particles per solute (i) is also called the van't Hoff factor. Assume molecular solutes do not dissociate, so i will equal 1. For ionic solutes, the number of particles formed by the solute can vary. For example, a $CaCl_2$ solute dissociates into 3 ions, so i = 3.

ΔT_f = change in freezing point
i = van't Hoff factor
K_f = freezing point depression constant for the solvent (°C/m)
K_b = boiling point elevation constant for the solvent (°C/m)
m = solution molality

change in freezing point
$\Delta T_f = i \times K_f \times m$

change in boiling point
$\Delta T_b = i \times K_b \times m$

Solved problem

The freezing point constant for water, K_f = 1.86 °C/m. Suppose you are instructed to make a 1.50 m sugar solution. How much will this solution reduce water's freezing point, and what is the solution's new freezing point?

Asked How much will water's freezing point decrease?

Given 1.5 m and K_f = 1.86 °C/m

Solve Sugar is a molecular substance so it will not dissociate; i = 1. Solve for ΔT_f:

$$\Delta T_f = 1 \times \frac{1.86\ °C}{m} \times 1.50\ m = 2.79\ °C$$

Water's normal freezing point is 0 °C. If adding the solute decreases water's freezing point by 2.79 °C, you must subtract this amount from the normal freezing point.

0 °C − 2.79 °C = -2.79 °C; the solution will freeze at -2.79 °C.

Answer Water's freezing point will be reduced by 2.79 °C. The sugar solution's freezing point is -2.79 °C.

Chapter 13

Section 4 review

The energy absorbed or released when a substance dissolves is called the heat of solution. When a solid dissolves in liquid the solute particles fit within the spaces between solvent molecules. This generally means that the density of a solution is greater than the density of the pure solvent. However, dissolving gases or liquids into a liquid solvent may increase or decrease the density compared to pure solvent. Colligative properties of solutions are ones that depend only on the relative amounts of solute and solvent, and not on the identity of the solute. Two colligative properties are freezing point depression and boiling point elevation. Molality is the number of moles of solute per kilogram of solvent. Remember that molarity is the number of moles of solute per liter of the total solution. Molality can be used to calculate changes in freezing point or boiling point.

Vocabulary words: heat of solution, colligative property, molality (m)

Key relationships
- Negative heat of solution indicates an exothermic dissolving process.
- Positive heat of solution indicates and endothermic dissolving process
- molality, $m = \dfrac{\text{moles solute}}{\text{kg solvent}}$
- Change in freezing point: $\Delta T_f = i \times K_f \times m$
- Change in boiling point: $\Delta T_b = i \times K_b \times m$

Review problems and questions

1. Substance AB is added to water and forms a solution. Is this dissolving process endothermic or exothermic? How can you tell?

    $AB(s) \xrightarrow{H_2O} A^+ (aq) + B^- (aq) + 39.9 \text{ kJ}$

2. Which way is energy moving when AB dissolves: from AB to the solution, or from the solution to AB? Explain your answer.

3. The freezing point constant for water, K_f equals 1.86 °C/m. How many grams of salt, NaCl, will you need to add to 0.50 kg of water to change its freezing point by 4.0 °C?

4. Calculate the changes in freezing point and boiling point of a solution made by dissolving 16 grams of NaCl in 118 grams of water. Use K_f = 1.86 °C/m; K_b = 0.520 °C/m.

5. Suppose you have a 4.60 m glucose solution that depressed the freezing point of the solution by 6.42 °C. What is the molal freezing point depression constant (K_f) of the solution?

6. 1.46×10^2 g of camphor ($C_{10}H_{16}O$) is dissolved in 8.24×10^2 g of dimethyl ether (C_2H_6O). What is the molality?

Don't Drink the Water

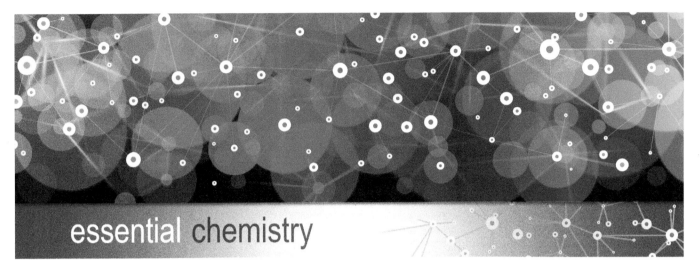

essential chemistry

Water is often called the universal solvent because it is capable of dissolving so many substances. Water is also ubiquitous. It is in the seas, in the air we breathe, and is crucial for our bodies. No matter where it goes, through the air, in the sea, or through our bodies, it carries solute particles with it. Because it is so good at dissolving substance and so crucial to life, contamination of water sources can be especially dangerous.

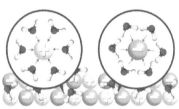

Water contamination can occur when a pollutant is added to a water source without adequate treatment. There are two different types of water pollution sources. Point source pollution occurs when there is a single, identifiable source of contaminants into the waterway. The source could be a discharge from a sewage treatment plant, from a factory, or even from a city's storm drain. Non-point source pollution, like runoff of agricultural fertilizers, is different because there is not a single identifiable source. Rather there are cumulative effects from small amounts of contamination over a large area. Some water pollution can occur even if the contaminant is not near a surface water source. In this case ground water can get contaminated if soil contamination eventually reaches the aquifer below.

Some contaminants like plastic bags, bottle caps, and soda cans are big enough to see. But much of the contamination is not visible to your eyes. Among the many possible contaminants, there are chemicals like acids from industrial processes, detergents, nitrates and phosphates from fertilizers, heavy metals, and hydrocarbons from fuels and oils. There could also be microorganisms, like bacteria, in the water. Many microorganisms are not detrimental, but some are and problems can arise if the water is not treated properly.

One notable example of water pollution occurred in Flint, Michigan. The issues began in 2014 when city officials decided to switch water suppliers in an effort to save money. To switch suppliers, a new pipeline had to be built. So in the interim, the city reverted to using water from the Flint River. This seemed like a sensible plan since the Flint River was the city's main water source until the 1960s.

Only a month into the switch residents began to complain that the water smelled and looked funny. It was also about 70 percent "harder" than the previous water source. Water is considered "hard" if it has high concentrations of calcium and magnesium ions dissolved in it.

Don't Drink the Water - 2

Much worse than being hard, E. coli and coliform bacteria were detected in the water. This prompted an advisory from the city to boil the water to kill the microorganisms. The city was going to address this situation systematically by adding more chlorine containing compounds to the water. However, this caused more problems. General Motors in Flint stopped using the water from the Flint River, deeming it too corrosive to its machines. It was determined later that the city did not have adequate corrosion testing or controls in place for the water.

one month exposure to Detroit water

one month exposure to Flint River water

Corrosion (rusting) can be an especially bad problem in cities that rely on older water lines. Water will travel from the source through miles and miles of pipes containing lead, iron and copper. Corrosion inhibitors, including phosphates (PO_4^{3-}), work by creating a passive layer on the inside of the pipes. This leaves a mineral scale that protects the metals from other chemicals in the water. Without the continuous addition of the inhibitor, the passive layer can dissolve back into the water. Dissolved chloride in the water will start to react with exposed iron from the metal pipes forming iron(II) ions (Fe^{2+}), and causing rust colored water. Dissolved oxygen is an essential solute in water – fish need it to breathe – but it can corrode pipes by reacting with exposed lead, increasing the concentration of lead(II) ions (Pb^{2+}).

pipe cross-sections

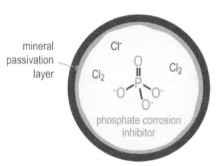
lead or iron pipe

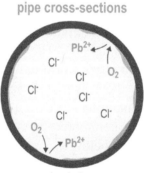
lead pipe

iron pipe

Not only did the corrosion affect GM's machines, it also affected the water pipes that lead to people's homes. The Environmental Protection Agency (EPA) considers a concentration of 15 parts per billion (ppb) to be safe for drinking water. A city test of the water at one resident's home found a lead concentration of 104 ppb. Three months later the child at this home was diagnosed with lead poisoning, and an independent test by researchers from Virginia Tech University found lead levels of 13,200 ppb. Water at anything greater than 5,000 ppb is considered toxic waste!

Months later, Flint switched back to their original water supply, fully optimized for corrosion control. The corrosion controls took some time to reach the homes in Flint. In order for the corrosion inhibitors to work, people had to run the water so the phosphates could "re-scale" the pipes. Even though lead concentration was reported at an acceptable 12 ppb in 2017, residents still do not trust the water and water suppliers.

For decades water quality in infrastructure was ignored in both developed and developing countries. Some countries made progress in controlling water pollution from domestic and industrial sources, but others have a long way to go to improve water quality. Being an active citizen and having an understanding of the essential science of water quality are the first steps in turning around this deterioration in the quality of our water.

Chapter 13 review

Vocabulary
Match each word to the sentence where it best fits.

Section 13.1

chromatography	conductivity
desalination	diffusion
dissociation	dissolution
dissolved	electrolyte
hydration	immiscible
miscible	osmosis
reverse osmosis	solute
solvation	solvent

1. The _____ makes up the largest part of the solution and is most often a liquid.
2. A(n) _____ substance is said to be completely and evenly dispersed throughout the solution.
3. The substance that is dissolved in a solution is called the _____.
4. _____ is the process where gas particles mix to form a solution.
5. The process by which water molecules in particular surround a charged ion or polar molecule is known as _____.
6. The process of _____ occurs when a solvent surrounds a solute.
7. The separation of an ionic compound into its ions in a solution is known as _____.
8. _____ is the name for the entire process of dissolving.
9. A charged particle capable of conducting electricity in a solution is called a(n) _____.
10. A solute and solvent are said to be _____ when they cannot mix to form a solution.
11. Solutes and solvents that do mix are _____.
12. The _____ of a solution is a measure of the ions present in a solution.
13. An ink solution can be separated using a process called _____.
14. The tendency of a solution to evenly distribute solutes within the the solvent is called _____.
15. _____ utilizes pressure to aid the process of separation between solute and solvent particles.
16. Salt water is converted into fresh water through a process called _____.

Section 13.2

aqueous equilibrium	concentrated
concentration	dilute
Henry's law	insoluble
saturated	solubility
soluble	supersaturated

17. A(n) _____ solution is holding the maximum amount of solute that it can hold.
18. A(n) _____ solution has very little solute dissolved in it.
19. A solution that has a large amount of dissolved solute is said to be _____.
20. A(n) _____ substance will not dissolve.
21. The _____ of solute in a particular solvent is the maximum amount that will dissolve at a given temperature and pressure.
22. When a substance is capable of being dissolved it is _____.
23. A solution is _____ when it contains more dissolved solute that it can typically hold at the temperature it is at.
24. Over time when the amount of dissolved solute becomes constant, a(n) _____ has been achieved.
25. The main idea of _____ is that for a given temperature the solubility of a gas in a liquid depends on the concentration and partial pressure of the gas directly above the liquid.
26. The _____ is the amount of dissolved solute compared to the total solution.

Section 13.3

Beer's law	colorimeter
molarity (M)	

27. A(n) _____ determines the concentration of a solution by measuring the intensity of light shined through the solution.
28. _____ is a unit of concentration measured as moles of solute divided by liters of solution.
29. The idea that absorbance and solution concentration are directly proportional is stated in _____.

Section 13.4

colligative property	heat of solution
molality (m)	

30. The energy released or absorbed when a solute dissolves in a solution is the _____.

Chapter 13

Chapter 13 review

Section 13.4

colligative property	heat of solution
molality (m)	

31. Being measured as moles of solute per kilogram of solvent, _____ is a concentration unit that does not vary with temperature.

32. A(n) _____ is a physical property that depends on the number of solute particles in solution but not the type of solute particle.

Conceptual questions

Section 13.1

33. Substance X will dissolve in water. What does this say about the substance's polarity?

34. Why is the major solvent in the human body water instead of a nonpolar solvent?

35. What is meant by the phrase like dissolves like?

36. Why do electrolyte solutions conduct electricity very well while molecular solutions do not?

37. What does a reading of high conductivity tell you about a solution?

38. Solids can act as the solvent in a solution. Explain how this type of solution would form.

39. The formula for hexane is C_6H_{14}.
    a. Draw the Lewis structure for mineral spirits.
    b. Is the molecule polar or nonpolar?
    c. Would mineral spirits be soluble in water? Why or why not?

40. Draw a picture of a sodium ion (Na^+) surrounded by 4 water molecules. Be sure to draw the water molecules in the correct position around the ion according to attractive forces.

41. Glucose ($C_6H_{12}O_6$) is mixed with water.
    a. Describe what happens to solute and solvent particles on the molecular level as the two combine to form a solution.
    b. Provide a sketch for your explanation. Make sure to label the solute and solvent in your drawing.

42. Table salt is an ionic compound that dissolves in water.
    a. Explain what happens to the solute and solvent on the atomic level.
    b. Provide a drawing for your explanation. Make sure to label the solute and solvent in your drawing.

43. How can a light source help you determine whether a mixture is a solution or a heterogeneous mixture?

44. When you make a solution of lime-flavored gelatin with a powdered mix,
    a. what is the solute, and what is the solvent? And,
    b. how can you distinguish the solute from the solvent?

Section 13.2

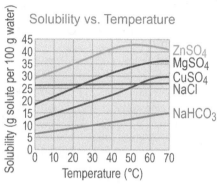

45. According to the graph above, how many grams of zinc sulfate ($ZnSO_4$) will dissolve per 100 grams water at:
    a. 20 °C
    b. 35 °C
    c. 50 °C

46. Use the graph in the previous question to answer the following:
    a. At 40 °C which substance has greater solubility: NaCl or $CuSO_4$?
    b. At 60 °C which substance has greater solubility: NaCl or $CuSO_4$?
    c. At what temperature will NaCl and $CuSO_4$ have equal solubility?

47. Suppose you are dissolving cocoa powder in milk to make chocolate milk. How will each scenario effect the rate of solution formation? Explain your answers.
    a. Warming up the milk
    b. Stirring the solution
    c. Increasing air pressure on the solution
    d. Using chocolate syrup instead of powder

48. Which has greater surface area: a solid 10 g chocolate bar or 10 g of chocolate chunks? Explain your answer.

49. Which will dissolve faster: 10 g of fine-grain salt or 10 g of rock salt? Explain your answer.

50. Explain why supersaturated solutions begin crystallizing when a seed crystal is added to the solution.

51. 5.0 mL of a liquid solute is added to 45.0 mL of a liquid solvent. The total solution volume equals 47 mL instead of 50 mL.
    a. Explain how it is possible for the total solution volume to be less than the sum of the individual solute and solvent volumes.
    b. Would you expect the total solution mass to be less than the sum of individual solute and solvent volumes? Why or why not?

442 Chapter 13: Solutions

Chapter 13 review

Section 13.2

52. Why are gas solutes less soluble in warmer solutions?

53. Suppose you have several 8-oz. glasses of water and a container of table salt. You use a stopwatch to help determine the rate of solution formation when 10.0 g of salt are added to an 8-oz glass of water at room temperature, 23 °C, and the solution is stirred. Decide whether the rate of solution formation will be faster or slower than the observed rate if each of the following are performed in a new glass of water:

    a. The water in the glass is chilled to 5 °C.
    b. 20.0 g of salt instead of 10.0 grams of salt are added to the water.
    c. Half the water is emptied from the glass.
    d. The water in the glass is heated up.

54. Suppose you have a 2.5 M solution of KCl and a 2.5 M solution of $AlCl_3$. Assuming both are water-soluble, which solution will likely conduct more electricity? Explain your answer.

55. Which solution would have the highest conductivity, a 1.0 M solution of LiCl or a 2.0 M solution of ethanol (C_2H_6O)? Explain your answer.

56. Apply solubility rules for each of the reactant pairs to write a balanced reaction with appropriate product states. Assume the charge on the metal does not change.

    a. $CuSO_4(aq) + Na_2S(aq) \rightarrow$
    b. $AgNO_3(aq) + NaCl(aq) \rightarrow$
    c. $Pb(NO_3)_2(aq) + K_2S(aq) \rightarrow$

Section 13.3

57. Research a past or current water, soil or air pollution issue near your home. Present your findings in a formal letter to your local congressional representative. Include an evidence-based argument about why the issue should be addressed. Acknowledge the perspective of the person(s) responsible for the pollution, but suggest a way that a person or persons can improve the pollution they created.

58. Environmental groups and corporations sometimes clash over the best way to address an industrial pollution issue. Research one such issue, and identify at least two strategies that each points of view support. What are the human, economic, and environmental costs and benefits associated with each strategy?

59. Use the accepted symbols to abbreviate the following phrase: A concentration of 0.65 molar sodium chloride.

60. Describe how you would increase the concentration of a solution of hot chocolate.

61. There are several ways to calculate concentration. Define 3 of them.

62. Why do chemists prefer a unit like molarity when working with aqueous solutions?

63. Under what conditions is it useful to report concentrations in units of ppb?

64. When making an aqueous solution of known molarity, what calculation(s) will you need to perform with:

    a. a solid solute
    b. a stock solution

65. What does Beer's law mean when it states that absorbance and solution concentration are directly proportional?

66. A high transmittance reading from a colorimeter tells you what about a solution?

67. What is the difference between a 1.0 M solution of potassium hydroxide and a 6.0 M solution of potassium hydroxide?

68. Instead of adding 40.0 mL of water as a dilution calls for, a student accidentally adds 41.8 mL of water. How does this affect the expected concentration of the dilute solution?

69. Why is important to measure the concentration of solutes like lead in a drinking water supply when the solutes are present in tiny amounts like parts per billion (ppb)?

70. Identify the symbol in the dilution equation,

$$M_1 \times V_1 = M_2 \times V_2,$$

that represents the following:

    a. Concentration of stock solution
    b. Concentration of dilute solution
    c. Volume of stock solution
    d. Volume of dilute solution

Section 13.4

71. Write a chemical equation that represents table salt, NaCl(s), dissolving in water. Make sure to include the appropriate state of matter for the products.

72. What does the ocean have in it that makes it not freeze as easily as a lake or pond. Explain your answer in terms of colligative effects.

73. Discuss how the density of a solution is affected by the addition of a solute?

74. When a solute is dissolved by a solvent, sometimes there is a change in solution temperature. Does energy flow towards or away from the solute when:

    a. The solution temperature increases
    b. The solution temperature decreases

75. On which side of the reaction equation is heat written when a salt dissolves and the solution absorbs heat (on the products side or the reactants side)? Explain your answer.

76. The water molecules either absorb heat or give heat to the solute when dissolving takes place in an aqueous solution. What are these two processes called?

77. If a solution has a positive heat of reaction when it forms, will the solution temperature increase or decrease during dissolution? Explain your answer.

Chapter 13 review

Section 13.4

78. Has the solute absorbed or released energy during the process of dissolving if the solution has a negative heat of reaction when it forms? Explain your answer.

79. Which has greater entropy: a solution or its pure solvent? Explain your answer.

80. Why does scattering salt on an icy sidewalk make the ice melt, even at temperatures below water's freezing point (0 °C)?

81. What would have to happen for the ice to re-form once salt melts ice on a sidewalk?

82. Why is molality used instead of molarity when applying the principles of colligative properties?

83. Which solution would reduce the freezing point more: a 1.0 M solution of NaCl or a 1.0 M solution of ethanol (C_2H_6O)? Explain your answer.

84. Which solution would reduce the freezing point more: a 0.5 M solution of a binary ionic compound or a 1.0 M solution of a covalent compound? Assume both substances are equally soluble. Explain your answer.

85. Which solution listed below would you expect to have the lowest freezing point, assuming all solutes are equally soluble? Explain your choice.

    a. 1 M LiCl
    b. 1 M Sugar, $C_{12}H_{22}O_{11}$
    c. 1 M $MgBr_2$
    d. 1 M KF

Quantitative problems

Section 13.3

86. Your lab partner has mixed 9.8 g of NaOH in 500.0 mL of water. What is the molarity of this solution?

87. How many moles of I⁻ are present in 60.0 mL of a 0.12 M CaI_2 solution?

88. What is the molarity of an acidic solution made by diluting 4.00 mL of a 12.1 M acid solution to a volume of 250.0 mL?

89. The directions on a "natural" single-serving drink mix state: dissolve 10.0 g of fruit sugar (fructose, $C_6H_{12}O_6$) in 150.0 g of water. Assume the density of water is 1.00 g/mL. Calculate the concentration of sugar in the "natural" solution as:

    a. Percent by mass
    b. Molarity
    c. Grams per liter

90. How many grams of HCl are in 65.0 mL of a 8.70 M solution?

91. What mass of KBr, in grams, is needed to prepare 250 mL of a 2.50 M solution?

92. What is the molarity of a stock acid solution if 250.0 mL of a 1.8 M solution is made by diluting 15 mL of the concentrated acid?

93. What is the final concentration when 50.0 mL of a 0.560 M glucose ($C_6H_{12}O_6$) solution is mixed with 130.0 mL of a 2.80 M glucose solution? Assume the volumes are additive.

94. The concentration of alcohol in an antiseptic solution is 0.10% by mass. How many grams of alcohol do you need to make 500 g of the antiseptic?

95. For a solution containing 47.0 mL of 6.68 M KOH:

    a. How many moles of KOH are in the solution?
    b. How many grams of KOH are in the solution?

96. Calculate the number of moles of ions in one liter of a 0.73 M potassium iodide(KI) solution.

97. How many grams of sweetener are dissolved in a bottle of soda if a 1 L bottle of soda has a sweetener concentration of 20 ppm?

98. The city of Flint, Michigan suffered a water quality crisis in 2015. Lead levels in the city's drinking water exceeded the 15 ppb safety limit to the point where a 1 L sample of tap water would contain as much as 0.104 mg of lead.

    a. What was the concentration of lead in the water (in ppb)? Assume the density of water is 1.00 g/mL.
    b. By how much does this lead concentration exceed the limit?

99. A recipe calls for 1 tablespoon of salt in 1 liter of water. Given that one tablespoon of salt has a mass of approximately 5.0 grams, what is the molarity of the solution?

100. If there are 35 g of $NaHCO_3$ dissolved in 2.5 L of solution, what is the percent mass of $NaHCO_3$ in the solution? Assume the density of water is 1.0 g/mL.

101. If you have 23.0 g of sugar how much water do you need to add to create a solution that has a concentration of 70 g/L of sugar?

102. You are preparing a dilution from 3.92 M to 0.0500 M. If you use 5.0 mL stock solution:

    a. What is the volume of the dilute solution being made?
    b. How much distilled water do you need?

103. If you add 51.4 g of $MgCl_2$ to 1 L of pure water, how concentrated will the solution be? Assume the density of water is 1.0 g/mL and calculate:

    a. The percent by mass of $MgCl_2$ in the solution.
    b. The concentration in g/L.
    c. The molarity of this solution.

Chapter 13 review

Section 13.3

104. Calculate the molarity of each of the following solutions.

    a. 22.5 g of ethanol (C_2H_5OH) in 466 mL of solution.
    b. 16.9 g of sucrose ($C_{12}H_{22}O_{11}$) in 77.6 mL of solution
    c. 11.0 g of sodium chloride (NaCl) in 97.2 mL of solution.

105. What is the volume in mL of the following solutions?

    a. A 0.450 M solution containing 3.50 g of KF.
    b. A 1.8 M solution containing 5.51 g of ethanol (C_2H_5OH).
    c. A 0.50 M solution containing 2.75 g of acetic acid (CH_3COOH).

106. A reaction calls for 150.0 mL of a 0.100 M solution. How much 4.56 M stock solution do you need to dilute to get the desired solution?

107. Certain fish thrive in rivers that contain at least an average of 6.0 ppm of dissolved oxygen (O_2) gas. Suppose you measured the O_2 concentration in a river during the wet season. You found on average there are 0.75 mg of O_2 per 250 mL water.

    a. Calculate the concentration of O_2 in the river in ppm.
    b. Is the river a healthy habitat for the fish? Why or why not?
    c. Rivers experience warming during drought years. What effect would a warmer river have on O_2 concentration? Hint: O_2 is dissolved as a gas!

Section 13.4

108. What is the molality of each of the following solutions?

    a. 50.0 g of solid $CuSO_4$ are dissolved in 1.5 kg of water.
    b. 70.0 g of LiI have been mixed into 300.0 g of water
    c. There are 26.5 g of $MgCl_2$ dissolved in 1,450 g of water.

109. The chemical name for antifreeze is ethylene glycol ($C_2H_6O_2$). What is the freezing point of a 3.50 m aqueous solution of antifreeze? For water, K_f = 1.86 °C/m.

110. A 2.82 m aqueous solution is made with sodium carbonate (Na_2CO_3) as the solute.

    a. Calculate the solution's boiling point. For water, K_b = 0.520 °C/m.
    b. Calculate the solution freezing point. For water, K_f = 1.86 °C/m.
    c. Can the freezing point of a solution be changed by adding solute without also changing the boiling point? Why or why not?

111. What is the boiling point of a 1.76 m aqueous solution of potassium chloride (KCl)? For water, K_b = 0.52 °C/m.

112. What mass of sulfuric acid (H_2SO_4) must be dissolved in 2,300 mL of water to make a 3.48 m solution?

113. The boiling point elevation constant (K_b) for water is 0.52 °C/m.

    a. How much will the boiling point change when 13.6 g of ammonia (NH_3) is dissolved in 300.5 g of water?
    b. What is the solution's new boiling point?

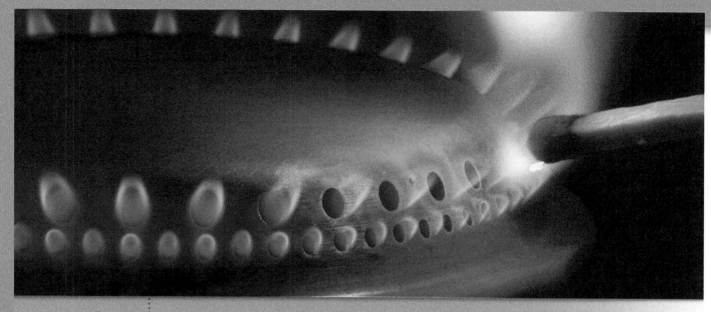

chapter 14 Reaction Rates

A scientifically important question is, "how long will it take?" Some reactions and physical processes are spontaneous, but they take a long time to occur. For example, under the right conditions carbon containing minerals will form diamonds. Unfortunately, the "right conditions" are very high temperatures, between 900-1,300 °C and pressures around 4.5×10^6 kPa. Even under these extreme conditions it still takes 1 billion years to complete the reaction. In contrast, rusting does not take billions of years, but is a process that happens over time. Usually we do not want rusting to occur, so we have devised ways to inhibit or slow down this reaction. One way we do this is by galvanizing metals.

Galvanization is a process in which zinc is used to coat materials that contain iron. Iron is very reactive and will rust quickly when it is exposed to air and moisture. Zinc is significantly less reactive and does not react with the atmosphere. Zinc is used to coat iron and steel structures so that they are corrosion resistant. Zinc can be painted on to the surface or applied in it molten form. Galvanization is used to slow down a reaction. Other chemicals are used to speed up reactions.

Sometimes a reaction needs a boost of energy or a little assistance from another substance to react fast enough. When you turn on a gas grill or the burner in your classroom, the oxygen and fuel are mixing, although initially nothing happens. These combustion reactions are spontaneous, but the rate is essentially zero until you add some energy in the form of a spark or a match. That little boost of energy gets some of the reactant molecules over an energy barrier and makes the reaction happen at a much faster rate. In our bodies and other biological systems reactions must happen at a fast rate, and we certainly cannot add sparks to make things happen more quickly. Instead, chemicals called enzymes are used to speed up the reaction. The presence of an enzyme changes the path from reactants to products with a new lower barrier.

The speed of a reaction is an important topic in chemistry. Understanding how to measure the speed of a reaction and the factors that can speed it up, or slow it down, is... essential.

Chapter 14 study guide

Chapter Preview

A balanced equation tells us what happens during a chemical reaction. But how do we know how quickly a reaction takes place? In this chapter you will learn about reaction rates and the factors that influence them. Reaction rates are measured as a change in concentration of reactants or products over time. The rate of a reaction isn't constant. It changes based on the concentration of reactants. Collision theory helps to explain what is necessary for a reaction to take place and why concentration, temperature and surface area affect reaction rates. Catalysts can be used to speed up reactions. Biological catalysts are known as enzymes. Up until now we've treated chemical reactions as a single step. But in reality there are often several steps in the reaction pathway with intermediate substances formed in between steps. The slowest step controls the overall speed of the reaction and is known as the rate-determining step.

Learning objectives

By the end of this chapter you should be able to:

- explain how certain factors influence reaction rate;
- calculate the rate of product appearance or rate of reactant disappearance for a given reaction;
- describe the activation energy (E_a) for a reaction and how it is affected by the presence of a catalyst;
- interpret a reaction profile and identify when intermediates are formed;
- describe the rate-determining step of a reaction and how it functions as a bottleneck for the reaction.

Investigations

14A: Optimum conditions

14B: Catalysts

Pages in this chapter

- 448 Reaction Rates
- 449 Rate of reaction
- 450 The collision theory
- 451 Factors affecting rate
- 452 Rate laws
- 453 Rate laws and units
- 454 Using the rate law
- 455 Zero, first and second order graphs
- 456 Using graphs to determine order
- 457 Section 1 Review
- 458 Catalysis
- 459 Enzymes
- 460 Catalysts and the environment
- 461 Catalysts and the world around us
- 462 Section 2 Review
- 463 Chemical Pathways
- 464 Two-step reaction mechanisms
- 465 Visualizing reactions with a curve
- 466 Reaction pathways
- 467 A biological pathway
- 468 Section 3 Review
- 469 Reaction Rates and Expiration Dates
- 470 Reaction Rates and Expiration Dates - 2
- 471 Reaction Rates and Expiration Dates - 3
- 472 Chapter review

Vocabulary

reaction rate
colligative property
enzymes
intermediate
rate law

activation energy
catalyst
elementary steps
unimolecular
rate determining step

activated complex
inhibitor
reaction mechanism
bimolecular

14.1 - Reaction Rates

Some reactions are fast, like a match burning. Some are slower, like milk spoiling when it is left out. The process of rusting occurs slowly over time when a metal object, such as a nail, is left out in the rain. The speeds of chemical reactions are affected by several factors, such as temperature, concentration, surface area and the presence of a catalyst.

Calculating rates

Speed and rates

You can measure how fast you are driving in a car by reading the speedometer. The speedometer tells us our speed in miles per hour (mph). When you travel at a speed of 70 mph, you will cover a distance of 70 miles in one hour. However, your speed is not 70 mph the entire time you are traveling. When you are preparing to enter a highway, you accelerate the car from 0 mph to 70 mph in a few seconds. When you exit the highway, you dramatically slow the car down. Let's say it took you half an hour to travel 26 miles. To find the average speed traveled in the car, divide the distance traveled by time:

$$\text{Speed (rate)} = \frac{\text{Distance}}{\text{Time}} = \frac{26 \text{ miles}}{30 \text{ min}} = 0.87 \frac{\text{miles}}{\text{min}} \text{ or } 52 \frac{\text{miles}}{\text{hr}}$$

The speed of the car varies during acceleration. Point A in the graph below shows that the car moves at a slower speed when first beginning to accelerate compared to point B, which represents the moment just before a constant speed (point C) is reached. Look at the slope of position A compared to position B. The slope indicates the car is speeding up more quickly at point B than at point A. The slope beyond point C remains the same because a constant speed has been reached. The rate of change is faster once the car is already moving fast compared to when the car first started accelerating from from zero.

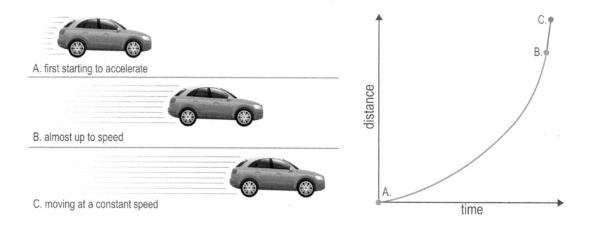

This example shows that the rate or speed can change while you are driving a car. The rate of a chemical reaction changes over time as well. The difference is that chemical reactions have a faster rate at the beginning and a slower rate at the end. Chemical reactions tend to slow down over time. We measure the rate of a chemical reaction by measuring a change in the amount of reactant or product per unit time.

$$\text{rate} = \frac{\text{change in concentration}}{\text{change in time}}$$

Rate of reaction

The rate of reactions

Chemists measure the change in concentration per unit time. The **reaction rate** is used to describe how reactions evolve over time. For a given generic reaction where A → C, we can measure the rate at which the reaction proceeds by measuring the decrease in the concentration of reactant A over time, or by measuring the increase in concentration of product C over time. As the reaction proceeds the amount of reactant A decreases, because it is being converted to the product C.

Reaction progress

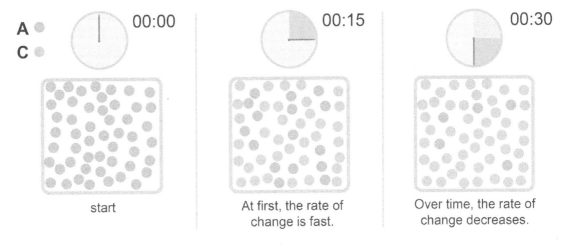

Reaction: A → C

Compare the amount of product made over time in each of the three diagrams above. Notice the time interval between each picture is constant at 15 seconds, but the rate of product formation decreases. Why? Because as time moves forward, there are fewer and fewer molecules of A to convert to C.

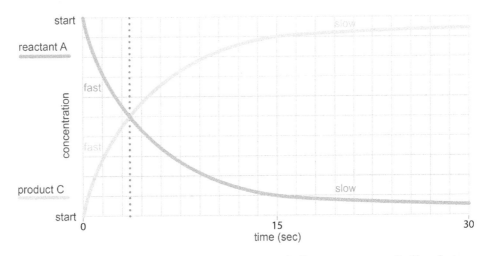

Reaction rate stoichiometry

The mole ratio relates the rates of appearance and disappearance of all substances in the reaction. If the rate of one substance is known, use stoichiometry to determine other rates.

Solved problem

According to the reaction below, how quickly will N_2 appear if the rate of disappearance of oxygen is 0.050 mol/s? $4NH_3(g) + 3O_2(g) \rightarrow 2N_2(g) + 6H_2O(g)$

Relationships: Multiply the rate by the mole ratio to solve for each unknown rate

Solve: Rate of N_2 appearance: $\dfrac{0.050 \text{ mol } O_2}{s} \times \dfrac{2 \text{ mol } N_2}{3 \text{ mol } O_2} = 0.033 \text{ mol } N_2/s$

Answer: Nitrogen gas, N_2, will appear at a rate of 0.033 mol/s.

Section 14.1: Reaction Rates

The collision theory

Molecular collisions cause reactions

Let's think about chemical reactions at the particle level. Reactants must collide with each other in order for a reaction to occur. If reactants never contacted one another, how could atoms be exchanged? As it turns out, not every collision between reactants results in a chemical change. Products only form when "successful" reactant collisions occur.

Successful collisions

In a pool game, you use a cue ball to roll a colored ball into a pocket. You must line up the cue ball with the colored ball to roll it towards a pocket. You also have to hit the cue ball at a proper angle to get the colored ball to roll in the desired direction. Another factor for success is how hard you hit the cue ball. If you do not hit the cue ball hard enough, it will not reach the pocket. You must hit the cue ball hard enough, at the correct angle, correctly lined up with the desired colored ball to successfully roll the ball into the pocket.

Collision theory

According to collision theory, a collision between reactants is considered successful only when both of the following conditions have been met:

1. Reactants must collide with enough energy to break reactant chemical bonds.
2. Each colliding reactant particle must be oriented in the proper position so once the bonds break they will form products.

Reactions must be positioned properly

Reactants have to hit each other to react. Bonds are only broken if the reactants hit each other with enough energy. Reacting particles must be aligned in such a way that when the bonds are broken, they are in contact with the correct atoms to form products. If they are misaligned, the reacting particles will re-form. Reactants must be aligned in such a way that once the bond is broken, the closest available atom is one that will form products.

The reaction energy profile

A reaction energy profile summarizes energy changes during a reaction. The **activation energy**, E_a, is the minimum amount of energy required for molecules to react. An **activated complex**, A_c, forms at the highest point of the reaction profile. The A_c is the product of successful collisions. When the A_c forms, it lasts for only a fraction of a second, and the colliding particles may form product or they may go back and reform reactants. The activated complex forms at such high energy that it is very unstable and it cannot be isolated in the lab. At the peak of the graph where the A_c sits, the likelihood that products will be formed equals the likelihood that reactants will reform.

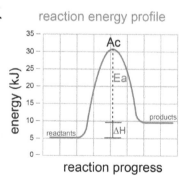

Positive and negative ΔH

The graph shows a positive enthalpy change, ΔH, for the reaction. The graph shows an endothermic reaction because product energy is higher than reactant energy. If the opposite were true, this would be an exothermic reaction with a -ΔH.

Factors affecting rate

Reaction speed

Changing the speed of a reaction is actually about changing the number of collisions. The greater the force and the more often particles collide, the faster they will react. What factors can be adjusted to control the rate at which a reaction takes place?

Concentration

1. **The concentration of the reacting chemicals.** Many chemical reactions happen more rapidly if the reactant concentrations are increased. The more reactants present the faster the reaction occurs because there are greater numbers of collisions.

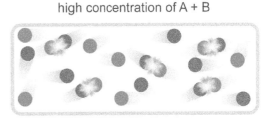

Temperature

2. **The temperature at which the reaction occurs.** For most reactions the higher the temperature the faster the reaction proceeds. Adding heat from a stove speeds up the chemical reactions that take place when cooking food. Temperature is a measure of the average kinetic energy of a substance. The higher the kinetic energy is, the greater the number of molecules that collide with the necessary activation energy. If we raise the temperature of a reaction mixture, more molecules will be moving fast enough to have effective collisions that provide the activation energy needed to from products.

Surface area

3. **Surface area—size of the reacting particles.** A reaction is most affected by surface area if the reactants are in solid form. The larger the surface area the faster the reaction will take place because there are more collisions of reacting particles. A medicine will be absorbed more quickly by our bodies if we ingest it in the form of a powder (held in a dissolvable capsule), rather than in the form of a solid pill. This is because our stomach can dissolve a powder more quickly than a pill. Consider making a sugar water solution with a sugar cube compared to loose granulated sugar. Loose sugar grains will contact more water than a sugar cube because the granulated sugar has a larger surface exposed to the water. More contact means more collisions.

Catalyst

4. **The addition of a catalyst.** A catalyst is a substance that facilitates a chemical reaction without being used up itself. It helps the chemicals to react, but it does not break down or become chemically changed. In a catalytic converter, a transition metal such as rhodium acts as a catalyst. It provides a surface for molecules to attach to, which lines the particles up so that they are in the right position for a successful collision. The rhodium metal itself does not participate in the chemical reaction, but it reduces the polluting gases that would be otherwise emitted in the exhaust.

Section 14.1: Reaction Rates

Rate laws

Rate law

The speed of reactions can be classified into categories or orders. Chemical reactions are typically either zero, first or second order. The reaction order is an exponent in a rate law. Rate laws have the following general form;

$$\text{Rate} = k[A]^m[B]^n$$

A and B are reactants and the exponents and the brackets represent concentration in molarity. The exponent's m and n are the orders with respect to A and B; k is the rate constant and changes with every reaction.

Zero order

For now, let's just consider just one reactant. When a reactant has a concentration of 2.0 M and is zero order, the rate law becomes, Rate= $k[2.0]^0$. Any number that has any exponent of zero becomes the number one. Therefore when the concentration is increased to 4.0M the reaction rate does not change. When a reactant is zero order, its concentration has no effect on the reaction rate.

First order

When a reactant has a concentration of 2.0 M and is first order, the rate law becomes, Rate= $k[2.0]^1$. An exponent of one does not change the base number. Therefore when the concentration is doubled and becomes 4.0M the reaction rate also doubles. When a reactant is first order, its concentration has a direct effect on reaction rate.

Second order

When a reactant has a concentration of 2.0 M and is second order, the rate law becomes, Rate= $k[2.0]^2$. An exponent of two exponentially changes the base number. Therefore when the concentration is doubled and becomes 4.0M the reaction rate increases by a factor of 16. When a reactant is second order, its concentration has an exponential effect on reaction rate.

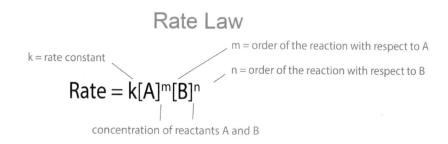

Two reactants

For a reaction such as, $2NO + O_2 \rightarrow 2NO_2$. The order with respect to NO does not have to be the same as the order with respect to O_2. If NO is second order and O_2 is first order the rate law becomes,

$$\text{Rate} = k[NO]^2[O_2]^1$$

The rate constant

The rate constant is dependent on experimental conditions and the order of the reacting species. The shape of the reacting container, the ways reactants flow into the container and micro fluctuations in heating all affect the rate constant. It has to be determined independently for every experimental situation.

Rate laws and units

The units on k

Notice the units on the rate constant. Liters squared should seem very strange. The units on k depend on the orders of the reacting species. Here is how they are calculated. Rate has units of M/s or another way to write it is mol/L·s. The time unit could be seconds, s, minutes, min, hours, hr, and so on. NO is second order so the units of mol/L have to be squared mol^2/L^2 and O_2 is a first order so the unit are mol/L. Putting just the units into the rate law looks like this;

$$\frac{mol}{L \times s} = k \frac{mol^2}{L^2} \times \frac{mol}{L}$$

$$\frac{mol}{L \times s} = k \frac{mol^3}{L^3}$$

$$\frac{mol}{L \times s} \times \frac{L^3}{mol^3} = k$$

$$\frac{L^2}{mol^2 \times s} = k$$

Putting it all together

The complete rate law is, Rate = 25 L^2/mol^2 s $[NO]^2[O_2]^1$. Using this form of the rate law, we can determine the rate of reaction for any concentration of NO and O_2.

Solved problem

What is the rate for Trial 4?

CO + NO$_2$ → CO$_2$ + NO			
Trial #	[CO]	[NO$_2$]	Rate mol/L·hr
1	1.0 x 10^{-3}	7.2 x 10^{-3}	3.4 x 10^{-8}
2	1.0 x 10^{-3}	3.6 x 10^{-3}	1.7 x 10^{-8}
3	2.0 x 10^{-3}	7.2 x 10^{-3}	6.8 x 10^{-8}
4	3.0 x 10^{-3}	1.4 x 10^{-2}	?

Relationships Rate = $k[CO]^m[NO_2]^n$

Solve Step 1: First determine the rate with respect to CO and NO_2

- When comparing Trials one and two, the concentration CO does not change but the concentration NO_2 is cut in half. The reaction rate also gets cut in half, so the reaction is first order with respect to NO_2.
- When comparing Trials one and three, the concentration of NO_2 does not change but the concentration CO doubles. The reaction rate also doubles, the reaction is also first order with respect to CO.
- For now, the rate law is Rate = $k[CO]^1[NO_2]^1$

Step 2: Determine the Rate Constant and Units

- Using Trial 3, substitute values into the rate law to find k.
  Rate = $k[CO]^1[NO_2]^1$
  6.8 x 10^{-8} mol/L·hr = k [2.0 x 10^{-3}]^1[7.2 x 10^{-3}]^1

$$k = \frac{6.8 \times 10^{-8} \frac{mol}{L \cdot hr}}{[2.0 \times 10^{-3} \frac{mol}{L}]^1 \times [7.2 \times 10^{-3} \frac{mol}{L}]^1} = \frac{6.8 \times 10^{-8} L}{1.44 \times 10^{-5} mol \cdot hr} = 4.7 \times 10^{-3} \frac{L}{mol \cdot hr}$$

Step 3: Determine the rate of Trial 4

- Put trial four values into the rate law.

$$\text{Rate} = 4.7 \times 10^{-3} \frac{L}{mol \cdot hr} \times \left[3.0 \times 10^{-3} \frac{mol}{L}\right] \times \left[1.4 \times 10^{-2} \frac{mol}{L}\right]^1 = 2.0 \times 10^{-7} \frac{mol}{L \cdot hr}$$

Answer Trial four will occur at 2.0 x 10^{-7} mol/L·hr.

Using the rate law

Using rate laws

Rate laws are used to determine how fast a reaction will run. Nitric oxide, NO, is a product of fossil fuel combustion. NO reacts with oxygen, O_2 in the air to produce nitrogen dioxide, NO_2, a brownish gas that contributes to smog. The rate law can be used to determine how quickly smog will form. This information is useful during emissions testing, to see how much pollution a vehicle or factory may contribute to air pollution.

Concentration data

The table below contains results for three experimental trials of the reaction,

$$2NO + O_2 \rightarrow 2NO_2.$$

A researcher varied the concentration of NO and O_2 and recorded the changes in reaction rate.

reaction rates for $2NO + O_2 \rightarrow 2NO_2$

Experimental Trial	Concentration NO (M)	Concentration O_2 (M)	Reaction Rate (M/s)
1	0.01	0.01	2.5×10^{-5}
2	0.02	0.01	1.0×10^{-4}
3	0.01	0.02	5.0×10^{-5}

Trials 1 and 2

Comparing trials 1 and 2 we see that the researcher doubled the concentration of NO and kept the concentration of O_2 constant. This made the rate increase from 2.5×10^{-5} M/s to 1.0×10^{-4} M/s. The rate increased by a factor of 4. This is an exponential increase. Therefore the reaction must be second order with respect to NO.

Trials 1 and 3

Comparing trials 1 and 3 we see that the researcher doubled the concentration of O_2 and kept the concentration of NO constant. This made the rate increase from 2.5×10^{-5} M/s to 5.0×10^{-5} M/s. The rate increased by a factor of 2. This is a direct increase. Therefore the reaction must be first order with respect to O_2. The rate law for this reaction becomes:

$$\text{Rate} = k[NO]^2[O_2]^1$$

Determining the rate constant

To determine the rate constant, we will substitute the data from any one of the trials into the rate expression. Let's arbitrarily chose trial 2.

$$\text{Rate} = k[NO]^2[O_2]^1$$
$$1.0 \times 10^{-4} \text{ M/s} = k[0.02 \text{ M}]^2[0.01 \text{ M}]^1$$
$$k = \frac{1.0 \times 10^{-4} \text{ M/s}}{[0.02 \text{ M}]^2[0.01 \text{ M}]^1}$$
$$k = \frac{0.0001 \text{ M/s}}{0.0004 \text{ M}^2 \times 0.01 \text{ M}} = 25 \text{ M}^{-2}\text{s}^{-1} \text{ or } 25 \text{ L}^2/\text{mol}^2\text{s}$$

Zero, first and second order graphs

Zero order The orders of reactions can also be determined graphically. Three plots need to be made: the concentration of the reacting species vs. time, the natural log of the concentration of the reacting species vs. time and inverse concentration of the reacting species vs. time. If the plot of concentration vs. time is the most linear of the three plots, the reaction is a zero order reaction.

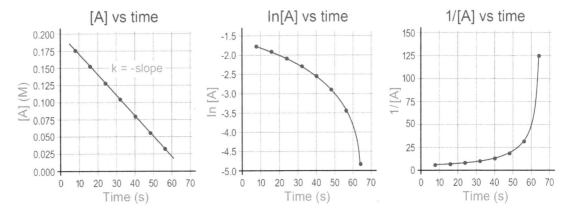

First order If the plot of the natural log of the concentration of the reacting species vs. time is the most linear of the three plots, the reaction is a first order reaction.

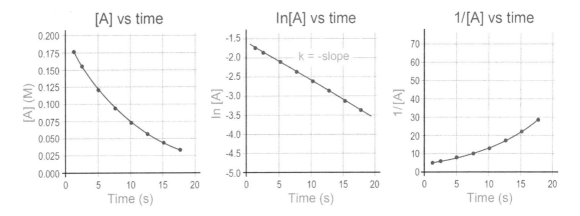

Second order If the plot of the inverse concentration of the reacting species vs. time is the most linear of the three plots, the reaction is a second order reaction.

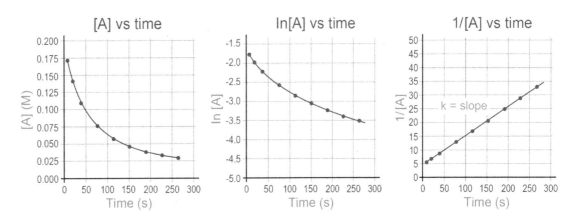

Using graphs to determine order

Solved problem

The following data was collected for NO_2. What is the order with respect to NO_2? Graph the following data and determine order of the reaction.

Given

$$2NO + O_2 \rightarrow 2NO_2$$

Time (s)	[NO_2]	ln[NO_2]	1/[NO_2]
0	1.00 x 10^{-2}	−4.605	100
60	6.83 x 10^{-3}	−4.986	146
120	5.18 x 10^{-3}	−5.263	193
180	4.18 x 10^{-3}	−5.477	239
240	3.50 x 10^{-3}	−5.655	286
300	3.01 x 10^{-3}	−5.806	332
360	2.64 x 10^{-3}	−5.937	379

Relationships The shape of the graph helps you determine reaction order.

- If the plot of Concentration vs Time is the most linear the reaction is zero order.
- If the plot of ln(Concentration) vs Time is the most linear the reaction is first order.
- If the plot of 1/(Concentration) vs Time is the most linear the reaction is second order.

Solve Graph each expression of concentration in the table:

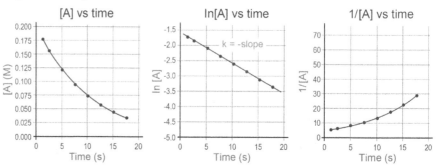

Answer The reaction is second order with respect to NO_2, because the plot of 1/[NO_2] is the most linear.

Overall Order

The overall order is found by adding the exponents in the rate law

$$\text{Rate} = 25 \, L^2/mol^2 \, s \, [NO]^2 [O_2]^1$$

2 + 1 = 3

In this reaction, the overall order is three.

Overall order

The overall order of a simple reaction is calculated by adding the reactant exponents in the rate law. The rate law for the reaction $2NO + O_2 \rightarrow 2NO_2$ is:

$$\text{Rate} = (25 \, L^2/mol^2 s)[NO]^2[O_2]^1$$

The overall order is 2 + 1 = 3 because the exponent on NO is 2 and the exponent on O_2 is 1.

Section 1 review

Chapter 14

Reaction rate is a measure of the speed at which a chemical reaction proceeds. Reaction rate can be calculated based on the change in concentration of a substance over time. The reaction rate generally slows as the reaction proceeds because there is less reactant left to convert. The mole ratio of a balanced equation determines the relative rates of formation of reactants and products. A successful reaction occurs when reactant molecules collide with enough energy to break chemical bonds, and in a way that leaves the particles aligned to form the products. The minimum amount of energy is known as the activation energy. Reaction rates are affected by the concentration of the reactants, the temperature, and by the surface area of the reacting particles. Reactions can also be sped up by the addition of a catalyst.

Vocabulary words   reaction rate, activation energy, activated complex, colligative property, catalyst

Review problems and questions

1. Calculate the average rate of disappearance of substance A at the beginning of the reaction: $2A \rightarrow 2B + C$.

Data for the Reaction $2A \rightarrow 2B + C$ (in a one liter container)			
Time (seconds)	Moles A	Moles B	Moles C
0.	0.0100	0.0000	0.0000
50.	0.0079	0.0021	0.0011
100.	0.0065	0.0035	0.0018
150.	0.0055	0.0045	0.0023
200.	0.0048	0.0052	0.0026
250.	0.0043	0.0057	0.0029
300.	0.0038	0.0062	0.0031
350.	0.0034	0.0066	0.0033
400.	0.0031	0.0069	0.0035

2. Calculate the average rate of disappearance of Substance A at the end of the reaction: $2A \rightarrow 2B + C$.

3. Is reaction rate slower at the beginning of a reaction or towards the end? Defend your answer.

4. There are two ways to calculate the rate of appearance of either Product B or C. Calculate the rate of appearance of Product B by:
    a. Performing an average reaction rate calculation, and
    b. Using stoichiometry and the rate of disappearance solved for in Question 1.

5. Raising the temperature causes most reactions to speed up. Use the collision theory to explain why.

14.2 - Catalysis

A catalyst is a substance that can speed up a chemical reaction without being changed or used up during the reaction. Catalysts provide a way for the reaction to start with lower activation energy (E_a). A catalyst is not a reactant, but it helps reactants to have effective collisions more easily.

Reaction profiles

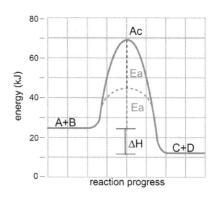

The reaction profile and catalysts

In the graph above, the reaction A + B → C + D is shown catalyzed and non-catalyzed. The activation energy of the non-catalyzed reaction shown in orange is (69 kJ - 25 kJ), which is equal to 44 kJ. The activation energy of the catalyzed reaction shown in green is (45 kJ - 25 kJ), which is equal to 20 kJ. Therefore, the catalyst reduced the activation energy by a factor of 24 kJ. With the lower activation energy, less energy input is needed to complete the reaction and more molecules have enough energy to react. The reaction now occurs at a faster rate overall.

Inhibitors

An **inhibitor** is a substance that slows a reaction down by increasing the activation energy. Penicillin and aspirin are inhibitors. Penicillin inhibits the production of cell walls in bacteria and aspirin reduces pain by interfering with molecules that cause swelling.

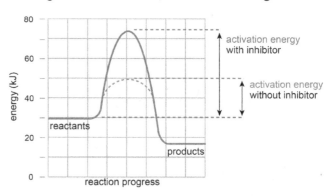

Enzymes

Catalysts in our bodies are called **enzymes**. We are alive right now because catalysts speed up complex reactions that would happen too slowly at normal body temperature.

Interactive simulation

Explore how catalysts change, or do not change during a reaction in the interactive simulation titled Catalysis. Investigate the impact of catalyst concentration on reaction rate. Look for evidence that catalysts are not reactants and observe how a catalyst helps bring reactants together to form products. Can you guess which substances are the "intermediates" that will be discussed in the next section?

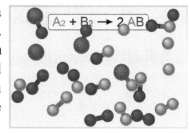

Enzymes

Enzymes

Carbonic anhydrase is an important enzyme in our bodies. It is responsible for the rapid conversion of carbon dioxide and water to hydrogen ion, H^+, and bicarbonate, HCO_3^-:

$$H_2O(l) + CO_2(g) \xrightarrow{\text{carbonic anhydrase}} H^+ + HCO_3^-$$

Carbonic anhydrase

Carbonic anhydrase helps to remove carbon dioxide, $CO_2(g)$ from your tissues by dissolving it in the form of HCO_3^- ion, which is carried to the lungs by our blood. Once it reaches our lungs, CO_2 converts back to its gaseous form and is exhaled. Without the carbonic anhydrase enzyme, we would not be able to effectively remove carbon dioxide from our tissues.

Soda tingle

If you take a sip of cold soda, you will feel a tingling on your tongue. How does this tingling sensation occur? Carbonated drinks contain dissolved carbon dioxide under high pressure. As you drink a soda, carbonic anhydrase in your saliva converts CO_2 into carbonic acid. The acid stimulates nerve endings on your tongue triggering a pain response that causes that tingling sensation.

Bromelain

Another natural enzyme, bromelain, is contained in certain fruits such as pineapple. Bromelain helps you digest specific protein molecules such as collagen. Collagen is found in many kinds of tissue such as skin and muscles. At the molecular level, bromelain breaks the bonds that hold the protein's structure together. Your mouth, lips, tongue and cheeks feel strange when you eat freshly-cut pineapple because bromelain is breaking down a small amount of collagen molecules in your mouth. When you eat canned or dried pineapple, your mouth does not feel as strange because enzymes themselves break down easily over time.

Experiment with bromelain

You can investigate the effect of bromelain on collagen with a simple experiment. Gelatin is derived from collagen. Make a batch of gelatin. Pour half of the liquid gelatin to set in one container. Add freshly-cut pineapple chunks to the remaining liquid and pour it in a second container. Allow both containers to set. What is the effect of bromelain on gelatin?

Catalysts and the environment

CFCs

The ozone (O_3) layer in the upper atmosphere absorbs up to 99% of the sun's harmful ultraviolet (UV) radiation before it reaches Earth's surface. Chemicals called chlorofluorocarbons (CFCs) were once widely used as propellants in aerosols, as refrigerants in air conditioners, and in foam production. Unfortunately, CFCs are also good at destroying ozone.

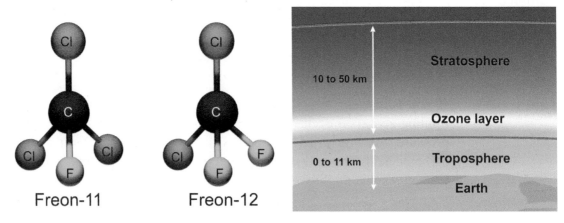

Ozone destruction

At Earth's surface, CFCs are mostly unreactive. CFCs are low-density gases that diffuse into the upper atmosphere where the sun's energy breaks off a chlorine atom. This forms a chlorine free radical, Cl•. Free radicals are short-lived, highly reactive chemicals that have unpaired electrons, indicated by the • symbol. Cl• free radicals act as catalysts in O_3 decomposition. Normally, O_3 is regenerated in the upper atmosphere as shown:

Natural Ozone Cycle (in the stratosphere)
Step 1: $O_3(g) + \text{UV (radiation)} \rightarrow \text{O•} + O_2(g)$
Step 2: $\text{O•} + O_2(g) \rightarrow O_3(g)$
Overall: $O_3(g) + \text{UV (radiation)} \rightarrow O_3(g)$

Although $O_3(g)$ is broken apart when it absorbs UV radiation, it is regenerated when O• combines with $O_2(g)$. Notice what happens when a chlorine free radical is present, shown below. You can see how ozone cannot be regenerated, leading to ozone depletion.

Ozone Depletion Mechanism
Step 1: $\text{Cl•} + O_3(g) \rightarrow \text{ClO} + O_2(g)$
Step 2: $\text{O•} + \text{ClO} \rightarrow \text{Cl•} + O_2(g)$
Overall: $\text{O•} + O_3 \rightarrow 2O_2$

Chlorine radicals

The Cl• free radical is a catalyst because it is present in the same form at the start of the reaction and at the end of the reaction. ClO is formed in the first reaction and then consumed by the second reaction, so it is a temporary molecule. The overall reaction shows that two O_2 molecules are formed instead of O_3. O_2 is unable to absorb harmful UV radiation, and the natural cycle to regenerate O_3 has been disrupted. One chlorine radical is estimated to catalyze the destruction of up to one million ozone molecules per second!

The ozone layer

The manufacture and use of most CFCs was banned by most countries internationally by 1996. Although the ozone layer has begun to recover, the effects of CFCs will linger for decades.

Catalysts and the world around us

Catalysts and reaction rate

Without a catalyst our only way to speed up a chemical reaction is to raise the temperature. This works for most chemical reactions, however, sometimes adding heat is not practical. Excess heat often causes the products to undergo unwanted reactions that decrease the percent yield. Heating a reaction can also be expensive or dangerous. Using a recyclable substance to facilitate the reaction can often be a better choice.

Industry and catalysts

Modern society relies on chemical process to make food, building materials and clothing. An industrial chemist's job is to find ways to speed up the production of chemicals that are in high demand. This is a very important factor for economic growth especially in developing countries. Most commercially produced chemicals use catalysts at some point in their production. Catalysts are also used to break down products like plastics into useful molecules for new substances.

Catalysts and the environment

An environmental chemist's job is to design ways to speed up reactions that reduce environmental pollution. The catalytic converter was invented in the 1930s by a chemist who was concerned about smog in Los Angeles, California. A catalytic converter is a device that uses a catalyst to speed up reactions that reduce the amount of harmful products of fossil fuel combustion in the air. Catalytic converters can be fitted to industrial and automobile exhaust systems. These devices are especially important to keep air clean in cities with a high population or large amounts of industrial manufacturing.

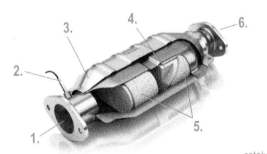

1. exhaust enters from engine
2. oxygen sensor
3. heat shield
4. catalyst for major reactions
5. catalyst for redox reactions
6. exhaust exits to muffler

car exhaust system with a catalytic converter

catalytic converter

Catalysts and nature

All living things need to use catalysts in the form of enzymes to speed up important metabolic reactions. We use enzymes in our bodies to help break down sugar molecules that provide energy and heat needed for everyday metabolism. The sugar-oxygen combustion reaction requires temperatures much higher than normal body temperature. Enzymes help these reactions take place at normal body temperatures.

Chapter 14
Section 2 review

Catalysts are substances that help speed up chemical reactions without being consumed in the process. Catalysts facilitate reactions by lowering the activation energy. Inhibitors do the opposite of catalysts. They slow reactions by increasing the activation energy required. Enzymes are biological catalysts and perform many critical functions in our bodies. Catalysts simply provide a way to efficiently speed up reactions. Sometimes that results in undesired effects such as ozone depletion, while in other cases it helps us offset damage done to the environment. Commercially, catalysts play a major role in producing important chemicals.

Vocabulary words inhibitor, enzymes

Review problems and questions

1. Use the graph to determine the activation energies of the catalyzed and uncatalyzed reaction of :
$$A + B \rightarrow C + D.$$

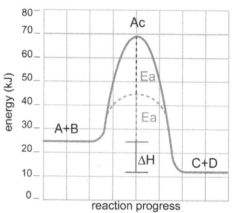

2. By how much does a catalyst reduce the activation energy for this reaction?

3. Describe at least one way you benefit from catalysts.

4. Describe at least one negative consequence of catalysts.

5. How do enzymes allow chemical reactions to happen faster?

5. Is an enzyme used up when it catalyzes a reaction?

6. Why are catalysts important in living things?

7. What are the importance of inhibitors?

7. Give two examples of inhibitors.

14.3 - Chemical Pathways

Chemical reactions do not always go straight from reactants to products. In fact, a balanced equation should be thought of as an abridged version of a reaction. The actual pathway is often several steps long.

Reaction mechanisms

Chemical reaction progression

Chemical reactions are usually written as though they go from reactant to product all in one step. Reactant atoms are rearranged to form new products in what seems like a single step. This is a major oversimplification. In reality, most chemical reactions occurs in a series of **elementary steps**. These smaller elementary steps combine to form a "pathway" for a chemical reaction.

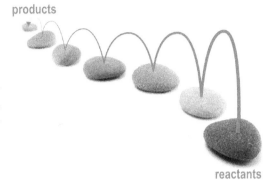

A series of steps

Think of a pathway with a series of stepping stones that connect the beginning and end of the path. It takes more than one step to convert reactants to products in many chemical reactions. The series of proposed steps a reaction takes is called a **reaction mechanism**. This proposed mechanism is based on experimental evidence supported observations in the laboratory. There can be more than one proposed reaction mechanism for a chemical reaction. Chemists do not know for sure which mechanism, or pathway, is correct because the high energy molecules formed along the steps of the reaction cannot be isolated. The high-energy molecules exist for only a small fraction of time.

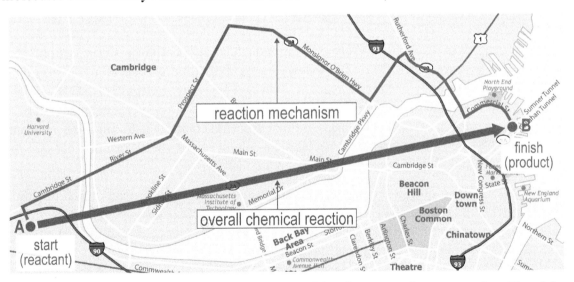

Reaction mechanism

A reaction mechanism is like following a map to drive from one place to another. You drive on a series of roads that take you to your destination. The reactants are like the starting point of your trip and the products are like your destination. The series of roads represents the elementary steps on the way from reactants to products. Together all of the elementary steps make up one single reaction mechanism. An overall chemical reaction only shows the start of the trip and the end of the trip, but it is impossible to go directly to your destination in a straight line. You have to follow whatever path the roads take. The reaction mechanism shows the roads you take on your trip, as well as where you start (reactants) and where you will finish (products).

Two-step reaction mechanisms

A reaction mechanism

Let's look at a reaction mechanism that has two elementary steps. We will investigate the reaction between nitric oxide gas, NO, and oxygen gas.

$$2NO(g) + O_2(g) \rightarrow 2NO_2(g)$$

Overall reaction

The reaction above suggests nitric oxide molecules collide with oxygen molecules, but that is not the case. Based on experimental evidence, scientists believe this reaction occurs in two steps:

Elementary step 1: $2NO(g) \rightarrow N_2O_2(g)$
Elementary step 2: $N_2O_2(g) + O_2(g) \rightarrow 2NO_2(g)$
Overall reaction: $2NO(g) + \cancel{N_2O_2(g)} + O_2(g) \rightarrow \cancel{N_2O_2(g)} + 2NO_2(g)$

The sum of the elementary steps must yield the overall balanced equation. All reactants are placed on the left side of the arrow, and all products are placed on the right side. Any molecules that appear as both a reactant and a product are canceled out of the reaction. In the first elementary step, two NO_2 molecules collide to form one N_2O_2 molecule.

step 1

This step is followed by the collision between N_2O_2 and an O_2 molecule.

step 2

Reaction Intermediates

It is much more likely that two molecules will collide with the correct energy and position in each step than three molecules, as shown by the overall reaction. The temporary N_2O_2 molecule is called a reaction intermediate. A reaction **intermediate** is a temporary product formed in an elementary step that is consumed in a later elementary step. Intermediates are the first new substances formed from reactants that become the reactant(s) for the next elementary step. Intermediates cancel out of the overall reaction because they are created in one step and are then used up in another step.

Why two steps?

Why does it take 2 steps for the reaction to occur, instead of just having 2 NO molecules react directly with O_2 molecule so the reaction can occur in just one step? Consider 3 people, each throwing a tennis ball so all 3 balls collide at the same time with the desired amount of energy and position. That would be difficult to do! It would be much easier if only 2 people were each throwing a tennis ball so both balls collide with one another with the desired energy and position. The chances of 3 tiny reactant molecules having an effective collision is practically zero compared to 2 reactant molecules.

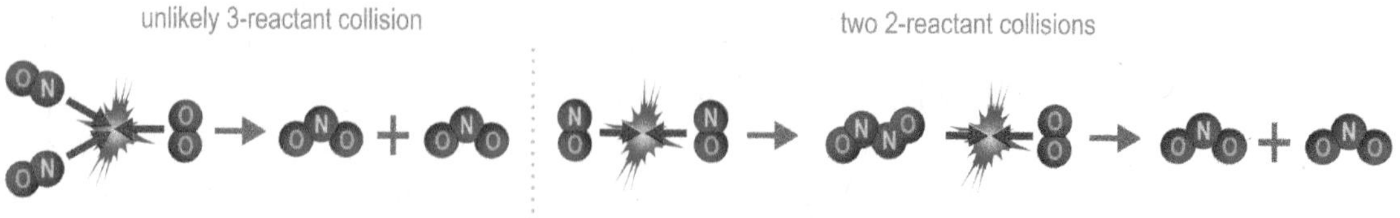

unlikely 3-reactant collision   two 2-reactant collisions

an effective 3-reactant collision is less likely to happen than two effective 2-molecule collisions

Visualizing reactions with a curve

Multiple elementary steps

The steps of the path that makes up a reaction mechanism can be graphed in the form of a reaction profile. The simplified reaction profile that you learned about earlier is only for one elementary step. Let's look at the reaction progress for a reaction with two elementary steps in the overall reaction.

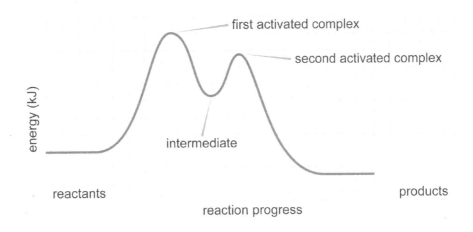

Graphing intermediates

The graph above shows that only one reaction intermediate has formed. You can locate the intermediate on the curve by looking for a dip (or valley) in the curve that has higher energy than the reactants. The reaction intermediate becomes a reactant in the next step of the reaction, which appears as the next peak.

You can see two peaks on the overall graph above. These peaks refer to the activation energies of the two activated complexes. There is one activated complex for each elementary step.

Analyzing graphs

Answer the following questions based on the graph below.

1. How many elementary steps do you see in this overall reaction?
2. How many intermediate products are formed?

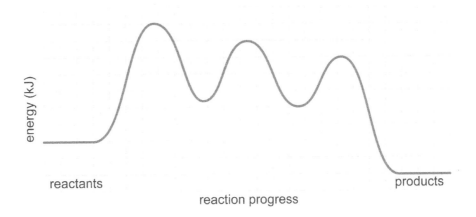

This reaction has 3 elementary steps, resulting in the formation of 2 intermediate molecules. You can think of this overall reaction having 3 smaller reactions. Like all reactions, each step in the overall reaction requires an energy input to proceed known as activation energy.

Section 14.3: Chemical Pathways

Reaction pathways

More complex reactions

As you saw in the graph on the previous page, some reactions have more than just two elementary steps. Here is an example of a proposed mechanism with three steps:

$$2HBr + O_2 \rightarrow H_2O_2 + Br_2$$

Elementary step 1: $HBr + O_2 \rightarrow HO_2Br$ (slow)
Elementary step 2: $HO_2Br + HBr \rightarrow 2HOBr$ (fast)
Elementary step 3: $2HOBr \rightarrow H_2O_2 + Br_2$ (fast)

Overall: $2HBr + O_2 + \cancel{HO_2Br} + \cancel{2HOBr} \rightarrow \cancel{HO_2Br} + \cancel{2HOBr} + H_2O_2 + Br_2$

Molecularity

Even though it seems like this 3-step reaction is complicated, notice the number of reactants in each step. The number of reactants in each step is called the molecularity. A **unimolecular** elementary step involves a single reactant dissociating or breaking apart. A **bimolecular** elementary step involves the collision of just two molecules. An overall reaction that involves three substances reacting, here two HBr molecules and one O_2 molecule, is highly unlikely! Having more steps with fewer simultaneous reactant collisions actually simplifies the overall reaction. Each step only requires one or two reactants to collide successfully, which is more probable based on collision theory.

The rate determining steps

Another aspect of a proposed reaction mechanism is that it must agree with the experimentally observed rate law. Recall that the rate law is an equation that predicts how the concentration of reactants will influence the speed of the chemical reaction. You can use the predicted reaction mechanism to predict the "slow" step or the **rate determining step**. This is the elementary process that takes the longest to complete, and therefore holds up progress in the overall reaction mechanism. The slow step involves the formation of a reaction intermediate.

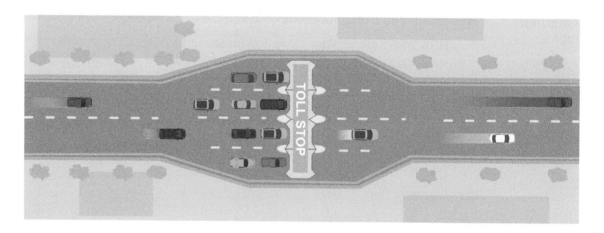

What makes a step "slow"?

The rate-determining step is like a tollbooth on a highway. The speed limit may be 55 mph on the highway, but all drivers must slow down and wait in line to pay for the toll. The tollbooth creates a bottleneck. Once you move past the tollbooth, you can resume your 55 mph speed. In a chemical reaction there is one step that has the highest activation energy, and it takes the longest to occur. In the three-step mechanism above, the first step is the slowest, and is therefore the rate-determining step. Remember the fact that even though the second step is fast, it cannot begin until the first step is complete.

A biological pathway

Biological pathways

All living things carry out complex chemical reactions as a series of steps, from tiny single-celled microorganisms to human beings. Biochemistry is the study of biological reaction pathways, most of which need many enzymes. Let's study the biochemical reaction mechanism that occurs during fermentation process. Fermentation is the chemical process microorganisms use to convert carbohydrates into smaller molecules instead of respiration.

Fermentation

Many microorganisms use the fermentation pathway to break down carbohydrates, and there is more than one type of fermentation. Yeasts use alcoholic fermentation to produce energy when oxygen is not available. Starting with a six carbon sugar such as glucose, the overall chemical process is:

$$\text{Overall reaction: } C_6H_{12}O_6 \rightarrow 2C_2H_5OH + 2CO_2 + \text{energy}$$

In words: the sugar glucose decomposes to produce ethanol, carbon dioxide, and energy.

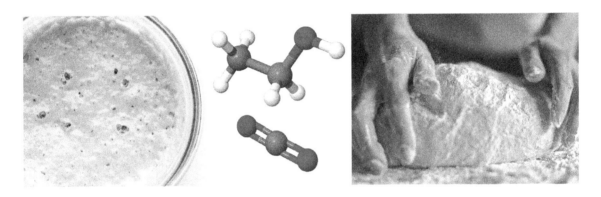

Fermentation and enzymes

During fermentation, many specific enzymes are required to modify a carbohydrate molecule over a series of several chemical reactions. Twelve different enzymes are required to break down one glucose molecule during fermentation.

Step 1

The first step of the fermentation reaction mechanism alone requires 10 different enzymes! In the first step, one glucose molecule is split into two pyruvic acid, CH_3COCO_2H, molecules.

$$\text{Step 1: } C_6H_{12}O_6 \rightarrow 2CH_3COCO_2H$$

This first step is in itself a series of elementary steps where the glucose molecule releases electrons. The electrons are accepted by another molecule that is used by cells as a form of energy for their metabolic processes.

Step 2

In the second step of this pathway, pyruvic acid is broken down into ethanol and carbon dioxide. This step requires 2 different enzymes.

$$\text{Step 2: } 2CH_3COCO_2H \rightarrow 2C_2H_5OH + 2CO_2$$

Who knew so many chemical changes were necessary just to prepare raw dough for baking!

Chapter 14

Section 3 review

While written as one simple and balanced equation, most reactions require multiple steps. This series of steps is known as a reaction mechanism. In reactions with multiple steps one or more intermediates are formed as the reaction progresses. Each step involves the collision of two reactants coming together. The rate law predicts how the concentration of reactants will influence the speed of a reaction. The rate-determining step is the slowest part of the reaction mechanism. For this reason it is the step that largely determines the overall speed of a reaction. Many biological pathways are complex and involve a series of steps. These reactions often depend on more than one enzyme to proceed.

Vocabulary words	elementary steps, reaction mechanism, intermediate, unimolecular, bimolecular, rate law, rate determining step

Review problems and questions

1. The following reaction occurs in two steps:
   Step 1: A + B → C + D
   Step 2: C + B → D + E
   Write the overall reaction.

2. Identify the intermediate(s) in the above reaction. How can you tell this is an intermediate, and is neither a reactant nor product?

3. Sketch a reaction profile graph for the reaction in Question 1.
   Your graph must:
   a. show the general shape of the reaction energy curve,
   b. have the correct number of activated complex "bumps",
   c. label where the intermediate(s) is/are formed
   d. label the energy of the reactants and the energy of the products
   e. label the point(s) where the activated complex(es) form(s)
   Label the x-axis "Reaction Progress" and label the y-axis "Energy." You do not need numbers on either axis.

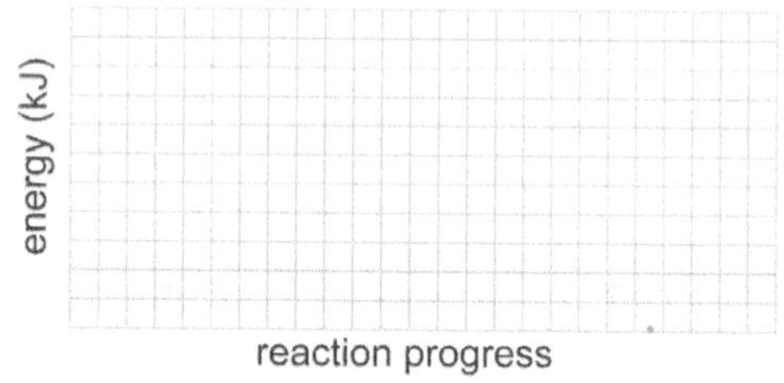

4. The reaction shown requires 4 steps. Which step is the rate-determining step? How can you tell? What effect does the rate-determining step have on the reaction?

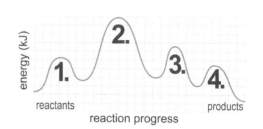

5. How many intermediates are formed in the above reaction? What feature on the graph did you use to find the intermediates?

Reaction Rates and Expiration Dates

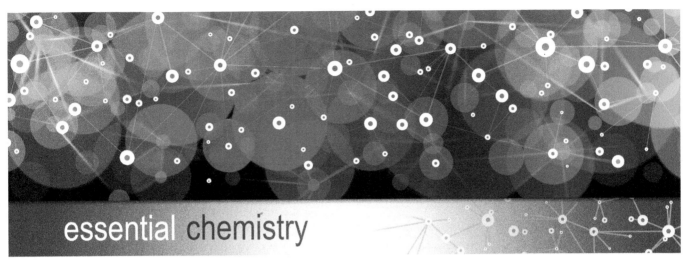

You have probably noticed that many foods, especially those that are packaged and refrigerated, have an expiration or "use by" date. These dates are an indication of how fresh the food or drink is based on the date of packaging. They also give an indication of the rate of a reaction that makes the food inedible or taste bad. In addition to a "use by" date, many food items also say, "keep refrigerated" or "refrigerate after opening." After learning about the factors that affect the rates of reactions, you can probably guess why this is true. It isn't just because it will taste better if it is cooler; it is because the lower temperature in the refrigerator will slow down the chemical reactions that could form unwanted by products.

Milk has a "use by" date that is typically 14 days after it has been pumped. After this date, there is a higher chance that your milk has soured. Milk is actually a complex mixture of thousands of compounds including calcium phosphate, proteins, vitamins, sugars and water. Milk contains many different sugars, but the primary sugar in milk is lactose. Lactic acid bacteria can end up in milk because it occurs naturally on plants like grasses. The lactic acid bacteria contain an enzyme called lactase that can take the lactose and turn it into glucose and galactose.

Lactose →(Lactase, H₂O)→ D-Galactose + D-Glucose

Other enzymes from the bacteria then take the glucose and galactose and convert them into lactic acid.

Glucose → 2 Lactic acid

Section 14.4: Connections

Reaction Rates and Expiration Dates - 2

When you say that milk has "soured" it is due to the increased amount of acid in the solution. This makes sense because acids have a sour taste. The increase in acid and decrease in pH change the structure of the protein in the milk. This structural change leads to clumps called "curds." The remaining liquid is called the "whey."

Refrigeration keeps the milk fresh by slowing down the reactions of the lactic acid bacteria. Most milk is pasteurized to kill most of the bacteria and has a shelf life of about 2 weeks. Once the container is opened, bacteria can get into the container. There is typically a week of time before it spoils. Milk that is "Ultra Pasteurized" will have a longer shelf life because 99.9% of the bacteria are killed during the high temperature pasteurization process. In chemistry terms, the ultra-pasteurization process lowers the concentration of enzymes, so milk has a shelf-life of about 70 days.

Milk isn't the only thing in your house with an expiration date. You have probably noticed that your medicines and over-the-counter supplements have an expiration date. When a new drug is being formulated, the manufacturer determines the expiration date. This date or "best before" date ensures that the product will still be effective for its intended use. For most products the expiration date is set to when the drug will lose 10% of its potency. This is sometimes noted as t90, or the time (t) when 90% of the drug is still effective.

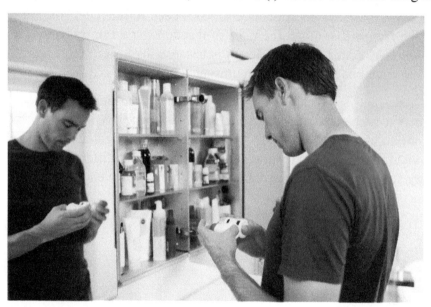

Drug stability will vary depending on the dosage form of the drug and environmental factors like the type of container, humidity, and temperature. Storing in a "cool, dry place" actually helps slow the rates of reactions! Liquid medicines typically have a shorter time before they expire. Depending on the type of medicine, you may need to store these in the refrigerator. Refrigeration can help, but if the time before expiration is short, the manufacturer might reformulate the liquid drug as a solid.

Many chemicals can undergo degradation when they encounter water. To slow these reactions down it is important to limit the amount of water that contacts the product. In these cases the solid dosage products are stored in tightly closed containers. Often vitamin or over-the-counter supplements are packed with small desiccants to continuously remove the moisture, even after the bottle has been opened. Some tablets are packed in blister-packs ensuring that the tablet is not exposed to any moisture until you are ready to use it.

Reaction Rates and Expiration Dates - 3

The actual degradation time for a drug could be weeks or months. Manufacturers don't always perform real-time stability studies to see how long it will take to lose 10% of their potency. Instead, studies at higher temperatures can be used to determine the stability of a drug at room temperature. By studying a reaction rate with the same concentrations of reactants at different elevated temperatures, scientists can calculate what the rate would be at lower temperatures.

A mathematical relationship known as the Arrhenius equation (named after a Nobel Prize winning chemist, Svante Arrhenius) was developed to explain the relationship between temperature and reaction rates. The Arrhenius equation also includes factors for the collision frequency of reactant molecules and activation energy. A good general rule of thumb is that for every 10 °C increase in temperature, the reaction rate will double.

Whether it's fresh milk or effective aspirin, you can now see that expiration dates are simply the result of applied chemistry with an understanding of collision theory. As we learn more about the chemicals and their reactions, we can modify the rate of a reaction to make sure we are getting effective products. This is not only important in chemistry class, it is essential to your everyday life.

Chapter 14 review

Vocabulary
Match each word to the sentence where it best fits.

Section 14.1

| activated complex | activation energy |
| catalyst | reaction rate |

1. The smallest amount of energy necessary for molecules to react is called the _____.

2. A chemical reaction is measured by the change in concentration of a reactant or product, per unit time. This is called the speed, or _____, of a chemical reaction.

3. A high energy state called the _____ is where bonds are broken and reformed.

4. A(n) _____ is a substance that increases the rate of a chemical reaction.

Section 14.2

| enzymes | inhibitor |

5. Biological catalysts called _____ speed up the chemical reactions in our bodies.

6. A substance that slows down a chemical reaction is called a(n) _____.

Section 14.3

bimolecular	elementary steps
intermediate	rate determining step
rate law	reaction mechanism
unimolecular	

7. A chemical species that is formed during the elementary steps but that is not present in the overall balanced equation is referred to as the reaction _____.

8. A(n) _____ collision is one that involves the collision of two molecules.

9. A series of simple reactions called _____ represent the overall progress of a chemical reaction.

10. A proposed pathway that must be supported by experimental evidence is termed the _____.

11. The _____ is the slowest elementary step in the reaction mechanism.

12. A(n) _____ elementary step involves only one reactant.

13. The _____ relates the speed of a chemical reaction with the concentrations or pressures of the reactants.

Conceptual questions

Section 14.1

14. If identical strips of aluminium metal are added to Beaker A containing 0.1 M HNO_3 and Beaker B containing 6.0 M HNO_3, which beaker will react faster? Explain.

15. What is collision theory?

16. How does increasing the concentration of the reactants affect reaction rate?

17. Describe how the act of decreasing the temperature will affect the rate of a chemical reaction. Give an example of this effect.

18. What is another word for the rate of a reaction and how is the rate of a chemical reaction measured?

19. Palladium can also be used in a catalytic converter as a catalyst. Describe the function of catalysts like palladium, Pd.

20. How does increasing the surface area of the reactants affect reaction rate?

21. What are the four factors that affect reaction rate? Describe how a change in each factor can lead to an increase in the speed of a reaction.

22. The rate of a chemical reaction changes over a period of time.

    a. How does the rate change with time?

    b. Why does this occur?

23. What is the formula used to calculate the "average" rate during a chemical reaction?

24. Use the general reaction below to answer the following:

    $$A + B \rightarrow 2C$$

    a. Describe the relationship between the rate of disappearance for substance A and that of substance B.

    b. Describe the relationship between the disappearance of substance A and the appearance of substance C.

25. What two conditions must be satisfied for a successful collision?

26. When molecules collide do they always form products? Why or why not?

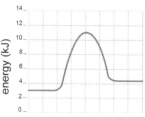

27. Sketch the above graph and label the following:
    a. activated complex
    b. reactants
    c. products
    d. the activation energy
    e. the enthalpy change, &DeltaH; f. explain whether the reaction is endothermic or exothermic

28. If a chemical reaction is proceeding to quickly, what are four things you can do to slow down the reaction.

29. In a chemical reaction, what occurs after the reactants have formed the activated complex?

Chapter 14 review

Section 14.1

30. How does increasing the temperature of the reactants affect reaction rate?

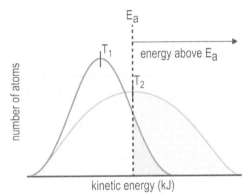

31. The graph above shows the amounts of atoms with a certain kinetic energy at two temperatures ($T_1 < T_2$). How does the relative size of the the yellow and blue areas demonstrate the relationship between reaction rates and temperature.

32. Use a molecular model set and a camera to create a short picture-based animation that explains how reactants form products according to the collision theory. Your animation must contain a minimum of 20 unique pictures/frames. Be prepared to share your animation with the class.

Section 14.2

33. What role does chlorine play in the ozone depletion mechanism? How do you know?

34. What does a catalyst alter that causes a change in the rate of a chemical reaction?

35. Our bodies use enzymes to assist with many biological reactions. What is the name and function of one of these enzymes?

36. Modern society relies on catalysts to speed up many chemical processes. What is one example of this?

37. The human body uses a specific variety of catalysts. What are the names of these catalysts?

38. Gelatin will not become firm and "gel" when fresh pineapple is added, why does this occur?

39. Sketch two reaction profiles. One for a general reaction and the other the same reaction but with a catalyst. Make sure to show the effects of the catalyst and to label your axes.

40. Sketch two reaction profiles. One for a general reaction and the other the same reaction but with an inhibitor. Make sure to show the effects of the inhibitor and to label your axes.

41. What is an example of an inhibitor that is useful in our everyday life. Explain how the inhibitor functions.

Section 14.3

42. Relate the rate determining step to a toll booth you encounter on a road trip.

43. Explain how a road trip can be related to a chemical reaction.

44. Explain why some reactions have more than two elementary steps.

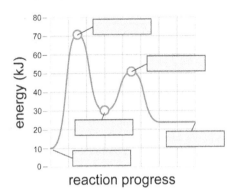

45. On the above graph, label the two activated complexes, the reaction intermediate, the reactants, and the products.

46. What are the two main steps of fermentation and how many enzymes does each step require?

47. Sketch a graph that shows 5 elementary steps and 4 intermediates. Label the axes, reactants, and products.

48. If you have seven elementary steps in an overall reaction, how many intermediates will be formed?

49. What is it about the following reaction that makes it unlikely that it will occur all in one step?

$$2NO_2^-(aq) + O_2(g) \rightarrow 2NO_3^-(aq)$$

50. What is the rate determining step and how do you identify it in a reaction mechanism?

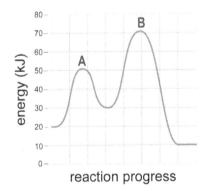

51. On the above graph, which is the rate determining step? How do you know?

Quantitative problems

Section 14.1

52. The formation of ammonia is shown below.

$$N_2(g) + 3H_2(g) \rightarrow 2NH_3(g)$$

The rate of formation of $NH_3(g)$ is measured to be 6.4×10^{-6} M/s. Assume the reaction proceeds via a single elementary step to calculate the rate of disappearance of $H_2(g)$?

Chapter 14 review

Section 14.1

53. Nitric oxide can react with itself in the reaction below to form laughing gas (N_2O) and nitrous oxide (NO_2).

    $$3NO(g) \rightarrow N_2O(g) + NO_2(g)$$

    If the reaction proceeded via a single elementary step, what would the rate of disappearance of nitric oxide (NO) when the rate of appearance of $NO_2(g)$ is measured to be 2.9×10^{-3} M/s?

54. Using the following equation: $2A \rightarrow 2B+C$
    a. If the disappearance of A over the period of 35 minutes to 58 minutes is .79 moles, what is the average rate of disappearance of A?
    b. Calculate the average rate of appearance of C during this period?

55. Nitrogen and oxygen can react to produce nitric oxide (NO) as shown below.

    $$N_2(g) + O_2(g) \rightarrow 2NO(g)$$

    The rate of disappearance of $O_2(g)$ is measured to be 3.8×10^{-7} M/s. Assuming that the mechanism occurs in one elementary step, what is the rate of formation of NO(g)?

56. Using the following equation: $A \rightarrow 2B+2C$
    a. If the disappearance of A over the period of 15 to 27 seconds is 0.052 moles, what is the average rate of disappearance of A?
    b. Calculate the average rate of appearance of C during this period?
    c. Would the average rate of appearance of B be the same as C? Why or Why not?

57. The reaction between the gases nitrogen dioxide (NO_2) and fluorine (F_2) gives:

    $$2NO_2(g) + F_2(g) \rightarrow 2NO_2F(g)$$

    If the reaction occurred in one step, what is the rate of formation of $NO_2F(g)$ if the rate of disappearance of $NO_2(g)$ is 4.3×10^{-5} M/s?

58. The table below shows data collected about a simple reaction: $A \rightarrow C$.

Time (sec)	Moles (A)	Moles (C)
0.0	5.00	0.00
2.0	4.00	1.00
4.0	3.20	1.80
6.0	2.60	2.40

    a. How is the formation of C related to the disappearance of A?
    b. What is the average rate for the appearance of C during first 2 seconds of the reaction (t=0 to t=2).
    c. What is the average rate for the appearance of from t=2 to t=4?
    d. Calculate the rate of disappearance of A during the last 2 seconds of the reaction (t=4 to t=6).
    e. How does the average rate at the start of the reaction compare to the rate at the end of the reaction? Explain why this trend occurs.

59. Given the data for reaction answer the questions below.

	$CO(g) + NO_2(g) \rightarrow CO_2(g) + NO(g)$		
Experiment	[CO] (M)	[NO_2] (M)	Initial rate (mol/L·h)
1	5.0×10^{-4}	3.6×10^{-5}	3.4×10^{-8}
2	5.0×10^{-4}	1.8×10^{-5}	1.7×10^{-8}
3	1.0×10^{-3}	3.6×10^{-5}	6.8×10^{-8}
4	1.5×10^{-3}	7.2×10^{-5}	?

    a. What is the order of the reaction with respect to [CO]?
    b. What is the order of the reaction with respect to [NO_2]?
    c. What is the value of k, including its units?
    d. What is the initial rate of the reaction in Experiment 4?

60. ⟪ Given the data for reaction answer the questions below.

	$2NO(g) + 2H_2(g) \rightarrow N_2(g) + 2H_2O(g)$		
Experiment	[NO]	[H_2]	Rate of N_2
1	1.680 M	0.488 M	0.544 M/s
2	0.840 M	0.488 M	0.136 M/s
3	0.840 M	0.976 M	0.272 M/s
4	0.421 M	1.952 M	0.136 M/s

    a. What is the order of the reaction with respect to [NO]?
    b. What is the order of the reaction with respect to [H_2]?
    c. What is the rate law?
    d. Determine the value of k.
    e. What is the overall order of the reaction?
    f. What is the rate of the reaction at the instant when [NO] = 0.920 M and [H_2] = 0.550 M?

61. ⟪ The results of several reactions of t-butylbromide, $(CH_3)_3CBr$, and hydroxide, OH^-, are shown below.

	$(CH_3)_3CBr + OH^-(aq) \rightarrow (CH_3)_3COH + Br^-$		
Experiment	[$(CH_3)_3CBr$] M	[OH^-] M	Rate (mol/L*h)
1	0.400	0.100	0.010
2	0.800	0.100	0.020
3	1.600	0.100	0.040
4	0.800	0.200	0.020
5	0.800	0.400	0.020

    a. What is the order of the reaction with respect to [$(CH_3)_3CBr$]?
    b. What is the order of the reaction with respect to [OH^-]?
    c. What is the rate law?
    d. What is the value of k, including its units?
    e. What is the overall order of the reaction?
    f. If the concentration of [OH^-] is quadrupled, what effect does it have on the reaction if the concentration of [$(CH_3)_3CBr$] remains at 0.800 M?

Chapter 14 review

Section 14.1

62. The reaction of $HgCl_2$ and $C_2O_4^{2-}$ is given below. Write out the rate law for the experiment:

$2HgCl_2(aq) + C_2O_4^{2-}(aq) \rightarrow$ $2Cl^-(aq) + 2CO_2(g) + Hg_2Cl_2(s)$			
Experiment	[$HgCl_2$]	[$C_2O_4^{2-}$]	Initial rate
1	0.210 M	0.300 M	1.8×10^{-5} M/s
2	0.210 M	0.600 M	7.1×10^{-5} M/s
3	0.104 M	0.600 M	3.5×10^{-5} M/s

63. What is the overall order for the given reaction?

$2A + B \rightarrow 2C$			
Experiment	[A]	[B]	Rate
1	0.240 M	0.100 M	0.020 M/s
2	0.120 M	0.100 M	0.020 M/s
3	0.120 M	0.300 M	0.060 M/s

Section 14.2

64. Using the graph provided, calculate the energy difference in kJ when a catalyst was added to the reaction.

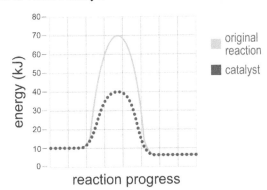

65. Using the graph provided, calculate the energy difference in kJ when an inhibitor was added to the reaction.

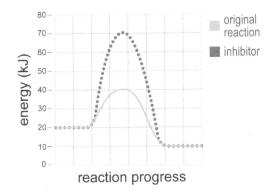

Section 14.3

66. Given the elementary steps below:
Step 1: $NO_2(g) + F_2(g) \rightarrow NO_2F(g) + F(g)$
Step 2: $NO_2(g) + F(g) \rightarrow NO_2F(g)$

   a. Write the balanced equation for the overall chemical reaction.
   b. If formed, what are the reaction intermediates?

67. What are the reaction intermediates and the overall balanced equation for the following elementary steps?

Step 1: $H_2O_2(aq) \rightarrow 2OH^-(aq)$

Step 2: $H_2O_2(aq) + OH^-(aq) \rightarrow H_2O(l) + HO_2^-(aq)$

Step 3: $HO_2^-(aq) + OH^-(aq) \rightarrow H_2O(l) + O_2(g)$

68. Consider the reaction that undergoes the following two elementary steps:

Step 1: $2NO(g) \rightarrow N_2O_2(g)$

Step 2: $N_2O_2(g) + O_2(g) \rightarrow 2NO_2(g)$

   a. Write the balanced equation for the overall chemical reaction.
   b. Are there reaction intermediates? How can you tell?

chapter 15: Equilibrium

When you were younger, did you ever try to play a game where you balanced multiple people on a seesaw? If there were six kids of about equal size on both sides of the seesaw, it would be balanced. They would have a force "equilibrium" on both sides of the seesaw. Equilibrium essentially means balance. If someone were to lift one side of the seesaw, then those six kids would have to adjust to rebalance. They might even have to adjust to the point where there are five kids on one side and one on the other to regain their equilibrium.

What does this have to do with chemistry? Well, it turns out that many physical processes and chemical reactions exist in equilibrium. They are ratios between reactants and products. Sometimes the amounts of reactants and products are about equal, like the six kids initially balanced on the seesaw. But external conditions often push the reaction heavily towards products or heavily towards reactants, until balance is achieved. The interesting thing about physical and chemical processes is that you can change the external conditions, just like pushing on a seesaw, to control the amounts of substances at equilibrium.

Equilibrium processes are all around you. For example, melting and freezing is a physical equilibrium between the solid and liquid states of matter, while dissolving and recrystallization is an equilibrium between the solid and dissolved solute. In both cases the equilibrium can be shifted easily by changing the temperature. The chemical systems in our bodies are in a delicate balance, called homeostasis. As you breathe you are continuously exchanging carbon dioxide and oxygen gases. During heavy exercise our metabolic rate increases creating excess carbon dioxide. Consequently, our breathing rate increases so we can remove the excess carbon dioxide and restore our body's proper balance. Perhaps it makes sense that if you truly want a proper "balance" in your life, you need to have an essential understanding of... equilibrium.

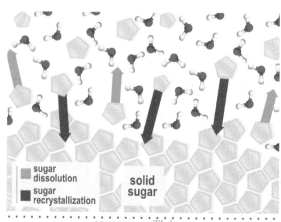

aqueous equilibrium:
rate of dissolving = rate of crystal formation

Chapter 15 study guide

Chapter Preview

Many chemical reactions are reversible and move not just from reactants to products but also from products to reactants. When initiated, a reaction will move more in one direction than the other. In time, however, equilibrium is reached and the forward and reverse reactions occur at the same average rate. In this chapter you will learn about chemical equilibrium and the factors that affect it. Equilibrium is dynamic. Reactions are still happening, they are simply balanced. The equilibrium constant is a way of quantifying the ratio of products to reactants when a given reaction reaches equilibrium at a given temperature. Le Châtelier's principle explains that varying the concentration, pressure or temperature can shift a reaction. The solubility product constant is a measure of how well a solid dissolves and can be used to predict whether or not a precipitate will appear in a given reaction.

Learning objectives

By the end of this chapter you should be able to:

- explain the concept of chemical equilibrium and what the equilibrium position represents;
- use equilibrium concentrations to calculate an equilibrium constant or use the equilibrium constant to determine an unknown concentration of a reactant or product;
- create an ICE chart to solve an equilibrium problem when given initial conditions;
- use Le Châtelier's principle to predict how a change of concentration, pressure/volume or temperature will shift a chemical reaction to maintain equilibrium; and
- calculate the solubility product constant and use it to predict when a precipitate will form as a result of a chemical reaction.

Investigations

15A: Reaction Equilibrium

15B: Le Châtelier's principle

Pages in this chapter

- 478 Chemical Equilibrium
- 479 Physical equilibrium
- 480 Equilibrium is dynamic
- 481 Chemical equilibrium
- 482 Chemical equilibrium in the air
- 483 Equilibrium position
- 484 Section 1 Review
- 485 The Equilibrium Expression
- 486 Creating the equilibrium expression
- 487 K_{eq} is constant
- 488 Using the equilibrium constant
- 489 Initial conditions
- 490 When to use an ICE chart
- 491 Section 2 Review
- 492 Le Châtelier's Principle
- 493 Conditions for chemical equilibrium
- 494 Using Le Châtelier's principle
- 495 Section 3 Review
- 496 Solubility Product Constant
- 497 Using K_{sp}
- 498 Precipitate formation
- 499 Section 4 Review
- 500 Nitrogen and the Nobel Prize
- 501 Nitrogen and the Nobel Prize - 2
- 502 Chapter review

Vocabulary

equilibrium
law of mass action
chemical species
solubility product constant

physical equilibrium
equilibrium expression
Le Châtelier's principle

equilibrium position
equilibrium constant
closed system

15.1 - Chemical Equilibrium

As you learned in a previous chapter, chemical reactions do not always occur in one step as written in a chemical equation. In this chapter, you will learn that some chemical reactions can go in the reverse direction. Reactions that are capable of going back and forth can achieve a balance between the rate at which reactants form products, and products re-form reactants. This balance is known as **equilibrium**.

Reversibility of chemical reactions

Chemical equilibrium

A chemical system that has established equilibrium experiences no overall change in the amounts of reactants and products at any given time. The reaction continues to move in two directions during equilibrium. Reactants are forming products, but products are also forming reactants at an equal rate. Imagine a chemical reaction running in place because while it is moving forward, it is moving backwards at the same speed! A symbol of two half-arrows pointing in opposite directions represents chemical equilibrium. The reaction $H_2 + I_2 \rightleftharpoons 2\ HI$ indicates H_2 and I_2 can make HI but HI also can decompose into H_2 and I_2.

A two way road

To understand equilibrium, let's look at a bridge with a steady traffic flow in both directions. It represents a chemical reaction. If the traffic flowing in both directions on the bridge is steady, the number of cars moving across the bridge into the city and out of the city stays relatively constant. This is like an established equilibrium.

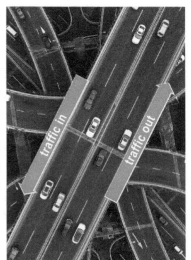

When the number of cars moving into and out of the city is constant, the system has established a balance. In contrast, a balance could not be established if the traffic flow is only in one direction, or if the traffic flow is continually changing.

At equilibrium concentrations are constant

Both forward and reverse reactions are happening simultaneously in a reaction at equilibrium. As a result, the amount of product and reactant remains constant over time, but the elements are continuously exchanged like the cars in and out of the city.

Equilibrium concentrations are not equal

Even though the traffic across the bridge is constant, that does not mean the number of cars going in both directions is equal. Chemical equilibriums rarely result in equal moles of products and reactants. Consider the equilibrium between vapor and liquid water at 90 °C and standard pressure. There are more molecules in the liquid phase than the gas phase. But if the rate at which the water is evaporating is equal to the rate at which water is condensing, the contents of the flask are at equilibrium.

Physical equilibrium

Evaporation and condensation at equal rates

When water in a closed container evaporates, it eventually reaches a **physical equilibrium**. It is a physical equilibrium because the equilibrium is between two phases of the same substance. Only physical changes take place. Think about what is happening to water molecules as they move between the vapor and liquid phase. If you fill a jar half way with water and then cover it, what will happen? Assume the jar will be kept at room temperature and standard pressure. Over time, will anything change inside the system?

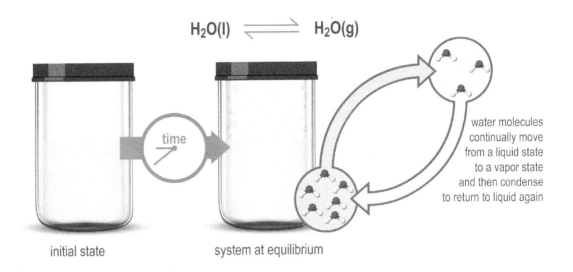

Balance and equilibrium

When the jar is first covered, water molecules that have already evaporated are trapped. But as more water evaporates, there will be so many molecules in the vapor phase that they will begin to condense on the inside of the jar and into the liquid phase. When the number of molecules evaporating is equal to the number of molecules condensing, the balance of a physical equilibrium is achieved. Water molecules are constantly condensing and evaporating during equilibrium.

Phases of water in balance

To the casual observer it may appear that nothing is happening because the water level in the covered jar remains the same. However, at the particle level water molecules are continuously being exchanged between the liquid and the gas phase. There is no noticeable change in the water level because the rates of the forward and reverse changes are equal. Once equilibrium is established, the water level appears to stay the same as long as the jar remains closed at a constant temperature. Notice that the amount of water in the vapor phase does not have to equal the amount of water in the liquid phase during equilibrium. The vast majority of water is in the liquid phase in the equilibrium system shown above.

Vapor pressure

Water molecules in the gas or vapor phase create pressure on the surface of the liquid. This is called vapor pressure. The air inside the jar is is also creating pressure on the liquid surface. When the vapor pressure is equal to the air pressure, boiling occurs. This is not an equilibrium situation unless you create conditions that will cause water in the gas phase to condense back to the liquid phase at the same rate at which the gas forms. During evaporation, the liquid-gas phase change only occurs at the surface of a liquid. During boiling, the liquid-gas phase change occurs throughout the entire liquid because vapor pressure is greater than or equal to atmospheric pressure.

Section 15.1: Chemical Equilibrium

Equilibrium is dynamic

An example of equilibrium

Consider a simple equilibrium between dissolved table sugar (sucrose) and water. At 25 °C the solubility of sucrose in water is an incredible 67%! That means 100 grams of saturated sugar solution contains 67 grams of sucrose and only 33 grams of water. The solubility of sugar increases with temperature - see the graph below. To ensure that a sugar solution is fully saturated, a good technique is to dissolve a higher concentration into hot water and let the solution slowly cool down. As it cools some of the sugar drops out of solution and forms large, solid crystals. Anyone who as ever eaten rock candy has tasted sucrose crystals that were historically prepared this way.

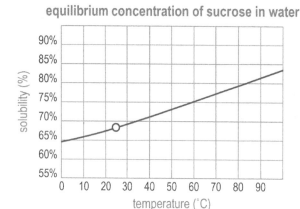

rock candy is crystalized sucrose

An example of dynamic equilibrium

Consider 100 grams of a sugar solution which starts at 75% sucrose at 90 °C. This is below the solubility limit so all the sugar stays dissolved. Now allow the solution to cool to 25 °C. At 25 °C the solubility is only 67% so some of the sugar forms solid crystals. The crystals grow until an equilibrium is reached between the dissolved sugar and the solid (un-dissolved) sugar. It looks like nothing is changing so how can this equilibrium be dynamic?

Particle-level view

At the particle level things are not static at all! Sugar molecules in the solution bump randomly into the solid sugar crystal surface and stick. These molecules become solid and drop out of solution. The opposite process happens at the same time. Water molecules bump the solid surface and dissolve solid sugar molecules back into solution.

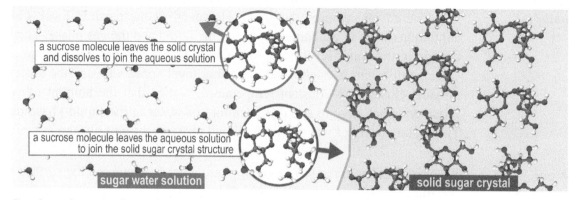

Equal rates

On the microscopic scale the meaning of dynamic equilibrium is NOT that either process (dissolving or un-dissolving) has stopped. The true meaning is that both processes are occurring at the same rate! At equilibrium, the rate of sugar molecules leaving the solid crystal and dissolving into the solution is the same as the rate of sugar molecules precipitating out of solution to re-form a solid crystal. On the macroscopic scale we see no change - but on the microscopic scale things are changing quite vigorously - but the rate of each of the two opposing processes are equal.

Chemical equilibrium

Reaction reversibility

In a closed system many chemical reactions tend to have a mixture of reactants and products present instead of only products. This is why we use a double arrow to symbolize equilibrium, which indicates that both the forward and reverse reactions are taking place.

Color change indicates a chemical change

Some reversible chemical reactions show observable evidence such as a color change. The reversible reaction $N_2O_4(g) \rightleftharpoons 2NO_2(g)$ is shown in the example below. $N_2O_4(g)$ is clear and colorless and $NO_2(g)$ is rust-colored. The test tube contents are clear when mostly $N_2O_4(g)$ is present and rust-colored when mostly $NO_2(g)$ is present. When the test tube is cooled, its contents are clear because mostly $N_2O_4(g)$ forms. When it warms up, the test tube contents are rust-colored because mostly $NO_2(g)$ forms. At an in-between temperature, the test tube has an intermediate color because the rate of formation of each substance occurs at an equal rate. This is an example of chemical equilibrium because the reaction shows a chemical change from reactant to product that reaches a balance, not just phase change of the same substance.

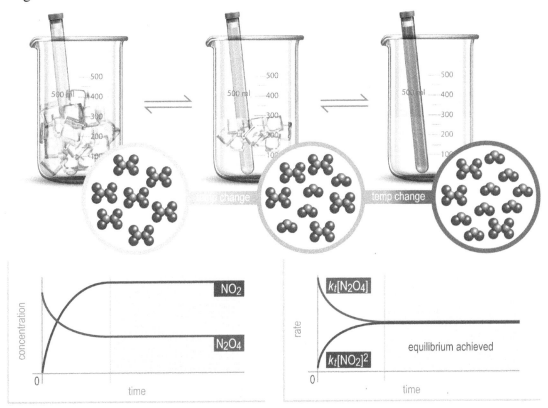

Equilibrium is established inside of each test tube, but the amount of product relative to reactant is different at different temperatures. Temperature is one of the variables that can affect an equilibrium system. We will investigate the reason why later in the chapter.

Chemical equilibrium graphs

Study the two graphs above. Both graphs indicate a chemical reaction has reached chemical equilibrium. You can tell this is the case for several reasons:

- The two data lines indicate there are two different chemicals changing amounts then coming to some kind of balance
- The graph on the left shows concentrations are not equal, however, you can see concentration has leveled off and is not changing over time
- The graph on the right shows that although the rates of the two chemicals did not start out the same, they eventually found a rate equal to one another

Chemical equilibrium in the air

Atmospheric equilibrium

An equilibrium of $N_2O_4(g) \rightleftharpoons 2NO_2(g)$ is occurring in the air around us. Nitrogen dioxide, $NO_2(g)$, is a poisonous gas produced in automobile exhaust emissions. You are exposed to different amounts of exhaust gases depending upon where you live. Areas with a high population tend to have more cars on the road, and therefore have higher levels of NO_2 in the air. The concentration of NO_2 in the air naturally reaches an equilibrium with N_2O_4 on its own.

$NO_2(g)$ produces smog

Nitrogen dioxide (NO_2) gas can cause health problems for people with breathing disorders such as asthma. Air quality is regularly tested for pollutants so sensitive groups can be warned to stay inside when air quality is poor. Smog can be worse in the summer because higher temperatures cause more NO_2 to form. NO_2 is one of the chemicals that gives smog its brown color.

Beijing, China struggles to control smog

Equilibrium reactions are two way

Equilibrium can be established with reversible reactions starting with reactant only or product only. If you take a closed flask of pure, frozen $N_2O_4(g)$ and warm it to room temperature, it will create some $NO_2(g)$. The $NO_2(g)$ will then re-form $N_2O_4(g)$. After some time, an equilibrium between both directions will be established.

The $N_2O_4(g) \rightleftharpoons 2NO_2(g)$ equilibrium reaction:

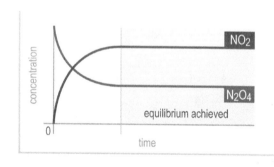
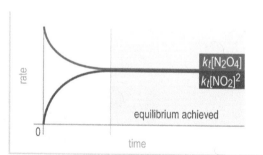

Equilibrium can also be established in reverse. If you start with just $NO_2(g)$ in the flask, the reaction will form N_2O_4 and eventually establish equilibrium. In each case, no matter what we start with, equilibrium will be established.

Reaction direction

Before moving on to more complex concepts, make sure you know which side of the equilibrium reaction is being described. The most common ways to say "to the left of the equilibrium arrow" are:

- towards the reactants
- favors the reactants
- in the reverse direction

Common ways to say "to the right of the equilibrium arrow" are similar:

- towards the products
- favors the products
- in the forward direction

Equilibrium position

Equilibrium concentrations are not equal

Reversible chemical reactions can favor or make more of either products or reactants, but it is not necessary for amounts of products and reactants to be equal. In fact, it is rare for a reaction in equilibrium to have similar amounts of reactants and products.

Equilibrium is a balance

A reaction has reached equilibrium when the rate of the forward reaction is equal to the rate of the reverse reaction. The **equilibrium position** indicates whether the products or reactants are favored at equilibrium. You can determine the equilibrium position for a reaction by monitoring the amount of reactants and products present once equilibrium has been established.

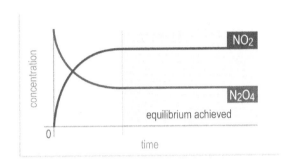 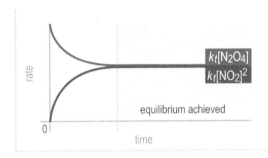

Favoring reactants

The graphs above show the equilibrium system $N_2O_4(g) \rightleftharpoons 2\ NO_2(g)$. Notice at the beginning of the reaction, the concentration of reactant, N_2O_4, is higher than the concentration of product, NO_2. But over time, the concentration of N_2O_4 decreases while NO_2 increases. When equilibrium is reached, the concentration of NO_2 is greater than N_2O_4. This tells us the products side of this equilibrium is "favored" under the conditions of 25 °C and 1 atm. The fact that the NO_2 concentration is higher indicates more of it is present at equilibrium.

Concentration, temperature and pressure

The amount of reactants compared to the amount products present at equilibrium for a given reaction depend on the starting concentrations, the temperature, and the pressure of the system. Depending on the conditions, it is possible that one direction of the reaction can be favored over the other. You can analyze concentration data over time to help determine which direction is favored under different conditions.

Graphs

Below are 2 graphs of Concentration vs. Time for two trials of the $2NO_2(g) + O_2(g) \rightleftharpoons 2NO_3(g)$ equilibrium system. The graph on the left is a trial that started with all reactant, and the graph on the right is a trial that started with all product. Equilibrium is established over time in both trials.

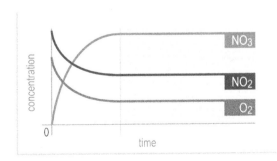 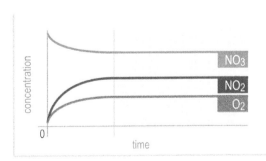

Analyzing graphs

Study the graphs. How can you tell which side of the reaction is favored? What does this tell you about the equilibrium position?

Chapter 15

Section 1 review

Equilibrium exists when amounts stay constant. Physical equilibrium is when a balance between two phases of the same substance is reached. Chemical equilibrium is reached when there is no further overall change in the amounts of reactants or products. Reactions still proceed in both directions, but they are balanced. Equilibrium does not mean that equal amounts of reactants and products exist. It simply means a balance has been reached. Equilibrium can be established from either direction of a chemical reaction. In either case, the equilibrium position of a reaction explains whether the reactants or products will be favored at equilibrium at a given temperature and pressure.

Vocabulary words equilibrium, physical equilibrium, equilibrium position

Review problems and questions

1. When reading a chemical equation, an arrow like this: ⇌, gives different information about a reaction than an arrow like this: →. Explain the difference between the two types of arrows.

2. Choose the letter that cannot correctly complete the sentence. "When a chemical reaction is at equilibrium, the _____ of reactants and products remain(s) unchanged."
    a. amount
    b. concentration
    c. ratio
    d. chemical bonds

3. Does the graph show a reaction at equilibrium? Why or why not?

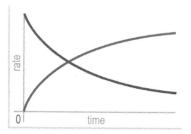

4. The graph shows a reaction at equilibrium. What exactly is at equilibrium; what should be written on the y-axis? Explain your answer.

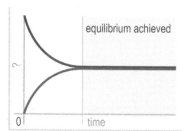

5. The graph shows the reaction $2SO_2(g) + O_2(g) \rightleftharpoons 2SO_3(g)$ as it reaches equilibrium. Which are favored: products or reactants? How can you tell?

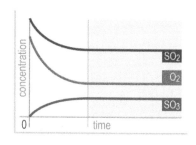

15.2 - The Equilibrium Expression

Can you predict which direction a reversible chemical system will favor? How does a change in reaction direction affect the equilibrium system? You can design a series of experiments under different conditions to answer those questions. The direction an equilibrium system favors depends on the reaction conditions, such as temperature.

What is an equilibrium expression?

Law of mass action

Our understanding of equilibrium chemistry is based on the work of two Norwegian chemists completed over many years in the second half of the 19th century. The **law of mass action** expresses the reaction rate of an equilibrium system as a product-to-reactant ratio. In the equilibrium reaction $aA + bB \rightleftharpoons cC + dD$, a moles of reactant A and b moles of reactant B yield c moles of product C and d moles of product D. You can use these amounts to set up an **equilibrium expression** which is a ratio of products to reactants.

$$K = \frac{[C]^c[D]^d}{[A]^a[B]^b} = \frac{[\text{products}]}{[\text{reactants}]}$$

Equilibrium expression

Reactant and product quantities are measured in molarity, M, which is the unit for concentration. Square brackets around reactants and products indicate concentrations are to be used. Each concentration is raised to the power of the corresponding coefficient from the balanced reaction. K or K_{eq} is the equilibrium constant, which represents the numeric value of the ratio. The **equilibrium constant** always has the same value at a given temperature for a specific chemical reaction. Most importantly, equilibrium data always support this ratio. You can imagine how useful the equilibrium expression is for chemists who need predictable ways of producing chemicals.

Applying equilibrium

Let's look at the equilibrium expression for the following reaction:

$$N_2(g) + 3\,H_2(g) \rightleftharpoons 2\,NH_3(g)$$

The first step in setting up an equilibrium expression is to write the concentration of each product multiplied together over the concentration of each reactant multiplied together.

$$K = \frac{[NH_3]}{[N_2][H_2]}$$

The concentration of each substance is raised to a power equal to the coefficient in the balanced equation. In other words, the coefficients become exponents:

$$K = \frac{[NH_3]^2}{[N_2][H_2]^3}$$

You would need to be given concentration data in order to solve for the equilibrium constant.

Creating the equilibrium expression

Writing the equilibrium expression

Consider the balanced equation $A + 2B \rightleftharpoons C + 4D$. Even though it is ridiculous, the expanded equation could be written as $A + B + B \rightleftharpoons C + D + D + D + D$ and the equilibrium expression can be written as:

$$K_{eq} = \frac{[C][D][D][D][D]}{[A][B][B]}$$

The equilibrium constant is a product constant. It is calculated by multiplying the products and multiplying the reactants then dividing them to find the ratio. This means the coefficients of the chemical equation become exponents in the equilibrium expression. Let's compare the equilibrium expression from the expanded equation with the simplified equation that has coefficients.

Expanded equation: $A + B + B \rightleftharpoons C + D + D + D + D$
Equation with coefficients: $A + 2B \rightleftharpoons C + 4D$

$$K_{eq} = \frac{[C][D][D][D][D]}{[A][B][B]} \text{ is equal to } K_{eq} = \frac{[C][D]^4}{[A][B]^2}$$

Notice how the expanded version and the simplified version of the expression are the same.

Solved problem

Calculate the equilibrium concentration of [HI] given the following information:

- $H_2(g) + I_2(g) \rightleftharpoons 2\,HI(g)$
- $K = 50.0$ at 450 °C
- The equilibrium concentrations of the reactants are $[H_2] = 0.22$ M and $[I_2] = 0.22$ M.

Asked Calculate the concentration of HI at equilibrium
Given $K = 50.0$ at 450 °C, and $[H_2] = [I_2] = 0.22$ M
Relationships Equilibrium expression: $K_{eq} = \frac{[HI]^2}{[H_2][I_2]}$

When known quantities are placed into the expression they do not have to be multiplied by the coefficient because they are at equilibrium and in the correct ratio.

Solve Write the expression:

$$K_{eq} = \frac{[HI]^2}{[H_2][I_2]}$$

Substite given values and solve for x:

$$50.0 = \frac{[x]^2}{[0.22\,M][0.22\,M]} \; ; \; 50.0 = \frac{x^2}{0.0484\,M^2}$$

$$50 \times (0.0484\,M^2) = x^2$$

$$2.42\,M^2 = x^2$$

$$\sqrt{2.42\,M^2} = x$$

$$1.6\,M = x$$

Answer The equilibrium concentration of [HI] = 1.6 M

K_{eq} is constant

K_{eq} is a constant

For a specific reaction at a given temperature the equilibrium constant K_{eq} will always be the same, or constant. This is verified by running many reactions with different starting concentrations at a given temperature. When the measured equilibrium concentrations are plugged into the equilibrium expression, the calculated K_{eq} is the same for all trials.

Using experimental data

Data for three trials of the reaction:
$$3H_2(g) + N_2(g) \rightleftharpoons 2NH_3(g) \; @470.0 \; °C$$

$$K_{eq} = \frac{[NH_3]^2}{[H_2][N_2]^3}$$

experiment	[H$_2$]	[N$_2$]	[NH$_3$]	K_{eq}
1	0.120	0.0400	0.00272	0.107
2	0.240	0.0800	0.0109	0.107
3	0.360	0.120	0.0244	0.106

Using the equilibrium expression

Here is how K_{eq} values are calculated from the measured equilibrium concentrations based on the experimental data above. Let's compare experiments 1 and 3.

Equilibrium expression calculation for data from experiment 1:

$$K_{eq} = \frac{[NH_3]^2}{[H_2][N_2]^3} = \frac{[0.00272]^2}{[0.120]^3[0.0400]} = 0.107$$

Equilibrium expression calculation for data from experiment 3:

$$K_{eq} = \frac{[NH_3]^2}{[H_2][N_2]^3} = \frac{[0.0244]^2}{[0.360]^3[0.120]} = 0.106$$

The values 0.107 and 0.106 for K_{eq} are close enough to be considered equal. This supports the fact that K_{eq} is constant at the same temperature. Equilibrium concentrations may be different for each experiment, but the ratio is the same in each case. K_{eq} would only be different if these three experiments were conducted at different temperatures.

Production of ammonia

One of the most important industrial chemicals produced today is ammonia (NH$_3$). Ammonia is used to manufacture fertilizer, plastics, explosives and many other products.

Ammonia production: $3H_2(g) + N_2(g) \rightleftharpoons 2NH_3(g) \; @470.0 \; °C$

Did you notice the temperature at which this reaction occurs? For this reaction, the increased temperature results in a lower K_{eq} compared to room temperature; remember, K_{eq} is based on a specific temperature. A lower K_{eq} would mean the ratio of ammonia to nitrogen and hydrogen has dropped. Why would a chemist run this reaction at such a high temperature if they are trying to produce more ammonia? It turns out temperature is not the only factor that influences equilibrium. Chemists can apply another equilibrium principle to cause the reaction to favor products instead of reactants. You will learn about this principle in the next section.

Section 15.2: The Equilibrium Expression

Using the equilibrium constant

The equilibrium expression

As you now know, you can use the equilibrium expression to determine a numeric value for the equilibrium constant. The equilibrium constant is used by chemists to find out which direction is favored by the equilibrium system and by how much. It can also be used to predict the concentration(s) of reactant(s) or product(s) at equilibrium. Sometimes, you may notice K_c sits in place of K_{eq} in an equilibrium expression. The "c" is a reference to the fact that equilibrium is in terms of concentration (M).

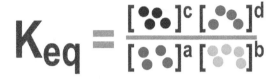

K_{eq} size determines which side is favored

The size of K_{eq} tells us to what extent the reaction forms products. Let's use the reaction $A(g) \rightleftharpoons B(g)$ to see how the size of K is interpreted. The expression is: $K_{eq} = \frac{[B]}{[A]}$

- When K_{eq} is very large ($K_{eq} > 10^4$), products are favored because a large K_{eq} occurs when [B] is much larger than [A].
- When K_{eq} is very small ($K_{eq} < 10^{-4}$), reactants are favored because a small K_{eq} occurs when [A] is much larger than [B].

Sometimes the value of K_{eq} is somewhere between very large and very small. In this case, there are significant amounts of both reactant [A] and product [B], or neither reaction direction is favored over the other. Let's use K to predict which direction is favored in the common equilibrium reactions shown below:

For $2H_2(g) + O_2(g) \rightleftharpoons 2H_2O(l)$

- $K_{eq} = 1.4 \times 10^{83}$ at 25.0 °C
- Large K favors the products

For $CaCO_3(s) \rightleftharpoons CaO(s) + CO_2(g)$

- $K_{eq} = 1.9 \times 10^{-23}$ at 25.0 °C
- Small K favors the reactants

Solved problem

The reaction $N_2 + 3H_2 \rightleftharpoons 2NH_3$ has a K_{eq} of 3.55×10^{-2} at 500 K and 7.76×10^{-5} at 700 K. Which temperature will make the greatest amount of products?

Given 3.55×10^{-2} at 500 K and 7.76×10^{-5} at 700 K

Relationships $K_{eq} = \frac{[NH_3]^2}{[N_2][H_2]^3}$ @ 500K = 3.55×10^{-2}

$K_{eq} = \frac{[NH_3]^2}{[N_2][H_2]^3}$ @ 700K = 7.76×10^{-5}

Solve Since the equilibrium expression is a ratio of products to reactants, the largest K_{eq} will produce the most products. 3.55×10^{-2} is larger than 7.76×10^{-5} and will make more products.

Answer 500 K will make more products.

Initial conditions

Initial conditions

It is useful to know how much product will be formed before the reaction starts. This was done with stoichiometry for one-way reactions. We will use the equilibrium constant and a new type of chart for two-way reactions.

The ICE chart

An ICE chart assists in calculating the concentration of all chemical species at equilibrium. ICE is an acronym for Initial, Change and Equilibrium. Consider the reaction, $N_2O_4(g) \rightleftharpoons 2NO_2(g)$, that starts with 0.1 M N_2O_4 and has an equilibrium constant of 5.76×10^{-3}. The following process describes how you calculate the equilibrium concentrations before the reaction starts.

1. The products and reactants are shown across the top row.

2. The reaction starts with 0.1 M N_2O_4 and there is no NO_2 because the reaction has not taken place yet. Those values go in the Initial row.

3. From the coefficients in the balanced equation we know that as the reaction starts to progress, one unit (x) of N_2O_4 will be used to create 2x NO_2. Place -x for N_2O_4 and +2x for NO_2 in the Change row. The - and + signs indicate "used up" and "produced."

4. The Equilibrium row is simply the combination of what you started with in the Initial row plus or minus and change that occurred in the Change row. Therefore 0.1-x goes in the Equilibrium row for N_2O_4 and 2x goes in for NO_2.

ICE chart

Species:	N_2O_4 $\rightleftharpoons$	$2NO_2$
Initial	0.1	0
Change	-x	+2x
Equilibrium	0.1-x	2x

5. The values in the Equilibrium row go to into the equilibrium expression.
$$K_{eq} = \frac{[NO_2]^2}{[N_2O_4]} = \frac{[2x]^2}{[0.1-x]} = 5.76 \times 10^{-3}$$

6. First square the 2x term and multiply both sides of the equation by 0.1-x.
$$\frac{4x^2}{0.1-x} = 5.76 \times 10^{-3}$$
$$4x^2 = 5.76 \times 10^{-3}(0.1-x)$$
$$4x^2 = 5.76 \times 10^{-4} - 5.76 \times 10^{-3}x$$

7. Set the equation equal to zero and arrange by decreasing powers of x.
$$4x^2 + 5.76 \times 10^{-3}x - 5.76 \times 10^{-4} = 0$$

8. This is a quadratic equation and we have to use the quadratic formula to solve it. To remind you, the quadratic formula is:

$$x = \frac{-b \pm \sqrt{b^2 - 4ac}}{2a}$$

Section 15.2: The Equilibrium Expression

When to use an ICE chart

Example ICE chart

9. After applying the quadratic formula you get two values for x. They are -1.27×10^{-2} and 1.13×10^{-2}. Only one of these values is valid. Each one needs to be tested by placing them into the equilibrium row of the ICE chart.

ICE chart

Species:	N_2O_4 ⇌	$2NO_2$
Initial	0.1	-
Change	-x	+2x
Equilibrium	0.1-x	2x

10. The negative number, -1.27×10^{-2}, makes sense when placed into the N_2O_4 Equilibrium row because when it is used in 0.1-x, it yields a positive number. However, when it is placed into the NO_2 Equilibrium row, it yields a negative number. Since concentration cannot be negative, the number -1.27×10^{-2} must be excluded and the other value of x, 1.13×10^{-2}, should be used exclusively.

11. Substituting 1.13×10^{-2} into the Equilibrium row we get 0.0887 M N_2O_4 and 0.0226 M NO_2. These are the amounts that we will find at equilibrium. As a double check, insert the equilibrium values into the equilibrium expression. You will know they are correct if they yield the original equilibrium constant.

When to use an ICE chart

summary for equilibrium initial concentration problems

1. make an ICE chart
2. fill in the initial, change and equilibrium rows
3. use the equilibrium row in the equilibrium expression
4. solve for X using the quadratic formula (if necessary)
5. determine which X is valid
6. place the valid X into the equilibrium row
7. find the equilibrium concentrations

For many equilibrium problems solved using an ICE table, you can simplify the process of solving it by assuming that 'x' is much smaller than the initial concentration of the reactants (so the 'x' drops out of terms like '0.1-x'). A good rule of thumb is that you can assume 'x' is very small when K_{eq} is smaller than 1.0×10^{-3}.

ICE chart with 3 species

The ICE Chart is not restricted to only two species. Here is an example with three species:

ICE Chart

	$I_2(g)$ +	$Cl_2(g)$ ⇌	$2ICl(g)$
Initial	1.00	1.00	0
Change	-x	-x	+2x
Equilibrium	1.00 - x	1.00 - x	2x

Section 2 review

Chapter 15

The favored direction of a reaction (toward the products or reactants) must be determined through experimental data. The equilibrium expression is a ratio of the concentration of the products divided by the concentration of the reactants. It leads to an equilibrium constant (K_{eq}) for a given reaction at a particular temperature and pressure. If the concentration of products and reactants at equilibrium are known, the equilibrium expression can be used to calculate the equilibrium constant. The value of K_{eq} also provides information as to whether the products or reactants are favored for a given reaction. When the initial conditions are given for an equilibrium problem, an ICE (Initial, Change and Equilibrium) chart will help in solving it.

Vocabulary words   law of mass action, equilibrium expression, equilibrium constant, chemical species

Review problems and questions

1. Write an equilibrium expression for the reaction: $COCl_2(g) \rightleftharpoons CO(g) + Cl_2(g)$.

2. Find the equilibrium concentration of $COCl_2$ when the equilibrium concentration of CO and Cl_2 are each 0.12M and K_{eq} is 1.29×10^{-2} at 600K. Use the reaction in Question 1.

3. Use the equilibrium constant given in Question 2 to identify whether products, reactants or neither side is favored for this reaction at 600K.

4. Create an ICE chart for the reaction in Question 1 starting with 0.85M of $COCl_2$. Assume there are no initial products.

5. What is the equilibrium expression for the following reaction?
   $2\ Cl_2(g) + 2\ H_2O(g) \rightleftharpoons 4\ HCl(g) + O_2(g)$

6. Write equilibrium expression, K_{eq} for:
   $CO(g) + H_2O(g) \rightleftharpoons CO_2(g) + H_2(g)$

7. At a given temperature, K = 62.0 for the reaction:
   $4\ HCl(g) + O_2(g) \rightleftharpoons 2\ H_2O(g) + Cl_2(g)$. At equilibrium, [HCl] = 0.150 M, [O_2] = 0.395 M and [H_2O] = 0.625 M. What is the concentration of Cl_2 at equilibrium?

8. For the reaction $CaCO_3(s) \rightleftharpoons CaO(s) + CO_2(g)$, at 25 °C, the following concentrations were obtained: [$CaCO_3$] = 0.207 M; [CaO] = 0.694 M; [CO_2] = 0.890 M. Find the equilibrium constant, K_{eq}.

9. Calculate the equilibrium concentrations of all species for the following reaction if the initial concentrations of H_2 and I_2 are both 1.00 M. Assume the initial concentration of HI is zero.
   $H_2(g) + I_2(g) \rightleftharpoons 2\ HI(g)$ $K_c = 70.2$

15.3 - Le Châtelier's Principle

You can imagine why a chemist would want an equilibrium reaction to favor one direction over another. For example, industrial chemists could boost the production of a desired product while environmental chemists could reduce production of an unwanted pollutant. What factors influence the direction of an equilibrium reaction? Four conditions affect the direction of equilibrium: concentration, temperature, pressure and volume. Let's investigate how changes in these conditions affect the equilibrium position.

Influencing directionality in equilibrium

Reactions shift to resist change

The French chemist, Henri Le Châtelier, studied how equilibrium reactions responded to change. **Le Châtelier's principle** states that when a system at equilibrium is changed or disturbed, the system will shift in a direction that reduces the change or the disturbance until equilibrium is reestablished. Changes or disturbances are also called stresses. The variables that place stress on a system at equilibrium are:

- concentration
- pressure
- volume
- temperature

Reactions shift to reestablish the K_{eq}

When a chemical system at equilibrium is changed, the products and reactants shift concentration back and forth until they reach a ratio that matches the K_{eq} for that reaction. For example, the reaction $H_2 + I_2 \rightleftharpoons 2HI$ has a K_{eq} close to 50. The table shows equilibrium concentrations established during an experiment. In trial two, extra HI was added. The products and reactants rearranged until the ratio returned to about 50. All species in the reaction increased because some of the added HI split into I_2 and H_2 molecules to maintain the K_{eq}. In trial three, HI was removed so the HI concentration went down, but so did I_2 and H_2. The K_{eq} for this trial still returned to about 50 when equilibrium was reestablished.

results of three experiments for:
$H_2(g) + I_2(g) \rightleftharpoons 2HI(g)$

	Equilibrium Concentrations			K_{eq} Calculation
Trial	$[I_2]$	$[H_2]$	$[HI]$	$Keq = \frac{[HI]^2}{[H_2][I_2]}$
1	1.52	1.58	10.90	49.47
2	1.81	1.84	13.01	50.82
3	1.20	1.18	8.45	50.43

K_{eq} at the particle level

Below you can see what happened in Trial 3, where HI product was removed. The system responded by using up reactants to restore lost HI product and reestablish equilibrium.

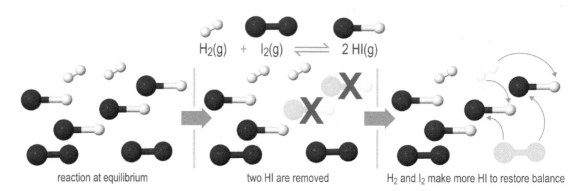

reaction at equilibrium — two HI are removed — H_2 and I_2 make more HI to restore balance

Conditions for chemical equilibrium

Closed system

In order for equilibrium to occur, the reaction flask must be sealed. It must be a **closed system**. A balance between reactants and products cannot be achieved if reactants and products are entering or leaving the reaction container.

Reversible reactions

Another condition for equilibrium is a reversible reaction. Many reactions, like combustion reactions, only go in one direction. They are not considered to be equilibrium reactions. Reversible reactions always have two-way arrows separating the reactants and products.

Constant pressure

Pressure affects the gas phase of an equilibrium. A reaction that has gaseous substances must be done at constant pressure. Increasing pressure favors the side of the reaction with the fewest number of gas particles. In the reaction below, which side is favored when pressure is increased?

$$3H_2(g) + N_2(g) \rightleftharpoons 2NH_3(g)$$

Using the coefficients, the reactant side has 3 moles of H_2 and 1 mole of N_2. The product side has 2 moles of NH_3. An increase in pressure will favor the product side because it has fewer total moles of gas. If there are equal moles of gas on either side of the reaction, or there are no gases in the reaction, pressure changes cannot disrupt equilibrium.

Constant concentration

If the concentration of either the products or reactants changes, the balance between reactants and products must also change. If CO were added to the flask containing the reaction below, how would the reaction shift to offset the change?

$$CO(g) + Cl_2(g) \rightleftharpoons COCl_2(g)$$

Considering the reaction was in balance before the CO was added, the materials in the closed flask will have to rearrange to restore the balance. This means that Cl_2 and the CO that was added will combine to form more phosgene, $COCl_2$. The rule is: adding reactants will shift the reaction to make more products. Adding products will shift the reaction to make more reactants.

Constant temperature

Unlike the other factors that shift equilibrium, changing temperature will change the value of the equilibrium constant. It is important to know whether the reaction is endothermic or exothermic when considering equilibrium shift. If energy is added to the reaction below how will the reaction shift?

$$CaO(aq) + H_2O(l) \rightleftharpoons Ca(OH)_2(aq) + energy$$

Adding energy will shift the reaction to the left. The reaction shifts to absorb the energy. Adding energy to an endothermic reaction will shift the reaction to the products.

Section 15.3: Le Châtelier's Principle

Using Le Châtelier's principle

Changing concentrations

Chemical engineers are tasked with making reactions work on an industrial scale. Often they are asked to adjust the conditions of a chemical reaction so that the maximum amount can be produced. They use the principles developed by Le Châtelier to create the optimal conditions that will cause the reaction to shift in the desired direction.

Le Châtelier's principle: Any change imposed on a system in dynamic equilibrium will cause it to shift in the direction that opposes the change

Solved problem

A scientist wants to make NH_3 from the reaction, $N_2(g) + 3 H_2(g) \rightleftharpoons 2 NH_3(g) + $ energy in a sealed container. What changes in pressure, concentration, and temperature will produce the maximum amount of NH_3?

Given $N_2(g) + 3 H_2(g) \rightleftharpoons 2 NH_3(g) + $ energy

Solve Use the ideas on the previous page to make the reaction shift in the desired direction. The reaction is a closed system and a reversible reaction.

Pressure: All of the reacting species are gases and affected by changes in pressure. Increasing pressure favors the side with the fewest particles and NH_3 is on that side. Increasing pressure will produce more NH_3.

Concentration: Adding any reactant will shift the equilibrium to the product side. Conversely, removing the product, NH_3, will also cause the reaction to replace what was taken out. This can be done by putting water on the bottom of the reaction container. N_2 and H_2 are both nonpolar and will not dissolve into the water. However, NH_3 is polar and it will dissolve into the water.

Temperature: You can think about temperature in the same terms as concentration. Consider energy and temperature to be the same thing. Adding temperature causes the reaction to shift to the left. Cooling the reaction will make the reaction shift to the right.

Answer To make the maximum amount of NH_3, the reaction should be done at high pressure, the concentration of both reactants increased, the concentration of the products decreased, and the temperature reduced.

Ammonia production

If you toured a factory that produced ammonia, you would see a complex system of pipes. The reactants and products are gases, so they are moved through the factory in pipes. Ammonia is cooled and stored under high pressure where it is compressed to a liquid. Liquid ammonia can either be used for manufacturing other products in the factory or it can be transferred to an insulated, pressurized ammonia tank for transportation elsewhere.

Section 3 review

Chapter 15

Earlier in this chapter you read that the equilibrium position of a reaction depends on the initial conditions. This is because Le Châtelier's principle states that for a reaction in dynamic equilibrium, changing concentration, temperature, or pressure/volume can influence the equilibrium position. If any of these conditions are changed, the reaction will shift in a way that opposes the change. For reactions with gases, increasing pressure favors the reaction that leads to fewer gas particles while decreasing pressure does the opposite. Increasing concentration on one side of the equation forces the reaction in the other direction. Changing temperature actually changes the value of the equilibrium constant, and its effect depends on whether the reaction is exothermic or endothermic.

Vocabulary words   Le Châtelier's principle, closed system

Review problems and questions

1. For the equilibrium system $2\ NO_2(g) \rightleftharpoons N_2O_4(g) + 57.5\ kJ$, indicate whether each of the following events is enough to disrupt equilibrium. If equilibrium is disrupted, apply Le Châtelier's principle to determine which side will be favored.

    a. Adding 1 mole each of $NO_2(g)$ and $N_2O_4(g)$.
    b. Decreasing the temperature.
    c. Increasing the pressure.
    d. Adding 2 moles of $NO_2(g)$ and 1 mole of $N_2O_4(g)$.
    e. Removing 1 mole of $NO_2(g)$.

2. Your teacher asks you how increasing the temperature on the equilibrium system, $COCl_2(g) \rightleftharpoons CO(g) + Cl_2(g)$, will affect its equilibrium. Explain why there is not enough information to answer your teacher's question.

3. What will be the effect of increasing the pressure on the equilibrium system, Energy $+ 2HI(g) \rightleftharpoons H_2(g) + I_2(g)$?

4. For the equilibrium system, Energy $+ 2HI(g) \rightleftharpoons H_2(g) + I_2(g)$, list two or more ways to increase the production of HI.

5. Suppose pressure is increased on the equilibrium system, $CaO(aq) + H_2O(l) \rightleftharpoons Ca(OH)_2(s) + $ Energy. What is the effect on equilibrium?

6. In the given reaction: $SO_2Cl_2(g) \rightleftharpoons SO_2(g) + Cl_2(g)$. What effect does the removal of Cl_2 have on equilibrium?

7. What effect would increasing the temperature of an exothermic reaction have on a chemical reaction at equilibrium?

15.4 - Solubility Product Constant

The **solubility product constant** K_{sp}, is a special case of an equilibrium constant. It is used to calculate how much of a solid will dissolve in a given amount of water. Solubility depends on a number of conditions including the identity of the substance, and solvent conditions such as temperature, pressure, and pH. The solubility product constant is applied in pharmaceutical chemistry. Less soluble substances may have medicinal value, so researchers must determine the minimum effective amount the body can safely dissolve.

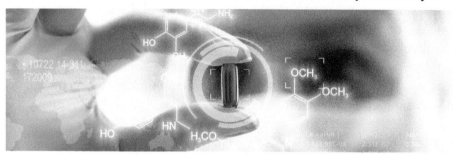

The solubility product constant, K_{sp}

K_{sp}

K_{sp} is calculated in a similar manner to any other equilibrium constant. For salts, the equilibrium is between the solid compound and its dissolved ions:

$$K_{sp} = \frac{[Products]}{[Reactants]}$$

It is used for solids dissolving, but solids cannot have a molarity. The reactants in a K_{sp} problem are always solid. So the equilibrium expression for K_{sp} becomes, K_{sp} = [Products]. For example, when aluminum carbonate, $Al_2(CO_3)_3$, dissolves the solution equilibrium is:

$$Al_2(CO_3)_3(s) \rightleftharpoons 2Al^{3+}(aq) + 3CO_3^{2-}(aq).$$

The K_{sp} expression for dissolving $Al_2(CO_3)_3$ is $K_{sp} = [Al^{3+}]^2[CO_3^{2-}]^3$.

Solved problem

A solution contains 2.25 grams of lead(II) chloride in 500. mL of water. What is the K_{sp} for lead(II) chloride?

Given 2.25 g of $PbCl_2$

Relationships The molar mass of lead(II) chloride is 278.106 g/mol and the $K_{sp} = [Pb^{2+}][Cl^-]^2$.

Solve First we find the molarity of $PbCl_2$.

$$\frac{2.25 \text{ g } PbCl_2}{1} \times \frac{1 \text{ mole } PbCl_2}{278.106 \text{ g } PbCl_2} = 8.09 \times 10^{-3} \text{ moles } PbCl_2$$

$$\frac{8.09 \times 10^{-3} \text{ moles } PbCl_2}{0.5 \text{ Liters}} = 1.62 \times 10^{-2} \text{ M } PbCl_2$$

Before we can use the K_{sp} we need to know the concentrations of the individual ions in the reaction, $PbCl_2 \rightleftharpoons Pb^{2+} + 2Cl^-$. Using stoichiometry we can see that 1 unit of $PbCl_2$ yields one unit of Pb^{2+} and 2 units of Cl^-. Therefore, we have 1.62×10^{-2} M Pb^{2+} and 3.24×10^{-2} M Cl^-. These values are placed into the K_{sp} expression:

$$K_{sp} = [Pb^{2+}][Cl^-]^2 = [1.62 \times 10^{-2}][3.24 \times 10^{-2}]^2 = 1.70 \times 10^{-5}$$

Answer The K_{sp} for lead(II) chloride is 1.70×10^{-5}.

Using K_{sp}

K_{sp}

Most often you use the K_{sp} to find out how much of a substance dissolves. A K_{sp} can be found on many reference tables. They are cross referenced with a corresponding temperature. The amount of a substance that dissolves is a function of temperature. Therefore the K_{sp} for a substance will change with temperature.

Solved problem

What is the concentration of ions when aluminum hydroxide dissolves at 25.0 °C? The K_{sp} is 1.26×10^{-33}.

Given   The K_{sp} at 25.0 °C for aluminum hydroxide is 1.26×10^{-33}.

Relationships $K_{sp} = [Al^{3+}][OH^-]^3$

Solve   First we find the concentration of $Al(OH)_3$.

Before we can use the K_{sp} expression we need to put the variable x into the equation. Using the coefficients of the balanced equation,

$$Al(OH)_3 \rightleftharpoons Al^{3+} + 3\,OH^-,$$

we will use x to represent Al^{3+} and 3x to represent OH^-.

$$K_{sp} = [Al^{3+}][OH^-]^3$$
$$1.26 \times 10^{-33} = [x][3x]^3$$
$$1.26 \times 10^{-33} = x(27x^3)$$
$$1.26 \times 10^{-33} = 27x^4$$
$$\frac{1.26 \times 10^{-33}}{27} = x^4$$
$$4.67 \times 10^{-35} = x^4$$
$$x = \sqrt[4]{4.67 \times 10^{-35}}$$
$$x = 2.61 \times 10^{-9}$$

The value for x represents the concentration of Al^{3+} in the solution and 3x represents the concentration of hydroxide.

Answer   The concentration of Al^{3+} is 2.61×10^{-9} M and the concentration of OH^- is 7.84×10^{-9} M.

The size of K_{sp}

The K_{sp} in the problem above is very small, 1.26×10^{-33}. K_{sp} is still considered a ratio of products to reactants even though we removed the reactants. Consider the concentration of reactants to be the number one. That means the ratio of products to reactants is 1.26×10^{-33} to 1. The concentration of products is very small when compared to the reactants, or in other words, very little dissolves. We can see this in the answer to the solved problem. The concentration of ions in the answer is very small. A general rule is, the smaller the K_{sp}, the smaller the amount that dissolves.

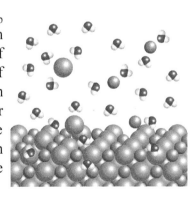

Section 15.4: Solubility Product Constant

Precipitate formation

Precipitation

K_{sp} can also be used to determine whether solute will precipitate out of a solution. K_{sp} is the ratio of products to reactants that saturate the solution. Generally, as a solution cools the amount of dissolved particles it can hold goes down. Once the ratio of dissolved particles falls below the K_{sp}, precipitation occurs.

Solved problem

Will a precipitate form if a solution that contains 1.99×10^{-4} M Ca^{2+} and 1.99×10^{-4} M CO_3^{2-} cools to 25.0 °C? The K_{sp} for $CaCO_3$ is 4.76×10^{-9}.

Given 1.99×10^{-4} M Ca^{2+}, 1.99×10^{-4} M CO_3^{2-}. The K_{sp} for $CaCO_3$ is 4.76×10^{-9}.

Relationships $CaCO_3 \rightleftharpoons Ca^{2+} + CO_3^{2-}$, $K_{sp} = [Ca^{2+}][CO_3^{2-}]$

Solve We first have to find the maximum concentration of Ca^{2+} and CO_3^{2-} that will dissolve at 25.0 °C.

$$K_{sp} = [Ca^{2+}][CO_3^{2-}]$$
$$4.76 \times 10^{-9} = [x][x]$$
$$4.76 \times 10^{-9} = x^2$$
$$x = \sqrt{4.76 \times 10^{-9}}$$
$$x = 6.90 \times 10^{-5}$$

The value for x represents the concentration of Ca^{2+} and CO_3^{2-}. This means that 6.90×10^{-5} M Ca^{2+} and CO_3^{2-} will saturate the solution, and any more of it will precipitate out of the solution. The given amount, 1.99×10^{-4} M, is much larger than the allowed amount of 6.90×10^{-5} M.

Answer A precipitate will form in the solution.

Lime

The chemical formula for lime is $CaCO_3$. As you can see from the previous problem the K_{sp} is very small and very little dissolves in water. However this little amount has big effects. When lime is dissolved in water it forms "hard water". Hard water gives you dry skin and keeps soap from lathering. It also precipitates out in pipes and in appliances like hot water heaters as scale and clogs them over time. To address this problem people in many parts of the world have to install water softeners in their homes. Water softeners exchange sodium ions for calcium ions. The amount of sodium in the water is very small and does not have negative effects on your body or the pipes in your house.

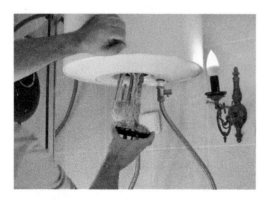

Section 4 review

Chapter 15

The solubility product constant (K_{sp}) is the equilibrium constant for a solid substance that is dissolved into an aqueous solution. The higher the K_{sp} value the more soluble a substance is. Because a solid substance doesn't have a molarity, the solubility product constant formula simplifies to a product of the concentration of the ions. K_{sp} is typically used to calculate the amount of a substance that dissolves. It can also be used to determine if a precipitate will form. Precipitation occurs when the ion product is greater than the solubility product constant.

Vocabulary words: solubility product constant

Review problems and questions

1. Write a K_{sp} expression for tin(II) hydroxide, $Sn(OH)_2$.

2. Find the concentration of ions when cobalt(II) hydroxide, $Co(OH)_2$, dissolves at 25.0 °C. The K_{sp} for $Co(OH)_2$ is 3.0×10^{16}.

3. The K_{sp} for LiF is 5×10^{-3}. Will a precipitate form if a solution with 1.11×10^{-9} M Li^+ and 2.22×10^{-9} M F^- cools?

4. What is the concentration of a saturated strontium sulfate solution? The K_{sp} for $SrSO_4$ is 3.44×10^{-7}.

5. What is the concentration of a saturated nickel(II) hydroxide solution? (K_{sp} for the solution is 5.48×10^{-16}).

6. Write K_{sp} expressions for the following solutions at 25 °C:
    a. Scandium hydroxide
    b. Strontium iodate
    c. Silver(I) chromate
    d. Lithium phosphate
    e. Lead(II) carbonate

7. A solution of $PbCl_2$ has 1.0×10^{-4} M of $[Pb^{2+}]$ and 1.5 M of $[Cl^-]$. Will it form a precipitate? K_{sp} for $PbCl_2 = 1.70 \times 10^{-5}$.

8. A solution contains 27.4 g of barium nitrate in 1000. mL of distilled water. Calculate the Ksp for barium nitrate?

Nitrogen and the Nobel Prize

essential chemistry

The complicated history behind the most famous equilibrium reaction

Ammonia (NH_3) is a critical component for many growing organisms, especially plants, but plants cannot directly use diatomic nitrogen gas from the air. Why would something so essential be so unusable? It turns out that nature has a solution to "fix" this problem. Biological nitrogen fixation is carried out by certain prokaryotic bacteria called diazotrophs. Many of the diazotrophs, like cyanobacteria, are free-living while other diazotrophs are symbiotic. These symbiotic diazotrophs grow and live in the roots of legumes like peas and beans. All the diazotrophs use an enzyme called nitrogenase to catalytically speed up the conversion of atmospheric nitrogen into ammonia. This ammonia is used by plants to create the biomolecules necessary for life. When the plants get eaten, they provide nitrogen to other living things. When those living things die, decomposition happens and the nitrogen containing compounds can end up in the soil to provide fertilizer for more plants, or eventually end up back in the atmosphere. From this description you can see there is a cycle of nitrogen as it moves through the atmosphere, the land and living things. You can learn more about the nitrogen cycle in the "Chemistry of the Earth" chapter.

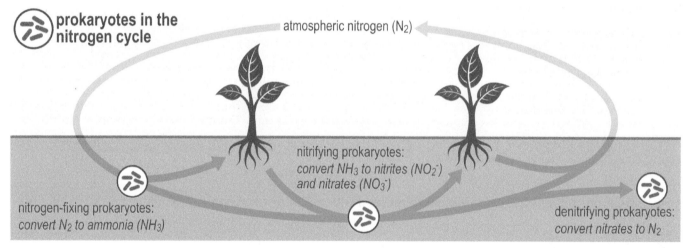

Nature has a good "fix" on the problem of nitrogen, but there is a catch. As the human population has increased there is greater and greater demand for food production. This food production means we need to grow more plants, and those plants need nitrogen. We use nitrogen fertilizer to enrich the soil. To grow more plants, we need more fertilizer. Here's where the catch comes in. Nature does not create fertilizer fast enough to keep up with population growth. As far back as the 1890s this was recognized as a problem.

Fritz Haber, a German chemist, turned his attention to the problem of artificially making fertilizer from atmospheric nitrogen. Nitrogen and hydrogen can react to turn into ammonia according to the following reaction.

$$N_2(g) + 3H_2(g) \rightleftharpoons 2NH_3(g) \quad \Delta H = -92 \text{ kJ/mol}$$

This is an interesting problem because it combines principles of both reaction rates and equilibrium. The equilibrium of this reaction under standard conditions lies on the product side of the reaction towards the formation of ammonia, but it happens at a very, very slow rate.

One way to increase the speed of a reaction is to add a catalyst. This makes sense since biologically the conversion of nitrogen to ammonia happens with an enzyme. Another way to speed up a reaction is to run it at higher temperatures. However, because this reaction is exothermic, an increase in temperature would create less products according to Le Châtelier's principle. Le Châtelier's principle provides another method for producing more products, especially when gases are involved. Looking at the equation there are four moles of gas on the reactants side, and two moles on the product side. An increase in pressure should shift the reaction to the side with less moles of gas. Higher pressures would lead to the production of more ammonia. In the early 1900s Fritz Haber came up with a process that uses catalysts, high pressure, and high temperature to create ammonia out of the nitrogen in the air. Later, through cooperation with an engineer, Carl Bosch, Haber created an industrial scale process to make ammonia at temperatures above 500 °C and pressures between 150 and 200 atm!

This creation of large amounts of ammonia from the air has had a tremendous impact on humanity. Fertilizers get incorporated into the plants that grow and the food we eat. It is estimated that 40% of the nitrogen in your body originated from a Haber-Bosch process! Ammonia and nitrogen containing compounds have uses outside of agriculture, and not all of them are beneficial. Compounds like ammonium nitrate are not only used for fertilizer, but also as explosives and munitions. The development of the Haber-Bosch process coincided with World War I. During the war a British naval blockade stopped shipments of ammonium nitrate from Chile to Germany to cut off their supplies. The Haber-Bosch process was modified to keep the Germans supplied with munitions, prolonging WWI for years.

Scientists do not live in a vacuum. Even brilliant scientific minds, like Fritz Haber, can have mixed contributions to society. The Haber-Bosch process has been estimated to be beneficial to billions of people. At the same time Haber supported the German war efforts. In addition to supplying Germans with munitions during WWI, he put his mind to weaponizing toxic compounds. In 1915, he was directly involved in the first large-scale deployment of chemical weapons in human history! As an example of how complicated his legacy is, Haber won a Nobel prize in Chemistry in 1918, and was named a WWI war criminal in 1919. Despite his loyalty to Germany, he was eventually forced to leave during the build-up to WWII because he was Jewish. Haber died in 1935, but to this day he remains one of the most influential and controversial scientists of the 20th century.

Chapter 15 review

Vocabulary
Match each word to the sentence where it best fits.

Section 15.1

> equilibrium   equilibrium position
> physical equilibrium

1. If the rate of the forward reaction is equal to the rate of the reverse reaction, the reaction is in _____ .

2. A system that is balanced between two or more phases of the same substance is in _____ .

3. The _____ tells us which direction is favored for reversible reactions.

Section 15.2

> chemical species   equilibrium constant
> equilibrium expression   law of mass action

4. The relationship of chemicals and their molar coefficients is shown by the _____, which is a general description of any equilibrium reaction.

5. The numerical ratio of products to reactants is the _____ .

6. A ratio that describes the relationship of the products relative to the reactants in a chemical equilibrium system is called the _____ .

7. A _____ is an atom, molecule, or ion taking part in a chemical process.

Section 15.3

> closed system   Le Châtelier's principle

8. A change in temperature, concentration, volume, or pressure will cause the system to shift in a direction that partially offsets the change is the definition of _____ .

9. Products and reactants in a chemical reaction often take place in a _____, which does not exchange matter with its surroundings, but does exchange energy.

Section 15.4

> solubility product constant

10. The equilibrium constant that refers to solids dissolving is called the _____ .

Conceptual questions

Section 15.1

11. Systems, much like reactions, strive towards a state of equilibrium. Using what you have learned in this section, describe an example of equilibrium that occurs in your own life and explain how it works.

12. For chemical reactions in equilibrium, a double arrow indicates that both the forward and reverse reactions are taking place. What is one method that can be used to observe the reversibility of a chemical reaction?

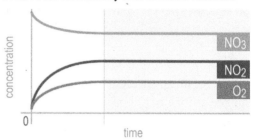

13. From the graph above, How can you tell whether the reactants are favored or if the products are favored in the equilibrium reaction below?

$$2NO_2(g) + O_2(g) \rightleftharpoons 2NO_3(g)$$

14. How does water in a physical system reach an equilibrium between molecules in two different phases? Explain.

15. Explain why chemical equilibrium can be established even when we start with only the products.

16. There is a common misconception that equilibrium is a static and unchanging process. Explain why is it more appropriate to describe equilibrium as dynamic.

17. Are the concentrations of the reactants and products equal in a system in equilibrium? Explain.

Section 15.2

18. When applying data from many experiments to the equilibrium expression, it is found that the value for the equilibrium constant K_{eq} was not the same. Name a factor that could have been changed in each experiment for the equilibrium expression to have a different value. Explain your reasoning.

19. A chemical reaction is at equilibrium. If you change the temperature, will the same reaction have the same K before and after the change? Explain.

20. Why would chemists need to know if the K_{eq} of a reaction is greater or less than 1? What does the K_{eq} value tell you about the reaction?

21. An ICE chart is very useful for calculating equilibrium concentrations. Why would you need an ICE chart in some equilibrium problems and not in others?

22. What is the relationship between products and reactants once a system reaches equilibrium?

23. Using what you have learned so far and in your own words, explain the equilibrium expression and the equilibrium constant.

24. Describe the process for calculating the equilibrium concentrations before a reaction starts. Remember to include the steps required to create an ICE chart in your answer.

Section 15.3

25. What does Le Châtelier's principle state?

Chapter 15 review

Section 15.3

26. HCl gas forms in the following reversible reaction.

$$Cl_2(g) + H_2(g) \rightleftharpoons 2HCl(g)$$
$$\Delta H = -92 \text{ kJ/mol}$$

Apply Le Châtelier's principle to find out which direction the above equilibrium will shift under the following conditions:

a. The temperature is increased.

b. $H_2(g)$ is added.

c. $Cl_2(g)$ is removed.

27. Consider the following endothermic reaction:

$$PCl_5(g) \rightleftharpoons PCl_3(g) + Cl_2(g)$$
$$\Delta H = 87.9 \text{ kJ}$$

Apply Le Châtelier's principle to find out which direction the above equilibrium will shift under the following conditions:

a. Pressure is decreased on the system

b. The temperature is lowered.

c. $PCl_5(g)$ is added.

d. $Cl_2(g)$ is removed.

28. Consider the following general equilibrium reaction:

$$2A(g) + B(g) \rightleftharpoons 2C(g)$$
$$\Delta H = -232 \text{ kJ}$$

Apply Le Châtelier's principle to find out which direction the above equilibrium will shift under the following conditions:

a. The temperature is increased.

b. B is added.

c. The pressure of the reaction mixture is decreased.

d. C is removed.

29. In order for equilibrium to occur, several conditions must be met. Explain why it is necessary for each of the following conditions to be met for equilibrium to be achieved. Explain which side is favored if it were to change.

a. It must be a closed system.

b. The reaction must be reversible.

c. There must be a constant pressure.

d. There must be a constant concentration.

e. There must be a constant temperature.

30. The industrial production of ammonia is described by the following reversible reaction.

$$N_2(g) + 3H_2(g) \rightleftharpoons 2NH_3(g)$$
$$\Delta H = -46 \text{ kJ/mol}$$

Apply Le Châtelier's principle to find out how the following effect the equilibrium.

a. The addition of heat.

b. An increase in pressure.

c. The addition of a catalyst.

d. Cooling the system.

e. The removal of $NH_3(g)$.

31. Use Le Châtelier's principle to explain why soft drinks go flat once they are opened.

32. A system at equilibrium is shifted to offset a change that has been introduced to it. What are some factors could cause a shift in an equilibrium system?

33. There are several conditions necessary for equilibrium to occur. What are they?

34. Chemical engineers are often tasked with making reactions work on an industrial scale. What is a chemical engineer able to do for large scale reactions using Le Châtelier's principle?

Section 15.4

35. The solubility product constant, K_{sp}, is a special case of an equilibrium constant. What does it represent?

36. The K_{sp} for silver(I) sulfate (Ag_2SO_3) is 1.20×10^{-5} at 25.0 °C.

a. At 50.0 °C, is the K_{sp} larger or smaller? Explain.

b. Which temperature, 25.0 °C or 50.0 °C, would have a higher concentration of dissolved silver ions? Explain.

37. Explain why reactants do not appear in the equilibrium expression for K_{sp} as they do for a general K_{eq} as seen below:

$$K_{eq} = \frac{[\text{Products}]}{[\text{Reactants}]}$$

What is the true equilibrium expression for K_{sp}?

38. How can K_{sp} be used to determine whether solute will precipitate out of a solution?

Quantitative problems

Section 15.2

39. Do the following reactions contain mainly reactants or products at equilibrium? Explain.

a. $2SO_3(g) \rightleftharpoons 2SO_2(g) + O_2(g)$ $K_{eq} = 3.91 \times 10^{-10}$

b. $N_2(g) + 3H_2(g) \rightleftharpoons 2NH_3(g)$ $K_{eq} = 75.3$

c. $2NO(g) + O_2(g) \rightleftharpoons 2NO_2(g)$ $K_{eq} = 5.14 \times 10^{12}$

Chapter 15 review

Section 15.2

40. Write the equilibrium expressions for the following reactions.

    a. $N_2(g) + 3H_2(g) \rightleftharpoons 2NH_3(g)$

    b. $HCO_3^-(aq) + NH_3(aq) \rightleftharpoons NH_4^+(aq) + CO_3^{2-}(aq)$

    c. $2SO_2(g) + O_2(g) \rightleftharpoons 2SO_3(g)$

    d. $Cl_2(g) + H_2(g) \rightleftharpoons 2HCl(g)$

    e. $N_2O_4(g) \rightleftharpoons 2NO_2(g)$

41. Write the equilibrium expressions for the following reactions.

    a. $I_2(g) + Cl_2(g) \rightleftharpoons 2ICl(g)$

    b. $2NO_2(g) \rightleftharpoons O_2(g) + 2NO(g)$

    c. $PCl_3(g) + Cl_2(g) \rightleftharpoons PCl_5(g)$

    d. $2N_2O_5(g) \rightleftharpoons 4NO_2(g) + O_2(g)$

42. Methanol is formed from the reaction of carbon monoxide and hydrogen gas.

    $$CO(g) + 2H_2(g) \rightleftharpoons CH_3OH(g)$$

    There are 0.260 moles of CH_3OH, 0.27 moles CO, and 0.45 moles H_2 present in a 1.5 L reaction vessel at equilibrium. What is the value of K_{eq} at a constant temperature of 227 °C.

43. What is the equilibrium concentration of $O_2(g)$ of the following reaction?

    $$2SO_2(g) + O_2(g) \rightleftharpoons 2SO_3(g)$$

    The equilibrium concentration of $SO_2(g)$ is 2.00 M and $SO_3(g)$ is 10.0 M? The equilibrium constant is 800.

44. The below reaction has a K_{eq} of 0.0410.

    $$PCl_5(g) \rightleftharpoons PCl_3(g) + Cl_2(g)$$

    What is the concentration of the species at equilibrium if 0.050 M PCl_5 is initially added to the reaction container?

45. The K_{eq} for the reaction below is 40 at 550.0 °C. What is the K_{eq} for the reverse reaction?

    $$H_2(g) + I_2(g) \rightleftharpoons 2HI(g)$$

46. The value of K_{eq} for the above reaction is 0.800.

    $$SO_3(g) + NO(g) \rightleftharpoons NO_2(g) + SO_2(g)$$

    A mixture had the following concentrations: $[SO_3]$ = 0.400 M, [NO] = 0.480 M, $[NO_2]$ = 0.600 M, and $[SO_2]$ = 0.450 M

    a. Show, by calculation, that this mixture is not at equilibrium.

    b. What will happen to the concentration of SO_3 and SO_2 as the system moves towards equilibrium?

47. The following reaction is the reversible decomposition of nitrous oxide (laughing gas).

    $$2N_2O(g) \rightleftharpoons 2N_2(g) + O_2(g)$$

    If the K_{eq} is 7.3 x 10^{34} and the concentration of $N_2O(g)$ is 1.4 x 10^{-2} M at equilibrium, what are the concentrations of the other species at equilibrium?

48. The following reaction was conducted in a 2.0 liter container.

    $$N_2(g) + O_2(g) \rightleftharpoons 2NO(g)$$

    At equilibrium, there are 0.25 moles of N_2, 0.25 moles of O_2 and 0.010 moles of NO. What is the K_{eq} for this reaction?

49. The following reaction takes place in a 1.00 liter container at 750.0 °C.

    $$H_2(g) + CO_2(g) \rightleftharpoons H_2O(g) + CO(g)$$

    At equilibrium there are 0.106 moles of H_2, 0.106 moles of CO_2, 0.094 moles of water and 0.094 moles of CO. What is the K_{eq} for this reaction?

50. Determine the equilibrium constant for the following reaction:

    $$CO(g) + Cl_2(g) \rightleftharpoons COCl_2(g)$$

    The equilibrium concentrations are 7.24 x 10^{-3} M $COCl_2$, 2.25 x 10^{-4} M Cl_2, and 1.84 x 10^{-3} M CO.

51. The at temperatures much higher than 550.0 °C the K_{eq} for the reaction below can reach 1.98 x 10^{-4}.

    $$2HI(g) \rightleftharpoons H_2(g) + I_2(g),$$

    Use the above values to find the concentrations of all the species in the reaction at equilibrium when 0.10 M HI(g) is initially placed in a container.

Section 15.4

52. Which will have a larger concentration of dissolved magnesium ions: a saturated solution of MgF_2 (K_{sp} = 5.16 x 10^{-11}) or a saturated solution of $Mg_3(PO_4)_2$ (K_{sp} = 1.04 x 10^{-24})?

53. Calcium carbonate is used as an antacid to relieve the symptoms of indigestion and heartburn. If you have 5.3 x 10^{-5} M Ca^{2+} and 5.3 x 10^{-5} M CO_3^{2-}, what is the K_{sp} of calcium carbonate at 25.0 °C?

54. Silver iodide has medical uses as an antiseptic. It is also used in cloud seeding to increase precipitation in an area or in fog suppression around airports. If 2.16 x 10^{-7} g of silver iodide dissolves in 100 ml of water, what is the K_{sp}?

55. Strontium fluoride, SrF_2, is used as optical coating on lenses. If 1.07 x 10^{-2} g of strontium fluoride dissolves in 100 ml of water, what is the K_{sp}?

56. Doctors use barium sulfate to coat the intestinal lining during x-rays so the intestines stand out. However, barium is poisonous to humans. The K_{sp} for barium sulfate is 1.08 x 10^{-10}. What is the concentration of barium ions that actually dissolve?

Chapter 15 review

Section 15.4

57. The K_{sp} for lithium phosphate is 3.2×10^{-9} and the K_{sp} for lithium carbonate is 2.5×10^{-2}.

    a. Which of the two compounds have a higher concentration of lithium ions?

    b. What is the difference in concentrations?

58. What is the concentration of dissolved cadmium if the K_{sp} for cadmium (II) phosphate is 2.53×10^{-33}?

59. Lead(II) chloride is an intermediate in refining bismuth ore and is used in the production of ornamental glass. If a solution of 3.00×10^{-2} M $PbCl_2$ at 89.0 °C cools to 25.0 °C, will a precipitate form if the K_{sp} at 25.0 °C for lead(ll) chloride is 1.62×10^{-5}?

60. Aluminum hydroxide can be used to reduce phosphate levels in people with certain kidney conditions. If a solution of 1.99×10^{-5} M $Al(OH)_2$ is cooled, will a precipitate form? The K_{sp} for aluminum hydroxide is 1.26×10^{-33}.

61. If silver ions are added dropwise to a solution contains 1.00×10^{-2} M Cl^- and 1.00×10^{-3} M I^-, which will completely precipitate out first, AgI or AgCl? The K_{sp} for AgI is 8.32×10^{-17} and the K_{sp} for AgCl is 1.81×10^{-10}.

62. Municipalities put fluoride into the water supply to improve the public's teeth. Tooth enamel contains the mineral hydroxyapatite, $Ca_5(PO_4)_3OH$, which has a K_{sp} of 6.8×10^{-37} and 2.7×10^{-5} mol/L will dissolve at 25.0 °C. Fluoride replaces hydroxide in hydroxyapatite and forms fluoroapatite, $Ca_5(PO_4)_3F$. This new mineral is harder and less likely to get cavities. The K_{sp} for fluoroapatite at 25.0 °C is 1.0×10^{-60}.

    a. How much fluoroapatite dissolves at 25.0 °C? Fluoroapatite dissolves into the ions, Ca^{2+}, PO_4^{3-} and F^-.

    b. Compare the solubility of fluoroapatite and hydroxyapatite and determine whether the addition of fluoride makes a difference.

63. One liter of saturated silver chloride solution contains 0.00192 g of dissolved AgCl at 25 °C.

    a. What is the equilibrium equation that illustrates this saturated solution?

    b. Calculate the K_{sp} for AgCl.

64. Calculate the molar solubility (in mol/L) of a saturated solution of the following substances.

    a. Copper(II) phosphate ($Cu_3(PO_4)_2$), $K_{sp}=1.93 \times 10^{-37}$

    b. Iron(III) sulfide (Fe_2S_3), $K_{sp}=1 \times 10^{-88}$

    c. Magnesium arsenate ($Mg_3(AsO_4)_2$), $K_{sp}=2.1 \times 10^{-20}$

    d. Nickel(II) phosphate ($Ni_3(PO_4)_2$), $K_{sp}=4.74 \times 10^{-32}$

chapter 16 | Acids and Bases

Have you ever bitten into something really sour? You might not have known it, but you were actually ingesting an acid! Lemons and limes cause your lips to pucker because they contain citric acid ($H_3C_6H_5O_7$). Citric acid is also used to make gums and candies "sour." In addition to citric acid, other fruits and vegetables contain acids such as malic acid ($H_2C_4H_4O_5$), oxalic acid ($H_2C_2O_4$), and ascorbic acid ($HC_6H_7O_6$) which is also known as Vitamin C. Vinegar gives foods a tart flavor because it is about 5% acetic acid ($HC_2H_3O_2$). Lactic acid ($HC_3H_5O_3$) is found in sour milk products like yogurt and kefir.

Of course, acids aren't just about taste. They are also an important part of your body chemistry. There are a series of reactions in your blood involving carbonic acid (H_2CO_3) that help you remove the carbon dioxide from your system after your body breaks down sugars for their energy. Your stomach contains hydrochloric acid (HCl), which not only digests food, but is strong enough to dissolve some metals! If you have an upset stomach, you can take an antacid to help you feel better. As the name implies, an antacid is the opposite of an acid. In chemistry, we call this type of substance a base. That antacid in the medicine cabinet is really just a base that will neutralize some excess acid in your stomach.

citric acid $C_6H_8O_7$

Acids and bases are integral to your everyday life. They are used in everything from foods and medicines to batteries and cleaning products. Acids and bases can also affect our environment by changing the quality of soil and water. This begs the question - what are acids and bases, and why are they so essential?

Chapter 16 study guide

Chapter Preview

You may know that the pH scale measures whether something is an acid or a base, but what are acids and bases? Arrhenius defined acids as the formation of hydrogen ions in solution and bases as the formation of hydroxide ions in solution. The Brønsted-Lowry definition refines this further, naming acids as proton donors and bases as proton acceptors. Acids and bases react to neutralize one another to form water and a salt. Strong acids and bases dissociate fully in water, while weak acids and bases only partially dissociate. In this chapter you'll also learn to calculate pH, investigate how titrations are performed, and explore buffers and how they work.

Learning objectives

By the end of this chapter you should be able to:

- define acids and bases using the Arrhenius and Brønsted-Lowry definitions;
- use the pH scale to determine whether a substance is an acid or a base;
- calculate pH and pOH using the logarithmic formula;
- solve for the pH of a given solution based on the concentration of H^+ or OH^- ions and/or the acid dissociation constant;
- calculate the amount or concentration of an acid or base needed for a neutralization reaction; and
- explain how a buffer is able to resist changes in pH when an acid or base is added.

Investigations

16A: What is pH?

16B: Titration of an unknown acid

16C: Antacids: an inquiry study

Pages in this chapter

- 508 Acids and Bases
- 509 Arrhenius acids and bases
- 510 Bases
- 511 The significance of the H^+ ion
- 512 Weak and strong acids and bases
- 513 Brønsted-Lowry acids and bases
- 514 Conjugate acid-base pairs
- 515 Section 1 Review
- 516 The pH Scale
- 517 pH is a logarithmic scale
- 518 Calculation of pH
- 519 Calculation of pH for bases
- 520 pH indicators
- 521 Section 2 Review
- 522 Acid - Base Equilibria
- 523 Weak acids
- 524 Weak bases
- 525 Section 3 Review
- 526 Acid - Base Reactions
- 527 Neutralization reactions
- 528 Titrations
- 529 Polyprotism
- 530 Normality
- 531 Salts and pH
- 532 Buffers
- 533 Buffer capacity
- 534 Section 4 Review
- 535 The Basics of Bicarbonate
- 536 The Basics of Bicarbonate - 2
- 537 Chapter review

Vocabulary

pH scale	Arrhenius acid	Arrhenius base
salt	neutral	acidic solution
base	hydronium ion	strong acid
strong base	weak acid	weak base
Brønsted-Lowry	conjugate acid-base pairs	amphoteric
logarithm	indicator	ion product constant
neutralization	titration	equivalence point
polyprotic acid	diprotic acid	triprotic acid
normality	equivalent	common ion
buffer	buffer capacity	

16.1 - Acids and Bases

Popular culture depicts acids as burning through metal and exploding. In reality, acids are not as exciting as the movies show. A few acids can cause harm when they are highly concentrated. However, the majority of acids are benign. Most of the food you eat and drink are acidic. In fact, much of your body is made from proteins that are formed from amino acids. Not only are most acids harmless, they are necessary for life. Many bases we encounter each day are not derived directly from nature. Common household cleaning chemicals are bases.

Acids and bases

The pH scale Because acids and bases are so important, they have a special measurement - the pH scale. The **pH scale** tells you whether a solution is acidic or basic. A pH less than 7 indicates an acid. An acid has a higher concentration of hydrogen ions, H^+, than pure water. A pH greater than 7 indicates a base. A base has an H^+ concentration less than pure water.

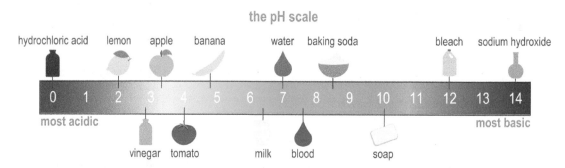

Properties of acids An acid is a compound that dissolves in water to make a solution that contains more H^+ ions than there are in pure water. Some properties of acids are listed below.

- Acids have a pH smaller than 7.
- Acids create the sour taste in food, such as lemons.
- Acids react with metals to produce hydrogen gas (H_2).
- Acids react with bases to produce a salt and water.
- Acids can corrode metals and burn skin through chemical action.
- Acids are conductive.
- Acids turn red litmus paper blue.

Properties of bases A base is a compound that dissolves in water to make a solution with more hydroxide, OH^- ions than there are in pure water. Some of the extra OH^- combines with H^+ to make water again, so another way to think about bases is that they reduce the concentration of H^+ ions. Some properties of bases are listed below:

- Bases have a pH greater than 7.
- Bases create a bitter taste.
- Bases react with acids to produce a salt and water.
- Bases can corrode metals and burn skin through chemical action.
- Bases have a slippery feel, like soap.
- Bases can neutralize acids.
- Bases are conductive.
- Bases turn blue litmus paper red.

Arrhenius acids and bases

Defining acids and bases

Svante Arrhenius first defined acids and bases in 1887. An **Arrhenius acid** is a substance that produces hydrogen ions, H^+ in water, and an **Arrhenius base** is a substance that produces hydroxide ions, OH^- in water. This definition is very easy to use but it is limited. During the course of this chapter, we will look at other definitions that expand the number of substances that are acids and bases.

Arrhenius acids and bases

HBr is an Arrhenius acid because when it dissolves in water it decomposes into H^+ and Br^-.

$$HBr(g) \rightarrow H^+(aq) + Br^-(aq)$$

KOH(s) is an Arrhenius base because when it dissolves in water it decomposes into K^+ and OH^-.

$$KOH(s) \rightarrow K^+(aq) + OH^-(aq)$$

Reacting acids and bases

According to the Arrhenius theory, when an acid and base are reacted, they form water and a salt. A **salt** is one of the products of an acid-base reaction. The other product is water. Salts are a large class of compounds and not always edible.

$$HBr(g) + KOH(s) \rightarrow H_2O(l) + KBr(aq)$$

In this example, KBr is the salt.

Neutral and acidic solutions

In a **neutral** solution, the concentrations of H^+ and OH^- are equal. When a H^+ ion and an OH^- ion react, they form a water molecule.

$$H^+ + OH^- \rightarrow H_2O$$

By definition, pure water is neutral. In the context of acidity, the word "neutral" means $[H^+] = [OH^-]$.

An **acidic solution** is a solution that contains more H^+ ions than OH^- ions.

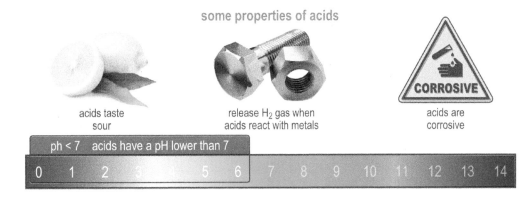

Making acids

Most acids are formed by dissolving nonmetals in water. Halogens and some polyatomic ions make some of the strongest acids known. For example, HCl, HBr, HNO_3 and H_2SO_4 are very strong acids. These acids can cause severe burns when they are concentrated.

Acids often start with an "H"

The chemical formula for an acid is usually written with an H first. However, the acid may be written several ways when it contains carbon. For example, acetic acid can be written as HCH_3COO, $C_2H_4O_2$ or CH_3COOH.

Bases

Bases have a pH greater than 7

A **base** is a substance that dissolves to produce a solution with more OH⁻ ions than H⁺ ions. Many properties of bases are the opposite of acids. For example, H⁺ and OH⁻ are opposite ions. However, bases do haves some things in common with acids. Both are corrosive, and both conduct electricity in solution.

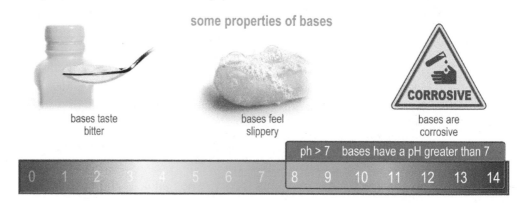

[Brackets] = concentration

Recall that brackets [ ] mean "concentration." When you see brackets around a formula in this chapter, assume the concentration is in moles per liter solution, or molarity (M). You would read "[H⁺] = 0.01 M" as "the hydrogen ion concentration equals 0.01 molar." High school students usually use dilute acid and base solutions with low molarities.

Solutions can be neutral, acidic, or basic

The balance between [H⁺] and [OH⁻] ions determines pH. If the concentration of H⁺ and OH⁻ ions is equal, pH equals 7.0 and the solution is neutral. If the concentration of H⁺ ions is greater than OH⁻ ions, the solution is acidic and pH is less than 7.0. A solution with a higher concentration of OH⁻ ions than H⁺ ions is basic, and pH is greater than 7.0.

Sodium hydroxide

Sodium hydroxide, NaOH, is a common base found in everyday chemicals in most homes and in industry. Sodium hydroxide is also known as lye. NaOH dissociates into sodium ions (Na⁺) and hydroxide ions (OH⁻) when added to water. It does not take much NaOH to damage body tissue, clothing, and a number of other materials. A 1.0 M NaOH solution has a pH of 14.

NaOH (lye) dissociates into sodium and hydroxide ions in water

$NaOH(aq) \longrightarrow Na^+(aq) + OH^-(aq)$

Alkaline means basic

Group I or II metals dissolve in water to make a base. The family names for group I and II are alkali and alkaline earth metals respectively. The word alkaline means "base". You may have heard of alkaline soil. This type of soil has a high concentration of sodium, calcium and magnesium ions that cause high soil pH. Potassium hydroxide (KOH) makes alkaline batteries basic. Potassium hydroxide is dissolved inside a battery to form conductive potassium ions (K⁺) and hydroxide ions (OH⁻).

potassium hydroxide (KOH) is used in **alkaline batteries**

free-floating K⁺ and OH⁻ ions

Recognizing bases

Chemical formulas for many bases contain hydroxide, for example, NaOH. An OH at the end of an ionic formula reminds you that it is a base. A molecular formula ending with OH is probably not a base, for example, formic acid (HCOOH). The OH at the end of the formula is not hydroxide because HCOOH is not an ionic compound. Formulas for some bases such as ammonia, NH₃, do not contain OH at all. You have to learn other ways to distinguish acids and bases, such as the Brønsted-Lowry definition.

The significance of the H⁺ ion

H⁺, the "naked proton"

Think about what a typical H⁺ ion is made of. Since it has neither has neutrons nor electrons, all it is made of is a nucleus with a single proton. No other element can exist or react as a bare nucleus. Hydrogen's nucleus has no neutrons and becomes only a single proton stripped of its electrons in ion form. We learned that chemical reactions are caused by interactions between electrons in unstable atoms. To state it more accurately, chemical behavior is driven by the electrical energy that links electrons and protons. The H⁺ ion is very small. Its positive charge is especially powerful because there is no electron cloud to shield its electrical force. All acid-base chemistry is directly related to the very potent electrical charge of the H⁺ ion. When you hear about protons involved in a chemical reaction, the reaction will involve the H⁺ ion.

H⁺ bonds with water to make hydronium

Recall that the oxygen-hydrogen bond in a water molecule is highly polar. The H⁺ ion is strongly attracted to the partially negative oxygen atom in a water molecule, forming H_3O^+. This is called the **hydronium ion**. The H⁺ ion becomes hydrated when it bonds to water molecules. When the H⁺ ion is dissolved in water, it immediately forms H_3O^+. The H⁺ ion is not found as a solitary ion in an aqueous solution.

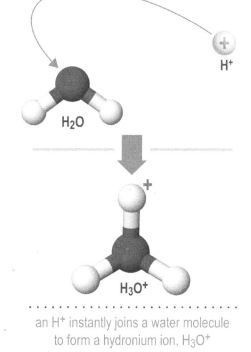

an H⁺ instantly joins a water molecule to form a hydronium ion, H_3O^+

H⁺ = H₃O⁺

It is common practice to refer to the hydronium ion as just H⁺ and not H_3O^+. This is because H⁺ is easier to write. However, always remember in an aqueous solution, H⁺ is just an abbreviation for the H_3O^+ ion. Protons, hydrogen ions and hydronium ions are all synonyms.

Water auto dissociates

Water dissociates into H⁺ and OH⁻.

$$H_2O \rightleftharpoons H^+ + OH^-$$

Water remains neutral because since one H_2O molecule yields one H⁺ and one OH⁻, the number of H⁺ and OH⁻ remains equal. A very small percentage of water molecules dissociate at any one time. In fact, the equilibrium constant for the dissociation of water, K_w, is 1×10^{-14}.

[H₂O]

Water cancels out of the equation and the brackets are a symbol that represents molarity. Since water is the solvent, writing [H₂O] is interpreted as the amount of water that dissolves in water. It is safe to say that all of the water dissolves in water. The [H₂O] is equal to the number one and can be dropped out of the equation.

$$K_w = \frac{[H^+][OH^-]}{[H_2O]} \quad \text{becomes} \quad K_w = [H^+][OH^-]$$

The dissociation of water has a large consequence for acid-base chemistry. The exponent on 10^{-14} is the reason that the pH scale goes from 0 to 14.

Weak and strong acids and bases

Strong and Weak

The terms "weak" and "strong" have special meaning in acid-base chemistry. A **strong acid** and a **strong base** fully dissociate in water. This means a strong acid or base that is placed in water fully dissolves into ions. Only small amounts of a **weak acid** and a **weak base** dissolve in water. Weak acids and bases only partially dissociate in water.

Acid-base strength and pH are not the same thing

Many people will incorrectly say an acid with a pH of 2 is a strong acid. Acid-base strength and pH are not the same thing. pH is related to the concentration of a solution and strength is a measure of how it dissociates. HCl is a strong acid and fully dissociates. Strong acids can form both concentrated and dilute solutions anywhere below a pH of 7. Stomach acid and the water in a swimming pool are solutions of hydrochloric acid. You think nothing of swimming in a dilute solution of HCl but a concentrated solution causes severe burns.

Identifying strong acids and bases

There are very few strong acids and bases, but thousands of weak ones. It is easier to identify the strong acids and bases. Assume that if it is not strong, it must be weak. Strong acids are made from halogens, except F^- and the polyatomic ions, SO_4^{2-}, NO_3^- and ClO_4^-. Strong bases are made from Ca^{2+}, Sr^{2+}, Ba^{2+} and the alkali metals. As you browse through the strong acids and bases listed below, look for patterns in the way an acid is named.

6 strong acids		6 strong bases	
HBr	hydrobromic acid	$Ba(OH)_2$	barium hydroxide
HCl	hydrochloric acid	$Ca(OH)_2$	calcium hydroxide
HI	hydroiodic acid	LiOH	lithium hydroxide
HNO_3	nitric acid	KOH	potassium hydroxide
$HClO_4$	perchloric acid	NaOH	sodium hydroxide
H_2SO_4	sulfuric acid	$Sr(OH)_2$	strontium hydroxide

Equilibrium constants

Because weak acids and bases only partially dissociate, they have equilibrium constants. Acid-base equilibrium constants are calculated in a manner similar to what you have seen earlier. The constants are product-to-reactant concentration ratios. We use the symbol K_a to represent the equilibrium constant for acids and K_b for bases. Strong acids and bases do not have equilibrium constants because all of the reactants become products.

$$K_a = \frac{[\text{products}]}{[\text{reactants}]}$$

Formulas for acids

The formula for an acid depends on the anion name, according to the following rules:
- Hydrogen + anion that ends in –ate: <u>(ion minus ending)</u> ic acid
    - Example: HNO_3 = nitric acid
- Hydrogen + anion that ends in –ite: <u>(ion minus ending)</u> ous acid
    - Example: HNO_2 = nitrous acid
- Hydrogen + anion that ends in –ide: <u>(ion minus ending)</u> ic acid
    - Example: H_2S = hydrosulfuric acid

Naming acids and bases

Acids can be binary or ternary. Binary acids have the form $H_nX(aq)$ and ternary acids have the form $H_nX_nO_n$ where X is a nonmetal. Use the flow chart to determine the name of an acid based on its chemical formula. Names for bases follow ionic compound naming rules as well as rules for writing ionic formulas with a few exceptions such as ammonia, NH_3.

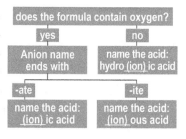

Brønsted-Lowry acids and bases

A base without hydroxide?

Not all bases contain the hydroxide ion. For example, ammonia (NH_3) is a weak base. When NH_3 dissolves in water, the ammonia molecule causes a water molecule to split into its ions. Ammonia attracts a proton from a water molecule to become NH_4^+, leaving behind an OH^- ion. Ammonia is a base because it creates OH^- ions in an aqueous solution. It doesn't matter that OH^- ions are only available because NH_3 takes protons (H^+) from water molecules. Since some of the ammonia molecules form NH_4^+ and OH^- ions, ammonia is a weak base. Most remain as NH_3 molecules in solution.

Proton donors and acceptors

Ammonia is a base because it accepts protons when dissolved in water. Another way to define a base is as a compound that accepts protons. When a compound accepts a proton from a water molecule, it leaves behind an OH^- ion. The idea of acids and bases as proton donors and acceptors is known as the **Brønsted-Lowry** definition of acids and bases. Review the process of accepting protons in the image below. Notice how the overall reaction arrow is reversible ($\rightleftharpoons$); this is a signal that indicates the base is weak.

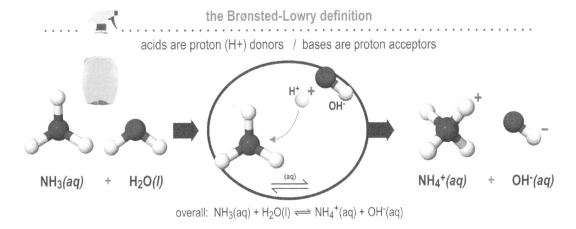

Why does NH_3 smell?

The main cleaning agent in glass cleaner is ammonia. Ammonia is a weak base that readily dissolves in water because it is a polar molecule just like water. In fact, airborne ammonia molecules easily dissolve in the water-based mucus inside your nose - that's why cleaners that contain ammonia smell so bad!

Brønsted-Lowry includes all acids and bases

The Brønsted-Lowry definition does not contradict the Arrhenius theory. For example, hydrochloric acid (HCl) fully dissociates in water to make H^+ ions and Cl^- ions. Notice the reaction arrow below is one-way ($\rightarrow$), indicating full dissociation of a strong acid. Hydrochloric acid molecules donate their protons (H^+) to the solution, which is why HCl is an acid according to both Arrhenius and Brønsted-Lowry definitions. However, some acids and bases do not count as acids and bases according to the Arrhenius definition. By contrast, all Arrhenius acids and bases can be defined by the Brønsted-Lowry theory.

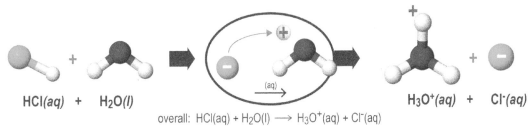

Conjugate acid-base pairs

Brønsted-Lowry acids and bases come in pairs

The Brønsted-Lowry definition states a molecule or ion can only donate a proton to act like an acid if another molecule or ion is available to act like a base by accepting the proton. The reverse is true for molecules and ions that act like bases. A substance can only act like an acid when another substance is present and able to act like a base, and vice versa. Acids and bases that exchange protons are called **conjugate acid-base pairs**.

Conjugate acid-base pairs

To identify conjugate acid-base pairs, start with reactants. Reactants are acids and bases, and their corresponding products are their conjugates. For example, in the dissociation reaction below, the ammonia (NH_3) molecule acts like a base by accepting a proton to become the ammonium ion, NH_4^+. The ammonium ion product is the conjugate acid for the reactant base, ammonia. In the reverse reaction the ammonium ion can act like an acid and donate its proton to the hydroxide ion. Water is the reactant that acts like an acid by giving up its proton. The hydroxide ion product, OH^-, is its conjugate base. The general reaction for conjugate acid-base pairs is: Acid + Base $\rightleftharpoons$ Conjugate Base + Conjugate Acid.

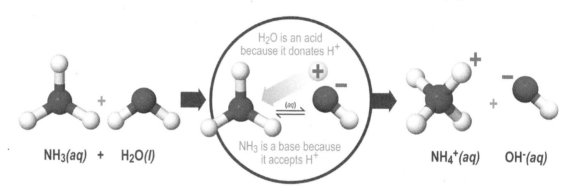

Water acts as either an acid or a base

This also works for strong acids and bases, because even though full dissociation occurs, the ion products can re-form reactants. When dissolved in water, nitric acid (HNO_3) is an acid because it donates its H^+ ion to water. Nitric acid's conjugate base is the nitrate ion. Water acts basic by accepting a proton and becoming H_3O^+, which is its conjugate acid. Water can act as either an acid or a base.

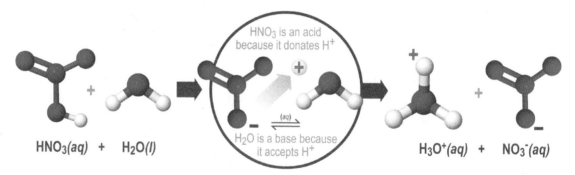

Amphoteric

When an acid is present, water acts like a base. When a base is present, water acts like an acid. An **amphoteric** substance can act as either an acid or a base. Water is an amphoteric compound because H_2O can either donate protons,

$$H_2O \rightleftharpoons H^+ + OH^-$$

or water can accept protons,

$$H_2O + H^+ \rightleftharpoons H_3O^+$$

Other substances such as the anions of weak acids are also amphoteric.

Section 1 review

Chapter 16

Using Arrhenius' definitions, an acid is a substance that produces hydrogen ions in water while a base is a substance that produces hydroxide ions. Water naturally dissociates at a low rate into both hydrogen ions and hydroxide ions. Acids and bases combine in a reaction to form water and a salt. Hydrogen ions are essentially protons that react with water to form hydronium ions. Strong acids and bases are those that fully dissolve into ions. By the more complete Brønsted-Lowry definition, acids are proton donors while bases are proton acceptors. Such pairs are known as conjugate acid-base pairs. It also explains that for something to act as an acid, something else must act as a base.

Vocabulary words: pH scale, Arrhenius acid, Arrhenius base, salt, neutral, acidic solution, base, hydronium ion, strong acid, strong base, weak acid, weak base, Brønsted-Lowry, conjugate acid-base pairs, amphoteric

Review problems and questions

1. Your friend asks you to take a sip of their drink to see whether you think it is acidic or basic. It tastes sour.
   a. Is the drink acidic, basic or neutral?
   b. What pH range would it have?
   c. Which ion has a greater concentration in the drink, H^+ or OH^-, or are the ions present in equal concentrations?

2. The following dissociation occurs in water: $NaOH \rightarrow Na^+ + OH^-$. According to the Arrhenius acid and base definitions, is the solution acidic, basic or neutral? Explain your answer.

3. What makes sulfuric acid (H_2SO_4) a strong acid and acetic acid (HCH_3COO) a weak acid?

4. The following conjugate acid-base pair forms in solution: $H_2CO_3 + H_2O \rightleftharpoons H_3O^+ + HCO_3^-$. According to the Brønsted-Lowry definition, is H_2CO_3 an acid or is it a base? Explain your answer.

5. a. Name the following compounds: H_2SO_3; H_3PO_4; HCN.
   b. Write formulas for the following: hydrofluoric acid; acetic acid; phosphorus acid.

6. Describe water, H_2O, as an Arrhenius acid and an Arrhenius base.

7. Describe water, H_2O, as a Brønsted-Lowry acid and a Brønsted-Lowry base.

16.2 - The pH Scale

Doctors, pool care technicians, and fish tank enthusiasts will tell you pH is very important. If the pH of your blood is above or below the normal range of 7.35 to 7.47, you may be seriously ill and you are at risk of damaging vital organs. If you have ever had uncomfortable burning of your eyes or skin in a pool, the pH of the water was probably outside of the ideal range of 7.4 to 7.6. There is a great deal of chemistry happening in an aquarium. Fish tank owners should keep an eye on the pH of their tank water because aquarium life can struggle or die if pH shifts abruptly or falls outside the ideal range.

pH is important

$[H_3O^+] > [OH^-]$ $[H_3O^+] = [OH^-]$ $[H_3O^+] < [OH^-]$

0 1 2 3 4 5 6 7 8 9 10 11 12 13 14

most acidic · neutral · most basic

The pH of most acids and bases

Pure water has a pH of 7.0 and is neutral. If you leave neutral water exposed to the air, it will actually become acidic over time. When carbon dioxide gas (CO_2) from the air dissolves in water, it forms a very dilute solution of carbonic acid, H_2CO_3, and the pH will slowly drop.

Concentrated acid and bases

Few individuals will work with concentrated acids and bases. Acids and bases are usually dilute so they are safer to work with. Most dilute acids and bases have a pH between 0 and 14. The farther away from 7 you get the more concentrated the acid or base gets.

Acid pH

Is a solution with a pH of 6.9 an acid? The answer is, yes! If pH is less than 7.0, a solution is an acid. Rain naturally has a pH between 5.5 and 6, which is slightly acidic, due to acids formed when certain atmospheric gases react with water. A typical lemon has a pH of 2.5. A 0.1 M hydrochloric acid solution has a pH of 1. Any substance with a pH less than 2 is dangerous and must be handled carefully.

Base pH

Baking soda, $NaHCO_3$, has a pH of 9 and antacids that contain magnesium hydroxide, $Mg(OH)_2$, may have a pH near 10.0 depending on the concentration. A concentrated solution of a strong base such as a 0.1 M sodium hydroxide solution has a pH of 13. Substances with a pH greater than 12 are dangerous and must be carefully handled.

Common substances

The substances in the table below will be familiar to you as everyday items. Compare the pH of foods with the pH of household cleaners. Foods tend to be acids; cleaners tend to be bases.

The pH of common household substances

HOUSEHOLD CHEMICAL	ACID or BASE	pH
citrus	acid	2
cola	acid	3
tomato sauce	acid	4.1
baking soda	base	8.5
soap	base	10
ammonia	base	11

The logarithmic scale

pH stands for "power of hydrogen." A **logarithm** is a power to which a number (the base) is raised to produce a given number. The pH scale is logarithmic because it is based on powers of 10. One unit of pH change represents a ten-fold change in the $[H^+]$. An acid at pH = 2 is 10 times more concentrated than an acid at pH = 3. An acid at pH = 5 is 100 times more concentrated than an acid at pH = 7. Small changes in pH equal large changes in the $[H^+]$.

pH is a logarithmic scale

The meaning of pH

A pH value tells you how concentrated an acid or base is based on the H^+ ion concentration, $[H^+]$ in a solution. A pH value is an expression of the molarity, M of H^+ ions. That means a molarity of H^+ ions can be used to calculate a pH value.

Why we use scientific notation

Water molecules can dissociate into H^+ and OH^- ions. However, you can only find about 0.0000001 mol of H^+ ions in 1 liter of pure, neutral H_2O molecules under ordinary conditions. This equals an $[H^+]$ of 0.0000001 M, best expressed as 1×10^{-7} M H^+.

the pH of neutral water

$$pH = -\log(H_3O^+)$$
$$-\log(1 \times 10^{-7})$$
$$= 7$$

H^+ in pH 7 vs. pH 4 solution

The $[H^+]$ in 1 L of an acidic water solution with a pH of 4 is 0.0001 M or 10^{-4} M H^+. Even though the number is small, that's a lot more hydrogen ions compared to a solution with a pH of 7! The concentration of H^+ ions for pH 7 is three powers of ten or 1,000 times smaller than pH 4. The image below may help you visualize the magnitude of difference between values on the pH scale.

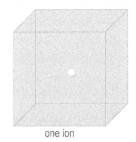
one ion

the number of H^+ ions in the box on the right is *1,000* times greater than the number of H^+ ions on the left

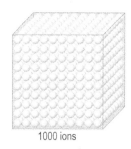
1000 ions

The pH equation

A pH value is -1 times the H^+ molarity expressed as a power of ten. Here is the same statement written as an equation:

$$pH = -\log[H^+]$$

$[H^+]$ other than 1

You can use the equation to find the pH of any given H^+ molarity. Suppose you have a solution with an $[H^+] = 0.08000$ M. If you take the logarithm of 0.08000, you get -1.097. The solution has a pH of 1.097 because pH is the logarithm of the $[H^+]$ multiplied by -1. A pH value has no units because it is simply a pure numbers of ions.

a mathematic definition

$$pH = -\log[H^+]$$

pH = -1 × the logarithm of the H^+ ion molarity

Low pH = high $[H^+]$

A strong acid like HNO_3 fully dissociates into H^+ and NO_3^- in water. A 1M solution of HNO_3 has a 1 M concentration of H^+ ions. In scientific notation, the number 1 is expressed as 1×10^0. If you take the negative log of 1, you get 0. This is why a 1M solution of HNO_3 has a pH of 0. A solution with a pH of 0 is a concentrated and strong acid.

$$HNO_3(aq) \rightarrow H^+(aq) + NO_3^-(aq)$$

Can pH be less than 0 or greater than 14?

Most common acids and bases have an H^+ molarity between 1 and 10^{-14}. The negative logarithm of 1 is zero, so when the molarity of H^+ is 1, the pH is zero. The negative logarithm of 10^{-14} is 14. If a solution has an H^+ concentration of 10^{-14} M, then pH is 14. pH can go below zero and be negative for very highly concentrated acids and pH can go higher than 14 for very highly concentrated bases. For example, a 12 M solution of concentrated hydrochloric acid has a pH of -1.08.

Section 16.2: The pH Scale

Calculation of pH

Working with logarithms

A range between 0 and 14 easier to work with than a range from 1 to 0.00000000000001, or a range from 1×10^0 to 1×10^{-14}. Logarithms make the 0-14 pH scale possible. Entering the equation pH = -log [H^+] into your calculator can be tricky. Some calculators require that you enter the concentration first, while on other calculators it is the last thing entered. Take a moment to learn how this works on your calculator with the following informaiton: the pH of a solution with a H^+ concentration of 1.3×10^{-3} M is 2.88. Try putting the concentration in your calculator then hit the log button and change the +/- sign. If that doesn't work try it in the reverse order.

Solved problem

You are working with a solution of sulfuric acid, H_2SO_4, that has an [H^+] of 2.1×10^{-6}. What is the pH of the solution?

Asked What is the pH?
Given Molarity of H^+ is 2.1×10^{-6}
Solve If pH = -1 x log[H^+], then pH = -log(2.1×10^{-6}) = 5.7
Answer The pH is 5.7.

Remember this rule: changing the pH by 1 changes the H^+ ion concentration by a factor of 10. A change from a pH of 5 to a pH of 3 is 2 pH units. This scales up to a change of 100 times in H^+ concentration. The same is true if pH changed from 5 to 3. When you add neutral water to dilute a solution by a factor of 10 to 1, the [H^+] decreases by 10 times, but the pH increases by 1. A dilution of 10 to 1 means you will add 1 part acid to 9 parts pure water, for example, 10 mL acid to 90 mL water. This works out to 10:1 because the dilute solution will have 10 ml of the original solution out of a total of 100 mL of solution.

changing the pH by 1 changes the concentration by a factor of 10

Solved problem

You have measured the pH of a hydrochloric acid solution, HCl, and found it is too low at a pH of 2.8. If you add 100 mL of the acid solution to 900 mL of neutral water, what will be the pH of the dilute solution?

Asked What is the pH of the new solution?
Given The original pH = 2.8; 100 mL of original acid is diluted with 900 mL water
Solve The new solution has a total of 1,000 mL. This is a 10:1 dilution because 1,000 mL of solution has 100 mL of original acid. The pH will increase by 1.
Answer The new solution has a pH of 3.8.

Determine molarity from pH

How can you use pH to determine the concentration of H^+ ions? This is easiest to determine when pH is a whole number. In that case, pH is the power of ten of the H^+ molarity. For example, if pH is 8, then the H^+ molarity is 10^{-8} or 0.00000001 M. When pH is a decimal number like the problem above, you will need to apply the opposite process of a logarithm. It is called anti-log and it looks like this: 10^x. Once again the order you put numbers in your calculator is important.

Solved problem

What is the concentration of H^+ in a solution that has a pH of 3.45?

Asked What is the concentration of the solution?
Given pH 3.45
Solve pH = -log[x], 3.45 = -log[x], $10^{-3.45}$ = x; x = 3.55×10^{-4}
Answer A solute with a pH of 3.45 has concentration of 3.55×10^{-4} M H^+.

Calculation of pH for bases

pOH

There is a scale similar to pH but for bases. It is called the pOH scale. It is the mirror image of the pH scale. The equation for pOH is also similar to the pH equation. Instead of stating the H⁺ ion concentration, pOH states the OH⁻ ion concentration.

$$pOH = -\log[OH^-]$$

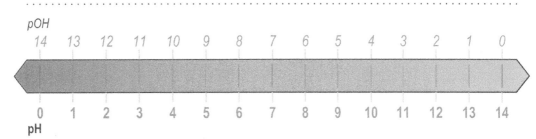

pH from pOH

Notice on the scales that when pH is added to the corresponding pOH they add up to 14.

$$pH + pOH = 14$$

We can use this to easily convert from the pH scale to the pOH scale.

Three equations for pH

The following three equations are used to calculate pH and pOH.

$$pH = -\log[H^+]$$
$$pOH = -\log[OH^-]$$
$$pH + pOH = 14$$

Often two of the three equations are used together.

Solved problem

The pOH of a clear, colorless solution is 6.1. What is its pH? Is the solution acidic, basic, or neutral?

    Asked What is the pH of the solution, and is it an acid, base, or neutral?
    Given pOH = 6.1
Relationships pH + pOH = 14
    Solve pH + 6.1 = 14;   pH = 7.9
   Answer The solution has a pH of 7.9, so it is a base.

Solved problem

What is the pH of a 0.00880 M solution of sodium hydroxide, NaOH?

    Asked What is the pH?
    Given An NaOH solution has a 0.00880 M concentration.
Relationships Since NaOH is a base, the molarity is for the [OH⁻]
             pOH = -log[OH⁻]   pH + pOH = 14
    Solve Use the concentration of base to find the pOH.
             pOH = -log[0.00880] = 2.06
             Then convert to the pH scale.
             pH + pOH = 14   pH + 2.06 = 14   pH = 11.9
   Answer The pH of the solution is 11.9.

pH indicators

Identifying acids and bases

If a chemist needs to identify an unknown clear solution, they can perform tests to see if it is an acid or a base. They can measure the conductivity of the solution. Acids and bases conduct electricity, but so do ionic solutions. For further testing they could combine the solution with another solution of known pH to see if there is an acid-base reaction. Or, the chemist could add a chemical called an **indicator** that responds to different pH values by changing color. Indicators can be extracted from plants like cabbage, or they can be produced in a lab setting.

Indicators

The color for a particular pH depends on the indicator used. Litmus dye is a pH indicator. Litmus paper gives you an idea of the pH range of a solution rather than a narrow numeric value. There are three types of paper: neutral, red, and blue. Light purple neutral paper turns red in an acid and turns blue in a base, but stays light purple in a neutral solution. Red paper turns blue in a base and stays red in an acid. Blue paper turns red in acid and stays blue in a base.

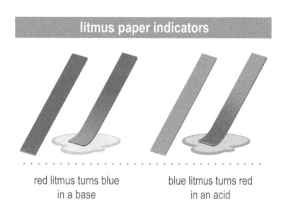

red litmus turns blue in a base

blue litmus turns red in an acid

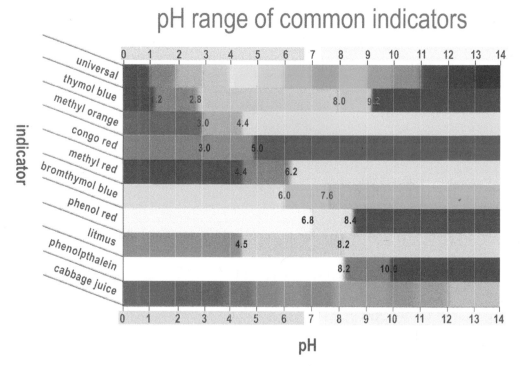

Choosing an indicator

As shown above, you can select a different indicator based on the pH of the substance you are working with. Universal indicator helps you narrow down a pH value if you are unsure of whether a substance is an acid, base, or neutral. If you know you are working with a base between pH 8 and 10, phenolphthalein is a good choice.

pH sensors

Instruments called pH sensors are used to measure pH with much greater accuracy than indicators. A pH sensor sends an electrical signal through a solution. The sensor measures the conductivity, or the ability of the solution to pass the electric signal through itself. Acids and bases conduct electricity because they dissociate into ions in aqueous solutions. The amount of H^+ ions present determines conductivity and pH.

Section 2 review

Chapter 16

The pH scale measures whether a solution is acidic, neutral, or basic. It is a logarithmic scale based on the concentration of H^+ ions in solution. The pH scale typically ranges from 0 to 14. A similar logarithmic scale exists for bases known as pOH. It measures the concentration of OH^- ions in solution. pH and pOH of a solution will always add up to 14. pH indicators are substances that are different colors at different pH levels. pH indicators are generally useful over a specific range of the pH scale. A direct measurement of a solution's pH can be made using a pH sensor or pH meter.

Vocabulary words: logarithm, indicator

Review problems and questions

1. A bar of soap has a pH of 10.2.
   a. Is the soap acidic or basic?
   b. What is its pOH?

pH 10.2

pOH?

2. A bar of soap has a pH of 10.2. What is the concentration of hydrogen and hydroxide ions?

3. A solution of hydrochloric acid, HCl, has a $[H^+] = 0.1$ M. What is its pH and pOH?

4. If you are working with a basic solution and want to use an indicator to estimate its pH, which is a better choice: methyl orange, or phenolphthalein? Explain your reasoning.

5. Why is a solution with H^+ concentration of 1.00×10^{-7} M said to be neutral?

6. What is the pH of a 3.55×10^{-2} M NaOH solution?

7. What are indicators?

8. For each of the following, determine pH, pOH, and state whether or not the solution is acidic, basic, or neutral.
   a. A sample of lemon juice with a concentration of 2.5×10^{-3} M.
   b. A sample of seawater with a concentration of 5.0×10^{-6} M.
   c. A sample of $Ca(OH)_2$ with a concentration of 0.210 g $Ca(OH)_2$ in 250.0 mL solution.

16.3 - Acid - Base Equilibria

In 1 liter of neutral water, there are are 1×10^{-7} moles of H^+ ions and 1×10^{-7} moles of OH^- ions. These ions combine to make water molecules at the same rate water molecules dissociate into their component ions. If a basic solution forms, the OH^- ion concentration increases and the H^+ ion concentration decreases. The opposite is true when an acidic solution forms.

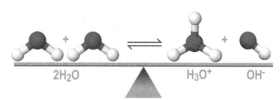

Neutralizing pH

The ion product constant

No matter what pH a dilute aqueous solution has, water molecules are constantly dissociating and recombining. That means the concentration of H^+ and OH^- ions will follow an equilibrium rule. For the dissociation ⇌ recombination reaction of water shown below, the **ion product constant**, K_w, equals 1.0×10^{-14} at 25 °C.

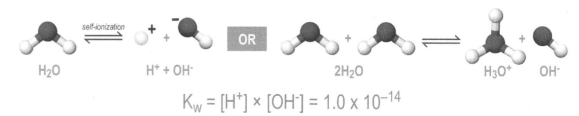

$$K_w = [H^+] \times [OH^-] = 1.0 \times 10^{-14}$$

Even if the addition of acids or bases causes the concentration of H^+ or OH^- ions to change, an equilibrium will be established. For this reason, you can use the concentration of OH^- ions to calculate pH. Solution pH depends on the H^+ concentration, but the ion product constant, K_w, defines the relationship between $[H^+]$ and $[OH^-]$.

Solved problem

The H^+ ion concentration of a solution is 0.000751 M. What is the OH^- ion concentration? Is the solution acidic or basic? Justify your answer with pH.

Asked What is $[OH^-]$; and based on pH, is the solution acidic or basic?
Given $[H^+] = 0.000751$ M
Relationships $[H^+] \times [OH^-] = 1.0 \times 10^{-14}$
pH = - logarithm $[H^+]$
Solve To calculate $[OH^-]$, plug the $[H^+]$ into the ion product constant expression.
$[0.000751] \times [OH^-] = 1.0 \times 10^{-14}$
$[OH^-] = 1.0 \times 10^{-14} / [0.000751]$
$[OH^-] = 1.3 \times 10^{-11}$
You can determine whether the solution is acidic or basic by its pH. Solve for pH with the equation pH = -log $[H^+]$.
pH = - log [0.000751] = 3.1
Answer The OH^- ion concentration is 1.3×10^{-11} M. The solution is an acid because its pH is 3.1.

The size of K

Sometimes you will see very small K values. The table shows how to interpret these values. The weakest acids and bases have the smallest K values. Very small K values indicate very few ions are available in solution at any given time. For water the K_w has an exponent of 10^{-14}. This small number reflects the fact that you will find very few H^+ and OH^- ions in pure, neutral water at any given time.

strength	K
very strong	> 0.1
moderately strong	$10^{-3} - 0.1$
weak	$10^{-5} - 10^{-3}$
very weak	$10^{-15} - 10^{-5}$
extremely weak	$< 10^{-15}$

Weak acids

Strong and weak acids

Weak acids or bases dissociate when added to water, but some ions go back to form whole molecules. Acetic acid ($HC_2H_3O_2$) is an example of a weak acid. A 0.050 M solution of $HC_2H_3O_2$ has a pH of 3.0. Less than 2%, or only 0.00093 M of H^+ ions out of the 0.050 M solution remain dissociated at any given time. However, a 0.050 M solution of hydrochloric acid (HCl) has a pH of 1.3. That's because HCl is a strong acid. When 0.050 M HCl is added to water, 100% of the HCl molecules dissociate into H^+ and Cl^- ions. For strong acids, the acid concentration equals the H^+ concentration. This is not true for weak acids.

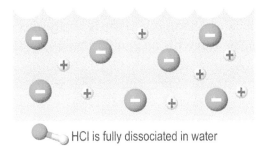
HCl is fully dissociated in water

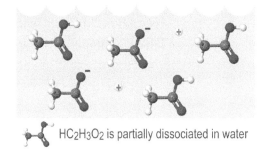
$HC_2H_3O_2$ is partially dissociated in water

Equilibrium expression

Acetic acid molecules find an equilibrium between dissociation and recombination. The balanced net ionic equilibrium equation and an equilibrium expression are written as:

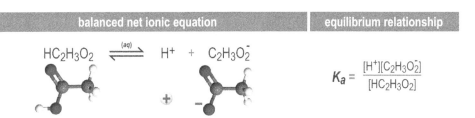

balanced net ionic equation	equilibrium relationship
$HC_2H_3O_2 \rightleftharpoons H^+ + C_2H_3O_2^-$	$K_a = \dfrac{[H^+][C_2H_3O_2^-]}{[HC_2H_3O_2]}$

What is K_a?

Since strong acids completely dissociate, you can use the acid concentration to calculate pH. By contrast, you must take the H^+ concentration at equilibrium into account for weak acid pH calculations. The acid dissociation constant, K_a is the value of the equilibrium expression. You can use the K_a to find the concentration of H^+ ions actually present in solution at equilibrium. Different weak acids reach equilibrium with different amounts of H^+ ions available in solution. For citric acid, $C_6H_8O_7$, $K_a = 7.4 \times 10^{-4}$. The K_a for acetic acid (1.8×10^{-5}) is smaller than the K_a for citric acid. That's because more citric acid molecules remain dissociated at equilibrium compared to acetic acid.

Solved problem

What is the pH of a solution that contains 1.67 M acetic acid ($K_a = 1.8 \times 10^{-5}$)?

Asked What is the pH of a weak acid?
Given 1.67 M $HC_2H_3O_2$ and $K_a = 1.8 \times 10^{-5}$
Relationships $K_a = \dfrac{[products]}{[reactants]}$
Solve $K_a = \dfrac{[H^+][C_2H_3O_2^-]}{[HC_2H_3O_2]}$

$1.8 \times 10^{-5} = \dfrac{[x][x]}{[1.67\ M]} = \dfrac{x^2}{1.67\ M}$

$\sqrt{1.8 \times 10^{-5} \times 1.67\ M} = x$; $x = 5.48 \times 10^{-3}$ M H^+

Now that you know the $[H^+]$, you can use pH = $-\log[H^+]$
pH = $-\log[5.48 \times 10^{-3}]$; pH = 2.26

Answer The pH of a 1.67 M solution of acetic acid is 2.26.

Weak bases

Weak bases produce few OH⁻ ions

Weak bases do not fully dissociate in solution, so they form a limited amount of OH⁻ ions. One common example of a weak base is ammonia, NH_3, which is found in many cleaning products. Ammonia has a distinct odor. You may have smelled ammonia if you have used glass cleaner, or if someone forgot to clean out the cat litter box.

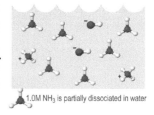

1.0M NH_3 is partially dissociated in water

Ammonia is a base

Ammonia is a base because it accepts a proton from water to produce an ammonium ion (NH_4^+) and a hydroxide ion (OH⁻). Ammonia is a weak base because the dissociated ions recombine to form whole molecules. An equilibrium is established between dissociation and recombination. This means you can write an equilibrium expression.

balanced net ionic equation	equilibrium relationship
$NH_3 + H_2O \rightleftharpoons^{(aq)} NH_4^+ + OH^-$	$K_b = \dfrac{[NH_4^+][OH^-]}{[NH_3]}$

Base equilibrium constant, K_b

You may have guessed that K_b is the equilibrium constant for bases. Since water drops out of the equilibrium relationship, the equation is similar to the weak acid equilibrium equation. Different weak bases have different values for K_b depending on how much OH⁻ ion remains dissociated at equilibrium. When given the "ion" concentration of a basic solution, the concentration refers to the OH⁻ ion instead of the H⁺ ion.

Solved problem

What is the pH of a solution that contains 1.67 M ammonia ($K_b = 1.8 \times 10^{-5}$)?

Asked What is the pH of a weak base?

Given 1.67 M NH_3

Relationships $pOH = -\log[OH^-]$, $K_b = \dfrac{[NH_4^+][OH^-]}{[NH_3]}$, and $K_b = 1.8 \times 10^{-5}$.

Solve Use K_b to determine the concentration of OH⁻ because NH_3 is a weak base.

$K_b = \dfrac{[NH_4^+][OH^-]}{[NH_3]}$

$1.8 \times 10^{-5} = \dfrac{[x][x]}{[1.67 M]} = \dfrac{x^2}{1.67 M}$; $x^2 = (1.8 \times 10^{-5} \times 1.67 M)$

$x = \sqrt{(1.8 \times 10^{-5} \times 1.67 M)}$

$x = 5.48270 \times 10^{-3}$ M OH⁻

Now use $pOH = -\log[OH^-]$ to find pOH:
$pOH = -\log[5.48270 \times 10^{-3}]$
$pOH = 2.26$

Use $pH + pOH = 14$ to find pH:
$pH + 2.26 = 14$
$pH = 11.74$

Answer The pH of a 1.67 M solution of ammonia is 11.74.

Chapter 16

Section 3 review

In dilute aqueous solutions, H^+ and OH^- ions follow an equilibrium rule. The ion product constant for water relates the concentrations of H^+ and OH^-. If the concentration of one goes up, the other goes down proportionately. Remember that strong acids completely dissociate while weak acids only partially dissociate. The acid dissociation constant is a measure of the strength of an acid in solution. Similar to acids, a weak base is one that does not fully dissociate. The base dissociation constant is a measure of the strength of a base in solution.

Vocabulary words   ion product constant

Review problems and questions

1. Why are some acid-base solutions considered to be in "equilibrium"?

2. Find the pH of a 0.95 M acetic acid solution: $HC_2H_3O_2 \rightleftharpoons H^+ + CH_3COO^-$ where K_a for acetic acid is 1.8×10^{-5}.

3. Find the pH and pOH of a 1.05 M ammonia solution: $NH_3 + H_2O \rightleftharpoons NH_4^+ + OH^-$ if $K_b = 1.8 \times 10^{-5}$.

4. Write equations for the following reactions associated with dissolving each of the following acids or bases in water. Make sure to indicate the appropriate reaction arrows ($\rightleftharpoons, \rightarrow$).

    a. Perchloric acid
    b. Magnesium hydroxide
    c. Ammonia
    d. Nitrous acid

5. For each of the following, determine $[H^+]$ and $[OH^-]$ concentrations.

    a. 0.018 M H_2SO_4
    b. 0.600 g $Sr(OH)_2$
    c. 2.05 M HF if $K_a = 6.6 \times 10^{-4}$

6. Determine the hydroxide ions concentrations of the following:

    a. A glass of pineapple juice with pH = 3.59
    b. A bowl of milk with pH = 6.61
    c. A bottle of lye with pH = 13.00
    d. A solution of milk of magnesia with pH = 10.55

7. Calculate the $[OH^-]$ concentration of a solution with a $[H^+]$ ion concentration of 8.65×10^{-4} M.

16.4 - Acid - Base Reactions

Scientists and careful observers have known about acid-base reactions for a long time. Even before the science of chemistry existed, acids were recognized as distinct substances. The word acid is derived from the Latin word acere, which means sour-tasting. Bases were known as substances that would neutralize acids. The term base arises from the fact that these substance provide the base for making a salt when acids and bases are mixed together. When an acid reacts with a base, a **neutralization** reaction occurs. The products are a salt and water, which have a neutral pH. The basic formula for a neutralization reaction is shown below.

$$\text{neutralization} \quad \underset{\text{acid}}{H_2SO_4} + \underset{\text{base}}{2KOH} \longrightarrow \underset{\text{a salt}}{K_2SO_4} + \underset{\text{water}}{2H_2O}$$

Acid - base reactions in everyday life

A neutralization reaction in your mouth

You may have heard that eating a lot of sweets will cause cavities in your teeth. Many people assume that sugar causes tooth decay, but that is not true. Cavities can develop no matter how many sweets you eat if you do not brush and floss your teeth. A thin film called plaque forms on your teeth every day. Bacteria live within plaque. These microscopic organisms digest food left sitting on your teeth and produce acid as a byproduct. Tooth damage from acids produced by bacteria and from eating acidic foods are the main causes of tooth decay. When you brush and floss your teeth, the toothbrush and floss physically remove plaque. Toothpaste is a base that neutralizes the acid from bacteria and from food.

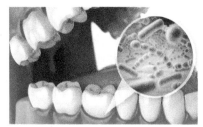
bacteria produce acid

toothpaste is a base that neutralizes acid

Acid + metal forms a salt + H_2 gas

Some, but not all metals react with acids to produce a salt and hydrogen gas:

$$\text{acid (aq)} + \text{metal (s)} \longrightarrow \text{a salt (aq)} + H_2(g)$$

This reaction is problematic in regions where acid rain has a particularly low pH. Sulfuric acid (H_2SO_4) is found in acid rain. Metals like steel are coated with zinc to prevent rusting, but the protective coating is lost when sulfuric acid reacts with zinc. Steel quickly rusts without its protective coating. An outdoor zinc-acid reaction happens too slowly to see evidence of hydrogen gas production. When a reactive metal like magnesium is added to concentrated hydrochloric acid, vigorous bubbling indicates hydrogen gas is present. Under the same conditions, less reactive zinc produces fewer H_2 gas bubbles.

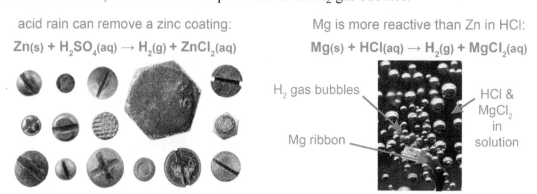

acid rain can remove a zinc coating:
$$Zn(s) + H_2SO_4(aq) \longrightarrow H_2(g) + ZnCl_2(aq)$$

Mg is more reactive than Zn in HCl:
$$Mg(s) + HCl(aq) \longrightarrow H_2(g) + MgCl_2(aq)$$

Neutralization reactions

Neutral on a molecular level

The pH of most solutions will decrease if you add an acid, and pH will increase if you add a base. The H^+ ions from the acid combine with ions from the base to produce neutral water. On the molecular level, neutralization is complete when the number of H^+ ions is equal to the number of OH^- ions. The reaction below shows an example of an ion other than hydroxide that participates in neutralization reactions: the bicarbonate ion, HCO_3^-.

Neutralizing and acid with a base

The vinegar-baking soda reaction is a familiar neutralization reaction. The chemical name for vinegar is acetic acid ($HC_2H_3O_2$). Baking soda is sodium bicarbonate ($NaHCO_3$). When an acid is mixed with a metal carbonate (or bicarbonate), a neutralization reaction occurs that produces a salt, water, and carbon dioxide gas – which causes bubbling. Look closely at the bicarbonate portion of sodium bicarbonate shown below. You can see how its structure provides O and H atoms that combine with a hydrogen ion to form water, releasing a CO_2 molecule. When all of the protons from acetic acid have been incorporated into a water molecule, the acid has been neutralized.

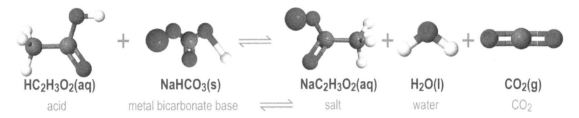

| $HC_2H_3O_2(aq)$ | $NaHCO_3(s)$ | $NaC_2H_3O_2(aq)$ | $H_2O(l)$ | $CO_2(g)$ |
| acid | metal bicarbonate base | salt | water | CO_2 |

Regulating the pH of your body

Bicarbonate can neutralize weak acids like acetic acid and partially neutralize strong acids like hydrochloric acid, HCl. Neutralization reactions occur in your body all the time. The chemical name for stomach acid is hydrochloric acid, HCl, a strong acid. As food and digestive fluids leave your stomach, your pancreas and liver produce bicarbonate to neutralize stomach acid. If your stomach produces too much acid, you may take an antacid such as milk of magnesia. The chemical name for this antacid is magnesium hydroxide, $Mg(OH)_2$, which is a strong base. Hydroxide ions (OH^-) neutralize excess hydrogen ions (H^+) to form water and magnesium chloride salt.

neutralization reaction between a strong acid and a strong base

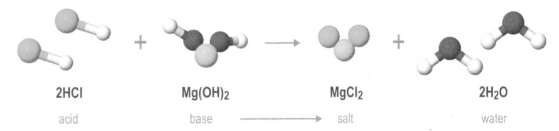

| 2HCl | $Mg(OH)_2$ | $MgCl_2$ | $2H_2O$ |
| acid | base | salt | water |

Changing the pH of soil

Sometimes soil is too acidic or too basic for homeowners to grow a healthy lawn or garden. Plants cannot properly absorb nutrients unless the soil pH is in the proper range. You can purchase a soil pH test kit from your local hardware store to see whether your soil is too acidic, too basic, or just right for growing grass at a pH around 6.5. If soil is too acidic, you can add lime to increase the pH. The chemical name for lime is calcium carbonate, $CaCO_3$, which is made from ground-up limestone. Carbonate ions combine with hydrogen ions which will increase pH. If the soil is too basic, you can add sulfur. When sulfur combines with water in soil, if forms sulfuric acid. Hydrogen ions from the acid will reduce pH.

Section 16.4: Acid - Base Reactions

Titrations

What is a titration?

A **titration** is a method used to measure the H⁺ or OH⁻ concentration within a solution. Titrations use a neutralization reaction to determine the concentration of an acid or base. An indicator provides a visual clue about when the titration is complete. You can get an accurate pH measurement when you combine a titration with an appropriate pH indicator.

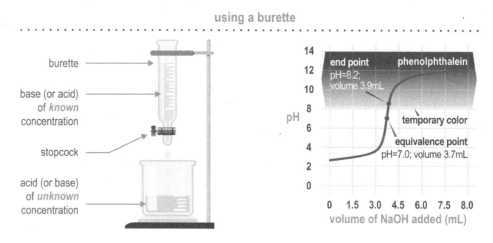

The equivalence point

Suppose you have 100 mL of a strong acid solution in a beaker and you use a burette to add a strong base one drop at a time. Each drop of base you add increases the pH by a small amount. At first, the solution has so much acid that adding a few drops of base does not change the pH much. As you approach the **equivalence point**, there is very little acid left and 1 drop of base can drastically change the pH. The equivalence point is reached when the moles of H⁺ in the acid are balanced by an equal number of moles of OH⁻. The pH is very close to neutral near the equivalence point.

Accuracy of titration is limited by indicators

An indicator changes color on the steep part of the curve near the equivalence point. Phenolphtalein is the indicator in the beaker above. It is colorless at pH values below 8.2 so the beaker stays clear as base is added from the burette until a pH of 8.2 is reached. The solution in the beaker turns pink at a pH of 8.2. Even though 8.2 is higher than the equivalence point of 7, the small volume difference between pH 7 and pH 8.2 falls within a reasonable amount of error. A small amount of base causes a large change in pH near the equivalence point, as shown by the steep of the curve.

Titration equation

You can use the known concentration of a base to determine the unknown concentration of an acid and vice versa. The dilution equation $M_1V_1 = M_2V_2$ is restated as $V_aM_a = V_bM_b$ when you use it for titrations assuming an acid:base mole ratio of 1:1. V_a is the acid volume and M_a is the acid molarity. V_b and M_b are the volume and molarity of the base.

Solved problem

5.0 mL of 0.1 M NaOH was used to titrate 50.0 mL of an unknown acid. What is the concentration of the acid?

Asked What is the concentration of an acid?
Given 5.0 mL of 0.1 M NaOH and 50.0 mL of an unknown acid.
Relationships $V_aM_a = V_bM_b$
Solve $V_aM_a = V_bM_b$
50.0 mL × M_a = 5.0 mL × 0.1 M
M_a = 0.01 M

Answer The molarity of the unknown acid is 0.01 M.

Polyprotism

Diprotic

A **polyprotic acid** has more than one proton in its chemical formula. Acids such as H_2SO_4 and H_2CO_3 have two protons per anion. These are both examples of a **diprotic acid** because they produce two moles of H^+ for every mole of acid. $H_2CO_3 \rightarrow 2H^+ + CO_3^{2-}$ The equation $M_aV_a = M_bV_b$ has to be modified to account for diprotic acids. The acid side of the equation has to be multiplied by two.

Solved problem

15.0 mL of 0.2 M NaOH was used to titrate 50.0 mL of an unknown diprotic acid. What is the concentration of the acid?

Asked   What is the concentration of an acid?
Given   15.0 mL of 0.2 M NaOH and 50.0 mL of an unknown diprotic acid.
Relationships $V_aM_a = V_bM_b$
Solve $(2 \times V_a)M_a = V_bM_b$
$(2 \times 50.0 \text{ mL}) \times M_a = 15.0 \text{ mL} \times 0.2 \text{ M}$
$M_a = \dfrac{15.0 \text{ mL} \times 0.2 \text{ M}}{100 \text{ mL}}$
$M_a = 0.03 \text{ M}$

Answer   The molarity of the unknown diprotic acid is 0.03 M.

Diprotic titration curve

The diprotic titration curve below has two distinct equivalence points. The lower area is where the first proton is removed in the reaction: $H_2CO_3 \rightarrow H^+ + HCO_3^-$. The second proton comes off in the upper area, where the reaction is: $HCO_3^- \rightarrow H^+ + CO_3^{2-}$. Since they are both weak acids, the upper and lower areas have 2 separate K_a expressions and values.

$$Ka_1 = \dfrac{[H^+][HCO_3^-]}{[H_2CO_3]} = 4.5 \times 10^{-7}$$

$$Ka_2 = \dfrac{[H^+][CO_3^{2-}]}{[HCO_3^-]} = 4.7 \times 10^{-11}$$

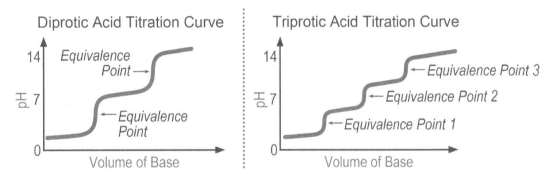

Triprotic acids

A **triprotic acid** yields three moles of protons for every mole of acid, as seen in phosphoric acid, H_3PO_4. Each proton comes off separately and each proton has its own K_a. Note three distinct equivalence points on the triprotic acid titration curve above.

Proton 1: $H_3PO_4 \rightarrow H^+ + H_2PO_4^-$ $Ka_1 = \dfrac{[H^+][H_2PO_4^-]}{[H_3PO_4]} = 6.9 \times 10^{-3}$

Proton 2: $H_2PO_4^- \rightarrow H^+ + HPO_4^{2-}$ $Ka_2 = \dfrac{[H^+][HPO_4^{2-}]}{[H_2PO_4^-]} = 6.2 \times 10^{-8}$

Proton 3: $HPO_4^{2-} \rightarrow H^+ + PO_4^{3-}$ $Ka_3 = \dfrac{[H^+][PO_4^{3-}]}{[HPO_4^{2-}]} = 4.8 \times 10^{-13}$

Normality

Equivalents

Normality is a concentration unit that accounts for the multiple protons in polyprotic acids. Normality is measured in equivalents per liter and is symbolized by a capital letter N. An **equivalent** is the mass of acid that produces one mole of protons in water. A diprotic acid such as oxalic acid produces two equivalents: $H_2C_2O_4 \rightarrow 2H^+ + C_2O_4^{-2}$. A triprotic acid such as phosphoric acid produces three equivalents: $H_3PO_4 \rightarrow 3H^+ + PO_4^{-3}$.

Solved problem

What is the normality of 156.34 g of oxalic acid, $H_2C_2O_4$, in 2.00 liters of water?

Asked What is the normality of oxalic acid?
Given 156.34 g of oxalic acid in 2 liters of water
Relationships One mole of oxalic acid has two equivalents.
The molar mass of oxalic acid is 90.035 g/mol.

Solve $\dfrac{156.34 \text{ g } H_2C_2O_4}{1} \times \dfrac{1 \text{ mol}}{90.035 \text{ g}} \times \dfrac{2 \text{ eq}}{1 \text{ mol}} = 3.47287$ equivalents

$\dfrac{3.47287 \text{ equivalents}}{2.00 \text{ liters}} = 1.736435 \xrightarrow{\text{round}} 1.74 \text{ N } H_2C_2O_4$

Answer The normality of oxalic acid is 1.74 N

Using normality

Bases such as $Al(OH)_3$ produce multiple hydroxide groups and in this case has three equivalents. When normality is used in titration ($V_aN_a = V_bN_b$) there is no need to multiply either side of the equation by the number of protons or the number of hydroxide groups. Normality accounts for the number of protons and hydroxide.

Solved problem

What volume of 4.50 N H_3PO_4 is required to titrate 53.67 grams of $Ca(OH)_2$ in 1.5 L of water?

Asked What is the volume needed?
Given 4.50 N H_3PO_4 and 53.67 grams of $Ca(OH)_2$ in 1.5 L of water
Relationships One mole of phosphoric acid has three equivalents.
One mole of calcium hydroxide has two equivalents.
The molar mass of calcium hydroxide is 74.093 g/mol.
Titration with normality: $V_aN_a = V_bN_b$

Solve First, find the normality of $Ca(OH)_2$:

$\dfrac{53.67 \text{ g}}{1} \times \dfrac{1 \text{ mol}}{74.093 \text{ g}} \times \dfrac{2 \text{ equivalents}}{1 \text{ mol}} = 1.448719$ equivalents

$\dfrac{1.448719 \text{ equivalents}}{1.5 \text{ L}} = 0.965812 \text{ N } Ca(OH)_2$

Next, use the normality to find the volume of H_3PO_4:

$V_aN_a = V_bN_b$
$1.5 \text{ L} \times 0.965812 \text{ N } Ca(OH)_2 = V_b \times 4.50 \text{ N } H_3PO_4$

$V_b = \dfrac{1.5 \text{ L} \times 0.965812 \text{ N}}{4.50 \text{ N}}$

$V_b = 0.321937 \xrightarrow{\text{round}} 0.32 \text{ L } H_3PO_4$

Answer 0.32 L of H_3PO_4 is needed to titrate 53.567 g of $Ca(OH)_2$ in 1.5 L of water.

Salts and pH

A salt that acts like a base

One of the products of an acid-base reaction is a salt. Some salts can affect solution pH if they are able to change the [H$^+$] and [OH$^-$] equilibrium. Salts of weak acids or bases are particularly effective due to partial dissociation. For example, when the weak acid hydrocyanic acid (HCN) reacts with a strong base such as sodium hydroxide (NaOH), sodium cyanide (NaCN) and water are produced. The NaCN salt dissociates in solution to form Na$^+$ and CN$^-$ ions. Since there are very few H$^+$ ions in water, the CN$^-$ ions tear H$^+$ away from water to form HCN, leaving behind excess OH$^-$ ions in the solution. NaCN acts like a base because its dissociation behavior makes the solution basic.

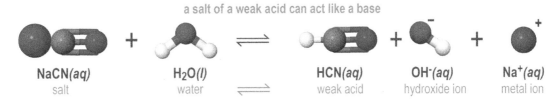

a salt of a weak acid can act like a base

NaCN(aq) salt + H$_2$O(l) water ⇌ HCN(aq) weak acid + OH$^-$(aq) hydroxide ion + Na$^+$(aq) metal ion

Salt that acts like an acid

Let's look at an example of a salt from a weak base that acts like an acid. The salt ammonium chloride, NH$_4$Cl, dissociates into its ions in water. The NH$_4^+$ ion acts like an acid because it donates a proton to H$_2$O. The products are a hydronium ion, H$_3$O$^+$, and a chloride ion, Cl$^-$. Excess protons make the solution acidic.

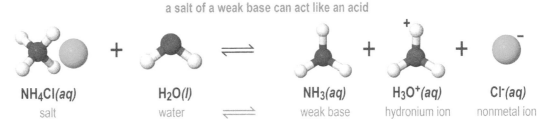

a salt of a weak base can act like an acid

NH$_4$Cl(aq) salt + H$_2$O(l) water ⇌ NH$_3$(aq) weak base + H$_3$O$^+$(aq) hydronium ion + Cl$^-$(aq) nonmetal ion

Anions and cations

All cations of weak bases undergo the previously described process. As a rule, anions of weak acids create basic solutions. Cations from weak bases also create equilibrium in water and cause solutions to become acidic.

Some salts can not affect pH

Not all salts resulting from acid-base reactions can affect pH. Strong acids and strong bases fully dissociate in water and do not undergo equilibrium because the reaction is not reversible. Potassium nitrate (KNO$_3$) is a salt that contains the cation of a strong base and the anion of a strong acid but it has no effect on pH. Potassium nitrate dissociates in water to create positive potassium ions, K$^+$ and negative nitrate ions, NO$_3^-$. Since there is no equilibrium, the ions do not tear water molecules apart. This does not allow excess H$^+$ or OH$^-$ to form in water. As a rule, a salt that is formed from a strong acid and a strong base has no effect on pH when the salt is dissolved in solution.

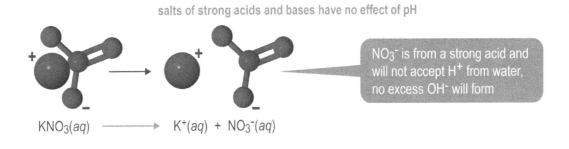

salts of strong acids and bases have no effect of pH

KNO$_3$(aq) → K$^+$(aq) + NO$_3^-$(aq)

NO$_3^-$ is from a strong acid and will not accept H$^+$ from water, no excess OH$^-$ will form

Section 16.4: Acid - Base Reactions

Buffers

Common ions

When solutions of weak acids and weak bases form, an equilibrium is established among [H⁺], [OH⁻], and salt dissociation. The salt and the weak acid or weak base have an ion in common. For example, hydrocyanic acid and its salt, sodium cyanide both incorporate the cyanide ion (CN⁻) in their structure. Cyanide is a **common ion** because two different chemicals in the same solution can release it when they dissociate.

Conjugate acid-base pairs

Earlier you saw that sodium cyanide (NaCN) acts like a base because it can form solutions with a pH greater than 7. The cyanide ion (CN⁻) is a conjugate base for hydrocyanic acid, HCN, as well as a common ion for NaCN and HCN. Hydrocyanic acid is a Brønsted-Lowry acid because it dissociates to donate a proton (H⁺ ions). The acid donates its proton to its conjugate base.

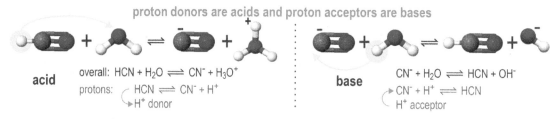

Buffer solutions resist pH change

A **buffer** is a solution made of a weak acid and its conjugate base. Buffer solutions resist small changes in pH by shifting ion equilibrium. If a small amount of acid is added to a hydrocyanic acid solution, you would expect the excess H⁺ ions to increase acidity and make the pH go lower. However, weak acids undergo equilibrium. For hydrocyanic acid,

$$K_a = \frac{[H^+][CN^-]}{[HCN]} = 6.7 \times 10^{-10}$$

The ratio of H⁺ to whole HCN molecules is a constant 6.7×10^{-10}. Whenever acid is added, the products and reactants will rearrange until they are in a ratio of 6.7×10^{-10} to one.

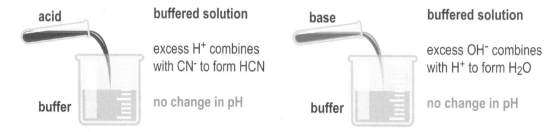

If a base is added to a buffer solution, the pH stays constant because the weak acid neutralizes the added base. Additional OH⁻ ions combine with H⁺ from the weak acid to make water molecules, which also rearrange until they are in a ratio of 6.7×10^{-10} to one.

Buffer capacity

Buffers have limits

A buffered solution resists pH changes because weak acids or weak bases shift equilibrium when excess H⁺ or OH⁻ ions are added. As shown below, the pH of a buffered solution does not change until a certain volume of base is added. The same would be true if an acid was added. All buffers have a limit to how much excess acid or base can be added before the pH changes. The curve below shows you how an indicator called bromthymol blue works to alert you to changing pH. Notice the solution stays yellow as the first 18 mL of base are added because pH remains below 6. The solution turns green when pH increases to 6, and then turns blue when pH exceeds 7.6. Up until 19 mL of base added, the common ion in the buffered solution combines with excess OH⁻ to keep pH stable.

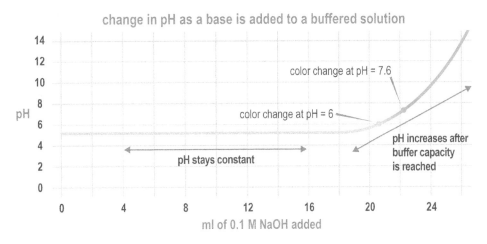

Buffer capacity

As more and more base is added, the common ions from the salt get used up and the solution runs out of ions to combine with extra OH⁻. If the common ions are used up, the solution cannot maintain equilibrium. The pH will change as if there was no buffer. The **buffer capacity** is the amount of excess H⁺ or OH⁻ ions a buffered solution can absorb without changing the pH. The buffer capacity of the system above is reached when about 19 mL of 0.1 M NaOH are added.

Ocean buffers

All forms of life have an ideal pH range for substances that interact with the inside and outside of their bodies. Everyday occurrences such as acid rain falling from the sky and consuming acidic foods could affect pH, so buffers are necessary to protect life. Ocean water maintains a pH of 8.3. It is buffered by calcium carbonate ($CaCO_3$), a weak base derived from the shells of ocean creatures, and carbonic acid (H_2CO_3), a weak acid that forms when CO_2 from the air dissolves into the water.

living things need buffer systems to survive changes in pH

Buffers in the human body

Your body has three major buffer systems: a carbonic acid-bicarbonate buffer system, a phosphate buffer system, and a protein buffer system. The carbonic acid-bicarbonate buffer system is the most important of these three systems because it maintains blood pH between 7.35-7.45. Critical tissues and molecules can be damaged outside of this pH range. Like the ocean, multiple body systems work together to maintain an appropriate ion ratio in order to keep pH stable.

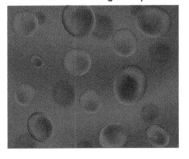

Chapter 16

Section 4 review

Mixing acids and bases is an example of a neutralization reaction. Acid-base neutralizations have many important commercial and biological applications. Titration is a technique used to determine the concentration of an unknown solution. An acid or base of known concentration is slowly added until the unknown solution is neutralized. The point where the moles of base is equal to the moles of acid is called the equivalence point. Salts of strong acids don't affect pH when added to solution but the salts of weak acids do. A mixture of a weak acid and its conjugate base creates a buffer. A buffer resists changes in pH by being able to neutralize small amounts of added acid or base. How much acid or base a buffer can neutralize is known as buffer capacity.

Vocabulary words: neutralization, titration, equivalence point, polyprotic acid, diprotic acid, triprotic acid, normality, equivalent, common ion, buffer, buffer capacity

Review problems and questions

1. If $[H^+] = 1.0 \times 10^{-7}$ M in a neutral solution, what is the $[OH^-]$?

2. A student used 7.20 mL of 0.10 M NaOH to neutralize 36.10 mL of an unknown monoprotic acid in a titration. What is the concentration of the acid?

3. The results of a titration can be used to determine pH. Use the molarity of the unknown acid above to find its pH. Assuming the acid is a strong acid.

4. Why should we be cautious about adding extra acids or bases to the ocean or our blood if those systems contain buffers that work by resisting a change in pH?

5. Determine if the salts formed from the following reactions will have a pH greater than, less than, or equal to seven?
    a. $HCl(aq) + NH_4OH(aq) \rightarrow NH_4Cl(aq) + H_2O(aq)$
    b. $HC_2H_3O_2(aq) + NaOH(aq) \rightarrow H_2O(aq) + Na^+(aq) + C_2H_3O_2^-(aq)$
    c. $HBr(aq) + KOH(aq) \rightarrow KBr(aq) + H_2O(aq)$

6. 24.20 mL of a 0.255 M HNO_2 neutralizes an unknown volume of 0.820 M CsOH solution. What is the volume of the base used?

7. How much 0.625 M propanoic acid is required to neutralize 100. mL of a 0.297 M of an unknown base?

8. Write a balanced chemical equation for each neutralization reaction:
    a. $H_2SO_4 + NaOH \rightarrow$?
    b. $HCl + Fe(OH)_2 \rightarrow$?
    c. $H_3PO_4 + Ni(OH)_2 \rightarrow$?

9. How many equivalents are in 1 mole of each of the following substances?
    i. Gallium(III) hydroxide, $Ga(OH)_3$
    ii. Phosphoric acid, H_3PO_4
    iii. Cesium hydroxide, CsOH
    iv. Acetic acid, $HC_2H_3O_2$

The Basics of Bicarbonate

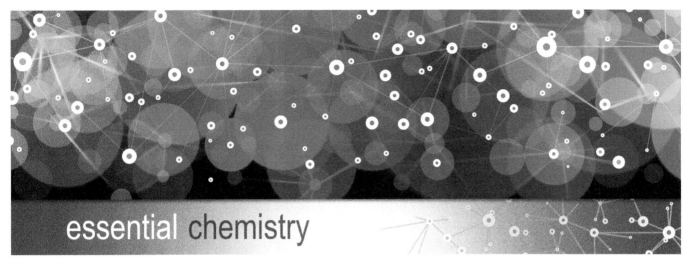

essential chemistry

Baking soda, technically sodium bicarbonate, may not look like much in your kitchen cabinet but the bicarbonate ion is a noteworthy substance that plays an important role for you and for the environment. As a white powdery solid, sodium bicarbonate is considered a base. When it is added to water, it will cause the pH of the solution to increase.

$$\text{Equation 1: } HCO_3^- + H_2O \rightleftharpoons H_2CO_3 + OH^-$$

In the kitchen, baking soda is typically used as a leavening agent. A leavening agent is something that causes the expansion of doughs and batters by the release of gases. When the recipe calls for baking soda, one of the other ingredients like vinegar, milk, yogurt, or cream of tartar is an acid. When the basic sodium bicarbonate reacts with the acidic ingredient, carbon dioxide gas bubbles are produced – leavening your batter.

$$\text{Equation 2: } HCO_3^- + H^+ \rightleftharpoons H_2CO_3 \rightleftharpoons H_2O + CO_2$$

If bicarbonate were just a base, it would not be a very interesting story. However, being a base is just half the story—literally. Bicarbonate can also act as an acid. Like water, bicarbonate is amphoteric. If sodium bicarbonate is added to an even stronger base, like a hydroxide, then it will act like an acid and form a carbonate.

$$\text{Equation 3: } HCO_3^- + OH^- \rightleftharpoons CO_3^{2-} + H_2O$$

Bicarbonate isn't just used for baking. It also plays a key role in your body. While your body is breaking down sugars for their energy, carbon dioxide is produced along with ATP and water. Blood transports the carbon dioxide from your tissues to your lungs. But to do so, most of the CO_2 gets converted into bicarbonate. The carbon dioxide and bicarbonate are part of a buffered equilibrium process with carbonic acid that occurs in your body.

$$\text{Equation 4: } H_2O + CO_2 \rightleftharpoons H_2CO_3 \rightleftharpoons H^+ + HCO_3^-$$

This buffer system maintains the pH in blood. The pH in your blood is typically between 7.35 and 7.45. Outside of this pH range proteins will get damaged, enzymes will lose their ability to function and your body is unable to sustain itself.

The Basics of Bicarbonate - 2

Your kidneys and lungs work together to maintain your pH by adjusting the amounts of the component in the equilibrium system. If your pH is too low, you have too much H^+ in your system. Your lungs help you take care of this problem. Short-term acid-base imbalances can be adjusted and overcome by changing the rate of ventilation. You might not consciously be aware of it, but increasing your breathing (especially after exercise) expels more CO_2. This causes a shift in the equilibrium. According to Le Châtelier's principle, if the amount of CO_2 is decreased, then the reaction will shift left to compensate.

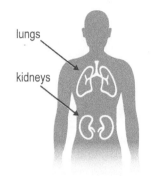

$$\text{Equation 4: } H_2O + CO_2 \rightleftharpoons H_2CO_3 \rightleftharpoons H^+ + HCO_3^-$$

This will reduce the amount of H^+ in your system and increase the pH. Kidneys also help maintain the acid-base equilibrium. If your pH is too high, the kidneys will remove bicarbonate ions from your system. This will cause an equilibrium shift to produce more H^+, lowering your pH.

Equilibrium between carbon dioxide, carbonates and bicarbonates helps to maintain ocean pH as well. Since oceans cover ¾ of the earth, this is a pretty big deal. Carbon dioxide from the atmosphere can get absorbed into the oceans. If there is more carbon dioxide in the atmosphere, then more will dissolve in the water. When carbon dioxide dissolves into water, it increases the acidity and reduces the pH of the water.

$$CO_2 + H_2O \rightleftharpoons H_2CO_3 \rightleftharpoons HCO_3^- + H^+$$

The bicarbonate can further dissociate in water to make more H^+.

$$HCO_3^- \rightleftharpoons CO_3^{2-} + H^+$$

Scientists estimate that since the industrial revolution, ocean pH has dropped from 8.2 to 8.1. This 0.1 drop may not seem like much – until you remember that pH is a logarithmic scale. That 0.1 drop is actually about a 30% increase in the concentration of H^+ in the ocean. Ocean acidification is a bit of a misnomer. The pH is becoming more acidic, but it is still not technically an acid. Just because it is not below 7, doesn't mean that the acidification cannot cause problems. Changes in ocean chemistry can have direct and indirect effects on marine life and their habitats.

Many marine animals use calcium carbonate for their skeletons. Coral reefs are a striking example of this. Coral are live animals that increase ocean biodiversity by building homes used by other creatures, but they only represent a small portion of ocean geography. Calcium carbonate is also the building block for the shells of marine organisms. These organisms live in areas of the ocean that have an abundant source of calcium carbonate dissolved in the water. An excess of dissolved carbon dioxide effectively removes the carbonate from the water.

$$CO_2 + CO_3^{2-} + H_2O \rightleftharpoons 2HCO_3^-$$

The net result of this reaction is a decrease in the amount of carbonate ions that are vital to marine life, and an increase in the amount of bicarbonate, which increases the amount of H^+ ions and decreases the pH. You might be able to see why ocean acidification has sometimes been called, "the other carbon dioxide problem."

So when you see the sodium bicarbonate in your cabinet, don't just think about making bread, or fluffy pancakes. Along with carbonates, carbonic acid and carbon dioxide, bicarbonates are an important part of acid-base equilibria that affect your health and the world around you.

Chapter 16 review

Vocabulary
Match each word to the sentence where it best fits.

Section 16.1

amphoteric	Arrhenius acid
Brønsted-Lowry	conjugate acid-base pairs
hydronium ion	neutral
salt	strong acid
strong base	weak acid

1. A chemical that dissolves in water to release hydrogen ions is a(n) _____.
2. A(n) _____ substance, such as water, can accept an H^+ ion or donate an H^+ ion.
3. An acid that dissociates completely in solution is called a(n) _____.
4. _____(s) ionize completely in water to produce hydroxide ions.
5. A(n) _____ only partially dissociates in a solution, yielding fewer hydrogen ions from less ionization.
6. The approach of _____ defines an acid and base by which substance is a hydrogen ion acceptor and which is a hydrogen ion donor.
7. Water bonds to a hydrogen ion in an aqueous solution to form _____(s).
8. A solution that has equal concentrations of H^+ and OH^- is called _____.
9. A(n) _____ is an ionic compound formed from the negative ion of a base and the positive ion of an acid.
10. A pair of molecules or ions act as _____ by performing a proton exchange.

Section 16.2

indicator	logarithm

11. The _____ of 1,000 is 3.
12. A(n) _____ is a substance like phenolphthalein that turns a different color as the pH changes.

Section 16.3

ion product constant

13. The _____ is the equilibrium constant for water at 25 °C and 1 atm and represents a value called K_w.

Section 16.4

buffer	buffer capacity
common ion	equivalence point
neutralization	normality
titration	

14. A laboratory method that is used to determine the concentration of an acid or a base is called _____.
15. A reaction where acids and bases react until the H^+ ions equal the OH^- ions and are in equilibrium is called _____.
16. In a titration, the _____ is where we can assume that the moles of acid is equal to the moles of base.
17. A(n) _____ is a solution that can resist small changes in pH.
18. Buffers work based on the principle of _____(s), which recognize that an ion is produced by two or more different chemicals in the same solution.
19. The amount of excess acid or base a solution can neutralize before changing pH is its _____.
20. _____ is a concentration unit that accounts for the multiple protons in polyprotic acids.

Conceptual questions

Section 16.1

21. What is an acid?
    a. Explain some of the chemical properties of acids.
    b. Give an example of an acid that you use regularly.
22. What is a base?
    a. Explain some of the chemical properties of bases.
    b. Give an example of a base that you use regularly.
23. How do the concentrations of H^+ ions and OH^- ions compare in a(n):
    a. acidic solution
    b. basic solution
    c. neutral solution
24. State whether each of the following compounds are an acid or a base.
    a. HCl
    b. $HC_2H_3O_2$
    c. $Ba(OH)_2$
    d. H_2SO_4
    e. LiOH
25. Explain the difference between the system used to name acids and the system used to name bases. Which system is more specialized?

Chapter 16

Chapter 16 review

Section 16.1

26. Even though it lacks a hydroxide ion, ammonia (NH_3) is a base. Use the Brønsted-Lowry theory of acids and bases to explain why NH_3 is a base. Support your explanation with a chemical equation.

27. Hydrochloric acid (HCl) dissociates in an aqueous solution. Write a chemical equation that demonstrates this.

28. Represent each of the following situations with a drawing. Explain what is occurring with each dissociation.
    a. A strong base ionizes in water.
    b. A weak base ionizes in water.
    c. A strong acid ionizes in water.
    d. A weak acid ionizes in water.

29. Answer the following questions regarding an acid dissolving in water:
    a. What ion is formed from the dissociation of an acid in water?
    b. How does this ion interact with water molecules?
    c. Why is this ion chemically unique?
    d. Draw the Lewis structure for this ion.

30. Name the following acids:
    a. $HClO_3$
    b. HCl
    c. HF
    d. H_2SO_4

31. Write formulas for the following acids:
    a. hydroiodic acid
    b. sulfuric acid
    c. oxalic acid
    d. chlorous acid

32. Which end of the pH scale will neutral water move towards when either Group I or Group II metals are dissolved in water? Explain your answer, and include a chemical equation to support your explanation.

33. The compounds below are in aqueous solutions. Write "YES" next to compounds that can act as an acid and "NO" to compounds that cannot act as an acid.
    a. NH_3
    b. Na_2SO_4
    c. PO_4^{3-}
    d. SO_4^{2-}
    e. HSO_4^-
    f. H_2SO_4
    g. KHP

34. For each of the following chemical reactions: which substance acts as the acid and which substance acts as the base? Identify the conjugate pairs.
    a. $HSO_4^- + NH_3 \rightleftharpoons SO_4^{2-} + NH_4^+$
    b. $HNO_3 + H_2O \rightleftharpoons NO_3^- + H_3O^+$
    c. $NH_3 + H_2O \rightleftharpoons NH_4^+ + OH^-$
    d. $HC_2H_3O_2 + NH_3 \rightleftharpoons C_2H_3O_2^- + NH_4^+$
    e. $C_2H_3O_2^- + HCl \rightleftharpoons HC_2H_3O_2 + Cl^-$

35. Compare the Brønsted-Lowry theory of acids and bases with the Arrhenius theory of acids and bases by answering the following.
    a. Which theory defines acids and bases in the context of ions formed in water? Identify the theory and re-state it in your own words.
    b. How does the other theory define acids and bases? Identify the theory and re-state it in your own words.

36. Fully translate the following symbol to words: $[H^+]$

Section 16.2

37. The pH scale is a logarithmic scale. How much does the $[H^+]$ concentration change for each unit of pH change?

38. When placed in an unknown solution, your pH sensor gives a reading of 7.2. Is this solution acidic, basic, or neutral? Explain your answer.

39. In order to estimate the pH of a solution, a litmus paper, methyl orange, or phenolphthalein indicator can be used. What is the value in having more than one kind of indicator available?

40. How many times more concentrated is a solution with a pH of 3 compared to a solution with a pH of 6?

41. If a solution changes from pH 7 to pH 5, what does this tell you about how the amount of $[H^+]$ is changing compared to how $[OH^-]$ is changing?

42. The scale for bases is the pOH scale, which is the mirror image of the pH scale, only it measures concentration of OH^- ions within solution. What are the three equations that show a mathematic relationship between pH an pOH?

43. There are very few strong acids and bases, but thousands of weak ones. If pH is a measurement of acidity, can the pH of a solution of a "strong" acid be as high as 6.5? Why or why not?

44. Explain what the acid dissociation constant is.

45. ⦗ Calculate the the pH of 1.0 M hydrochloric acid and the pH of 1.0 M hydrocyanic acid. Compare the pH values and explain what makes the pH different for these two acids even though they have the same concentration. ($K_a = 6.2 \times 10^{-10}$)

46. A student determined the pH for a 12.0 M perchloric acid to be 4.06 using a K_a value of 6.2×10^{-10}. What is the accuracy of thier calculation? Explain.

47. Plants require a specific pH of the soil for proper growth, organisms require different pH in order to aid with digestion or prevent tooth decay. What does pH measure in an aqueous solution?

48. Predict the pH sensor readings for following substances. Explain your reasoning for each prediction.
    a. water
    b. glass/window cleaner
    c. liquid laundry detergent
    d. cola
    e. fruit punch

Chapter 16 review

Section 16.3

49. How is calculating the pH of a weak acid different from the way you calculate the pH for a strong acid when given a concentration?

50. You are given a 100-mL sample of a strong base and a 100-mL sample of a weak base. Both have a concentration of 0.50 M. Which will have a higher pH value, and what are some ways to confirm your answer? Explain.

51. Each of the following are either a weak acid or a weak base that will partially ionize to form 2 ions. Complete the ionization equation for each.

    a. $HCN + H_2O \rightleftharpoons$
    b. $HF + H_2O \rightleftharpoons$
    c. $NH_4OH + H_2O \rightleftharpoons$
    d. $C_5H_5N + H_2O \rightleftharpoons$

52. Answer the following about the dissociation of water.

    a. What is the chemical equation that illustrates the dissociation of water?
    b. What is the equilibrium constant, K_w, for this dissociation reaction under normal conditions?
    c. What does the size of K_w indicate?

53. Answer the following questions about dissociation reactions. Be sure to use the correct reaction arrow for strong or weak acids and bases.

    a. Write the dissociation equation for the ionization of hydrochloric acid, HCl (aq), in an aqueous solution.
    b. Write the dissociation equation for the ionization of acetic acid, $HC_2H_3O_2$ (aq), in an aqueous solution.
    c. Which dissociation equation above has a greater potential to change the pH of a solution? Explain.

54. Which of the following conditions indicate a basic solution at 25 °C?

    a. pOH = 12.25
    b. $[OH^-] > [H^+]$
    c. $[OH^-] > 1.0 \times 10^{-7}$ M
    d. $[OH^-] < [H^+]$
    e. pOH = 10.70

55. Which of the following conditions indicate an acidic solution at 25 °C?

    a. pOH = 9.66
    b. pH = 4.25
    c. $[H^+] > 1.0 \times 10^{-7}$ M
    d. $[H^+] > [OH^-]$
    e. $[OH^-] > [H^+]$

56. For each of the following substances, state whether it is a strong or weak acid or base. Write equilibrium expressions for weak bases.

    a. $HClO_4$
    b. $Ca(OH)_2$
    c. HCOOH
    d. LiOH
    e. CH_3NH_2

57. Write conjugates for the following:

    a. OH^-
    b. HCN
    c. $NaHCO_3$
    d. HF
    e. H_2O
    f. NH_3

58. Use the table below to help you answer the following questions.

    interpreting K values

strength	K
very strong	> 0.1
moderately strong	10^{-3} – 0.1
weak	10^{-5} – 10^{-3}
very weak	10^{-15} – 10^{-5}
extremely weak	< 10^{-15}

    a. Hydrofluoric acid has a K_a of 5.9×10^{-2} while hydrocyanic acid has a K_a of 6.17×10^{-10}. Which acid is weaker?
    b. Which acid will produce more H^+ ions when acid concentrations are equal?
    c. Which base is most likely to have a K_b with a positive exponent, ammonia (NH_3) or sodium hydroxide (NaOH)? Explain your answer.

59. Is an aqueous solution of $NaHSO_4$ acidic, basic or neutral? Support your answer with a chemical equation.

Section 16.4

60. Complete the generic neutralization reaction:
    strong acid(aq) + strong base(aq) → _____ + _____.

61. Write a complete, balanced equation for each of the following reactions that involve an acid.

    a. NaOH(aq) + HCl(aq) → _____ + _____
    b. HNO_3(aq) + Ca(s) → _____ + _____
    c. H_2SO_4(aq) + $CuCO_3$(s) → _____ + _____ + _____
    d. Which reaction(s) will produce bubbles? Explain.

62. Suppose you accidentally spilled acid in the lab. You used baking soda (sodium bicarbonate) to neutralize the acid. What products are left behind after neturalization?

63. Answer the following questions about buffers.

    a. What does a buffer solution do to pH as acid is added to solution?
    b. What ion makes acetic acid act as a buffer?

64. A student adds a few grams of solid ammonium chloride (NH_4Cl) to a neutral sample of pure water. How does this affect the pH? Explain.

65. If solid potassium chloride (KCl) is added to a neutral solution, how would the overall pH be affected? Explain.

66. A weak acid is titrated with a strong base. What happens when the concentration of the acid is equal to the concentration of the conjugate base?

67. In a neutralization reaction, why isn't the equivalence point always at pH = 7? Explain your answer.

Chapter 16

Chapter 16 review

Quantitative problems

Section 16.2

68. A solution has a pH of 6.25. What is the hydrogen ion concentration of the solution?

69. The hydrogen ion concentrations for solutions a-e are given below. Calculate the pH of each solution and identify the solution as an acid, base, or neutral.
    a. $[H^+] = 8.57 \times 10^{-12}$ M
    b. $[H^+] = 3.19 \times 10^{-10}$ M
    c. $[H^+] = 1.00 \times 10^{-7}$ M
    d. $[H^+] = 5.06 \times 10^{-7}$ M
    e. $[H^+] = 3.09 \times 10^{-3}$ M

70. What is the pH of a solution if it has an $[H^+]$ of 4.2×10^{-4}?

71. Calculate the pH and pOH for each of the solutions listed below, and classify each as acidic, basic, or neutral.
    a. $[OH^-] = 9.30 \times 10^{-10}$ M
    b. $[OH^-] = 3.42 \times 10^{-1}$ M
    c. $[OH^-] = 9.87 \times 10^{-9}$ M
    d. $[OH^-] = 9.87 \times 10^{-7}$ M

72. Calculate the pH and pOH of the following types of drinks. Based on your calculations, is the drink with the lowest $[H^+]$ also the drink with the lowest pH? Explain your answer.
    a. lemon-lime soda, $[H^+]$ 2.51×10^{-3} M
    b. fruit punch, $[H^+] = 1.91 \times 10^{-4}$ M
    c. root beer, $[H^+] = 4.47 \times 10^{-5}$ M
    d. milk, $[H^+] = 1.99 \times 10^{-7}$ M

73. Fill in the missing information below.

	pH	pOH	$[H^+]$	$[OH^-]$
Solution I	5.88			
Solution II			6.2×10^{-11} M	
Solution III		3.00		
Solution IV				1.7×10^{-11} M
Solution V			1.0×10^{-7} M	

74. Calculate the $[H^+]$ concentration and pH for each of the following $[OH^-]$ concentrations.
    a. $[OH^-] = 7.17 \times 10^{-2}$ M
    b. $[OH^-] = 6.86 \times 10^{-8}$ M
    c. $[OH^-] = 4.51 \times 10^{-6}$ M

75. Calculate the pH for each of the solutions of strong acids.
    a. 6.7×10^{-5} M HCl
    b. 0.0075 M H_2SO_4
    c. 8.87×10^{-4} M HNO_3
    d. 4.6×10^{-7} M HCl

76. What is the $[OH^-]$ concentration for each of the following $[H^+]$ concentrations?
    a. $[H^+] = 9.25 \times 10^{-4}$ M
    b. $[H^+] = 3.14 \times 10^{-9}$ M
    c. $[H^+] = 4.87 \times 10^{-7}$ M

Section 16.3

77. The K_a for acetic acid is 1.8×10^{-5}. What is the pH of a 1.27 M solution of acetic acid?

78. The K_a for lactic acid is 1.4×10^{-4}. Calculate the pH of a 0.50 M solution of lactic acid.

79. A 6.0-M concentration of a weak acid, HF, was determined to have a K_a value of 6.3×10^{-4}. What is the pH of the acid?

80. Ammonia has a K_b of 1.8×10^{-5}. Calculate the pH of a 0.30 M solution of ammonia.

81. Aniline, $C_6H_5NH_2$, is used to manufacture dyes and plastics. Aniline has a K_b of 3.8×10^{-10}. Calculate the pH of a 0.30 M solution of aniline.

82. Use your answers for the aniline and ammonia calculations in the two previous problems to calculate the pOH of those solutions. Then use at least one calculated quantity from these two sample solutions to explain how pH is related to K_b.

83. (((Complete the table for each of the following solutions given the following values:

K_a for $HC_3H_5O_3 = 1.4 \times 10^{-4}$

K_b for $C_2H_5NH_2 = 5.6 \times 10^{-4}$

	pH	pOH	$[H^+]$	$[OH^-]$
0.0070 M HNO_3				
3.0 M KOH				
1.5×10^{-2} M HCl				
5.0×10^{-5} M $Mg(OH)_2$				
0.67 M $C_2H_5NH_2$				
0.90 M $HC_3H_5O_3$				

84. Drinks that contain caffeine, $C_8H_{10}N_4O_2$, are acidic but the caffeine molecule itself is weakly basic in an aqueous solution. What is the pH of a caffeine solution with a concentration of 1.4×10^{-3} M? K_b for caffeine = 4.1×10^{-4}.

85. A mystery solution has a pH of 3.61.
    a. Is this solution an acid, base, or neutral?
    b. What is the pOH of the solution?
    c. What is the hydrogen ion concentration of this solution, in molarity, M?
    d. What is the hydroxide ion concentration of this solution, in molarity, M?

Chapter 16 review

Section 16.4

86. Citric acid, $C_6H_8O_7$, gives sour candy its intense flavor. A sample of 30.0 mL of citric acid is titrated with a 1.00 M solution of NaOH. If the sample required 2.65 mL of NaOH to neutralize it, what is the concentration of the citric acid?

87. If it takes 25.5 mL of an NaOH solution to neutralize 16.4 mL of a 0.25 M HCl solution, what is the molarity of the NaOH solution?

88. Solid magnesium hydroxide, $Mg(OH)_2$, and aqueous hydrochloric acid, HCl, undergo an acid-base neutralization.

    a. Write the complete, balanced neutralization reaction.
    b. What volume of 0.755 M HCl solution is required to neutralize 4.50 g of $Mg(OH)_2$?

89. Find the normality of 0.684g of sulfuric acid in a 500-mL solution.

90. Determine the normality of the following solutions.

    a. 1.98 M H_3PO_4.
    b. 208 g $Ba(OH)_2$ in 300 mL solution.
    c. 2.0 M $Al(OH)_3$

91. « A chemist mixed 230 g of the following compounds in 455 mL of water. What are their concentrations in molarity, molality and normality?

    a. indium(III) hydroxide
    b. strontium hydroxide
    c. tin(II) hydroxide
    d. carbonic acid

92. How many milliliters of 0.085 M nitric acid (HNO_3) are needed to completely neutralize 32.50 mL of 0.15 M sodium hydroxide (NaOH)? Write out the balanced chemical reaction.

93. A 0.750 M HNO_3 solution is used to neutralize both of the following solutions. Calculate the volume of HNO_3 that will be required for each solution.

    a. 15.00 mL of a 0.525 M LiOH solution.
    b. 24.00 mL of a 0.120 M KOH solution.

94. A 1.5 M solution of a strong base is used to titrate 50.00 mL of a 4.50 M HCl solution. What volume of base do you expect to add?

95. A 1.5 M NaOH solution is used to titrate 10.0 mL of an unknown acid with a known concentration of 6.30 M. Calculate the volume of base needed. Insert a monoprotic strong acid as the acidic substance in your calculation.

96. A 10.0 g sample of an unknown acid is dissolved in 25.0 mL of pure, neutral water. If the solution requires 40.0 mL of a 1.60 M NaOH solution to titrate, what is the molar mass of the acid?

97. If 30.00 mL of acetic acid solution is titrated with 39.18 mL of 0.1812 N lithium hydroxide, what is the concentration of acetic acid?

98. 41.25 mL of an acid solution is titrated with 0.662 N base, if it takes 21.05 mL of the base to reach the end point, what is the normality of the acid?

99. « What is the normality of the following compounds in 3.5×10^2 mL of water?

    a. 188 g of benzoic acid, HOC_6H_4COOH
    b. 154.65 g of phosphoric acid, H_3PO_4
    c. 18.0 g ammonium hydroxide, NH_4OH
    d. 241 g of citric acid, $C_3H_5O(COOH)_3$
    e. 128 g of salicyclic acid, HOC_6H_4COOH

100. « What is the normality of:

    a. 2.5 M gallium(III) hydroxide
    b. 6.0 M barium hydroxide
    c. 4.5 M sulfuric acid
    d. 0.54 M iron(II) hydroxide

chapter 17 Oxidation and Reduction

What do you think of when you hear the words oxidation and reduction? If the word oxidation makes you think of oxygen, then you are not mistaken. In fact the word oxidation has its historical roots in reactions involving oxygen-producing oxides that release energy. Combustion of a fuel is oxidation because oxygen is combined with a hydrocarbon. For example, methane reacts with oxygen, to produce carbon dioxide, water and energy. But not all oxidation reactions involve burning. The rusting of iron is also an oxidation reaction because iron is combined with oxygen to form iron oxides. Oxidation also happens in your body as food is broken down by combining with the oxygen you breathe. The term reduction also has historical roots that are related to oxygen. In its earliest instances reduction meant that the amount of something, usually a metal oxide, was getting smaller. Upon heating or chemical processing, the compound was losing mass as oxygen was released and the metal was reduced to its elemental form.

During oxidation, an atom such as iron loses electrons becomes an ion, for example: $Fe \rightarrow Fe^{2+}$. When an atom like chlorine is reduced, it gains electrons. The electrons that are lost by solid iron are gained by chlorine. Neither of these processes can occur without the other. When iron is in a separate container from chlorine and they are connected with a wire, the electrons are transferred from iron to chlorine. This is an electric current. The movement of the electrons from the oxidized element to the reduced element creates electricity.

As our understanding has progressed, scientists came to realize that oxidation and reduction processes need each other. There can be no oxidation without reduction, and no reduction without oxidation. The two terms are so intertwined that a new word, "redox," is often used. Not only that, it turns out that redox reactions do not even need to involve oxygen at all! Like many chemistry topics, redox reactions can be explained by looking at electrons. In this chapter, you will learn how to keep track of electrons and how their movement is essential to redox reactions.

Chapter 17 study guide

Chapter Preview

Electrochemistry is the study of chemical reactions that generate electricity. In this chapter you'll learn about oxidation-reduction, or redox reactions. These reactions involve a transfer of electrons between chemical species and are central in the study of electrochemistry. In a redox reaction, one chemical species is oxidized while another is reduced. Oxidation numbers provide a way to both identify redox reactions and track the movement of electrons in such reactions. When balancing redox reactions it is important to balance both mass and charge. While oxidation and reduction always occur together, it sometimes helps to write half reactions that just tracks the oxidized or reduced elements individually to more easily track the electron transfers.

Learning objectives

By the end of this chapter you should be able to:

- explain what a redox reaction is and what it means for an element or compound to be oxidized or reduced;
- determine oxidation numbers for elements in compounds;
- identify a reaction as a redox reaction and state which elements or compounds are oxidized and which ones are reduced;
- identify the oxidizing agent and reducing agent in a redox reaction; and
- balance redox reactions using the half reactions method.

Investigations

17A: Vitamin C titration

Pages in this chapter

- 544 Oxidation and Reduction
- 545 Electron transfer
- 546 Oxidation numbers
- 547 Section 1 Review
- 548 Determining Oxidation Numbers
- 549 Applying oxidation number rules
- 550 Conflicting rules
- 551 Oxidation numbers of polyatomic ions
- 552 Oxidation states of transition metals
- 553 Section 2 Review
- 554 Oxidation Numbers in Redox Reactions
- 555 Oxidizing and reducing agents
- 556 Conservation of mass and charge
- 557 Balancing: The oxidation number method
- 558 Section 3 Review
- 559 What is a Half-Reaction?
- 560 The half-reactions balancing method
- 561 Half-reactions method (continued)
- 562 Balance an acidic redox reaction
- 563 A redox reaction in a basic solution
- 564 Balance a basic redox reaction
- 565 Section 4 Review
- 566 Antioxidants
- 567 Antioxidants - 2
- 568 Chapter review

Vocabulary

electrochemistry
free radical
redox
reducing agent
oxidation
antioxidant
oxidation number
half-reactions
electroneutrality
reduction
oxidizing agent

17.1 - Oxidation and Reduction

Living organisms need thousands of molecules to work together to keep them healthy. These molecules can be damaged when they undergo harmful chemical reactions that undermine health in the same way a bridge can fail if its critical pieces are weakened from rust. Multiple systems in your body work together to deactivate compounds that promote harmful reactions. Whether in the body or on a rusting piece of metal, these chemical processes involve electron transfer from one substance to another. However, electron transfers do not always have a negative outcome. For example if we control where transferred electrons go by directing their movement through a wire, we can harness electricity. **Electrochemistry** is the study of chemical processes that produce or use a flow of electricity.

Oxidation, free radicals, and antioxidants

Oxidized substances have lost electrons

Rusting is a synthesis reaction where a metal such as iron reacts with oxygen in the air to produce a metal oxide. Iron rusts when it loses electrons to oxygen in a chemical process called **oxidation**. Rust is the common name for iron oxides, most commonly iron(III) oxide, Fe_2O_3. Although many oxidation reactions involve oxygen, the loss of electrons defines oxidation—not the presence of oxygen as was originally thought. Many chemical reactions you see every day are oxidation reactions such as burning fuel, photosynthesis in plants and respiration in animals.

Free radicals in the body

Molecules in our bodies undergo oxidation. An oxidized molecule takes electrons from nearby molecules to regain its neutral charge. **Electroneutrality** is the property of charge balance. A highly reactive, electron-taking ion is called a **free radical**. Small amounts of free radicals are generated during normal body processes and are necessary for good health. High levels of free radicals are unhealthy. Your body creates excess free radicals when you experience high levels of stress. Toxins and pollutants in the environment can also increase the amount of free radicals in your body. A free radical can set off a chain reaction by removing electrons from stable molecules, turning them into free radicals. This can damage cells.

antioxidants stabilize free radicals

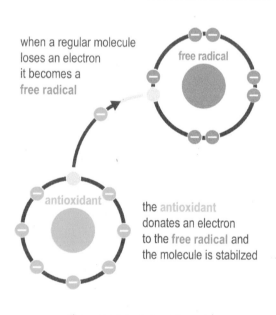

when a regular molecule loses an electron it becomes a free radical

the antioxidant donates an electron to the free radical and the molecule is stabilzed

the antioxidant doesn't react further and the destructive chain reaction simply stops

Antioxidants

Your body needs antioxidants to protect itself from the activity of free radicals. **Antioxidant** molecules are chemical compounds that undergo oxidation by donating electrons to free radicals. Free radical production and introduction in our bodies is a continuous process. The body's supply of antioxidants needs to be replenished by regular exercise and a healthy diet.

Electron transfer

Metal refinement

Refinement of pure metallic elements such as iron, copper, aluminum and tin drove the development of early chemical knowledge. Many metals are refined from mineral compounds such as copper sulfides. Alchemists coined the term "reduction" to describe purification of metals into their elemental form before the chemistry of electron transfer was understood. To early chemists, reducing meant a compound was treated until its mass was reduced to its lowest value. This meant the metal reached its pure form.

Oxidized vs. reduced

Chemists have discovered that when aluminum oxide is reduced to elemental aluminum and oxygen, electrons are transferred from oxygen to aluminum. Oxygen is oxidized because it loses electrons and aluminum is reduced because it gains electrons. **Reduction** is the opposite of oxidation where an element gains electrons and its charge becomes more negative. The charge on a reduced element is reduced because the charge goes down. For example, Al^{3+} ions gain 3 electrons and are reduced to Al atoms with no charge. Electrochemical processes are driven by electron transfers between elements trying to become more stable.

OIL RIG or LEO says GER

Oxidation occurs when an element loses electrons and its charge becomes more positive. Reduction and oxidation occur together within a reaction because when one element loses its electrons, another element gains those electrons. Either of the following phrases can help you remember both processes: OIL RIG (Oxidation Is Loss, Reduction Is Gain) or LEO the lion says "GER" (Losing Electrons is Oxidation, Gaining Electrons is Reduction).

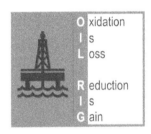

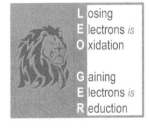

Reduction - oxidation reactions

Reduction-oxidation reactions or **redox** reactions are important chemical reactions that are also very common. Reduction cannot happen without oxidation and oxidation cannot happen without reduction; the two processes occur simultaneously. Electrochemistry is founded upon the concept of charge conservation as the driving force behind reduction-oxidation processes.

Evidence of electron transfer

There is no way to directly observe the process of electron transfer, but we can see evidence of an electron exchange. A copper(II) sulfate ($CuSO_4$) solution will plate or cover a piece of galvanized metal. Something that is galvanized means it has a zinc (Zn) coating. The zinc metal becomes plated with a coating of copper metal when electrons are transferred between Zn and Cu.

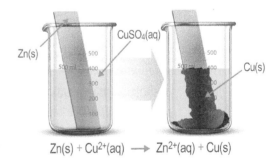

$Zn(s) + Cu^{2+}(aq) \rightarrow Zn^{2+}(aq) + Cu(s)$

Copper is reduced and zinc is oxidized

The blue $CuSO_4$ solution contains Cu^{2+} and SO_4^{2-} ions. The Cu^{2+} ions look blue when they are in solution. Once the Cu^{2+} ions from the solution are reduced to pure copper metal atoms, they form a solid copper-colored deposit on the strip when they gain electrons from the zinc. Notice the solid copper clumps forming on the strip shown inside the beaker in the diagram. The zinc atoms on the strip are oxidized when they lose electrons and become Zn^{2+} ions in the solution: $Zn(s) + CuSO_4(aq) \rightarrow ZnSO_4(aq) + Cu(s)$.

Oxidation numbers

Redox is common

You are familiar with chemical processes such as iron (Fe) rusting and the combustion of fuel. Both of these are redox reactions. Redox reactions happen all around you every day.

$4Fe(s) + 3O_2(g) \longrightarrow 2Fe_2O_3(s)$

$C_4H_{12} + 7O_2 \longrightarrow 6H_2O + 4CO_2$

iron rusting

propane burning

Elements combining to form NaCl is redox

The formation of sodium chloride (NaCl) from its constituent elements is a redox reaction. Sodium has one valence electron it is willing to lose, and chlorine has one valence electron to gain. When sodium and chlorine react, sodium loses its valence electron to chlorine. This electron transfer is what designates the reaction as redox.

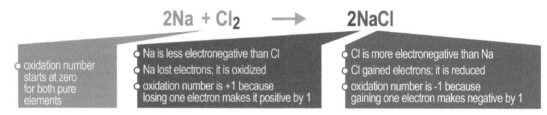

$2Na + Cl_2 \longrightarrow 2NaCl$

- oxidation number starts at zero for both pure elements
- Na is less electronegative than Cl
- Na lost electrons; it is oxidized
- oxidation number is +1 because losing one electron makes it positive by 1
- Cl is more electronegative than Na
- Cl gained electrons; it is reduced
- oxidation number is -1 because gaining one electron makes negative by 1

Oxidation numbers indicate electron loss or gain

The **oxidation number** indicates electron transfer. Think of it as the charge an element has if electrons are transferred to the more electronegative element. The more electronegative chlorine atom accepts an electron from sodium and its charge becomes -1. Therefore chlorine's oxidation number in the compound NaCl is -1. Both Na and Cl started with an oxidation number of 0 when they were pure elements. During the reaction chlorine gained an electron and it was reduced. Its oxidation number decreased (became more negative). NaCl is neutral, so the oxidation number of sodium must be +1. Sodium is oxidized because it lost an electron and its oxidation number increased (became more positive).

Electron "loss" in molecular substances

Molecular compounds also undergo redox reactions even though electrons are shared in covalent bonds. If an element starts out with a non-polar bond where electrons are equally shared, then ends up with a polar bond where electrons are unequally shared, the more electronegative element "gains" the shared electron for a longer time than the less electronegative element. Consider the electron as "lost" most of the time by the element with lower electronegativity. If the oxidation numbers change, it is a redox reaction.

electrons spend more time closer to Cl

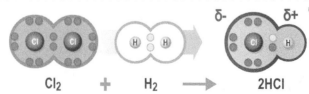

These reactants are pure elements with an oxidation number of 0

$Cl_2 + H_2 \longrightarrow 2HCl$

Cl's oxidation number is -1 since it "gains" an electron; H "loses" an electron so its oxidation number is +1.

Section 1 review

Chapter 17

While oxygen is a part of many oxidation reactions, the term is defined more broadly. Oxidation is the loss of electrons during a reaction. Some oxidation processes form highly reactive free radicals. Antioxidants donate electrons to free radicals to stop them from reacting further. Reduction is the gain of electrons in a reaction and is the opposite of oxidation. Oxidation and reduction always happen concurrently. A reaction can be determined to be a redox reaction when there is electron loss and electron gain taking place. The oxidation numbers are used to keep track of electron transfer.

Vocabulary words: electrochemistry, oxidation, electroneutrality, free radical, antioxidant, reduction, redox, oxidation number

Key relationships: Use OIL RIG or LEO says GER to remember how electrons are exchanged in reduction and oxidation:

OIL RIG:
- OIL: Oxidation is loss.
- RIG: Reduction is gain.

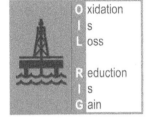

LEO says GER:
- LEO: Lose electrons is oxidation.
- GER: Gaining electrons is reduction.

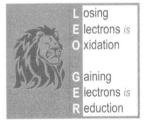

Review problems and questions

1. What must happen during a chemical reaction in order for it to be a "redox" reaction?

2. An element forms an ion when it loses two electrons. Has it undergone oxidation, or reduction?

3. An element changes its oxidation number from -1 to -2. Has it undergone oxidation, or reduction? Explain your answer.

4. Which type of chemical reaction is always a redox reaction?

5. Each of the reactions below is a redox reaction. Identify the element that has lost electrons and identify the element that has gained electrons.
   a. $4Fe + 3O_2 \rightarrow 2Fe_2O_3$
   b. $Ca + F_2 \rightarrow CaF_2$
   c. $2Na + Cl_2 \rightarrow 2NaCl$

Section 17.1: Oxidation and Reduction

17.2 - Determining Oxidation Numbers

The oxidation number is commonly—but not always—the same as the charge of the atom or ion. For example, oxygen in OF_2 is assigned a +2 charge, but it is debatable whether that is a reality. Oxidation numbers are assigned to make calculations easier. Also, the oxidation number of an element can differ depending on the compound it is found in. The purpose of an oxidation number is to help you keep track of electrons associated with atoms. An oxidation number is written with a sign followed by a number such as -1, +1, -3, and so on. This is opposite of the way we write ionic charge where the sign follows the number as in 1-, 2-, 3+, and so on.

| oxidation number is denoted number *follows* the sign | -2 |
| ionic charge is denoted sign *follows* the number | 2- |

Assigning oxidation numbers

The rules Chemists have created a set of rules to help you assign an oxidation number to each atom in a reaction. The list of rules is based on properties of atoms and the way they bond with other atoms to form various compounds. This table is also available in the Appendix.

rules for assigning oxidation numbers

Rule	Example	Oxidation #
1. The oxidation number of an atom in a pure element is zero.	Fe Cl Cl_2	0 0 0
2. The sum of the oxidation numbers in a neutral molecule is zero. Subscripts become coefficients in the equation: $0 = n_{element} + n_{element}$	CCl_4 $0 = n_C + 4(-1)$	Cl: -1 (Rule 5) C: +4 (Rule 2)
3. The sum of all atoms' oxidation numbers for an ion is equal to the charge of the ion. For polyatomic ions, Charge = $n_{element} + n_{element}$	Cu^{2+} SO_4^{2-} $-2 = n_S + 4(-2)$	+2 O: -2 (Rule 5) S: +6 (Rule 3)
4. Group 1 and 2 metals in compounds have oxidation numbers according to group number. Group 1 metals = +1. Group 2 metals = +2.	$CaCl_2$ $0 = 2 + 2(n_{Cl})$	Ca: +2 (Rule 4) Cl: -1 (Rule 2)
5. The oxidation number of a nonmetal in a compound is as follows; if the compound is made of two different nonmetals, assign the oxidation number to the nonmetal that appears first in this list and calculate the other number with the appropriate equation. - Fluorine (F): -1 - Hydrogen (H): +1 except when combined with a metal - Oxygen (O): -2 except in peroxides - Group 17 (Cl, Br, I): -1 - Group 16 (S, Se, Te): -2 - Group 15 (N, P, As): -3	F_2O $0 = 2(-1) + n_O$ Pb_2O_3 $0 = 2n_{Pb} + 3(-2)$ CuS $0 = n_{Cu} + (-2)$ LiH $0 = 1 + (n_H)$	F: -1 O: +2 (Rule 2) O: -2 Pb: +3 (Rule 2) S: -2 (Rule 5) Cu: +2 (Rule 2) Li: +1 (Rule 4) H: -1 (Rule 2)

Applying oxidation number rules

Individual elements, molecular elements and monatomic ions

When you are asked to find the oxidation number of a single element or ion, only rules one and three will apply.

Rule #1: The oxidation number of an atom in a pure element is zero.

- Single elements like K and Pb will have oxidation numbers of zero. This rule also applies to diatomic molecules made of the same element such as H_2 and Br_2. When only one kind of element is present, its oxidation number equals zero.

Rule #3: The oxidation number for an ion is equal to its charge.

- The oxidation number for Fe^{2+} is +2 while the oxidation number for Fe^{3+} is +3.

Compounds or ions with 2 elements

Most of the time an element's oxidation number in a compound is the same as the charge you would expect for the ion. The rules for assigning oxidation numbers help you get oxidation numbers correct when they are not the same as ionic charge. Follow the steps below to find oxidation numbers for compounds or ions with 2 different elements.

1. Choose an equation based on whether the compound is charged or neutral.

    - For neutral compounds such as H_2O and CO, use the equation in Rule #2.
    - For charged substances such as H_3O^+ and NO_2^-, use the equation in Rule #3.

2. Find the first rule that applies to assign an oxidation number to one element. Your goal is to assign the first rule that applies to an element (just one) in the compound.

3. Plug the known oxidation number into the equation and solve for the unknown oxidation number. Use the appropriate equation identified in Step 1.

4. Check your answer. Verify your answer by plugging all oxidation numbers into the equation. Make sure the oxidation numbers validate the equation.

Solved problem

Find the oxidation numbers for the elements in hydrogen bromide, HBr.

Asked Find the oxidation numbers for H and for Br

Given The compound HBr

Relationships HBr is a neutral, 2-element compound

Solve
1. HBr is neutral; use the equation from Rule #2: $0 = (n_H) + (n_{Br})$
2. According to Rule #5, the oxidation number of H is +1.
3. Substitute hydrogen's oxidation number in the equation and solve for n_{Br}.
   $0 = 1 + (n_{Br})$; $-1 = n_{Br}$
4. Check your answer. Plug both oxidation numbers into the equation.
   $0 = (+1) + (-1)$ is valid because $0 = 0$

Answer In HBr the oxidation number for H is +1 and the oxidation number for Br is -1.

Conflicting rules

Conflicting rules

You will not need to worry about conflicting rules for assigning oxidation numbers if you follow the four steps given on the previous page. The rule highest in the list has priority if more than one element in the compound is associated with a rule. You are most likely to see conflicting rules for oxygen or hydrogen.

Solved problem

Find oxidation numbers for K and O in the compound K_2O_2, potassium peroxide.

Given K_2O_2, potassium peroxide
Relationships K_2O_2 is a neutral
Solve
1. K_2O_2 is neutral.
   Use the equation from Rule #2: $0 = 2(n_K) + 2(n_O)$.

2. According to Rule #4 the oxidation number of K is +1.

3. Substitute potassium's oxidation number in the equation and solve.
   $0 = 2(+1) + 2(n_O); \quad 0 = +2 + 2(n_O); \quad -2 = 2(n_O); \quad -1 = n_O$

4. Check your answer. Plug both oxidation numbers into the equation
   $0 = 2(+1) + 2(-1)$ is valid because $0 = 0$

Answer The oxidation number for K is +1 and the oxidation number for O is -1.

Apply rules in order

As seen above, if you assumed oxygen's oxidation number was the same as its ion charge, you might have accidentally assigned -2 as oxidation's number instead of -1, and potassium would have received an erroneous +2 oxidation number. The same mistake would have happened if you used Rule #5 first instead of Rule #4.

Charged substances

When determining oxidation states for charged compounds or polyatomic ions, you will still use the $(n_{element}) + (n_{element})$ equation but you cannot set it equal to zero. The equation must be set equal to the overall charge of the ion. For example, a polyatomic ion with a 3- charge will use the equation: $-3 = (n_{element}) + (n_{element})$.

Solved problem

Find the oxidation number for each element in the sulfite ion, SO_3^{2-}.

Given The sulfite ion, SO_3^{2-}
Relationships The sulfite ion is a charged, 2-element substance
Solve
1. SO_3^{2-} is charged.
   Use the equation from Rule #3: $-2 = n_S + 3(n_O)$.

2. According to Rule #5 the oxidation number of O is -2.

3. Substitute oxygen's oxidation number in the equation and solve.
   $-2 = n_S + 3(-2); \quad -2 = n_S - 6; \quad +4 = n_S$

4. Check your answer. Plug both oxidation numbers into the equation.
   $-2 = (+4) + 3(-2)$ is valid because $-2 = -2$.

Answer In SO_3^{2-} the oxidation number for S is +4 and the oxidation number for O is -2.

Oxidation numbers of polyatomic ions

Compounds or ions with 3 or more elements

Compounds with three or more atoms use the same rules for oxidation numbers, with a slight modification to the solving process. You will need to split the compound into its ions and solve for oxidation numbers according to each ion. The solving process is:

Step 1: Choose an equation based on whether the compound is charged or neutral.

- For neutral compounds such as K_3PO_3 and NH_4OH, use the equation in Rule #2.
- For charged substances such as HCO_3^-, use the equation in Rule #3.

Step 2: Apply Rule #3 to both ions.

Step 3: Write an equation to solve for individual atoms in the polyatomic ion(s). Use the equation: Charge = $n_{element}$ + $n_{element}$ for each polyatomic ion.

Step 4: Check your answer. Verify your answer by plugging all oxidation numbers into the equation identified in Step 1. Make sure the oxidation numbers validate the equation.

Solved problem

Find oxidation numbers for K, Cr and O in the compound $K_2Cr_2O_7$, potassium dichromate.

Asked Assign oxidation numbers to the elements listed.

Solve
1. $K_2Cr_2O_7$ is neutral.
   Use the equation from Rule #2: $0 = 2(n_K) + 2(n_{Cr}) + 7(n_O)$ for the overall compound and split the compound into its ions.
   K^+ $Cr_2O_7^{2-}$

2. According to Rule #3, the oxidation number of K is +1 and the overall oxidation number for $Cr_2O_7^{2-}$ is -2.

3. Write an equation to solve for individual atoms in the polyatomic ion.
   $Cr_2O_7^{2-}$ is charged. According to Rule #3, $-2 = 2(n_{Cr}) + 7(n_O)$.

   According to Rule #5, the oxidation number of O is -2. Substitute and solve.

   $-2 = 2(n_{Cr}) + 7(-2)$
   $-2 = 2(n_{Cr}) - 14$
   $+12 = 2(n_{Cr})$
   $+6 = n_{Cr}$

4. Check your answer. Plug oxidation numbers into the equation from Step 1.
   $0 = 2(+1) + 2(+6) + 7(-2)$ is valid because $0 = 0$.

Answer In $K_2Cr_2O_7$ the oxidation number for K is +1, Cr is +6 and the oxidation number for O is -2.

Oxidation states of transition metals

Multiple oxidation states

One element can have different oxidation numbers from one compound to the next. For example, iron has a +2 and +3 oxidation state. Most of the transition metals change their oxidation state to become more stable.

Hund's rule and unpaired d electrons

According to Hund's Rule the electrons in the five d orbitals single fill before doubling up. Because there are 5 d orbitals to fill, it is possible to have many unpaired electrons. Unpaired electrons make atoms unstable, so they add or remove electrons from their valence to increase stability. The large number of d orbitals provides several options for shuffling electrons around.

Multiple oxidation states

The more unpaired d electrons a transition metal has, the more options it has for losing electrons. Iron has two unpaired d electrons and commonly forms either +2 or +3 oxidation states. Manganese has five unpaired d electrons and can form +2, +3, +4, +6, and +7 oxidation states. Transition metals like manganese, nearest the middle of the d-block, will have the most unpaired electrons and therefore the most oxidation states. A metal's oxidation state impacts its chemical and physical properties. Differences between the two most common oxidation states of iron (Fe) are shown below.

periodic table d-block

21 Sc	22 Ti	23 V	24 Cr	25 Mn	26 Fe	27 Co	28 Ni	29 Cu	30 Zn
39 Y	40 Zr	41 Nb	42 Mo	43 Tc	44 Ru	45 Rh	46 Pd	47 Ag	48 Cd
57-71 +	72 Hf	73 Ta	74 W	75 Re	76 Os	77 Ir	78 Pt	79 Au	80 Hg
89-103 +	104 Rf	105 Db	106 Sg	107 Bh	108 Hs	109 Mt	110 Ds	111 Rg	112 Cn

comparison of Fe^{+2} & Fe^{+3}

PROPERTY	Fe^{+2}	Fe^{+3}
solubility	higher	lower
color in chloride solution	light green	red-brown
attracted to magnets	usually	always
ion formed	Fe^{2+}	Fe^{3+}
stability	less	more

Cr^{3+} and Cr^{6+}

Chromium is a transition metal that can have several oxidation states. Chromium(III) or Cr^{3+} is an essential element our bodies need in small amounts to function properly. It also gives rubies their red color. Chromium(VI), or Cr^{6+}, is known as hexavalent chromium. Chromium(VI) is easily reduced. Since reduction and oxidation occur together, the presence of Cr^{6+} ions can cause unwanted redox reactions in living tissue. When Cr^{6+} is reduced, another substance is oxidized. Repeated exposure to oxidizing agents like chromium(VI) can cause serious health problems including cancer.

Section 2 review

Chapter 17

Oxidation numbers are a way to keep track the number of electrons lost and gained in a chemical reaction. The oxidation numbers aren't necessarily the same as actual charges on molecules. There are several rules or steps for assigning oxidation numbers. Following the steps in order will help avoid mistakes. Elements and compounds are neutral. The oxidation numbers of monatomic and polyatomic ions is determined by setting the sum of the oxidation numbers equal to the charge.

Key relationships

- An oxidation number is written as: -1 while a charge is written as: 1-.
- Reference in the Appendix: Rules for assigning oxidation numbers.

Review problems and questions

1. Assigning oxidation numbers are based on a set of rules. What are the list of rules based on?

2. Assign oxidation numbers to each element in the compound Al_2O_3.

3. Assign oxidation numbers to each element in the ion H_3O^+.

4. Why is oxygen's oxidation number -1 and not -2 in the compound H_2O_2?

5. Assign an oxidation number to each atom in each ion:
    a. Sr^{2+}
    b. PO_4^{3-}
    c. ClO_4^-
    d. SO_4^{2-}
    e. Al^{3+}

6. Find oxidation numbers for the following compounds:
    a. K_2O_2
    b. CO_2
    c. HF
    d. Sr
    e. Mg_3N_2
    f. $KMnO_4$
    g. BaO_2

7. Can an oxidation reaction occur without a reduction? Explain.

8. Which of the following reactions always involves the transfer electrons?
    a. Neutralization reaction
    b. Decomposition reaction
    c. Redox reaction
    d. Synthesis

Section 17.2: Determining Oxidation Numbers

17.3 - Oxidation Numbers in Redox Reactions

Oxidation numbers help you keep track of how electrons move between each atom or ion during a redox reaction. The first step to identifying a redox reaction is to determine oxidation numbers for elements. The next step is to see if the oxidation numbers have changed for any elements in the reaction. If the oxidation numbers change from the reactant side to the product side, then a redox reaction has occurred. If the oxidation numbers of the reactants and products do not change, it is not a redox reaction. The reducing agent is distinguished from the oxidizing agent when the the reaction is balanced with respt to mass and charge.

Identifying redox reactions

Solved problem

Aluminum oxide forms as follows: $4Al(s) + 3O_2(g) \rightarrow 2Al_2O_3(s)$. Does the reaction of aluminum with oxygen result in a redox reaction? Why or why not?

Given The balanced reaction
Relationships If the oxidation number for an element has changed, it is redox.
Solve Begin with reactants.

1. Al and O_2 are both zero because both are pure elements.
   $$4Al(s) + 3O_2(g) \rightarrow 2Al_2O_3(s)$$
   $\quad Al^0 \quad\quad O^0$

2. Move on to products. The sum of the oxidation numbers for Al and O must add up to 0 because Al_2O_3 is a neutral compound.
   - Set up the equation: $2(n_{Al}) + 3(n_O) = 0$
   - The oxidation number of O is -2. Substituting for oxygen's oxidation number:

   $2(n_{Al}) + 3(-2) = 0$
   $2(n_{Al}) - 6 = 0$
   $2(n_{Al}) = 6$
   $n_{Al} = 3$

3. Oxygen decreased because it started at 0 on the reactant side and ended at -2 on the product side. Aluminum increased because it started at 0 and ended at +3.
   $$4Al(s) + 3O_2(g) \rightarrow 2Al_2O_3(s)$$
   $\quad Al^0 \quad\quad O^0 \quad\quad Al^{+3} \; O^{-2}$

4. Use the change in oxidation state to determine which element has gained electrons and which has lost electrons. Oxygen became more negative so it gained electrons and was therefore reduced. Aluminum became more positive so it has lost electrons and was therefore oxidized.

Answer This is a redox reaction. Aluminum is oxidized because it lost electrons and its oxidation number increased from 0 to +3. Oxygen is reduced because it gained electrons and its oxidation number decreased from 0 to -2.

Oxidizing and reducing agents

Electrons lost = electrons gained

When an element is oxidized, the number of electrons it has given up is equal to the number of electrons that are gained by the reduced element. A neutral charge balance in the overall redox reaction is maintained because the amount of electrons lost is equal to electrons gained.

An oxidizing agent is reduced, a reducing agent is oxidized

During a rusting reaction iron is oxidized and oxygen is reduced. You can see that iron (Fe) is oxidized by oxygen (O). Another way to think about this is oxygen helps Fe become oxidized by taking electrons from iron. The reduced element (O) helps oxidize another element (Fe) so it is called the **oxidizing agent**. In this case, oxygen is the oxidizing agent. The oxidized element (Fe) helps reduce another element (O) so it is called the **reducing agent**. In this example Fe is oxidized and it is the reducing agent that assists in the reduction of O_2.

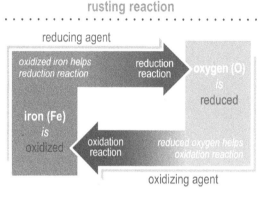

rusting reaction

$4Fe(s) + 3O_2(g) \rightarrow 2Fe_2O_3(s)$

Solved problem

Identify the reducing and oxidizing agents in: $Ca(s) + Cl_2(g) \rightarrow CaCl_2(s)$.
Explain your answers.

Asked Identify which reactant is oxidized in order to find the reducing agent, and identify the reduced reactant to find the oxidizing agent.

Solve Use the rules for assigning oxidation numbers in the Appendix.

1. Using rule number 1, assign oxidation numbers to reactants. Ca and Cl_2 both begin at zero charge.

   $Ca(s) + Cl_2(g) \rightarrow CaCl_2$
   $Ca^0 \quad Cl^0$

2. Using rule number 2, assign oxidation numbers to the neutral product MgO.
   Ca + 2Cl = 0
   Ca is +2 and each Cl is -1.

3. Review before-and-after oxidation states.

   $Ca(s) + Cl_2(g) \rightarrow CaCl_2$
   $Ca^0 \quad Cl^0 \quad\quad Ca^{+2} \; Cl^{-1}$

   - Calcium increased from 0 to +2, therefore it was oxidized.
   - Chlorine decreased from 0 to -1, therefore it was reduced.

4. The reduced element helps oxidize another element so it is an oxidizing agent. The oxidized element assists in reduction so it is a reducing agent.

Answer Chlorine is reduced, so it is an oxidizing agent that helps Ca undergo oxidation. Since Ca is the oxidized element, it is a reducing agent because it assists in reducing Cl.

Conservation of mass and charge

Mass and charge conservation

The law of conservation of mass applies to redox reactions like any other chemical reaction. The amounts and identities of atoms in the reactants are equal to the amounts and identities of atoms in the products. Charge is also conserved in redox reactions. This is true whether the reactants and products are neutral substances or charged ions. Conservation of charge is checked by keeping track of the electrons exchanged during oxidation and reduction. In other words, the number of electrons gained has to be equal to the number of electrons lost. Redox reactions are balanced when mass and charge are both conserved.

balancing redox reactions

first | count atoms of each element
balance mass with coefficients if necessary

next | use oxidation numbers to count electrons
balance charge by adding electrons where needed

A balanced redox reaction requires the balance of both mass and charge. It takes more steps to balance redox reactions compared to non-redox reactions.

Using the inspection method

Sometimes it is easy to balance a redox reaction by the inspection method, which is the balancing method taught previously. When balancing redox reactions that take place in aqueous (water-based) solutions, the redox reaction is not usually able to be balanced with the inspection method. Let's determine whether the reaction below is a redox reaction before balancing it. Begin by finding oxidation numbers for reactants and products.

$$Fe(s) + HCl(g) \rightarrow FeCl_3(s) + H_2(g)$$
$$Fe^0 \ \ + H^+ \ Cl^- \rightarrow Fe^{+3} \ Cl^- + H^0$$

This is a redox reaction because iron is oxidized (increases from 0 to +3) and hydrogen is reduced (decreases from + to 0). You can use the inspection method because this reaction does not involve aqueous solutions. The oxidation number of chlorine does not change in this reaction, so you can focus on balancing Fe and H first.

$$2Fe(s) + 6HCl(g) \rightarrow 2FeCl_3(s) + 3H_2(g)$$

Redox reactions purify metals

An important iron purification redox reaction takes place in a high-temperature vessel called a blast furnace:

$$3C + 2Fe_2O_3 \rightarrow 4Fe + 3CO_2$$

Iron(III) oxide (Fe_2O_3) is also known as iron ore or hematite. Under very high temperatures, hematite melts and reacts with carbon. This process is known as smelting. When hematite is reduced, it separates from oxygen and pure iron is produced. The leftover impurities collect together as slag. Purified liquid iron can be further processed and mixed with other elements to make steel. Humans have relied on redox reactions to purify metals for nearly 10,000 years. Ancient humans used hardened ceramic pottery over a very hot fire as a blast furnace. Temperatures can reach up to 2,000 °C in a modern blast furnace, shown to the right.

Balancing: The oxidation number method

Conserving mass and charge

Charge balance is the guiding principle when using the oxidation number method. The increase in oxidation number for oxidized atoms must equal the decrease in oxidation number for reduced atoms. A redox reaction used to isolate elemental sulfur is given below as an example of how to use the oxidation number method. The final, balanced equation should satisfy both mass and charge conservation.

the oxidation number method

step		example
1	Find the oxidation number for each atom in the reaction.	$\overset{+1\ -2}{H_2S} + \overset{+1\ +5\ -2}{HNO_3} \longrightarrow \overset{0}{S} + \overset{+2\ -2}{NO} + \overset{+1\ -2}{H_2O}$
2	Identify atoms that have changed their oxidation state. Determine the number of electrons lost and gained. In the example, S lost 2 electrons and N gained 3 electrons.	$\overset{-2}{H_2S} + \overset{+5}{HNO_3} \longrightarrow \overset{0}{S} + \overset{+2}{NO} + H_2O$ $-2 \to 0$ = lose 2e⁻ $+5 \to +2$ = gain 3e⁻
3	Write number of electrons lost as a positive number. Write number of electrons gained as a negative number. S: +2; N: -3	$\overset{-2}{H_2S} + \overset{+5}{HNO_3} \longrightarrow \overset{0}{S} + \overset{+2}{NO} + H_2O$ (+2 above S, -3 below N)
4	Use coefficients to make the total number of electrons lost equal the total number of electrons gained. Apply coefficients to the reaction. +6 is cancelled by -6	S: +2 x 3 = +6 $3\overset{-2}{H_2S} + 2\overset{+5}{HNO_3} \longrightarrow 3\overset{0}{S} + 2\overset{+2}{NO} + H_2O$ N: -3 x 2 = -6
5	Balance the rest of the atoms by the inspection method.	$3H_2S + 2HNO_3 \longrightarrow 3S + 2NO + 4H_2O$

Mass and charge balance

You could have successfully used the inspection method to balance the reaction above without worrying about the charge balance. However, redox reactions that occur in acidic or basic solutions often leave out water molecules or hydroxide ions. The only way to balance atoms in these reactions is to balance charges. Let's practice counting charges.

Solved problem

Verify that charge is balanced in the reaction: $2Fe(s) + 6HCl(g) \to 2FeCl_3(s) + 3H_2(g)$.

Asked Make sure charges balance; the total number of electrons lost by oxidation should equal the total number of electrons gained by reduction.

Given The reaction above, showing mass balance.

Solve

$\overset{0}{2Fe} + \overset{+1\ -1}{6HCl} \to \overset{+3\ -1}{2FeCl_3} + \overset{0}{3H_2}$

$0 \to +3$ = lose 3 e⁻ per Fe = 6 e⁻ $+1 \to 0$ = gain 1 e⁻ per H = 6 e⁻

Answer The charge is balanced: 6 e⁻ lost by iron = 6 e⁻ gained by hydrogen.

Section 17.3: Oxidation Numbers in Redox Reactions

Chapter 17

Section 3 review

The first step in identifying a redox reaction is to determine the oxidation numbers. If oxidation numbers change in the reaction, it is a redox reaction. The reducing agent in a redox reaction is an element or compound that reduces another element or compound. The oxidizing agent is the element or compound that oxidizes another element or compound. In a redox reaction the oxidizing agent is the chemical species that gets reduced while the reducing agent is the chemical species that gets oxidized. A proper redox reaction has to be balanced both in terms of the type and number of atoms in the reactants and products as well as in terms of charge. The oxidation number method can be used to balance mass and charge of some redox reactions.

Vocabulary words: oxidizing agent, reducing agent

Key relationships:
- An increased oxidation state means electrons were lost (oxidation). Oxidized substances act as reducing agents.
- A decreased oxidation state means electrons were gained (reduction). Reduced substances act as oxidizing agents.
- Redox reactions show mass balance with coefficients and charge balance by verifying the total number of electrons lost by oxidized atoms is equal to the total number of electrons gained by reduced atoms.

step	balancing redox reactions: the oxidation number method
1	Find the oxidation number for each atom in the reaction.
2	Identify atoms that have changed their oxidation state. Determine the number of electrons lost and gained.
3	Write number of electrons lost as a positive number. Write number of electrons gained as a negative number.
4	Use coefficients to make the total number of electrons lost equal the total number of electrons gained. Apply coefficients to the reaction.
5	Balance the rest of the atoms by the inspection method.

Review problems and questions

1. Show the oxidation state for each element in the reaction: $FeS + 2HCl \rightarrow FeCl_2 + H_2S$

2. Use your answer to Question #1 to determine whether the given reaction is a redox reaction or not. Explain your answer.

3. Identify the reducing agent and the oxidizing agent in the reaction: $4Fe + 3O_2 \rightarrow 2\ Fe_2O_3$

4. Verify that charge is balanced in the redox reaction: $4Fe + 3O_2 \rightarrow 2\ Fe_2O_3$

5. Use the oxidation number method to balance the following reaction:
   $Fe_2O_3(s) + CO(g) \rightarrow Fe(s) + CO_2(g)$

17.4 - What is a Half-Reaction?

The complete reaction in an aqueous solution can be separated into two half-reactions. The two **half-reactions** involve the oxidized elements and the reduced elements. It's easier to track electron transfers when the oxidation reaction is written separately from the reduction reaction. The reaction shown occurs when a piece of zinc-coated metal is placed in copper(II) sulfate ($CuSO_4$) solution. Follow the steps given in the table to learn how to write mass and charge balanced half-reactions for the redox reaction:

$$Zn(s) + CuSO_4(aq) \rightarrow ZnSO_4(aq) + Cu(s)$$

Balancing half-reactions

Half-reactions

Some redox reactions can be particularly difficult to balance. You must learn how to write half-reactions before attempting to balance those challenging reactions. Remember, a redox reaction involves oxidation and reduction occurring together. When you write half-reactions, you are writing each half of the redox reaction: one oxidation reaction, and one reduction reaction. Mass and electrons, or charge, must balance between the two half-reactions.

how to write balanced half-reactions

step	example
1. Rewrite the reaction with compounds split into ions with charges.	$Zn(s) + Cu^{2+}(aq) + SO_4^{2-}(aq) \rightarrow Zn^{2+}(aq) + SO_4^{2-}(aq) + Cu(s)$
2. Assign oxidation numbers to each element. You can keep polyatomic ions together and use the ion charge as the oxidation number.	$Zn(s) + Cu^{2+}(aq) + SO_4^{2-}(aq) \rightarrow Zn^{2+}(aq) + SO_4^{2-}(aq) + Cu(s)$ 0 +2 -2 +2 -2 0

3. Ignore spectator ions until the last step. Substances that have not changed oxidation state are spectator ions.	$Zn(s) + Cu^{2+}(aq) + \boxed{SO_4^{2-}(aq)} \rightarrow Zn^{2+}(aq) + \boxed{SO_4^{2-}(aq)} + Cu(s)$
	  0 +2 -2 +2 -2 0
	                          spectator

4. Identify oxidation (e⁻ loss) and reduction gain (e⁻ gain).	$\boxed{Zn(s)} + \boxed{Cu^{2+}(aq)} + SO_4^{2-}(aq) \rightarrow \boxed{Zn^{2+}(aq)} + SO_4^{2-}(aq) + \boxed{Cu(s)}$
	  0 +2 -2 +2 -2 0
	   oxidation → / reduction → / spectator

5. Write a half-reaction for oxidation and a half-reaction for reduction. Balance atom mass if necessary.	oxidation: $Zn(s) \rightarrow Zn^{2+}(aq)$ reduction: $Cu^{2+}(aq) \rightarrow Cu(s)$
6. Add enough electrons to the most positive side to balance charge in each half-reaction. Lost electrons are products, gained electrons are reactants. Adjust coefficients on atoms if needed.	oxidation: $Zn(s) \rightarrow Zn^{2+}(aq) + 2\ e^-$ reduction: $2e^- + Cu^{2+}(aq) \rightarrow Cu(s)$

The half-reactions balancing method

Keeping track of electrons

What is the purpose of writing half-reactions to balance reaction like the zinc-copper redox reaction, $Zn(s) + CuSO_4(aq) \rightarrow ZnSO_4(aq) + Cu(s)$? You could write the final balanced equation with little effort, but the only way you would know two electrons were transferred from zinc to copper is to write the half-reactions. If this reaction occurred inside a battery or inside a living thing, it would be important to know that 2 electrons were exchanged. Moving electrons can create or contribute to an electrical current that can be used to perform work. In that case, it is important to know exactly how many electrons are moving and in which direction they are flowing.

Half-reactions method

As you know, many different conditions can affect the speed or rate at which a chemical reaction will take place. In aqueous redox reactions where the reactants and products are in ion form, the solution pH can have a big impact on reaction rate. Redox reaction equations that take place in neutral, acidic or basic solutions can be balanced using half-reactions in a process called the half-reactions method. Let's balance the following redox reaction, which occurs in an acidic solution: $Cl_2 + S_2O_3^{2-} \rightarrow Cl^- + SO_4^{2-}$.

half reactions method to balance a redox reaction

Step	Example
1. Write separate half-reactions.	Reduction: $Cl_2 \rightarrow Cl^-$ Oxidation: $S_2O_3^{2-} \rightarrow SO_4^{2-}$
2. Add electrons to balance oxidation numbers.	Reduction: $e^- + Cl_2 \rightarrow Cl^-$ Oxidation: $S_2O_3^{2-} \rightarrow SO_4^{2-} + 4e^-$
3. Balance the atoms in each half-reaction as follows: a. Use coefficients to balance atoms except for hydrogen and oxygen. Apply coefficient to associated electrons. b. Add enough water (H_2O) to balance the side that needs oxygen c. Add enough hydrogen ions (H^+) to the side that needs hydrogen to balance hydrogen	Reduction: $2e^- + Cl_2 \rightarrow 2Cl^-$ Oxidation: $S_2O_3^{2-} \rightarrow 2SO_4^{2-} + 8e^-$ Reduction: N/A Oxidation: $S_2O_3^{2-} + 5H_2O \rightarrow 2SO_4^{2-} + 8e^-$ Reduction: N/A Oxidation: $S_2O_3^{2-} + 5H_2O \rightarrow 2SO_4^{2-} + 8e^- + 10H^+$
4. Use coefficients to balance electrons across both half-reactions. Multiply the entire half-reaction by the coefficient.	Reduction: $8e^- + 4Cl_2 \rightarrow 8Cl^-$ Oxidation: $S_2O_3^{2-} + 5H_2O \rightarrow 2SO_4^{2-} + 8e^- + 10H^+$
5. Combine the two half-reactions. Cancel duplicate items on either side of the equation.	$\cancel{8e^-} + 4Cl_2 + S_2O_3^{2-} + 5H_2O \rightarrow 8Cl^- + 2SO_4^{2-} + \cancel{8e^-} + 10H^+$ $4Cl_2 + S_2O_3^{2-} + 5H_2O \rightarrow 8Cl^- + 2SO_4^{2-} + 10H^+$ Charges and atoms are balanced!
For basic equations: 6. Neutralize remaining H^+ ions with OH^- ions to form water. Add equal amounts of OH^- to both sides. Combine multiple water molecules on the same side.	N/A; this reaction occurs in an acidic solution

Half-reactions method (continued)

When to use the half-reaction method

You were given two clues that you would need to use the half-reaction method to balance the example redox reaction:
- this redox reaction occurs in an acidic solution.
- $Cl_2(g) + S_2O_3^{2-} \rightarrow Cl^- + SO_4^{2-}$ is impossible to balance by the inspection method.

Methods of balancing

You are now aware of the inspection method, the oxidation number method, and the half-reaction method for balancing chemical reactions. All of these methods share one thing in common: the principle of mass conservation. Matter can neither be created nor destroyed. The difference between redox and non-redox reactions is a redox reaction involves a transfer of electrons from one atom to another, resulting in a change in oxidation numbers. Charge must be balanced in addition to mass to satisfy the law of conservation.

Solved problem

Use the half-reactions method to balance the redox reaction: $Fe^{2+} + Al \rightarrow Fe + Al^{3+}$

Asked Balance the given redox reaction.

Given The unbalanced reaction.

Solve Use only the steps of the half-reactions method that apply.

- Write the half-reactions. Add electrons to balance oxidation numbers.

$$Fe^{2+} + 2e^- \rightarrow Fe$$
$$Al \rightarrow Al^{3+} + 3e^-$$

- Balance atoms except H and O. To balance oxygen, add H_2O. To balance hydrogen, add H^+.

N/A

- Balance the electrons across both reactions. Multiply by coefficient used.

$$3 \times (Fe^{2+} + 2e^- \rightarrow Fe) = 3Fe^{2+} + 6e^- \rightarrow 3Fe$$
$$2 \times (Al \rightarrow Al^{3+} + 3e^-) = 2Al \rightarrow 2Al^{3+} + 6e^-$$

- Combine the half-reactions. Cancel duplicate terms.

$$3Fe^{2+} + \cancel{6e^-} + 2Al \rightarrow 2Al^{3+} + \cancel{6e^-} + 3Fe$$
$$3Fe^{2+} + 2Al \rightarrow 3Fe + 2Al^{3+}$$

- For basic equations: add equal amounts of OH^- to both sides of the reaction to neutralize H^+ ions. Combine water molecules on the same side.

N/A

Answer The balanced reaction is: $3Fe^{2+} + 2Al \rightarrow 3Fe + 2Al^{3+}$

Balance an acidic redox reaction

Follow the steps Following the steps of the half-reactions method will lead you to the correctly balanced redox equation as long as you check for charge and mass balance as you work through the steps.

Solved problem Balance the following acidic redox reaction:
$Sn(s) + NO_3^-(aq) \rightarrow Sn^{2+}(aq) + NO(g)$

- Asked Balance the given redox reaction.
- Given The unbalanced reaction, which occurs in an acidic solution.
- Solve Use the half-reactions method.

- Write the half-reactions. Add electrons to balance oxidation numbers.

$$Sn \rightarrow Sn^{2+} + 2e^-$$
$$NO_3^- \rightarrow NO + e^-$$

To balance oxygen, add H_2O.

$$NO_3^- \rightarrow NO + 2H_2O + e^-$$

To balance hydrogen, add H^+.

$$4H^+ + 4e^- + NO_3^- \rightarrow NO + 2H_2O + e^-$$

Cross out the electrons that appear on both sides of the equation
$$4H^+ + 3e^- + NO_3^- \rightarrow NO + 2H_2O$$

- Balance the electrons across both reactions. Multiply by coefficient used.

$$3 \times (Sn \rightarrow Sn^{2+} + 2e^-) = 3Sn \rightarrow 3Sn^{2+} + 6e^-$$

$$2 \times (4H^+ + 3e^- + NO_3^- \rightarrow NO + 2H_2O) =$$
$$8H^+ + 6e^- + 2NO_3^- \rightarrow 2NO + 4H_2O$$

- Combine the half-reactions. Cancel duplicate terms.

$$3Sn + \cancel{6e^-} + 8H^+ + 2NO_3^- \rightarrow 3Sn^{2+} + \cancel{6e^-} + 2NO + 4H_2O$$

$$3Sn + 8H^+ + 2NO_3^- \rightarrow 3Sn^{2+} + 2NO + 4H_2O$$

- For basic equations: Add equal amounts of OH^- to both sides of the reaction to neutralize H^+ ions. Combine water molecules on the same side.

N/A

Answer The balanced reaction is: $3Sn + 8H^+ + 2NO_3^- \rightarrow 3Sn^{2+} + 2NO + 4H_2O$

A redox reaction in a basic solution

A basic redox reaction

Let's walk through an example of how to balance a redox reaction in a basic solution using the half-reaction method with the reaction: $Cr(OH)_3 + ClO_3^- \rightarrow CrO_4^{2-} + Cl^-$

1. Write the half-reactions.
   $ClO_3^- \rightarrow Cl^-$
   $Cr(OH)_3 \rightarrow CrO_4^{2-}$

2. Add electrons to balance oxidation numbers.
   $ClO_3^- \rightarrow Cl^-$
   $+ 2e^- + Cr(OH)_3 \rightarrow CrO_4^{2-}$

3. Balance atoms. Cl and Cr are already balanced so you can move on to O and H.
   Balance O by adding H_2O:
   $ClO_3^- \rightarrow Cl^- + 3H_2O$
   $H_2O + 2e^- + Cr(OH)_3 \rightarrow CrO_4^{2-}$

   Balance H by adding H^+.
   $6H^+ + 6e^- + ClO_3^- \rightarrow Cl^- + 3H_2O$
   $H_2O + Cr(OH)_3 \rightarrow CrO_4^{2-} + 5H^+ + 3e^-$

4. Balance electrons across both reactions. Multiply the entire half-reaction by the coefficient used.
   $6H^+ + ClO_3^- + 6e^- \rightarrow Cl^- + 3H_2O$
   $2 \times (Cr(OH)_3 + H_2O \rightarrow CrO_4^{2-} + 5H^+ + 3e^-) =$
   $2H_2 + Cr(OH)_3 \rightarrow 2CrO_4^{2-} + 10H^+ + 6e^-$

5. Combine half reactions. Cancel duplicate items.
   $\cancel{6H^+} + ClO_3^- + \cancel{6e^-} + 2Cr(OH)_3 + \underbrace{2H_2O}_{} \rightarrow Cl^- + \underbrace{3H_2O}_{H_2O} + 2CrO_4^{2-} + \underbrace{\cancel{10H^+}}_{4H^+} + \cancel{6e^-}$

   $ClO_3^- + 2Cr(OH)_3 \rightarrow Cl^- + H_2O + 2CrO_4^{2-} + 4H^+$

6. This is a basic solution, so you must neutralize remaining H^+ ions with OH^- ions to form water.

   - Add equal amounts of OH^- to both sides.
     $4OH^- + ClO_3^- + 2Cr(OH)_3 \rightarrow Cl^- + H_2O + 2CrO_4^{2-} + 4H^+ + 4OH^-$

   - When they appear on the same side, combine H^+ and OH^- to form H_2O.
     $4OH^- + ClO_3^- + 2Cr(OH)_3 \rightarrow Cl^- + H_2O + 2CrO_4^{2-} + 4H_2O$

   - Combine multiple water molecules if they are on the same side.
     $4OH^- + ClO_3^- + 2Cr(OH)_3 \rightarrow Cl^- + 5H_2O + 2CrO_4^{2-}$

The reaction is balanced as: $4OH^- + ClO_3^- + 2Cr(OH)_3 \rightarrow Cl^- + 5H_2O + 2CrO_4^{2-}$.

Balance a basic redox reaction

Solved problem

Balance the following redox reaction that occurs in a basic solution:
$Cl_2(g) + S_2O_3^{2-}(aq) \rightarrow Cl^-(aq) + SO_4^{2-}(aq)$.

Given The unbalanced redox reaction occurs in a basic solution.

Solve

1. Write the half-reactions. Add electrons to balance oxidation numbers.
   $Cl_2 + e^- \rightarrow Cl^-$
   $S_2O_3^{2-} \rightarrow SO_4^{2-} + 4e^-$

2. Balance atoms and associated electrons.
   $Cl_2 + 2e^- \rightarrow 2Cl^-$
   $S_2O_3^{2-} \rightarrow 2SO_4^{2-} + 8e^-$

   - Balance O by adding water.
     $5H_2O + S_2O_3^{2-} \rightarrow 2SO_4^{2-} + 8e^-$

   - Balance H by adding hydrogen ions.
     $5H_2O + S_2O_3^{2-} \rightarrow 2SO_4^{2-} + 8e^- + 10H^+$

3. Balance electrons across both reactions. Multiply out the half-reaction that receives a coefficient.
   $4 \times (Cl_2 + 2e^- \rightarrow 2Cl^-) = 4Cl_2 + 8e^- \rightarrow 8Cl^-$
   $5H_2O + S_2O_3^{2-} \rightarrow 2SO_4^{2-} + 8e^- + 10H^+$

4. Combine half reactions. Cancel duplicate items.
   $4Cl_2 + \cancel{8e^-} + 5H_2O + S_2O_3^{2-} \rightarrow 8Cl^- + 2SO_4^{2-} + \cancel{8e^-} + 10H^+$

   $4Cl_2 + 5H_2O + S_2O_3^{2-} \rightarrow 8Cl^- + 2SO_4^{2-} + 10H^+$

5. Neutralize remaining H^+ ions with OH^- ions to form water.
   - Add equal amounts of OH^- to both sides.
     $10OH^- + 4Cl_2 + 5H_2O + S_2O_3^{2-} \rightarrow 8Cl^- + 2SO_4^{2-} + 10H^+ + 10OH^-$

   - Combine hydroxide and hydrogen ions into water when they are on the same side.
     $10OH^- + 4Cl_2 + 5H_2O + S_2O_3^{2-} \rightarrow 8Cl^- + 2SO_4^{2-} + 10H_2O$

   - Reduce water molecules across the equation.
     $10OH^- + 4Cl_2 + \cancel{5H_2O} + S_2O_3^{2-} \rightarrow 8Cl^- + 2SO_4^{2-} + \underbrace{\cancel{10H_2O}}_{5H_2O}$

     $10OH^- + 4Cl_2 + S_2O_3^{2-} \rightarrow 8Cl^- + 2SO_4^{2-} + 5H_2O$

Answer The reaction is balanced as: $10OH^- + 4Cl_2 + S_2O_3^{2-} \rightarrow 8Cl^- + 2SO_4^{2-} + 5H_2O$

Section 4 review

Chapter 17

Oxidation and reduction reactions take place simultaneously. However, we can write just the oxidation reaction or just the reduction reaction. These are called half-reactions. Half-reactions are useful for tracking charges in a redox reaction. Half-reactions can be used to explain what goes on inside a battery or other reactions that create electrical current. Half-reactions can also be used as a way to balance certain redox reactions.

Vocabulary words

half-reactions

Key relationships

how to write balanced half-reactions

step

1. Rewrite the reaction with compounds split into ions with charges.
2. Assign oxidation numbers to each element. You can keep polyatomic ions together and use the ion charge as the oxidation number.
3. Ignore spectator ions until the last step. Substances that have not changed oxidation state are spectator ions.
4. Identify oxidation (e⁻ loss) and reduction gain (e⁻ gain).
5. Write a half-reaction for oxidation and a half-reaction for reduction. Balance atom mass if necessary.
6. Add enough electrons to the most positive side to balance charge in each half-reaction. Lost electrons are products, gained electrons are reactants. Adjust coefficients on atoms if needed.

half reactions method to balance a redox reaction

step

1. Write separate half-reactions.
2. Add electrons to balance oxidation numbers.
3. Balance the atoms in each half-reaction as follows:
   a. Use coeficients to balance atoms except for hydrogen and oxygen. Apply coefficient to associated electrons.
   b. Add enough water (H_2O) to balance the side that needs oxygen.
   c. Add enough hydrogen ions (H^+) to the side that needs hydrogen to balance hydrogen.
4. Use coefficients to balance electrons across both half-reactions. Multiply the entire half-reaction by the coefficient.
5. Combine the two half-reactions. Cancel duplicate items on either side of the equation.

For basic equations:

6. Neutralize remaining H^+ ions with OH^- ions to form water. Add equal amounts of OH^- to both sides. Combine multiple water molecules on the same side.

Review problems and questions

1. Write balanced half-reactions for the unbalanced redox reaction.
   $Cu + AgNO_3 \rightarrow Ag + Cu(NO_3)_2$

2. When using the half-reactions method to balance a reaction, what is the purpose of adding water molecules or hydroxide ions to the reaction?

3. Balance the reaction in which hydrazine, N_2H_4 and copper(II) hydroxide are combined in a basic solution: $Cu(OH)_2 + N_2H_4 \rightarrow Cu + N_2$

Antioxidants

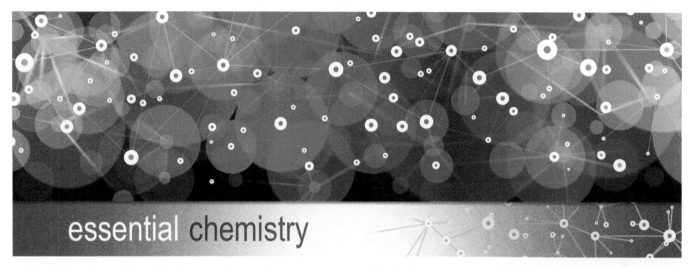

essential chemistry

Antioxidants get a lot of hype as being an important part of healthy foods – including so-called superfoods. But what really are antioxidants, and why are they important? As the name implies, antioxidants are chemical compounds that counteract oxidants. An oxidant, or oxidizing agent, is something that will cause oxidation by gaining an electron. An antioxidant is something that can stop oxidation.

Oxidation occurs naturally in your body all the time. This is a regular process, and there are systems in place to balance out oxidation. Problems can arise when there is an imbalance between the oxidation and your body's ability to detox. This imbalance is called oxidative stress and it can be caused normally by metabolism in the body. Oxidative stress is influenced by outside factors like smoking or environmental toxins. Oxidative stress creates reactive oxygen-containing species, including free radicals. Free radicals are relatively short-lived molecules that contain an unpaired electron. The electron-hungry free radical molecules react with other molecules by stealing an electron. In doing so they can set off a chain reaction of molecules seeking extra electrons. In your body this is especially dangerous because free radicals will affect important biomolecules and cause cell injury. Antioxidants stop the free radical chain reaction by acting as electron donors. The key to their antioxidant power is their ability to donate an electron and still remain stable!

Your body produces some antioxidants as a natural defense against too much oxidation. Glutathione is an important antioxidant produced by your body. It scavenges cells looking for free radicals and other reactive oxygen species. The sulfur group (yellow atom) on the compound is a reducing agent. When glutathione donates an electron to a free radical, a stable glutathione disulfide compound is formed. This stops the chain reaction and prevents cellular injury. The disulfide gets converted back to glutathione by another compound in the body to regenerate this important antioxidant. All cells contain glutathione, and because the conversion occurs constantly, the ratio of glutathione to its disulfide is used as a measure of oxidative stress.

Antioxidants - 2

Fruits and vegetables are an excellent source of many phytonutrients (plant chemicals), including antioxidants. You might have heard that carrots are good for your eyes. This is because carrots contain beta-carotene. Beta-carotene is a pigment that gives the orange color to carrots, as well as pumpkins and sweet potatoes. Beta-carotene is in a class of compounds called carotenoids, and they are all effective antioxidants. The key to their antioxidant powers lies in their molecular formula. Carotenoids contain 40 carbon atoms with an alternating double bond structure. This arrangement makes these molecules very electron-rich. In general, the more double bonds, the greater the antioxidant activity. In addition to being an antioxidant, beta-carotene gets metabolized into Vitamin A in your body. Vitamin A is essential for our immune system, healthy skin and, or course, vision.

Lycopene is another carotenoid that is found in some fruits and vegetables. It is the red plant pigment that gives tomatoes, watermelons and guava their color. The molecular structure gives it its color, and makes it insoluble in water. If you have ever noticed a reddish stain on plastic containers that held tomato sauce – even after you washed them - you now know that you can blame lycopene. You should be able to see from its structure that lycopene is very similar to beta-carotene except it does not have ring groups at the end of the molecule. Like beta carotene, it has an alternating double bond structure making an effective antioxidant. Unlike beta-carotene, it cannot metabolize vitamin A.

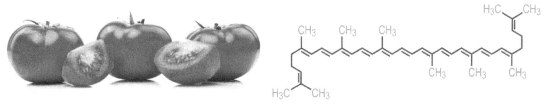

Antioxidants do not have to be carotenoids. Vitamins E and C are also antioxidants. Vitamin C, ascorbic acid, is found naturally in citrus fruits, as well as spinach, kale and broccoli. Vitamin C is water-soluble and it can reduce reactive oxygen species, including hydrogen peroxide. Vitamin E is an important antioxidant that is not water soluble. It is found in sunflower oil, almonds, peanuts and avocados. Evidence suggests that it protects the lipid membranes of cells from free radicals.

Antioxidants get promoted as a cure for many health issues, including preventing cancer and fighting the effects of aging. There are many factors involved in antioxidants' effectiveness, including the ability of your body to absorb the antioxidants and how they interact with other molecules in your body. A variety of antioxidant supplements are available in stores, but should you take them? And if so, how much? So far studies on individual antioxidants as supplements are inconclusive. In fact, some of the studies suggest that too much antioxidant intake can be harmful! The best advice is to make sure you get your antioxidants from healthy servings of natural sources. Your parents were right all along – it is essential that you eat your fruits and vegetables.

Chapter 17 review

Vocabulary
Match each word to the sentence where it best fits.

Section 17.1

antioxidant	electrochemistry
electroneutrality	free radical
oxidation	oxidation number
redox	reduction

1. _____ (s) are reactive substances that seek electrons to replace lost electrons.

2. _____ studies the change in oxidation state as a science combining chemistry and the flow of electricity.

3. A process where an element gains electrons is _____.

4. The _____ indicates the amount of electrons gained or lost by an element.

5. A molecule called a(n) _____ protects us from the harmful effects of free radicals.

6. A chemical reaction that increases the charge of an atom or ion by removing electrons is called _____.

7. _____ reactions are identified by chemical reactions in which elements are oxidized and reduced.

8. Redox reactions obey _____, the property of charges to balance each other's loss and gain of electrons to become neutral.

Section 17.3

oxidizing agent	reducing agent

9. A(n) _____ is an element that is reduced as it oxidized another element.

10. An element called a(n) _____ reduces another element as it gets oxidized.

Section 17.4

half-reactions

11. _____ show the oxidation and reduction parts of a redox reaction.

Conceptual questions

Section 17.1

12. What is an example of an electrochemical process that is accomplished by a non-living thing?

13. When a neutral molecule loses an electron, it becomes a free radical.
    a. What chemical behavior makes free radicals dangerous?
    b. Provide an example of how free radicals can be harmful to the human body.

14. Antioxidants are chemical compounds that fight oxidation.
    a. How do antioxidants stabilize free radicals?
    b. What happens to the antioxidant during this process?

15. Electrochemical processes are driven by electron transfers between elements trying to become more stable.
    a. Explain what happens to an element when it becomes reduced, and how the oxidation number is affected by the change.
    b. Explain what happens to an element when it becomes oxidized, and how the oxidation number is affected by the change.

16. There are many examples of how electrochemistry applies to life. Describe an electrochemical process that is accomplished by a living organism.

17. Oxidation does not happen without reduction and reduction does not happen without oxidation. How does this lead to the concept of electroneutrality?

18. In a neutral state, an element has an equal number of protons (+) and electrons (-). Consider an element that has lost two electrons.
    a. What was the oxidation number of the element before it lost two electrons?
    b. Is the element oxidized or reduced?
    c. Will its oxidation number increase, or will it decrease? What is the new oxidation number?
    d. Will its charge increase or will it decrease? What is the new charge?

19. Electrochemistry is the study of chemical processes that are capable of producing a flow of electricity. How are reactions in electrochemistry different from other types of reactions?

Section 17.2

20. The oxidation number of nitrogen in the N_2 molecule is zero. What rule does this follow?

21. The oxidation number of barium is $^{+2}$. What rules for assigning oxidation numbers proves this?

22. Redox reactions are among the most common and most important chemical reactions in everyday life. What distinguishes a redox reaction from a non-redox reaction?

23. What should all of the oxidation numbers in a neutral molecule add up to? What rule for assigning oxidation numbers does this equate to?

24. How is the charge of an ion related to its oxidation number? What equation can be used to determine the oxidation numbers for polyatomic ions?

25. What formula should be used when determining the oxidation numbers for an ion with a charge?

Chapter 17 review

Section 17.2

26. How does the process change when determining oxidation numbers for neutral compounds with 3 elements as opposed to neutral compounds with 2 elements?

27. Transition metals like iron (Fe) and chromium (Cr) regularly form ions with different charges. What feature of transition metals allow them to have multiple different oxidation states?

28. You are tasked with finding an element's oxidation number in a compound whose charge is not what you would expect for the ion. What steps should be followed in order to assign oxidation numbers for the compound when they are not the same as their ionic charge.

29. When assigning oxidation numbers to a compound where two conflicting rules apply for the same element, which rule should be applied?

Section 17.3

30. How is a neutral charge balance maintained between oxidizing agents and reducing agents?

31. You are assigning oxidation numbers to elements in a reaction and notice that the element has changed its oxidation number from 0 to $^{+1}$. Was this element oxidized or reduced? Explain.

32. After determining the oxidation number of the element in the previous problem, is the element an oxidizing agent or a reducing agent? Explain.

33. Most redox reactions cannot be balanced with the inspection method. An alternative to consider is the oxidation number method of balancing redox reactions.

    a. What is the basic rule that drives this method of balancing reactions?

    b. Are gained electrons written as a positive or negative number?

    c. Are lost electrons written as positive or negative?

34. What needs to be in equal quantities on both sides of a redox reaction in order for it to be considered balanced?

35. When identifying whether a reaction is an oxidation-reduction reaction or not:

    a. What steps should be taken?

    b. Once it has been identified as redox, what should be done to establish a balanced equation?

Section 17.4

36. A complete chemical reaction in an aqueous solution can be separated into two half-reactions. What does a half-reaction show that a full reaction does not?

37. When should the half-reaction method be used instead of the oxidation number method when balancing a redox reaction?

38. Why are spectator ions left out when writing half-reactions?

39. When placing electrons on the reactant side or product side in a half-reaction, what should be considered to determine which side they should be placed? What steps should be taken to ensure a balanced equation?

40. The dioxovanadium(V) ion appears yellow in an acidic solution. After being introduced to zinc, the ion becomes reduced and gradually changes color to purple according to the following unbalanced equation.

$$VO_2^+(aq) + Zn(s) \rightarrow VO^{2+}(aq) + Zn^{2+}(aq)$$

   a. What must be added to balance oxygen when a redox reaction takes place in an acidic solution?

   b. Which side of the reaction will this be added to?

   c. What must be added to balance hydrogen in a redox reaction that takes place in an acidic solution?

   d. Which side of the reaction will this be added?

41. Aluminum metal is oxidized in an aqueous base, with water serving as its oxidizing agent according to the following unbalanced reaction.

$$Al(s) + H_2O(l) \rightarrow [Al(OH)_4]^-(aq) + H_2(g)$$

   a. What must be added to balance oxygen in a redox reaction that takes place in a basic solution?

   b. Which side of the reaction will this be added to?

   c. What must be added to balance hydrogen in a redox reaction that takes place in a basic solution?

   d. Which side of the reaction will this be added to?

42. What are the differences between balancing redox reactions in a basic solution as compared to balancing a redox reaction in an acidic solution?

Quantitative problems

Section 17.2

43. What are the oxidation numbers for each element in the following compounds?

    a. SiO_2

    b. Br_2

    c. KCl

    d. $NaCl$

    e. HNO_2

44. Find the oxidation number of sulfur in each of the following compounds.

    a. SO_4^{2-}

    b. H_2S

    c. SO_3^{2-}

Chapter 17 review

Section 17.2

45. Assign an oxidation number for each element in the following substances.
    a. C_2H_4
    b. H_2CO_3
    c. NO_3^-
    d. ClO_4^-
    e. ClO_2^-
    f. NH_4^+
    g. Ca^{2+}
    h. CO

46. Determine the oxidation number of each element in the following compounds.
    a. Carbon disulfide, CS_2
    b. Sodium cobaltinitrite, $Na_3Co(NO_2)_6$
    c. Ammonium orthomolybdate, $(NH_4)_2MoO_4$
    d. Barium peroxide, BaO_2

Section 17.3

47. Determine whether each of the following reactions are redox or not. Explain your answers.
    a. $MnO_2 + 4HCl \rightarrow MnCl_2 + Cl_2 + 4H_2O$
    b. $Na_2SO_4 + CaCl_2 \rightarrow CaSO_4 + 2NaCl$
    c. $Cl_2 + 2KI \rightarrow I_2 + 2KCl$
    d. $NaOH(l) + HNO_3(l) \rightarrow NaNO_3(l) + 2H_2O(l)$

48. For each of the following redox reactions,
    i. Identify which elements are oxidized and which are reduced; and,
    ii. Identify the oxidizing agent and the reducing agent.
    a. $2NO(g) + 5H_2(g) \rightarrow 2NH_3(g) + 2H_2O(g)$
    b. $2Na(s) + 2H_2O(l) \rightarrow 2NaOH(aq) + H_2(g)$
    c. $2H_2O(l) \rightarrow 2H_2(g) + O_2(g)$
    d. $4Li(s) + O_2(g) \rightarrow 2Li_2O(s)$

49. Use the oxidation number method to balance the following reactions, whether the inspection method is easier or not.
    a. $Cu(s) + AgI(aq) \rightarrow CuI_2(aq) + Ag(s)$
    b. $HNO_3(aq) + Cu_2O(s) \rightarrow Cu(NO_3)_2(aq) + NO(g) + H_2O(l)$
    c. $Zn(s) + HCl(aq) \rightarrow ZnCl_2(aq) + H_2(g)$

50. The chemical reaction below shows the redox reaction of sulfurous acid and permanganate ions:

    $H^+(aq) + MnO_4^-(aq) + SO_3^{2-}(aq) \rightarrow$
    $MnO_2(aq) + SO_4^{2-}(aq) + H_2O(l)$

    a. Determine the oxidation number for each element in the chemical equation.
    b. Identify the elements that are oxidized and reduced.
    c. What are the oxidizing and reducing agents?
    d. Balance the redox reaction using the oxidation number method.

51. During photosynthesis, plants use energy from sunlight to power a redox reaction between carbon dioxide and water to make their food source, glucose. The by-product of photosynthesis is oxygen gas. The reaction for photosynthesis is shown below:

    $CO_2 + H_2O \rightarrow C_6H_{12}O_6 + O_2$

    a. Determine the oxidation number for each element in the chemical equation.
    b. Identify the elements that are oxidized and reduced.
    c. What are the oxidizing and reducing agents?
    d. Balance the redox reaction using the oxidation number method.

Section 17.4

52. Copper reacts with silver nitrate in solution to form copper nitrate solution and silver metal.

    $Cu(s) + AgNO_3(aq) \rightarrow Cu(NO_3)_2(aq) + Ag(s)$

    a. Determine the oxidation number for each element in the chemical equation.
    b. Identify the elements that are oxidized and reduced.
    c. What are the oxidizing and reducing agents?
    d. Identify spectator ions, if any.
    e. What are the half-reactions for this redox reaction?
    f. Balance the redox reaction with the half-reaction method in a neutral solution.

53. The determination of carbon monoxide content in blood is performed by reacting extracted carbon monoxide with iodine pentoxide at 150°, allowing the following to take place:

    $CO + I_2O_5 \rightarrow CO_2 + I_2$

    a. Determine the oxidation number for each element in the chemical equation.
    b. Identify the elements that are oxidized and reduced.
    c. What are the oxidizing and reducing agents?
    d. What are the half-reactions for this redox reaction?
    e. Using the half-reaction method in a basic solution: balance the redox reaction.

Chapter 17 review

Section 17.4

54. Niobium dioxide reacts with tungsten:

$$NbO_2(aq) + W(s) \rightarrow Nb(s) + WO_4^{2-}(aq)$$

   a. Determine the oxidation number for each element in the chemical equation.
   b. Identify the elements that are oxidized and reduced.
   c. What are the oxidizing and reducing agents?
   d. What are the half-reactions for this redox reaction?
   e. Balance the redox reaction with the half-reaction method in an acidic solution.

55. The following reaction occurs in a basic solution:

$$NiO_2 + Fe \rightarrow Ni^{2+} + 2OH^- + Fe^{2+} + 2OH^-$$

   a. Determine the oxidation number for each element in the chemical equation.
   b. Identify the elements that are oxidized and reduced.
   c. What are the oxidizing and reducing agents?
   d. What are the half-reactions for this redox reaction?
   e. Using the half-reaction method, balance the redox reaction.

56. Ammonia reacts with hypochlorite ions:

$$NH_3 + ClO^- \rightarrow N_2H_4 + Cl^-$$

   a. Determine the oxidation number for each element in the chemical equation.
   b. Identify the elements that are oxidized and reduced.
   c. What are the oxidizing and reducing agents?
   d. What are the half-reactions for this redox reaction?
   e. Balance the redox reaction with the half-reaction method in an basic solution.

57. The following reaction occurs in an acidic solution:

$$ClO_3^- + SO_2 \rightarrow SO_4^{2-} + Cl^-$$

   a. Determine the oxidation number for each element in the chemical equation.
   b. Identify the elements that are oxidized and reduced.
   c. What are the oxidizing and reducing agents?
   d. What are the half-reactions for this redox reaction?
   e. Using the half-reaction method, balance the redox reaction.

58. Use the half-reaction method in a neutral solution to balance the following redox reactions.

   a. $Ni^{2+}(aq) + Al(s) \rightarrow Al^{3+}(aq) + Ni(s)$
   b. $Zn(s) + Pb^{2+}(aq) \rightarrow Zn^{2+}(aq) + Pb(s)$
   c. $Mg(s) + Cu^+(aq) \rightarrow Mg^{2+}(aq) + Cu(s)$

59. Formaldehyde is reduced by aluminium hydride ions in a basic solution:

$$AlH_4^- + H_2CO \rightarrow Al^{3+} + CH_3COH$$

   a. Determine the oxidation number for each element in the chemical equation.
   b. Identify the elements that are oxidized and reduced.
   c. What are the oxidizing and reducing agents?
   d. What are the half-reactions for this redox reaction?
   e. Using the half-reaction method in an basic solution: balance the redox reaction.

60. Iron (III) oxide reacts with carbon monoxide to form molten iron and carbon dioxide.

$$Fe_2O_3(s) + CO(g) \rightarrow Fe(s) + CO_2(g)$$

   a. Determine the oxidation number for each element in the chemical equation.
   b. Identify the elements that are oxidized and reduced.
   c. What are the oxidizing and reducing agents?
   d. What are the half-reactions for this redox reaction?
   e. Using the half-reaction method in an acidic solution: balance the redox reaction.

61. Ethanol reacts with permanganate ions in an acidic solution:

$$MnO_4^-(aq) + C_2H_5OH(aq) \rightarrow Mn^{2+}(aq) + CH_3COOH(aq)$$

   a. Determine the oxidation number for each element in the chemical equation.
   b. Identify the elements that are oxidized and reduced.
   c. What are the oxidizing and reducing agents?
   d. What are the half-reactions for this redox reaction?
   e. Using the half-reaction method, balance the redox reaction.

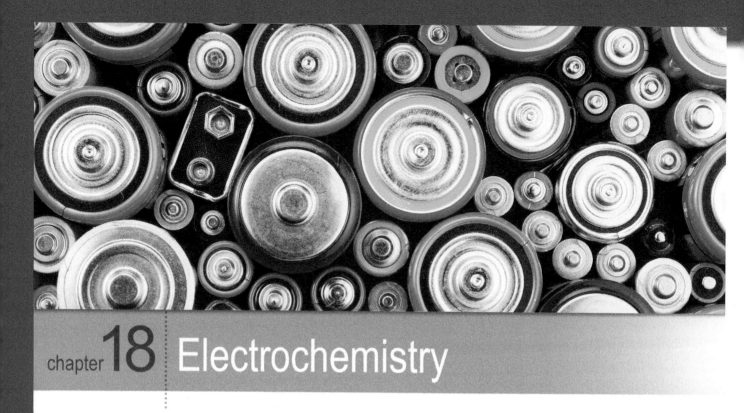

chapter 18 Electrochemistry

Redox reactions involve changing oxidation states and the movement of electrons. What if you could harness the movement of electrons and use that energy to do work? It turns out that we can, and you use that energy every day. Every single type of battery uses a redox reaction to power devices like cell phones, computers, flashlights, and even cars!

The key to any battery lies in the fact that oxidation and reduction happen in different places. You may have noticed that there is a positive (+) and a negative (-) terminal on a battery. The positive and negative indicate that reduction and oxidation happen at those terminal ends. Many redox reactions happen within a single location or solution – electrons are exchanged between the oxidation and reduction half-reactions that are in direct contact with each other. With a battery, the half-reactions are separated so they occur in two different places. The movement of electrons will still happen, but only if a wire connects the positive and negative ends. The spontaneous movement of electrons through the wire is what powers your devices.

Batteries use a spontaneous redox reaction to generate electricity. Sometimes the reverse is true. Electricity is used to drive a nonspontaneous redox reaction. A rechargeable battery is actually a great example of both of these processes. A rechargeable battery uses a spontaneous reaction to power your phone, then electricity is used to make that reaction happen "backwards" and recharge the battery.

Electrochemistry is the study of the relationship between electricity and chemical processes. If you look at the electrical devices in the world around you, you will see why understanding electrochemistry is so essential.

Chapter 18 study guide

Chapter Preview

As discussed briefly in the last chapter, electrochemistry explains how chemical reactions can generate electricity as well as how electricity can cause chemical change. But what is electricity? Fundamentally it is a type of energy created by charged particles. Protons and electrons carry electrical charge. When made to flow, electrons create an electrical current. Redox reactions are ones in which charges are exchanged. These reactions play a major role in electrochemistry. In this chapter you will learn about the fundamentals of electricity, how separating oxidation and reduction reactions can be used to generate electricity, and how galvanic and electrolytic cells are constructed as well as how they function.

Learning objectives

By the end of this chapter you should be able to:

- describe how ions produce an electric current;
- explain what voltage is a measurement of;
- identify the parts of an electrochemical cell;
- explain the difference between galvanic and electrolytic cells;
- write the half-reactions that occur at the anode and cathode of a galvanic cell;
- calculate cell voltage;
- determine if a reaction will occur spontaneously; and
- calculate the total free energy available when a metal reacts.

Investigations

18A: Electrochemical cells
18B: Electroplating
18C: Lemon battery
18D: Project: Design a Galvanic Cell
18E: Writing and Discussion Enhancement: Galvanic Cell

Pages in this chapter

- 574 Fundamentals of Electricity
- 575 Charges flow in solutions
- 576 Voltage is a force that pushes electrons
- 577 Section 1 Review
- 578 Electrochemical Cells and Electrolysis
- 579 Galvanic and electrolytic cells
- 580 Electrolysis with an electrolytic cell
- 581 Electroplating with an electrolytic cell
- 582 Chemistry in a galvanic cell
- 583 The salt bridge
- 584 Section 2 Review
- 585 Electricity From Electrochemical Cells
- 586 Calculating energy from cell voltage
- 587 Standard reduction potentials
- 588 Calculating cell voltage
- 589 Cell emf spontaneity
- 590 Activity series
- 591 Nernst equation
- 592 Section 3 Review
- 593 Batteries of All Shapes and Sizes
- 594 Batteries of All Shapes and Sizes - 2
- 595 Batteries of All Shapes and Sizes -3
- 596 Chapter review

Vocabulary

- coulomb (C)
- voltage
- ohm (Ω)
- electrode
- electrolyte
- voltaic cell
- electrolysis
- corrosion
- standard reduction potential
- electric current
- volt
- Ohm's law
- anode
- galvanic cell
- electrolytic cell
- electroplating
- electromotive force (emf)
- activity series
- ampere
- resistance
- electrochemical cell
- cathode
- spontaneous
- nonspontaneous
- salt bridge
- cell emf
- Nernst equation

Chapter 18 Summary

18.1 - Fundamentals of Electricity

Did you know a lemon and two different kinds of metal can generate electricity? Electrochemistry is the study of the chemical reactions in the lemon that generate electricity. Electrochemistry is also the study of chemical changes caused by electricity. Let's review the fundamentals of electricity.

Electric charge and its flow

The coulomb

A proton's charge is equal and opposite to an electron's charge. The unit of charge is the **coulomb (C)**. The charge of a single proton is $+1.602 \times 10^{-19}$ C. The charge of single electron is -1.602×10^{-19} C. Equal numbers of protons and electrons in an atom produce no overall charge.

Conservation of charges

Like mass, electric charge obeys the law of conservation. The electric charge lost by one part of a closed system is gained somewhere else in a system, therefore the total electric charge is constant. Since charge is conserved you can say the charge of a single electron is equal to 1.602×10^{-19} C. If you have one mole or an Avogadro's number of electrons the total charge equals 96,440 C. The more electrons there are, the greater the charge.

Metallic bonds allow electrons to flow freely

Metals such as copper form metallic bonds between atoms. As presented in the sea of electrons metallic bonding model, some of the valence electrons are free to move from one copper atom to the next. In a solid copper wire, free electrons travel randomly among positive ions which are fixed in their positions. Free electrons have very little mass and move so quickly that they move energy very effectively. For this reason copper and other metals are good conductors of heat and electricity.

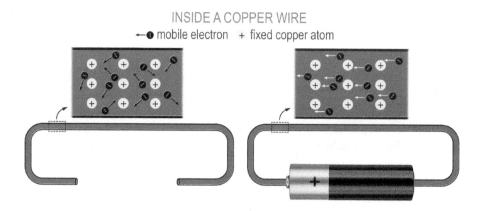

Electron movement produces current

When a battery provides energy for portable devices, it organizes the motion of electrons. A battery's energy causes electrons to move in one single direction instead of random paths. The arrows in the diagram show how energy from the battery will change each electron's movement. When no battery is present, electrons move in any direction as shown by the black arrows within a copper wire. When a battery is connected to the copper wire, the white arrows show the direction of the electrons will begin flowing to the left towards the positive battery terminal. When the free electrons move together as a group in the same direction, the charge also moves in that direction. This movement of charge is what produces an **electric current**. The SI unit for electric current is the **ampere** (A), or amp. One ampere is the rate of flow of one coulomb of electric charge per second.

Charges flow in solutions

Electrolyte solutions contain mobile ions

An electrolyte is a mobile ion with either a positive or negative charge. Ions are mobile in liquids in aqueous solutions. An example of an electrolyte solution is salt water where sodium chloride (NaCl) is dissolved in water. Dissolved salt crystals produce Na^+ and Cl^- ions that move around freely in solution. When positive sodium cations (Na^+) and/or negative chloride anions (Cl^-) move freely they can create an electric current. Acidic solutions can also create current because acids form ions in solution. When electrically charged particles are free to move they may have the ability to create an electric current.

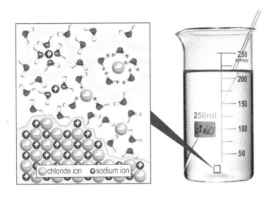
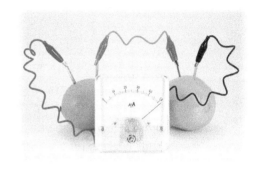

Chemistry generates current

You might have built a lemon battery in the past. The diagram shows a battery made of oranges. You can use any fruit, vegetable or solution that contains free-flowing charged particles to build a battery. Electric current is created by chemical reactions that occur where the orange juice, a piece of copper, and a piece of magnesium meet. Orange juice is an electrolyte solution that contains ions formed from citric acid. The atoms in the magnesium metal lose electrons to form ions in the orange juice. Magnesium's lost electrons flow into the wire, through the LED and back to the orange juice through a piece of copper. Electric charge is conserved because the total number of electrons in the system does not change. The system is neutral because the total number of negative charges equals the total number of positive charges.

Mobile charged particles and different metals

Electric current can only flow if the two pieces of metal are different and the substance they are placed in is capable of carrying an electric charge. A fresh potato contains plenty of sodium and potassium ions that can carry an electric charge. However the potato battery shown will not work because the two metals inserted in the potato are identical. Current will flow once one of the metals is replaced with a different metal. On the other hand if electric charges cannot move, then no electric current can flow. For example, even though sugar molecules are made of atoms that have protons and electrons, the molecules are electrically neutral because each molecule has an equal number of electrons and protons that cannot move independently from each other to produce current. Furthermore, sugar molecules do not dissociate into ions when dissolved in water. Fixed particles with no net charge cannot produce an electric current.

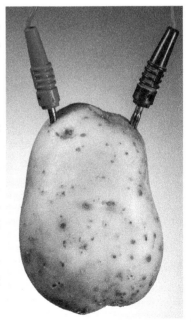

Section 18.1: Fundamentals of Electricity

Voltage is a force that pushes electrons

Electrons flow when pushed

Water follows the easiest path. Water in a pipe is pushed by a pressure difference between two points in the pipe. Water pushes itself out where pressure is lowest. Think of a wire as a pipe full of electrons. Free electrons in a copper wire will move randomly in a sea of electrons unless they are forced to move in one direction because of a pressure difference between two points.

Charge difference creates electrical potential

If electrons flow through a wire in one direction, an electric current is present. Electrons can only flow through a wire if there is a difference in electrical pressure between two points. If one point has more electrons than another, there is a pressure difference. This difference in electrical pressure created by different amounts of electrons is a force called **voltage** measured with the **volt** (V) unit. Electric current can only flow if there is a voltage to push it. Higher voltages generate bigger pushes. A difference in charge between two locations creates the potential for a current to flow between them. Voltage is also called electric potential because it is a measure of electrical potential energy. One volt is measured as one unit of electrical potential energy (joule, J) per unit charge (coulomb, C).

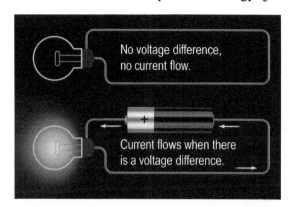

$$volt\ (V) = \frac{joule\ (J)}{coulomb\ (C)}$$

The potential energy of electricity

The coulomb indicates how strong a force or push is between charged particles. Charged particles that have the same charge will repel each other, but they will also push each other. If 10 coulombs of charge moves through a difference of 10 volts, the charge gains or loses 100 joules of potential energy. A charge loses energy when it goes from a higher voltage to a lower voltage and a charge gains energy when it goes from a lower to a higher voltage.

Collisions create resistance

Electricity is like a running river. Water molecules are like electrons, voltage is the force pushing the water so it can flow, and current is the rate at which the water is flowing. If you added pebbles to the river, you would slow down the flow of water. The pebbles cause resistance. **Resistance** impacts the amount of current that can flow. High resistance results in low current and low resistance results in high current. When electrons flow in solid copper they collide with fixed copper ions. The continuous collisions absorb energy and cause resistance. Resistance is measured in the unit **ohm** (Ω). A resistance of one ohm means 1 amp of current flows with 1 volt. The Greek letter omega (Ω) represents ohms.

Ohm's law

Ohm's law relates the current, the voltage and the resistance in the equation $V = I \times R$. The current, I, depends on the voltage, V, and the resistance, R.

Section 1 review

Chapter 18

Electric charge is a fundamental property of matter. Protons are said to have a positive charge and electrons have a negative charge that is equal in magnitude. When the net movement of electrons is in one direction, that flow of charge is called an electric current. When charged particles are free to move, an electric current can be produced. Metals have delocalized electrons that are free to move. Ions in solution are charged particles that are free to move. When electrons are not free to move, a current cannot flow. Voltage is the difference in electric potential between two points that causes charges to flow. Resistance is a measure of the hindrance to the flow of charge. Ohm's law relates current, voltage, and resistance.

Vocabulary words: coulomb (C), electric current, ampere, voltage, volt, resistance, ohm (Ω), Ohm's law

Key relationships

- Charges of protons and electrons are equal and opposite.
- Charge of a proton = +1.602 x 10^{-19} C.
- Charge of an electron = -1.602 x 10^{-19} C.
- Ohm's law: voltage (V) = current (amps) × resistance (Ω)   or $V = I \times R$
- One volt is one unit potential energy per one unit charge: 1 volt (V) = $\frac{1 J}{1 C}$

Review problems and questions

1. Free electrons are moving inside a wire, but in random directions. Is this wire conducting an electric current? Why or why not?

2. If there is a wire that is not conducting an electric current, what will it take to get current to move through it?

3. Is sugar water an electrolyte solution? Why or why not?

4. The compound shown is a solid capable of being dissolved by water.

    a. Will a solution of this substance conduct electricity?

    b. Explain your answer.

5. Under what circumstances are ions able to produce an electric current?

6. When you measure voltage, what is being measured?

Section 18.1: Fundamentals of Electricity

18.2 - Electrochemical Cells and Electrolysis

If you are able to separate oxidation and reduction reactions, you can force the exchanged electrons to move through a wire. An electrical current is made up of electrons moving through a wire. A device that separates oxidation and reduction reactions to access current is an electrochemical cell.

Electrical energy from redox reactions

Components of an electrochemical cell

An **electrochemical cell** is a device that uses redox reactions either to generate electrical energy or to use electrical energy to initiate a chemical reaction. The various parts of the electrochemical cell are:

1. Two **electrode**(s), where the oxidation and reduction reactions occur. The electrodes in the diagram below are aluminum and copper metals.
   - The electrode at which oxidation occurs is called the **anode**.
   - The electrode at which reduction occurs is called the **cathode**.

2. The **electrolyte** is a conductive solution where the electrodes are immersed. Standard electrolyte solutions have a 1M concentration.
   - The electrolyte contains mobile ions that are free to move anywhere in the solution.
   - The electrolyte conducts electricity because ions are charged particles.

3. A conductive path or wire that connects the electrodes outside of the solution.
   - Electrons flow from one electrode to another through this path.

4. The salt bridge which is in one particular kind of electrochemical cell called a galvanic cell. The picture below shows a galvanic cell.

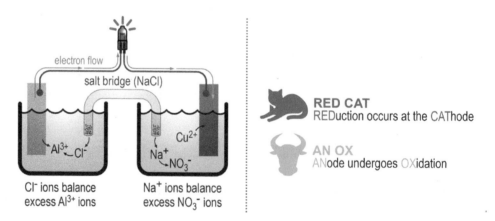

Reduction at cathode, oxidation at anode

To help you remember which electrode experiences reduction and which electrode experiences oxidation, picture a red cat and an ox like the ones shown. A RED CAT helps you remember that REDuction occurs at the CAThode, and AN OX helps you remember that the ANode undergoes OXidation.

Electrons flow from anode to cathode

It can be confusing to remember which way electrons move in an electrochemical cell. Electrons always flow from anode to cathode. This is true no matter which type of electrochemical cell you are studying. If you remember that the anode undergoes oxidation, and oxidation means electrons are lost, it makes sense that electrons leave the anode to move towards the cathode. Reduction occurs at the cathode which is receiving or gaining electrons.

Galvanic and electrolytic cells

Transforming chemical and electrical energy

There are two kinds of electrochemical cells: electrolytic and galvanic cells. Galvanic cells are also called voltaic cells. Both kinds of cells generate voltage from a potential difference in charge between an anode and cathode, where current flows between them. The difference is galvanic cells convert chemical energy into electrical energy and electrolytic cells perform the reverse.

Galvanic cells

In a **galvanic cell**, spontaneous chemical reactions occur at the electrodes allowing electrons to flow from the anode to the cathode. These reactions produce an electrical current. A **spontaneous** reaction is one that happens in the expected forward direction without energy input. A galvanic cell converts chemical energy into electric energy. The galvanic cell is named after Luigi Galvani, known as the father of electrochemistry. Another name for a galvanic cell is the **voltaic cell** named after Alessandro Volta, who invented the battery. An alkaline battery is a galvanic cell.

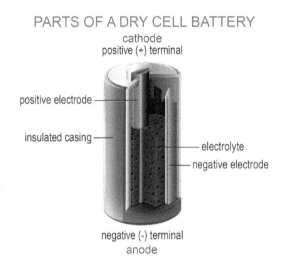

Electrolytic cells

An **electrolytic cell** works when an externally applied electrical current drives non-spontaneous chemical reactions. A **nonspontaneous** reaction is one that requires an energy input and proceeds in the reverse direction. This is where electrical energy is converted to chemical energy. The anode is the positive electrode and the cathode is the negative electrode, which is the reverse of a galvanic cell. The direction of electron flow is still from anode to cathode. Both galvanic and electrolytic cells contain two half-cells. In either type of cell, one half-cell is the anode where an oxidation half-reaction occurs. The other half-cell is the cathode where a reduction half-reaction occurs.

Electron flow is balanced by ion flow

The battery shown below is a galvanic cell because the cathode is identified as the positive terminal. This cell can produce 3 V to move electrons; the positive voltage indicates the reaction is spontaneous. When an external circuit such as a light bulb is connected between the electrodes of the battery, electrons flow from the anode to the cathode outside of the battery through the wire. Positive and negative ions from the electrolyte interact with the appropriate electrode to maintain charge neutrality.

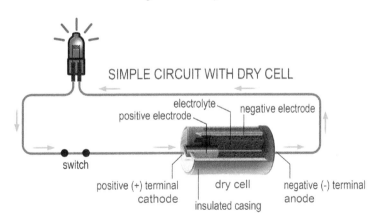

Electrolysis with an electrolytic cell

Electrolysis

Electrolysis is a nonspontaneous, induced redox reaction in an electrolytic cell. Aluminum, chlorine and other useful substances are commercially produced using electrolysis. Notice the battery between electrodes in the diagram. This indicates an energy input is required for the electrolysis or decomposition of water molecules:

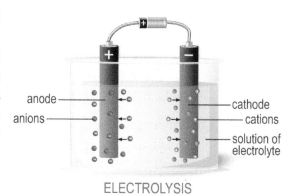

$$2H_2O(l) \xrightarrow{energy} 2H_2(g) + O_2(g).$$

Half-reactions for electrolysis of water

The reaction will not move forward unless you add energy. Hydrogen has an oxidation number of +1 in H_2O but in H_2 the oxidation number 0. The oxidation number decreased, therefore hydrogen has been reduced. The oxidation number of oxygen in H_2O is -2 but it is 0 in O_2, therefore oxygen is oxidized. For this redox reaction, the oxidation and reduction half-reactions and their associated voltages (from a reference chart) are:

$$\text{Reduction: } 2H_2O(l) + 2e^- \rightarrow H_2(g) + 2OH^-(aq) = -0.83 \text{ V}$$

$$\text{Oxidation: } 2H_2O(l) \rightarrow O_2(g) + 4H^+(aq) + 4e^- = -1.23 \text{ V}$$

Negative means non-spontaneous

If you add the reduction and oxidation potentials together you get -2.06 V. The negative total voltage indicates the reaction is nonspontaneous. You need to add energy to move electrons in a direction that breaks up water into hydrogen gas and oxygen gas.

Hydrogen fuel cells

Hydrogen gas is a potential future energy source. It does not exist in abundance on Earth so it must be produced commercially from water, fossil fuels, or biomass. Generating hydrogen gas from the electrolysis of water is one of the cleanest ways to produce fuel, but it is also among the most expensive. There is no inexpensive way to add enough energy to get the reaction to happen. When hydrogen gas is produced by the reaction, it can be stored in a hydrogen fuel cell. When the fuel cell is used, molecular hydrogen and molecular oxygen undergo a spontaneous redox reaction to form water. Oxygen is reduced and hydrogen is oxidized with the help of an electrolyte and a catalyst.

Brine vs. molten NaCl

Chlorine gas is used as a disinfectant for water and cleaning products. Chlorine gas and pure sodium metal are retrieved through the electrically-induced decomposition of molten sodium chloride, NaCl. The half-reactions total -4.07 V when added together.

$$\text{Reduction: } Na^+(l) + e^- \rightarrow Na(l) = -2.71 \text{ V}$$

$$\text{Oxidation: } 2Cl^-(l) \rightarrow Cl_2(g) + 2e^- = -1.36 \text{ V}$$

This energy-expensive reaction is the standard way to produce pure sodium. When this reaction is attempted in a salt water solution, also known as brine, chlorine is oxidized as expected but sodium is not reduced. Water is reduced instead because its reduction potential is -0.83 V, far less than sodium. The products of brine electrolysis are hydrogen gas from the reduction of water, chlorine gas from the oxidation of chlorine, and the remaining Na^+ and OH^- ions form aqueous sodium hydroxide (NaOH).

Electroplating with an electrolytic cell

Electro-plating

Just a few years after Alessandro Volta invented the battery, Luigi Brugnatelli used an electrolytic cell to invent electroplating. **Electroplating** is a process that uses electricity to oxidize anions or reduce cations, resulting in a thin metal coating on an object. The result is a pleasant and protective shiny metal covering. Examples of electroplated or plated goods include jewelry, coins, wheel rims, and anything made of brass.

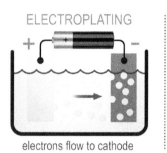

Benefits

The most familiar electroplating processes involve reducing a metal cation so it gets deposited on the surface of the cathode. When the cathode is coated with a layer of the reduced metal, it gains a number of new properties. The added metal coating thickens the cathode, and changes its conductivity. High-quality electronic components are plated with gold (Au) on their surface to improve their electrical conductivity. The metal beneath the gold layer is commonly copper or silver. "White gold" jewelry is made by a double-electroplating process. The jewelry begins as silver (Ag), then a layer of nickel (Ni) is plated over the silver, and the outer coating is a layer of rhodium (Rh). Galvanized nails, screws, nuts and bolts are able to resist corrosion because a zinc (Zn) coating prevents oxidation of the metal below.

How rechargeable batteries work

By now you can see how much you rely on electrolytic and electrochemical processes for everyday things. Most high school students feel lost when their phone battery dies; luckily, the battery is rechargeable. A rechargeable battery uses a reversible redox reaction to act like an electrolytic cell when it is being charged, and a galvanic cell when it is being used. During charging, electricity is used to drive a nonspontaneous redox reaction to move electrons from anode to cathode. The positive terminal is oxidized and the negative terminal is reduced. When the battery is being used, chemical energy spontaneously generates electrical current as electrons move from anode to cathode. The positive terminal is reduced and the negative terminal is oxidized.

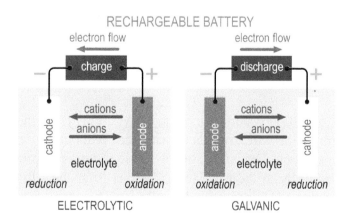

Chemistry in a galvanic cell

The galvanic cell

A galvanic cell made with two half-cells can be used to generate an electric current. Two different metals must be used for the electrodes because one metal must be more willing to give up its electrons than the other metal. This is what creates a potential charge difference. Electrons are released at the anode where oxidation occurs. Electrons collect at the cathode where reduction happens. The wire that connects the two electrodes allows you to access the electric current generated by the redox reaction.

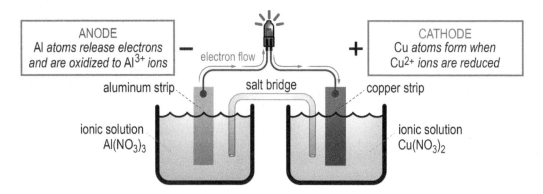

Al-Cu voltaic cell

Let's look at a galvanic cell using electrodes made of aluminum (Al) atoms and copper (Cu) atoms. The aluminum electrode is immersed in an aluminum nitrate, $Al(NO_3)_3$ solution that contains Al^{3+} and NO_3^- ions. The copper electrode is immersed in a copper(II) nitrate, $Cu(NO_3)_2$ solution that contains Cu^{2+} and NO_3^- ions. The half-reactions for the oxidation of Al and the reduction of Cu are:

Oxidation at the anode: $Al(s) \rightarrow Al^{3+}(aq) + 3e^-$ (Aluminum atoms become ions)

Reduction at the cathode: $Cu^{2+}(aq) + 2e^- \rightarrow Cu(s)$ (Copper ions become atoms)

Current is produced

Aluminum is more reactive than Cu, so Al loses electrons more easily. When the two electrodes are connected by a wire, electrons released by Al at the anode move through the wire to the Cu cathode. This produces an electrical current. But this reaction will not last very long if the cell is set up exactly as shown in the picture below. The "bridge" between the two electrolyte solutions is missing!

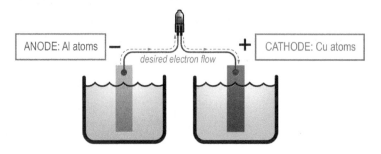

There is a charge imbalance

As oxidation occurs at the anode, Al^{3+} ions enter the solution while their electrons move through the wire towards the cathode. Without the "bridge" positive Al^{3+} ions build up in the solution. Oxidation stops because it can only proceed when there is charge neutrality. Meanwhile at the cathode, Cu^{2+} ions from the solution deposit on the cathode, leaving negative NO_3^- ions behind to pile up. This stops reduction. If the oxidation and reduction reactions stop, so does the electric current. Something is needed to balance ions so excess charges do not build up on either side of the cell.

The salt bridge

Balancing charges

A very important component of a galvanic cell is the salt bridge. A **salt bridge** contains a solution of positive and negative ions whose job is to balance charges in the cell solution. The salt bridge connects the oxidation and reduction half-cells without letting the half-cell solutions mix together.

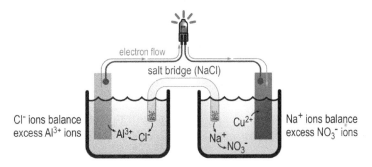

Parts of a salt bridge

A typical salt bridge is a tube that contains an electrolyte such as sodium chloride (NaCl). The ends of the tube are fitted with a porous material, such as a cotton ball, so ions can move in and out of the tube. You can make a salt bridge out of simple materials like a paper towel or a filter paper that retains most of the salt solution but allows ions such as Na^+ and Cl^- to pass through it. Positive and negative ions in the salt bridge can move to either cell as needed to balance charges.

Written form of cell setup

There is a written convention to represent an electrochemical cell setup instead of drawing it. The written form of the aluminum-copper galvanic cell is:

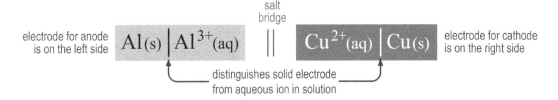

How to arrange the written form

The anode or the oxidation half-cell is always written first. The double-line in the middle represents the salt bridge. The cathode or reduction half-cell is written to the right of the salt bridge. The solid electrodes are written on the left (anode) and right (cathode) of the expression. The single vertical line that separates the atom from its ion represents the phase boundary where the solid electrode and electrolyte solution meet.

What happens at the anode?

Let's walk through how the complete Al-Cu galvanic cell works. The anode is negative because it is the source of electrons through oxidation. The Al^{3+} ions formed continue to move from the electrode to the $Al(NO_3)_3$ solution as long as the salt bridge is there. Negative chloride ions from the salt bridge balance the excess positive charges. The Al electrode loses mass because its atoms form ions that move into the electrolyte solution. This is known as **corrosion**.

What happens at the cathode?

Meanwhile, electrons given up by Al travel through the wire to the cathode. When Cu^{2+} ions from the solution contact the Cu electrode, they receive electrons and get reduced to Cu atoms. In atom form, Cu stays or deposits on the electrode while NO_3^- ions stay in the solution. This is why the cathode mass increases and the solution changes color. The salt bridge provides Na^+ ions to balance the negative charge in the solution from NO_3^- ions.

Chapter 18

Section 2 review

Electrochemical cells use redox reactions to produce electrical current. It is composed of two electrodes made of two different metals, an electrolyte and a conducting path between the electrodes. A salt bridge connects the half-cells without letting the solutions mix. Oxidation occurs at the anode, releasing electrons. Reduction occurs at the cathode, consuming electrons. In a galvanic cell the energy from a spontaneous redox reaction creates electrical current.

Alkaline batteries are an example of a galvanic cell. In an electrolytic cell, energy is added to drive a nonspontaneous redox reaction. Rechargeable batteries use an electrolytic process to restore charge. Electrolysis in an electrolytic cell is used to produce many useful substances, including the decomposition of water into hydrogen gas and oxygen gas. Electroplating is the process of using hydrolysis to apply a coating of one metal on another.

Vocabulary words	electrochemical cell, electrode, anode, cathode, electrolyte, galvanic cell, spontaneous, voltaic cell, electrolytic cell, nonspontaneous, electrolysis, electroplating, salt bridge, corrosion

Key relationships
- Oxidation = loss of electrons; occurs at the anode
- Reduction = gain of electrons; occurs at the cathode

- Diagram of a galvanic cell:

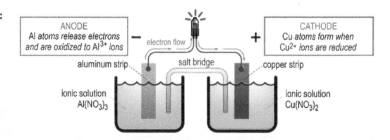

Review problems and questions

1. Which type of electrochemical cell generates electricity without needing to add any extra energy to it?

2. The chemical reaction shown occurs in an electrochemical cell: $A + BX \xrightarrow{energy} AX + B$. What kind of electrochemical cell is this: galvanic or electrolytic? How can you tell?

3. The half reaction shown occurs in an electrochemical cell: $X \rightarrow X + e^-$. Has element X undergone oxidation or reduction? How can you tell?

4. A cation becomes an atom. Is this oxidation or reduction? How can you tell?

5. The net ionic equation for a reaction for a galvanic cell is: $2Ag^+(aq) + Cd(s) \rightarrow Cd^{2+}(aq) + 2Ag(s)$. Convert this reaction into electrochemical cell notation.

18.3 - Electricity From Electrochemical Cells

Water flows from high pressure to low pressure, like water flowing from a pipe out of a faucet in your home. Similarly, in a galvanic cell electrons flow from an area of high pressure where there are a lot of electrons to an area where there are fewer electrons and less pressure. As long as there is a difference in the number of electrons between two points, electrons can flow in an electric current.

Electromotive force

Emf is the cell driving force in volts (V)

Electrical charge is equivalent to the amount of electrons available. This is like the amount of water available to move in a water system. An electrical charge can only move from one point to another when there is a difference in potential energy between the anode and cathode. A potential difference is required for electric charge to be able to move from a point of higher potential to a point of lower potential. In an electrochemical cell the electrons that move from the anode to the cathode are pushed by the voltage. Recall that voltage is the force that is generated by the potential difference between two points, in this case, the anode and cathode. This potential difference between the electrodes is called the **electromotive force (emf)** of the cell, also called **cell emf**. The symbol for cell emf is E^0_{cell}.

ΔG is maximum cell energy

You can use the emf to calculate the maximum amount of energy and work that an electrochemical cell can provide when a certain amount of reactants are consumed. A Gibbs free energy calculation quantifies this maximum cell energy. The equation is:

$$\Delta G = -n \times F \times E^0_{cell}$$

Use the oxidation half-reaction to determine moles, n

Gibbs free energy (ΔG) is the maximum amount of energy produced by an electrochemical cell. You need to know the half-reaction for the oxidized reactant to find the number of moles of electrons (n) that are released by the anode as the electrode is used up. For the Al-Cu cell, the Al half-reaction shown below indicates 3 moles of electrons are released for every mole of Al atoms oxidized.

$$Al\,(s) \rightarrow Al^{3+} + 3e^-$$

1 mole of solid aluminum ⟶ yields 3 moles of electrons

$$\Delta G = -(3)F\,E^0_{cell} \quad n = 3$$

If ΔG is (-) the reaction is spontaneous

The Faraday constant, F equals a charge of 96,000 coulombs per mole of electrons or 96,000 C/mol. The cell emf (E^0_{cell}) in volts (V) is either given or it can be determined by a simple calculation from a table of known standard reduction potential values. The cell emf gives us a measure of the driving force of the cell reaction. When the ΔG calculation is complete, a negative ΔG means energy is released. Therefore the redox reaction is spontaneous and it represents a galvanic cell. If ΔG is positive, energy was consumed. Thus the reaction is nonspontaneous and it represents an electrolytic cell.

Calculating energy from cell voltage

Reaction spontaneity

A spontaneous reaction proceeds in a direction that favors products while a nonspontaneous reaction occurs in the reverse direction which favors reactants. You can use the Gibbs free energy calculation to determine whether a reaction is spontaneous or not.

Solved problem

A Zn/Cu cell reacts according to: $Zn(s) + Cu^{2+}(aq) \rightarrow Zn^{2+}(aq) + Cu(s)$. The reaction has an $E^°_{cell}$ of 1.10 V. Calculate the total energy that can be obtained from this cell if 4.78 g of Zn are consumed in the reaction, and determine the reaction spontaneity.

Given The reaction, emf voltage and the amount of reactant consumed.

Relationships
- $\Delta G = -n \times F \times E^°_{cell}$; a $-\Delta G$ is spontaneous and a $+\Delta G$ is nonspontaneous
- The half-reactions are: Oxidation: $Zn(s) \rightarrow Zn^{2+}(aq) + 2e^-$ and Reduction: $Cu^{2+}(aq) + 2e^- \rightarrow Cu(s)$
- Faraday constant, F = 96,000 C/mol
- For this calculation, the unit V is best expressed as J/C
  $E^°_{cell} = 1.10$ V is equivalent to 1.10 J/C

Solve The half reactions show 1 mole of Zn generates 2 mol of electrons.

- One mole of Zn has a mass of 65.38 g.

$$\frac{4.78 \text{ g Zn}}{1} \times \frac{1 \text{ mol Zn}}{65.38 \text{ g Zn}} = 0.07311 \text{ mol Zn} \longrightarrow \text{round } 00.0731 \text{ mol Zn}$$

- Each mole of Zn oxidized produces 2 moles of electrons.

$$\frac{0.0731 \text{ mol Zn}}{1} \times \frac{2 \text{ mol electrons}}{1 \text{ mol Zn}} = 0.146 \text{ mol electrons}$$

- Substitute into the energy equation.

$$G = -0.146 \text{ mol} \times 96,000 \frac{C}{\text{mol}} \times 1.10 \frac{J}{C} = -15,417.6 \text{ J} \longrightarrow \text{round } -15,400 \text{ J}$$

Answer 4.78 g of Zn release -15,400 J of energy in a spontaneous reaction.

How to find the cell voltage

Total cell voltage $E^°_{cell}$ equals the combined potentials of each half-cell. You can calculate $E^°_{cell}$ by adding reduction potential and oxidation potential:

$$E^°_{cell} = E_{ox} + E_{red}$$

Look up the standard reduction value for each electrode in the table on the next page. All values are given for reduction, so you must change the sign of a metal that is oxidized. For example, if the voltage for the zinc reduction half-reaction is

$$Zn^{2+}(aq) + 2e^- \rightarrow Zn(s) \text{ is } -0.76 \text{ V,}$$

then the voltage for the oxidation reaction, $Zn(s) \rightarrow Zn^{2+}(aq) + 2e^-$, is -(-0.76) or +0.76 V. The $E^°_{cell}$ given for the solved problem above was calculated in this manner. Because Zn is oxidized its (-) sign in the table of reduction values was changed to (+). Copper is reduced in the reaction so its sign did not change. Using the equation $E^°_{cell} = E_{ox} + E_{red}$,

$$E^°_{cell} = 0.76 \text{ V} + 0.34 \text{ V} = 1.10 \text{ V}.$$

Standard reduction potentials

Choosing the right anode

You can use the table of standard reduction potentials to predict the output of a cell. Suppose you have a silver (Ag) cathode and you were given a choice to use either a magnesium (Mg) or an aluminum (Al) anode. When you calculate the cell potential with the equation $E^0_{cell} = E_{ox} + E_{red}$, you can see the Ag-Mg cell voltage is 3.18 V while the Ag-Al cell voltage is 2.46 V. An E^0_{cell} calculation helps you decide which metal to choose for the anode.

standard reduction potentials

reduction half-reaction	E^0 (V)	
$Li^+(aq) + e^- \rightarrow Li(s)$	-3.04	Strong reducing agent, strong tendency to be oxidized, strong tendency to give up electrons, weak oxidizing agent, tendency to occur in the reverse direction.
$Na^+(aq) + e^- \rightarrow Na(s)$	-2.71	
$Mg^{2+}(aq) + 2e^- \rightarrow Mg(s)$	-2.38	
$Al^{3+}(aq) + 3e^- \rightarrow Al(s)$	-1.66	
$2H_2O(l) + 2e^- \rightarrow H_2(g) + 2OH^-(aq)$	-0.83	
$Zn^{2+}(aq) + 2e^- \rightarrow Zn(s)$	-0.76	
$Cr^{2+}(aq) + 2e^- \rightarrow Cr(s)$	-0.74	
$Fe^{2+}(aq) + 2e^- \rightarrow Fe(s)$	-0.41	
$Ni^{2+}(aq) + 2e^- \rightarrow Ni(s)$	-0.23	
$Sn^{2+}(aq) + 2e^- \rightarrow Sn(s)$	-0.14	
$Pb^{2+}(aq) + 2e^- \rightarrow Pb(s)$	-0.13	
$Fe^{3+}(aq) + 3e^- \rightarrow Fe(s)$	-0.04	
$2H^+(aq) + 2e^- \rightarrow H_2(g)$	0.00	Reference half-cell
$Cu^{2+}(aq) + 2e^- \rightarrow Cu(s)$	0.34	Strong oxidizing agent, strong tendency to be reduced, strong tendency to attract electrons, weak reducing agent, tendency to occur in the forward direction.
$O_2(g) + 2H_2O(l) + 4e^- \rightarrow 4OH^-$	0.40	
$Cu^+(aq) + e^- \rightarrow Cu(s)$	0.52	
$Ag^+(aq) + e^- \rightarrow Ag(s)$	0.80	
$ClO_2(g) + e^- \rightarrow ClO_2^-(aq)$	0.95	
$O_2(g) + 4H^+(aq) + 4e^- \rightarrow 2H_2O(l)$	1.23	
$Cl_2(g) + 2e^- \rightarrow 2Cl^-O(aq)$	1.36	
$PbO_2(s) + 4H^+(aq) + 4e^- \rightarrow Pb^{2+} + 2H_2O(l)$	1.46	
$Au^{3+}(aq) + 3e^- \rightarrow Au(s)$	1.50	
$H_2O_2(aq) + 2H^+(aq) + 2e^- \rightarrow 2H_2O(l)$	1.78	
$F_2(g) + 2e^- \rightarrow 2F^-(aq)$	2.87	

Calculating cell voltage

Standard conditions for reduction potentials

The standard reduction potential table only lists the E^0_{red} voltages for single metals. That is because for a single metal, $E^0_{ox} = -E^0_{red}$. If you need an oxidation voltage, you can simply switch the sign on the reduction voltage. As you know, rate of chemical reactions depends on conditions such as temperature, pressure and concentration. Reduction potentials on the table are recorded under the following set of standard conditions:

- Temperature: 25 °C
- Pressure: 1 atmosphere (for gases)
- Concentration: 1 M (for solutions)

Standard reduction potential

The **standard reduction potential** labeled as $E°_{cell}$ is the cell voltage at 1.0 M and standard conditions. It is not possible to measure the potential of single electrodes, so the standard reduction potentials are based on the half-reaction: $2H^+ (aq) + 2e^- \rightarrow H_2(g)$. The standard potential assigned to this half-reaction is 0 V and the potential of every other cell is measured in comparison to hydrogen's value. Let's use the standard reduction potential table to determine the cell voltage of the zinc-copper galvanic cell shown.

Solved problem

Look up values for the standard reduction potentials to calculate the voltage of the galvanic cell: $Mg(s) | Mg^{2+}(aq) \| Cu^{2+}(aq) | Cu(s)$.

Asked Find E^0_{cell} of the reaction $Mg(s) + Cu^{2+}(aq) \rightarrow Mg^{2+}(aq) + Cu(s)$.

Given The cell reaction and the standard reduction potentials.

Solve Magnesium is oxidized and copper is reduced.

- Since magnesium is oxidized and copper is reduced, you must change the sign on the voltage for Mg from the Standard Reduction Potential table.

- From the table the standard potential of the reduction half reaction is:
  $Mg^{2+}(aq) + 2e^- \rightarrow Mg(s)$, with a voltage of -2.38 V.

  Therefore, the corresponding oxidation reaction is the reverse:
  $Mg(s) \rightarrow Mg^{2+}(aq) + 2e^-$, with a voltage of -(2.38 V) or +2.38.

  Therefore, $E^0_{ox} = +2.38V$.

- Copper is reduced: $Cu^{2+}(aq) + 2e^- \rightarrow Cu(s)$
  The half-reaction reduction value is: $E^0_{red} = +0.34V$.

- Use the equation: $E^0_{cell} = E^0_{ox} + E^0_{red}$

  Substitute the values for $E^0_{ox} = 2.38$ V and $E^0_{red} = 0.34$ V.

- $E^0_{cell} = 2.38$ V $+ 0.34$ V $= +2.72$ V

Answer The cell voltage is +2.72 V.

Cell emf spontaneity

Spontaneous and non-spontaneous cell emf

You can calculate the cell emf (E^0_{cell}) of a redox reaction using the standard potentials given in the standard reduction potential table. The sign of the calculated E^0_{cell} tells you whether the reaction is spontaneous or not. A positive E^0_{cell} indicates a spontaneous, product-favored reaction proceeds without additional energy input. A nonspontaneous reaction is one that favors reverse direction and requires additional energy input to proceed. A negative E^0_{cell} indicates a nonspontaneous, reactant-favored reaction.

$$E^0_{cell} > 0: \text{ spontaneous reaction}$$
$$E^0_{cell} < 0: \text{ nonspontaneous reaction}$$

Solved problem

Is the reaction $Cr(s) + Ni^{2+}(aq) \rightarrow Cr^{2+}(aq) + Ni(s)$ spontaneous under standard conditions?

Given The reaction and the table of standard reduction potentials.

Relationships Use the equation: $E^0_{cell} = E^0_{ox} + E^0_{red}$

- According to the table, the half-cell potentials are:

  Oxidation: $Cr(s) \rightarrow Cr^{2+}(aq) + 2e^-$ $E^0_{ox} = +0.74V$
  Reduction: $Ni^{2+}(aq) + 2e^- \rightarrow Ni(s)$ $E^0_{red} = -0.23V$

Solve Substitute values in the equation.

- $E^0_{cell} = 0.74 \text{ V} + (-0.23 \text{ V}) = 0.74 \text{ V} - 0.23 \text{ V} = 0.51 \text{ V}$

Answer Because the total cell voltage is a positive number, the reaction is spontaneous and it proceeds in the forward direction.

LITHIUM ION button cell batteries

The higher metal is oxidized and the lower metal is reduced

If you want to build a better battery, look no further than the Standard Reduction Potential table. The position of the half-reactions in the table tells us if the redox reaction will be spontaneous or not. The reactions at the top of the table have a tendency to occur in the reverse direction. The metals in these half-reactions tend to undergo oxidation, and are strong reducing agents. The reactions at the bottom of the table tend to occur in the forward direction. These metals have a propensity to attract electrons so they tend to undergo reduction and are strong oxidizing agents. A reduction half-reaction in the table will be spontaneous when it is paired with the reverse of the half-reaction above it.

Activity series

Activity series is listed by oxidation

You used the standard reduction potential table to determine cell emf. But if you only want to know whether a single replacement redox reaction will occur spontaneously or not, you can use the **activity series**. Sometimes called the reactivity series, the activity series lists metals in order of how easily the metal will lose electrons or become oxidized. If you want to know whether sodium atoms can replace aluminum in a compound by a spontaneous redox reaction, look for the two metals in the activity series. If the single metal is higher up in the list than the metal within the compound, a single replacement redox reaction will occur spontaneously. If the single metal is lower than the compound-bound metal, the reaction will not occur spontaneously. For example, a neutral iron atom will spontaneously replace a tin atom bound within a tin compound because iron appears higher up than tin in the activity series.

activity series of metals in aqueous solution

metal	oxidation reaction
lithium	$Li(s) \rightarrow Li^+(aq) + e^-$
potassium	$K(s) \rightarrow K^+(aq) + e^-$
barium	$Ba(s) \rightarrow Ba^{2+}(aq) + 2e^-$
calcium	$Ca(s) \rightarrow Ca^{2+}(aq) + 2e^-$
sodium	$Na(s) \rightarrow Na^+(aq) + e^-$
magnesium	$Mg(s) \rightarrow Mg^{2+}(aq) + 2e^-$
aluminum	$Al(s) \rightarrow Al^{3+}(aq) + 3e^-$
manganese(II)	$Mn(s) \rightarrow Mn^{2+}(aq) + 2e^-$
zinc	$Zn(s) \rightarrow Zn^{2+}(aq) + 2e^-$
chromium(III)	$Cr(s) \rightarrow Cr^{3+}(aq) + 3e^-$
iron(II)	$Fe(s) \rightarrow Fe^{2+}(aq) + 2e^-$
cobalt(II)	$Co(s) \rightarrow Co^{2+}(aq) + 2e^-$
nickel(II)	$Ni(s) \rightarrow Ni^{2+}(aq) + 2e^-$
tin(II)	$Sn(s) \rightarrow Sn^{2+}(aq) + 2e^-$
lead(II)	$Pb(s) \rightarrow Pb^{2+}(aq) + 2e^-$
hydrogen gas	$H_2(g) \rightarrow 2H^+(aq) + 2e^-$
copper(II)	$Cu(s) \rightarrow Cu^{2+}(aq) + 2e^-$
silver	$Ag(s) \rightarrow Ag^+(aq) + e^-$
mercury(II)	$Hg(l) \rightarrow Hg^{2+}(aq) + 2e^-$
platinum	$Pt(s) \rightarrow Pt^{2+}(aq) + 2e^-$
gold	$Au(s) \rightarrow Au^{3+}(aq) + 3e^-$

Solved problem

Determine if the reaction $3Na(s) + AlCl_3(s) \rightarrow 3NaCl(s) + Al(s)$ will occur spontaneously under standard conditions.

Given The reaction and the activity series

Relationships
- The reaction is spontaneous if the single reactant metal is higher than the element in the compound. Na is the single reactant metal, and Al is in a compound.

Solve
- Na is higher in the activity series than Al.

Answer Because the single metal is higher than the metal in a compound, sodium will replace aluminum in a spontaneous single replacement redox reaction.

More reactive metals replace less reactive metals

Think of the activity series as a list of individuals you are attracted to. Put the individual you are most attracted to at the top of the list, and add names beneath that individual in order of highest attraction to lowest attraction. Let's say you are at prom dancing with someone in the middle of the list when someone towards the bottom of the list asks to cut in. You may not want to switch dancing partners. But if someone towards the top of the list asks to cut in, you will switch dancing partners. The activity series is in order of reactivity with the most reactive metal at the top. A metal in a compound will be replaced by a more reactive metal above it.

Solved problem

Will zinc spontaneously replace calcium in calcium chloride?

Given The reaction and the activity series

Relationships
- The reaction is spontaneous if the single reactant metal is higher than the element in the compound. Zn is the single reactant metal, and Ca is in a compound.

Solve
- Ca is higher in the activity series than Zn.

Answer Because the single metal is lower than the metal in a compound, zinc will not spontaneously replace calcium.

Nernst equation

Battery consumption

Battery voltage is a function of its cell emf. As you know, batteries stop working after some time. This happens because the chemical energy stored in the battery eventually runs out. The more you use a battery, the more electrical energy you use. A battery uses redox reactions to convert chemical energy to electrical energy. As the redox reactions proceed in an electrochemical cell, the relative amounts of reactants and products changes.

Voltage output depends on Q

Redox reactions that occur in batteries are like any other chemical reaction when you consider the reactants are used up as products are formed during the reaction. This means the concentration of each substance in the cell changes as long as the reaction continues. As reactants get used up, their concentration decreases while the concentration of the products increases. The voltage output of a battery, E_{cell}, depends on the concentration of reactants and products. The relationship that describes this dependency is the **Nernst equation**:

$$E_{cell} = E^0_{cell} - \frac{0.592}{n} \log(Q) \qquad Q = \frac{\text{product concentration}}{\text{reactant concentration}}$$

where E^0_{cell} is the standard cell potential and n is the number of moles.

Voltage output is cell emf

In the cell voltage equation shown above, the quantity n is the number of moles of electrons produced by the redox reaction. The quantity Q is the ratio of product to reactant concentrations. Under standard conditions, $Q = 1$. If $\log(1) = 0$, then $E_{cell} = E^0_{cell}$.

Discharging a battery

The concentration of the products increases and the concentration of reactants decreases when using a battery. This results in Q greater than 1 and $E_{cell} < E^0_{cell}$. This will decrease the battery voltage. As the battery is discharged, it eventually runs out of energy.

Charging a battery

The concentration of reactants increases and the concentration of products decreases when recharging a battery. This causes Q to decrease and drives E_{cell} towards E^0_{cell}. The battery becomes fully recharged when $Q = 1$ and $E_{cell} = E^0_{cell}$.

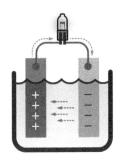

DISCHARGING
reactants decrease
products increase

Q increases
log(Q) increases
E_{cell} decreases

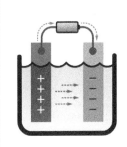

CHARGING
reactants increase
products decrease

Q decreases
log(Q) decreases
E_{cell} increases

Chapter 18

Section 3 review

The difference in electrical potential between the anode and cathode in an electrochemical cell is called the electromotive force, or emf. The emf of an electrochemical cell can be used in a Gibbs free energy calculation to determine the maximum amount of work and energy the cell can provide. The total cell voltage is found by combining the potential at each half-cell. Those values for half-cells can be looked up in a Table of Standard Reduction Potentials. The sign of the cell emf determines whether the redox reaction is spontaneous or nonspontaneous. Another way to determine if a given single replacement redox reaction will occur spontaneously is to compare how easily the two metals involved become oxidized. This information is found in the activity series of metals. The voltage output of a battery decreases as the redox reactions proceed. The Nernst equation is used to calculate the output of a battery based on the concentration of reactants and products.

Vocabulary words	electromotive force (emf), cell emf, standard reduction potential, activity series, Nernst equation

Key relationships

- Gibbs free energy, $\Delta G = -n \times F \times E^0_{cell}$
- Faraday constant, $F = 96{,}000 \; \frac{C}{mol}$
- Cell voltage, $E^0_{cell} = E_{ox} + E_{red}$
- The unit volts, V is equivalent to $\frac{J}{C}$
- Oxidized species change their sign when retrieving a value from the Table of Standard Reduction Potentials (in the Appendix).
- $E^0_{cell} \; 0$ = Spontaneoous reaction
- $E^0_{cell} > 0$ = Nonspontaneous reaction
- Nernst equation, $E_{cell} = E^0_{cell} - \frac{0.0592}{n} \log(Q)$
- When a battery is used up, [products] increase while [reactants] decrease: $E_{cell} < E^0_{cell}$ and $Q > 1$
- When a battery is recharged, [products] decrease while [reactants] increase: $E_{cell} = E^0_{cell}$ and $Q = 1$ when the battery is fully charged

Review problems and questions

1. Predict the cell voltage produced by the reaction in an electrochemical cell:
   $Cr^{2+}(aq) + Cu(s) \rightarrow Cu^{2+}(aq) + Cr(s)$.

2. Explain whether the above reaction is spontaneous or not.

3. Calculate the total free energy available when 32.0 g Cu react.

4. Use the Activity Series (in the Chapter) to determine whether this reaction will occur spontaneously or not: $CaCl_2(s) + Zn(s) \rightarrow Ca(s) + ZnCl_2(s)$.

Batteries of All Shapes and Sizes

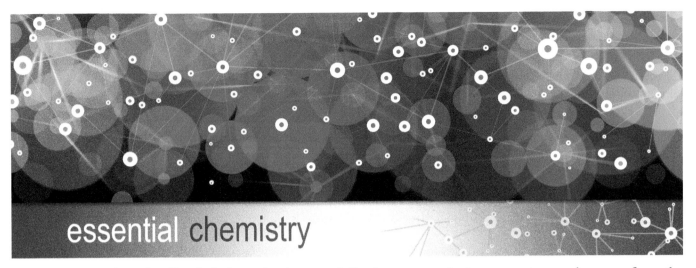

essential chemistry

The electrochemistry in all voltaic batteries is essentially the same – electrons spontaneously move from the anode to the cathode to power a device and complete a redox reaction. If all batteries are basically the same, why are there so many different shapes and sizes of batteries that you can buy?

The design of batteries has changed significantly over the years. Alexander Volta designed and built the first electrochemical battery called the voltaic pile. It was simply a series of alternating copper and zinc plates separated by salt-water soaked disks. This simple design lead to more complex explorations in electrochemistry. But it was not a commercially applicable design.

Voltaic Pile   Daniell Cell

The first battery to be a practical source of electricity was a Daniell cell invented in 1836 by John Daniell. This voltaic cell also used zinc and copper metals. At the cathode, copper is immersed in a solution of copper(II) sulfate. At the anode, zinc is immersed in a solution of zinc sulfate or sulfuric acid. The overall reaction for the Daniell cell is:

$$Zn(s) + Cu^{2+}(aq) \rightarrow Zn^{2+}(aq) + Cu(s)$$

Over time the physical design of the Daniell cell was modified to reduce cost and generate stronger currents. One variation of the Daniell cell was called a gravity cell. A copper electrode was placed on the bottom of a large glass jar, and a zinc electrode was suspended near the top. The bottom of the jar was scattered with solid copper(II) sulfate, and then it was filled with water. When the circuit was connected, the zinc oxidized into zinc sulfate and the copper(II) sulfate reduced to copper. As long as a current was drawn, the copper and zinc solutions would stay separate based on their density. This became a battery of choice for some industry applications but it could only be used if it was kept stationary.

Batteries of All Shapes and Sizes - 2

When you use a battery, you probably think about a portable or mobile device. A gravity cell would definitely not work in your flashlight! Some of the more common batteries now are alkaline batteries. Alkaline batteries get their name from the basic alkaline paste that is used in their construction. This type of battery uses zinc and manganese dioxide for the oxidation and reduction reactions. At the anode, zinc powder is dispersed in a gel containing potassium hydroxide (the basic, alkaline electrolyte). During the reactions the zinc is oxidized to zinc oxide according to the following equation:

$$Zn(s) + 2OH^-(aq) \rightarrow ZnO(s) + H_2O(l) + 2e^-$$

At the positive electrode, the manganese(IV) oxide is formed into a paste with carbon powder to increase conductivity. It gets reduced to manganese(III) oxide according to the following equation:

$$2MnO_2(s) + H_2O(l) + 2e^- \rightarrow Mn_2O_3(s) + 2OH^-(aq)$$

This happens inside a steel can with a button top (metal top cover) to collect the current and be the positive end of the battery. The overall reaction for an alkaline battery is:

$$Zn + 2MnO_2 \rightarrow ZnO + Mn_2O_3$$

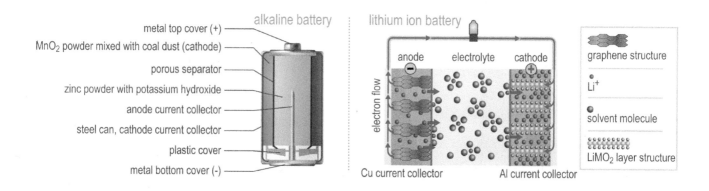

Rechargeable batteries are becoming more popular as they are used in phones and computers. Lithium-ion chemistry is common for rechargeable batteries. Lithium ions are shuffled back and forth between the anode and the cathode as the battery is discharged and recharged. As the lithium ions move, a stoichiometric number of electrons move externally through your device. In a typical lithium ion battery, graphite is used at the anode and lithium cobalt oxide (LiCoO$_2$) is used at the cathode. The anode and cathode are separated by an electrolytic solvent, gel, or polymer that allows the lithium ions to move back and forth.

There are advantages and disadvantages to using lithium-ion batteries. The first advantage is pretty obvious – they are rechargeable! Can you imagine if you had to replace your phone battery every day? In addition to saving you the expense of having to buy new batteries, rechargeable batteries generate less waste. However, lithium–ion batteries are not perfect. You may have noticed that your phone holds less charge over time. This is because an internal resistance can develop which will reduce the ability to deliver current. If there is an internal malfunction, lithium-ion batteries can short circuit and catch fire. This problem is solved with a current interrupting device, and by taking appropriate safety precautions.

Batteries of All Shapes and Sizes - 3

The oxidation and reduction reactions in a small, light lithium-ion battery can generate 3.6 V. This means that these batteries provide a high amount of energy relative to their volume, or a high energy density. Their energy density is much higher than alkaline batteries, older rechargeable batteries made from nickel, or lead acid batteries used to start cars. But the energy density of a battery is still less than the energy density of food or a fuel, like gasoline.

Energy density, safety, and impacts on natural resources are all considerations for battery developers. An essential understanding of electrochemistry can help as we move toward a future with more and more electric devices and vehicles.

Chapter 18 review

Vocabulary
Match each word to the sentence where it best fits.

Section 18.1

ampere	coulomb (C)
electric current	ohm (Ω)
Ohm's law	resistance
volt	voltage

1. The _____ is the unit of electrical charge.
2. A(n) _____ is produced when electrical charges are moved.
3. The unit of voltage is the _____.
4. _____ is the difference in electrical pressure or potential between two objects.
5. The _____ is used to measure electrical resistance.
6. As electrical charges move in a conductor they experience _____ because they participate in many collisions with other particles.
7. The SI unit of electrical current is the _____.
8. The relationship between current, voltage and resistance is described as _____.

Section 18.2

anode	cathode
corrosion	electrochemical cell
electrode	electrolysis
electrolyte	electrolytic cell
electroplating	galvanic cell
nonspontaneous	salt bridge
spontaneous	voltaic cell

9. Reduction occurs at the _____ of the electrochemical cell.
10. A(n) _____ is the site of energy conversion from electrical energy to chemical energy.
11. Oxidation occurs at the _____ of the electrochemical cell.
12. The connection between the oxidation and the reduction half cells called the _____ allows ions to move back and forth between electrolytes.
13. The _____ is the solution where the cell electrodes are immersed.
14. _____ causes a metal to lose mass during a redox reaction.
15. The anode and cathode are examples of a(n) _____, the site of oxidation and reduction reactions.
16. Spontaneous chemical reactions generate electricity in a(n) _____.
17. A device called a(n) _____ is also known as a voltaic cell.
18. A redox reaction that requires no energy input is said to be _____.
19. A nonspontaneous, induced redox reaction termed _____ occurs in an electrolytic cell.
20. A(n) _____ reaction requires an energy input in order to proceed in a direction that favors reactants.
21. A(n) _____ is a device in which electrical energy is used to produce a chemical reaction.
22. The process that uses electricity to oxidize anions or reduce cations called _____ results in a thin metal coating on an object.

Section 18.3

activity series	cell emf
electromotive force (emf)	Nernst equation
standard reduction potential	

23. The _____ is the difference in the electrical potential of the anode and cathode.
24. The cell voltage at 25 °C, 1 atmosphere, and 1M is called the _____.
25. A list of metals that helps you determine whether a single replacement redox reaction will occur spontaneously or not is called a(n) _____.
26. The abbreviation for the electromotive force of a cell is the _____.
27. The output of a battery based on the concentration of reactants and products is calculated using the _____.

Conceptual questions

Section 18.1

28. What causes an electric current?
29. Explain how electrons are pushed through a circuit and name the force that aids in the process.
30. Voltage is sometimes referred to as potential. Explain why.
31. Provide the correct units for each of the following:
    a. charge
    b. current
    c. voltage
    d. resistance
32. Explain how a lemon, a piece of copper, and a piece of magnesium are able to function like a battery.
33. How is the current affected if the voltage is decreased when resistance is high? Explain.
34. Relate the components of a running river to the components of an electrical system.
35. Does sugar water conduct electrical current? Why or why not?

Chapter 18 review

Section 18.1

36. Does salt water conduct electrical current? Why or why not?

37. Do acids conduct electricity in solution? Why or why not? What about bases?

38. If a charge goes from a higher voltage to a lower voltage, does it lose or gain energy? Explain your answer.

Section 18.2

39. Describe what occurs at the cathode with an example of a chemical process.

40. Which type of cell undergoes a redox reaction without the addition of energy? Explain.

41. Describe what occurs at the anode with an example of a chemical process.

42. How is an electrolytic cell similar to a galvanic cell? Explain your answer.

43. What are the benefits of using a salt bridge in an electrochemical cell?

44. Sketch a battery and label the parts that allow it to be considered a galvanic cell. Include the anode, cathode, electrolyte, and the positive and negative electrode.

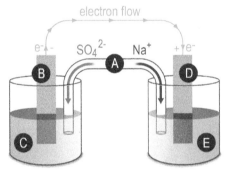

45. Using the given diagram, answer the following questions (letters may be used more than once).

    a. What is the name and role of A in the diagram?

    b. Does the diagram represent an electrolytic cell or a galvanic cell? How do you know?

    c. Which two letters represent electrodes?

    d. Which two letters represent electrolyte solution?

    e. Which letter in the diagram represents the site of reduction?

    f. Which letter in the diagram represents the site of oxidation?

    g. Which letter in the diagram is experiencing electron gain?

    h. Which letter represents a transfer of mass to the electrode from its solution?

    i. Which electrode undergoes corrosion?

46. Which of the following terms is most appropriate to describe electroplating: spontaneous, or nonspontaneous? Explain your reasoning.

47. What are the two names for the cells that convert chemical energy into electrical energy?

48. Describe the reason that the cathode is positive in a galvanic cell and and negative in an electrolytic cell.

49. Describe electroplating and why it is used for high quality electronic components.

50. Hydrogen gas is said to be the energy source of the future. Why is it not widely used today?

51. Describe electrolysis and how the process is used to produce chlorine gas.

52. What direction (right or left) do the electrons flow in the electrolytic cell shown below? How do you know?

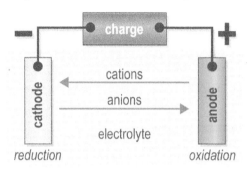

53. How can a rechargeable battery act as both an electrolytic and a galvanic cell?

54. According to the following half reaction of an electrochemical cell,

$$Na(s) \rightarrow Na^+(aq) + e^-,$$

has sodium undergone oxidation or reduction? How do you know?

55. According to the following half reaction of an electrochemical cell,

$$F_2(g) + 2e^- \rightarrow 2F^-(aq),$$

has flourine undergone oxidation or reduction? How do you know?

Section 18.3

56. What does the Gibbs free energy calculate for an electrochemical cell?

57. When you are calculating the Gibbs free energy, what units should $E°_{cell}$ be measured in?

58. What is the formula for determining the total cell voltage, $E°_{cell}$?

59. Using the standard reduction potential table, how can you tell which metals will be reduced?

Chapter 18 review

Section 18.3

60. When using the standard reduction potential table, why do you have to change the sign in front of the numerical value for a metal that is being oxidized?

61. How do variations in reactant concentration and product concentration affect the voltage of a galvanic cell? Explain your answer.

62. Using the activity series table, will potassium spontaneously replace barium in barium chloride?

63. Using the activity series table, determine if the reaction $Co(s) + ZnCl_2(s) \rightarrow CoCl_2(s) + Zn(s)$ will occur spontaneously under standard conditions.

64. In the standard reduction potential table, which metals are strong oxidizing agents? Those at the top, middle, or bottom?

65. In the standard reduction potential table, where can you find the chemicals that have a strong tendency to give up electrons? Those at the top, middle, or bottom?

66. When is Q equal to one in the Nernst equation? Explain your answer.

67. Describe the relationship between cell voltage and spontaneous reactions.

Quantitative problems

Section 18.2

68. Draw a diagram of a complete Ag/Mg galvanic cell.

    a. Label the cathode, anode, and salt bridge.

    b. Label the metal electrodes and their ions. Indicate which metal is being oxidized and which is being reduced.

    c. What voltage will a standard Ag/Mg galvanic cell produce (Section 18.3)?

69. Write the half reactions that occur at the cathode and at the anode for the following cells.

    a. $Zn(s) | Zn^{2+} \| Ag^{2+}(aq) | Ag(s)$

    b. $Ni(s) | Ni^{2+} \| Au^{3+}(aq) | Au(s)$

    c. $Fe(s) | Fe^{2+} \| Cu^{2+}(aq) | Cu(s)$

70. Sketch an electrochemical cell with the following reaction equation: $Zn(s) + Ni^{2+} \rightarrow Zn^{2+}(aq) + Ni(s)$

    a. Label the anode and the cathode and identify the metal on each.

    b. Use either NaCl or KCl as the salt bridge. Write the ions that form in the salt bridge and label the direction each ion moves.

    c. Use an arrow to indicate the direction of electron flow through a wire.

71. An electrode in a galvanic cell has changed mass from 0.32 g to 0.46 g after several minutes.

    a. Is this electrode a cathode or an anode?

    b. How do you know?

72. Given the following net ionic equation:

$$Fe(s) + Au^{3+}(aq) \rightarrow Fe^{3+}(aq) + Au(s)$$

determine which element is oxidized and which element is reduced. In addition, convert the reaction into electrochemical cell notation.

73. Given the following net ionic equation:

$$Pb(s) + 2Cu^{+}(aq) \rightarrow Pb^{2+}(aq) + 2Cu(s)$$

determine which element is oxidized and which element is reduced. In addition, convert the reaction into electrochemical cell notation.

Section 18.3

74. You are building a lithium/gold electrochemical cell. Use the standard reduction potential table to determine which element is most likely to be oxidized and which element will be reduced, then write the electrochemical cell notation.

75. You are building an aluminium/silver electrochemical cell. Use the standard reduction potential table to determine which element is most likely to be oxidized and which element will be reduced, then write the electrochemical cell notation.

76. Using standard conditions, calculate the cell voltage for each of the following; you may have already completed the half-reactions in a previous problem:

    a. $Zn(s) | Zn^{2+} \| Ag^{2+}(aq) | Ag(s)$

    b. $Ni(s) | Ni^{2+} \| Au^{2+}(aq) | Au(s)$

    c. $Fe(s) | Fe^{2+} \| Cu^{2+}(aq) | Cu(s)$

77. Calculate the voltage of the following cell and determine whether it is spontaneous or nonspontaneous. Explain your answer.

$$Zn(s) | Zn^{2+}(aq) \| Cl^{-}(aq) | Cl_2(g)$$

78. Calculate the cell voltage of the following cell, and determine whether the reaction is spontaneous or nonspontaneous. Explain your answer.

$$Al^{3+}(aq) + Au(s) \rightarrow Au^{3+}(aq) + Al(s)$$

79. Calculate the cell voltage of the following cell, and determine whether the reaction is spontaneous or nonspontaneous. Explain your answer.

$$2Ag^{+}(aq) + Zn(s) \rightarrow Zn^{2+}(aq) + 2Ag(s)$$

Chapter 18 review

Section 18.3

80. In an electrochemical cell, the following reaction occurs at the anode:

$$Na(s) \rightarrow Na^+(aq) + e^-$$

and at the cathode, the following reaction occurs:

$$Ag^+(aq) + e^- \rightarrow Ag(s)$$

What is the total energy that can be obtained from 6.71 g of silver ions in this cell?

81. Calculate the total energy that can be obtained from 4.15 g of copper ions in an Sn/Cu cell. The reaction is:

$$Sn(s) + Cu^{2+}(aq) \rightarrow Sn^{2+}(aq) + Cu(s).$$

82. Calculate the total energy that can be obtained from 6.49 g of gold ions in a Mg/Au cell. The reaction is:

$$3Mg(s) + 2Au^{3+}(aq) \rightarrow 3Mg^{2+}(aq) + 2Au(s).$$

83. Of the following reduction reactions occurring at the cathode, which will produce the lowest cell voltage under standard conditions?

   a. $Ag^+ (aq) + 1e^- \rightarrow Ag(s)$

   b. $Ni^{2+} (aq) + 2e^- \rightarrow Ni(s)$

   c. $Au^{3+} (aq) + 3e^- \rightarrow Au(s)$

   d. $Zn^{2+} (aq) + 2e^- \rightarrow Zn(s)$

84. Given the following information, use the Nernst equation to solve for E_{cell}.

   a. $E°_{cell} = 0.93$ V

   b. n = 2

   c. Q = 2.5

85. Given the following information, use the Nernst equation to solve for $E°_{cell}$.

   a. $E_{cell} = 2.93$ V

   b. n = 3

   c. Q = 4.72

86. If the product concentration is equal to 15 M, and the reactant concentration is equal to 3.6 M, what is the value of Q?

87. If Q is equal to 0.75, and the reactant concentration is equal to 9.8 M, what is the product concentration?

88. If the product concentration is equal to 21 M, and Q is equal to 2.6, what is the value of the reactant concentration?

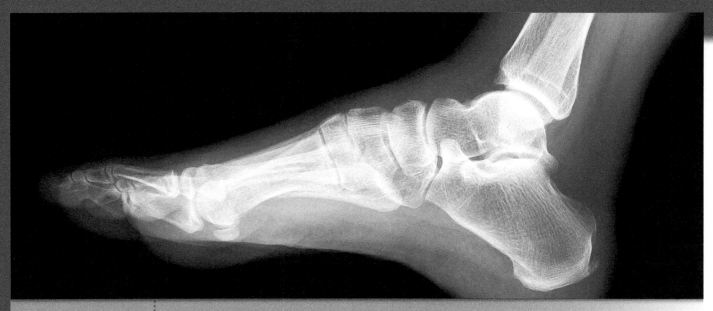

chapter 19 Nuclear Chemistry

Most of the chemistry you have studied involves the outer electrons of atoms. Electrons are lost, gained, or shared to form bonds. The electrostatic attraction of electrons and the strength of bonds affect the physical and chemical properties of the matter that we observe. We don't think about the nuclei of atoms during normal chemical reactions because they stay intact. Let's imagine we could magnify one atom to the size of a soccer stadium. The nucleus would be a little, dense marble-sized mass filled with positive charge in the middle of the stadium. The electrons would be whirling around in the bleachers, far from the nucleus.

But the nucleus of an atom can change! And when it does, it can have profound impacts. In fact, nuclear chemistry and nuclear processes are used by many products and devices that we take for granted in our daily lives. The smoke detector in your home uses a small nuclear reaction to keep you safe from a fire. Nuclear radiation is used in medical techniques that help us make medical diagnoses, treat illnesses, and study drug effectiveness. Even our understanding of history and the human condition is aided by nuclear chemistry. Archaeologists date ancient objects by using nuclear chemistry techniques.

The first nuclear weapons were developed in Los Alamos, New Mexico. This occurred during World War Two and was part of a top secret mission called the Manhattan Project. Many of the world's greatest scientists gathered to split an atom and release the immense energy inside of an atomic nucleus.

Nuclear reactions generate much more energy per mol than a normal chemical reaction. This energy is a great source of power, and potentially a great risk. Life on earth is only possible because the energy from nuclear reactions in the Sun drives photosynthesis. Nuclear energy can also be harnessed in the form of nuclear weapons.

Development of nuclear technology presents us with major considerations of the trade-offs between benefit and risk. It is important for you to learn about the essentials of nuclear chemistry to help you accurately evaluate its impact on our lives.

Chapter 19 study guide

Chapter Preview

Nuclear energy and radiation appear in popular culture generally as the source of a super hero's powers or a catastrophe that threatens many lives. But what are nuclear energy and radiation? Unlike chemical reactions, nuclear reactions are changes to the nucleus that can change one element into another. A nucleus can decay in several different ways, emitting different types of radiation. In this chapter you will learn about what makes some nuclei more stable than others, and how Einstein's famous equation calculates the energy released in nuclear reactions. You will also explore the difference between nuclear fission and nuclear fusion and the actual dangers radiation can pose to living things.

Learning objectives

By the end of this chapter you should be able to:

- compare properties of stable and unstable nuclei;
- explain why nuclear reactions exchange far more energy than ordinary chemical reactions;
- write complete nuclear equations for the decay of isotopes;
- describe the half-life of a radioactive isotope and how it can be used to date matter;
- determine the amount of radioactive sample remaining after a certain amount of time;
- distinguish between nuclear fission and nuclear fusion; and
- explain how nuclear radiation affects biological matter.

Investigations

19A: Half-lives

19B: Isotopic composition

Pages in this chapter

- 602 Nuclear Chemistry
- 603 Isotopes and the periodic table
- 604 Writing nuclear reactions
- 605 The stable nuclei and radioactivity
- 606 Nuclear binding energy
- 607 Nuclear reactions involve new particles
- 608 Mass and energy in nuclear reactions
- 609 Section 1 Review
- 610 Radioactivity
- 611 Alpha decay
- 612 Beta decay
- 613 Gamma decay and decay chains
- 614 Half life and the decay equation
- 615 Solving radioactive decay problems
- 616 Carbon dating
- 617 Choosing an isotope for dating
- 618 Nuclear decay graphs
- 619 Section 2 Review
- 620 Nuclear Energy
- 621 Fission reactions
- 622 Chain reactions
- 623 Nuclear energy
- 624 Fusion reactions
- 625 Fusion power
- 626 Section 3 Review
- 627 Radiation
- 628 The intensity of radiation
- 629 Radiation exposure
- 630 X-rays and CAT scans
- 631 Section 4 Review
- 632 Radioactive Tracers
- 633 Radioactive Tracers - 2
- 634 Radioactive Tracers - 3
- 635 Chapter review

Vocabulary

nucleons	atomic number (Z)	isotopes
mass number (A)	radioactivity	binding energy
positron	antimatter	mass defect
radiation	alpha decay	beta decay
beta radiation	positron emission	gamma decay
half-life	rate constant	rate of decay
carbon dating	carbon-14	fission reaction
chain reaction	critical mass	fusion reaction
ionization	ionizing radiation	non-ionizing radiation
intensity	inverse square law	watt
radiation dose		

19.1 - Nuclear Chemistry

Every other reaction in this book involves electrons - with the exception of the H⁺ ion which is a bare proton. Because they involve electrons chemical reactions do not change the identity of any element involved. Chemical reactions rearrange elements into new compounds leaving the elements themselves unchanged. Nuclear reactions can change an element into a different element. Nuclear reactions fundamentally involve the **nucleons**, and electrons are only participants in the sense that they balance the electric charges.

Comparing chemical and nuclear reactions

Chemical reactions

Chemical reactions change chemical bonds by moving electrons around. Aerobic respiration is a good example of a chemical reaction that rearranges the chemical bonds between carbon, oxygen, and hydrogen atoms. If you react one kilogram of glucose and oxygen in the correct proportion you get one kilogram of water and carbon dioxide, and release 6,126 kJ of energy. The same carbon, oxygen, and hydrogen atoms exist before and after the reaction.

Comparing nuclear and chemical reactions

		Energy release
Aerobic respiration (chemical)	$C_6H_{12}O_6 + 6O_2 \longleftrightarrow 6CO_2 + 6H_2O$ A chemical reaction rearranges chemical bonds changing compounds into other compounds	6,126 kJ/kg
Deuterium-tritium fusion (nuclear)	$^2_1H + ^3_1H \longrightarrow ^4_2He + ^1_0n$ A nuclear reaction rearranges particles in the nucleus - changing elements into other elements	337,445,000,000 kJ/kg

Nuclear reactions

An example of nuclear reaction is the fusion of two isotopes of hydrogen, deuterium (^2_1H) and tritium (^3_1H), to make helium (^4_2He). The elements on the reactant side of the equation (hydrogen) are different than the elements on the product side of the reaction (helium). A nuclear particle, a neutron, is also released. Nuclear reactions change elements and may involve individual particles such as protons, neutrons, neutrinos, and even antimatter particles. Fusion of one kilogram of deuterium and tritium releases 337 billion kilojoules!

Energy

The forces that hold the nucleus together are much stronger than the electromagnetic forces that act between electrons. Stronger forces mean higher energy must be expended to overcome them, or higher energy is released when the forces are allowed to act. Notice that the nuclear energy release per kilogram is 55 million times greater compared to the chemical reaction with the same quantity of matter.

Nuclear reactions convert mass into energy

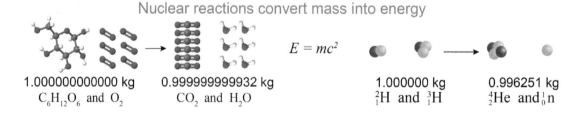

1.000000000000 kg	0.999999999932 kg		1.000000 kg	0.996251 kg
$C_6H_{12}O_6$ and O_2	CO_2 and H_2O	$E = mc^2$	^2_1H and ^3_1H	^4_2He and ^1_0n

Mass - energy conversion

Nuclear reactions convert a significant fraction of mass to energy. Chemical reactions also convert mass to energy but the amount of mass converted is so small it cannot be detected. The deuterium-tritium fusion reaction converts 3.749 grams to energy.

Isotopes and the periodic table

Isotopes

Isotopes are atoms of the same element with the same number of protons and different numbers of neutrons. For example, all carbon atoms have six protons but there are two stable isotopes of carbon. One isotope, carbon-12, has six neutrons and the other isotope, carbon-13, has seven neutrons. Isotopes behave the same in chemical reactions but undergo different nuclear reactions. A third (rare) isotope, carbon-14, has eight neutrons and is radioactive.

Writing isotopes

There are several ways to write an isotope. One is to use a dash and the **mass number (A)**. The mass number is the total number of protons and neutrons in a nucleus. Carbon-14 tells you there are 14 particles in the nucleus, and since carbon has six protons, there must be 8 neutrons. A more compact notation uses one or two numbers added to the usual chemical symbol to indicate the isotope. For example, $^{14}_{6}C$ means carbon-14 which can also be written with both the **atomic number (Z)** and mass number as below.

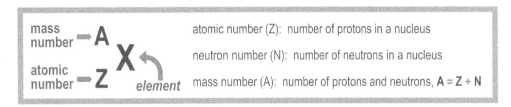

Naturally occurring isotopes

The chart below shows the isotopes that occur naturally for the first 54 elements. Note that some elements such as beryllium have only a single isotope $^{9}_{4}Be$. Other elements have more isotopes; magnesium has three isotopes ($^{24}_{12}Mg$, $^{25}_{12}Mg$, $^{26}_{12}Mg$) and calcium has five isotopes ($^{40}_{20}Ca$, $^{42}_{20}Ca$, $^{43}_{20}Ca$, $^{44}_{20}Ca$, $^{46}_{20}Ca$). Technetium (Tc) has no stable naturally-occuring isotopes.

Solved problem

How many neutrons are in an atom of the isotope $^{61}_{28}Ni$?

Relationships   Ni has atomic number 28. The mass number is A + Z

Solve   Given the mass number is 61, there are 61-28 = 33 neutrons.

Answer   33 neutrons

Stable isotopes for the first 54 elements
stable isotopes occur naturally and do not decay (not radioactive)

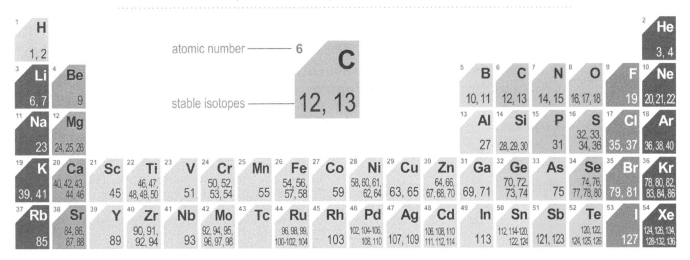

Section 19.1: Nuclear Reactions

Writing nuclear reactions

Nuclear equations

Nuclear reactions are described by equations similar to chemical equations. The difference is that the elements may not be the same on both sides of the reaction. Two hydrogen atoms on the reactant side can become one helium atom on the product side of a nuclear reaction. This is one of the reactions that occurs in the Sun to produce energy. The diagram below shows several nuclear reactions that occur in the Sun.

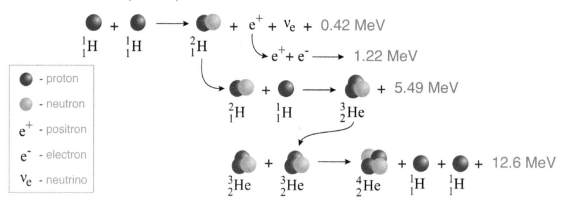

The proton-proton chain of nuclear reactions in the sun

Writing isotopes in equations

Nuclear reactions involve isotopes, and the chemical symbols are written with the atomic number as a subscript and the mass number as a superscript before the element symbol. For example, the above reaction involves two isotopes of helium: helium-3, written $^{3}_{2}He$, and helium-4, written $^{4}_{2}He$. It is also common to put both the atomic number and mass number preceding the element symbol. In this book we will just use the mass number since the element symbol already implies the atomic number.

Writing isotopes in equations for nuclear reactions

mass number
element symbol
atomic number

helium-3
^{3}He or $^{3}_{2}He$

helium-4
^{4}He or $^{4}_{2}He$

Conservation rules

Just as chemical equations conserve mass and atoms, nuclear equations also obey conservation laws. However, the properties that are conserved are different.

1. The total charge of the reactants must equal the total charge of the products.
2. The total mass + energy of the reactants must equal the total mass + energy of the products.
3. The sum of the atomic numbers must be the same before and after the reaction.

For example, when boron-10 is bombarded with a neutron it yields lithium-7 and an alpha particle.

Conservation in Nuclear Reactions

reactants
Mass = 11
Atomic Number = 5

$^{10}_{5}B + ^{1}_{0}n \rightarrow ^{7}_{3}Li + ^{4}_{2}He$

products
Mass = 11
Atomic Number = 5

The stable nuclei and radioactivity

Stability and instability

Inside the nucleus there are two competing forces. The protons repel each other with the electromagnetic force. Protons and neutrons attract each other with the strong nuclear force. If there are enough neutrons, the attractive forces are collectively stronger than the repulsive forces and the nucleus stays together. We use the word stable to describe a nucleus that stays together. On the opposite side of the balance, if there are too many protons compared to neutrons, the repulsive forces eventually win and the nucleus comes apart. Marie Curie coined the word **radioactivity** to describe elements for which the nucleus is unstable, and spontaneously changes to become more stable.

Radioactivity
An unstable nucleus spontaneously changes - and may emit particles and/or energy

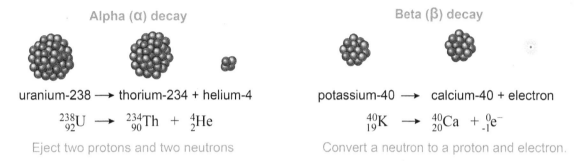

Alpha (α) decay

uranium-238 → thorium-234 + helium-4

$^{238}_{92}U \rightarrow \, ^{234}_{90}Th + \, ^{4}_{2}He$

Eject two protons and two neutrons

Beta (β) decay

potassium-40 → calcium-40 + electron

$^{40}_{19}K \rightarrow \, ^{40}_{20}Ca + \, ^{0}_{-1}e^{-}$

Convert a neutron to a proton and electron.

Neutron to proton ratio

The graph below shows the number of protons on the vertical axis and the number of neutrons on the horizontal axis. As the atomic number increases, it takes more neutrons to maintain stability. All the stable nuclei are shown with green dots. You can see that the green dots follow a curve in which the neutron-to-proton ratio is nearly 1:1 for light elements. After element 16 the curve of stable nuclei bends down toward a neutron ratio of 1.5:1. To maintain stability, there are more neutrons than protons as the elements get heavier.

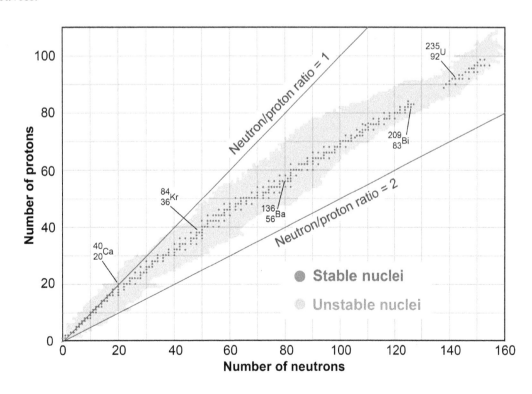

Section 19.1: Nuclear Reactions

Nuclear binding energy

Binding energy

Protons and neutrons stick together in the nuclei of atoms because a nucleus is a much lower energy arrangement than separate protons and neutrons. The **binding energy** is the energy difference between the nucleus compared to its constituent parts. For example, two protons and two neutrons have far higher energy as individual particles compared to when they are bound together in a helium nucleus. The energy that holds the nucleus together comes from the strong nuclear force. You can think of the binding energy as the height of a hill that the particle has to climb to escape the nucleus. A large binding energy means each proton and neutron is tightly held to the nucleus.

The nuclear binding energy curve

Protons and neutrons are referred to as nucleons because both particles are in the nucleus. The graph below shows the binding energy per nucleon. Since each element has a different number of nucleons, the binding energy per nucleon is a better measure of how tightly each particle is held to the nucleus. Notice how the graph slopes up steeply. Find iron-56 at the highest point, which has 26 protons and 30 neutrons.

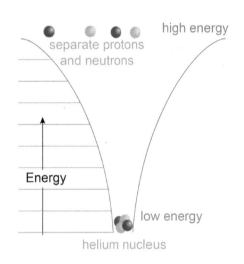
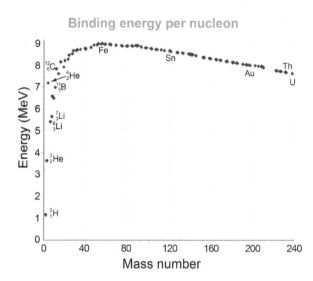

The most stable element is iron

As you can see in the graph, nucleons are bound most tightly together near iron. Elements with the highest binding energy per nucleon (BE/n) have the most stable nuclei. Iron, which has the highest binding energy per nucleon, is the most stable element. Iron can be easily found all over the planet because of its stability.

stable nuclei have high binding energy

The iron nucleus

Iron's stability is not beneficial when it comes to nuclear energy. This is because iron's nucleons (protons and neutrons) have the highest binding energy. As you can see on the graph above, other nuclei are less tightly bound. We can only get energy from nuclei that appear to the left or to the right of iron on the graph.

iron is a poor candidate for nuclear energy due to its tightly bound nucleus and high binding energy

Nuclear reactions involve new particles

The "other" particles

Ordinary atoms consist of protons, electrons, and neutrons. Once scientists discovered that atoms were able to split and change into other elements, it was also discovered that there are other particles besides electrons, protons, and neutrons. We do not normally experience most of the direct evidence for these particles, but they exist, and to understand nuclear reactions we need to introduce them.

Neutrinos

The lightest known particle, the neutrino, has no charge and almost never interacts with ordinary matter. In fact, more than 65 billion neutrinos pass through every square centimeter of Earth, including you, every second. Neutrinos are represented with the Greek letter gnu, ν. Neutrinos come in three types: electron neutrinos, muon neutrinos, and tau neutrinos. Electron neutrinos are produced in the radioactive nuclear reaction called beta decay.

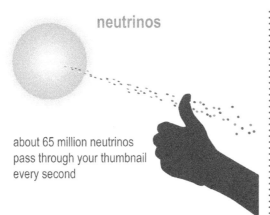

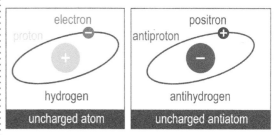

Neutrinos help us study the universe

Most neutrinos came into existence at the same time the universe formed but new ones continue to be formed as a product of nuclear reactions. Because neutrinos travel at the speed of light, lack a charge, and are only affected by gravity, they are perfect candidates to help us understand the universe. However, neutrinos are hard to detect because their tiny size makes it unlikely they will interact with matter.

Antimatter

Ordinary electrons have negative charge but there are electrons with positive charge. An electron with positive charge, called a positron, is an example of antimatter. Antimatter is exactly the same as ordinary matter except the electric charge is reversed. A positron has the same mass as an ordinary electron. Certain high-energy nuclear reactions may create anti-protons. An antiproton has the same mass as a normal (positive) proton except the charge is negative.

Gamma rays

Gamma rays are not really particles in the sense that they have no mass. Gamma rays are very high energy photons, similar to visible light but with millions of times more energy. We learned that atoms give off visible light photons when electrons change energy levels. The protons and neutrons in the nucleus also have energy levels but the energies are billions of times greater. Gamma rays are emitted when a nucleus loses energy. Gamma rays are also created when matter and antimatter meet. For example, if a normal electron meets a positron, both particles are instantly annihilated and turned into pure energy in the form of two gamma ray photons.

Mass and energy in nuclear reactions

Einstein's formula

Nuclear reactions do not conserve mass and energy separately however, nuclear reactions conserve mass-energy together. The explanation relies on Einstein's formula, $E = mc^2$. In this equation, c represents the speed of light or 3×10^8 m/s. The value of c^2 is very large so a small amount of mass equates to a very large amount of energy. A good illustration is combining two neutrons and two protons to make one helium-4 nucleus.

Conservation of mass-energy

The total mass of 2 moles of protons and 2 moles of neutrons is 4.01388 g. But, the mass of one mole of helium-4 nuclei which has a total of 2 moles of protons and 2 moles of neutrons is only 4.00260 grams. It would seem that mass is not conserved because the product mass is less than the reactant mass, however, we must consider the enormous amount of energy associated with nuclear formation. The nuclear reaction that forms the helium nucleus gives off 2.635×10^{11} joules of energy. According to Einstein's formula, mass is energy divided by c^2 therefore 2.635×10^{11} J is equivalent to 0.002928 grams. If we count mass and energy together then the total mass-energy on the reactant side equals the total mass-energy on the product side. The difference between expected nuclei mass and actual nuclei mass due to mass-energy conversion is called **mass defect**.

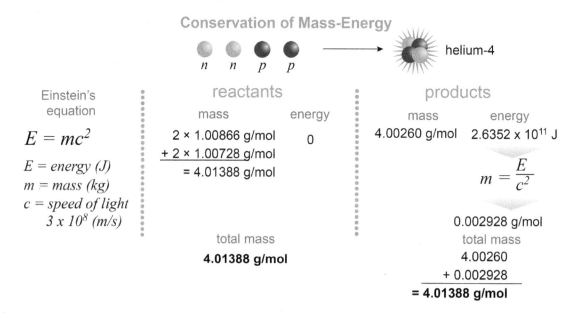

Mass is not conserved in nuclear reactions

Mass is not conserved separately in nuclear reactions. The forces are so strong in nuclear reactions that a significant fraction of mass is converted to energy. The total of mass-energy together is conserved however. Einstein's formula is the relationship between mass and energy. One gram of mass is equivalent to 9×10^{13} J of energy. To put this in perspective, one gram of gasoline reacts with oxygen to release 46,400 J of energy. This seems like a lot of energy, however, converting one gram of matter to pure energy releases 9×10^{13} J, or two billion times as much energy!

Solved problem

The U.S. uses 1.0×10^{20} J of energy in one year. If we could convert mass to energy directly, how much mass would be required? Note: the kg is equal to one J·s²/m²

Given 1.0×10^{20} J of energy

Relationships $E=mc^2$ and kg = J·s²/m²

Solve $m = \dfrac{E}{c^2} = \dfrac{1.0 \times 10^{20} \text{ J}}{(3 \times 10^8 \text{ m/s})^2} = 1,111.11111$ kg $\xrightarrow{\text{round}}$ 1,100 kg

Answer It would take only 1,100 kg to power the whole U.S. for 1 year.

Section 1 review

Chapter 19

Isotopes are elements with the same number of protons but a different number of neutrons. Some isotopes are unstable, leading them to decompose and release energy in the form of radiation. The atomic mass on the periodic table is a weighted average of the existing isotopes.

Nuclear reactions occur when a nucleus is rearranged, releasing a large amount of energy according to the equation $E = mc^2$. The mass of a nucleus is always less than the sum of the individual protons and neutrons that comprise it. The energy released due to this change in mass is the binding energy. If you plot a graph showing combinations of protons and neutrons that create stable nuclei, you will see a band of stable nuclei that fall in a neutron-to-proton ratio of 1:1 or 1.5:1. Nuclei outside of this band tend to be unstable.

Vocabulary words: nucleons, atomic number (Z), isotopes, mass number (A), radioactivity, binding energy, positron, antimatter, mass defect

Key relationships
- Isotopic notation & abbreviation: $^A_Z X$ and X-A (A = mass number, Z = atomic number)
- Energy-mass relationship, $\Delta E = \Delta m c^2$ where $c = 3.0 \times 10^8$ m/s
- Expected mass of a single isotope, $m_{nucleons} = m_{protons} + m_{neutrons}$
- Masses of nucleons: 1 proton = 1.0073 amu and 1 neutron = 1.0087 amu (1 amu = 1.6606×10^{-27} kg)
- 1 J = 1 kg·m²/s²

Review problems and questions

1. An isotope forms as a result of radioactive decay. It has 8 protons and 10 neutrons. Write the symbol for this isotope using isotopic notation and abbreviation.

2. How does the ratio of neutrons to protons change as the nucleus gets larger?

3. How is the ratio of neutrons to protons related to nucleus stability?

4. Why do nuclear reactions release so much more energy than ordinary chemical reactions?

5. Identify all stable isotopes for the following elements. Make sure to include atomic numbers.
    a. Strontium
    b. Molybdenum
    c. Selenium
    d. Antimony
    e. Copper

19.2 - Radioactivity

On a cloudy March day in 1896, French physicist Antoine Henri Becquerel put some potassium uranyl sulfate $K_2UO_2(SO_4)_2$ crystals in a dark drawer with some fresh photographic plates. He had been doing experiments with sunlight, but since the day was cloudy, Becquerel put his supplies away to wait for the sun. The next day, when he opened the drawer, on a hunch, he developed the photographic images and to his great surprise the uranium compound had created a photographic image of itself without any sunlight - in total darkness. Becquerel knew it took energy to expose a photographic plate and somehow the $K_2UO_2(SO_4)_2$ crystals made energy out of nothing! Becquerel had discovered radioactivity, and he would share the 1906 Nobel Prize with Pierre and Marie Curie for the discovery.

The nature of Becquerel's discovery

Stability and radioactivity

There is an almost infinite number of ways to combine protons and neutrons however, only a very few combinations are stable. If there are too many protons, the repulsive force rapidly breaks the nucleus up into separate parts. If there are too many neutrons, a process called beta decay turns them into protons and electrons. Carbon is a good example. There are two stable natural isotopes of carbon, ^{12}C and ^{13}C. The other isotopes are radioactive and either break up or turn into something other than carbon. The process of nuclear reactions occurring spontaneously is the explanation for **radioactivity**.

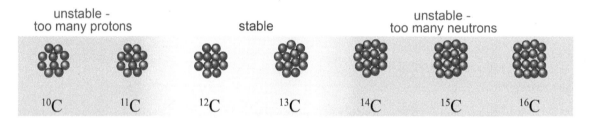

Radioactivity

A spontaneous nuclear reaction is called a nuclear decay. The four most common types of nuclear decay are alpha (α) decay, beta (β) decay, positron emission (β^+) and gamma (γ) decay. Alpha and beta decays change the nucleus into another element, releasing both particles and energy. Gamma decay releases pure energy and does not change the nucleus into another element.

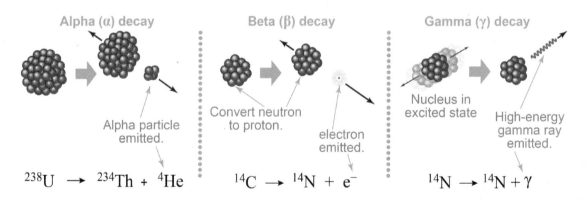

Radiation

The transmission of energy, matter, or waves through space is called **radiation** Some radiation is harmless, such as sunlight. The alpha particles, electrons, or gamma ray photons emitted by a nuclear decay are other examples of radiation. Alpha, beta, and gamma radiation can be dangerous if the energy is high enough to break chemical bonds. Ultraviolet (UV) light from the sun is another example of radiation that can be harmful.

Alpha decay

Alpha decay emits a helium nucleus

When people buy a home they may inspect for radon gas. The isotope radon-222, $^{222}_{86}\text{Rn}$, is radioactive and a common example of **alpha decay**. An alpha decay occurs when the nucleus emits two protons and two neutrons bound together in an alpha particle. An alpha particle is the nucleus of a helium atom with two protons and two neutrons. In a nuclear reaction, we write an alpha particle as $^{4}_{2}\text{He}$. The alpha decay of radon-222 produces polonium-218, $^{218}_{84}\text{Po}$. Polonium emits another alpha particle to become lead-214, $^{214}_{82}\text{Pb}$. Radon-222 is produced by the alpha decay of natural uranium in rocks and soil.

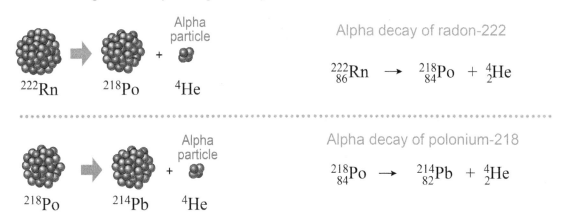

Alpha decay of radon-222

$$^{222}_{86}\text{Rn} \rightarrow {}^{218}_{84}\text{Po} + {}^{4}_{2}\text{He}$$

Alpha decay of polonium-218

$$^{218}_{84}\text{Po} \rightarrow {}^{214}_{82}\text{Pb} + {}^{4}_{2}\text{He}$$

Change the atomic number and mass number

An alpha decay carries away two protons and two neutrons and changes both the mass number and the atomic number.

1. The atomic number is reduced by two. For example, radon is atomic number 86 and polonium is atomic number 84.
2. The mass number is reduced by four. $^{222}_{86}\text{Rn}$ becomes $^{218}_{84}\text{Po}$.

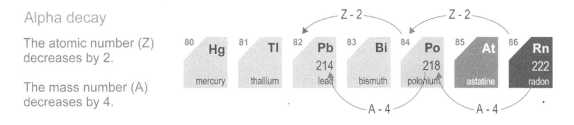

Alpha decay

The atomic number (Z) decreases by 2.

The mass number (A) decreases by 4.

Solved problem

Write the nuclear equation for the α-decay of U-238.

Asked Write the nuclear equation.

Relationships for α decay the atomic number, Z, decreases by 2 and mass number, A, decreases by 4.

Solve
1. For $^{238}_{92}\text{U}$, A = 92 and Z = 238.
2. Alpha decay decreases atomic number by two and mass number by four. After the α-decay, A = 90 and Z = 234.
3. Atomic number 90 is thorium, Th, and the isotope is $^{234}_{90}\text{Th}$.
4. Uranium-238 becomes thorium-234 and an alpha particle (helium-4)

Answer U-238 decays according to the equation:

$$^{238}_{92}\text{U} \rightarrow {}^{234}_{90}\text{Th} + {}^{4}_{2}\text{He}$$

Beta decay

A beta decay emits an electron

A neutron has zero electric charge; however, a neutron is made of even smaller particles that do have a charge. A neutron can change into an electron and a proton. Charge is conserved because there is 0 total charge before the breakup and 0 total charge after the breakup. This kind of nuclear decay is called **beta decay**. Radiation released during beta decay is called **beta radiation** although beta radiation is actually moving electrons.

Beta decay increases atomic number

Ordinary beta decay is also called β^- decay. The beta (β^-) particle is a negative electron that is ejected from the nucleus. The positive proton stays in the nucleus. The β^- decay reaction increases the atomic number by one because one neutron changes to a proton. The mass number stays the same because the total number of protons plus neutrons does not change. For quantum reasons the β^- decay also emits an electron anti-neutrino, $\bar{\nu}_e$.

Beta decay - β^-

The atomic number (Z) increases by 1.

The mass number (A) stays the same.

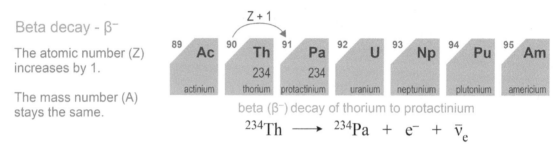

beta (β^-) decay of thorium to protactinium

$$^{234}\text{Th} \longrightarrow {}^{234}\text{Pa} + e^- + \bar{\nu}_e$$

Positron, β^+ decay

Another beta decay, called **positron emission** (β^+), results in emission of a positron rather than an electron. A positron is equal to an electron in its mass but opposite in its charge. β^+ decay occurs when a proton changes into a neutron and the nucleus emits a positron (e^+). The atomic number drops by one because one proton becomes a neutron. The mass number stays the same. β^+ decay creates an electron neutrino ν_e.

Beta plus decay - β^+
(positron decay)

The atomic number (Z) decreases by 1.

The mass number (A) stays the same.

beta plus (β^+) decay of fluorine-18 to oxygen-18

$$^{18}\text{F} \longrightarrow {}^{18}\text{O} + e^+ + \nu_e$$

Solved problem

Write the nuclear equation for the β^- decay of C-14.

Relationships Atomic number, Z, increases by 1 and mass number, A, stays the same.

Solve
1. For $^{14}_{6}\text{C}$, A = 6 and Z = 14.
2. β^- decay increases atomic number by one therefore after the β^- decay, A = 7 and Z = 14.
3. Atomic number 7 is nitrogen, N, and the isotope is $^{14}_{7}\text{N}$.
4. The electron anti-neutrino, $\bar{\nu}_e$, does not need to be included.

Answer $^{14}_{6}\text{C} \rightarrow {}^{14}_{7}\text{N} + {}^{0}_{-1}e$

Solved problem

Write the nuclear equation for the positron emission of C-11.

Relationships Atomic number, Z, decreases by 1 and mass number, A, stays the same.

Solve
1. For $^{11}_{6}\text{C}$, A = 6 and Z = 11. A decreases by one so A = 5 and Z = 11.
2. The isotope is $^{11}_{5}\text{B}$; an electron neutrino, ν_e is created but not included.

Answer $^{11}_{6}\text{C} \rightarrow {}^{11}_{5}\text{B} + {}^{0}_{+1}e$

Gamma decay and decay chains

Gamma decay releases electromagnetic energy

Gamma decay, abbreviated γ-decay, is the emission of energy by a nucleus. Recall that atoms can absorb or emit energy by moving electrons up and down energy levels. A similar process works in the nucleus when protons and neutrons change energy levels. However, the amount of energy is much, much greater with nuclear energy levels. A nucleus with excess energy often releases the energy with a gamma ray photon. Gamma rays have much higher energy than x-rays.

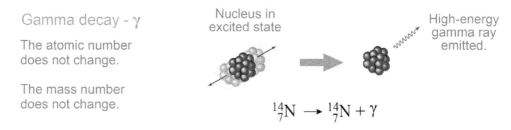

Gamma decay - γ

The atomic number does not change.

The mass number does not change.

Nucleus in excited state

High-energy gamma ray emitted.

$${}^{14}_{7}N \rightarrow {}^{14}_{7}N + \gamma$$

γ radiation

Because only energy is emitted, the number of protons and neutrons does not change during γ-decay. However, gamma decay often follows another radioactive decay, such as β-decay or α-decay. Both α and β decay release energy. Some of this energy is carried away as kinetic energy of the particles moving away from the nucleus. A portion of the energy may be emitted as a gamma ray.

Gamma rays have high energy

Gamma ray photons have high enough energy to break apart other atoms, making them dangerous to living organisms. Because they have no mass and move at the speed of light, gamma rays have high penetrating power and can pass through most material. The best way to protect yourself from γ-rays is to use a thick shielding material made of lead or concrete.

The ^{238}U decay chain

Many nuclear decays start with a radioactive nucleus and produce another radioactive nucleus, which decays into another radioactive nucleus until a stable nucleus is reached. This is called a nuclear decay chain. The diagram below shows the nuclear decay chain for uranium-238. The chain starts in the upper right with $^{238}_{92}$U and ends with stable $^{208}_{82}$Pb in the lower left. Can you find all three types of decay reactions (α, β, γ)?

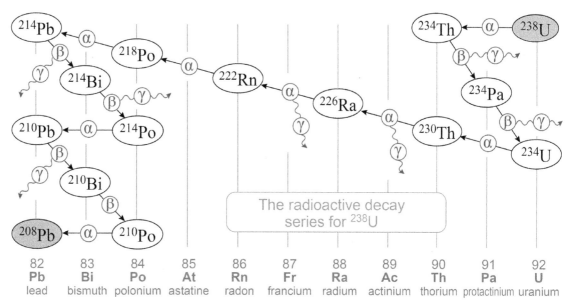

The radioactive decay series for ^{238}U

Section 19.2: Radioactivity

Half life and the decay equation

Radioactvity and chance

Radioactive isotopes decay spontaneously. That is, the decay reaction occurs randomly without outside influences. We do not know when the nucleus of a radioactive isotope will undergo radioactive decay. It could be in the next second, a few minutes from now, or in days, years, centuries—or even in a few millennia. The decay of radioactive atoms is a chance process. We cannot predict when any single atom will decay, but given a very large number of atoms we can predict how the whole group of atoms will behave. This is one way in which our everyday world and the quantum world are fundamentally different.

Half-life
In each half-life, one half of the remaining atoms decay (on average).

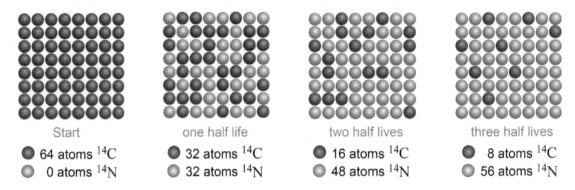

Half life and rate of decay

The rate at which radioactive atoms decay is described by the **half-life**. In one half-life, half of the radioactive atoms will decay. The half-life is denoted by the symbol $t_{1/2}$. Imagine that you have 64 atoms of the radioactive isotope ^{14}C. This isotope has a half-life of 5,700 years. After one half-life, half of the 64 atoms will have decayed into ^{14}N. Wait another half life and half of the remaining 32 ^{14}C atoms will decay. In another half-life, half of the remaining 16 ^{14}C atoms will decay. That is how half-life works: after the time period of one half-life, half of the atoms will have decayed radioactively. The decay process is random so with only 64 atoms the numbers would not neatly decrease as 64, 32, 16, 8. However, with 10^{23} atoms the half life is a very accurate predictor. For a given radioactive isotope, the number of radioactive decays per unit of time is known as the **rate of decay**. The rate of decay decreases over time as there are fewer atoms remaining that have not yet decayed. For example, the ^{14}C sample above experienced 32 decays during the first 5,700 years but only 16 decays during the next 5,700 years, so the rate of decay decreased.

The Decay Equation

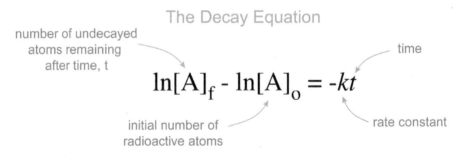

$$\ln[A]_f - \ln[A]_o = -kt$$

Rate constant

The **rate constant**, k, describes the probability of a radioactive decay occurring per unit of time. A large k value means each atom has a high probability of decaying and therefore decays very fast. For ^{21}F, k = 0.016 s^{-1} and a half-life of only 4.16 seconds. By comparison, ^{14}C has a k of 0.00012 years^{-1} and a half-life of 5,700 years. The rate constant k has units of 1/time. If time is in years then the units of the rate constant are 1/years which is usually written years^{-1}. Half-life, $t_{1/2}$, is related to the rate constant, k in the decay equation above.

Solving radioactive decay problems

Using the rate equation

Radioactivity problems typically involve the half-life or the rate constant. The two are related by the equation below. The larger the rate constant, the shorter the half-life. These are some things to be careful of:

1. The units of the rate constant and half life should match the units for time. For example, if time is in seconds then the half-life should also be in seconds and the rate constant shoud be in 1/seconds (s^{-1}).
2. The initial and final quantities must always be in the same units (atoms, mass or moles). The number of atoms is proportional to both mass and moles so the equation works the same for any units.
3. The symbol e is the base of the natural logarithm, e ≈ 2.718. Scientific calculators have this function built in as e^x.

The half-life and the rate constant

$$t_{1/2} = \frac{0.693}{k} \qquad k = \frac{0.693}{t_{1/2}}$$

(half-life) (rate constant) (rate constant) (half-life)

Logarithm

The parameter e is a constant equal to the number 2.718, which is the base of the natural logarithm. This mathematical constant is used across many areas of science and engineering.

Solved problem

Fluorine-18 is a radioactive isotope used in medicine. ^{18}F has a half-life of 109.8 minutes and usually decays by positron emission (β^+). If a 1.25 g sample of ^{18}F is left for 8 hours, how much ^{18}F is left?

Asked Calculate the amount of ^{18}F left after 8 hours.
Given The half-life, $t_{1/2}$ = 109.8 min, t = 8 hours, a_0 = 1 gram
Relationships $\ln[A]_f - \ln[A]_0 = -kt$, $k = 0.693/t_{1/2}$, and e = 2.718
Solve

Calculate the rate constant:

$$k = \frac{0.693}{t_{1/2}} = \frac{0.693}{109.8 \text{ min}} = 0.006311 \text{ min}^{-1}$$

Convert hours to minutes:

$$\frac{8 \text{ hours}}{1} \times \frac{60 \text{ mins}}{1 \text{ hours}} = 480 \text{ minutes}$$

Use the rate equation, $\ln[A]_f - \ln[A]_0 = -kt$

$\ln[A]_f - \ln[1.25g] = -0.006311 \text{ min}^{-1} \times 480 \text{ min}$

$\ln[A]_f = -0.006311 \times 480 + 0.223$

$\ln[A]_f = -2.80613$

$[A]_f = e^{-2.80613}$

$[A]_f = 2.718^{-2.80613}$

$[A]_f = 0.0604$ g

Answer After 8 hours the initial 1.25 g of ^{18}F is reduced to 0.0604 g.

Carbon dating

Carbon-14

The isotope, ^{14}C is radioactive and decays to ^{14}N (β^-) with a half-life of 5,730 years. Since the Earth is 4.6 billion years old, any carbon-14 initially present has long since decayed to stable nitrogen-14. However, ^{14}C is produced in the atmosphere by cosmic rays interacting with nitrogen atoms. Since the intensity of cosmic rays is fairly stable, the rate of production of ^{14}C is roughly constant over time leading to a concentration of 1.5 carbon-14 atoms per 10^{12} atoms of carbon-12 and carbon-13.

Carbon exchange in living organisms

While they are alive, living organisms constantly exchange carbon with the environment maintaining the same ratio of 1.5 carbon-14 atoms per 10^{12} atoms of carbon. When an organism dies, active carbon exchange with the environment stops. **Carbon-14** already within the organism decays radioactively with a half-life of 5,730 years. Without any new source, the remaining amount of ^{14}C provides a natural clock which starts ticking the moment an organism dies.

Egyptian papyrus can be carbon dated because it was made from living plant matter.

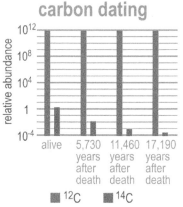

Paleolithic campfires can be carbon dated because the wood was once alive.

Carbon dating

Carbon dating is the technique of comparing the concentrations of carbon-12 and carbon-14 to determine the age of a biological material. For example, carbon dating can determine the age of ancient Egyptian papyrus samples and the charcoal from prehistoric campfires. Carbon dating is reliable for samples up to 57,300 years old, which is about 10 half-lives. After 10 half-lives, in most samples there is not enough carbon-14 remaining to measure accurately.

Limits to carbon dating

The technique of carbon dating only works with matter that was once alive and actively exchanging carbon with the environment. Egyptian papyrus can be carbon dated but most ancient pottery cannot be dated because the inorganic clays in pots were never living matter. Carbon dating has been successfully used to measure the age of relics up to 50,000 years old. After 50,000 years only 0.24% of the original ^{14}C remains.

Beta decay (β^-) of ^{14}C

$$^{14}C \longrightarrow {}^{14}N + e^- + \bar{\nu}_e \qquad t_{1/2} = 5{,}730 \text{ years}$$

Detecting ^{14}C

Carbon dating works because the beta particle emitted by ^{14}C has a known energy, and beta detectors are very sensitive. A typical technique is to count beta decays for 250 minutes, which is statistically sufficient to give a date with an error of ± 80 years, with 68% confidence.

Choosing an isotope for dating

Useful range

Some isotopes are better than other for dating very old things. Uranium-235 has a half life of 7.04×10^8 years. Uranium-235 has half life that is so large it would be useless for measuring items less than 704 million years old. If you wanted to know the age of a bottle of wine found in a very old home, U-235 would not be an appropriate choice. The isotope cesium-137 has a half life of 30.17 years. Cesium-137 is a better choice than U-235 because the bottle of wine is likely to be hundreds of years old or less, not millions.

radioisotopes and their useful range

Parent	Daughter	Half-Life of Parent (years)	Useful Range (years)
carbon-14	nitrogen-14	5,730	100 - 30,000
potassium-40	argon-40	1.3 billion	100,000 - 4.5 billion
rubidium-87	strontium-87	47 billion	10 million - 4.6 billion
uranium-238	lead-206	4.5 billion	10 million - 4.5 billion
uranium-235	lead-207	710 million	10 million - 4.6 billion

Solved problem

The half life of carbon-14 is 5,730 years. A bone is found that contains 2.50 mg of nitrogen-14. It is known that nitrogen-14 comes from carbon-14 radioactive decay. The bone is calculated to have had 4.5 mg while it was alive. How old is the sample?

Given: 4.5 mg of C-14, 2.50 mg of N-14 and C-14 has a half life of 5,730 years

Relationships: $\ln[A]_f - \ln[A]_0 = -kt$, $k = 0.693 / t_{1/2}$

Solve: First, find the rate constant:
$$k = \frac{0.693}{t_{1/2}} = \frac{0.693}{5730} = 1.29050 \times 10^{-4}$$

Next, find the amount of Carbon-14 remaining by subtraction:
Original amount of C-14 − Amount of N-14 = C-14 Remaining
4.5 mg − 2.5 mg = 2.0 mg

Finally, use the decay equation to find the age of the sample:
$\ln[A]_f - \ln[A]_0 = -kt$
$\ln[2.0] - \ln[4.5] = -1.29050 \times 10^{-4} \times t$
$0.693147 - 1.50477 = -1.29050 \times 10^{-4} \times t$

$$\frac{0.693147 - 1.50477}{-1.29050 \times 10^{-4}} = t; \quad t = 6{,}289.26 \longrightarrow \text{round} \longrightarrow 6{,}300 \text{ years}$$

Answer: The bone is 6,300 years old.

Nuclear decay graphs

Interpreting decay graphs

You can use a radioactive decay graph to find the age of a radioactive material or predict how much will remain after a period of time. As shown on the graph, when 50% of a substance remains, one half-life has passed. When the second half life has passed, 75% of the substance has decayed and only 25% remains. The remaining sample is divided in half with each half-life that passes. The % remaining never reaches zero because the rate of decay slows down over time; and, even a tiny sample has trillions upon trillions of atoms.

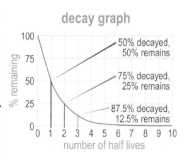

How decay graphs are set up

A decay graph indicates time on the x-axis, usually in number of half-lives or with the half life time specified. The y-axis shows the amount of radioactive material usually as a percentage, as a mass, or as a concentration. All three graphs below have the same shape because for each graph, time begins at zero with 100% of the sample intact.

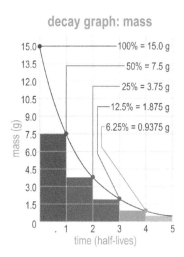
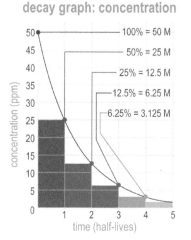
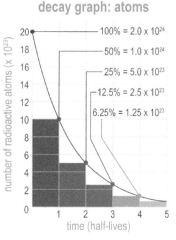

Using the graph

Let's use a decay graph to solve a radioactive decay problem very similar to the one on the previous page. On an exam you would be given the graph shown to the right, minus the blue arrows labeled A and B. Graphs are best used for estimates, so if you need to know accurate amounts, the half-life and rate constant equations are a better choice. Notice how the question is worded somewhat differently to emphasize the fact that your answer is an estimate.

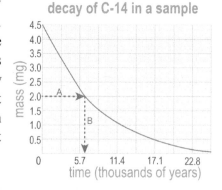

Solved problem

C-14 has a half life of about 5,700 years. The bone has 2.50 mg of nitrogen-14, but the bone had 4.5 mg while its owner was alive. Use the graph above to estimate its age.

Given 4.5 mg of C-14, 2.50 mg of N-14 and C-14 has a half life of about 5,700 years

Solve Find the mass of C-14 remaining by subtraction:
Original amount of C-14 − Amount of N-14 = C-14 Remaining
4.5 mg − 2.5 mg = 2.0 mg
Draw a line from 2.0 mg to the decay line (line A), then draw a line from the decay line to the x-axis (line B). The line intersects the axis at about 6,300 years.

Answer The bone is about 6,300 years old.

Section 2 review

Chapter 19

Nuclear decay reactions are classified as α, β, γ and β⁺ and release different types of radiation. In alpha decay, a nucleus ejects an alpha particle, which is comprised of two protons and two neutrons. In beta decay, a nucleus emits a high-energy electron formed when a neutron decays into a proton and electron. In gamma decay, a nucleus emits electromagnetic radiation. Positron emission occurs when an unstable nucleus emits a positron, converting a proton into a neutron.

The half-life of an isotope is a measure of the time it takes for half the sample to decay. The rate of decay is directly proportional to the number of nuclei that are present to begin with as well as the rate constant for that particular isotope. Carbon dating is a way of determining the age of organic material by comparing the ratio of carbon-14 to carbon-12 in the sample versus the ratio found in the environment. Several other isotopes are also useful for radioactive dating. The ratio of carbon-18 to carbon-16 in ice cores or calcite is a key indicator of climate change while the long half-life of Uranium-238 is ideal for geological dating.

Vocabulary words: radioactivity, radiation, alpha decay, beta decay, beta radiation, positron emission, gamma decay, half-life, rate constant, rate of decay, carbon dating, carbon-14

Key relationships

- Rate constant, $k = \dfrac{0.693}{t_{1/2}}$ where $t_{1/2}$ = half-life
- Alpha decay: $^{A}_{Z}X \rightarrow {}^{(A-4)}_{(Z-2)}\text{NewElement} + {}^{4}_{2}\text{He}$ or $^{A}_{Z}X \rightarrow {}^{(A-4)}_{(Z-2)}\text{NewElement} + {}^{4}_{2}\alpha$
- Beta decay: $^{(A)}_{(Z+1)}\text{NewElement} + {}^{0}_{-1}e$ or $^{A}_{Z}X \rightarrow {}^{(A)}_{(Z+1)}\text{NewElement} + {}^{0}_{-1}\beta$
- Gamma decay: $^{A}_{Z}X \rightarrow {}^{A}_{Z}X + {}^{0}_{0}\gamma$
- Positron emission: $^{A}_{Z}X \rightarrow {}^{(A)}_{(Z-1)}\text{NewElement} + {}^{0}_{+1}e$
- Decay equation: $\ln[A]_f - \ln[A]_0 = -kt$ where:
  - $[A]_f$ = number of undecayed atoms remaining after time, t
  - $[A]_0$ = initial number of radioactive atoms
  - k = rate constant
  - t = time
- Half-life and half-life constant:
  $t_{1/2} = \dfrac{0.693}{k}$ or $k = \dfrac{0.693}{t_{1/2}}$ where k = rate constant and $t_{1/2}$ = half-life

Review problems and questions

1. Identify the type of decay each isotope undergoes in the following, and complete each equation.

   a. ?-95 is used in smoke detectors in your home: $^{241}_{95}? \rightarrow {}_{93}? + ?$

   b. Th-230 can be used to determine the age of ancient ocean corals: $^{230}_{90}\text{Th} \rightarrow {}^{230}_{90}? + ?$

   c. Radioactive "seeds" can be implanted in the body to treat non-aggressive prostate cancer in men: $^{192}_{?}\text{Ir} \rightarrow {}^{188}_{?}? + ?$

   d. ?-131 is used both to diagnose and treat thyroid cancer: $^{131}_{53}? \rightarrow {}^{(?)}_{(?)}? + {}^{0}_{-1}e$

2. In the decay graph shown, there are 80.0 g of a substance at t = 0 days. What is the half-life of this substance?

3. How many grams of Sample X will remain after 3 half-lives have passed? Use the graph, then verify your answer with a calculation.

4. How long will it take for 35.0 g to remain? Use the graph, then verify your answer with a calculation.

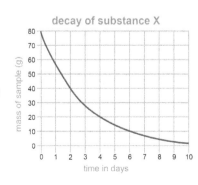

decay of substance X

Section 19.2: Radioactivity

19.3 - Nuclear Energy

Chemical reactions can release the energy of chemical bonds. Nuclear reactions can release the energy in the far stronger nuclear bonds between protons and neutrons. Virtually all the available energy on Earth comes from nuclear reactions in the Sun. Even fossil fuels are technically stored solar energy that was captured during the carboniferous era 300 million years ago. In the core of the Sun, nuclear fusion reactions convert 600 billion kilograms of hydrogen into helium every second. Fortunately, the Sun has an enormous mass of 2×10^{30} kg and there is enough hydrogen in the Sun's core to last another four billion years!

Nuclear reactions and energy

The source of nuclear energy

The energy for nuclear reactions is shown on the nuclear binding energy curve. In an earlier chapter we discussed that different molecules have different amounts of energy even though they may contain the same atoms. For example, one glucose molecule and six oxygen molecules have more energy than six carbon dioxide molecules and six water molecules. This energy difference is what sustains living organisms such as yourself. A similar situation occurs with protons and neutrons. Different nuclei have different amounts of binding energy. Rearranging protons and neutrons from one nucleus into another can either absorb more energy or release energy.

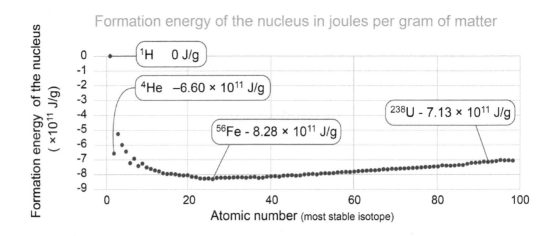

The formation energy of the nucleus

The graph above shows the energy of formation of the nucleus expressed in joules of energy per gram of matter. The energy is relative to hydrogen, which is assigned a value of zero. Helium has an energy of formation of -6.60×10^{11} J/gram. Compared to hydrogen, the graph tells us one gram of separate protons and neutrons that come together to form one gram of helium nuclei release an astounding 660 billion joules (6.60×10^{11}) of energy, in one gram! The chain of nuclear reactions that convert hydrogen to helium is the energy source of the Sun (see the chapter on The Universe).

Why the curve has a minimum

The nucleus with the lowest formation energy per gram is ^{56}Fe at −828 billion joules per gram. Nuclei that are lighter or heavier than iron release less energy when they form from separate protons and neutrons. The minimum in the curve is a result of the competition between the repulsive electromagnetic force between protons and the attractive nuclear force between protons and neutrons. The nuclear force has a very short range, about the radius of an iron nucleus. For nuclei heavier than iron, the outermost protons and neutrons do not feel the full strength of the attraction. That makes the nucleus less tightly bound and therefore it takes less energy to break it apart.

Fission reactions

Fission breaks up the nucleus

In a **fission reaction** a larger nucleus splits into two smaller nuclei. For example, when ^{235}U is struck by a neutron, the uranium nucleus breaks up into two nuclei plus leftover neutrons. In the reaction below, one uranium-235 ($^{235}_{92}$U) nucleus breaks up into a cesium-140 ($^{140}_{55}$Cs) nucleus, a rubidium-93 ($^{93}_{37}$Rb) nucleus, and two neutrons. These two neutrons can impact other $^{235}_{92}$U nuclei and induce more fission reactions.

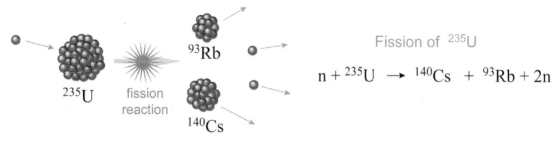

Fission of ^{235}U

$$n + {}^{235}U \rightarrow {}^{140}Cs + {}^{93}Rb + 2n$$

Why fission releases energy

Fission reactions are exothermic when heavy nuclei break up into nuclei that are lower on the nuclear energy curve. For example, when ^{235}U fissions, both the cesium and rubidium daughter isotopes have a lower energy of formation. Protons and neutrons have lower total energy in the form of the rubidium and cesium nuclei than they do in the form of a uranium nucleus. The energy difference is the energy given off by the fission reaction.

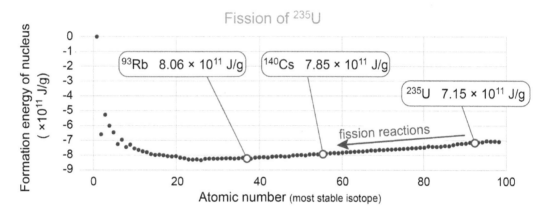

Why don't all heavy nuclei fission?

If iron is the lowest energy, why don't all heavier nuclei decay? The answer is similar to why glucose doesn't spontaneously transform to water and carbon dioxide. Once a nucleus is formed, it takes a substantial amount of energy to break the existing bonds and "loosen-up" the protons and neutrons in order to allow them to re-form into a lower energy nucleus. Virtually all nuclei heavier than helium were initially formed in the intense energy at the core of stars or in cataclysmic supernova explosions. For all but a very few nuclei, the activation energy needed to initiate a fission reaction far exceeds the energy available in the environment. However, fission reactions of all elements are routinely produced in accelerator experiments where high energy nuclei or particles are slammed together.

Fission products

The heavier a nucleus gets, the closer it gets to the boundary at which the strong force cannot hold the nucleus together against the repulsion of the protons. Bismuth, with 83 protons, is the heaviest stable nucleus. Every element heavier than bismuth is radioactive. Very few isotopes are so close to the boundary of stability that they can undergo a fission reaction with only a small amount of "trigger" energy.

Section 19.3: Nuclear Energy

Chain reactions

Uranium

Uranium is a radioactive, silvery-white metal. Natural uranium is found in an ore called pitchblende, which is a combination of UO_2 and U_3O_5. Five uranium isotopes exist naturally. The most abundant (99.275%) is ^{238}U with 92 protons and 146 neutrons. All of Earth's natural uranium was forged in stars and supernovae between five and ten billion years ago. Today's uranium has been decaying ever since. Uranium-238 has a very long half-life of 4.5 billion years, which explains why some still exists today.

^{235}U

The isotope ^{235}U with 143 neutrons is the next longest-lived with a half life of 704 million years. Only 0.72% of natural uranium is ^{235}U. However, this isotope is special because it is the only naturally occurring isotope that is fissile. This means it is close enough to the stability boundary that ^{235}U nuclei undergo fission reactions when struck by a neutron with thermal energy. For this reason, ^{235}U can support a nuclear chain reaction which is why this isotope is the foundation of all nuclear power, for both peaceful and destructive purposes.

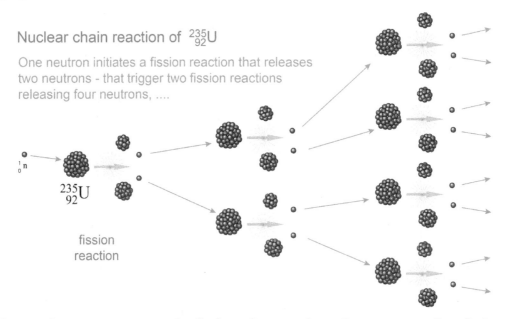

Chain reaction

In a nuclear **chain reaction**, the fission of one nucleus triggers a cascading fission of many other nuclei. This is possible because when ^{235}U undergoes fission, two neutrons are released on average. If these two neutrons hit other ^{235}U nuclei, two more fission reactions are triggered. This releases four neutrons, that trigger four fissions releasing eight neutrons, and so on. If the density of ^{235}U nuclei is high, this can occur very fast. Within microseconds, one neutron can precipitate 10^{20} fission reactions releasing an enormous amount of energy very quickly.

Critical mass

For a chain reaction to occur, there must be a high probability of each neutron hitting another fissile nucleus before being absorbed, or slowing down. This requires a **critical mass** of fissionable matter. The critical mass is the minimum amount of fissionable material, usually Uranium-235, within a relatively small volume so the chain reaction can occur. Once a critical mass is assembled, it immediately begins the chain reaction. Unless the reaction is interrupted, it proceeds until all the fissile material is either gone, or has been dispersed by the resulting explosion. The control of nuclear energy is based on the many ways to absorb neutrons to slow the chain reaction.

Nuclear energy

Fission energy release

The fission of one gram of ^{235}U releases 75 billion joules of energy. The calculation below is similar to the calculation for the enthalpy release of a chemical reaction, but using the nuclei's formation energy. The total mass of products is less than the mass of reactants (one neutron is a catalyst). This reaction converts 0.51% of the mass of uranium into energy.

Energy released by fission of one gram of ^{235}U

$$n + {}^{235}U \rightarrow {}^{140}Cs + {}^{93}Rb + 2n$$

		mass	Energy of formation of nucleus	
reactants	^{235}U	1.0000 g	-7.15×10^{11} J/g	
products	^{140}Cs	0.59528 g	$+4.70 \times 10^{11}$ J/g	
	^{93}Rb	0.39534 g	$+3.20 \times 10^{11}$ J/g	
	n	0.0042915 g	0 J/g	
0.005090 g converted to energy ~ 0.51%		0.99491 g	-0.750×10^{11} J/g	75.0 billion joules of energy released

How a reactor works

All nuclear power in the world today is based on fission of ^{235}U. Uranium containing an enriched percentage of ^{235}U is contained in fuel rods suspended in water in a sealed and pressurized reactor vessel. Control rods containing a strong neutron absorber, such as boron, are moved in and out of the fuel rods. The farther the control rods are inserted into the reactor, the slower the chain reaction becomes as more neutrons are absorbed before they can cause fissions. The energy from fission heats the fuel rods which heats the water that is flowing around them. To move the energy to the outside world, this superheated radioactive water is piped into the lower half of a steam generator. The water in the steam generator boils and creates high-pressure steam which, like a jet engine in reverse, spins a turbine that is attached to a generator.

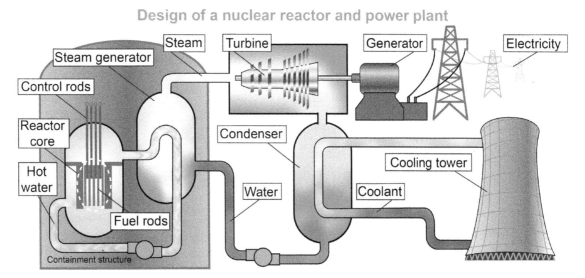

Design of a nuclear reactor and power plant

Interactive simulation

The Control Rod simulation allows you to control the fission rate by raising and lowering the control rods. Try to keep the heat output in the green range. Actual reactors do this automatically and the simulation will switch to automatic control if you get "too hot!"

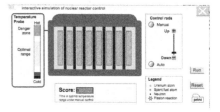

Interactive simulation of the control rods in a nuclear reactor

Fusion reactions

Combining nuclei

A nuclear **fusion reaction** combines or "fuses" two smaller nuclei into a larger nucleus. A good example is the fusion of ^{3}He and ^{4}He to produce ^{7}Be. This fusion reaction occurs in the Sun, although it is a minor energy producer compared to the proton-proton reaction on the next page.

Fusion reactions - combine smaller nuclei into larger nuclei

$$^3He + {}^4He \longrightarrow {}^7Be + e^-$$

Energy from fusion reactions

Fusion reactions follow the energy curve from left to right, from lighter nuclei to heavier nuclei. This is opposite from fission reactions. Fusion reactions can release a very large amount of energy, even more than fission reactions because light nuclei, such as deuterium (^{2}H) and tritium (^{3}H), are far up the energy curve from iron. The graph below shows the energy of formation of the nucleus for several light isotopes.

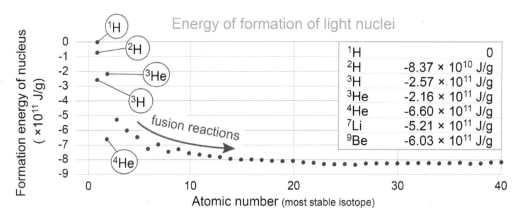

Extreme heat allows fusion reactions to progress

For a fusion reaction to occur, the two nuclei must get close enough for the strong force to attract them before the electromagnetic force repels them. Imagine two balls of sticky clay flying through the air towards one another. They can pass by very close to each other yet still fly apart. Clay balls have to actually touch to stick together. The strong nuclear force is like clay, protons have to almost touch before they stick together. Protons need to be moving fast to get close before flying apart and fast means high temperature. It takes temperatures above 10 million degrees Celsius to get nuclei close enough enough together for fusion to occur.

Nucleo-synthesis

The universe began with only hydrogen and helium. All the heavier elements were created by fusion reactions in the cores of stars. Stars fuse hydrogen into helium, and helium into heavier elements to produce energy. Elements heavier then hydrogen are created this way. When a star exhausts its hydrogen fuel, depending on the star's mass, it may explode and spew heavier elements out into the universe. These may condense into new stars, like our Sun.

Synthesis of carbon through nuclear fusion

Fusion power

Harnessing fusion

Fission power has the advantage of not using fossil fuels, and being relatively simple. The disadvantage is that fission creates long-lived radioactive waste and the technology can be used destructively. Fusion reactions that combine hydrogen in water into helium could produce limitless energy and theoretically have no radioactive waste. However, the temperature and pressure at which fusion energy is practical is daunting. A fusion scientist famously remarked that "taming the nuclear fire of a star is perhaps the most ambitious technical challenge ever attempted by humans." Nonetheless, the pictures below show an experimental tokamak fusion reactor in which the plasma temperature exceeds 85 million degrees Celsius, five times hotter than the core of the Sun.

The Alcator C-MOD experimental fusion reactor at MIT

Nuclear energy of tomorrow

The easiest fusion reaction to reproduce on Earth uses two isotopes of hydrogen: deuterium, ^{2}H, and tritium, ^{3}H. Deuterium is found naturally in seawater and its abundance is one out of every 6,420 hydrogen atoms. Notice that the only by-products of this reaction are helium and a neutron. The reaction also uses tritium, a radioactive but short-lived isotope ($t_{1/2}$ = 12.3 yr). Tritium does not occur naturally but can be generated within the reactor by the extra neutron interacting with lithium, which is a very common element.

D-T fusion reaction

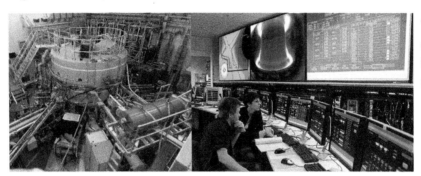

$$^2H + {}^3H \longrightarrow {}^4He + {}^1_0n$$

	reactants				products		
	mass	moles	energy		mass	moles	energy
^{2}H	1.0000 g	0.49650	8.37 × 10^{10} J	^{4}He	1.9872 g	0.49650	-13.1 × 10^{11} J
^{3}H	1.4975 g	0.49650	3.85 × 10^{11} J	$^1_0$n	0.50338 g	0.49650	0 J
	2.4975 g		+4.69 × 10^{11} J		2.48811g		-13.1 × 10^{11} J

Energy released per gram of deuterium fuel

4.69 × 10^{11} J - 13.1 × 10^{11} J = 8.43 × 10^{11} J

The challenge of nuclear fusion

The energy released by one gram of deuterium fuel is 843 billion joules. This is 20 million times the energy from burning one gram of gasoline. Even though nuclear fusion is an ideal energy source, the technology of building a fusion reactor is still under development. Part of the challenge faced by engineers is how to safely heat the deuterium and tritium fuel to more than 50 million degrees Celsius. Another problem is that the hot fusion plasma must be contained in a magnetic force field and kept isolated from solid materials. The tokamak in the illustration is a step in a seventy-year research program to make fusion power practical.

Chapter 19

Section 3 review

Fission reactions split larger nuclei into smaller ones, releasing the difference in binding energy. Fission of U-235 can be used to produce electricity or create nuclear explosions. Fusion reactions join two smaller nuclei into a larger nucleus. Fusion energy is produced in the interior of stars at extremely high temperatures. Research is ongoing to create fusion reactions on Earth.

Vocabulary words	fission reaction, chain reaction, critical mass, fusion reaction
Key relationships	• Fission: A heavy nucleus splits into smaller nuclei. • Fusion: Lighter nuclei combine to form a single nucleus.

Review problems and questions

1. List the types of changes in order from most energy exchanged to least energy exchanged: fission reaction, chemical reaction, phase change, fusion reaction.

2. How are fusion and fission reactions similar, and how are they different?

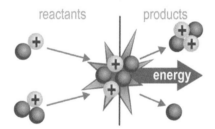

3. Does the picture above show a fission reaction, or a fusion reaction? How can you tell?

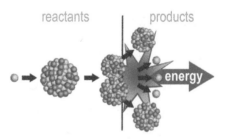

4. Does the picture above show a fission reaction, or a fusion reaction? How can you tell?

19.4 - Radiation

Radiation is often associated with danger but technically, sunlight is radiation, and so is the heat from a campfire. Radiation is the transmission of energy through space by waves or particles, or both. Light is electromagnetic radiation, and so are microwaves, x-rays and gamma rays. Beta rays are fast-moving electrons, which are particles. What all these different forms of radiation have in common is the amount of energy transferred per second per unit of area.

Ionizing and non-ionizing radiation

Ionizing and non-ionizing radiation

Whether a form of radiation is dangerous or not depends on two factors. One is the intensity, or power, of the radiation. The second factor is the energy of the radiation. The important scale for energy is the energy required to eject an electron from an atom. Removing an electron from an atom is called **ionization**. Radiation that has enough energy to cause ionization is called **ionizing radiation**. X-rays are a good example of ionizing radiation. Visible sunlight and radio waves are low-energy and cannot eject an electron from an atom. Sunlight and radio waves are examples of **non-ionizing radiation**.

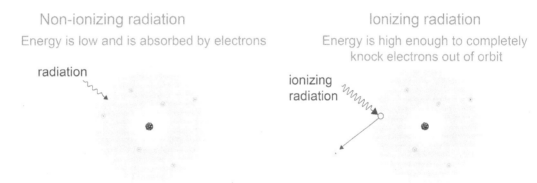

Breaking chemical bonds and sunburn

The reason that ionizing radiation is potentially harmful is that the energy in chemical bonds is similar to the ionization energy. Ionizing radiation can break chemical bonds and that is the essence of the danger. The lowest energy radiation which can be ionizing is ultraviolet light. Ultraviolet, or UV light, causes sunburns because the radiation breaks bonds in molecules in your skin. Skin cells are damaged and that is the underlying cause of a sun burn. Fortunately our bodies have evolved to heal minor damage such as sunburns.

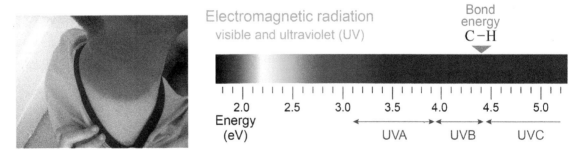

Sunscreens

The diagram shows the energy of different types of light, or electromagnetic radiation, in units of electron-volts (eV). The carbon-hydrogen bond breaks at 4.4 eV. Ultraviolet light breaks bonds in skin molecules starting at around 3.5 eV in the UVA band. Sunscreens are chemicals that have high absorption of UV. The best sunscreens absorb both UVA and UVB from 3.1 eV to more than 4.5 eV.

The intensity of radiation

Units of intensity

The difference between a dim light and a bright light is the **intensity**. Intensity describes how much energy flows per unit of area per second. Energy flow (or power) is expressed in watts. One **watt** (W) is one joule per second. The unit for intensity is a rate of energy per unit time, energy per unit area, or power per unit area (W/m²). For example, on a sunny day between 400 and 1,000 watts of radiant energy fall on each square meter. A high-quality flashlight might have an intensity of 100 W/m² in the brightest part of the beam. The light in a shadow might have an intensity of only 1 W/m².

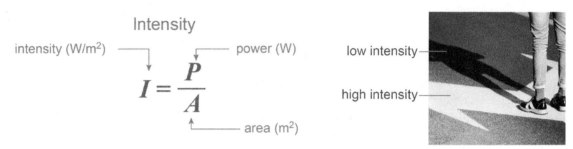

Inverse square law

The intensity of light from a small source decreases as the inverse square of the distance. Going twice as far decreases the intensity by $2^2 = 4$. This is a result of geometry. Consider a light source that emits a total power P equally in all directions. At any distance r, the power is spread out over a sphere of surface area $4\pi r^2$. The intensity of the light is $I = P/(4\pi r^2)$. A relationship in which a quantity decreases as $1/r^2$ is called an **inverse square law**.

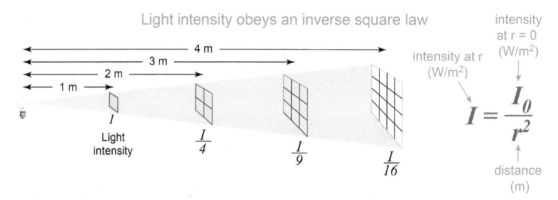

Radiation and distance

When radioactive particles are accidentally released, scientists use the inverse square law to define safe and unsafe distances from the site of the accident. The distance from the radioactive source, amount of radioactive material, and type of radioactivity all influence what is considered a "safe" distance. For example, you can protect yourself from alpha particles with a piece of paper. Thick cardboard or a thin sheet of metal can protect you from beta particles. You need a thick layer of lead to protect you from gamma radiation!

Solved problem

A source of x-rays is planned for a lab. How far away do you need to be for the intensity to be reduced to 1% (100x less) of the intensity at 0.5 meters?

Asked Distance for intensity to drop to 1%
Given The reference distance is 0.5 meters.
Relationships Intensity decreases as $1/r^2$.
Solve $I = \dfrac{I_0}{r^2} \rightarrow \dfrac{I_r}{I_{0.5m}} = \dfrac{(0.5m)^2}{r^2}$

$r = 0.5m \times \sqrt{\dfrac{I_{0.5m}}{I_r}} = 0.5m \times \sqrt{100} = 5m$

Answer Increasing the distance by 10x decreases the intensity by $10^2 = 100$.

Radiation exposure

Radiation dose

The science of radiation exposure concerns the effect of radiation on living tissue. The **radiation dose** is the amount of radiation absorbed by body tissue. Different kinds and energies of radiation have different biological effects. To take account of the differences there are different units used to measure dose. The units for radiation dose are listed below.

Units of radiation dose

1. **Gray (Gy):** an absorbed energy of one joule per kilogram of tissue.
2. **Sievert (Sv):** a biologically weighted dose taking the type of radiation and the type of tissue into account. The units are also joules per kilogram, (same as Gy) but one Sv can be more or less than one gray depending on the radiation type and the part of the body that is exposed. One sievert equals 100 rem.
3. **Radiation dose (rad):** an older unit equal to 0.01 gray.
4. **Radiation equivalent man (rem):** an older unit equal to 0.01 sievert.

Grays vs. sieverts

To illustrate the difference between one gray and one sievert, in one hour at the beach a 50 kg person might absorb a dose of 7,000 joules per kilogram, or 7,000 gray, of visible and infrared light. Because visible light is non-ionizing, the dose in sieverts is zero. Yet, a dose of one gray (1 J/kg) of gamma rays can cause serious injury and would be assigned a far higher value in sieverts. The same dose in grays creates a different dose in sieverts depending on the type of radiation and the part of the body affected.

Radiation doses in your environment

Source	Approximate dose
chest x-ray	100 µSv
Being in a stone or brick house for year	70 µSv
One average flight from US to Europe	40 µSv
One day of average background radiation	10 µSv
One dental x-ray	5 µSv
Eating one banana (from the potassium-40)	0.1 µSv

Average radiation exposure

On average, a person should only be exposed to about 0.0035 Sv (0.35 rem) per year. Most of this radiation comes from natural sources, but your exposure can depend on your job. Pilots that fly for 10 hours at 10,000 meters are exposed to the same amount of radiation as a person on the ground would be exposed to over 200 days. People that are routinely exposed to radiation sources at work carry devices that help them estimate their exposure. These devices are checked regularly to make sure that the total amount of radiation exposure stays in a safe range. A worker's total radiation dose should be below 0.05 Sv (5 rem) per year regardless of their profession.

Solved problem

A 5.0 kg sample exposed to a source emitting beta radiation absorbs 0.60 J of energy. What is the dose in grays?

Given : 5.0 kg sample and 0.60 J of energy
Relationships : 1 gray = 1 J/kg body tissue
Solve : $\frac{0.60 \text{ J}}{5.0 \text{ kg}} = 0.12 \text{ Gy}$
Answer : The dose is 0.12 Gy.

X-rays and CAT scans

x-rays

Today it is routine to diagnose a broken bone, or a tooth problem, or even serious internal injuries with an x-ray or CAT scan. A hundred years ago a doctor would have to cut you open to diagnose a fracture or internal injury! The "treatment" could easily be worse than the injury. The development of x-rays has saved innumerable lives and trauma by allowing doctors to see through the body.

How x-rays work

Like visible light, X-rays are electromagnetic radiation, but with much higher energy. Calcium is a strong absorber of x-rays which is why the image of a bone on an x-ray film is lighter than the surroundings. The high calcium content of bones absorbs x-rays and fewer reach the detector camera so those areas are light. Soft tissues absorb less of the x-ray energy, more reaches the detector, and those areas are darker.

How medical x-rays work

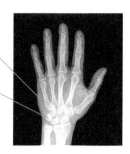

CAT scans

Computer Aided Tomography (CAT) uses multiple x-rays to construct a three-dimensional model of a volume. By rotating the x-ray source 360 degrees around and taking exposures from different angles, the computer is able to reconstruct a detailed three-dimensional model of tissues and bones. CAT scans use multiple x-rays that release a significantly higher radiation dose than regular x-rays. For this reason they are used on young children and pregnant women only when medically justified.

CAT scan uses x-rays from many angles to compute a 3-D model

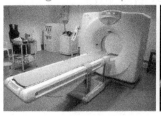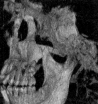

MRI technology uses magnetism and radio waves to image in 3-D

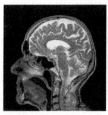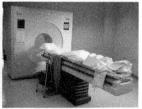

MRI imaging

The disadvantage of x-rays is that they do not image soft tissues well, and that there is a small but measurable health risk from radiation. The magnetic resonance imaging (MRI) technology can image any element in the body, not just calcium, and uses non-ionizing radiation with essentially no health risk. The nuclei of each atom act like very tiny magnets and each element has a different nuclear magnetic strength. An MRI scanner tunes radio waves to the exact frequency at which the magnetic axis of a specific element flips back and forth in an applied magnetic field. Because the frequency is different for every element, MRI can create more detailed images of soft tissue by looking at carbon, or phosphorus, or other elements.

Section 4 review

Chapter 19

Radioactivity can have damaging effects on living things. When the radiation energy is high enough it can free electrons from atoms or molecules. This affects chemical bonds and can change the function of biological molecules. For this reason it is important to limit exposure to radioactivity and shield ourselves from it whenever possible. The amount of radiation absorbed by a body is measured using one of the following units: gray, sievert, rad, or rem.

Vocabulary words: ionization, ionizing radiation, non-ionizing radiation, intensity, inverse square law, watt, radiation dose

Key relationships
- Intensity of radiation, $I = \dfrac{\text{power (W)}}{\text{area (m}^2\text{)}}$
- 1 Gy = 1 J/kg body tissue
- 1 Sv = 100 rem
- 1 rad = 0.010 J/kg body tissue
- 1 rem = 0.01 Sv

Review problems and questions

1. Given that a thin sheet of metal can protect you from β-particles, a sheet of paper will protect you from α-particles, and a thick shield of lead is required to protect you from γ-emission, which of these forms of radiation do you think is most damaging to tissue? Which is the least damaging? Explain your argument.

2. A space vessel does not have a magnetic field around it like Earth to protect astronauts and tourists from cosmic radiation. The safe limit of exposure is calculated at 25 rads per trip or mission. How much energy is this in J/kg body tissue?

3. A space tourist that weighs 130 lbs. has a mass of about 59 kg. If the space tourist has been exposed to 0.25 J of beta radiation on a trip, is the dose safe? Support your answer with a radiation dose in rads.

Section 19.4: Radiation

Radioactive Tracers

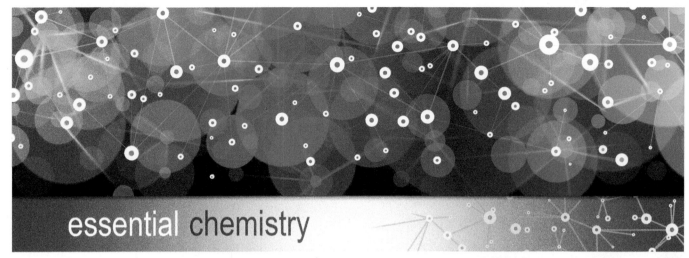

essential chemistry

Radioactive substances can give off ionizing radiation. So why would we want to purposefully use something that is radioactive? Well, it turns out that radiation is used as a chemical tracer to follow chemicals as they move through a system. Tracers are either radioactive isotopes or chemical compounds that have been marked by substituting an atom with its radioactive isotope. Radioactive tracers are used to help us understand chemical, biological, and medical processes. The next time you are in a hospital, dental office or other medical facility look around for the radiation symbol – or look for a sign warning you about radiation!

There are a few factors that make an isotope useful as a tracer. To be safe, isotopes are selected that expose the body to only small doses of radiation and work for only a short amount of time. Effective isotopes give off radiation that can be detected outside of the body. The most commonly used radioisotope is technetium-99m. The "m" after the mass number 99 indicates that this isotope of technetium is metastable and is not susceptible to uncontrolled nuclear decay. It is well suited to the role of a tracer because it emits gamma radiation that is easily detected and has a half-life of only 6.01 hours. The short half-life ensures that the concentration becomes negligible in a few days and is harmless. Technetium-99m is the most common radioisotope for diagnosis, accounting for about 80% of all scans.

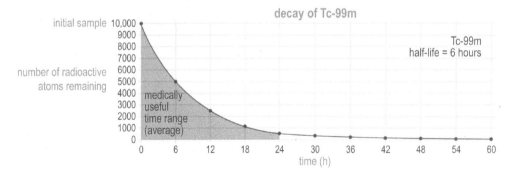

One medical study using a radioactive tracer, but not technetium-99, involves the thyroid gland. Your thyroid gland is in your neck and produces hormones that control your metabolism rate. The thyroid hormones are triiodothyronine and thyroxine. Looking at the molecular structure of the hormones, you can see that their compositions contain iodine. We get the iodine needed to make these hormones from foods, including dairy and seafood, and from "iodized" table salt.

Radioactive Tracers - 2

Thyroid issues affect the production of these hormones. Too much hormone production and your metabolism is too fast, too little production and you feel tired, cold and sluggish.

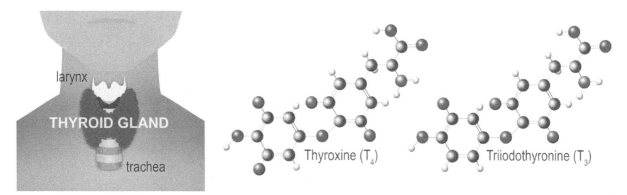

To detect and diagnose thyroid issues, a patient is given some sodium iodide or potassium iodide. But these are not normal salts! Instead of the stable and naturally abundant I-127 isotope, these tracers contain the radioactive isotope I-123. The isotope is relatively short-lived, with a half-life of only 13 hours. The iodine containing compounds are then taken up by the thyroid gland. As the isotope decays the emitted radiation is used to monitor thyroid health. A different iodine radioisotope can be used to not only identify but also treat thyroid diseases. The iodine-131 isotope is comparatively energetic. It undergoes beta emission and has a half-life of eight days. Instead of just tracing a process, this radioisotope is used to destroy cells. As it accumulates in the thyroid gland area it can destroy over-active thyroid tissues and cancerous cells. Iodine-131 is one of the most successful cancer treatments.

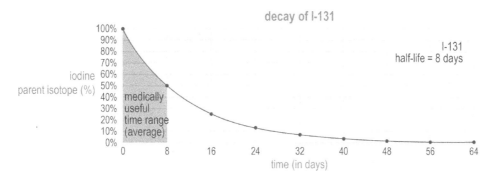

Phosphorous, like iodine, is another important essential element for biochemical systems. You might remember a specific phosphorous-containing molecule called adenine triphosphate (ATP) from your biology class. You will learn much more about the chemistry of ATP in the Biochemistry chapter of this book. Since phosphorous is so vital to biological systems, a radioisotope of phosphorous can be used to trace these systems.

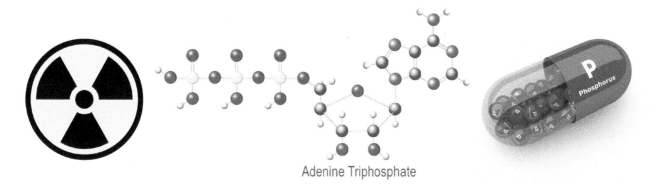

Adenine Triphosphate

Radioactive Tracers - 3

Phosphorous-32 is a beta emitter with a half-life of 14.29 days and has found uses in medicine, biochemistry, and even plant sciences. Medically, phosphorous-32 is particularly useful in identifying tumors. Cancer cells in tumors tend to accumulate more phosphorous than regular cells. When phosphorous-32 is used as a tracer, external scans of the body will reveal where the phosphorous has accumulated and helps to identify potentially malignant cells. Since it is a beta-emitter, the radiation given off by phosphorous-32 is diagnostic and can also be potentially therapeutic.

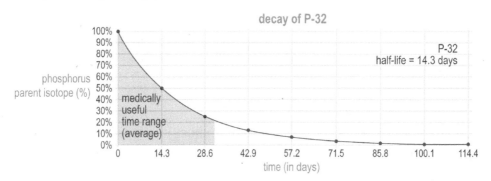

Biochemistry also benefits from the use of phosphorous-32. We can understand the chemical changes that happen in a cell as well as the movement of molecules by tracing the progress and movement of the phosphorous-32 radioisotope. This is even true in plant science when phosphorous-32 labeled fertilizer is "fed" to the plant via the soil or water.

Radioactive tracers have relatively short lived half-lives. This is great when they are being used, especially for medical purposes. But it also means you cannot just store a sample on a shelf and use it next month. Technetium-99m is produced from a parent nuclei of molybdenum-99, which itself is produced and extracted from the fission of uranium. The molybdenum-99 is then shielded and transported to hospitals to create Tc-99m. The molybdenum-99 has a half-life of 66 hours so can only generate Tc-99m for about one week. Other medically useful radioactive tracers must be created in a nearby nuclear reactor or in a cyclotron. A cyclotron is a particle accelerator that is useful in creating isotopes rich in protons. Cyclotrons are used around the world to create medically useful isotopes like fluorine-18 and carbon-11.

It was their essential understanding of the interactions between radiation and matter that gave scientists the foresight and ability to create radioactive tracers.

Chapter 19 review

Vocabulary
Match each word to the sentence where it best fits.

Section 19.1

antimatter	atomic number (Z)
binding energy	isotopes
mass defect	mass number (A)
nucleons	positron
radioactivity	

1. The _____ represents the number of protons in a nucleus and is unique to each element.

2. The total number of protons and neutrons in a nucleus is measured by its _____.

3. The energy that is required to keep nucleons inside a nucleus together is called _____.

4. _____(s) have the same mass as electrons and have a positive charge.

5. Atoms or elements called _____ have the same atomic number but have different masses.

6. The particles that make up the nucleus of an atom are known as _____.

7. The property of some elements to spontaneously change by releasing particles or energy is called _____.

8. Particles with the opposite charge of normal matter are called _____ particles.

9. The _____ between the reactants and the products results in the energy released during a nuclear reaction.

Section 19.2

alpha decay	beta decay
beta radiation	carbon-14
carbon dating	gamma decay
half-life	positron emission
radiation	radioactivity
rate constant	rate of decay

10. A nucleus that decreases its mass number by four and its atomic number by two is undergoing _____.

11. _____ is when an unstable nucleus releases a high-energy electron and its atomic number increases by one.

12. The _____ is a measure of the number of radioactive decays per unit time.

13. _____ is the transmission of energy through space.

14. When an excited nucleus emits high-energy electromagnetic radiation, but does not change atomic number or mass it is undergoing _____.

15. _____ is a radioactive isotope of carbon.

16. The _____ is the time it takes for half of the atoms in a pure radioactive substance to decrease by one half.

17. _____ is the process by which a nucleus of an atom changes by emitting particles or energy.

18. _____ is the result of β+ decay.

19. The _____ is a parameter that describes the speed at which a radioactive nuclide is decayed.

20. The radiation resulting from the emission of an electron or a positron is called _____.

21. A technique used to determine the age of organic materials by a ratio of isotopes is _____.

Section 19.3

chain reaction	critical mass
fission reaction	fusion reaction

22. During a(n) _____, a nucleus is split.

23. The minimal mass required to sustain a nuclear chain reaction is called the _____.

24. A nuclear _____ causes the neutrons produced in one fission reaction to cause more fission reactions to take place.

25. The type of nuclear reaction occurring in our Sun and stars is a _____.

Section 19.4

intensity	inverse square law
ionization	ionizing radiation
non-ionizing radiation	radiation dose
watt	

26. _____ does not have enough energy to induce ionization.

27. The _____ is the amount of radiation that is absorbed by the body.

28. When radiation ejects electrons out of their orbits the process is called _____.

29. The high energy of _____ is enough to eject electrons from their orbitals.

30. The _____ states that the intensity of radiation emitted from a radioactive source is decreased by a factor of four as you move twice times the distance away.

31. A measure of _____ describes how much energy flows per unit area per second.

32. 1 J/s is equivalent to one _____.

Conceptual questions

Section 19.1

33. With nuclear binding energy in mind, explain how when a star starts producing iron, it begins to collapse.

34. What are the main differences between nuclear reactions and chemical reactions?

Chapter 19 review

Section 19.1

35. What are two ways to symbolize an isotope of oxygen that has a mass number of 18?

36. How is binding energy related to the energy that is released in a nuclear reaction?

37. When calculating the mass of an atom, why aren't electrons included?

38. Why are molar masses on the periodic table shown as decimals instead of being whole numbers?

39. What is the "mass defect" and why is it important?

40. How does the energy released during nuclear reactions compare to that released from chemical reactions? What is responsible for the difference?

41. Write and label each symbol in Einstein's equation. Explain how nuclear changes relate to each symbol.

Section 19.2

42. What are the four most common types of decay reactions?

nickel-63

43. The decay of nickel-63 is shown in the image above. What type of decay is this, and what is the missing particle?

44. Explain how to balance a nuclear equation.

thorium-230

45. What type of decay is shown in the image above, and what is the missing particle?

46. What happens during alpha (α) decay, and what is the general equation for this reaction?

47. What happens during beta (β or β⁻) decay, and what is the general equation for this reaction?

48. What happens during gamma (γ) decay, and how does it change a nucleus?

49. What happens during positron (β⁺) emission decay, and what is the general equation for this reaction?

50. What happens to a radioactive sample after one half-life?

51. Identify the nuclear decay reactions that release high-energy particles.

52. Identify the nuclear decay reactions that release energy as electromagnetic radiation.

53. Smoke detectors contain the radioactive isotope Am-235, which emits α-particles. Assess the extent to which specific evidence provided on the Environmental Protection Agency (EPA) website supports this claim.

54. Henri Becquerel is credited as having discovered radioactivity. Research his work and write a couple sentences about it and what radioactivity is. Compare your research to how radioactivity is portrayed in movies, TV, shows, and novels.

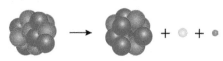

- neutron
- proton
- positron
- beta particle
- neutrino
- anti-neutrino

55. Study the nuclear reaction shown above.

a. What type of decay is shown?

b. In what way will the mass number of the original nucleus change? Explain your answer.

c. In what way will the atomic number of the original nucleus change? Explain your answer.

d. Does the atom in the products have the same identity as the atom in the reactants? Why or why not?

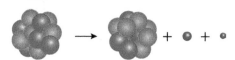

- neutron
- proton
- positron
- beta particle
- neutrino
- anti-neutrino

56. Study the nuclear reaction shown above.

a. What type of decay is shown?

b. In what way will the mass number of the original nucleus change? Explain your answer.

c. In what way will the atomic number of the original nucleus change? Explain your answer.

d. Does the atom in the products have the same identity as the atom in the reactants? Why or why not?

57. Use a graphing program to determine which of the following functions best describes the shape of a nuclear decay graph.

a. $y = x^4 - 4x^3 + 6x^2 - 4x + 1$
b. $y = \sqrt[3]{x}$
c. $y = -\sqrt{x}$
d. $y = x^2 + x - 1$
e. $y = 1/x$
f. $y = x^{-2}$

Chapter 19 review

Section 19.2

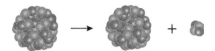

● neutron ◉ beta particle
● proton   ◉ neutrino
○ positron ◉ anti-neutrino

58. Study the nuclear reaction shown above.

    a. What type of decay is shown?

    b. How will the mass number of the original nucleus change? Explain your answer.

    c. In what way will the atomic number of the original nucleus change? Explain your answer.

    d. Does the atom in the products have the same identity as the atom in the reactants? Why or why not?

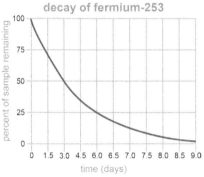

59. Use the graph above to estimate answers to questions a-c.

    a. What is the approximate half-life of Fm-253?

    b. If you begin with 10 g of a sample of Fm-253, how much mass will remain after 6 days?

    c. How long will it take for about 75% of the sample to decay?

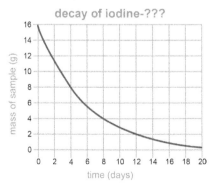

60. The half-life for I-124 is 4.18 days; half-life for I-125 is 59.40 days; half-life for I-131 is 8.02 days. Which of the 3 iodine isotopes is shown in the graph above? Explain.

61. Describe rate of decay and give an example.

62. How is carbon-14 used for dating once-living archaeological artifacts?

63. Draw diagrams that show alpha, beta, and gamma decay. Indicate how they produce more stable nuclei.

Section 19.3

64. There are two major types of nuclear reactions. What are they and what happens during each of them?

65. Why is it difficult to initiate and sustain fusion reactions?

66. Explain how a nuclear power plant generates electricity from nuclear changes.

67. How is a nuclear power plant similar to a coal power plant?

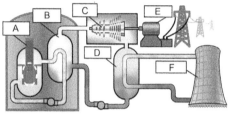

68. Choose any structure labeled A through F from the nuclear power plant diagram above.

    a. Identify the structure.

    b. If this structure were to fail, how would the power plant be affected? Write a 1-sentence summary of what you learned.

69. Synthesize information about nuclear fission from multiple sources in order to draft a report on whether the country should produce a larger fraction of electricity using nuclear power or not. Provide proper citations and support your opinion with at least two sources.

70. Research scientists may find themselves caught up in ethical dilemmas. During World War II, many of the world's leading scientists were recruited by the United States to develop the first atomic bomb. Write a 2-page discussion on the ethics for the United States to develop nuclear weapons and whether the research helped to advance society. Consider the following questions faced by scientists at that time:

    a. What were some reasons why the US chose to pursue nuclear weapons?

    b. What did scientists who worked on the bomb say about its creation? What were the consequences for those who spoke out for and against the continued development of nuclear weapons?

    c. What were the impacts of the people who lived where nuclear bombs were dropped in Hiroshima and Nagasaki in Japan?

    d. Could the technology be useful beyond the scope of military use?

Chapter 19 review

Section 19.4

71. Exposure to a large amount of radiation can damage biological tissues. Describe the process that takes place.

72. What determines if radiation will be harmful?

73. Identify three ways a scientist can protect themselves when studying radioactive samples.

74. Compare rads with rems. How is each unit used to keep track of radiation absorbed by our bodies?

75. What law does the intensity of radiation follow? According to this law, what is the relationship between the intensity of radiation and the distance from its source or point of origin?

Quantitative problems

Section 19.1

76. Identify the number of protons and neutrons in each of the following isotopes.

    a. $^{2}_{1}H$

    b. $^{7}_{3}Li$

    c. $^{16}_{8}O$

    d. $^{58}_{28}Ni$

    e. $^{129}_{54}Xe$

    f. $^{232}_{90}Th$

77. Convert each of the following into energy in joules. (1 amu = 1.6606×10^{-27} kg)

    a. The mass of one neutron (m = 1.0087 amu).

    b. The mass of one proton (m = 1.0073 amu).

78. The sun releases 3.86×10^{26} joules of energy per second. How much mass is converted to energy per second?

79. The Earth receives 1.800×10^{17} J/s of solar energy.

    a. How much energy is received by Earth in one day?

    b. What mass of coal (in kg) would have to be burned to provide the same amount of energy? (Coal produces 32 kJ of energy per gram when burned)

    c. The nuclear reaction taking place in the Sun produces far more energy per kg of fuel than the chemical recreation of burning coal. How many times more mass of coal must be burned than hydrogen fuel reacted in the Sun to produce the same amount of energy?

    d. There are approximately 1.1 trillion tonnes of coal reserves on Earth. If all the energy received by Earth from the Sun had to be produced by burning coal, how long would the coal reserves last?

80. According to Einstein's equation for mass-energy equivalence, how much energy would be released if 10.0 g of mass was converted into energy?

Section 19.2

81. Write balanced nuclear equations for each of the following:

    a. $^{38}_{19}K \rightarrow \underline{\quad} + ^{0}_{+1}e$

    b. $^{205}_{83}Bi \rightarrow ^{201}_{81}Tl + \underline{\quad}$

    c. $^{18}_{9}F \rightarrow ^{18}_{8}O + \underline{\quad}$

    d. $^{6}_{2}He \rightarrow \underline{\quad} + ^{0}_{-1}e$

82. Write balanced nuclear equations for the following:

    a. $\underline{\quad} \rightarrow ^{236}_{92}U + ^{4}_{2}He$

    b. $\underline{\quad} \rightarrow ^{40}_{20}Ca + ^{0}_{-1}e$

    c. $\underline{\quad} \rightarrow ^{40}_{18}Ar + ^{0}_{+1}e$

83. Iodine-123 is a synthetic, unstable isotope used for medical imaging and diagnosis. Its decay constant equals 0.05775 hours^{-1}.

    a. What is the half-life of I-123?

    b. If a patient drinks a potassium iodide (KI) solution containing iodine-123 how many hours will it take before the concentration of iodine-123 reaches 21.0% of its initial concentration?

84. Iodine-131 is a radioactive isotope used to treat thyroid cancer. I-131 has a half-life of 8.02 days.

    a. If a sample of Iodine-131 contains 20.0 g, how many will remain after 24.06 days?

    b. What is Iodine-131's rate of decay per day?

85. Radiocarbon dating is an effective method of dating organic matter up to a maximum of approximately 50,000 years. What percentage of the original carbon-14 exists after this time? The half-life of carbon-14 is 5,730 years.

86. A paleontologist finds a completely preserved wooly mammoth, but only needs a small sample of bone to determine its age by C-14 analysis.

    a. Find the percentage of C-14 remaining in the bone if it has 1/5 of the amount that living things have.

    b. The half-life of C-14 is 5,730 years. How old is the bone?

87. How many successive α-decays occur in the decay of the thorium isotope ^{228}Th into ^{212}Pb?

88. Uranium isotopes are used to date very old non-living things due to their long half-lives. U-238 has a half-life of 4.47×10^9 years. If a rock today has 0.5000 g of U-238 in it, how much U-238 will be in the rock 2.5 billion years from now?

Chapter 19: Nuclear Chemistry

Chapter 19 review

Section 19.2

89. Carbon-14 has a half-life of 5,730 years.

    a. What percentage of a C-14 sample remains after 4,250 years?

    b. What percentage remains after 24,000 years?

90. Tritium is an isotope of hydrogen that contains two neutrons and one proton. Tritium is beta-radioactive and emits an electron.

    a. Write the abbreviation for the tritium isotope.

    b. What does tritium become after undergoing β-decay?

    c. The rate constant for tritium is 0.05544/years. What is the half-life of tritium?

    d. How much of a 5.0 g sample will remain after 25 years?

91. Cobalt-60 is a useful gamma ray source that is used in medicine and to preserve and sterilize foods. The Food and Drug Administration (FDA) has confirmed Co-60 food irradiation is safe for humans, but is lethal for viruses, bacteria, and insects. The half-life of Co-60 is 5.2714 years.

    a. What is the decay constant of Co-60?

    b. A food processing plant has 165.0 g of a Co-60 source. How much of the sample will still be effective for irradiation in 10 years?

92. Potassium-40 has a half-life of 1.25×10^9 years. If a sample of K-40 has a mass of 8.50498×10^{-11} g, how many atoms will decay in one week?

Section 19.4

93. The exposure at 2.0 meters from a source of radiation is 15 rads, what will the exposure be at 5 meters?

94. If a source of radiation is releasing 120 J per second, what is the intensity experienced by a person standing 3.0 meters away from the point of radiation?

95. The dosage of Co-60 irradiation depends on the purpose.

    a. Fresh fruit sterilization and ripening delay requires low-dose Co-60 irradiation up to 1 kGy. How much is this in rads?

    b. Medium-dose Co-60 irradiation is used to delay meat spoilage and kill pathogens in meat and spices. If a dose of 7.0 kGy is applied to 5.0 kg of meat, how many joules are absorbed by the meat?

    c. High-dose applications of Co-60 are used to sterilize packaged meat. If 35.0 kGy has been applied to a 2.0-pound package of ground beef patties, how much radiation energy has been absorbed? One pound = 453.592 g.

96. While waiting in line at the airport, the person in front of you tells you they are afraid of radiation from the X-ray scanner at the security checkpoint.

    a. If a person receives a dose of 0.001 mrem from the full-body X-ray scanner, how much is this in sieverts?

    b. The maximum radiation a person should receive in one year is 0.05 Sv. How many trips through the scanner in one year would it take to receive this dose?

97. A 20-g piece of fruit absorbs 0.520 J of energy.

    a. What is the dose in grays?

    b. What is the dose in rads?

98. Using the inverse square law, by what factor would you have to increase the distance by to reduce the exposure from a point of radiation by 99%?

99. Find the intensity of a light source at a distance of 12.5 m that produces an illumination of 1.93 lux. (1 lux = 1 lumen/m²)

100. A source of radiation releases 1,400 W from a point. Use the inverse square law to answer the following:

    a. What is the intensity of radiation 5.0 meters away?

    b. What is the intensity of radiation 10.0 meters away?

101. Radiation doses of over 1,000 rads are almost invariably fatal. How many joules of energy is absorbed by a person's body if the person has 62 kg body mass?

chapter 20 Organic Chemistry

When you look inside your refrigerator or go to the supermarket, you may see the word "organic" on many food labels. In this case, the label "organic" implies that the products were grown naturally without the use of synthetic pesticides, fertilizers or additives. Properly labeled "organic" foods are generally considered healthier, even though there is not sufficient evidence to scientifically support those claims.

In chemistry, the word organic originally meant something similar. It was believed that organic chemicals could only be synthesized by living things. In 1828 Friedrich Wohler disproved this idea when he synthesized an organic compound from inorganic reactants. Wohler converted inorganic ammonium cyanate into organic urea. Urea is produced by mammals during protein breakdown and is eliminated in urine.

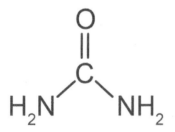
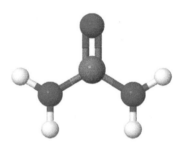

Now we understand that organic chemicals don't have to come from plants or animals, but are simply compounds that contain carbon. Because of the versatility of the carbon atom, there are millions of organic compounds that can vary greatly in size and complexity. Organic molecules can be as simple as methane (CH_4) which is a fuel for Bunsen burners. They can also be larger and more complicated like plastics or proteins.

So, the next time you pick up a gallon of "Organic" milk, try to remember that the plastic jug is made of organic chemicals too!

Chapter 20 study guide

Chapter Preview

Organic chemistry began as the study of compounds produced by living organisms. It has since been expanded to the study of carbon-containing compounds. In this chapter you will learn about the composition, properties, structure and reactions of carbon compounds. The simplest of these are hydrocarbons called alkanes. They are composed only of carbon and hydrogen atoms connected by single bonds. Branched chains, double bonds, triple bonds and ring structures can add complexity to hydrocarbons. Functional groups are important groups of atoms that add special functionality to hydrocarbon chains. There are many functional groups but we will focus on the most important ones in this chapter. Finally, organic compounds can take part in many biologically and commercially important reactions including polymerization, combustion, and hydrogenation.

Learning objectives

By the end of this chapter you should be able to:

- write chemical formulas and draw structural formulas for saturated hydrocarbons;
- write chemical formulas and draw structural formulas for unsaturated hydrocarbons;
- distinguish between non-isomers, structural isomers, and stereoisomers;
- name straight chain, branched and ringed hydrocarbons as using the naming rules;
- identify functional groups and their properties; and
- identify common applications of organic reactions.

Investigations

20A: Bonding and organic chemistry

20B: Distilling aromatic compounds

20C: Fragrant esters

Pages in this chapter

- 642 Carbon Chemistry
- 643 Alkanes
- 644 Drawing and naming alkanes
- 645 Branched alkanes
- 646 Naming branched alkanes
- 647 Characteristics of hydrocarbons
- 648 Everyday uses of hydrocarbons
- 649 Section 1 Review
- 650 Other Organic Compounds
- 651 Naming unsaturated hydrocarbons
- 652 Structural isomers
- 653 Stereoisomers
- 654 Drawing and naming ring structures
- 655 Ring structures and aromatic hydrocarbons
- 656 Section 2 Review
- 657 Functional Groups
- 658 Alcohols and carboxylic acids
- 659 Ketones and aldehydes
- 660 Familiar ketones and aldehydes
- 661 Esters and ethers
- 662 Amines and alkyl halides
- 663 Summary of organic names
- 664 Section 3 Review
- 665 Organic Reactions
- 666 Reactions of alkanes: Dehydrogenation
- 667 Reactions of alkanes: Substitution
- 668 Reactions of alkanes: Cracking
- 669 Reactions of alkenes and alkynes
- 670 Organic polymerization reactions
- 671 Section 4 Review
- 672 Plastics: What's in a Number?
- 673 Plastics: What's in a Number? - 2
- 674 Chapter review

Vocabulary

organic chemistry	hydrocarbon	saturated hydrocarbon
unsaturated hydrocarbon	alkane	parent compound
R group	alkene	alkyne
structural isomer	geometric isomer	optical isomers
aromatic hydrocarbon	resonance	functional group
alcohol	carboxylic acid	aldehyde
ketone	carbonyl group	carbohydrates
ester	ether	amine
alkyl halide	dehydrogenation reaction	substitution reaction
free radical	petroleum	cracking
addition reaction	hydrogenation reaction	partial hydrogenation
addition polymerization	condensation polymerization	

Chapter 20 Summary

20.1 - Carbon Chemistry

Organic chemistry is the study of compounds that contain carbon. Carbon atoms are able to generate an infinite number of chemical compounds with different properties thanks to their atomic structure. Carbon makes four bonds and can bond easily with itself and with other atoms. Carbon has the ability to make double and triple bonds with itself and other nonmetals.

Organic chemistry

Organic chemistry is carbon chemistry

Organic chemistry is the study of the structure and properties of compounds that contain carbon, except for salts like carbonates. Carbon is one of the most special elements on the periodic table because its bonding abilities are unmatched by any other element. Carbon has an electron configuration that makes it easy for the atom to bond to itself and other elements in what seems like an unlimited number of combinations. Organic compounds take on many shapes, including long chains, flat sheets, loops, rings and three-dimensional networks. As you learned in a previous chapter, carbon makes a tetrahedral structure when it has all single bonds. Diamonds get their hardness and strength from the tetrahedral lattice of carbon-carbon bonds as shown.

Carbon forms hybrid orbitals

A carbon atom has six electrons with an electron configuration of $1s^2 2s^2 2p^2$. As you remember, the four valence electrons hybridize to create four sp^3 orbitals that spread out to reach the maximum distance from one another. The four bonds are 109.5° away from each other, which forms a very stable and strong tetrahedral shape.

Tetrahederal   Pentagonal

Carbon can bond from 1 through 6 atoms

Carbon can bond to fewer than four atoms if it makes a double or triple bond. When carbon makes double and triple bonds with itself or other atoms, it forms different hybrid orbitals with different geometries. Under very specific conditions, carbon can bond to more than four atoms! Although it has not been found in nature, scientists have created a synthetic organic compound called hexamethylbenzene dication, $C_6(CH_3)_6^{2+}$ in which carbon makes six bonds as shown above. This six-bond structure is unstable unless its environment is highly acidic. Under these conditions carbon forms a five-sided pyramid called a pentagonal pyramid. The bonds in a pentagonal pyramid are weak compared to a tetrahedral pyramid.

Hydro-carbons

The least complex organic compound is called a **hydrocarbon** because it contains only carbon and hydrogen. Hydrocarbons are either saturated or unsaturated. A **saturated hydrocarbon** contains only single carbon-carbon bonds. An **unsaturated hydrocarbon** has at least one double or triple bond between two carbon atoms.

Alkanes

Alkanes are saturated

Organic reactions tend to occur at the site of a double or triple bond. An **alkane** is a saturated hydrocarbon that is not very reactive because it contains only single bonds. Alkanes range in complexity from one carbon atom to a series of carbon atoms linked together in a branched chain. The smallest hydrocarbons are named with a prefix that indicates how many carbons are joined together to form the main "parent" carbon chain. The prefixes and the formulas for the simplest alkane chains are listed below.

prefixes for small hydrocarbons

prefix	number of carbon atoms	formula for an alkane
meth-	1	CH_4
eth-	2	C_2H_6
prop-	3	C_3H_8
but-	4	C_4H_{10}
pent-	5	C_5H_{12}
hex-	6	C_6H_{14}
hept-	7	C_7H_{16}
oct-	8	C_8H_{18}
non-	9	C_9H_{20}
dec-	10	$C_{10}H_{22}$

example of an alkane

chemical name: **pentane**

chemical formula: C_5H_{12}

structural formula:

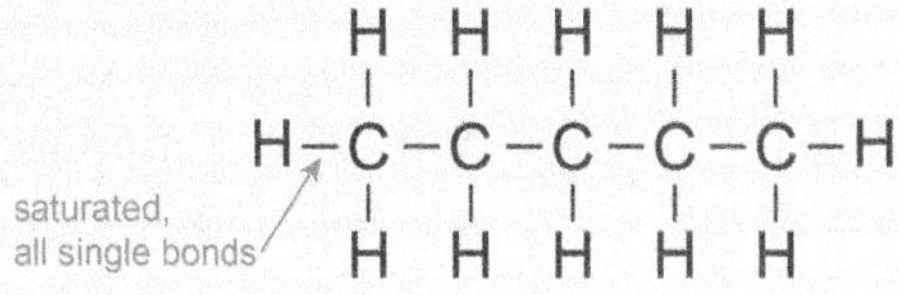

```
    H  H  H  H  H
    |  |  |  |  |
H - C -C -C -C -C - H
    |  |  |  |  |
    H  H  H  H  H
```
saturated, all single bonds

condensed structural formula:

$CH_3CH_2CH_2CH_2CH_3$

Simple alkanes

Look in the table for the pattern in carbon-to-hydrogen bonding in simple alkanes. For each carbon atom (N), there are two times plus two hydrogen atoms (2N + 2). The general formula for a simple alkane is $C_NH_{(2N+2)}$ where N represents the number of carbon atoms.

Carbon chains are not straight

The prefix pent- in a molecule named pentane has 5 carbon atoms bonded to each other in a parent chain. The suffix -ane indicates the molecule is an alkane and it has only single bonds. Build a pentane molecule to understand how a structural formula is represented in a drawing compared its the three-dimensional structure. Start by forming the parent chain of carbon-to-carbon bonds. Connect 5 carbon atoms with single bonds in a straight chain, then add 12 hydrogen atoms. When you observe the structure of pentane, you will see that a straight chain is really not straight or linear. Why is this?

pentane

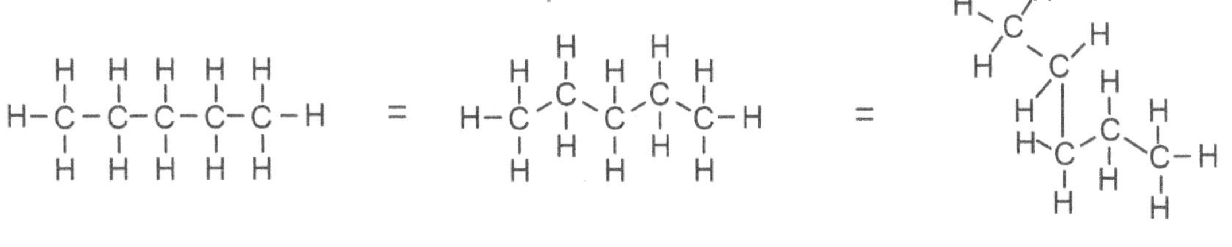

Carbon makes a three dimensional tetrahedral structure when it forms four single bonds, therefore a linear chain is not possible. Spin each carbon atom around one of its bonds. Notice how hydrogen atoms end up in different places as the carbon atoms rotate. If you rotate more than one carbon atom at a time, you can twist the chain like the one shown. No matter how you rotate the carbon and hydrogen atoms, as long as you do not break any bonds you still have pentane because the formula stays the same. All three molecules shown above are pentane because all three contain five carbons in the parent chain with twelve hydrogen atoms attached with single bonds.

Drawing and naming alkanes

Alkane naming rules

Apply the following rules to name an alkane:

1. If the compound is not pictured, draw it to verify there are only single bonds. Check your drawing by counting H and C atoms. Your H and C count should match the given chemical formula.

2. Determine the number of carbons in the longest carbon chain. Choose the appropriate Greek prefix from the table based on this number.

3. Combine the Greek prefix with the -ane suffix.

Number of Carbons	Greek Prefix
1	Meth
2	Eth
3	Prop
4	But
5	Pent
6	Hex
7	Hept
8	Oct
9	Non
10	Dec

Solved problem

A student is asked to retrieve C_7H_{16} from the chemical storage area, but all chemicals are labeled by their name. What is C_7H_{16} named?

Asked: Name the compound C_7H_{16}

Given: C_7H_{16}

Relationships: The number of carbons indicates the prefix. Alkanes receive the –ane suffix.

Solve:

1. Draw the molecule to verify this is an alkane. Start by connecting 7 carbon atoms together to form the parent chain. Give each carbon atom 4 bonds.

$$-\overset{|}{\underset{|}{C}}-\overset{|}{\underset{|}{C}}-\overset{|}{\underset{|}{C}}-\overset{|}{\underset{|}{C}}-\overset{|}{\underset{|}{C}}-\overset{|}{\underset{|}{C}}-\overset{|}{\underset{|}{C}}-$$

Add a hydrogen atom to each empty carbon bond.

$$H-\overset{H}{\underset{H}{C}}-\overset{H}{\underset{H}{C}}-\overset{H}{\underset{H}{C}}-\overset{H}{\underset{H}{C}}-\overset{H}{\underset{H}{C}}-\overset{H}{\underset{H}{C}}-\overset{H}{\underset{H}{C}}-H$$

Compare the carbon and hydrogen total against the given formula. There are 7 carbon atoms and 16 hydrogen atoms. The drawing is correct.

2. The parent chain is 7 carbons long; choose hept- for the prefix.

3. Combine the prefix hept- with the suffix -ane.

Answer: The molecule above is called heptane.

Branched alkanes

Find the longest continuous carbon chain

Branched chain alkanes contain carbons that have atoms or groups of atoms other than hydrogen bonded to the carbon chain. The longest continuous chain of carbons is called the **parent compound**. The parent compound helps you name the alkane. When you are asked to name an organic compound given a drawing, look for the highest number of carbons linked together in a chain. Use a pencil to draw lines through carbon atoms as you count to help you find the longest carbon chain. Explore all options by starting and stopping the carbon count from all possible ends of the molecule. Find the longest parent compound in the molecule represented below. The molecule is shown with its regular structural formula and its condensed structural formula.

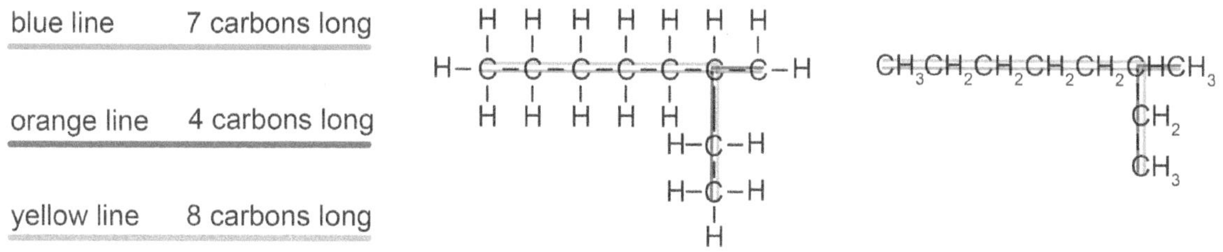

blue line — 7 carbons long

orange line — 4 carbons long

yellow line — 8 carbons long

Name the parent

There are three different carbon chains in this molecule. The longest carbon chain has eight carbon atoms. The parent compound is called octane. The hydrogen atoms are not part of the parent chain, they are just condensed in the diagram to the right to save space.

Name the R-group with the –yl suffix

There is a side chain coming off the parent chain. The one-carbon -CH_3 group attached to the parent is called a methyl group. Meth- represents one carbon and –yl means it is a branch attached to the parent compound. The methyl group is also called an **R group**. An R-group is an abbreviation for a side chain attached to a parent carbon compound.

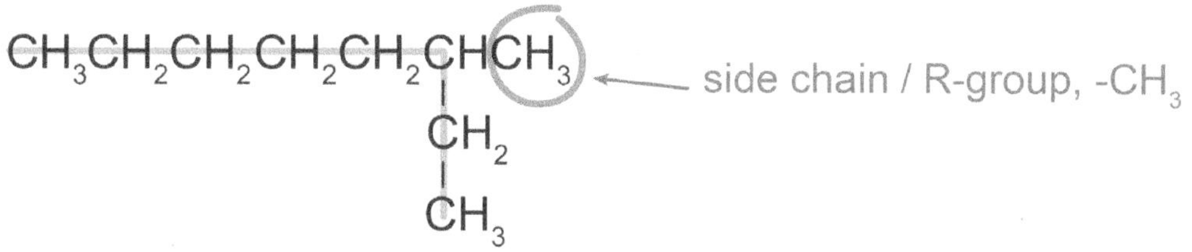

Give the branch a number

Identify which parent carbon contains the methyl group. Number the parent in a direction that results in the lowest number possible for the branch. Here it is on the third parent carbon when you number the parent from right to left. If the methyl group was located on the fourth carbon instead of the third, the compound would have a different name along with different chemical and physical properties. Always check numbers in both directions!

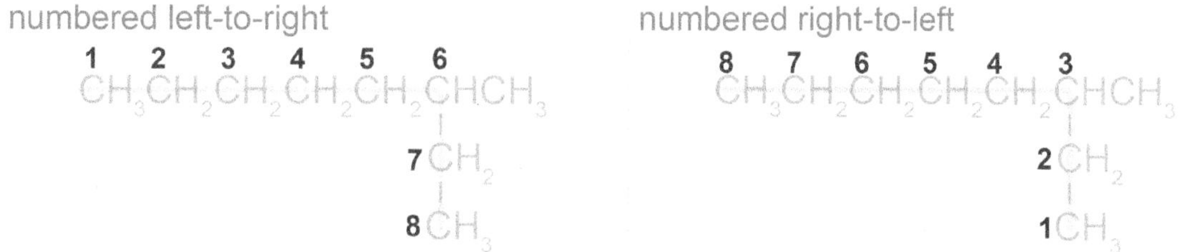

numbered left-to-right

numbered right-to-left

Punctuate and combine names

The molecule is called 3–methyloctane. The R-group location number and name are placed at the front. Notice the dash connecting the group number and name. Group and parent names become one word. The name does not contain spaces.

Naming branched alkanes

Naming and drawing Let's practice naming and drawing branched alkanes. The first problem is an example of how to name a branched alkane, and the second problem shows the steps to draw a branched alkane.

Solved problem Name the molecule shown

Relationships Use the prefix that matches parent chain length and add the -ane ending.

Solve
1. Find the longest continuous chain to name the parent. In this case the 5-carbon saturated chain is called pentane

2. Name the R-group based on the number of carbons it has, and add the –yl suffix - in this example "methyl"

3. Number the parent in a direction that gives the R-group the lowest possible number…

4. Put the name together in this order:
   first: R-group number, **then** dash, **then** R-group name
   last: parent name.

Answer 2-methylpentane

Solved problem Draw 2-methylbutane.

Relationships The number indicates the position of the methyl group on the parent chain.

Solve
1. Draw the parent chain first; give each carbon 4 bonds.
2. Add the 1-carbon methyl group to either the top or bottom of carbon #2; give all carbons 4 bonds.
3. Add hydrogens to open bonds.

Answer (Pictured to the right)

Chapter 20: Organic Chemistry

Characteristics of hydrocarbons

The C-H bond

Methane, CH_4, is a non-polar molecule. Carbon and hydrogen have a small difference in electronegativity, meaning electrons will mostly be shared evenly between the nuclei of the two atoms. This is why the methane molecule is considered to be non-polar. In fact, most alkanes are non-polar. Methane molecules form liquids and solids at very low temperatures, and the particles are only weakly held together by London dispersion forces.

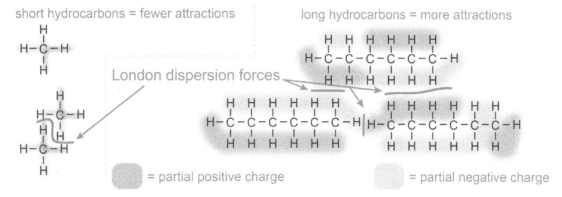

Attractive forces influence properties

Hydrocarbons with low molecular weights, such as methane, are gases at room temperature. The particles are not attracted enough to form a liquid. A particle needs less kinetic energy to escape into the gas phase if intermolecular attractive forces are weak. Hydrocarbons with larger molecular weights, such as hexane and octane, show a trend of increasing boiling point as molecule size increases. As shown above, as a hydrocarbon chain gets longer the London dispersion forces become more and more effective. Stronger intermolecular attractions keep molecules nearer to each other. Increased contact holds the hydrocarbon chains together closely enough to form liquids at lower temperatures. Hydrocarbons with low boiling points are flammable at room temperature.

boiling points for small hydrocarbons

hydrocarbon	formula	boiling point (°C)
methane	CH_4	-164
ethane	C_2H_6	-89
propane	C_3H_8	-42
butane	C_4H_{10}	-0.5
pentane	C_5H_{12}	36
hexane	C_6H_{14}	69
heptane	C_7H_{16}	98
octane	C_8H_{18}	125

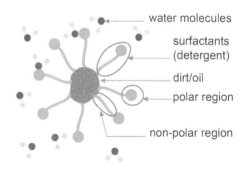

how surfactants work

Non-polar molecules with polar side chains

The non-polar nature of hydrocarbons makes them less soluble in polar solvents like water. However, a hydrocarbon can have a polar side chain attached to it. This gives some hydrocarbons the flexibility to dissolve in either polar or non-polar solvents. A surfactant is a molecule that has polar and non-polar regions. For example, the polar ends of laundry detergent molecules reduce water's surface tension so it can spread into fabric more easily. The non-polar ends are attracted to non-polar dirt and oil molecules on your clothes. The non-polar ends of the detergent break up oil and surround it like an envelope. Polar regions face outside and non-polar regions face inside. These "envelopes" carry oil particles from your clothes to the water in a washing machine. Surfactants are also used to disperse oil from oil spills into water where bacteria can further break down the oil.

Section 20.1: Carbon Chemistry

Everyday uses of hydrocarbons

Petroleum isomers

Petroleum is a mixture of liquid hydrocarbon chains with different lengths. Short chain hydrocarbons from methane through butane are useful fuels because they are flammable, compressible gases. Chains of five to seven carbons are useful as non-polar solvents. Chains seven through twelve carbons long are used as liquid fuels. In oil refineries, petroleum can be separated into different hydrocarbons through distillation based on each compound's boiling point. Hydrocarbons are further processed when they are exposed to high temperatures and catalysts to create new products with a variety of structures. Gasoline is an example of a hydrocarbon that is modified in oil refineries. You probably recognize "octane" from a gas station, but gasoline is not actually made of linear octane molecules. A branched hydrocarbon called isooctane makes up "octane" at the gas pump.

Octane ratings for gasoline

Gasoline is a mixture of hydrocarbons. Octane ratings at the fuel pump are based on a ratio of branched isooctane to linear heptane. The number 87 at the gas pump indicates a ratio of 87% isooctane to 13% heptane. A higher octane rating has a greater percentage of isooctane. Isooctane is a more efficient fuel than linear octane because the branches help the fuel burn at a rate that produces optimal power for an engine.

Other uses of hydrocarbons

Hydrocarbons are abundant on planet Earth and in other places inside and outside our solar system. Titan, one of Saturn's moons, has lakes of liquid ethane on its surface. Ring-shaped hydrocarbons have been detected in distant nebulae. Most people are familiar with hydrocarbons as fuels and oils, but other common uses include glues, paints, antiseptics, numbing agents, plastics, and solvents such as paint thinner and dry cleaning fluid. There are countless practical uses for hydrocarbons as well as fun uses.

Small hydrocarbons have low boiling points

A drinking bird toy uses the low boiling point of a hydrocarbon called dichloromethane, shown above. The chemical name tells you the parent compound is a single carbon, and it is saturated because it ends in -ane. You can guess from the dichloro- prefix that there are two chlorine atoms on the molecule. When the bird's hollow body is built, all air is removed from it and is replaced with some liquid dichloromethane and dye. A little dichloromethane evaporates because its boiling point drops in the absence of air and it has a small molecular size. If you wet the bird's head, the water evaporates and cools the head down. Dichloromethane gas in the head condenses to a liquid, setting off a chain reaction of pressure changes. If the bird's beak touches water and the air temperature is above dichloromethane's boiling point, the bird will appear to "drink water" from a glass.

Section 1 review

Chapter 20

Organic chemistry is the study of the structure and properties of carbon compounds. Carbon has four valence electrons and can form 4 bonds. Hydrocarbons are organic compounds containing only carbon and hydrogen. Alkanes are hydrocarbons containing only a single bond. The formula for alkanes is $C_NH_{(2N+2)}$ and are named using Greek prefixes depending on the number of carbons. Branched alkanes are named using the longest chain, and then by numbering and naming the side chain. Hydrocarbons are non-polar molecules held together by weak London dispersion forces. This explains why the smallest hydrocarbons are gases while larger ones are liquids at normal temperatures.

Vocabulary words: organic chemistry, hydrocarbon, saturated hydrocarbon, unsaturated hydrocarbon, alkane, parent compound, R group

Key relationships
- Alkanes, also known as saturated hydrocarbons, contain only single bonds.
- Alkanes are named with the prefix shown in the chart based on the number of carbons in the parent chain and receive the –ane suffix.
- Number the parent chain in a direction that results in the lowest number combination for multiple bonds and branching groups.

Review problems and questions

1. For the following organic compounds:

   a.   b.   c.   d.

   a. Determine whether each is saturated or unsaturated.
   b. Name only the saturated compounds.

2. Draw 4-ethylnonane.

3. What part of a hydrocarbon's name tells you whether it is saturated or not?

4. What part of a group's name tells you it is a group and not part of the saturated parent chain?

Section 20.1: Carbon Chemistry

20.2 - Other Organic Compounds

Some hydrocarbons contain either double or triple carbon to carbon bonds in their parent chain. The naming system for these types of hydrocarbons is very similar to the process for naming saturated hydrocarbon chains. Because carbon has so many opportunities to form bonds, sometimes one chemical formula can be represented by more than one chemical structure.

Alkenes and alkynes

Unsaturated molecules contain multiple bonds

Alkanes are hydrocarbons that are saturated with the maximum number of bonds with atoms other than carbon. Carbon can bond to itself or other atoms by forming double or triple bonds. Unsaturated hydrocarbons have multiple bonds. Each carbon atom will still form four bonds, but two of the four may be found in a double bond or three of the four may be located in a triple bond.

energy needed to break C-C bonds

bond type	example	bond energy (kJ)
single –	ethane: CH_3-CH_3	348
double =	ethene: $CH_2=CH_2$	614
triple ≡	ethyne: $CH\equiv CH$	839

Bond strength

Double bonds take more energy to break than single bonds and triple bonds require even more energy. As more bonds are added, the nuclei between atoms get closer to each other. More energy is needed to pull apart atoms joined by multiple bonds.

Alkenes have double bonds

An **alkene** is an unsaturated hydrocarbon that contains one or more double bonds. Alkenes use the same prefix system for alkanes, except the -ene suffix indicates double bonds are present. Ethene is a two-carbon alkene used to make a plastic called polyethylene, PE. Look for the letters PE at the bottom of a recyclable plastic container. Many products, from small sandwich bags to artificial knee replacement joints are made with PE. Ethene is a naturally-occurring plant hormone that is responsible for many chemical changes. For example, ethene builds up in fruit as its seeds mature. Higher levels of ethene causes fruit to ripen. Ethene can be used commercially to ripen fruits that are harvested early.

ethene

Alkynes have triple bonds

Ethyne is an example of an **alkyne**, a hydrocarbon that contains one or more triple bonds. Eth- means the parent compound contains two carbon atoms and -yne indicates there is a triple bond. Ethyne can also be used to manufacture plastics and it is also used as a fuel in welding torches.

Naming unsaturated hydrocarbons

Double and triple bonds

To name a hydrocarbon with multiple bonds, first determine whether the compound is an alkene or alkyne. Then choose a prefix that matches the length of the parent compound. If the chain has a double bond, add the suffix –ene. If you see a triple bond, use the –yne suffix. The three examples below each show a different rule for a 5-carbon chain.

The first molecule is called 2-pentene. The prefix pent- indicates the parent is 5 carbons in length. The suffix -ene indicates there is a double bond. The number 2- is used to identify the location of the multiple bond. Since the double bond is located between the second and third carbon, choose the lowest carbon number, 2. No location number is needed in the pentene molecule because the double bond is located on carbon number 1. The number 1 is invisible just like a chemical formula. You might have noticed the bonds on carbon #1 in pentene form a triangle. This is because bonds in structural diagrams are drawn to show maximum repulsion. The last molecule is called 2-pentyne because there is a triple bond between the second and third carbon atoms in a five-carbon chain.

Solved problem

Name the compound shown.

Relationships Combine a prefix, the -yne suffix, and the triple bond location

Solve
1. Name the parent chain; hept- for 7 carbons and -yne for the triple bond. The parent is named heptyne.
2. Number the parent carbons in a direction that results in the lowest number pair for the triple bond.

   In this example, numbering from right to left locates the triple bond between carbons number 3 and 4. Three is the lowest number.
3. Add the location of the triple bond in front of the parent prefix followed by a dash: 3-hept

Answer The molecule is named 3-heptyne.

Hydrogen atoms and multiple bonds

Did you notice fewer hydrogen atoms are bonded to carbons with double and triple bonds? This is because carbon can only make a total of 4 bonds. A carbon has hydrogen atoms if there are less than 4 carbon-to-carbon bonds. Recall the formula pattern for alkanes: $C_nH_{(2n+2)}$. The alkene pattern has two less hydrogens: $C_nH_{(2n)}$ and alkynes have four less hydrogens: $C_nH_{(2n-2)}$. Count the carbon and hydrogen atoms in 3–heptyne to verify the formula pattern for alkynes.

Structural isomers

What is a structural isomer?

If asked to draw C_8H_{18}, you might draw octane. But there are 24 unique ways to draw structural formulas for C_8H_{18}! Just 4 of the 24 structural formulas are shown below. Notice each drawing shows a different structure and chemical name with the same chemical formula of C_8H_{18}. Each name is based on where atoms are located on the compound. A **structural isomer** is a chemical formula that can take on more than one unique structural formula. Isomers have identical chemical formulas, but have different chemical and physical properties due to different bonding patterns.

C_8H_{18} / 4 of 24 structural isomers

$CH_3-CH_2-CH_2-CH_2-CH_2-CH_2-CH_2-CH_3$
octane

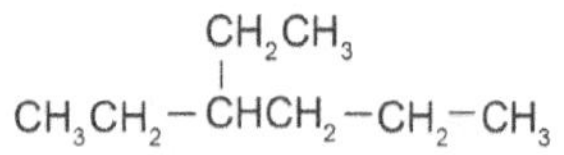

$CH_3CH_2-\underset{\underset{CH_2CH_3}{|}}{CH}CH_2-CH_2-CH_3$
3-ethylhexane

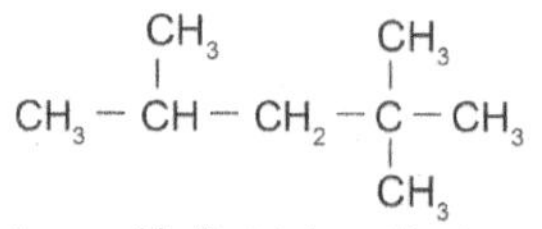

$CH_3-\underset{\underset{}{|}}{\overset{\overset{CH_3}{|}}{CH}}-CH_2-\underset{\underset{CH_3}{|}}{\overset{\overset{CH_3}{|}}{C}}-CH_3$
isooctane (2,2,4 trimethylpentane)

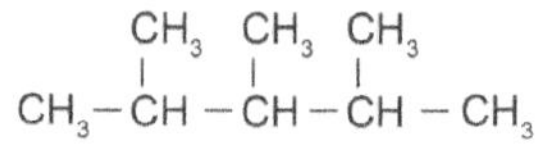

$CH_3-\overset{\overset{CH_3}{|}}{CH}-\overset{\overset{CH_3}{|}}{CH}-\overset{\overset{CH_3}{|}}{CH}-CH_3$
(2,3,4 trimethylpentane)

As you might guess, "tri" is added to the name "trimethylpentane" to state there are three methyl groups on the parent compound. The carbon location numbers are placed from low to high. For example, 2,2,4- is correct but 4,2,2- and 2,4,2- are incorrect.

How to distinguish isomers from non-isomers

The key to identifying and drawing isomers is understanding when structural formulas are actually different. The image below shows three ways to draw the structural formula for C_5H_8. Even though the chains are arranged in a different way, all three structures are identical. Look for the parent chain. In all three drawings the parent is a saturated 5-carbon chain with no attached groups, therefore all three structures are named pentane. If you trace the parent chain in the structures above, you will see that you get different parent lengths, different kinds of groups branching from the parent, and different branch locations.

C_5H_8 / non-isomers

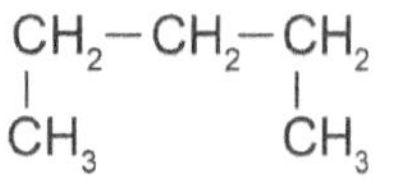

$\underset{\underset{CH_3}{|}}{CH_2}-CH_2-\underset{\underset{CH_3}{|}}{CH_2}$
pentane

$CH_3-CH_2-CH_2-CH_2-CH_3$
pentane

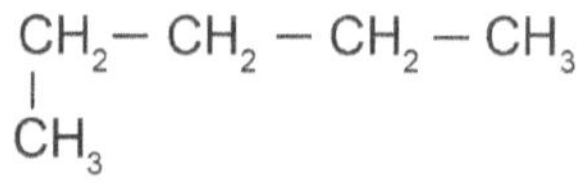

$\underset{\underset{CH_3}{|}}{CH_2}-CH_2-CH_2-CH_3$
pentane

More isomers of C_8H_{18}

Decide whether the molecule shown to the right is another isomer of the 8-carbon molecules not already shown in the diagram at the top of the page. First, make sure the chemical formula is the same. Is it C_8H_{18}? Yes, it is. Next, name the compound. Does it have the same name as one of the isomers already shown?

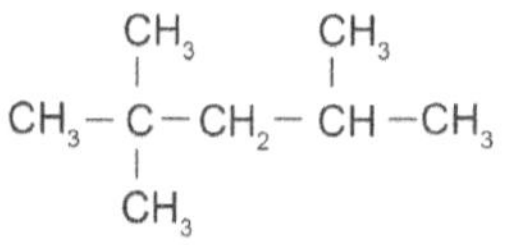

$CH_3-\underset{\underset{CH_3}{|}}{\overset{\overset{CH_3}{|}}{C}}-CH_2-\overset{\overset{CH_3}{|}}{CH}-CH_3$

Try drawing more isomers

This molecule is called 2,2,4-trimethylpentane, which has already been included in the diagram above. Try drawing more isomers for C_8H_{18}. Can you find all 24? Here are some hints: you can either draw isomers or use a molecular model set to build them. Experiment with different branched group lengths. Try adding more than one ethyl- group; see how just one methyl- group works out. Can you fit four methyl groups on the parent compound (tetramethyl-)? As you work, constantly check your chemical formula to make sure you still have 8 carbon atoms and 18 hydrogen atoms.

Stereoisomers

Different layouts

In another group of isomers called stereoisomers, atoms are bonded in the same sequence but the spatial, three-dimensional layout of the atoms is different. There are two kinds of stereoisomers: geometric and optical.

Geometric isomers

Build a 2-butene molecule with your molecular model set. Notice how the carbon atoms at both ends of the molecule can rotate freely around the carbon-carbon bond. Now try to rotate carbon number two or number three. You can see how a double bond prevents a carbon atom from rotating freely. A **geometric isomer** forms when a double or triple bond forces atoms or groups to stay in the same three-dimensional position. Look at the 2-butene isomers below. The two structures are different, but if you tried to name the isomers with the standard system, both will have the same name. In this case, you need to use the cis- and trans- prefixes to distinguish one configuration from another.

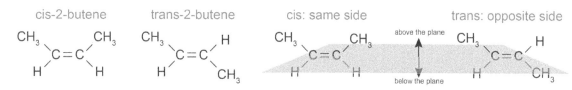

Different arrangement

Imagine a flat plane that cuts through the middle of the 2-butene molecule. When both $-CH_3$ ends of the parent compound are both on the same side of the plane, the molecule has a cis- configuration and is called cis-2-butene. When the $-CH_3$ ends are on opposite sides of the plane, the molecule has a trans- configuration and is named trans-2-butene.

Optical isomers

Superimpose your hands by placing your left hand over your right hand and line up your thumbs. It's impossible! The only way to get your thumbs to line up is to place your palms together. Your hands are a mirror image of one another but you cannot superimpose them. Like your hands, **optical isomers** are non-superimposable mirror images of one another.

d- and l-

The center carbon atom in the lactic acid molecule shown below has four different atoms or groups attached to it. This allows two structural configurations which results in different chemical properties. The chemical name identifies which isomer is present based on how the molecule rotates polarized light. The "d-" form rotates light clockwise (to the right), and the "l-" form rotates light counterclockwise (to the left).

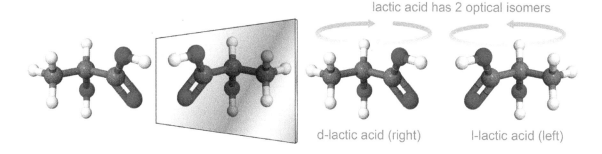

Drawing and naming ring structures

Use the prefix cyclo-

When a ring structure is present, the molecule's name will contain the prefix cyclo-. Use the number of carbons that make up the ring structure to name the parent. Choose a suffix that matches the saturated or unsaturated state of the structure.

Solved problem

Name the molecule shown below.

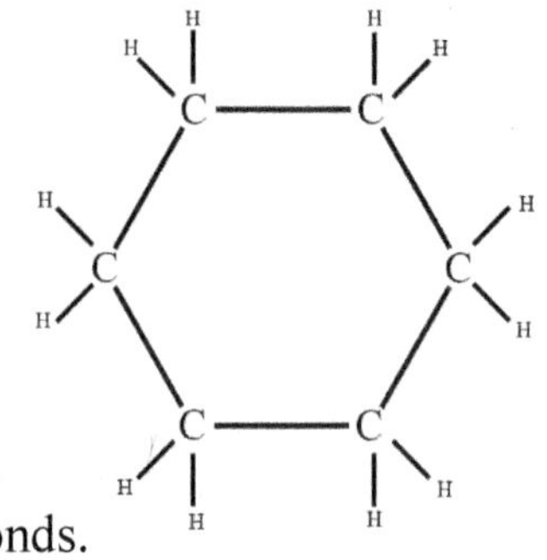

Asked: Apply the rules for naming ring structures.

Solve:
1. Begin with the name "cyclo-"
   cyclo-
2. Name the parent by counting carbons that make up the ring.
   cyclohex-
3. Add the suffix; is this an -ane, -ene, or -yne?
   The alkane molecule is saturated with all single bonds.
   Use the suffix -ane.

Answer: The molecule is named cyclohexane.

Cycloalkenes and cycloalkynes

Multiple bonds are indicated by the suffix in the molecule's name and are shown as additional lines in the ring structure. As you can see, cyclopentene has one extra line in its structure. This indicates a double bond. Cyclopentyne has two extra lines in its structure. This indicates a triple bond. Notice the number of hydrogen atoms accommodates carbon's four-bond maximum as multiple bonds form.

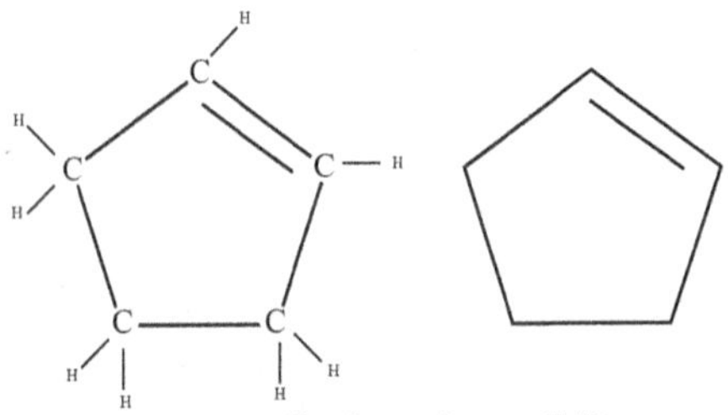

Cyclopentene, C_5H_8 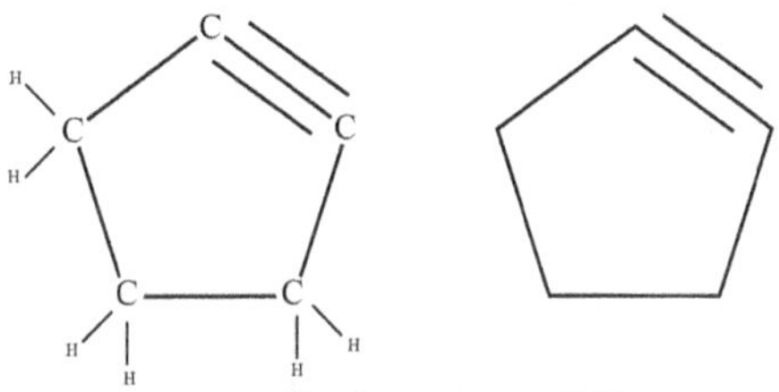   Cyclopentyne, C_5H_6

Skeletal structural diagrams

One type of structural diagram above has all the symbols for all atoms in the molecule, while the other structural diagram contains no atomic symbols at all. Hydrocarbons are sometimes drawn with skeletal structural formulas to communicate only molecular shape and important features like multiple bonds or bonded groups branching from the parent compound. You can assume all corners and chain ends represent a carbon atom. Also assume each carbon atom has enough invisible H atoms bonded to it to fill four bonds.

Examples

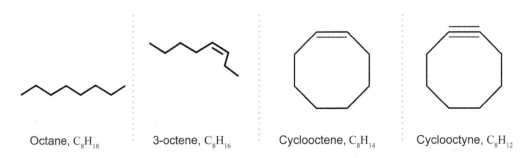

Octane, C_8H_{18} 3-octene, C_8H_{16} Cyclooctene, C_8H_{14} Cyclooctyne, C_8H_{12}

Ring structures and aromatic hydrocarbons

Aromatic molecules produce strong smells

For centuries, scientists have been interested in determining what gives certain forms of matter their characteristic aromas and flavors. Substances that have flavors or aromas contain molecules with a six-carbon ring structure known as an **aromatic hydrocarbon**. The chemical nature of tastes and smells can be difficult to pinpoint. Some odors and flavors require a complex combination of thousands of different kinds of aromatic hydrocarbons!

2-phenylethanol

ß-damascone

ß-ionone

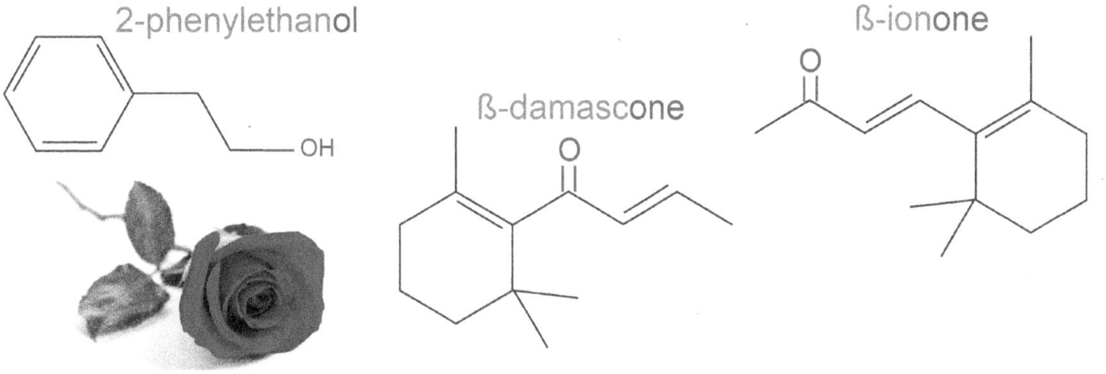

Aromatic molecules contain ring structures

Just a few of the many aromatic hydrocarbons needed to produce the smell of a red rose are pictured above. A ring structure forms when a linear carbon chain joins its ends together to form a circular structure. Ring structures are also referred to as cyclic structures. An example of a ring structure is a six-carbon cyclic alkene with alternating double bonds called benzene, C_6H_6. Many aromatic hydrocarbons contain one or more benzene rings. Because electrons are always moving, benzene's double and single bonds are never fixed in place. Delocalized electrons create structures with alternating double bonds known as **resonance** structures. Two resonance structures for benzene are shown to the right. In reality, benzene never exists as either of the resonance structures. The electrons in the carbon-to-carbon bonds do not belong to any one carbon in the benzene ring. They spread out evenly within the ring.

Benzene's Resonance Structures

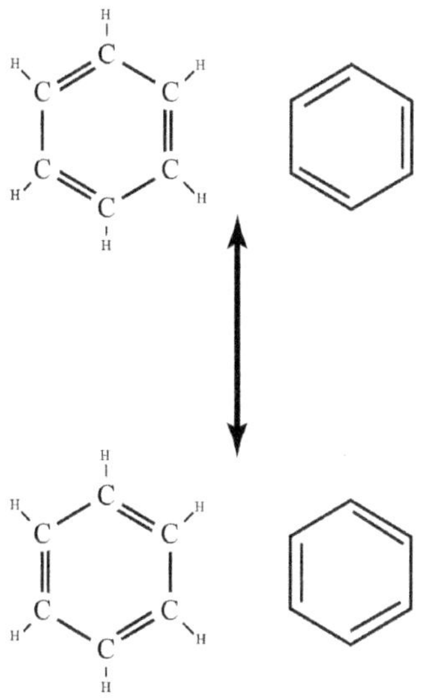

Benzene's structural diagram

A better structural diagram for benzene is a six-carbon ring with a circle inside it. The circle represents the shared electrons evenly distributed among the carbon atoms in the parent compound. This form of electron distribution makes benzene an especially stable molecule.

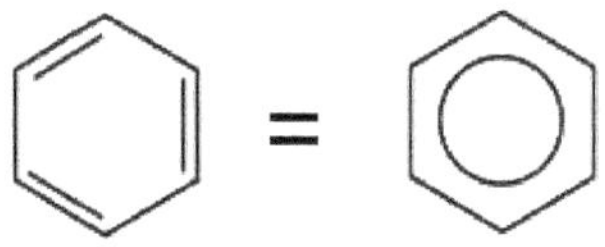

Benzene in and out of compounds

Benzene is a natural molecule with a pleasant smell. It makes a great non-polar organic solvent with many industrial uses such as the production of plastics. However, benzene is very dangerous! It is highly flammable and it causes cancer if safety precautions are not followed. When embedded in an aromatic compound, benzene produces smells and flavors that can be wonderful... or not-so-wonderful.

2-quinolinemethanthiol

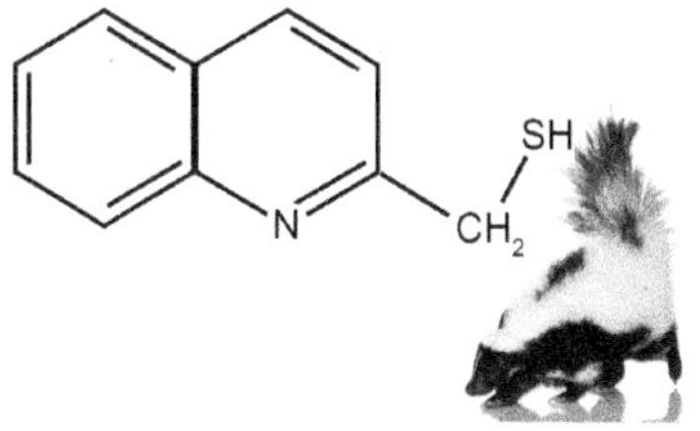

Section 20.2: Other Organic Compounds

Chapter 20

Section 2 review

Hydrocarbons containing one or more double bonds are called alkenes. Hydrocarbons containing one or more triple bonds are called alkynes. Alkenes with one double bond have the formula C_NH_{2N} while alkynes with one triple bond have the formula $C_NH_{(2N-2)}$. Two branched hydrocarbons can have the same chemical formula but have the branch at a different point. These structural isomers generally have different chemical and physical properties. Atoms in stereoisomers are bonded in the same order but differ in spatial arrangement.

Carbon compounds can have ring structures. Cycloalkanes are rings with only single bonds. Cycloalkenes and cycloalkynes contain double and triple bonds respectively. In benzene, delocalized electrons are shared among all the bonds in the ring. This is known as resonance.

Vocabulary words	alkene, alkyne, structural isomer, geometric isomer, optical isomers, aromatic hydrocarbon, resonance
Key relationships	• Alkenes, also known as unsaturated hydrocarbons, contain double bonds and are named with a prefix based on the parent chain and receive the –ene suffix. • Unsaturated alkynes contain triple bonds, are named with a prefix based on the parent chain, and receive the –yne suffix. • Ring structures receive the "cyclo-" prefix.

Review problems and questions

1. Draw the following molecules.

    a. 4-propyldecane
    b. 3-heptyne
    c. cyclohexyne
    d. benzene

2. Name the molecules shown.

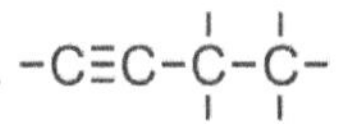

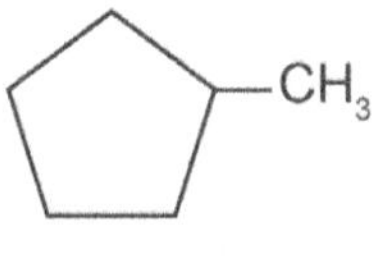

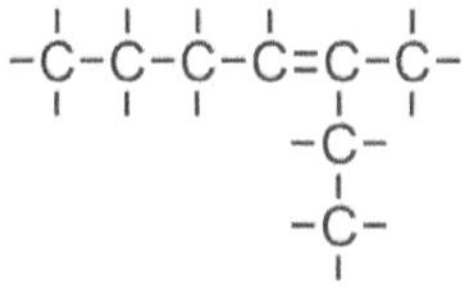

3. Are the two structures shown isomers of one another? Why or why not?

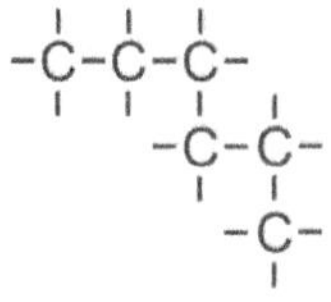

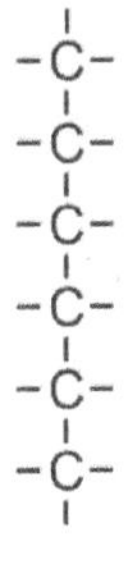

20.3 - Functional Groups

The carbon atom is able to bond in such a versatile way that an almost infinite number of organic molecules can be created. How do we make sense of all the possibilities? Scientists look for patterns in the chemical properties of organic compounds and group them by their chemical composition and behavior. Organic compounds take on specific properties depending on which functional groups are attached to the parent.

Functional groups determine properties

Branches of elements

A *functional group* is a branch of elements that give an organic compound its chemical and physical properties. Much of the chemistry an organic compound experiences is located at its functional group(s). When you remove a hydrogen atom from a benzene molecule and replace it with a hydroxyl functional group, -OH, you get a compound called phenol. Phenol is the substance in a throat spray that soothes a sore throat. If you have ever used throat spray, you will recognize phenol as an aromatic hydrocarbon with an unpleasant flavor.

benzene, C_6H_6

phenol, C_6H_5OH

Functional groups change properties

All alcohols contain a hydroxyl functional group. Benzene is hazardous to your health but it becomes a medicinal alcohol when you add a hydroxyl group to it. All hydrocarbon chains and rings gain new properties when functional groups are added. Hydrocarbon-based groups like methyl- and ethyl- are called alkyl functional groups. Observe the functional groups in the table below. All of the structural formulas in the table have the symbol R− attached to the functional group. R− stands for a saturated or unsaturated hydrocarbon ring or chain of any length. Look for a pattern in the way a molecule containing a functional group is named.

table of common functional groups

functional group	structural formula	sample substance	substance name
alcohol	R-OH	CH_3OH	methanol
ether	R-O-R	CH_3CH_2-O-CH_2CH_3	diethyl ether
alkyl halide	R-X (X = a halogen)	CH_3-CHCH$_3$ with Cl	2-chloropropane
amine*	R-NH$_2$	CH_3-NH$_2$	methylamine
ketone	R-CO-R'	CH_3-CO-CH_3	acetone
aldehyde	R-CHO	HCHO (H-CO-H, HCH$_3$)	formaldehyde
carboxylic acid	R-COOH	CH_3-COOH	acetic acid
ester	R-CO-O-R'	CH_3CH_2-CO-O-CH_3	ethylethanoate

*structure for a primary amine; secondary amines have two R− groups and tertiary amines have three R− groups

Section 20.3: Functional Groups

Alcohols and carboxylic acids

A hydroxyl is found in alcohols and carboxylic acids

Organic compounds often have more than one functional group attached to the parent compound. We will look only at organic compounds that have a single functional group at a time. Hydrocarbons that have a hydroxyl group, −OH, at any end belong to the **alcohol** class of compounds. If you see a hydroxyl at the end of a hydrocarbon combined with a C=O bond immediately in front of it, it is a **carboxylic acid**.

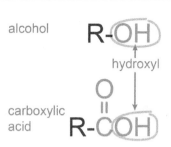

Region of polarity

Recall that the large electronegativity difference between O and H creates a polar bond. Molecules are slightly polar in the region where a hydroxyl group is located even when the molecule is non-polar overall. This gives alcohol and carboxylic acid molecules the ability to dissolve in polar and non-polar solvents. The −OH group can form hydrogen bonds which raises the molecule's boiling point. Compare the structure of ethane in the form of an alcohol and a carboxylic acid. Look for differences in three-dimensional shape and boiling point. Also note the difference between a molecular, chemical and structural formula. The structural formula is the most helpful when drawing hydrocarbons with functional groups.

molecule name:	ethane	ethanol	ethanoic acid
boiling point:	-88.6 °C	78.3 °C	117.9 °C
functional group:	none	alcohol (R-OH)	carboxylic acid (R-COOH)
chemical formula:	C_2H_6	C_2H_5OH	C_2H_3COOH
structural formula:	CH_3CH_3	CH_3CH_2OH	CH_3COOH
molecular formula:	C_2H_6	C_2H_6O	$C_2H_4O_2$

Properties of alcohols

Smaller alcohol molecules like ethanol and isopropyl alcohol dry out your skin and make acne worse. These alcohols damage cells above a certain concentration. This property of alcohols makes them great disinfectants for bacteria, fungi, and viruses. Ethanol is found in antibacterial wipes, hand sanitizers, and as a fuel source or additive. It is also found in wine, spirits and beer. Alcohols found in cosmetics, lotions and facial cleansers that are "good" for your skin include cetyl and stearyl alcohol. These alcohols have long chains and belong to a sub-class called fatty alcohols.

stearyl alcohol

Properties of carboxylic acids

Carboxylic acids form hydrogen bonds that raise melting and boiling points. They are found in all amino acids which are the building blocks of proteins. The hydrogen from a hydroxyl end can be pulled off by water to form H_3O^+ ions, but only partially, making this class of molecules weak acids. Ethanoic acid shown above is also called acetic acid, or vinegar. Acetic acid and citric acid are among the acids that give food a sour taste. Look for three carboxylic acid groups and one alcohol group in the citric acid molecule shown.

citric acid

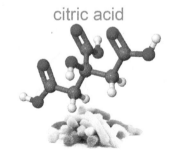

Ketones and aldehydes

A carbonyl is found in aldehydes and ketones

Organic molecules that have a carbonyl group, −C=O, located between carbons belong to the **ketone** class. Ketones are found on an inner part of a molecule because both sides of the carbon are available to join a carbon chain. If you see a carbonyl group at the end of a molecule, is is an **aldehyde**. An aldehyde is found at the end of a molecule because 3 of its 4 bonds are already occupied with an oxygen atom and a hydrogen atom.

molecule name:	hexane	2-hexanone	hexanaldehyde (or hexanal)
boiling point:	68.5 °C	127.6 °C	131.0 °C
functional group:	none	ketone (R-CO-R)	aldehyde (R-CHO)

From odorless to fragrant

Aldehydes and ketones give polarity to organic molecules. Smaller aldehydes and ketones are highly water-soluble but as molecular size increases, solubility decreases. The boiling point of hexane increases when ketone and aldehyde groups are added. Pure hexane is an odorless solvent used to produce glues and cooking oils. When you add a carbonyl to the second carbon, the ketone 2-hexanone is produced. This compound is also used as a solvent but it has a sharp smell that might remind you of nail polish remover. When an aldehyde is added to the end of a hexane molecule, the aldehyde hexanal is produced. This compound smells like fresh-cut grass. There are a large variety of aldehydes and ketones found in nature but many are synthesized in the lab to use as less expensive artificial food flavorings. These compounds produce smells and flavors that range from sweet to acrid.

Properties of ketones

A ketone called 2-heptanone contributes to the pungent smell of blue cheese. If you replaced the ketone with an aldehyde, you would get a fruity smell used in perfumes. Ketones are more stable than aldehydes because they lack the hydrogen atom seen on an aldehyde group. Many important biological molecules, such as sugars, are ketones. A condition called ketosis occurs in the body when blood contains a high level of ketones. Ketosis has received a lot of attention from the diet fad community because elevated levels of ketones promote the release and metabolism of fat.

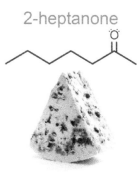

Properties of aldehydes

Aldehydes are easy to oxidize because the exposed O-H bond on the end of a molecule is reactive. Aldehydes make excellent reducing agents. Smaller aldehyde molecules with low molecular weights tend to smell like rotten fruit. Higher-weight aldehydes have pleasant smells you would recognize as oranges or roses in essential oils. Heptanal, pictured left, is among the most commonly used aldehydes in the fragrance industry. Aldehydes eventually lose their scent when they are oxidized, which is why a perfume's fragrance loses potency.

Familiar ketones and aldehydes

Acetone

Acetone is the smallest ketone molecule. The smell of nail polish remover comes from acetone molecules. It is easy for you to detect the smell of acetone because its low boiling point allows it to evaporate easily at room temperature. Acetone is an important solvent in laboratories, industry and at home because the polar and non-polar regions of the molecule allow it to dissolve either polar or non-polar solutes. Acetone is also used in industry as a precursor, or temporary building block, to create larger molecules through organic reactions.

acetone

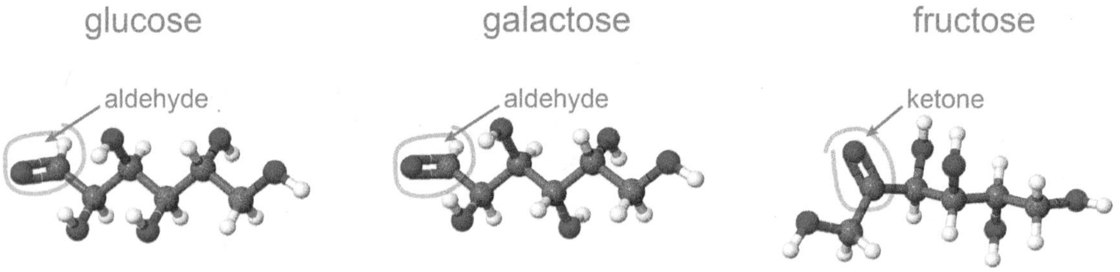

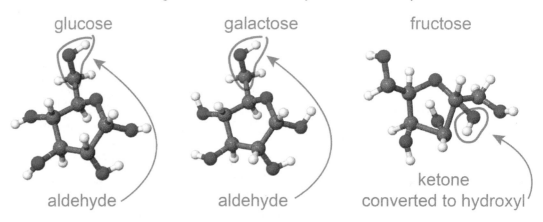

Functional groups of carbohydrates

Carbohydrates like sugar, starch and cellulose are among the most important examples of aldehydes and ketones in nature because they can be broken down by living things to release energy. Let's look at the smallest building blocks that make up larger carbohydrate molecules. Glucose, $C_6H_{12}O_6$, is an aldehyde that contains several hydroxyl groups. Galactose, an isomer of glucose, is also an aldehyde sugar. Fructose is a third isomer that has a ketone group instead of an aldehyde. The structures above show the simple sugars in their linear form, but these sugars are most commonly found in their cyclic form.

Linear vs. cyclic sugars

Notice the chemical formulas remain the same whether the molecules are in their linear or cyclic form. Glucose and galactose are still aldehydes, but fructose is no longer a ketone. The ketone is converted to a hydroxyl group when a double bond is replaced with a hydrogen atom. The hydrogen atom came from a hydroxyl group at the end of the chain when the oxygen joined the ring.

Functional group linkages

Pay close attention to the hydroxyl, aldehyde and ketone functional groups on sugars like glucose, galactose and fructose. These groups may be rearranged when the molecules are linked together to make larger carbohydrate molecules like starch..

Esters and ethers

Esters have a carbonyl and an ether

You saw an example of an **ether** in the cyclic sugar structures on the previous page where an oxygen atom is placed between carbon atoms. An **ester** group is made of a carbonyl group with an adjacent oxygen atom. Esters and ethers provide a polar site on a molecule and increase its boiling point. Solubility in water decreases as molecule size increases. Esters are derived from carboxylic acids and look similar, but carboxylic acids are a combination of adjacent carbonyl and hydroxyl groups.

Properties of esters

Plants and animals produce esters that range from small chains to large fat molecules. Esters produce flavors and aromas in fruits and flowers. There are usually a great number of aromatic compounds responsible for specific flavors and aromas but sometimes one molecule stands out. For example, ethyl butanoate is an ester that smells and tastes like pineapple. This chemical is inexpensive to produce artifically. If your food or drink contains artificial pineapple flavor, it contains ethyl butanoate.

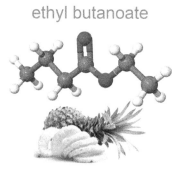

ethyl butanoate

Synthetic esters

The synthetic material called polyester means many esters. Polyesters are made by using ester groups to link molecules together in a long hydrocarbon chain. Your clothes likely have some amount of polyester added to them. Ester groups give fabric stretchy, strong, wrinkle-free and stain resistant qualities. Polyethylene terephthalate, or PET as it is abbreviated on recyclable plastic goods, is one of the most common synthetic fibers. Notice how the benzene rings in the parent chain are connected with esters.

polyethylene terephthalate / PET

Properties of ethers

Many people think anise and fennel have a similar sweet-spicy fragrance. An ether called methoxybenzene is common between the two spices. Ethers are found in natural and artificially produced flavors, dyes, perfumes, fuel additives, oils and waxes. The ether functional group is among the least reactive of the functional groups. The C-O bond of an ether is fairly stable and slightly polar. Ethers do not have any O-H bonds, so they cannot form hydrogen bonds. The reduced intermolecular forces among ethers causes them to have lower boiling points compared to other functional groups.

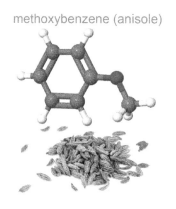

methoxybenzene (anisole)

Diethyl ether

Diethyl ether, commonly called ether, is one of the most industrially important organic solvents. It was once used for general anesthesia to address pain during surgery in the 1800s. Diethyl ether is highly flammable and has been replaced by safer anesthetics in hospitals.

diethyl ether

Section 20.3: Functional Groups

Amines and alkyl halides

Amine groups

An **amine** is a functional group that has a nitrogen atom at its center. There are three kinds of amine groups: primary (1'), secondary (2'), and tertiary (3'). The three kinds of amines differ in the number of hydrogen atoms attached to the central nitrogen. Since a N-H bond can form hydrogen bonds and is more reactive than a N-R bond, primary amines are the most reactive. Amine groups are derived from the ammonia (NH_3) molecule. The lone pair of electrons on the nitrogen atom influences molecular shape, polarity, and pH. Amines are weak bases because the lone pair can accept an additional H^+ ion. This causes pH to increase, which could influence the chemical behavior of the molecule or other compounds.

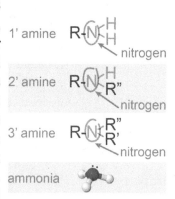

Biogenic amines

Monosodium glutamate (MSG) is an amine that gives protein-rich foods a savory flavor. An amine group is found in all amino acids, which are the building blocks of proteins. Proteins make up many important structures and organs in the body, such as muscles and skin. A sub-class of amines called biogenic amines are made by all plants and animals. Natural foods that have proteins have biogenic amines. Your body uses biogenic amines to make larger molecules with specific functions such as hormones.

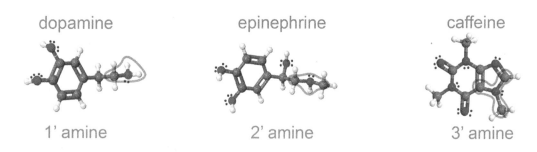

Familiar amines

Many molecules that regulate your body's functions are amines. This is why it is common to find amines in medicines. Dopamine is a primary amine that your nerve cells release at times when you feel good, like when you laugh or after a big accomplishment. Epinephrine is a secondary amine also known as adrenaline, which makes you feel exhilarated - or panicked - depending on the situation. Caffeine is a tertiary amine that stimulates your nervous system and can keep you from sleeping if you drink it too late at night. Notice how each of these molecules also contain benzene rings, and hydroxyl and ketone groups.

Alkyl halides

An **alkyl halide** is composed of a halogen connected to an R− group. Alkyl halides are also known as haloalkanes. Fluorine-containing alkyl halides are commonly used in the pharmaceutical industry because the C-F bond is strong and stable. The C-Cl and C-Br bonds are reactive. A C-I bond does not form because too much energy input is required. Alkyl halides are largely used as solvents and refrigerants. They can be added to flammable compounds like diethyl ether to increase boiling point and reduce flammability. Some alkyl halides, such as chlorofluoromethane, have damaging environmental effects that will be discussed in the next section.

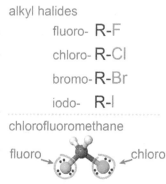

Summary of organic names

Summary

This page provides a summary of the naming rules and functional groups you are most likely to encounter when first studying organic chemistry.

Prefixes for hydro-carbons

The following prefix system is used to name simple hydrocarbons whose parent contains 10 carbon atoms or less. These prefixes can also be used to name small carbon chain groups extending from the parent hydrocarbon.

prefixes for small hydrocarbons

prefix	number of carbon atoms	formula for an alkane
meth-	1	CH_4
eth-	2	C_2H_6
prop-	3	C_3H_8
but-	4	C_4H_{10}
pent-	5	C_5H_{12}
hex-	6	C_6H_{14}
hept-	7	C_7H_{16}
oct-	8	C_8H_{18}
non-	9	C_9H_{20}
dec-	10	$C_{10}H_{22}$

Naming parent hydro-carbons

Here is a summary of which suffix to use based on whether you are naming an alkane, alkene, or alkyne:

type of bond	suffix
small hydrocarbons with only single bonds between carbons end with:	-ane
small hydrocarbons with at least one double bond between carbons end with:	-ene
small hydrocarbons with at least one triple bond between carbons end with:	-yne

Functional groups

table of common functional groups

functional group	alcohol	ether	alkyl halide	amine*	ketone	aldehyde	carboxylic acid	ester
structural formula	R-OH	R-O-R'	R-X (X = a halogen)	R-NH$_2$*	$\underset{R-C-R'}{\overset{O}{\parallel}}$	$\underset{R-C-H}{\overset{O}{\parallel}}$	$\underset{R-C-OH}{\overset{O}{\parallel}}$	$\underset{R-C-O-R}{\overset{O}{\parallel}}$

*structure for a primary amine; secondary amines have two R– groups and tertiary amines have three R– groups

Ring structures

Here is a summary of how to name ring structures:

begin with the name "cyclo-"	cyclo-
name the parent by counting carbons that make up the ring	cyclohexane

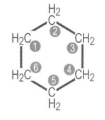

Section 20.3: Functional Groups

Chapter 20

Section 3 review

Functional groups are certain groups of atoms that give organic molecules specific properties and characteristics. An –OH group creates an alcohol. They have low boiling points and the hydroxide group allows them to form hydrogen bonds. Ethers include a single bonded oxygen atom. They have weak intermolecular forces and vaporize easily. Aldehydes and ketones both have an oxygen atom double bonded to a carbon atom. With aldehydes this functional group is at the end of a chain, while in ketones it is in the middle of the chain. A carboxyl functional group creates a weak acid known as a carboxylic acid. The carboxyl group can form strong hydrogen bonds. Amines are organic derivatives of ammonia where one or more hydrogen atoms are replaced by organic groups. Amines can form strong hydrogen bonds and can accept protons, making them bases using the Brønsted-Lowry definition. Esters are derived from acids, often carboxylic acids. They often have distinct smells and flavors.

Vocabulary words: functional group, alcohol, carboxylic acid, aldehyde, ketone, carbonyl group, carbohydrates, ester, ether, amine, alkyl halide

Review problems and questions

1. Identify the functional group shown on each hydrocarbon parent.

    a. 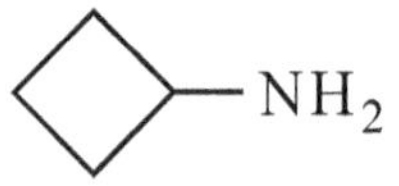 ⬡—NH₂

    b. 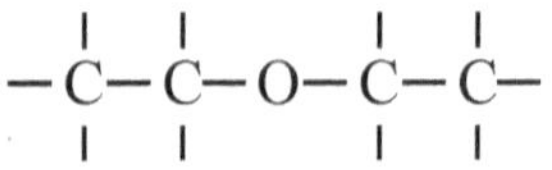 —C—C—O—C—C—

    c. H—O—C—C—C—C—
              ‖
              O

    c. —C—C—O—C—C—
           ‖
           O

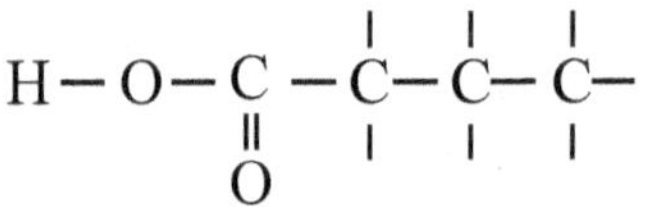

    e. 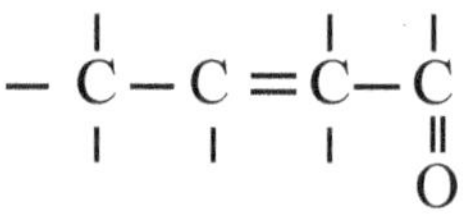 —C—C=C—C
              ‖
              O

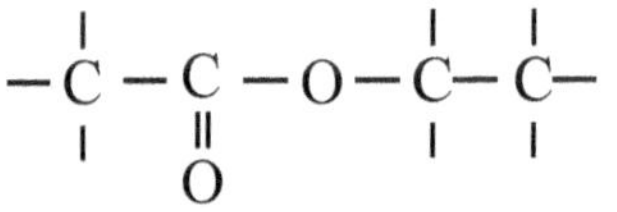

    f. 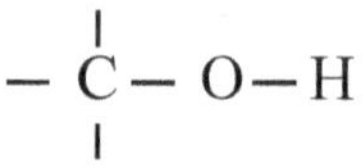 —C—O—H

    g. 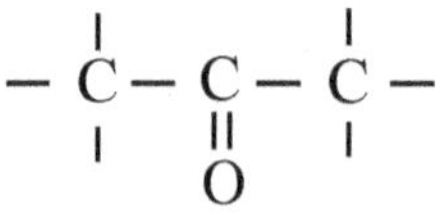 —C—C—C—
           ‖
           O

2. What property of alcohols allows them to easily dissolve in water?

3. What is the most important role of a functional group?

4. The two molecules listed below are real chemicals that have been named by chemists. Identify the functional group present in each chemical based on its name.

    a. Traumatic acid is a healing compound produced by plants when their tissues get injured.
    b. Penguinone is being researched in the drug industry, and has a structural diagram that looks like a penguin.

20.4 - Organic Reactions

You learned about the different types of inorganic reactions such as single replacement, decomposition, and synthesis. You can also classify organic reactions into different types and predict their products if you learn the patterns associated with each type.

Combustion/oxidation of hydrocarbons

Combustion of alkanes

Organic reactions have long been important to humans. The first chemical reaction controlled by humans is most likely fire, which is the combustion of hydrocarbons. The combustion of hydrocarbons is also a redox reaction because the hydrocarbon is oxidized by atmospheric oxygen. When a hydrocarbon burns, it reacts with oxygen to produce carbon dioxide and water. This reaction is particularly useful because it is exothermic. We use the energy released to keep us warm, cook our food, and power mechanical and electronic devices. Examples of organic combustion reactions that go to completion are shown below.

combustion of octane
$$2C_8H_{18} + 25O_2 \rightarrow 16CO_2 + 18H_2O$$

digestion of food
$$C_6H_{12}O_6 + 6O_2 \rightarrow 6CO_2 + 6H_2O$$

Combustion produces pollutants

Ordinary combustion of hydrocarbons never goes to completion because there is always some unburned fuel, smoke, soot or ash remaining. The compounds left behind after incomplete combustion are toxic pollutants. Cars and industries that burn organic compounds for energy are required to capture the majority of those pollutants before they are sent into the air. A smoldering "fire" has no flame, but combustion continues to occurs at a lower temperature and slower rate. Lower-temperature smoldering combustion usually results in incomplete combustion, leaving behind more pollutants than flaming combustion. However, smoldering combustion can be used to clean contaminated soil or to reduce the risk of wildfires in a controlled burn.

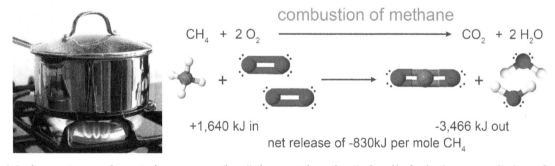

combustion of methane
$$CH_4 + 2O_2 \longrightarrow CO_2 + 2H_2O$$

+1,640 kJ in   −3,466 kJ out
net release of −830kJ per mole CH_4

Energy is released

Methane (natural gas) is among the "cleanest-burning" fossil fuels because it has fewer impurities than other fuels. As you know, you must first add energy to get a combustion reaction started. The products have less total bond energy than the reactants, so there is a net release of energy. This released energy is more than enough to keep the combustion reaction going as long as there are enough reactants. The combustion of methane results in a net release of 830 kJ per mole, as shown above.

Reactions of alkanes: Dehydrogenation

Alkanes are not highly reactive

Although you have seen many examples of combustion reactions with alkanes, this group is among the least reactive class of organic compounds. Alkane bonds are stable because they do not easily attract other groups or compounds. The C-C bonds in parent compounds are non-polar, and C-H bonds are only slightly polar. Alkanes tend to be more reactive when they are highly branched or when they are exposed to alkyl halide groups and ultraviolet (UV) light. In industry, catalysts and heat are added to alkanes to increase their reactivity.

Dehydrogenation creates multiple bonds

Dehydrogenation reactions occur in the presence of heat and a catalyst. A **dehydrogenation reaction** removes two hydrogen atoms from the alkane for each new carbon-to-carbon bond formed. A single C-C bond forms a new C=C double bond and a molecule of hydrogen, H_2. Alkenes can also undergo dehydrogenation to form alkynes. Dehydrogenation causes areas of unsaturation along a hydrocarbon chain. One of the most common dehydrogenation reactions in industry is the production of styrene molecules. Ethylbenzene gains a double bond when exposed to an iron(III) oxide catalyst, as shown below. Styrene molecules are important precursors for plastics.

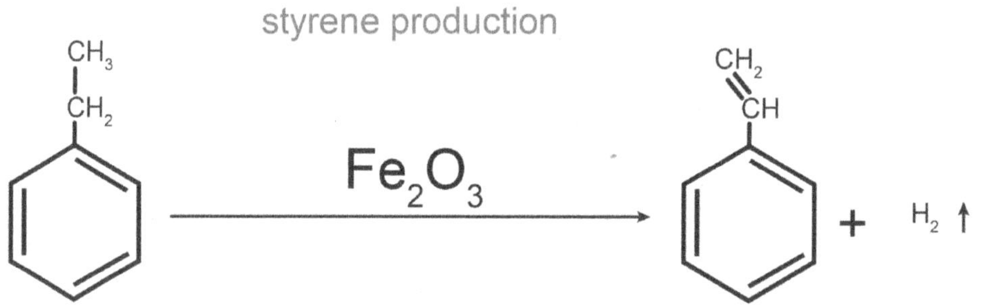

styrene production

Reactivity of unsaturated molecules

The alkene products of dehydrogenation are often more reactive than the original alkane reactants. Dehydrogenation reactions are used to produce many different kinds of molecules in industry, including aromatic hydrocarbons and alcohols. This is also the reaction used to convert saturated fats to unsaturated fats, and is a step in petroleum refinement to increase the octane rating of a fuel. You might be wondering how an octane rating can be over 100. The fuel industry allows octane ratings over 100 when substances like alcohols have been added to improve fuel performance in high-powered engines.

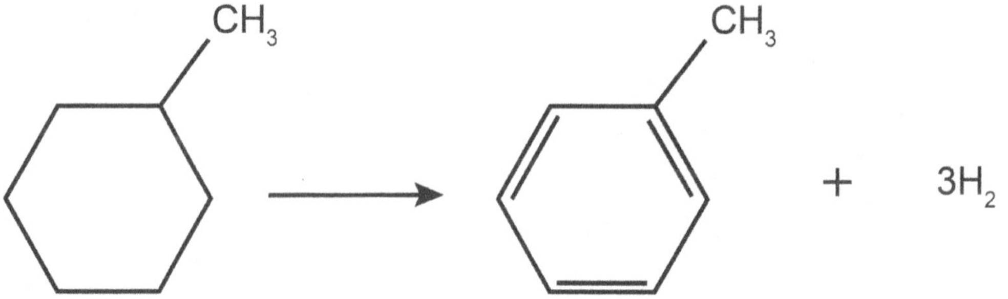

Octane number 75   Octane number 124

High temperature reactions

Converting stable alkanes to more reactive alkenes is an energy-intensive process. Dehydrogenation reactions often occur at temperatures above 500 °C! If the reaction used to convert a saturated hydrocarbon to an unsaturated hydrocarbon is called dehydrogenation, can you guess what the opposite reaction is called?

Reactions of alkanes: Substitution

Substitution reactions

In a **substitution reaction**, one or more hydrogen atoms on a molecule are removed and replaced by different atoms. Alkanes and cycloalkanes react with alkyl halides, particularly chlorine and bromine. For example, when a halogen such as bromine reacts with methane, a bromine atom will take a hydrogen atom's place. This substitution occurs in the presence of light at room temperature.

Photo-chemical reaction

Before an alkyl halide can substitute a hydrogen atom, a diatomic halogen molecule must be split apart. Let's look at a bromine molecule. Light energy breaks the Br–Br bond. The hv symbol above the arrow in the decomposition of the bromine molecule represents light energy. This is also called a photochemical reaction because light energy is required. Two free bromine radicals are produced. A **free radical** is a very reactive molecule, ion, or atom with an unpaired electron represented with a dot (such as Br· or Cl·).

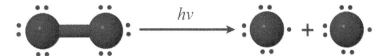

One substitution at a time

Even though two free radicals are available, substitution reactions occur one at a time. A bromine radical breaks a C–H bond on the methane molecule and takes the place of a hydrogen atom. The other bromine radical bonds to the newly-freed hydrogen to form strong hydrobromic acid HBr.

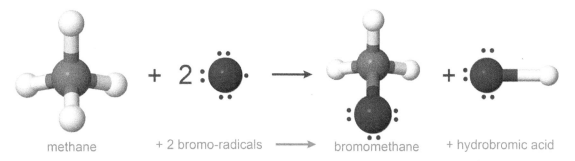

methane + 2 bromo-radicals ⟶ bromomethane + hydrobromic acid

All H atoms can be substituted

Then, another substitution reaction can happen to add a second bromine atom to the molecule. Several or all of the hydrogen atoms can be replaced on the hydrocarbon, resulting in a mixture of bromomethane gas, the liquids dibromomethane, tribromomethane, tetrabromomethane and several hydrobromic acid molecules.

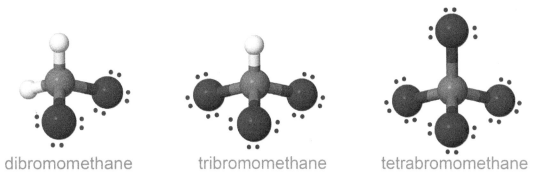

dibromomethane   tribromomethane   tetrabromomethane

Products of substitution are reactive

Alkanes with alkyl halide groups increase their reactivity. The carbon-halide bond is not as stable as a C-H bond, and it is more polar. Some alkane substitution reactions can occur in the absence light. Cyclopropane undergoes bromine substitution in the dark.

Reactions of alkanes: Cracking

Crude oil is mostly plankton

We drill into Earth's crust to extract the chemically modified organic remains of ancient plants and animals. When these ancient organisms died, their remains sank to the bottom of their watery home and became covered with sediment. Over millions of years and exposure to high temperature and pressure, the organic matter chemically changed into crude oil, or **petroleum**. Most oil is made up of a type of microscopic plankton called diatoms. Diatoms are similar to plants because they are capable of converting the sun's energy to chemical energy stored in organic molecules.

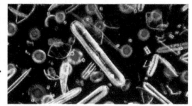
fossilized plankton

Petroleum is a mixture of hydrocarbons

Petroleum is made of many different hydrocarbons mixed together, primarily straight and branched alkane chains. The first hydrocarbon to be released when drilling for oil is methane. Methane is a gas at room temperature and it is usually released into the air when a drill reaches a petroleum deposit. The hydrocarbons mixed in petroleum can be used for many purposes such as fuels, solvents or raw materials to make new organic molecules.

Alkylation of small fractions and cracking of large fractions

The hydrocarbon mixture needs to be separated into groups called fractions. Fractions are hydrocarbon molecules of similar chain length with similar properties, such as molecular weight and boiling point. Petroleum is sent into a tall tower heated to over 600 °C where fractions separate by density and boiling point. The fractions with the longest chains and highest boiling points sink to the bottom, and the least dense fractions with the lowest boiling points float to the top. Each level in the tower can be accessed so the separate fractions can be removed for further processing. Smaller hydrocarbons can undergo an alkylation reaction where an alkyl halide reacts with a hydrocarbon chain to saturate it, branch it, or lengthen the parent chain. The alkyl halide transfers a carbon-based group to the hydrocarbon chain to modify it.

Cracking is the process used to break larger hydrocarbons into smaller chains. Cracking requires very high temperatures and the use of catalysts. Fractions with longer hydrocarbon chains are broken down, or cracked, to form shorter-chain hydrocarbons like octane and kerosene. The smaller hydrocarbon molecules produced during the cracking process can be used as building blocks to create products such as plastics. Products created by the cracking process include industrially useful alkenes and hydrogen gas.

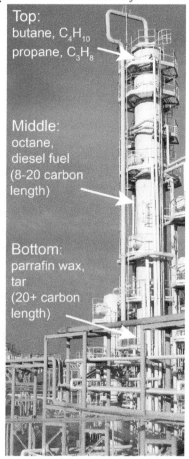
oil refinery

Top: butane, C_4H_{10} propane, C_3H_8

Middle: octane, diesel fuel (8-20 carbon length)

Bottom: parrafin wax, tar (20+ carbon length)

Reactions of alkenes and alkynes

Double and triple bond sites are more reactive

The double and triple bonds in alkenes and alkynes make them more reactive than alkanes. Industrial chemists perform organic reactions that involve using the site of a multiple bond on an alkene or alkyne to modify a molecule. Chemists make plastics and synthetic fibers by linking modified molecules together. In an **addition reaction** atoms are added to two carbon atoms that share a multiple bond, as shown below. Hydrogenation is among the most common types of addition reactions. In a **hydrogenation reaction**, hydrogen atoms are added to the unsaturated part of a carbon chain. The process of hydrogenation is used to make solid shortenings for baking. Hydrocarbon molecules can pack tighter when multiple bonds are removed. Tighter molecule arrangements allow intermolecular forces to become more effective. This causes a liquid oil to form a solid at room temperature.

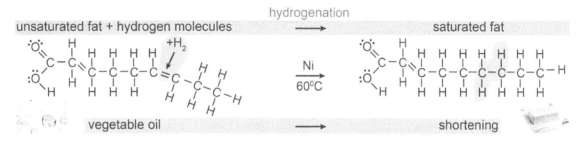

Solved problem

When 2-butene, C_4H_8, undergoes hydrogenation, what will the product be?

Given The molecule is called 2-butene

Relationships
- The prefix but- indicates the parent chain is 4 carbons long; the suffix -ene indicates a double bond is present and the number 2- indicates the double bond is located between carbon #2 and carbon #3.
- In a hydrogenation reaction, a double bond between 2 carbons is removed and replaced by one H atom on each carbon.
- One reactant is 2-butene and the other is H_2 since this is a hydrogenation reaction.

Solve
1. First, draw 2-butene:

   H–C(H)(H)–C(H)=C(H)–C(H)(H)–H

2. Next, use the hydrogen molecule (H_2) to remove the double bond. Add one H to each C that once shared the double bond.

   H–C(H)(H)–C(H)=C(H)–C(H)(H)–H + H_2 → H–C(H)(H)–C(H)(H)–C(H)(H)–C(H)(H)–H

Answer The hydrocarbon is now saturated because it contains only single bonds. The molecule is now called butane, C_4H_{10}, a saturated alkane.

Partial hydrogenation

Trans fats are not typically found in nature. They were created by accident through the **partial hydrogenation** of unsaturated fats. Some double bonds still remain in the fat structure, so it is not completely hydrogenated. Partial hydrogenation leaves a trans orientation on either side of the remaining double bonds. You will learn more about natural and artificial fats in the next chapter.

Section 20.4: Organic Reactions

Organic polymerization reactions

Polymer chemistry

Many of the polymers you rely on every day are organic molecules. Just over 100 years ago, Leo Baekeland invented a synthetic polymer called Bakelite. Prior to Bakelite, synthetic plastics were flimsy or rubbery compounds at room temperature. Solid objects were not made of plastic for that reason. Bakelite was the first synthetic plastic that could be molded into any shape to create a durable solid at room temperature. Bakelite is made by heating phenol to a high temperature with formaldehyde. Phenol contains a hydroxyl group and formaldehyde has an aldehyde group. When the substances combine at high temperatures, they form a polymer chain. Bakelite was instantly popular, giving brilliant color to the solid objects of the world in the 1920s.

2 phenol + formaldehyde → bakelite polymer + water

Condensation polymers

The reaction above shows two phenol monomers combining with one formaldehyde monomer producing a partial chain and water. If there are enough reactants, the Bakelite polymer chain can continue to grow longer. You can see the open bond on the left and right side of the polymer to indicate the chain can can be lengthened in either direction. Plastic, polyester and foam are polymers that are produced through a **condensation polymerization** reaction. As the name condensation implies, water is removed in order for two monomers to join. In the example below, the hydroxyl from a carboxylic acid group on one monomer bonds with a hydrogen from a hydroxyl group on a neighboring monomer to form a water molecule. As a result, a new bond linking the two monomers is formed.

terethalic acid + ethylene glycol → polyethylene terepthalate (PET) + water

Addition polymers

In an **addition polymerization** reaction, atoms or groups are added to existing molecules. Hydrogenation is one type of addition reaction, but addition polymerization reactions allow multiple monomers to continue to add to a polymer chain. The reaction below shows an abbreviated version of how polystyrene is formed. Styrene molecules are stable, but if they are exposed to a free radical, it causes the double bonds to be eliminated, allowing phenol monomers to form a polymer chain. Polystyrene is a synthetic polymer used to make transparent plastics, fabric and packing peanuts. Polystyrene's stability is beneficial for industrial and consumer purposes, but it is harmful to the environment.

styrene + styrene + styrene → polystyrene (plastics, foam)

Section 4 review

Chapter 20

Organic compounds are involved in several important types of reactions. Combustion of hydrocarbons releases carbon dioxide, water and heat. Dehydrogenation reactions use a catalyst to remove hydrogen atoms as hydrogen gas and convert a carbon-carbon single bond into a double bond. Many important products can be formed through dehydrogenation. In substitution reactions one or more hydrogen atoms of a hydrocarbon are replaced, creating a more reactive compound. The opposite of dehydrogenation is hydrogenation, a type of addition reaction. In these reactions, a carbon-carbon double bond is removed and hydrogen atoms attach to each carbon. Hydrogenation is often used for polymerization reactions, creating long chains of repeating monomers. Petroleum refinement is a process of separating a mixture of hydrocarbons.

Vocabulary words: dehydrogenation reaction, substitution reaction, free radical, petroleum, cracking, addition reaction, hydrogenation reaction, partial hydrogenation, addition polymerization, condensation polymerization

Key relationships
- Combustion: $C_xH_x + O_2 \rightarrow CO_2 + H_2O +$ heat
- Dehydrogenation: $R\text{-}CH_2\text{-}CH_2\text{-}R \xrightarrow[\text{cat}]{\text{heat}} R\text{-}CH=CH\text{-}R + H_2$ (can also form alkynes)
- Substitution: $R\text{-}CH_2\text{-}CH_2\text{-}R + X \rightarrow R\text{-}CH_2\text{-}CHX\text{-}R + H$ (also in alkenes and alkynes)
- Addition: $R\text{-}CH=CH\text{-}R + X_2 \rightarrow R\text{-}CX\text{-}CX\text{-}R + R$ (also in alkynes)
   - Hydrogenation: $R\text{-}CH=CH\text{-}R + H_2 \xrightarrow{\text{cat}} R\text{-}CH_2\text{-}CH_2\text{-}R$ (also in alkynes)

Review problems and questions

1. A fat is solid at room temperature.
   a. Is it a saturated or unsaturated fat?
   b. If you wanted the fat to become a liquid at room temperature, what type of organic reaction would the fat need to undergo?

2. What kind of organic reaction occurred in the reaction shown? How can you tell?

3. What kind of organic reaction occurred in the reaction shown? How can you tell?

4. Describe at least one cost and one benefit of organic combustion reactions.

Plastics: What's in a Number?

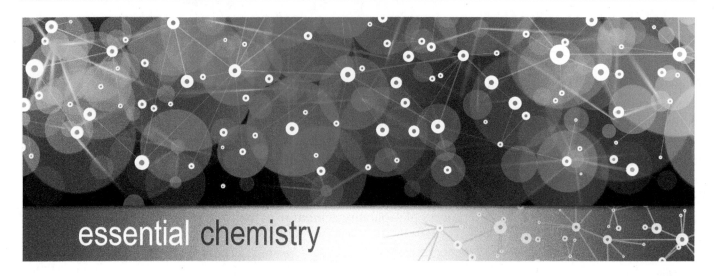

essential chemistry

"Plastic" is not any one material. It is more like a family of materials that can be engineered to meet an assortment of needs. All plastics are organic chemicals and are derivatives of petroleum. They are also all polymers, but the monomer from which they are built makes the plastics very different. The different types of polymers can then be shaped and molded into different forms to suit a variety of needs.

Plastic with the symbol #1 is PET or PETE. PET is short for polyethylene terephthalate. You can probably guess from the name that the monomer is ethylene terephthalate. The subscript "n" on the ethylene terephthalate picture indicates that this is the repeating unit monomer. "n" can be any number and for polymers, there could be thousands of monomers bonded together. Can you see the repeating unit monomer in the picture of the polymer?

PET is typically used to make bottles for soaps, detergents, soft drinks, and peanut butter. Most curbside recycling programs will pick up #1 containers. When it is recycled, it can be used to make fibers for carpets or fleece jackets.

Plastic #2 is HDPE, or high density polyethylene. HDPE is a straight chain version of polyethylene. These chains can stack together well because their shape maximizes intermolecular forces. This efficient stacking is what makes the polyethylene "dense." HPDE is typically opaque. It can be made into cereal box liners and some grocery bags. When it is in the form of milk jugs or narrow-necked bottles, it can be picked up by most curbside recycling programs. Those items can then be processed into shampoo bottles, floor tiles and flower pots.

Like plastic #2, plastic #4 is also made from a monomer of ethylene. Plastic #4 is LDPE, or low density polyethylene. LDPE is a branched version of polyethylene. Branches interrupt the molecules' ability to closely interact with other molecules. Reduced interaction lowers the strength of intermolecular forces. The branching makes the polyethylene less dense and more flexible. LDPE is found in squeezable bottles like the wash bottles in your lab and plastic grocery bags. Most curbside programs do not recycle LDPE bottles, but many food stores do collect LDPE bags. When it is recycled, it can be made into garbage bags, floor tile and shipping envelopes.

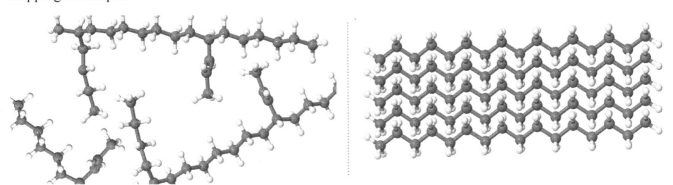

Plastic #5 is polypropylene, or PP. It is generally considered safe for containing foods and beverages. It is a hard but flexible polymer of propylene that is commonly used for ice cream and yogurt containers, and ketchup and syrup bottles. It is becoming more accepted for curbside recycling. Once recycled, you can find polypropylene in car battery cases, brooms or ice scrapers.

Plastics #3 is PVC, or polyvinyl chloride. The monomer is vinyl chloride, as called chloroethylene. As you can see from the picture, it is like ethylene, but one of the hydrogens has been replaced by a chlorine atom. As you might have guessed, old-fashioned vinyl records that are popular among audiophiles are made from PVC that is colored black. Fake leather clothing is vinyl made from PVC with some additives that make it soft and flexible. Most PVC has construction-related uses like pipes, home siding, and fences.

If the chlorine in PVC is replaced with an aromatic ring, then you have the structure of polystyrene, or PS. Polystyrene is plastic #6. It is naturally transparent and can be found in disposable plates and utensils. In its foam form, polystyrene is used to make packing peanuts, insulating cups, or protective cases like egg cartons. Polystyrene is controversial due to health and environmental concerns. Like all plastics, polystyrene takes a long time to break down.

Plastic number #7 is for "Other" plastics. These plastics are either not one of the other 6 listed, or the product is a combination of the different plastics. The most common #7 plastic is polycarbonate. Polycarbonates are engineered to be strong, tough, and transparent so it is useful as a windscreen on a motorcycle, swimming googles, and television displays. There is some controversy around the use of polycarbonate for food storage since it is suspected of releasing a potentially toxic synthetic chemical called bisphenol A, or BPA. Plastic #7 is not curbside recyclable even when it is made into 5-gallon water jugs.

Now that you have a better understanding of some of the types of plastics and their uses you can make more informed decisions about the containers you use and recycle. So the next time you throw a #2 plastic into the recycle bin, remember that it might come back to you as a flower pot!

Chapter 20

Chapter 20 review

Vocabulary
Match each word to the sentence where it best fits.

Section 20.1

alkane	hydrocarbon
organic chemistry	parent compound
R group	saturated hydrocarbon
unsaturated hydrocarbon	

1. A(n) _____ is the longest continuous chain of carbons.

2. If an organic compound contains hydrogen and carbon it is called a(n) _____ .

3. A compound that only contains single carbon-to-carbon bonds is called a(n) _____ .

4. _____ is the shorthand notation that is sometimes used to represent a hydrocarbon side chain.

5. A hydrocarbon compound that contains multiple bonds is called a(n) _____ .

6. _____ is another name for an alkane.

7. _____ is the study of carbon-containing molecular compounds.

Section 20.2

alkene	alkyne
aromatic hydrocarbon	benzene
geometric isomer	optical isomers
resonance	structural isomer

8. If an aromatic ring structure has a formula of C_6H_6, it is called a(n) _____ .

9. Propene is an example of a(n) _____ .

10. Atoms that are arranged differently with respect to their double bonds are referred to as Cis and Trans isomers and also _____ .

11. _____ contains at least one benzene ring in its structure and has a distinct odor.

12. Molecules that rotate light differently with respect to one another are called _____ .

13. Butyne is an example of a(n) _____ .

14. Molecules that have the same chemical formula but different bonding pattern are referred to as _____ .

15. Benzene exhibits _____ because it contains delocalized electrons that create alternating double bonds.

Section 20.3

alcohol	aldehyde
amine	carbohydrates
carbonyl group	carboxylic acid
ester	ether
ketone	

16. Ethanol is an example of the _____ functional group.

17. Vanilla and cinnamon give us familiar smells and flavors and have the _____ functional group.

18. A(n) _____ is the functional group that contains a carbon to oxygen double bond with hydrocarbon groups on both sides.

19. Acetic acid is an example of a(n) _____ , which is an organic acid containing -COOH.

20. A(n) _____ functional group is an important part of amino acids and can also act as a weak base in aqueous solutions.

21. The _____ functional group is responsible for the fragrance of flowers and ripened fruit.

22. An organic molecule where oxygen is bonded in the middle, with single bond hydrocarbon chains on either sides refers to the functional group of _____ .

23. A carbon double bonded to an oxygen is referred to as a _____ .

24. An example of aldehydes and ketones that provide energy in living organisms are _____ .

Section 20.4

addition polymerization	addition reaction
condensation polymerization	cracking
dehydrogenation reaction	free radical
hydrogenation reaction	partial hydrogenation
petroleum	substitution reaction

25. The process in which hydrogens are removed and a double bond is created in the structure of a saturated alkane is called a _____ .

26. When long chain hydrocarbons break into smaller fragments, the process is called _____ .

27. A blend of hydrocarbons that was formed from prehistoric organic material is called _____ .

28. The _____ adds hydrogen atoms to the double bonded areas of an unsaturated hydrocarbon molecule.

29. The process of _____ creates trans fats and allows foods that contain oils to last longer on the shelf at the grocery store.

Chapter 20 review

Section 20.4

addition polymerization	addition reaction
condensation polymerization	cracking
dehydrogenation reaction	free radical
hydrogenation reaction	partial hydrogenation
petroleum	substitution reaction

30. A(n) _____ forms multiple bond(s) in a molecule by adding atoms to carbon atoms.

31. A highly reactive molecule or atom that has an unpaired electron is known as a(n) _____.

32. _____ is the reaction where hydrogen atoms are removed from an alkane and replaced with different atoms, usually a halogen.

33. A polymer can be formed by adding monomers together using their double bonds through the process of _____.

34. Polymers can be formed by linking monomers through the loss of a small molecule by the process of _____.

Conceptual questions

Section 20.1

35. If a sample is made up of very short hydrocarbon chains, is it most likely to be a solid, liquid or gas at room temperature? Explain.

36. What is the name of the following compound?

$$CH_3CH_2CH_2CHCH_3$$
$$\quad\quad\quad\quad\quad | $$
$$\quad\quad\quad\quad\quad CH_3$$

37. Sometimes you see the parent chain of a saturated hydrocarbon drawn with a straight chain, and sometimes you see the parent chain drawn in a zigzag pattern.

    a. Which drawing is more correct: straight, or zigzag?
    b. What is responsible for this pattern?

38. Write the chemical formula for an alkane that is used as a rating for gasoline.

39. What do the numbers 87, 89, and 93 indicate about gasoline? Explain.

40. Use the generic formula for a straight-chain alkane, $C_nH_{(2n+2)}$, to predict the formula for a 12-carbon compound.

41. Write the chemical formula and name for a straight chain alkane that has 9 carbons.

42. Answer the following for an alkane with a 6-carbon parent chain:

    a. Draw this compound.
    b. Write the chemical formula for it.
    c. Name this compound.

43. A substance dissolves easily in liquid pentane.

    a. Is pentane polar or nonpolar? Explain your answer.
    b. Is the dissolved substance polar or nonpolar? How do you know?
    c. Will the substance dissolve in water? Why or why not?

44. Why is there a big difference in boiling points of small hydrocarbons compared to longer hydrocarbons? Explain.

45. Choose the correct name for the molecule shown:

$$\quad\quad\quad\quad\quad CH_2CH_3$$
$$\quad\quad\quad\quad\quad | $$
$$CH_3CH_2CH_2CHCH_2CH_3$$

    a. Ethylhexane
    b. 3-ethylhexane
    c. 4-ethylhexane

46. Does the $C_nH_{(2n+2)}$ rule apply to saturated ring structures? Why or why not?

Section 20.2

47. What is the name of the following compound?

$$CH_3CH_2C\equiv CCH_2CH_2CH_3$$

48. What is the name of the following compound?

$$CH_3CH=CHCH_2CH_3$$

49. What feature must be present in a hydrocarbon that is an alkene?

50. How many hydrogen atoms should be added to carbon #2 in the structure shown?

$$C-C\equiv C-C-C-C$$

51. Should there be a single, double, or triple bond in-between carbon #3 and carbon #4 in the structure shown below?

$$CH_3CH_2CCHCH_2CH_3$$
$$\quad\quad\quad\quad | $$
$$\quad\quad\quad\quad CH_2$$
$$\quad\quad\quad\quad CH_3$$

52. Should there be a single, double, or triple bond in-between carbon #1 and carbon #2 in the structure shown below?

$$CH_2CHCH_2CH_2CH_3$$

53. Draw the structure for each of the following compounds.

    a. 2-methylbutane
    b. 1-propene
    c. 3-heptyne
    d. 1-butene

54. What is the name of the skeletal structure that is pictured above?

Chapter 20 review

Section 20.2

55. What is the minimum number of carbons an alkane can have to make a structural isomer? Explain your answer.

56. Draw the cis- and trans- geometric isomers for 2-butene.

57. One type of isomer is an optical isomer.
    a. Define optical isomer.
    b. Draw a simple tetrahedral optical isomer.
    c. What is the difference between an optical and structural isomer?

58. Which of the following isomers shows the cis-configuration?

$$
\begin{array}{ccc}
\text{CH}_3 & \text{CH}_3 & \text{CH}_3 & \text{H} \\
| & | & | & | \\
\text{CHCH=CHCH} & & \text{CHCH=CHCH} \\
| & | & | & | \\
\text{H} & \text{H} & \text{H} & \text{CH}_3
\end{array}
$$

59. Draw two structural isomers of butane.

60. What is the difference between trans and cis? Explain why the difference is important.

61. Explain what the circle inside a hexagonal benzene ring represents.

62. What are two properties that distinguish aromatic compounds from other hydrocarbons?

Section 20.3

63. What is the name of the parent compound in the structure shown?

$$\text{CH}_3\text{CH}_2\text{CH}_2\text{CH}_2\overset{\overset{\text{O}}{\|}}{\text{C}}-\text{OH}$$

64. Which functional group is present in the compound in the previous question?

65. Vanillin, pictured below, is an organic compound that gives vanilla beans their distinct odor and flavor.

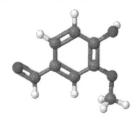

   a. What is the name of the parent on this molecule?
   b. Name as many functional groups as you can find attached to the parent.

66. Formic acid contains which functional group?

67. Diphenylamine is a yellow-colored organic compound used as an insecticide and fungicide. Which functional group does the molecule contain?

68. Compare ketones and aldehydes.
    a. What is similar?
    b. What is different?

69. Compare esters and carboxylic acids.
    a. What is similar?
    b. What is different?

70. Are ethers and esters water-soluble? Explain your answer.

71. Why do carboxylic acids, such as acetic acid, form weak acids in an aqueous solution?

72. Explain how amines act as bases in an aqueous solution.

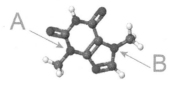

73. Theobromine, shown above, is an aromatic compound found in chocolate and tea. Letters A and B point to similar-looking functional groups, but letter A points out a specific group called an amide group. Letter B cannot also be called an amide.

    a. What type of functional group is indicated by letter B?
    b. Re-draw the molecule and circle the part you believe makes the amide group different from the group indicated by letter B. Explain your reasoning.

74. Give an example of what an aldehyde is commonly used for.

75. Identify the functional group shown in each molecule below.

   a. $\text{CH}_3\text{CH}_2-\text{O}-\text{CH}_2\text{CH}_3$

   b. $\text{CH}_3\text{CH}_2\text{CH}_2\text{NH}_2$

   c. $\text{CH}_3\text{CH}_2\text{CH}_2\overset{\overset{\text{O}}{\|}}{\text{C}}-\text{OH}$

   d. $\text{CH}_3-\overset{\overset{\text{O}}{\|}}{\text{C}}-\text{CH}_2\text{CH}_2\text{CH}_3$

   e. $\text{CH}_3\text{CH}_2\text{OH}$

76. Draw the generic structural formula that represents each of the following functional groups.
    a. Alkyl halide
    b. Ester
    c. Ketone
    d. Amine
    e. Alcohol
    f. Ether
    g. Carboxylic Acid
    h. Aldehyde

Chapter 20 review

Section 20.4

77. Polystyrene is made by addition polymerization. State the benefits and disadvantages of this polymer.

78. In your own words, explain what happens in each of the following types of organic reactions.

    a. Substitution

    b. Hydrogenation

    c. Dehydrogenation

79. How does cracking help to produce raw materials?

80. Fractions are produced during crude petroleum processing. What are fractions, and how are they distinguished from one another?

81. What must be added to reactants to make the cracking process work?

82. What is done to petroleum to make it more useful in our life?

83. What chemical process would you use to convert a fat that is a liquid at room temperature into a solid form? What happens during this process?

84. Transfats were created by accident.

    a. What is the name of the process that formed transfats?

    b. How is this process different than the standard processes?

85. What type of organic reaction is represented by the following?

    a. When hydrogen gas is added to an alkene, it becomes an alkane.

    b. A saturated hydrocarbon becomes unsaturated.

    c. Butane fuel is reacted with oxygen gas at high temperature.

    d. A chlorine atom takes the place of a hydrogen atom in a hydrocarbon.

86. $CH_4 + O_2(g) \rightarrow CO_2(g) + 2H_2O(g) + Heat$

    a. What type of alkane reaction is shown above?

    b. What can this reaction be used for?

87. What do open bonds on the left and right sides of a polymer indicate, as seen on the monomer drawing above?

88. What everyday products are produced through condensation polymerization?

89. Over 100 years ago, Leo Baekeland invented a groundbreaking polymer.

    a. Why was his invention different than what already existed?

    b. What was his invention called?

    c. How is the polymer created?

90. What types of reactions are shown below?

    a. $CH_3CH_2CH_2CH_2CH_3 \rightarrow CH_3CH_2CH_2CH=CH_2 + H_2$

    b. $CH_3CH_2CH_2CH_2CH_3(g) + Br_2(g) \rightarrow CH_3CH_2CH_2CH_2CH_2Br + HBr$

    c. $CH_3CH=CHCH=CH_2 + H_2(g) \rightarrow CH_3CH_2CH_2CH_2CH_3$

    d. $2C_8H_{18} + 25O_2 \rightarrow 16CO_2 + 18H_2O$

91. The following substances will undergo a reaction. Write a complete, balanced reaction for each of the following.

    a. 2-pentene is hydrogenated.

    b. Hexane is burned.

    c. Butane is dehydrogenated and forms an alkene with one double bond.

92. Find a nutrition label on any food item and answer the following.

    a. Identify the brand and type of food.

    b. What is the serving size in grams?

    c. What is the amount of total fat per serving?

    d. How much saturated fat is in one serving?

    e. How much trans fat is in one serving?

chapter 21: Molecular Biology

Chemistry isn't only something that occurs out in the world; it exists inside your body too! In fact, you and all living systems are a series of molecular processes that work together in a very specialized way. The chemicals responsible for driving living systems are commonly called biological macromolecules, also known as biomolecules. Biomolecules are large and often complex organic molecules that include fats, carbohydrates, proteins and nucleic acids.

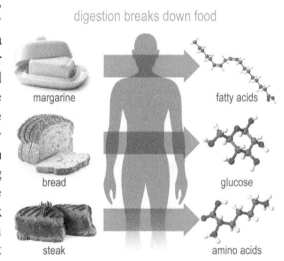

If the words fats, carbohydrates, and proteins sound familiar, they should. These complex organic molecules are typically itemized on nutrition labels. "You are what you eat" is not just a saying concocted by your parents to get you to eat your vegetables when you were a child. Our daily diet of fats and carbohydrates provides us with energy. We can look at the structure of macromolecules to see why some fats are considered healthy while others are not, and we can see why complex carbohydrates provide more sustained energy than simple sugars. We need proteins because they are the building blocks of the body. Without them our bodies could not make muscles, bones, and enzymes. We consume foods that are a mix of fats, carbohydrates and proteins, then break them down through digestion, and use these smaller molecules to support growth and keep us healthy.

Some macromolecules are not fuel, but are important for replication. You probably learned about DNA and RNA in your biology class. These molecules are important because they carry important information coding for every cell function. We can now look at them through the eyes of a chemist and see how things like acids and bases, and hydrogen bonding play a role in the structure and function of these macromolecules.

Let's see how we can apply science concepts learned throughout the year to understand why biochemistry is so essential.

Chapter 21 study guide

Chapter Preview

When you read nutrition labels you can see that calories for foods are broken down into carbohydrates, fats and proteins. But what are those compounds and what is their role in our bodies? In this chapter you will learn about the structure and function of carbohydrates, lipids and proteins, as well as DNA. Carbohydrates are polymers of sugar molecules that are the main energy source for plants and animals. Lipids are fats, waxes and oils. One type of fat molecule is a triglyceride made up of glycerol and three fatty acid chains. Proteins are large chains made up of amino acids. Both fats and proteins have various roles within our bodies. DNA holds the instructions for an organism with each gene coding for a particular protein.

Learning objectives

By the end of this chapter you should be able to:

- describe the production and structure of carbohydrates;
- explain what is meant by amphipathic and why phospholipids are important for living things;
- distinguish between the primary, secondary, tertiary and quaternary structure of proteins;
- explain the basic structure of DNA and how it is copied; and
- describe the process for a gene to code for a protein.

Investigations

21A: Polymers
21B: Amino acids

Pages in this chapter

- 680 Carbohydrates
- 681 Simple and complex carbohydrates
- 682 Polymerization
- 683 Carbohydrates form isomers
- 684 Consuming carbohydrates
- 685 Section 1 Review
- 686 Lipids
- 687 Fats
- 688 Phospholipids
- 689 Food and fats
- 690 Section 2 Review
- 691 Proteins
- 692 Amino acids
- 693 Amino acid table
- 694 Chiral proteins
- 695 The role of proteins in the body
- 696 Primary protein structure
- 697 Secondary protein structure
- 698 Tertiary and quaternary protein structure
- 699 Enzymes
- 700 Section 3 Review
- 701 Nucleic Acids
- 702 The DNA molecule
- 703 Replication
- 704 Transcription and translation
- 705 Section 4 Review
- 706 Semi-Synthetic Life
- 707 Semi-Synthetic Life - 2
- 708 Chapter review

Vocabulary

carbohydrates	simple carbohydrates	complex carbohydrates
monosaccharide	disaccharides	oligosaccharides
polysaccharide	polymer	monomer
polymerization	condensation polymerization	chiral carbon
lipid	adipose tissue	fat
triglyceride	lipoproteins	phospholipid
amphipathic	hydrophilic	hydrophobic
micelle	protein	amino acid
enzymes	inhibitor	denatured
peptide bond	primary structure	secondary structure
alpha helix	beta pleated sheet	tertiary structure
quaternary structure	substrate	active site
optimum range	DNA	nucleotide
nitrogenous base	gene	transcription
RNA	translation	

21.1 - Carbohydrates

Food provides the atoms and molecules your body needs to build everything from brain cells to toenails. Food also delivers energy for the countless chemical processes that keep living things alive. We get most of our daily energy from fat, protein and carbohydrate molecules that we eat. **Carbohydrates** are made of both short and long chain sugar molecule polymers. All organisms including plants release energy from carbohydrate molecules. Our bodies can either break apart these molecules when we need energy or create more of these molecules when our bodies need building materials. So you are what you eat!

Carbohydrates

Molecules for energy and structure

Carbohydrates are the primary source of energy for all living things. Let's look at carbohydrates found in food. Starch is a carbohydrate the body uses to make the sugar, glucose. Any glucose that is not immediately used for energy can be stored as a different molecule called glycogen in the liver and in muscles. Athletes often eat starchy foods like pasta before a big race to boost glycogen storage, which allows them to perform longer. Plants use a different carbohydrate called cellulose for structural purposes. Cellulose found in plant cell walls allows trees and plants to grow tall without falling over. Cellulose is also a dietary fiber that is essential to your health. Our bodies cannot digest cellulose, but the fibers help keep digested food moving along through the digestive tract. When individual glucose molecules (shown below) form a polymer, they form starch molecules.

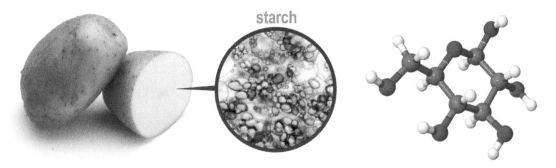

Variety of structures

What do gummy bears, spaghetti and wood have in common? All three contain carbohydrates. The function of a carbohydrate depends largely on its structure.

Structure is related to function

Gummy bears are high in sugars. Nature made sure our bodies respond favorably to foods that are high in carbohydrates because they provide energy for our bodies. Spaghetti is high in starch, another energy-containing carbohydrate. Wood gets its tough structure from cellulose, the most abundant carbohydrate on Earth. The cellulose molecule is a long polymer chain with plenty of hydroxyl groups that are capable of forming a great deal of hydrogen bonds with neighboring chains. The strong attractive forces between carbohydrate chains makes wood and other plant structures very strong.

Simple and complex carbohydrates

Carbohydrate composition

Simple carbohydrates are smaller sugar molecules and **complex carbohydrates** are larger starch or cellulose molecules, commonly known as grain and fiber. Simple carbohydrates are sugar molecules that are only made of one or two subunits. Sucrose is a simple 2-unit carbohydrate known as table sugar. Complex carbohydrates are large polymers that are anywhere from three or more sugar units long. These polymers serve as energy storage.

Single molecule carbohydrates

A **monosaccharide** is a single sugar molecule that is the smallest repeating unit in a carbohydrate. Fructose, glucose and galactose are examples of monosaccharide isomers that all have the formula $C_6H_{12}O_6$. Fructose gives fruits their sweet taste. Glucose is found in fruits and honey. Galactose is found in dairy products and in vegetables like sugar beets. These sugar molecules are different from one another, yet they have similar chemical properties. Many monosaccharides have a sweet taste and form colorless network covalent crystals that dissolve in water.

Short-chain carbohydrates

Small carbohydrate chains are formed by linking 2 to 10 monosaccharides together. Two-monomer sugars are called **disaccharides** while chains between 2 and 10 monosaccharide units are called **oligosaccharides**. Lactose is an example of an oligosaccharide that is also a disaccharide. When glucose and galactose combine to form a single molecule, they form the sugar lactose, which is found in dairy products.

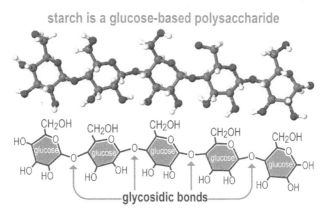

Long-chain carbohydrates

A **polysaccharide** is a monosaccharide chain that is longer than 10 monosaccharide units. These complex carbohydrates give living things either a source of energy or materials to build structural molecules. Starch, glycogen and cellulose are polysaccharides. Starch is the energy storage molecule found in rice, potatoes and corn. Glycogen is the energy storage molecule found in animals. This molecule is stored in the liver and in muscle tissue so it can be used as a fuel source when needed. Plants use cellulose to give rigidity to their structure. Many mono-, di-, and oligosaccharides have a sweet taste, but polysaccharides do not.

Polymerization

Polymers form through chemical reactions

A **polymer** is a large molecule made by linking a number of smaller **monomer** molecules together with chemical bonds. Many biologically important molecules are polymers. When you look at polymer molecules, look for the functional groups that make covalent bond linkages between molecules possible. A **polymerization** reaction occurs when monomers are linked together to form polymers.

Natural and artificial polymers

Polysaccharides are made of repeating sugar monomers. Polysaccharide polymers formed in nature make up the majority of polymers you are exposed to in your daily life. Starch, cotton, silk, proteins and rubber are natural polymers. Synthetic polymers have been around since the 1800s but they only began to be mass-produced during World War II when natural polymers like silk became scarce. Polyester fabrics and nylon are among the first mass-produced synthetic polymers. Today artificial polymers dominate multiple industries.

Condensed polymers

One reaction used by nature and industry to synthesize polymers is called **condensation polymerization**. A polymer is formed by eliminating a smaller molecule, such as water, when two monomers are joined. This is how carbohydrate polymers form. Condensation polymerization can occur between any two monomers.

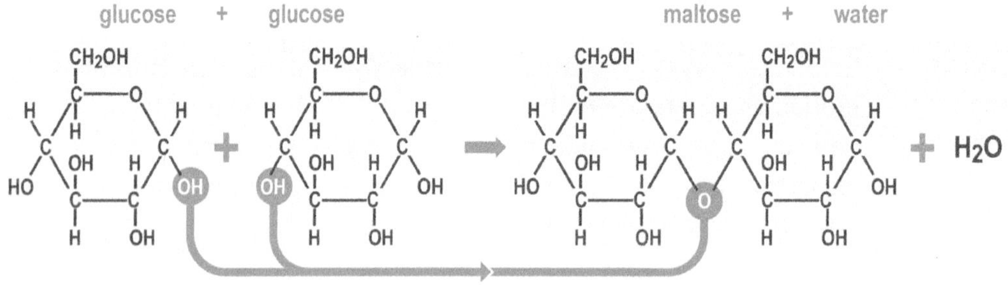

Water is released when monomers join hydroxyl groups

In the condensation polymerization reaction shown, a water molecule is lost when two glucose monomers are joined. The water molecule is produced when the hydroxyl group from the first glucose unit combines with a hydrogen atom from the hydroxyl group on the second glucose unit. Once the water molecule is eliminated, the glucose units become joined by the leftover oxygen. The overall products of the reaction are a disaccharide called maltose and a water molecule.

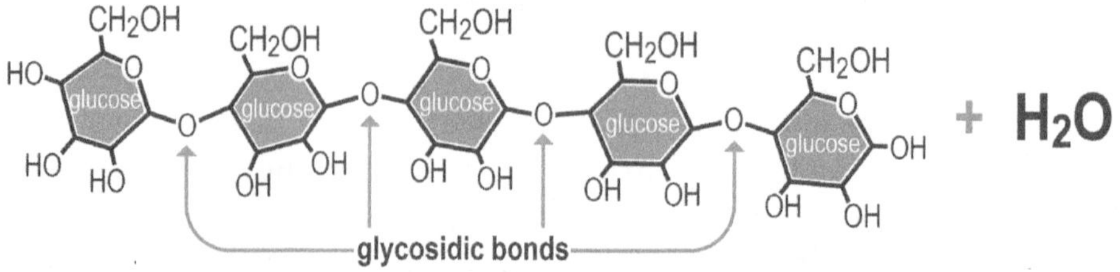

How polysaccharides form

If the condensation reactions continued, the chain would continue to grow by adding more glucose monomers to either end of the molecule. You would end up with the starch polysaccharide shown above along with more water molecules. Notice how the chain begins to twist as it gets longer.

Carbohydrates form isomers

Chiral carbons

Carbohydrates are made by linking monosaccharides in their cyclic or ring form. These ring structures contain at least one **chiral carbon** that have four different groups or atoms attached. The chiral carbons are circled green in the glucose isomers shown. Notice how each atom or group bonded to the chiral carbon are not repeats of one another. The attached groups or atoms are highlighted yellow.

D-glucose and L-glucose are isomers

You can see the chiral carbon ends up on different sides of the D-glucose and L-glucose isomers. D-glucose is produced by plants through photosynthesis and is used by organisms to fuel respiration or fermentation reactions. L-glucose is also produced by plants, but in much smaller amounts. Living things lack the enzyme needed to start the respiration reaction with the L-glucose molecule's specific shape. For that reason, L-glucose is not commonly found in nature. Although L-glucose is not considered biologically important, it has similar properties to D-glucose, such as a sweet taste.

Biologically important D isomers

D and L are abbreviations for dextrorotatory and levorotatory which are fancy ways of saying "turns to the right" and "turns to the left." Both forms of glucose are optical isomers, meaning their different atom arrangements cause them to rotate light in opposite directions. D-glucose rotates light clockwise, or to the right, so it is D-rotary. L-glucose rotates light counterclockwise, or to the left, so it is L-rotary. In most living things the D- carbohydrate forms are the ones that are utilized. When you read or hear about "glucose," assume it is D-glucose which rotates light to the right.

α-D- and β-D-glucose form different structures

Find the chiral carbons highlighted green on alpha-D-glucose (α-D) and beta-D-glucose (β-D) above. Locate the hydroxyl, -OH groups on the chiral carbons. In α-D-glucose the hydroxyl group points down and in β-D-glucose the hydroxyl group points up. This single difference makes the two molecules isomers and impacts the way bonds form during polymerization. When α-D-glucose monomers form starch, the polymer ends up with ethyl group linkages all pointing on the same side of the molecule. When β-D-glucose monomers join, ethyl group linkages are on alternating sides of the cellulose molecule. Starch forms tasty spiral-like polysaccharide chains that our bodies can use for energy. Cellulose forms linear chains that allow multiple hydrogen bonds to form. This results in chain-to-chain attractions that form strong fibers plants use for building structures that our bodies cannot digest.

Section 21.1: Carbohydrates

Consuming carbohydrates

Storage

Fatty foods alone do not cause excess weight gain. When we eat food, our body converts carbohydrates to energy. If we eat more carbohydrates than needed, our bodies store the leftover energy as glycogen polymers in muscle and liver tissues. However, muscles and the liver cannot store very much extra carbohydrates. Any excess carbohydrates that cannot be stored by muscle and liver cells are converted to fat and stored in fat cells in our bodies. This is how eating too many carbohydrates increases body fat.

Molecular structure influences digestion

When you eat food, you take in two types of carbohydrates: simple and complex. Simple carbohydrates like fructose are smaller molecules that are digested quicker. They do not contain much cellulose fiber from plants. Our digestive system has evolved to get glucose molecules into the bloodstream quickly. Examples of complex carbohydrates include whole wheat bread, brown rice, pasta and vegetables. These foods have large amounts of cellulose, vitamins and minerals. Cellulose molecules take longer to digest because they tend to have long chains and branched molecular structures. Incompletely digested complex carbohydrates leave us feeling full for longer periods of time. Vegetables are also high in water content, adding to the feeling of fullness.

Choose carbs carefully

Some diets encourage you to avoid carbohydrates altogether, but carbohydrates are necessary for a healthy diet. You have to choose your carbohydrates wisely. Avoid foods high in simple carbohydrates, like candy. They have a lot of calories and very few nutrients, and they leave us feeling hungry again in a short period of time. Unsatisfied hunger causes us to consume more calories than we planned on and we could end up overeating. Fruits and vegetables high in complex carbohydrates like cellulose keep us full longer as cellulose fiber does not get digested. The full feeling helps keep us from overeating.

health impact	simple carbohydrates	complex carbohydrates
healthy	• honey • fruit • dairy (milk, yogurt) • root vegetables	• vegetables and beans • whole grain bread, pasta, cereal • brown rice
less healthy	• cakes, pies, cookies • table sugar, candy • ice cream	• processed grains like bread and pasta made with white flour • white rice

Carbs are important for health

Complex carbohydrates are important for our overall health. Nutritionists recommend just over half of our daily calorie intake should come from carbohydrates. We are advised to eat a variety of carbohydrates to make sure we get a healthy combination of nutrients.

Section 1 review

Chapter 21

Carbohydrates are polymers of sugar molecules of varying lengths. Carbohydrates are a primary energy source for animals and plants. Cellulose is a carbohydrate polymer used widely by plants for structural purposes. A monosaccharide is a single sugar molecule or monomer. Disaccharides are two monomer sugars. Polysaccharides contain 10 or more sugar monomers. Carbohydrate polymers are formed in a condensation polymerization where water is released and monomers are joined. Carbohydrates can form isomers with different properties. Carbohydrates in our diet provide quick energy but often have no other nutritional value. Excess carbohydrates can be converted to fat.

Vocabulary words: carbohydrates, simple carbohydrates, complex carbohydrates, monosaccharide, disaccharides, oligosaccharides, polysaccharide, polymer, monomer, polymerization, condensation polymerization, chiral carbon

Review problems and questions

1. Which type of molecule is the main energy source for living things?

2. How do simple carbohydrates form complex carbohydrates?

3. Which part of a glucose molecule is altered during dehydration synthesis? What compound is formed as a by-product after the molecule is altered?

4. A student is investigating the direction in which two forms of a carbohydrate rotate light. Carbohydrate form A rotates light to the left (counter-clockwise) and carbohydrate form B rotates light to the right. Which carbohydrate form is probably the most biologically useful? Defend your answer.

Section 21.1: Carbohydrates

21.2 - Lipids

A **lipid** is a fat, wax, oil or steroid molecule. Both plants and animals are capable of producing lipids. For example, milk is a water-based dispersion of fat molecules made by animals, and olive oil comes from olives. Lipids store energy, act as signals in biological processes, and are important building blocks of cell membranes. There are many biologically important types of lipids, but we will focus on fats.

Lipids

Fats vs. oils Fats are solids at room temperature while oils are liquids. The main difference between a fat and an oil is parent chain length and the number of saturated carbon-carbon bonds in the parent carbon chains. Molecules with a greater number of multiple bonds are less saturated and are likely to be liquid at room temperature. Oils are simply liquid fats. Fats store more energy per gram than carbohydrates.

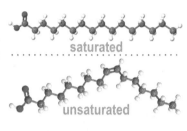

Cis unsaturated fats You may know that some fats are better for you than others. Fried foods are 'unhealthy' but room temperature olive oil is very 'healthy'. Fat is essential to your health, but the type of fat you eat matters. Like carbohydrates, some fats are easier for your body to break up than others. The number of double bonds between parent carbon atoms and where those bonds are located determine how healthy a fat is. Saturated fats contain only single bonds and are usually solid at room temperature. Although necessary for good health, saturated fats are not good for our bodies in excess amounts because they are harder to split apart for energy. Unsaturated fats are much easier for our bodies to metabolize because double bonds change a straight chain into a bent chain. Straighter fat molecules are harder to digest than bent ones. This is why configuration around the double bond matters. Trans fats are more difficult to digest because their chains are straighter than cis fats. Nearly identical fats with different configurations around the double bond in the middle of the chain are shown below.

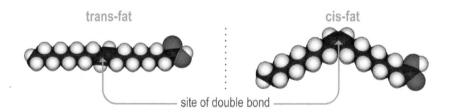

Trans fats may be in oily foods Trans fats are added to oily foods like cookies to increase shelf life in a grocery store. Since research has shown that trans fats are harmful to health, scientists have come up with ways to process oils that do not produce harmful trans fats. In fact, most manufacturers now advertise "zero trans fats" on product labels so that consumers can choose a healthier option.

Why we need fat Fat molecules keep your body functioning properly. For example, the layer of fat under skin provides protection from disease. The fatty coating over your nerve cells and spinal cord speeds up communication between nerve cells. Every cell in your body has a fat-based membrane that forms a barrier between the inside and outside of each cell. Fats make up **adipose tissue** that provides energy storage, insulates your body from temperature changes, and protects vital organs.

Fats

How fats form

Let's look at how a fat molecule is formed. Locate the glycerol molecule and three fatty acid chains in the diagram below. Notice the three hydroxyl groups on the glycerol molecule and the hydrogen atom on each fatty acid chain highlighted in yellow.

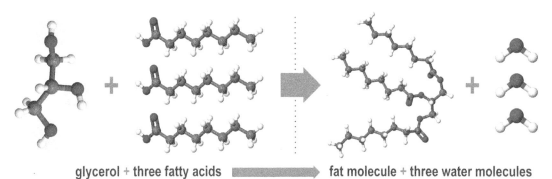

glycerol + three fatty acids ⟶ fat molecule + three water molecules

Fats are attracted to one another

Hydrogen atoms from the fatty acid chains combine with the hydroxyl groups on glycerol to form a **fat** molecule and three water molecules. This is a condensation reaction because water is a product. When a non-polar glycerol molecule attaches to 3 fatty acids, a **triglyceride** is formed. Fats tend to clump together because London dispersion forces along the long non-polar fatty acid chains attract other fatty acid chains.

Fatty acids determine the type of fat

Fats come in different varieties depending on the makeup of the fatty acid molecules. All 3 fatty acids can be the same molecule, or they can be different in saturation or length. Fatty acid molecules are usually between 4 and 24 carbons in length. Fatty acid chains 16 to 18 carbon atoms long are the most common in humans. Fatty acid chains can also be saturated or unsaturated. Unsaturated omega-3 fatty acids are well-known for their health benefits.

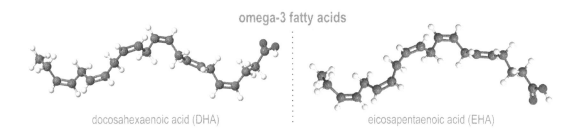

omega-3 fatty acids

docosahexaenoic acid (DHA) : eicosapentaenoic acid (EHA)

Saturated = solid

Fats are solid at room temperature when they contain more saturated bonds. Oils are liquid instead of solid at room temperature because they have more unsaturated bonds. Foods with unsaturated double bonds are easier for our bodies to digest.

Non-polar solvents dissolve fats

Because like dissolves like, fats dissolve in non-polar solvents such as benzene or vegetable oil. But our blood and body fluids are mostly water. A non-polar fat molecule can dissolve in water if it has charged areas, either from polar regions on the molecule or from ions.

Lipoproteins have polar and non-polar surfaces

Our bodies use **lipoproteins** to carry fat molecules in our bloodstream. Lipoproteins are small, hollow spheres with polar groups on the outside and non-polar groups on the inside. The polar outside lets the molecule dissolve in blood. The non-polar inner lining allows fat to be carried inside the lipoprotein. Fat-soluble molecules like Vitamin D get transported in our blood by traveling inside lipoproteins.

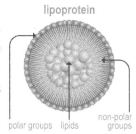

lipoprotein

polar groups | lipids | non-polar groups

Phospholipids

Polar heads and non-polar tails

A type of lipid molecule called a **phospholipid** is incorporated into every membrane in every cell of every living thing. Phospholipids are **amphipathic** meaning they have polar and non-polar regions. A phospholipid has two parts: the "head" is a polar phosphate group, and the "tail" is made of two long non-polar hydrocarbon chains. The difference in polarity lets these molecules strategically arrange themselves in water. The polar phosphate head is attracted to water. This portion is **hydrophilic**, or "water-loving." The non-polar hydrocarbon tail repels water molecules so it is **hydrophobic**, or "water-hating."

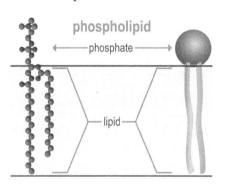
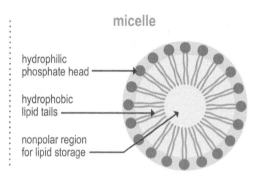

Micelles

Amphipathic molecules organize themselves into a sphere called a **micelle**. The sphere forms due to intermolecular attractive forces. Polar phosphate heads orient outwards to contact water while Van der Waals attractions help non-polar hydrocarbon tails to line up on the inside. The result is a spherical phospholipid shell with a polar outer surface and non-polar inner surface. Lipoproteins are examples of micelles in your body.

Interactive simulation

Soaps use micelles to dissolve non-polar oil or dirt and remove it when rinsed in water. In the interactive simulation titled Polar and Non-polar Interface, watch amphipathic molecules self-align in an interface between oil and water according to their polarity. Can you determine which end of the molecule is hydrophobic and which end is hydrophilic?

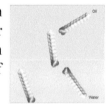

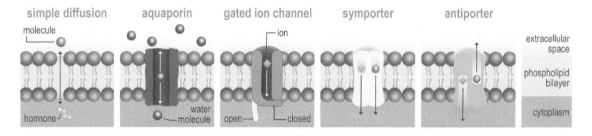

Two-layer membranes

In living things, micelles form a two-layer membrane called a phospholipid bilayer. Living things rely on membranes to regulate the highly complex flow of energy and matter in and out of cells. Complex forms of life with multiple organ systems would be impossible without a phospholipid bilayer. The bilayer's most basic function is to act as a barrier between the cell and its external environment. It also brings needed molecules inside while allowing wastes and cell products to move outside the cell. Fats and proteins embedded in the membrane provide identification tags that help cells exchange messages with one another, allow molecule transport in and out of the cell, and provide structural support. Our cell membranes can rearrange themselves to form the variety of structures above because phospholipid molecules can move or migrate within a layer to accommodate a cell's needs.

Food and fats

Energy storage in humans

Triglyceride molecules are the most common type of fat found in the human body. Our bloodstream transports triglyceride fats inside lipoproteins. Our cells use triglycerides for energy, but cell membranes do not allow the molecules to enter freely. Enzymes must first break triglycerides down into their fatty acid and glycerol molecules. The fatty acid molecule shown below is palmitic acid. Palmitic acid is the most common saturated fatty acid produced by plants, animals and microorganisms.

Fatty acids impact food properties

Natural triglyceride fats often contain different fatty acid chains on one molecule. This gives manufacturers the ability to vary the texture and melting point of foods. For example, ice cream and frozen yogurt are both dairy-based foods you would expect to contain triglycerides with palmitic acid chains. As you know, ice cream and frozen yogurt have different textures and melting points. One reason is because ice cream has a larger amount of lecithin added. Lecithin is a phospholipid that helps triglycerides form large blobs, which helps ice cream form a more firm structure than frozen yogurt. Similarly, different chocolate candies have different melting points and textures based on the variety of fatty acid ingredients. Chocolates with a higher degree of saturated fatty acid molecules typically have higher melting points than chocolates with unsaturated fatty acid chains.

Longer fatty acid chains yield more ATP

Lipids provide more energy per gram than carbohydrates or proteins by far. Long chain fatty acids yield large amounts of adenosine triphosphate (ATP) molecules when they are oxidized. For example a 12-carbon fatty acid chain yields nearly 80 ATP molecules! A greater number of carbon-to-carbon bonds available to break results in a greater number of ATP molecules formed.

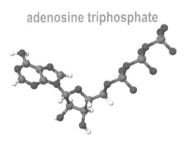

Excess triglycerides can lead to heart disease

Triglycerides are essential for good health. If we consume more fat than our body needs for energy, some of the excess is stored in our bodies in adipose (or fat) tissue. Most fat in our bodies comes from our diet, but the body can make triglycerides if needed. When we consume foods that have triglycerides, our digestive system breaks them apart into smaller molecules. Some of these smaller molecules can re-form triglyceride molecules. This is why a high fat diet results in a larger amount of triglycerides floating around in the blood. A high amount of fat molecules floating in the blood can lead to blocked blood vessels and arteries. To avoid heart disease, we can eat a balanced diet and test our blood to keep an eye on triglyceride levels during yearly check-ups with a doctor.

Chapter 21

Section 2 review

Lipids are an important class of biological molecules. Fats are one type of lipid. Fat molecules are used to protect, insulate and cushion our bodies. They also provide a way to store energy. In terms of diet, unsaturated fats (without double bonds) are easier for our bodies to break down and are thus healthier. A triglyceride is made up of a glycerol molecule and three fatty acid chains. Fats are non-polar molecules so lipoproteins are used to carry them in the bloodstream. Triglycerides can provide a lot of energy for our bodies, but lead to health problems when eaten in excess. Phospholipids are lipids with both polar and non-polar regions. They form into micelles and are important in forming biological membranes.

| Vocabulary words | lipid, adipose tissue, fat, triglyceride, lipoproteins, phospholipid, amphipathic, hydrophilic, hydrophobic, micelle |

Review problems and questions

1. Most people assume "fat" is unhealthy, but lipids serve many important biological functions. Describe at least three functions of lipid molecules for living things.

2. What is a phospholipid?

3. Explain how the phospholipid bi-layer that makes up cell membranes have the perfect chemistry to transport polar and non-polar substances in and out of cells.

4. Explain at least two dietetic reasons why some fats are better to eat than others.

5. What is the chemical difference between cis- and trans- fats?

21.3 - Proteins

A **protein** is a large biomolecule that performs numerous functions in all living things. Proteins build our bodies and keep them running smoothly. They are involved in every active cell process and almost every chemical reaction that keeps us alive. Our bodies break apart the proteins we eat and are used to build and repair tissue such as skin and muscle. Every protein that makes up your body was once made by your own body.

Polymers of amino acids

Proteins are made from amino acids

A protein is a polymer made of **amino acid** molecules. Amino acids molecules have an amine (-NH₂) group on one end, a carboxyl group (-COOH) on the other, and a side chain or R-group that gives the amino acid its unique identity. More than 100 amino acids link together to form a protein polymer. Different combinations of over 20 naturally occurring amino acids give rise to a great diversity of biological molecules. A large protein chain length and a variety of amino acid building blocks allows for over 22,100 different combinations.

examples of amino acids

asparagic acid
asp

glutamic acid
glu

tyrosine
tyr

Proteins change reaction rates in our bodies

Many proteins found in the body are biological catalysts otherwise known as **enzymes**. Catalysts speed up the rate of chemical reactions. If we did not have these protein-based enzymes, reactions in our body would happen too slowly to keep us alive! Sometimes reactions in our bodies need to be slowed down by a protein-based molecule called an **inhibitor** so all of the reactant molecules are not used up too quickly.

hemoglobin molecule

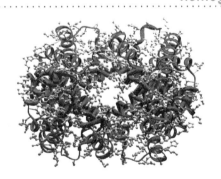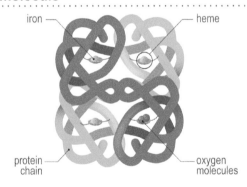

Proteins change their shape based on their environment

The physical structures in our bodies are made with proteins. Fingernails, hair, skin, muscles and bones are protein-based. Proteins also have special roles like transporting oxygen and nutrients to our tissues. Proteins can have a fixed three-dimensional shape or they can have a changeable 3D shape when their function demands it. For example, a hemoglobin molecule is made of four proteins that transport oxygen from our lungs to our tissues. Hemoglobin binds oxygen molecules and holds on to them while traveling in blood. When it reaches tissues with low oxygen, hemoglobin changes its shape and releases oxygen molecules to allow the tissues to function normally. When the hemoglobin protein returns to the oxygen-rich capillaries in the lungs, it once again changes its shape so it can pick up as many as four oxygen molecules. There are over 250,000,000 hemoglobin molecules for every red blood cell in the body. Every red blood cell can carry over 1 billion oxygen molecules!

Amino acids

Amine, carboxyl and R-group

There are hundreds of amino acids occurring in nature, but most living things create only 20 on their own. Amino acids produced by living things all have an amine (NH_2) and a carboxylic acid (COOH) group bonded to a central carbon on one side of the molecule. The central carbon also contains one hydrogen atom and one R-group. The R-group is what is unique among the 20 most common biologically produced amino acids.

R-group polarity

The central carbon atom has four different groups attached to it, so it forms an optical isomer. The R-group side chain significantly affects the amino acid's chemical properties. An R-group can be either polar or non-polar.

Polarity determines shape

Proteins can only do their job when their 3D shape is correct. A protein carefully folds itself into a 3D shape through a series of loops, sheets, and zigzags to accommodate the polarity of R-groups on its amino acids. Amino acids such as alanine contain non-polar R-groups. Non-polar R-groups are hydrophobic, so they curl to the inside of the 3D protein molecule in an aqueous environment. Other amino acids such as serine contain polar R-groups that are hydrophilic. Polar R-groups curl to the outside of an amino acid because they are hydrophilic and can form hydrogen bonds with water molecules in an aqueous solution.

R-groups are either acidic or basic

Amino acids can have an overall neutral, acidic or basic pH. The "acid" part of the name amino acid comes from the carboxylic acid functional group, but the R-groups are what determine pH. Side chains are exposed to a roughly neutral solution in the body with a pH of about 7.35. Amino acids are very pH-sensitive. Proteins can be made to come apart and denature under conditions of high or low pH. A **denatured** protein suffers damage to its 3D structure and can no longer work correctly. If pH is too high or too low in your body, your enzymes could denature and could cause serious illness or death. That's why your blood contains a buffering system to keep pH stable. Biochemists take advantage of this when they want to isolate specific proteins. High temperatures can also cause proteins to denature, and that's why fevers above 104 °F can be dangerous.

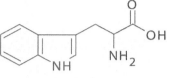

common amino acids found in nature

non-polar

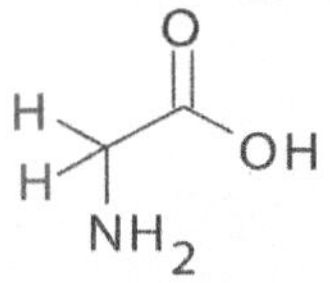

glycine

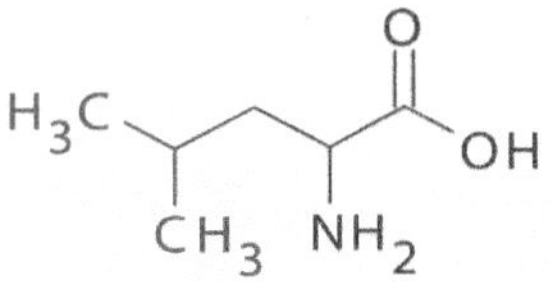

leucine

tryptophan

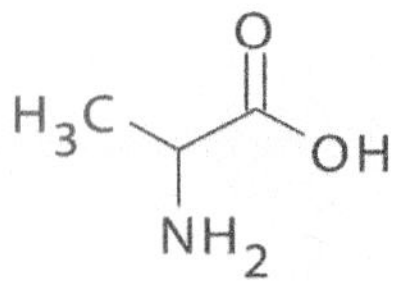

alanine

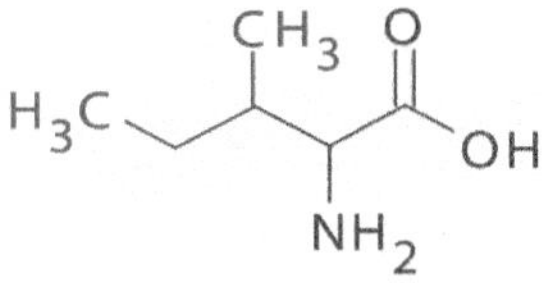

isoleucine

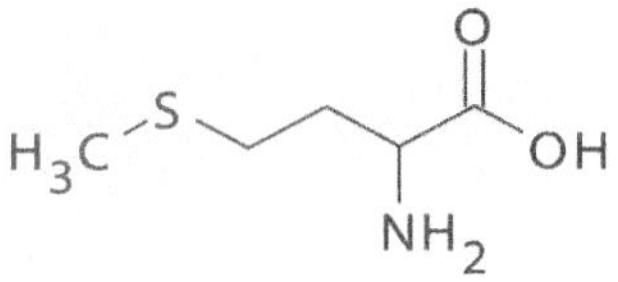

methionine

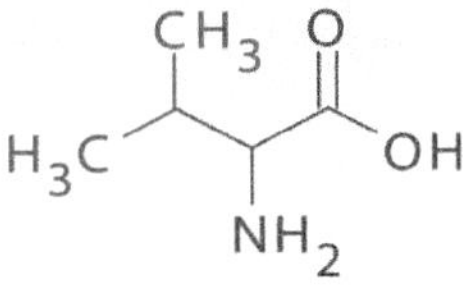

valine

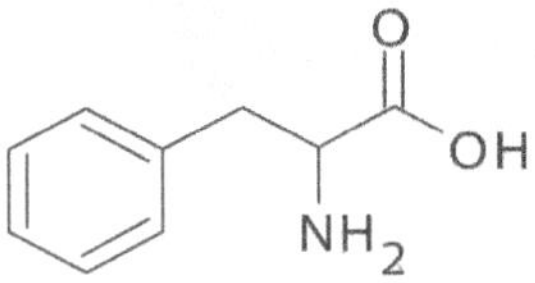

phenylalanine

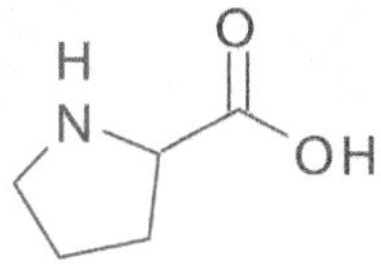

proline

acidic

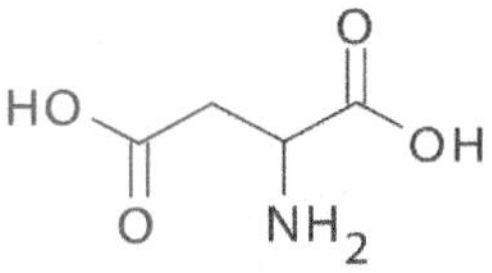

aspartic acid

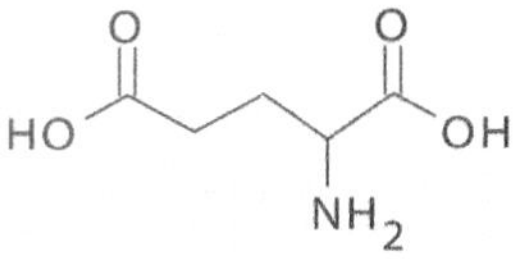

glutamic acid

basic

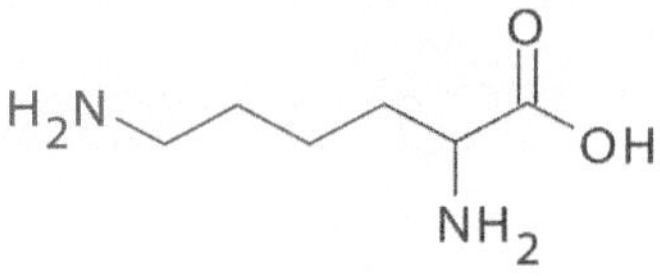

lysine

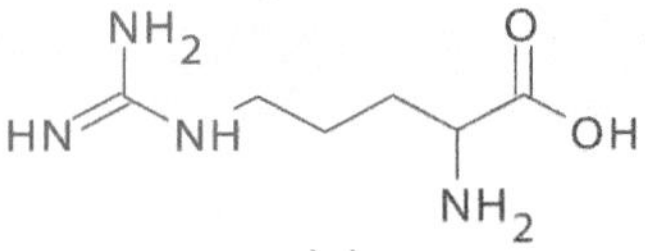

arginine

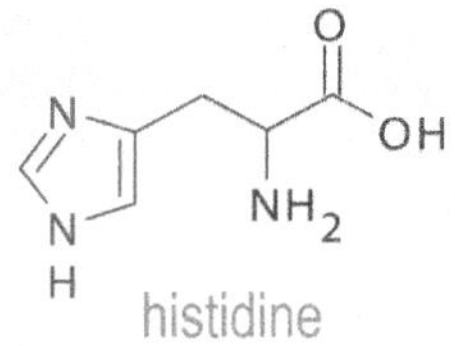

histidine

polar

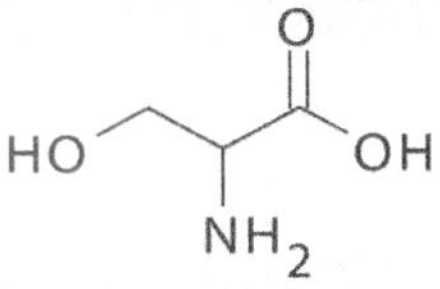

serine

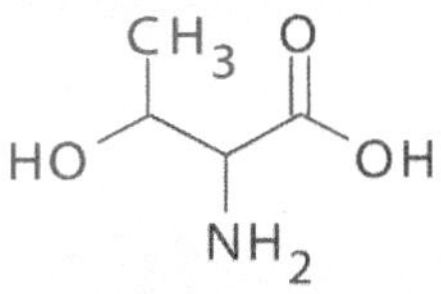

threonine

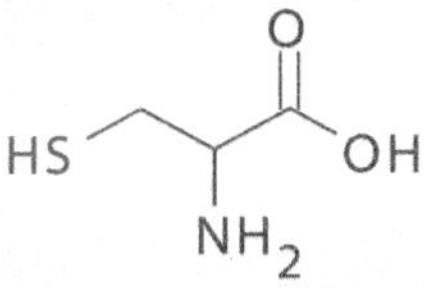

cysteine

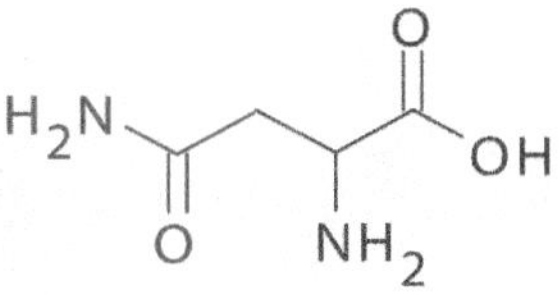

asparagine

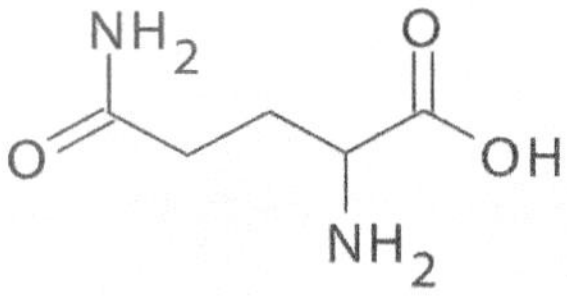

glutamine

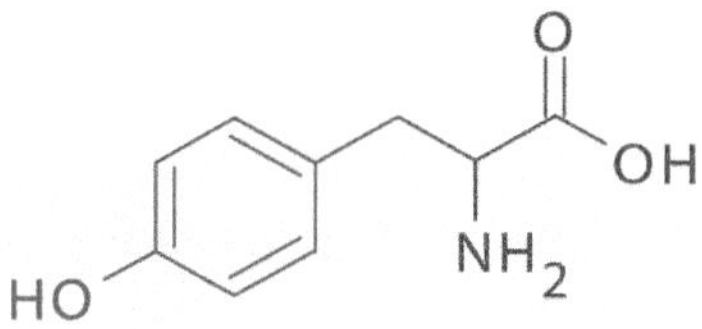

tyrosine

Section 21.3: Proteins

Chiral proteins

L-rotation is dominant

Living things use D-rotary carbohydrates but for amino acids, L-rotary is dominant. Many hypotheses attempt to explain why one form is more common. One hypothesis points out L-rotary amino acids are favored when dissolved in water. This would mean they are easy to distribute in aqueous environments, which is where the earliest forms of life likely evolved. Another explanation suggests meteorites brought L-rotary amino acids to earth. The Murchison meteorite that landed in Australia in 1969 and other meteorites contain L-amino acids. Some scientists think L-rotation is the result of circular light polarization by stars. If one chiral form of light was available back when life was first developing, perhaps nature adapted itself to use the same chiral form. Perhaps the earliest life forms developed the L-rotary form by chance and nature remained consistent to conserve energy.

Polarized light

Reptiles, fish, birds and insects are among the animals that can detect polarized light. Animals use polarized light to navigate, find water and communicate with one another. Animals sensitive to polarized light have photoreceptors, which are light-sensitive cells in the eye that have a particular geometric arrangement. Humans lack sensitivity to polarized light, but when you put on a pair of polarized sunglasses, you enjoy enhanced contrast that makes everything look sharper.

Shape-specific reactions

All amino acids found in proteins have at least one stereoisomer, except for glycine. Our bodies have receptors for one stereoisomer: the R-chiral D-rotary form for carbohydrates and the L-chiral L-rotary form for amino acids. Each chiral form has a unique 3-dimensional shape. If a shape does not match by as little as one atom, a reaction may not proceed because the exact shape is necessary for proper chemical function. This is why only one chiral form works.

Engineering medicine from isomers

More than half of manufactured drugs are chiral, and more than 25% of prescription medicines used each year in the United States contain chiral compounds from plants and animals. It is economical and environmentally responsible for scientists to create drugs in a laboratory instead of harvesting them from nature. When a drug is discovered in nature, scientists isolate different optically active molecules to identify which one is useful. The drug used to treat childhood leukemia comes from a periwinkle plant you may

have in your backyard. When scientists tried to isolate the compound that acts on cancer, they found a mix of isomers that was expensive to separate. Once the biologically active isomer was found, they synthesized the compound in the lab. Typically, lab-generated isomers create a problem of mixed isomers that is not seen in nature. Nature usually only produces L-rotary isomers while synthetic processes in the lab produce a mix of L- and D-rotary isomers that have to be chemically separated.

Different biological effects

Isomers have been shown to have very different effects in the body. Molecules with different chirality produce different biological activity. For example, in the 1950s a drug was marketed to help expectant mothers find relief from nausea. The D-form did provide relief, but the L-form caused severe birth defects. This example taught us some medications must be carefully controlled to exclude harmful isomers.

The role of proteins in the body

Protein and muscles

You very much are what you eat. A healthy diet includes plenty of protein because our bodies break proteins down into amino acids. These amino acids are used to make new cells such as muscle cells. Muscles are mainly made up of protein. Without protein, muscles would not be able to contract. Athletes and body builders who want to build muscle consume protein shakes to help build muscle mass.

Creatine is a protein supplement

Creatine is a popular body-building supplement taken by athletes. It is a natural protein found in meat that helps provide energy and builds muscle mass faster without the many negative side effects of steroids. Scientists are uncertain how effective the supplement is for enhancing athletic performance and side effects are unknown. However, individuals with health conditions such as heart or kidney problems are urged to avoid the supplement.

Structure, storage and transport

Think about all of the different things that give the inside and outside of your body its structure. Fingernails, skin, bones, tendons, and hair are all made primarily of protein. You may have heard that collagen is one of the things that keeps your skin looking young, and keratin makes up your hair and nails. Both collagen and keratin are protein molecules. The way the proteins are arranged determine how tough or stretchy each structure is. Proteins also make up the portable structures in our bodies. All kinds of proteins are dissolved in our blood. Some of these proteins store minerals needed by the body and release them when they are needed. Other proteins carry important molecules from one place to another in the body, such as hemoglobin.

Enzymes and energy

Protein-based catalysts called enzymes speed up reactions in your body. All living things need enzymes to keep the body functioning normally. Proteins help move electrons through a series of redox reactions called metabolism. The goal of metabolism is to provide energy for the body. Proteins help improve the efficiency of our metabolism.

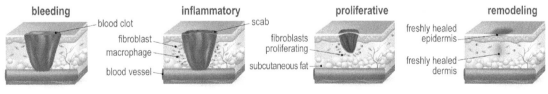

Protection and signaling

Did you know your cells communicate with one another? There are countless coordinated activities happening in your body at any given time. Proteins are always at the center of those activities. When you get cut, cells send the signal that it is time to repair a wound by activating blood-clotting proteins. Proteins are involved at every step of wound healing. If bacteria or other foreign invaders enter your body either through a cut or by other means, proteins in your immune system identify them and destroy them. Hormones are protein-based chemical signals that constantly regulate your body systems.

Primary protein structure

Peptide bonds

A series of chemical reactions occur when your body builds proteins. Deoxyribonucleic acid (DNA) provides instructions to your body on how to assemble amino acids into proteins. Amino acids are chemically linked together by a peptide bond. A **peptide bond** forms between the carboxyl and amine groups of adjacent amino acids. A water molecule is released when a peptide bond forms, therefore proteins are formed through condensation reactions.

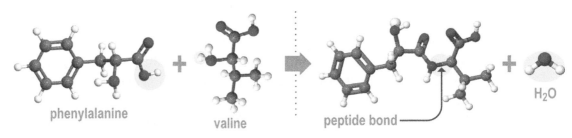

phenylalanine   valine   peptide bond H_2O

Forming the polymer

The condensation polymerization reaction begins when a bond forms between two amino acid monomers. A polymer is formed when more amino acids are added to the sequence. Amino acids are added to the right end of the polymer as it continues to grow. Each of the 20 biologically significant amino acids has the two groups needed to form a peptide bond: a carboxyl and an amine group. In the diagram above, you can see how a peptide bond forms when the carboxyl group from phenylalanine combines with a hydrogen atom from an amine group in valine. Note the water molecule released as a condensation product.

1' structure is the amino acid sequence

The amino acid sequence in a protein chain is called the **primary structure**, abbreviated as: 1' structure. If amino acids are missing or in the wrong order, the protein may not work the way it is supposed to. Most proteins contain a sequence of at least 100 amino acids. Imagine having 100 different ingredients in a recipe. When grouped in a different order, the same ingredients can make completely different meals, snacks, or desserts.

Interactive simulation

Explore how a sequence of amino acids forms a protein in the interactive simulation titled Exploring Proteins in 3D. There are six different amino acids that you can highlight in the interactive, but the protein is made of more than six kinds of amino acids. Toggle the visual style, colored and highlighted sections to get different views of the protein and how it is built.

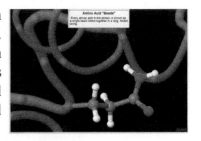

Essential amino acids

Our bodies can synthesize some, but not all of the amino acids called for in a protein's primary structure. In humans there are 9 essential amino acids that our DNA does not code for. Essential amino acids come from foods in our diet, such as meats, dairy, and plant-produced foods. When we consume these foods, their proteins are broken up into amino acids. Our bodies recycle some of these amino acids when we assemble the primary structure of proteins. A balanced diet provides you with a greater chance of eating foods with the essential amino acids your body cannot synthesize.

sources of proteins = sources of amino acids

Secondary protein structure

2' structure is the chain shape

Have you ever been frustrated by an extra-long cable that is twisted in some places and flat in others? This is similar to the shape a protein's amino acid sequence makes. A protein's **secondary structure**, or 2' structure, refers to the way parts of its amino acid chain are shaped. Like a long cable, there are two spatial arrangements that amino acid sequences take on: an **alpha helix**, α-helix, or a **beta pleated sheet**, β-pleated sheet. The α-helix resembles a spring. The β-pleated sheet resembles a folded paper fan or an accordion. As you can see in the catalase enzyme, proteins commonly use a combination of these two structures to form the right shape.

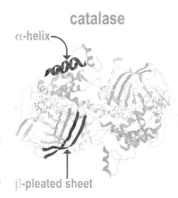
catalase

H-bonds keep coils together

The coil shape of an α-helix is held in place by regularly-spaced hydrogen bonds between the hydrogen of an amine and the oxygen of a carbonyl on a single amino acid chain. The diagram shows how hydrogen bonds (highlighted green) help form the coil. The H-bonds allow an α-helix to stretch and return to its coiled shape like a spring. The ability to stretch influences protein function. Natural fibers like wool, hair and tendons have the α-helix shape. The twist causes the side chains or R-groups to stick out on the outside of the helix, where they can interact with their environment.

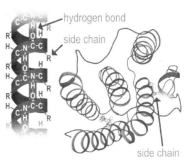
α-helices in a protein

H-bonds across a β-sheet

Large amino acids like tyrosine and tryptophan have bulky side chains that cannot fit into an α-helix. They instead form β-pleated sheets. Imagine drawing a polymer chain along the left and right margins of a sheet of paper, then folding it into an accordion-like zigzag shape. This is what a β-pleated sheet looks like. A β-pleated sheet is made up of two or more protein chains located side by side. Hydrogen bonds form between the chains. This gives the protein strength and flexibility, but it can't stretch.

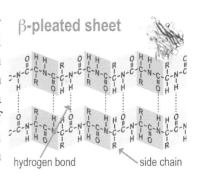
β-pleated sheet

β-sheets are strong

Pleated sheet proteins are found in muscle tissue and in natural fibers like silk. Like an α-helix, the hydrogen bonds form between a hydrogen on an amine and an oxygen from a carbonyl. The accordion shape appears as R-groups alternate above and below the sheet. Look at the diagrams below. Imagine the H-bonds holding each type of structure together. Imagine pulling α-helices like a spring, and twisting the β-pleated sheets without stretching them, like a flexible plastic sheet.

amino acid R-groups extending from α-helices and β-pleated sheets

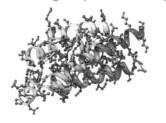 

Tertiary and quaternary protein structure

3' structure is the 3D shape

Think about everything you have learned about bonding and attractive forces. Hydrogen bonds, attractions between charged or polar molecules, hydrophobic and hydrophilic interactions — you must apply all of these concepts to truly understand tertiary protein structure. **Tertiary structure**, or 3' structure, is the protein's 3D shape that plays a major role in how a protein will function. Secondary α-helix and β-pleated sheet structures fold or bend into what is their globular tertiary structure. When you see proteins symbolized like the enzyme shown as a solid blue shape, realize the true structure is created by a series of helices and sheets.

two views of catalase

Globular shapes

The term globular is appropriate because the protein looks like a glob of amino acids clumped together. A globular protein has a complex tertiary structure that makes it able to complete a very specific role. The protein pictured above is an enzyme called catalase that is present in all living things. Catalase helps break down hydrogen peroxide, which is released as a by-product during normal cell activity. If catalase became denatured, hydrogen peroxide could build up to toxic levels in cells and tissues.

Inter-molecular forces and shape

R-groups or side chains sticking out of α-helix and β-sheet structures dictate the 3' structure of the protein. Most proteins in living systems are in aqueous water environments. Non-polar side chains tend to tuck inside the 3' structure away from polar water molecules. Polar side chains are on the outside so they can interact with the aqueous environment. This balance of intermolecular forces helps to stabilize a protein's tertiary structure.

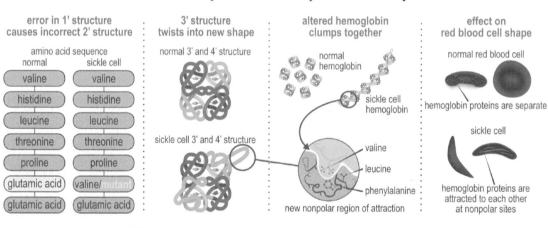

When 4' structure is wrong

Quaternary structure (4') develops when single protein molecules interact to form a larger 3D structure. Hemoglobin forms a 4' structure when its four units join to make one large protein. An error in the amino acid sequence for hemoglobin shown above can drastically change the shape of the 4' structure. When valine is expressed instead of glutamic acid, a non-polar region develops in the protein. This change in structure and attractive forces causes hemoglobin molecules to stick together. Sickle cell anemia is the result.

Interactive simulation

Observe a number of proteins with a hydrophobic core and hydrophilic surface in the interactive simulation titled Exploring a Hydrophobic Core. Take advantage of the tools in the simulation that allow you to rotate the molecule and focus on specific regions of different polarity. Look for the spaces in the molecule that are more friendly to non-polar substances than polar substances.

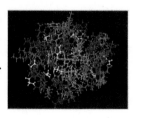

Enzymes

Enzymes and their substrates

Enzymes are large proteins that act as biological catalysts in living systems. They speed up chemical reactions by reducing the activation energy required for a reaction to take place. Enzymes bind **substrate** molecules, which are reactants in a chemical reaction. The substrate binds to a small region on the enzyme called the **active site**. The active site is only about 3 to 4 amino acids in length. Enzyme molecules have very high specificity for substrates. In other words, an enzyme will only allow one type of substrate molecule to fit in its active site. The active site shape is dictated by the tertiary structure of the protein. This specific fit is often called the lock-and-key model. As with a lock, the enzyme takes a substrate key with a very specific shape to unlock the reaction. When substrate molecules bind to the active site, the enzyme and the substrate change shape and mold to each other. The human body has thousands of kinds of enzymes, each catalyzing a specific reaction.

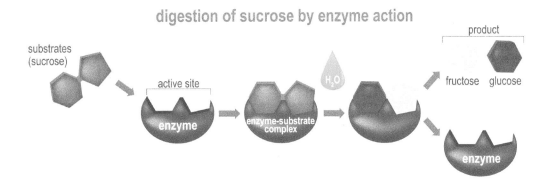

Regulatory enzymes

Enzymes serve a variety of roles in living systems. Living systems have a complex array of integrated chemical reactions that must be timed just right. Enzymes orchestrate these complex reactions by turning molecules "on" and "off" at appropriate times. This type of enzyme is called a regulatory enzyme.

Optimum pH and temperature ranges

An enzyme works best in a specific pH and temperature range depending on where it works in the body. This is known as the **optimum range**. Stomach acid, 0.1M hydrochloric acid, has a pH range between 1.5 and 3.5. Digestive enzymes in your stomach work best in that pH range. The same pH range would denature other digestive enzymes found in your small intestine where pH is closer to 7. Most enzymes in your body work best right around normal body temperature, but they can tolerate somewhat lower or higher temperatures. Significantly lower temperatures slow enzyme activity down too much and higher temperatures could denature enzymes.

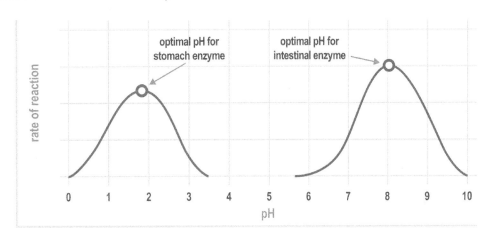

Section 21.3: Proteins

Chapter 21

Section 3 review

Proteins are a type of biological polymer. They are made of long chains of amino acids. Each of the 22 different amino acids has a different side chain. Some side chains are polar while others are non-polar. Some are acidic and some are basic. Together all the many amino acids of a protein help determine the three dimensional shape and functionality of that protein. Proteins serve many important roles in the body, including comprising muscle, and catalyzing reactions as enzymes. Peptide bonds connect amino acids in proteins. A protein's sequence of amino acids is called the primary structure. How chains of amino acids form three-dimensional structures such as helices or sheets is called the secondary structure. Interactions and bonds of side chains determine a protein's tertiary structure. A quaternary structure is determined by individual protein molecules interacting with one another to form more complex structures.

| Vocabulary words | protein, amino acid, enzymes, inhibitor, denatured, peptide bond, primary structure, secondary structure, alpha helix, beta pleated sheet, tertiary structure, quaternary structure, substrate, active site, optimum range |

Review problems and questions

1. Describe at least three functions of protein molecules for living things.

2. Re-organize the following list in order from primary protein structure to quaternary protein structure: 3D structure; α-helices and β-sheets; amino acid sequence; connected proteins

3. When a protein is denatured, which structure is affected first: primary, secondary, tertiary or quaternary? Explain your answer.

4. What determines whether a protein region will fold into an α-helix or β-sheet?

5. A student is investigating two forms of a protein in a lab. Protein form A rotates light to the left (counter-clockwise) and protein B rotates light to the right (clockwise). Which form of the protein is probably more biologically useful? Defend your answer.

21.4 - Nucleic Acids

How can a molecule carry enough information to code all of the directions to build a living organism and keep it alive? **DNA** is a set of biological polymers that store and transmit genetic information. Single-celled bacteria have only one DNA molecule per cell, but humans have 46 DNA molecules in every cell (except blood cells).

Nucleic acids

Nucleotide monomers

DNA is made of monomers called nucleotides. A **nucleotide** has three parts: a 5-carbon cyclic sugar called deoxyribose, a phosphate group, and an organic molecule called a nitrogenous base. DNA nucleotides are arranged in specific sequences that carry all of the genetic information an organism needs to grow, survive and reproduce. Human DNA molecules store over 6 billion nucleotides per cell nucleus!

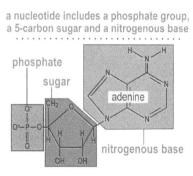

a nucleotide includes a phosphate group, a 5-carbon sugar and a nitrogenous base

Your DNA

Less than 1% of your DNA is responsible for causing the physical, chemical, and behavioral differences that make you unique among the billions of people on Earth. The other 99% of your DNA is identical to the billions of people on Earth! Although you were born with a set of DNA from your parents, your DNA can change throughout your life. Viruses, the environment inside and outside your body, and random mutations can change your DNA. These changes in DNA can be harmless, harmful or helpful.

Nucleotides encode our genes

We take for granted that cells can replicate and pass on genetic material to new cells. This ability is responsible for life itself. The genetic material in living things is encoded by only 4 different nucleotides. You may not immediately think 4 molecules are enough to provide the diversity we see across the kingdoms of life. The way DNA nucleotide sequences are arranged is what makes so much diversity in living things possible.

Nitrogenous bases

A **nitrogenous base** prominently contains nitrogen in its ring structure as you can see below. Nucleotides differ only by the nitrogenous base they have. The bases are cytosine, thymine, adenine, and guanine. You can see that all 4 nitrogenous bases have a 6-carbon ring, but adenine and guanine have an additional 5-carbon ring. The image of a nucleotide monomer at the top of the page shows the guanine nitrogenous base attached to a deoxyribose sugar.

the four nitrogenous bases found in DNA

cytosine   thymine   adenine   guanine

Section 21.4: Nucleic Acids

The DNA molecule

Sugar-phosphate backbone

The vertical part of a DNA "ladder" forms when the phosphate group from one nucleotide bonds with the deoxyribose sugar from the next nucleotide, releasing a water molecule. This is a condensation polymerization reaction. Below you can see how the repeated sugar and phosphate groups make up the "backbone" that holds 2 DNA strands together. Notice how the nitrogenous bases are positioned inside the molecule. The DNA helix is a 3D structure that makes a corkscrew shape. DNA is a double helix because two individual DNA molecules twist around one another like two corkscrews.

DNA's shape

At first glance you may think DNA looks like an α-helix that makes up a protein's secondary structure. Unlike an α-helix in a protein, there are no vertical hydrogen bonds holding the DNA coil together, and DNA is made of two polymer chains. For that reason, DNA is like a β-pleated sheet because hydrogen bonding occurs between groups through the molecule center. If you untwisted a DNA molecule, it would look similar to a β-sheet.

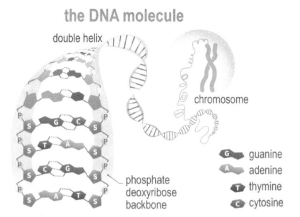

DNA strands are joined by attractions

The two strands are not actually bonded to one another through a covalent or ionic bond. They are primarily attracted to one another through hydrogen bonding between nitrogenous bases. Each nitrogenous base is found paired up with another base determined by the number of hydrogen bonds the bases make with each other. Adenine, A, always pairs with thymine, T, with two hydrogen bonds. Guanine, G, pairs with cytosine, C, with three hydrogen bonds. The spacing between the turns of the helix is just large enough for the specific A=T and C≡G base pairings. One complete turn in the helix is 10 base pairs long. Pairings like A to G do not fit properly in the helix, and G would be missing its third hydrogen bond because A only forms two hydrogen bonds.

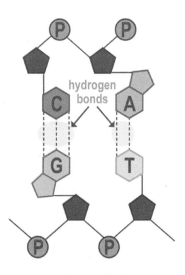

Interactive simulation

Explore the structure of DNA in detail in the interactive simulation titled DNA: The Double Helix. Look for the deoxyribose sugars and phosphate groups that make up the molecule's backbone. Experiment with the controls and see if you can identify the nucleotides based on hydrogen bonding between base pairs.

Replication

DNA coils upon itself

Almost every cell in the human body carries DNA in its nucleus. We might expect DNA molecules to be very long because they are carrying so much information. The 3 billion nitrogenous base pairs found in one set of DNA would measure over 1 meter long! How could something that long fit into a tiny cell? Fortunately, DNA twists itself up many times over to form compact chromosomes so it can fit into a cell's nucleus.

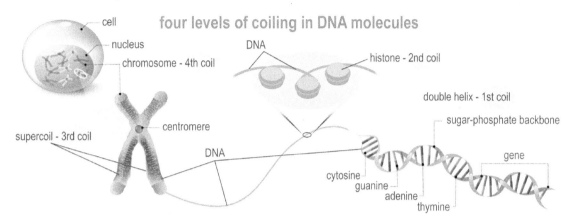

4 levels of coiling

If you had a very long string, you could make it seem shorter by coiling it up. Human DNA makes its first coil at the helix level. Next, the helix coils around proteins called histones that are linked together like pearls on a string (2nd coil). The string of histones curls up into a supercoil (3rd coil), forming a chromosome (4th coil). DNA molecules are wound into 23 pairs of chromosomes in the human genome. So, there are 46 DNA molecules in a cell's nucleus. But before a cell can divide, it must replicate, or make a copy of, all the DNA molecules in the nucleus. After replication, there are 92 DNA molecules in the nucleus!

DNA polymerase opens the helix

Before cells can divide, the double helix unwinds so that new DNA strands can be formed from the original strands. An enzyme called DNA polymerase breaks hydrogen bonds between the base pairs to "unzip" the double helix. New base pairs are added to each side as the original strand unzips. Because base pairing follows the A=T and C≡G rules, we can predict what the sequence of nucleotides will be on the new strand. After DNA replication, cell division can occur. The new cell and the original cell will both have a copy of the original DNA molecule. The genetic information is coded in a base pair sequence three letters long. Every 3-letter sequence codes for an amino acid, but there can be more than one three-letter code for each amino acid. For example, there are four codes for the amino acid alanine: CGA, CGG, CGC and CGT. Scientists think this allows for some flexibility if there are mistakes made during DNA replication.

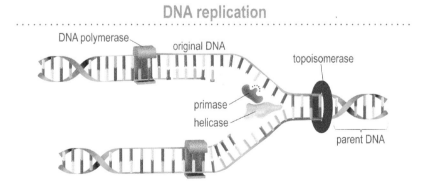

Transcription and translation

Genes code for amino acids

Proteins are involved in every chemical reaction in your body. The proteins you eat cannot be used directly by your body because they get broken down into amino acids during digestion. Your cells use specific instructions from DNA to make the proteins you need to live. A segment of DNA called a **gene** codes for an amino acid sequence that forms a specific protein. Each human chromosome contains over 20,000 genes.

mRNA delivers code outside the nucleus

Humans store DNA in the cell nucleus, but amino acids and proteins are built outside the nucleus in the cytoplasm. When a cell needs to access a gene to make a protein, it needs to copy a specific section of DNA. **Transcription** is the process where only the portion of DNA that codes a specific gene is copied. During transcription, DNA is unzipped by the enzyme polymerase at the beginning of a gene. The gene's nucleotide sequence is copied into a new molecule called **RNA** (ribonucleic acid). RNA is similar to DNA except it contains a different 5-carbon sugar called ribose. Ribose can only form a single strand instead of a double-helix, and it uses one different nitrogenous base called uracil, U. Uracil is used in place of thymine, T on an RNA strand. Human RNA molecules are much smaller than DNA molecules because they only need to code fragments of the genome instead of the entire genome. The RNA strand created in the nucleus is called messenger RNA, or mRNA. Its role is to deliver the nucleotide sequence from DNA in the nucleus to the cytoplasm outside the nucleus.

Ribosomes work in the cytoplasm

Ribosomes are like factories that help put proteins together. After the transcribed mRNA strand leaves the nucleus, it enters the cytoplasm where a ribosome attaches to it. Another type of RNA called transfer RNA or tRNA carries amino acids to the ribosome. tRNA molecules bind to a specific 3-nucleotide mRNA sequence based on RNA's pairing rules: A=U and C≡G. When peptide bonds form between amino acids, the protein's primary structure begins to take shape.

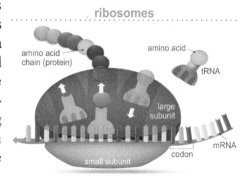

2' and 3' folding

The process where mRNA and tRNA convert a nucleotide sequence into a protein's primary structure is called **translation**. After translation, the protein detaches from the ribosome and automatically begins folding into its secondary and tertiary structures on its own. There are thousands of ribosomes in a single cell's cytoplasm, so you can imagine how many proteins are made at once in a cell. Helper molecules called chaperones can attach to proteins to help them fold correctly. Chaperones can also fix an incorrectly folded protein shape, whether newly synthesized or not, so the protein can function normally.

Interactive simulation

In the DNA to Protein interactive simulation, follow the series of events that occur when DNA's chemical code is used to create a 3D protein. The simulation helps you review the events that occur from the beginning of transcription through translation. Take a moment to edit the DNA molecule. Make the nucleotide sequence either shorter or longer, or mutate the sequence by changing one single nucleotide or more to produce different amino acids.

Section 4 review

Chapter 21

DNA molecules carry all the hereditary information for an organism. DNA is made up of repeating nucleotides. Each nucleotide contains a phosphate group, a deoxyribose sugar molecule and one of four nitrogen bases. The four bases are adenine, guanine, cytosine and thymine. The structure of DNA is two strands of repeating phosphate/sugar backbone wound around as a double helix. This structure is held together through hydrogen bonds between matching nucleotides. Adenine pairs with Thymine and Guanine with Cytosine. DNA is compacted by winding around histones, which are then coiled into chromosomes. DNA can be unwound and copied since the base pairs provide a template to build a new strand. DNA molecules hold the instructions for building all the proteins an organism requires. Through the use of RNA as an intermediary, segments of DNA called genes are translated into the primary structure of a given protein.

Vocabulary words DNA, nucleotide, nitrogenous base, gene, transcription, RNA, translation

Review problems and questions

1. What is a "nitrogenous base?"

2. Explain how a series of nitrogenous bases in a DNA molecule ultimately codes for a protein.

3. How does hydrogen bonding both help a DNA molecule twist into a double-helix and also break open the helix into single strands when it needs to replicate?

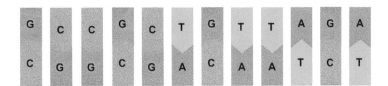

4. Use the diagram abone to answer the following questions.

   a. How many amino acids does this sequence of DNA code for?

   b. What is the sequence of amino acids that would result from a cell interpreting this DNA code?

   c. There are two ways to "spell" one of the amino acids in coded for in the diagram. What amino acid is spelled two ways and what other ways are there to code for the same amino acid?

Section 21.4: Nucleic Acids

Semi-Synthetic Life

essential chemistry

The code of life is found in DNA and for as long as we have known, it is written in 2 sets of molecule pairs. Adenine and thymine (A-T) is one pair, while cytosine and guanine (C-G) is the other pair. The two sets of pairs build the 22 naturally occurring amino acids and these twenty two amino acids build all the proteins that are responsible for life.

In a feat of synthetic biology, scientists have found a way to expand the DNA alphabet from A-T and C-G to include X-Y also. The fuller but still abbreviated name for X and Y are d5SICS and dNaM. When you look at a C-G pair compared to an X-Y pair you will notice that the size of the pairs and distance between the molecule backbones are about the same. This is important because the X-Y pair needs to be able to fit into the DNA structure without interrupting the overall shape of the molecule. The other important aspect of the X-Y pair is that unlike other pairs, it is not held together by hydrogen bonding. X and Y do not contain nitrogen, and consequently are held together by hydrophobic interactions.

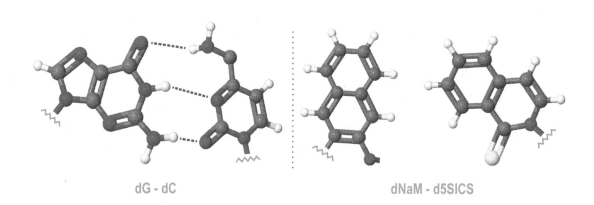

dG - dC   dNaM - d5SICS

How is it possible to add a synthetic base pair into DNA? Scientists engineered E. coli bacteria to include a transporter that would specifically bring in the X and Y molecules. DNA containing the new X-Y pair was added to the cell. When the cell is placed in an environment that contains X and Y it is essentially being fed these unnatural base pairs (UBPs). Once in the cell, the enzyme polymerase will seamlessly incorporate the UBPs into their genetic material when the cell replicates. Scientists essentially made modified E. coli microbes that can carry DNA with 3 base pairs!

Semi-Synthetic Life - 2

No need to worry about a Jurassic Park event with these semi-synthetic organisms. Scientists built a failsafe into the mechanism. Unlike A, T, C and G, cells cannot make X or Y. These modified cells can only add X and Y to their DNA as long as they are being fed X and Y. Once the cells are removed from the environment with X and Y, the natural processes in the cell will eliminate X and Y from the DNA and return to its natural state. Could E. coli evolve to make X and Y naturally? Scientists don't think so. For evolution to occur, nature needs a starting material and then it makes small changes. X and Y are so unlike natural DNA, nature doesn't have the starting material to work with.

Scientists maintain that these new cells represent a semi-synthetic life which lays the foundation for achieving the central goal of synthetic biology—the creation of new life forms and functions. Biochemists first announced their findings in 2014. At the time, the microbes with the expanded DNA were sickly and soon died out. In just a couple of years since then, they refined their processes resulting in heartier microbes. One of the major improvements came when they modified the microbe's immune system to attack any DNA without the synthetic bases. The result is a stable microbe that will faithfully hold and replicate its three-base pair genetic material.

Why would scientists want to make this Frankenstein E. coli? The short answer is… to see if they could. Right now, the synthetic base pairs do nothing other than survive in the microbes. But there is tremendous potential in this achievement. Part of the potential lies in the ability to make proteins. Currently the 2-base pair system can build over 20 amino acids, but with 3 base pairs, there are 172 potential amino acids! The proteins from this expanded source of amino acids could be used to create new and novel drugs and materials.

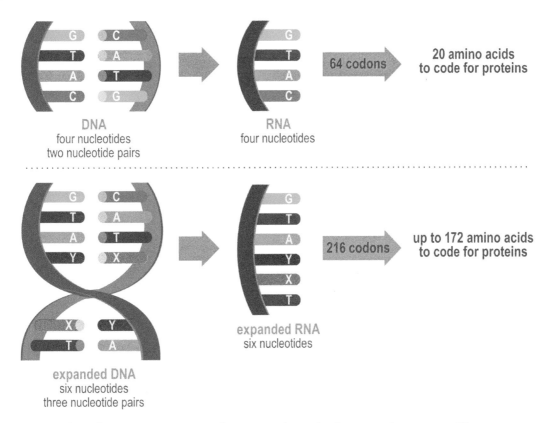

At this point in the discovery process these semi-synthetic organisms are like any new scientific breakthrough. There is excitement and tremendous promise, but there will have to be a lot of research and experimentation to see where this potential leads.

Chapter 21 review

Vocabulary
Match each word to the sentence where it best fits.

Section 21.1

carbohydrates	chiral carbon
complex carbohydrates	condensation polymerization
monomer	monosaccharide
oligosaccharides	polymer
polymerization	polysaccharide
simple carbohydrates	

1. A single sugar unit is called a(n) _____.
2. _____(s) are long-chain molecules made up of many smaller, repeating subunits.
3. Glucose and sucrose are both examples of _____, small sugar molecules composed of one or two monosaccharides.
4. _____(s) are the building blocks of polymers.
5. Cellulose and starch are common examples of _____, larger molecules of three or more monosaccharides.
6. This is the class of molecules that includes two to ten monosaccharide units linked together in a chain.
7. _____ are carbon atoms in a ring structure with four different groups or atoms attached to it.
8. _____ is a reaction that joins together many monomers.
9. A reaction called _____ links monomers by eliminating a small molecule such as water.
10. The primary source of energy for all organisms is _____.
11. A long chain of monosaccharides connected together form the _____ type of molecule such as starch, cellulose, or glycogen.

Section 21.2

adipose tissue	amphipathic
hydrophilic	hydrophobic
lipid	lipoproteins
micelle	phospholipid
triglyceride	

12. Examples of _____ molecules include fats, waxes, oils, and steroids.
13. A _____ is a fat molecule used for energy storage and is composed of glycerol and three fatty acids.
14. _____(s) have a polar head phosphate group attached to a non-polar hydrocarbon tail.
15. A water-loving, polar molecule is considered _____.
16. _____ molecules do not dissolve in water due to being non-polar.
17. Phospholipids assembled in a sphere where the polar head line the exterior are called _____.
18. Fat is stored in _____ for energy storage, insulation, and cushioning.
19. A molecule is considered _____ when it has a polar hydrophilic region and a non-polar hydrophobic region.
20. The dual polar nature of fatty acids force them to form a _____ in aqueous solutions.

Section 21.3

active site	alpha helix
amino acid	beta pleated sheet
denatured	enzymes
peptide bond	primary structure
protein	quaternary structure
secondary structure	substrate
tertiary structure	

21. The sequence of amino acids of a protein is called its _____.
22. A _____ forms between the amino group and the carboxyl group of two amino acids.
23. α-helices or a β-pleated sheets are forms of a proteins _____.
24. The _____ of an enzyme changes shape when a particular substrate molecule binds to it.
25. _____(s) are reactant molecules that bind to an enzyme.
26. The folding or bending of a proteins secondary structure creates a _____ as the result of molecular interactions.
27. Long chain polymers of amino acids used in every cell process in our bodies are called a _____.
28. There are twenty different naturally occurring _____(s).
29. Biological catalysts called _____(s) assist in speeding up the chemical reactions inside our bodies.
30. When a protein or amino acid is _____, function is compromised as a result of damage to its 3D structure.

Chapter 21 review

Section 21.3

active site	alpha helix
amino acid	beta pleated sheet
denatured	enzymes
peptide bond	primary structure
protein	quaternary structure
secondary structure	substrate
tertiary structure	

31. A secondary structure of a protein that forms a spiral or twist called a(n) _____ in which a single protein chain is held together by hydrogen bonds.

32. A secondary structure of a protein that forms a folded pattern of a _____ which is held side-by-side protein chains by hydrogen bonding.

33. The structure formed by complete proteins when they bond or interact with another is called their _____.

Section 21.4

DNA	gene
nitrogenous base	nucleotide
transcription	translation

34. A _____ is a component of a nucleotide that is linked by hydrogen bonds in different pairs.

35. A long polymer chain of nucleotides formed as a double helix that transmits genetic information is called _____.

36. A _____ is a molecular unit composed of a five-carbon sugar, a phosphate group, and a nitrogenous base.

37. A segment of DNA called a _____ contains the code for a specific protein.

38. The stage of gene expression where DNA sequences are copied and edited in the nucleus is called _____.

39. The stage of gene expression where amino acid sequences are decoded and build into proteins is called _____.

Conceptual questions

Section 21.1

40. Sugar molecules are broken up into categories of simple and complex. What is the difference between a monosaccharide, a disaccharide, and an oligosaccharide? Which one(s) are considered complex?

41. Compare the formulas for and uses of glucose and fructose.

42. What do plants do with the cellulose they make?

43. Carbohydrates are polymers of sugar molecules of varying lengths. Why are carbohydrates important? Provide some examples of how living things use carbohydrates.

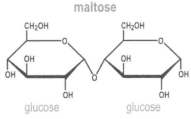

44. Maltose is composed of two glucose monomers. Is it a mono-, di- or oligosaccharide?

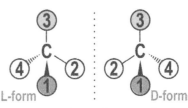

45. The isomers shown above are simple carbon compounds. If the above isomers were carbohydrates, which form would be utilized by living things?

46. Glucose molecules are bonded together to form a polysaccharide chain through condensation polymerization. Which groups or atoms are used to synthesize these polymers?

47. What disaccharide is formed from the condensation polymerization of the monosaccharides glucose and galactose?

48. Carbohydrates are important for overall health. We are advised to eat a variety of carbohydrates to ensure that we get adequate nutrients.

    a. What are some examples of simple carbohydrates that are considered to be healthy?

    b. What are some examples of complex carbohydrates that are considered to be healthy?

49. What happens to the extra carbohydrates that your body does not need when consumed?

50. Examples of common complex carbohydrates include large starch and cellulose molecules.

    a. What is the common name for cellulose that you will see printed on nutrition labels?

    b. Why is a complex carbohydrate like cellulose important for our diet?

51. What happens to the extra carbohydrates that your body does not need when consumed?

52. How much of our daily diet do nutritionists recommend come from carbohydrates?

53. Complex carbohydrates are considered healthier than simple carbohydrates. Why is this?

Chapter 21 review

Section 21.2

54. If fats, oils, waxes, and steroids are all lipids, how are fat and oil molecules different?

55. Cell membranes have two phospholipid layers that form a bi-layer.

    a. What is the basic function of the phospholipid bi-layer?

    b. Compare the structure and function of the phospholipid head with the tail.

56. When we eat food, our body uses the carbohydrates it needs and stores the excess as glycerol attached to variable length fatty acid chains. Why do long fatty acid chains provide more energy than short fatty acid chains?

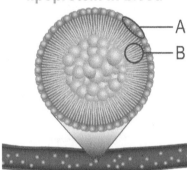

lipoprotein in blood

57. Our bodies use lipoproteins to collect fat molecules from the body's cells and tissues and transport them through our bloodstream.

    a. What are lipoproteins made of, and how do they achieve this function?

    b. Compare region "A" with region "B"; what are the differences in their chemical makeup, polarity, and attraction to water?

58. Why does the body need lipoproteins to transport fats in blood?

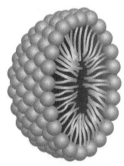

59. Micelles are a collection of molecules that form a sphere, allowing non-polar tails to be in the center of the sphere away from the aqueous solvent. An example of this would be lipoproteins. What type of lipid molecules are micelles made up of in the human body?

60. Why does the phospholipid molecule attract water on one end but not on the other?

61. Lipids store energy, act as signals in biological processes, and are important building blocks of cell membranes. What smaller molecules make up a typical fat molecule?

62. Fat molecules are important for keeping our bodies functioning properly. What are the purposes of fats in our bodies? Explain with some specific examples.

63. Plants store their energy as carbohydrates in the form of starch. Why do animals store most of their energy in fat molecules instead of carbohydrates?

64. The body needs fat for a number of reasons. If fat is essential to your health, why are saturated fats considered less healthy than unsaturated fats?

65. How are the structures of a saturated fat and an unsaturated fat different?

66. What kinds of natural foods contain unsaturated fatty acids?

67. Oils are generally healthier for us to consume and are liquid at room temperature. What primarily determines whether a fat will be solid or liquid at room temperature?

Section 21.3

68. Proteins are involved in every active cell process and are necessary for almost every chemical reaction that keeps living organisms alive. Where do animals get the proteins they need to either repair or build new cell and body structures?

69. Identify the seven different functions of proteins in our bodies, and provide at least one specific example for each function.

70. Sketch the condensation reaction between two amino acid molecules. Draw the 2 reacting amino acids along with the 2 new products. Circle the atoms that leave the reactant molecules to form water and label the peptide bond in the newly formed polymer.

71. What is the 1' or primary structure of a protein composed of? Describe the 1' structure of a typical protein.

72. What is the 2' or secondary structure of a protein? Describe the components that could make up a 2' structure.

73. In relation to proteins, why is it necessary to keep pH and temperature stable in the human body?

74. What are some theories why nature prefers amino acids with L-chirality?

75. Why is chirality important to living things?

76. How many different naturally occurring amino acids do living things incorporate into their proteins?

77. Draw and label the three essential components that all amino acids have.

78. Sketch any amino acid. Circle the part of the molecule that determines whether it will be acidic, basic or neutral in solution.

Chapter 21 review

Section 21.3

79. Compare the properties that an α-helix gives to a protein with those of a β-pleated sheet.

80. Fireflies use an enzyme to oxidize a chemical in order to achieve bioluminescence. What is the enzyme?

81. What happens if an enzyme's tertiary structure is damaged or denatured?

82. What happens structurally and functionally when a substrate binds to the active site of an enzyme?

83. What role do enzymes play for living systems?

84. What is the purpose of an active site on an enzyme?

Section 21.4

85. Caffeine is in the same class of molecules as DNA and RNA. Write a 1-page report on whether caffeine is good for you or bad for you. Be sure to:

    a. Use at least 2 sources from a scientific text or a reliable news source;

    b. Use your own words and cite your sources according to your teacher's directions;

    c. Point out inconsistent statements among authors, or if all authors' statements agree, paraphrase the statements on which they most strongly agree;

    d. Explain how you know your sources are reliable;

    e. Identify they type of person or source do you wish you could interview to help you answer this question;

    f. Explain why you would choose that source.

86. How is RNA different from DNA?

87. What keeps two DNA strands attached together?

88. What four nitrogenous bases encode the genetic material in all living things?

89. Which nitrogenous base pairs with cytosine and how many hydrogen bonds are formed between the two bases?

90. What molecules are bound together to form the "backbone" of DNA?

91. How many DNA molecules compose the human genome?

92. How is DNA replicated before cell division? What breaks the hydrogen bonds between strands?

93. What is the difference between DNA replication and DNA transcription?

94. Explain the relationship between transcription and translation and the products of each.

95. How are such large quantities of DNA able to fit inside a cell's nucleus?

96. Identify the three molecules that make up a nucleotide.

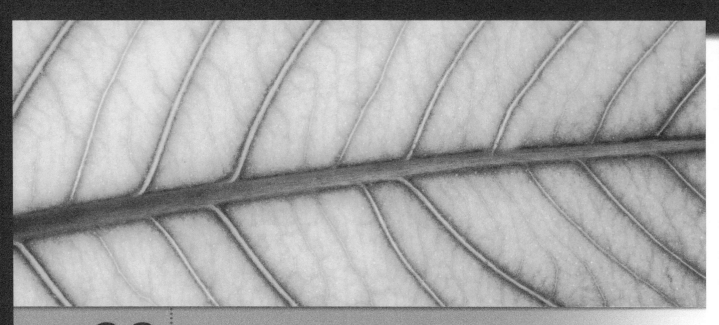

chapter 22 Biochemistry

Energy transfers are all around us. The toaster you used this morning converts electrical energy into thermal energy. The battery in your cell phone converts chemical energy into electrical energy to power your device. To recharge your phone, the reverse process happens—electrical energy is converted into chemical energy and stored in the battery. At a hydroelectric plant, the gravitational potential energy of water at great heights is converted to electrical energy. But how do living things move energy around in their cells? It turns out that ATP (adenosine triphosphate) is the crucial energy carrier for energy transfer within cells. ATP (adenosine triphosphate) is so critical that it is sometimes called the energy currency of life.

The production and use of ATP as an energy carrier is itself a series of energy and matter transformations. Organisms that can undergo photosynthesis and produce ATP from carbon dioxide, water and the energy from sunlight. In another set of reactions during photosynthesis, the ATP is broken down to provide the energy to produce the end product of glucose. All living things (even plants!) undergo respiration. Glucose provides the fuel for respiration which involves a series of reactions that result in the formation of carbon dioxide, water and even more ATP.

The key to the power in ATP is in the tri of the triphosphate. This third phosphate group is easily broken producing ADP (adenosine diphosphate). ATP is considered a "high energy" molecule not because its bonds are strong, but because when its third phosphate bond is broken, free energy is released as new bonds form. This free energy then drives cellular processes like reproduction, transportation across cell membranes, or contracting muscles, allowing us to perform basic bodily functions. Your body constantly cycles through energy and matter, including the ATP-ADP cycle. As you read this chapter you will see that these cycles are essential to the chemistry of life.

Chapter 22 study guide

Chapter Preview

Biochemistry deals with the chemical processes within and related to living things. In this chapter you will learn how plants use solar energy to create food, how your body creates the energy needed for chemical reactions, and how signals are carried throughout your body to regulate temperature, pH and more. Photosynthesis is a chemical reaction that converts carbon dioxide and water into glucose and oxygen gas with the use of light energy to excite electrons in chlorophyll. Cellular respiration is fundamentally the reverse reaction, using glucose and oxygen to create carbon dioxide, water and ATP. ATP is used by your body to transport molecules, cause muscle contractions, create impulses in nerves and more. A goal of many of your body's processes is to maintain a stable environment even as conditions internally or externally are changed.

Learning objectives

By the end of this chapter you should be able to:

- describe the chemical light and dark reactions of photosynthesis, including the Calvin cycle;
- list the steps of cellular respiration;
- write the balanced reactions for photosynthesis and cellular respiration;
- explain how cells in the human body use ATP;
- explain how carbonic acid is used to regulate the pH of blood; and
- describe the different methods your body uses to communicate signals in order to regulate conditions.

Investigations

22A: Chlorophyll extraction

22B: Respiration and energy

Pages in this chapter

- 714 Carbohydrate Synthesis in Autotrophs
- 715 Photosynthesis
- 716 Pigments in chloroplasts
- 717 Photosynthesis Part 1: Light-dependent reactions
- 718 Photosynthesis Part 2: Dark reactions
- 719 Energy transfer summary
- 720 Section 1 Review
- 721 Cellular Respiration
- 722 Respiration: Glycolysis
- 723 Respiration: The Krebs cycle and electron transport
- 724 Energy transfer during respiration reactions
- 725 Section 2 Review
- 726 Energy Use in Cells
- 727 ATP-ADP energy cycle
- 728 How cells use ATP
- 729 Chemical work: Cell signaling
- 730 The endocrine system and chemical signaling
- 731 Transport work powered by ATP
- 732 Mechanical work powered by ATP
- 733 Section 3 Review
- 734 Maintaining Homeostasis
- 735 pH range in the body
- 736 Buffering system in the body
- 737 Temperature equilibrium
- 738 Nervous system signals
- 739 Section 4 Review
- 740 Energy Transfers in Living Things
- 741 Energy Transfer in Living Things - 2
- 742 Chapter review

Vocabulary

autotroph
oxidation
chlorophyll
ATP
heterotroph
NADH
metabolism
work
diffusion
buffer
electrochemical

phototroph
photosynthesis
chloroplast
ADP
cellular respiration
Krebs cycle
catabolism
homeostasis
Le Châtelier's principle
isoenzymes

chemotroph
pigment
NADPH
Calvin cycle
glycolysis
electron transport chain
anabolism
transport
optimum range
thermoregulation

22.1 - Carbohydrate Synthesis in Autotrophs

Energy cannot be created or destroyed, but it can be converted to different forms. Where does the energy in food come from? No matter what kind of food you eat, you can trace its energy back to an autotroph. A living thing that creates its own organic molecules from simple organic or inorganic chemicals is called an **autotroph** or producer. Autotrophs are the foundation of a food chain. All living things depend on autotrophs to enable the flow of energy and matter through their bodies.

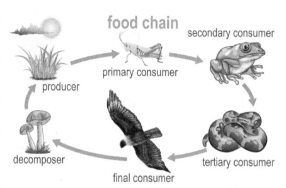

Autotrophs

Phototrophs require light, chemotrophs do not

An autotroph that uses light energy to drive food molecule production is a **phototroph**. Plants and algae are examples of phototrophs. Like a phototroph, a **chemotroph** creates carbohydrates and other organic molecules from inorganic chemicals, although the energy source is not light. Chemotrophs use **oxidation** reactions as their energy source. For example, some chemotrophic bacteria near iron-rich lava oxidize iron(II) to iron(III) in the presence of oxygen. The energy released when iron loses an electron is used to power the reaction that converts organic or inorganic substances to carbohydrates.

Where are chemotrophs found?

Chemotrophs can also oxidize inorganic manganese, sulfur, ammonium, methane or hydrogen sulfide to convert inorganic CO_2 into organic carbohydrates. Chemotrophs are found in extreme environments where it seems life is not possible, often in dark, extremely hot conditions. Most chemotrophs are bacteria found near volcanic features such as those found in Yellowstone National Park's 150 °F hot springs, pictured below (left). The colors come from different bacteria growing at different temperatures. A volcanic park in New Zealand is pictured to the right.

Chemo-autotrophs support communities in harsh environments

Some volcanic vents like hot springs are at Earth's surface and others are on the deep, dark ocean floor. Hydrothermal vents on the ocean floor heat surrounding water as high as 800 °F. Entire communities of worms, snails, fish and other marine creatures can live in deep sea zones because chemotrophic bacteria produce organic food. Scientists are interested in chemotrophs particularly because they provide clues about how life may exist in other places in the universe. Here is an example of how one species of chemotrophic bacteria found at hydrothermal vents produces carbohydrates without sunlight on the ocean floor:

$$CO_2 + 4\,H_2S + O_2 \rightarrow CH_2O + 3\,H_2O + 4\,S$$

Carbon dioxide, water, hydrogen sulfide and oxygen combine to produce a carbohydrate and sulfuric acid. The carbohydrates produced are consumed by larger organisms thereby supporting an ecosystem in an otherwise lifeless environment.

Photosynthesis

Plants convert sunlight into chemical energy

The glucose sugar molecule is the most important energy-yielding molecule in living systems. **Photosynthesis** is a process in which plants harness light energy from the sun and convert it to chemical energy. Chemical energy is stored in chemical bonds within the sugar product. Sunlight is used to power reactions that rearrange water and carbon dioxide molecules into sugars and starches. Photosynthesis is the first step in the pathway. Carbon atoms from carbon dioxide are used as building blocks to make amino acids, proteins, and other complex organic molecules. Plants use the chemical energy found in these molecules to grow and move nutrients. Animals that consume plants also benefit from the chemical energy made available by photosynthesis. Animals use chemical energy stored in glucose molecules to fuel life-sustaining reactions. The photosynthesis reaction may look familiar from a past biology class: six moles of carbon dioxide gas combine with six moles of liquid water in the presence of sunlight to form one mole of solid glucose molecules and six moles of oxygen gas.

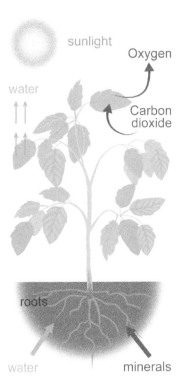

$$6CO_2(g) + 6H_2O(l) \xrightarrow{light} C_6H_{12}O_6(s) + 6O_2(g)$$

The photosynthetic reactants

Living things that use sunlight to make glucose include plants, algae, and some bacteria. These organisms use water and carbon dioxide reactants to produce sugar molecules inside their colored leaves. Oxygen is also produced as waste and is released to the atmosphere through leaves. Most of the oxygen in our atmosphere is produced by photosynthetic organisms. Land plants take in the CO_2 reactant from the air and aquatic plants take in the CO_2 dissolved in water.

Chlorophyll absorbs light energy

Sugars do not spontaneously form from CO_2 and H_2O molecules. Energy must be added for this to occur. Plants use **pigment** molecules to collect the energy needed to kick off a photosynthesis reaction. Pigments give color to plants. A pigment called chlorophyll helps a plant transfer the sun's light energy into chemical energy stored in a six-carbon molecule. **Chlorophyll** is a large molecule capable of absorbing photons in the blue and red wavelengths of the visible spectrum. Chlorophyll A absorbs red photons and chlorophyll B absorbs blue photons. Chlorophyll does not absorb green light, so it is reflected, giving plants their green color.

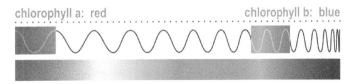

Excited electrons

Electrons in the atoms that make up the chlorophyll molecules absorb light energy, and these electrons become excited to a higher energy state. The energy from these electrons is used to break the chemical bonds in the reactants and form new bonds in the products.

Section 22.1: Carbohydrate Synthesis

Pigments in chloroplasts

Chloroplasts

Plants keep their pigments inside their cells in different structures depending on the pigment function. Photosynthetic pigments are found inside plant cells within a structure called a **chloroplast**. Chlorophyll is found inside the stack of membrane-bound discs seen in the diagram. Photosynthesis consists of two major phases in different locations within the chloroplast. Only the first part of photosynthesis needs light and chlorophyll.

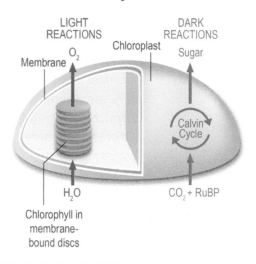

Photosynthesis

How do leaves change color in fall?

Not all pigments absorb light in the same part of the visible spectrum, and not all are photosynthetic. A pigment molecule class called xanthophyll gives plants yellow colors, and the carotenoid class color plants orange, yellow and brown. Plants usually have a mix of more than one kind of pigment to maximize their absorption of light energy. Leaves change color in fall because cues like less sunlight and cooler temperatures prompt chlorophyll pigment molecules to undergo a decomposition reaction. When most of the green pigment molecules have broken down, the yellow, red, and orange pigments that were present all along in the leaves become visible. When all pigments have decomposed, photosynthesis no longer occurs. The tree drops its brown leaves because they are no longer able to produce sugar.

Other pigments

Other pigment functions include assistance in various plant reactions, attracting insects or animals to help with pollination or seed dispersal, and protection from the sun's damaging UV rays. Anthocyanins are a class of pigments commonly seen in red, purple and blue leaves and fruit that are popular for their antioxidant properties. These molecules prevent potentially harmful oxidation reactions that could damage cells. The more anthocyanins a plant part has, the deeper its color. For example, blueberries are promoted as a healthy antioxidant fruit because they have high anthocyanin levels.

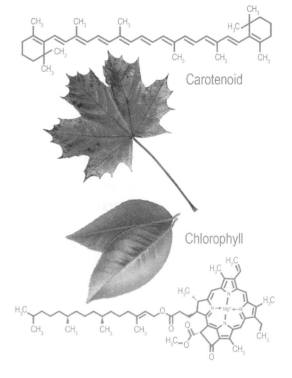

Pigments

Photosynthesis Part 1: Light-dependent reactions

Water is oxidized or split to release electrons

A plant rearranges carbon dioxide and water molecules into glucose and oxygen in a series of enzyme-catalyzed redox reactions. The first reactions use the energy absorbed from sunlight to split water molecules. This part of photosynthesis is called the light reactions because the reactions depend on sunlight. Chlorophyll molecules absorb photons of light. Photon energy is used to split water molecules apart in an oxidation reaction. The products are 2 H^+ ions, 2 electrons, and 1 oxygen atom. Oxygen atoms combine to form O_2 molecules, which are released into the air. H^+ ions and electrons stay in the chloroplast to support the next part of photosynthesis.

$$2\ H_2O(l) \xrightarrow{\text{light}} 4\ H^+(aq) + O_2(g) + 4e^-$$

Energy is transferred through a NADP+ and NADPH redox reaction cycle

The electrons produced are added to molecules whose only job is to transfer energy. A molecule called nicotinamide adenine dinucleotide phosphate ($NADP^+$) is able to accept high energy electrons and pass them along a series of redox reactions. Each $NADP^+$ molecule is reduced when it receives 2 high energy electrons and a H^+ ion. The reduced form is called **NADPH**. NADPH carries the sun's light energy as chemical energy in its bonds. This energy is carried to the next reaction when NADPH returns to its oxidized $NADP^+$ form by losing electrons. It is then ready to carry energy by gaining electrons, and the energy transfer cycle continues.

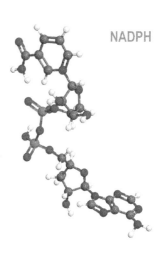

NADPH

Adenosine triphosphate

Another energy carrier molecule called adenosine triphosphate or **ATP** carries energy in bonds between each phosphate, PO_4^{-3}. Find the PO_4^{-3} groups in the picture. Water molecules break bonds between the PO_4^{-3} groups. The bond energy released fuels chemical reactions within cells. With one less PO_4^{-3} group, the ATP molecule is now called adenosine diphosphate, or **ADP**. ADP can be recharged to ATP with the addition of another PO_4^{-3} group.

ATP and ADP

adenosine triphosphate

adenosine diphosphate

Energy carriers

The importance of ATP and NADPH lies in their special ability to carry energy from spontaneous light-driven reactions to non-spontaneous reactions that require chemical energy. ATP and NADPH provide chemical energy for the next step of photosynthesis.

Plants cycle energy and matter

Photosynthetic organisms have the most important place in the food chain as energy molecule producers. Their role in transforming energy from the sun into a form non-photosynthetic organisms can use is irreplaceable. Plants are important for cycling energy and for cycling carbon. The next phase of photosynthesis highlights photosynthetic organisms' role in the carbon cycle. Plants take carbon dioxide out of the air and "fix" it into glucose molecules, helping carbon atoms move quickly through the carbon cycle.

Photosynthesis Part 2: Dark reactions

NADPH and ATP carry energy to the dark reactions

The remaining reactions in photosynthesis are called the dark or light-independent reactions because they do not use chlorophyll. This does not mean the dark reactions cannot happen when there is daylight! In fact, the dark reactions occur only when light is available because these reactions use energy from molecules formed in the light reactions. Another name for the dark reactions is the **Calvin cycle**. During the Calvin cycle, CO_2 is added to a 5-carbon sugar molecule called RuBP (ribulose-1,5-bisphosphate). After a series of redox reactions, the 6-carbon glucose molecule is formed.

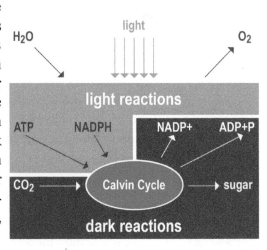

The Calvin cycle uses energy from NADPH & ATP to make glucose

As you can see, ATP and NADPH produced from the light reactions are fed into the Calvin cycle. The Calvin cycle takes in CO_2 and RuBP to make the sugar glucose, and when the energy carriers transfer their energy, they become ADP and $NADP^+$. The Calvin cycle returns the spent molecules back to the light reactions to restore them as energy carriers. ATP is regenerated when ADP and a phosphate ion react during the light reactions. The light and dark reactions need to take in six CO_2 molecules to make one glucose molecule. Plants continuously remove CO_2 from our atmosphere during daylight to make energy-rich sugars.

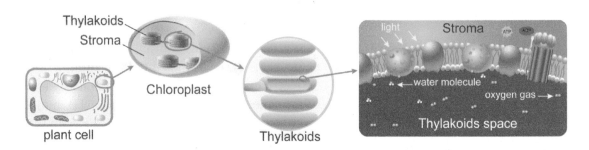

Calvin cycle happens in the stroma

The diagram above shows stacked discs in the chloroplast called thylakoids surrounded by a fluid region called the stroma. Light reactions occur in the thylakoid membrane where chlorophyll molecules are located. Products from the light reactions are released to the stroma where the Calvin cycle takes place.

All living things rely on similar reactions to rearrange energy and matter

Plants use the atoms and energy in glucose to supply energy for other reactions and to make new molecules like starch and cellulose. These molecules are incorporated into plant tissue mainly as new growth or as storage molecules. Similarly, animals that eat plants build molecules from glucose. Animals must break down starch into simpler carbohydrates before they are useful for energy or structures. Living things depend on chemical reactions to reactions rearrange energy or elements to produce the substances needed to sustain life. The molecules and chemical reactions that keep living things alive are remarkably similar, whether you are a single-celled bacterium or a multicellular human. None of these reactions would be possible without photosynthesis. Scientists believe the most abundant enzyme on Earth is the one that aids in the formation of the RuBP molecule.

Energy transfer summary

Light energy is transferred to chlorophyll electrons, then to ATP and NADPH

We can trace the flow of energy during photosynthesis in an energy bar chart. Plants convert electromagnetic light energy from the sun into chemical energy, E_{ch} which is eventually stored in carbohydrate bonds. The two main stages of photosynthesis occur in different parts of the chloroplast. The light reactions kick off when light enters chlorophyll molecules. Chlorophyll absorbs light and uses it to split water into oxygen atoms, hydrogen ions, and excited electrons. Oxygen atoms combine to form O_2 molecules, which are released to the atmosphere. Hydrogen ions and electrons help transfer energy to ATP and NADPH. These molecules now store the initial light energy as chemical energy, E_{ch}, in their chemical bonds.

ATP, NADPH carry chemical energy

Not all of the light energy absorbed is converted to chemical energy. No system can convert energy with 100% efficiency. Some energy is lost, but there is a net gain. ATP and NADPH molecules carry chemical energy needed to power the next phase of photosynthesis: the Calvin cycle otherwise known as the dark reactions.

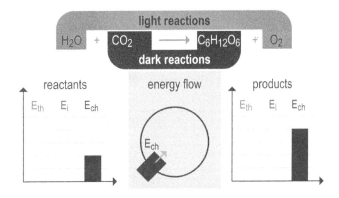

ATP, NADPH transfer their energy to glucose

Chemical energy carried within ATP and NADPH molecules power the Calvin cycle. Their chemical energy is used to bind inorganic CO_2 molecules and RuBP to form organic sugar molecules. Light energy from the sun is finally transformed to chemical energy in carbohydrate bonds.

Formation of glucose is non-spontaneous

Photosynthesis obeys the first law of thermodynamics: energy can neither be created nor destroyed, but it can be converted from one form to another. You can see the reaction is endothermic because the chemical energy in the products is greater than the reactants. The light reactions begin as a spontaneous process when light energy excites electrons in chlorophyll molecules. This energy drives non-spontaneous reactions and produces the glucose molecule which stores energy. Dozens of molecules and ions work together during the light and dark reactions. They shuffle electrons around until the series of interactions we call photosynthesis completes a cycle. Although the overall reaction is simplified below, you are now aware that the process of converting electromagnetic light energy into chemical energy does not occur in one simple stage.

$$6\,CO_2(g) + 6\,H_2O(l) \xrightarrow{\text{Light}} C_6H_{12}O_6(s) + 6\,O_2(g)$$

Section 22.1: Carbohydrate Synthesis

Chapter 22
Section 1 review

Autotrophs are organisms that can produce nutritional organic substances from inorganic substances. Phototrophs use light as an energy source while chemotrophs use oxidation reactions. Photosynthesis uses energy from the sun to convert water and carbon dioxide into glucose and oxygen gas. Chlorophyll is a molecule that can absorb photons in the physical spectrum. The decomposition of chlorophyll is responsible for the color of autumn leaves. Once chlorophyll has absorbed photons, that energy is used to split water molecules and release electrons that are passed along in reactions. NADPH and ATP are formed in this process. The Calvin cycle uses the energy from these compounds to form glucose from a five-carbon sugar and carbon dioxide.

Vocabulary words	autotroph, phototroph, chemotroph, oxidation, photosynthesis, pigment, chlorophyll, chloroplast, NADPH, ATP, ADP, Calvin cycle

Review problems and questions

1. Explain the role of chlorophyll in photosynthesis.

2. Summarize what happens in Part 1 of photosynthesis.

3. Summarize what happens in Part 2 of photosynthesis.

4. Consider the "light" and "dark" photosynthesis reactions. Which reactions can occur in the light, and which can occur in the dark?

5. In your own words, briefly describe the process of photosynthesis. Where does it occur? How many stages are involved in this process?

6. Write the complete equation associated with the end-product of photosynthesis, include the phase of each molecule.

7. What does ATP stand for and what does it do?

8. What does ADP stand for and what does it do?

9. Specifically, in what region does the Calvin cycle take place?

10. During the Calvin cycle, glucose is formed from carbon dioxide and a five-carbon sugar. What is the five-carbon sugar called?

22.2 - Cellular Respiration

A **heterotroph** is an organism that cannot make its own food. Heterotrophs get their energy from the six-carbon sugar, glucose produced by autotrophs. Heterotrophs obtain energy from the glucose molecule through **cellular respiration**.

Energy for life

Respiration is the reverse of photosynthesis

Glucose and other food molecules undergo oxidation reactions during cellular respiration. Oxygen reacts with food molecules and as chemical bonds are broken apart, energy is released. The cellular respiration reaction is the reverse of photosynthesis. You might recognize it as a combustion reaction. But instead of allowing most of the energy to quickly escape as heat and light, the goal is to carefully store it within chemical bonds of molecules like ATP. Another reason to keep the reaction rate in control is to avoid protein damage from heat. The equation is over-simplified because it makes the respiration appear to happen all at once. Like photosynthesis, respiration occurs in a series of enzyme-catalyzed redox reactions. Respiration involves three main reactions that slowly release energy from bonds in the glucose molecule: glycolysis, the Krebs cycle, and the electron transport chain.

$$6O_2(g) + C_6H_{12}O_6(s) \rightarrow 6CO_2(g) + 6H_2O(l) + Energy(ATP)$$

Cells break down glucose slowly

If bond energy in the glucose molecule is released too fast, much of the energy will be lost as heat. As the glucose bonds are slowly broken the energy is captured by molecules like ADP. ADP molecules use the energy released by glucose to bond another phosphate group and form ATP. The body's chemical reactions need energy that is immediately available, which ATP can easily provide.

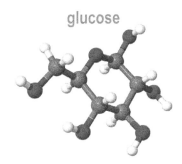
glucose

Photosynthesis and respiration cycle energy and carbon

Photosynthesis and cellular respiration work together to cycle carbon through the air, soil, water, and living things. Carbon-based molecules work their way up the food chain from autotrophs to heterotrophs. Some molecules are used only for their energy, some molecules are stored in new cells and tissues, and many molecules are discarded as waste materials. None of the matter is truly wasted because atoms are constantly recombined in different ways as they pass through food webs. Matter and energy are converted and conserved at every step in the process.

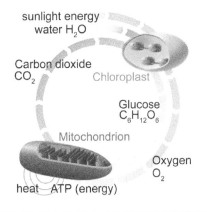

Respiration: Glycolysis

Glycolysis occurs in the cell's cytoplasm

How does glucose get to all the cells in your body that need it? The circulatory system helps deliver glucose to cells. Receptor proteins on cell membranes help cells recognize glucose molecules and bring them into the cell's cytoplasm where **glycolysis**, the first phase of respiration, begins.

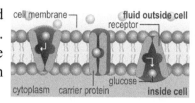

Glucose splits when it gets oxidized

During glycolysis, energy stored in the glucose molecule is transferred to energy carrier molecules through a series of redox reactions. In the first part of glycolysis, glucose is oxidized and split into two temporary 3-carbon molecules which are further modified to form two molecules called pyruvate.

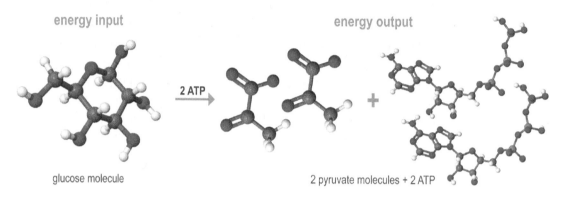

ATP output exceeds input

Two ATP molecules are needed to split the glucose molecule at the beginning of glycolysis. Activation energy needs to be added to start the reaction, but by the end of glycolysis the energy produced is greater than the energy input with a net gain of 2 ATP. In the second part of glycolysis, four ATP molecules are produced. The reaction above is simplified; glycolysis is a multi-step reaction catalyzed by many enzymes.

NAD^+ oxidizes the intermediate molecules and yields 4 ATP and 2 pyruvate molecules

In the second phase of the glycolysis reaction, two nicotinamide adenine dinucleotide, **NADH** molecules are formed when the temporary 3-C molecules are reduced. NADH is the molecule that carries chemical energy during respiration reactions. As the dinucleotide part of the name implies, NADH is made of two nucleotides: adenine and nicotinamide. The nucleotides are attached to one another with phosphate groups. Look for the familiar adenine portion at the top of the molecule and find the nicotinamide portion towards the bottom.

NADH is shown as NAD^+ in its oxidized form. NAD^+ molecules are reduced when they gain electrons during the oxidation of temporary 3-C molecules. NADH molecules are used to make ATP molecules in a different respiration phase.

Glycolysis has a small energy yield

After accounting for the input energy, glycolysis only yields 2 ATP molecules. The energy payoff is realized in next two cellular respiration reactions. Pyruvate molecules yield more ATP when they are processed within mitochondria. In your Biology class, you probably learned that the mitochondria are the powerhouses of the cell. The "power" produced in mitochondria is chemical energy released from the breakdown of the pyruvate molecules.

Respiration: The Krebs cycle and electron transport

After glycolysis pyruvate moves into mitochondria

The first step in cellular respiration, glycolysis, harvests 10% or less of the bond energy available in from glucose. After glycolysis, the remainder of the energy can be harvested by breaking down pyruvate molecules. Pyruvate and oxygen molecules enter mitochondria where the **Krebs cycle** continues the respiration reaction. The Krebs cycle occurs within the gel-like fluid called the mitochondrial matrix

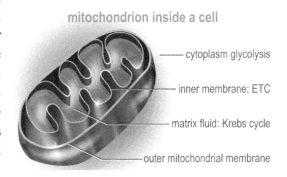

The Krebs cycle breaks down pyruvate to form ATP and CO_2

The Krebs cycle is an enzyme-catalyzed sequence of reactions that transfer energy when pyruvate breaks down. During the Krebs cycle, the carbons from pyruvate are released as carbon dioxide and some energy is stored in 2 more ATP molecules. The remaining energy is stored as energized electrons in NADH and an energy carrier called flavin adenine dinucleotide, $FADH_2$. Both energy carriers bring high-energy electrons to the last phase of respiration, the **electron transport chain** (ETC). The ETC uses the high-energy electrons from NADH to make more ATP molecules than glycolysis and the Krebs cycle combined.

The ETC is made of proteins

The "chain" part of the ETC is a series of proteins called cytochromes embedded within the inner mitochondrial membrane. These proteins help NADH and $FADH_2$ give up electrons. The electrons given up travel from protein to protein in the ETC and push H^+ ions to the opposite side of the membrane. The H^+ ion buildup is relieved through a protein channel called ATP synthase. When the H^+ ions move through the channel, they provide the energy ADP needs to attach another phosphate group and form ATP.

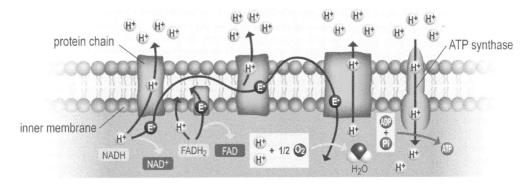

The ETC needs oxygen to keep electrons flowing down the chain

Molecular oxygen (O_2) sits at the end of the ETC, ready to get split into O^{2-} ions to form water. Without O_2, electrons would stay in the membrane and pile up instead of regenerating the NAD^+ molecules needed in the Krebs cycle. Oxygen has a high electronegativity, so electrons are attracted away from the last chain protein towards the oxygen molecule. H^+ ions that travel through the ATP synthase channel combine with O^{2-} ions to form water. Oxygen's role in picking up electrons and H^+ ions at the end of the ETC is the primary reason why you need a constant supply of O_2 molecules in your body. Cell metabolism cannot occur for very long in your body without ATP. As you now know, ATP production through cellular respiration cannot occur efficiently without oxygen. As shown below, the overall reaction within the ETC produces up to 34 ATP molecules!

$$4H^+(aq) + 4e^- + O_2(g) \xrightarrow{\text{NADH \& FADH}_2} 2H_2O(l) + 34ATP$$

Energy transfer during respiration reactions

Flow of energy

Let's follow the flow of chemical energy from glucose into chemical energy stored in ATP at each of the three main phases of respiration. Before we review each part, let's remember animal and plant cells undergo respiration. Like animals, plants need energy at all times to keep systems properly functioning. At times when photosynthesis is not possible, plants undergo cellular respiration to get energy.

Part 1: Glycolysis

The first phase of respiration is called glycolysis. Glycolysis starts within the cytoplasm (highlighted yellow) as soon as glucose diffuses into a cell. Glucose molecules lose some chemical energy as heat when it is split apart and converted to pyruvate molecules. ATP is produced and NAD^+ is converted to NADH. The pyruvate molecules are sent to the next step in respiration.

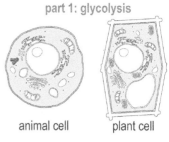

part 1: glycolysis

animal cell   plant cell

Part 2: The Krebs cycle

The second phase in respiration is called the Krebs cycle. Pyruvate molecules diffuse from the cytoplasm into the mitochondrial matrix (highlighted yellow) to undergo oxidation. Depending on the type of cell, there may be zero to thousands of mitochondria per cell. During the Krebs cycle pyruvate is converted into CO_2, ATP, NADH & $FADH_2$ and some energy is lost as heat. Oxygen (O_2) is required for the Krebs cycle and the electron transport chain.

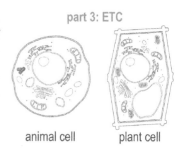

part 2: the Krebs cycle

animal cell   plant cell

Part 3: The ETC

The third phase of respiration overlaps with the second phase. The electron transport chain (ETC) enables a series of oxidation reactions. NADH and $FADH_2$ are oxidized by proteins in the inner mitochondrial membrane (highlighted yellow). Lost electrons are sent down the chain and H^+ ions are sent to the opposite side of the membrane. H_2O and ATP are produced, and heat is lost again.

part 3: ETC

animal cell   plant cell

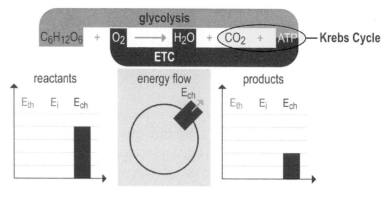

Second law of thermodynamics

Respiration obeys the second law of thermodynamics. Energy is lost at each step of the reaction, and every step requires an energy input from a molecule like ATP, NADH or $FADH_2$. Only 38% of the energy available in a glucose molecule is transferred to ATP molecules. The remaining 62% is lost as heat which adds to your normal body temperature. The respiration reaction sometimes shows 34 ATP molecules are produced. Because the conversion process is not 100% efficient, fewer than 34 ATP are produced. That's why you see the reaction above with an unspecified number of ATP molecules in the products.

Section 2 review

Chapter 22

Heterotrophs use the glucose produced by autotrophs for energy through cellular respiration. The net reaction of cellular respiration is the reverse of photosynthesis. Cellular respiration starts with glycolysis. Glycolysis is the breakdown of glucose into pyruvate which produces two molecules of ATP. After glycolysis, the pyruvate and oxygen enter the mitochondria and the Krebs cycle converts pyruvate to carbon dioxide and two more molecules of ATP. The last phase of respiration is the electron transport chain (ETC). The ETC is a stepwise movement of electrons from high to low energy that produces up to 34 molecules of ATP. In all, up to 38 molecules of ATP are created per molecule of glucose during cellular respiration.

Vocabulary words heterotroph, cellular respiration, glycolysis, NADH, Krebs cycle, electron transport chain

Review problems and questions

1. What are the main phases of respiration called?

2. Summarize what happens during the glycolysis phase of respiration.

3. Summarize what happens during the Krebs cycle of respiration.

4. Summarize what happens during the electron transport chain (ETC) phase of respiration.

5. Write balanced reactions for photosynthesis and respiration. Indicate whether an energy input is needed or an energy output is produced for each reaction.

6. During cellular respiration, what happens to the energy stored in glucose?

7. What two products are released during respiration?

8. Where does glycolysis take place?

9. How many stages does glycolysis undergo?

10. Where does the Krebs cycle take place?

Section 22.2: Cellular Respiration

22.3 - Energy Use in Cells

Your body is an open system that exchanges matter and energy with the environment. The sun's energy is passed down to you with the help of chloroplasts in plants. Food energy drives matter cycling in the human body. Like any system, human body functions are limited unless enough energy is available. Cells are the fundamental unit in any living thing that make energy and matter transfer possible. The chemical reactions that keep you alive occur in cells. The materials to build structures in your body are assembled by cells. The instructions to perpetuate life are stored in cells.

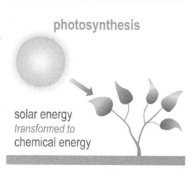

photosynthesis

solar energy *transformed to* chemical energy

Energy in cells

Catabolic and anabolic processes

The collection of chemical reactions that work together to keep you alive is called your **metabolism**. Metabolic reactions are categorized by whether the reaction breaks down or builds new molecules. **Catabolism** is the breakdown of molecules while **anabolism** is the construction of new molecules. Catabolic processes are usually exothermic reactions while anabolic processes are endothermic. Cellular respiration is a catabolic reaction while healing skin is anabolic. Both processes work together with the help of enzymes to give cells the energy and materials they need to get their work done.

Individual cells respond to changing chemical and physical conditions

Your body is a collection of systems called organs. Organs are made of tissues, and tissues are made of specialized cells. Individual cells allow the body to respond to changing internal and external conditions. For example your internal chemistry changes a great deal when you exercise. The greater energy demand on your body requires an increased rate of cellular respiration so more ATP can be produced. But more ATP production also results in more CO_2 and body heat production along with a greater O_2 demand. Individual cells need to be organized so they can communicate with one another and respond to the need to reduce body temperature and CO_2 concentration while bringing more O_2 to cells that need it.

Solved problem

Give one example of anabolism and one example of catabolism in your body.

Relationships   Anabolism builds biological molecules and absorbs energy; catabolism is the reverse.

Solve   What tissues and cells does your body need to either use or create for it to function properly? Also think about the four major classes of molecules: proteins, nucleic acids, lipids and carbohydrates. Is energy absorbed or released when the molecules are processed to complete their functions?

Answer   A couple examples of anabolism are DNA synthesis and protein formation from amino acids. Examples of catabolism include digestion of proteins to get amino acids and breakdown of carbohydrates to get energy.

ATP-ADP energy cycle

Biological molecules have basic building blocks

When you played with plastic bricks as a child, chances are you had a favorite piece that you used for everything you built. Like plastic bricks, many biological molecules are built around a basic molecule. Pieces added to or removed from the molecule change its chemical and physical properties. ATP is one such molecule.

Adenosine has multiple roles

All living things produce ATP. Every cell in every living thing relies on ATP as a fuel source. Recall that ATP is an abbreviation for adenosine triphosphate. Adenosine is composed of adenine and an organic 5-carbon sugar called ribose. You might recognize adenine as a nitrogenous base found in DNA and RNA, and the sugar ribose associated with the DNA and RNA backbone. When a phosphate group binds to adenosine, it is called cyclic adenosine monophosphate, cAMP, and its role is to regulate biological processes. Adenosine bound to two phosphates is called adenosine diphosphate, ADP. When three phosphate groups are present, it is adenosine triphosphate, ATP. ADP and ATP are energy carrier molecules that work like an energy tag team during metabolism.

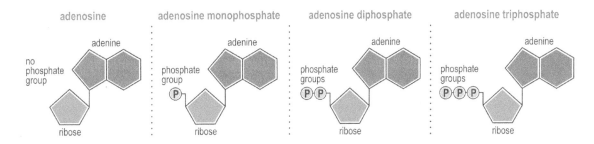

Phosphate bonds store energy

The molecules differ by the number of phosphate groups attached to the ribose. ATP carries energy in the bonds between phosphate groups. You might have guessed that ATP has the most energy available, followed by ADP and lastly cAMP because more phosphate bonds hold more energy. When a cell needs energy, an enzyme-catalyzed reaction causes ATP to release a phosphate group and form ADP. The energy that was once stored in the phosphate bond is transferred to the reactant in the cell that needs it.

ATP is used as soon as it is made

Unlike glucose, ATP does not have a storage form. As soon as an ATP molecule is produced, it is consumed. The body must produce ATP quickly as it is needed. ADP molecules are recycled during any of the three phases of cellular respiration to form ATP. A single active cell in your body can produce, use, and recycle over 10 million ATP molecules every second!

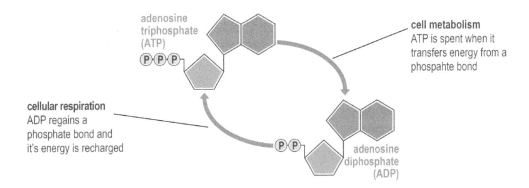

How cells use ATP

Body functions require ATP

Different kinds of cells take advantage of the energy released by ATP's phosphate bond. You can think of cell activities as reactions that recycle and maneuver matter through the body. These activities require energy in the form of ATP. Basic body functions that constantly demand energy include digestion, breathing, blood circulation, brain activity and other organ system functions.

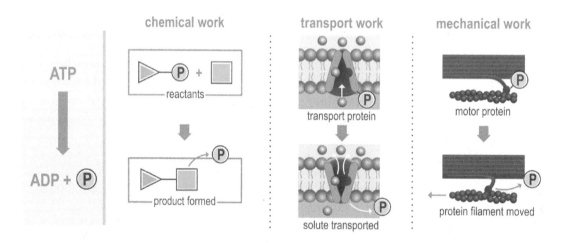

ATP is used to perform work

Work within living cells is the transfer of energy from one system to another to produce a change. Biological work falls into three broad categories: chemical work such as skin cells reproducing to repair a cut on your skin; transportation work such as an ATP-powered channel allowing an ion to pass through a cell membrane; and mechanical work such as contracting a finger muscle to scroll down a computer screen.

Daily energy needs

For an average adult, about 90% of the body's daily energy use is needed for background body activities we are often unaware of, and the remaining 10% is used for physical activities. There are trillions of cells in your body, each one needing to make and use ATP for ordinary activities. There are some cells that use more ATP than others. These include muscle cells and cells that constantly replace themselves such as skin cells.

Cells consume energy

A cell must duplicate its DNA before it can divide to make new cells. New cells also need to activate instructions in their DNA to build new proteins as they grow and work. Growing, making proteins, contracting—all of these processes demand high levels of energy at all times. But the cells that consume the most energy in your body are your brain cells. Some of the energy your brain uses is for ordinary maintenance, but your brain is like a muscle. The more you use your brain, the more ATP it demands and the better it will function!

Chemical work: Cell signaling

Cells communicate with chemical signals

Did you know your cells are constantly talking about you? Cells share messages like "break down glucose faster to keep up with exercise!" or "make more skin cells to fix a cut!" Cells need to respond to changes inside and outside your body because they only survive when they maintain a narrow range of temperature, pH, and oxygen-level conditions called **homeostasis**. Cells use different kinds of chemical messages to communicate with other body systems to maintain homeostasis. A message might tell a cell to begin or end growth, protein production, or muscle movement. Other messages direct your body to attack a cell that might make you sick. Another may increase or decrease body temperature or breathing rate. All messages have two things in common: ATP is required, and the goal is to bring about some kind of change in a cell.

Cells use lipids and proteins to "see" and to "be seen"

Different cell types need to recognize similar or different cells, and they also need to be recognizable by other cells. Recognition is critical for normal cell processes like growth, immunity response, and hormone response. Carbohydrates and molecules called glycoproteins and glycolipids are prominently displayed on the outside of cell membranes so they can "see" other cells and molecules to react with, and also "be seen." The glyco prefix means combined with one or more carbohydrate groups.

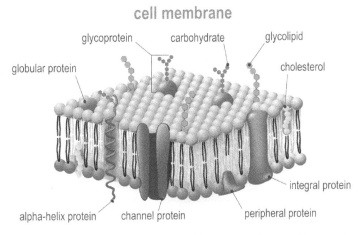

Short- and long-distance signaling

Some signals are designed to work over short distances when cells are in contact with one another. Other signals affect cells that are somewhere else in the body. Long-distance communication generally relies on hormones that enter the bloodstream so the whole body receives the signal. The communication system works when a cell sending a message releases a chemical hormone. Only cells with the correct receptor in their membranes called target cells can receive the signal. A target cell then interprets the signal and responds accordingly.

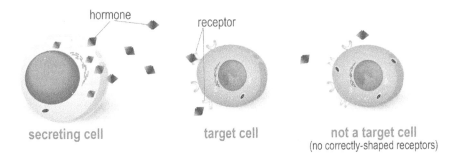

The endocrine system and chemical signaling

High glucose concentration around β-cells causes diffusion

Your digestive system breaks carbohydrates down into glucose molecules which enter the bloodstream from the small intestine. Your circulatory system distributes glucose molecules all over your body. The pancreas contains specialized cells called beta (β) cells that store a hormone called insulin. Insulin affects sugar, fat, and protein metabolism throughout the body. If the concentration of glucose around a β-cell is high, glucose enters the cell through a plasma membrane channel. Glucose channels are like gatekeepers embedded in a cell's membrane, allowing only glucose molecules to enter.

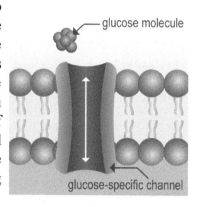

Respiration leads to insulin release

Once inside a β-cell, glucose undergoes cellular respiration. When ATP is produced, the β-cell membrane becomes more positively charged. As a result, vesicles containing the hormone insulin are attracted to the cell membrane. When the vesicles reach the β-cell membrane, they fuse with it and insulin gets released into the bloodstream.

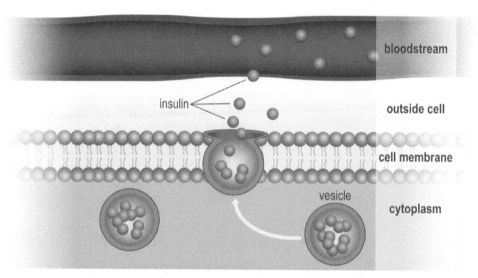

Insulin tells other cells to make a glucose receptor

Insulin is like a key that unlocks a cell's ability to take in glucose molecules. When insulin is in the bloodstream, other cells besides β-cells can absorb glucose. Insulin binds to a receptor that sits on the outside of a cell membrane. Once insulin attaches to the receptor it can deliver a message that tells the cell to build a glucose receptor protein or channel into its membrane. The newly-built glucose membrane channel protein can now bring glucose into the cell.

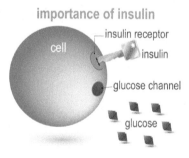

Insulin and diabetes

Glucose undergoes respiration inside the cell to produce ATP. As glucose gets used up, the drop in glucose concentration stops pancreatic β-cells from releasing insulin. Insulin controls blood sugar by limiting the amount of glucose cells can take up. When there is too much glucose in your body, insulin signals your liver to store the excess as glycogen. Diabetes is related to insulin production. If there isn't enough insulin in your body, cells cannot receive the signal to take in glucose so they start using proteins and fats for energy while blood sugar gets too high. If your body makes enough insulin but cells stop responding to its signal, β-cells overwork themselves in an attempt to make more insulin.

Transport work powered by ATP

CO_2, O_2, and H_2O don't need ATP to move in and out of cells

Movement of solutes in and out of cells through membranes is called **transport**. Like bouncers, membranes control when, how much, and where solutes go in and out of a cell. Even structures within cells like the mitochondria and nucleus are enclosed within a membrane to control which substances go in or stay out. Only a few small, nonpolar molecules like CO_2 and O_2 get past membranes by simple diffusion. Recall that **diffusion** is the flow of solutes from high to low concentration. Diffusion allows CO_2 waste from respiration to flow out of the mitochondria, out of the cell, and into the bloodstream. O_2 molecules diffuse into cells as they get used up. Water tends to flow into cells because there are more solutes inside a cell compared to outside a cell. Simple diffusion is also called passive transport because it works without adding ATP energy.

Interactive simulation

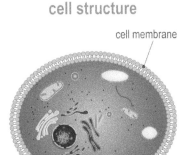

Explore a model of how substances like CO_2 and O_2 diffuse across a semi-permeable membrane in the interactive simulation titled Diffusion Across a Membrane. Experiment with different pore sizes to see how a membrane can be more or less selective in which substances it allows to diffuse in and out of a cell.

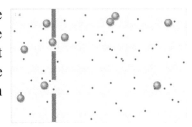

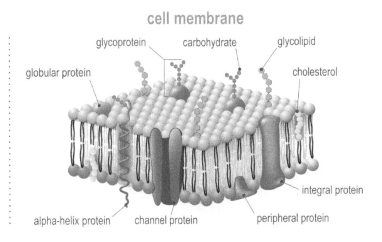

Facilitated diffusion through channels

Cell membranes are made of a double phospholipid layer. Polar heads line the outer and inner cell surfaces while nonpolar tails point inward. Notice proteins and carbohydrates embedded in the membrane. Some proteins span across the entire membrane and have a carbohydrate attached. These are called channels. They perform facilitated diffusion of larger molecules like glucose that are too large to squeeze though the phospholipids in the membrane on their own. Channels only let specific molecules in without using ATP.

Active transport through channels requires ATP

Other solutes like ions, proteins and amino acids need to be kept inside a cell no matter what the concentration is outside the cell. Ions are often transported into cells from low to high concentration, working against the natural flow. In this case, solutes enter the cell through protein channels in the cell membrane. These special channels need energy from ATP to work against the concentration gradient, so the movement through channels is considered active transport. The energy released when ATP loses a phosphate group causes the protein channel to change its shape just enough to let specific substances such as potassium ions into the cell. Without ATP-powered active transport, cells would not be able to specialize their functions.

Section 22.3: Energy Use in Cells

Mechanical work powered by ATP

A muscle's job is to contract

Molecules in your muscle cells convert chemical energy to mechanical energy with the help of ATP. There are three kinds of muscle cells in the human body. Each one has a different function, but all three types behave similarly on the molecular level. Whether located around your skeleton, in your heart, or in your digestive tract, a muscle's job is to create movement by contracting. Muscle tissue is made of bundles of fibers, and each fiber is made of many muscle cells.

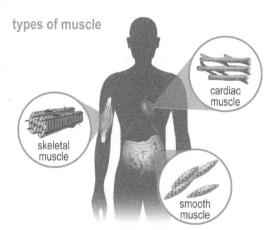

ADP is bound to myosin at rest

Muscles use electricity, chemical bonds, ions, ATP and several molecules including two proteins called actin and myosin to accomplish a contraction. ADP and a single phosphate are bound to myosin when a muscle is resting. When a muscle contraction occurs, myosin releases the phosphate and the protein changes shape.

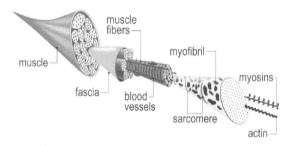

Myosin pulls actin towards itself during contraction

This new shape helps myosin attach to actin and pull it towards itself, similar to an adjustable bracelet. The pieces that make up the bracelet do not change length, but their position to each other can make the bracelet bigger or smaller. When myosin pulls actin towards itself, the muscle fiber shortens and spent ADP is released. ATP then attaches to myosin, causing it to release actin. ATP breaks up into ADP and a phosphate group on the myosin protein, and the muscle fibers continue to contract as long as ATP and other required ions and proteins are present.

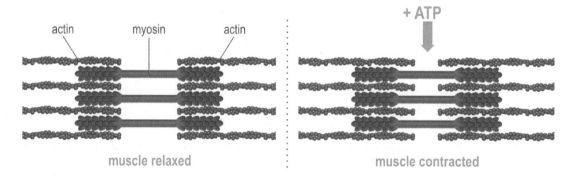

Muscles at rest

The process of myosin pulling actin towards itself occurs very quickly and many times during a muscle contraction. When the muscle contraction is finished, myosin releases actin to its original position, lengthening the muscle fiber at rest. A muscle at rest means exactly that - myosin is resting and is not actively pulling actin towards itself.

Section 3 review

Chapter 22

Metabolism refers to the biochemical processes that occur in living things to maintain life. Catabolism is the breakdown of molecules, and anabolism is the formation of new molecules. The sum of these reactions must change as conditions change. Adenosine triphosphate (ATP) is an important molecule in the metabolic pathway. The three phosphate bonds carry energy that is used in many reactions. The majority of ATP in our bodies is used for regulating the body to keep conditions optimal, even when changes occur. The endocrine system uses hormones to communicate the need for changes. ATP is used to move large molecules through cell membranes and to perform active transport, which moves solutes against a concentration gradient. Additionally, the energy of ATP is required to produce muscle contractions.

Vocabulary words: metabolism, catabolism, anabolism, work, homeostasis, transport, diffusion

Review problems and questions

1. Why does ATP have more energy than ADP or AMP? What is the energy source in these molecules?

2. How does the cell membrane make cell signaling possible?

3. Identify at least three ways the energy from ATP is used by cells.

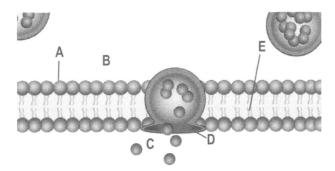

4. Use the diagram to answer the following:
   Which letter identifies an exclusively non-polar region?

5. Use the diagram to answer the following:
   Is B located inside a cell or outside a cell? How do you know?

6. Use the diagram to answer the following:
   How do the substances shown at point C interact with other cells?

22.4 - Maintaining Homeostasis

When a substance enters your body by eating, breathing, or diffusion through your skin, your body needs to recognize it as useful, useless, or dangerous. Will the substance provide energy for metabolism? Will it damage DNA? Is it a harmful bacterium that should be destroyed by the immune system? Should it be stored for later use?

Maintaining Homeostasis

Body systems as feedback loops

Feedback mechanisms in living things use physical and chemical signals to either increase or decrease a process. Appetite and body fat are controlled by protein chemical messages called hormones. An empty stomach releases a hormone called ghrelin into the blood. When ghrelin reaches your brain, you feel hunger. If you eat enough food to satisfy your body's energy needs, another hormone called leptin is released from fat cells into the blood. Your brain translates leptin into the feeling of fullness. This prompts you to stop eating. Leptin also signals starvation in the body. If you are not eating enough food, leptin levels get too low. Your brain senses low leptin levels and interprets it as starvation, thus stimulating appetite. The goal is to maintain homeostasis.

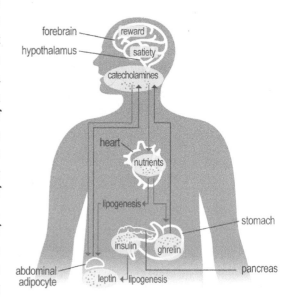

Equilibrium and homeostasis

Homeostasis is possible because our bodies have an interconnected feedback loop system. Homeostasis is **Le Châtelier's principle** in action: when a system at equilibrium faces a change, the system will shift in a direction to counteract the change and restore equilibrium.

Common equilibrium disruptors

Your body faces challenges to equilibrium regularly. The environment outside the body can have rapid changes in temperature and humidity, and gases in the air you breathe may change composition or concentration. Inside the body, acids and bases are constantly presenting the problem of keeping pH stable. The internal body environment responds to changes outside your body as well as inside. Every activity, no matter how small, changes conditions inside the body. Even when you spend all day sitting on the couch watching TV, the metabolic reactions in your body require reactants obtained from food. The products of metabolism must be distributed where they are needed or disposed of if they are waste. Let's look at a few feedback loops in the human body, all of which rely on chemical work powered by ATP

pH range in the body

pH in different parts of the body

Some chemical reactions work better at pH 5 and others at pH 7. How does your body create more acid where it is needed, for example, in your stomach? Active transport proteins in cell membranes are designed to create whatever hydrogen ion concentration is needed for proper enzyme function in the cell. When hydrogen is in control, so is pH.

Enzymes work best at an optimum range

Enzymes are protein-based so they are sensitive to pH and temperature. An enzyme's *optimum range* is the set of conditions where the enzyme works best. Your skin, mouth, stomach, intestines, blood and body fluids all have a different pH. The enzymes that work in each of those areas will have a different optimum pH range.

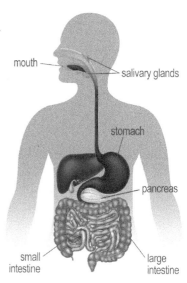

Thousands of enzymes

Your body relies on thousands of enzymes to complete metabolic processes and maintain homeostasis. When you eat something, digestion begins in your mouth. Your saliva contains a digestive enzyme called amylase. Salivary amylase breaks down starch into smaller units for faster digestion in the small intestine. Salivary amylase has an optimum pH range of 4.6-5.2. Most food you eat is acidic and is found in this pH range.

Isoenzymes

Amylase is in the family of *isoenzymes*, an enzyme that has more than one form. Another form of amylase is found in your pancreas, which secretes enzymes into the stomach and small intestine. Pancreatic amylase has an optimum pH range of 6.7-7.0, and it works to break down complex carbohydrates. After food is largely digested in the stomach using enzymes, it moves through the intestines for final digestion with even more enzymes. Your circulatory system collects the useful products of digestion from the small intestine: amino acids, glucose, and other molecular building blocks.

Bicarbonate blood buffer

Blood pH must be between 7.35 and 7.45. You will feel ill if pH is outside of this range. How does the body handle constant threats to pH? Your blood has *buffer* systems that resist changes in pH from carbon dioxide (CO_2) when it is carried into your blood as a waste product of cellular respiration. When it reacts with water, CO_2 forms the weak acid carbonic acid, H_2CO_3. The conjugate base for carbonic acid is the bicarbonate ion, HCO_3^-. The pH change is stable in a buffered system because either the weak acid or its conjugate base can neutralize added acids and bases.

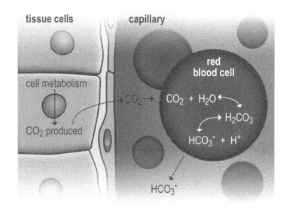

bicarbonate buffer system

$$H_2CO_3 \rightleftharpoons HCO_3^- + H^+$$

carbonic acid — bicarbonate ion — hydrogen ion
a weak acid — *its conjugate base*

Buffering system in the body

Breathing rate is controlled by CO_2, not O_2

Many people think breathing rate increases during exercise because their body needs more oxygen. Your breathing rate is determined by the amount of carbon dioxide, not oxygen, in your blood. The primary reason your body changes the way you breathe is to keep pH stable. Exercise increases the rate of cellular respiration to meet ATP demand, which increases CO_2 production. Study the equilibrium expression below and think about Le Châtelier's Principle. Why does an increase in CO_2 lead to a pH change?

$$CO_2(g) + H_2O(l) \rightleftharpoons H_2CO_3(aq) \rightleftharpoons HCO_3^-(aq) + H^+(aq)$$

CO_2 affects H^+ ion concentration in the blood

The equilibrium system above is called the bicarbonate buffering system. Carbonic acid reversibly dissociates. The bicarbonate ion, HCO_3^- either binds or releases hydrogen ions. According to Le Châtelier's Principle, increases in CO_2 will shift equilibrium to the right, increasing hydrogen ions, H^+. More H^+ leads to a lower, more acidic pH. Decreases in CO_2 will shift equilibrium to the left, which removes H^+ and increases pH.

Multiple organs work together to maintain pH

The lungs, kidney and brain play critical roles in the body's bicarbonate buffering system. Receptors in the brain stem sense the CO_2 and H^+ concentration in your blood and control lung activity with chemical signals sent through the circulatory system. Increased metabolic activity increases the rate of CO_2 produced, shifting equilibrium to the right. Receptors in your brain stem bind H^+ and CO_2 and recognize when they increase. The brain sends signals to your lungs to increase breathing rate to eliminate excess CO_2 before H^+ ions build up.

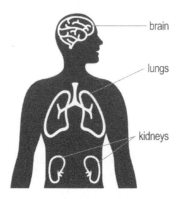

How kidneys regulate pH

Kidney cells are sensitive to changes in CO_2, HCO_3^-, and H^+ concentrations. Kidneys can either increase pH by releasing HCO_3^- ions into your blood so less free H^+ ions are present, or when pH goes too high they absorb HCO_3^- to allow excess H^+ ions in the blood. These ion exchanges occur in the structure shown, a nephron. Kidneys contain many nephrons because they are the main functional unit.

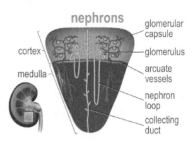

Other buffers in the body

The bicarbonate buffering system is just one of a few major buffering systems in the body, and it works in fluid outside of cells. A phosphate ion buffering system works inside cells to maintain pH, and many proteins like hemoglobin in the body act as buffers inside and outside cells. Proteins with acid and base groups can either give up or gain H^+ ions. These buffering systems rely on chemical equilibrium and Le Châtelier's principle to deal with constant challenges to maintaining homeostasis.

Temperature equilibrium

Thermo-regulation and enzyme activity

Why does your body temperature stay around 98.6 °F? There are at least two reasons: because cellular respiration is exothermic, and there are feedback mechanisms in the body that keep internal conditions consistent. But a third reason why your body temperature is able to stay constant is because about 60% of your body's mass is water. The percentage of water inside individual cells is even higher at about 85%. Water's high specific heat means it has the ability to store a lot of heat allowing it to resist changes in temperature. Maintaining body temperature is important because enzymes operate within an optimum temperature range. If body temperature dips too low, the enzymes cannot catalyze life-sustaining chemical reactions quickly enough to keep you alive. If body temperature rises too high, enzyme shapes can denature, making them ineffective. The ability of an organism to maintain a constant body temperature is called **thermoregulation**, and is accomplished through physical, chemical or behavioral means.

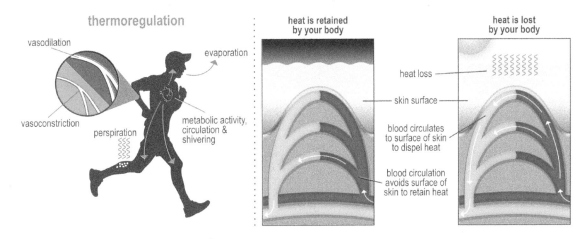

Cold weather temperature regulation

Different animals have different thermoregulation methods. Arctic animals keep warm and nourished by storing large amounts of fat under their fur-covered skin. Their large size also retains a large volume of water, which retains heat very well. Some animals migrate or store a lot of excess fat in fall and then hibernate in winter when food is scarce. Humans and other mammals get goosebumps and shiver to reduce heat loss.

Sweating and panting

Humans sweat to transfer heat from the inside of their bodies to the outside, while animals that cannot sweat pant. Sweating and panting move water from inside the body to the outside to take advantage of evaporative cooling. When liquids evaporate, they must absorb energy from their surroundings—in this case, your skin—and they carry the heat away when they change to the vapor phase.

Desert animals

Desert animals tend to have small bodies to make it easier to keep cool, and to keep energy needs to a minimum when food is hard to find. Larger desert animals have bodies that are designed to trap and release heat as efficiently as possible. Many people believe a camel's hump is filled with water, but that is a myth. The hump is a desert adaptation that concentrates fat in one part of the body instead of the whole body. By concentrating fat in only a few areas, heat is also trapped in only a few areas. As an additional benefit, the stored fat provides energy when food is scarce.

Nervous system signals

The nervous system is an electrical system

Reactions to changing conditions in the environment rely largely on the nervous system and how it transmits signals throughout the body. Your nervous system uses electrochemical forces within and between atoms. An **electrochemical** change involves the conversion of chemical energy to electrical energy or vice versa. Electrochemical changes in the nervous system depend on ion concentrations and protein channels in cell membranes. A common type of cell found in the nervous system called a neuron has a shape and behavior that is different from typical cells. A neuron cell has typical structures like a nucleus with DNA and mitochondria to make ATP, but it is long and branched.

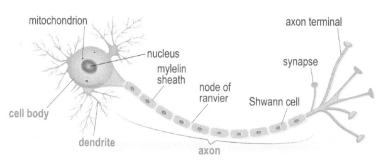

Neurons at rest

Neural cell membranes have ATP-powered protein channels or "pumps" that force an ion imbalance inside the cell to create an electrical potential difference or voltage. Neurons at rest are "polarized" with a positive charge outside the cell. Membrane pumps keep potassium ions (K^+) inside the cell and push sodium ions (Na^+) out of the cell. Negatively charged proteins stay in the cell and chloride ions stay outside. This creates a potential voltage. The inside of the neuron is 70 mV (millivolts) less than the outside.

Neurons at work

When a neuron receives a signal, Na^+ rushes into the cell followed by a rush of K^+ out of the cell. This sends an electrical impulse down to the nerve's end where a small gap called a synapse is found. The synapse is where a chemical change will take place.

Neurotransmitters carry the signal across the synapse

When the signal reaches the axon, ATP is used to open cell membrane channels that let in calcium (Ca^{2+}). This allows vesicles that store proteins called neurotransmitters to bind to the neuron cell membrane and release neurotransmitters into the synapse. The neurotransmitters chemically bond to receptors on the next nerve cell causing it to send the electrical message down the nerve, and the process continues.

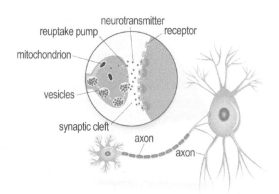

Life organizes energy & matter

Think about all the chemistry that is happening in your body right now. Even simple living things as small as bacteria use chemistry to organize energy and matter in a way that is beneficial to them. All of your body systems work as a team to produce the physical, chemical, electrical, and biological changes necessary to keep you alive.

Section 4 review

Chapter 22

Homeostasis refers to any process that living organisms use to maintain a stable internal environment. This is accomplished through many feedback loops that regulate processes when changes occur. Different parts of the body (and the enzymes found there) work best within a specific pH range. As an example, blood must be kept within a fairly narrow pH range. Carbonic acid and its conjugate base serve as a buffer, while the lungs and kidneys are signaled to help regulate changes. Maintaining body temperature or thermoregulation is accomplished through several means including shivering, the use of stored fat, and sweating. Many reactions in the body depend on electrochemical changes. A potential voltage creates an electrical impulse that releases neurotransmitters that signal changes in the body.

Vocabulary words Le Châtelier's principle, optimum range, buffer, isoenzymes, thermoregulation, electrochemical

Review problems and questions

1. Explain what an optimum range is.

How does running affect blood pH?

2. A student athlete is almost finished with her daily jog. Apply Le Châtelier's principle to predict whether the blood's bicarbonate buffer system will shift to produce more CO_2 or more HCO_3^-. Explain what will happen to pH as a result of this shift.

$$CO_2(g) + H_2O(l) \rightleftharpoons H_2CO_3(aq) \rightleftharpoons HCO_3^-(aq) + H^+(aq)$$

3. Which body system is capable of converting chemical energy to electrical energy?

4. Choose any body system and describe how that system uses physical and or chemical methods to maintain homeostasis when conditions inside or outside the body change.

Energy Transfers in Living Things

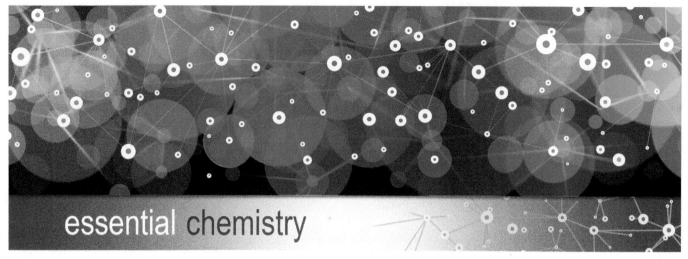

essential chemistry

You probably don't think about the sun too much except for occasionally remembering to put on sunglasses, or to wear sunscreen in the summer. But, when you dig a little deeper, you see why the sun is considered the ultimate energy source for life on Earth. Plants, algae, and cyanobacteria are primary producers - they take energy from the sun and convert it into organic compounds like carbohydrates. If you think about the food web, those primary producers are eaten by other animals called primary consumers, who are then eaten by secondary consumers. The food that these higher and higher consumers eat can be traced back to the sun.

Plants collect energy from the sun during photosynthesis. Photosynthesis can be broken into two main stages. In the first stage, energy from light is captured by photosynthetic pigments like the chlorophyll that gives plants their green color. The energy from the light is used to split water into oxygen, hydrogen ions, and electrons. NADPH and ATP are also created in this stage and these carry the energy from the sun's light to the next stage. At the second stage, energy-carrying molecules drive CO_2 through the Calvin cycle to ultimately produce a carbohydrate. Meanwhile, discharged energy-carrying molecules are converted back to their original forms to start the process all over again. The overall equation looks like the reverse of a combustion reaction.

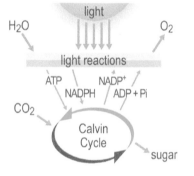

$$6CO_2(g) + 6H_2O \rightarrow C_6H_{12}O_6(s) + 6O_2(g)$$

The two stages can be summed up as: collect light energy to split water to release oxygen and hydrogen, and capture carbon dioxide to make a sugar. You can see that plants and trees are essential to our survival. Powered by the sun's energy, they can generate oxygen, sequester carbon dioxide, and are the primary producers in a food web.

Similarly, we have tried to harness the sun's power to generate heat and electricity. Solar power collection comes in two major forms: thermal energy and light energy. Solar thermal energy is exactly what the name implies. Using disks or mirrors the sun's radiant energy is converted into thermal energy to warm objects or create steam. The simplest example of this is a solar cooker that you might use when camping. The more common form of solar energy collection is through solar cells. In fact, when you see solar panels on the roof of a house, you are seeing an array of solar cells. Solar cells, also called photovoltaic cells, convert light energy directly into electrical energy.

Energy Transfer in Living Things - 2

Solar cells consist of a material that exhibits a photoelectric effect. When a photon of light hits photoelectric material, an electron can be knocked loose. If these free electrons can flow through a circuit, an electric current will result generating electricity. At any given moment, the sun generates about 3.86×10^{26} watts of power, and about 1.74×10^{17} watts of that power strikes the earth. While a massive amount of energy strikes the earth from the sun, currently we use solar energy to satisfy less than 1% of our electricity needs.

What if we could make energy collectors that could perform the same function as leaves? That question has pushed scientists into the field of artificial photosynthesis. Scientists have taken different approaches toward artificial photosynthesis but their goal is the same: take solar power to transform water and carbon dioxide into fuels and other useful organic chemicals. By tackling artificial photosynthesis, research groups are finding new ways to address growing energy production and carbon dioxide issues.

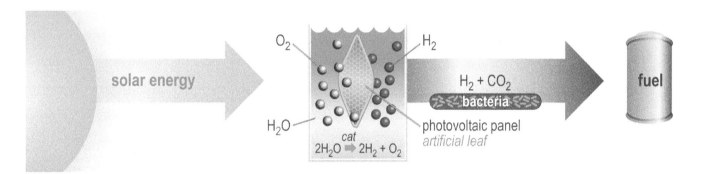

Chemist Daniel Nocera and synthetic biologist Pamela Silver have looked to nature for inspiration and created what is called a "bionic leaf". Silver is quoted as saying "plants are the best chemists there are." So it makes sense to not only incorporate reactions that mimic plants, but also include living things in the artificial photosynthesis process. Starting in 2015 their team reported success in converting sunlight, carbon dioxide and water into liquid fuel. Their living battery uses a photovoltaic panel to collect solar energy. The solar energy powers a catalyzed chemical reaction that splits water into hydrogen and oxygen. Then the biology takes over – microbes feed on the hydrogen that is produced from the chemical reaction and combine it with carbon dioxide from the air to make fuels.

The Nocera and Silver team have modified their designs to improve performance. Their initial device in 2015 was able to create 216 milligrams of alcohol fuel per liter of water. This isn't a great efficiency, and the nickel-molybdenum-zinc catalyst that they chose had the inadvertent effect of poisoning the microbes!

In 2016 the Nocera and Silver team used a new catalyst made from cobalt phosphide. The added advantage is that the phosphorous, in the form of phosphate in solution, is actually good for the R. eutropha bacteria in the device they dubbed "bionic leaf 2.0." The team reported that for every kilowatt-hour of electricity used, 130 grams of CO_2 is taken out of 230,000 liters of air to make 60 grams of isopropanol fuel. With advances in synthetic biology, it is possible to modify the R. eutropha to generate other organic products. At these levels, bionic leaf 2.0 has an efficiency of 10%. Not only is this a big leap from their previous design, it is more efficient that natural photosynthesis!

There is much more research to do with artificial photosynthesis. But in theory, this platform makes it possible to take any carbon-based molecule using water, carbon dioxide and some energy from the sun!

Chapter 22 review

Vocabulary
Match each word to the sentence where it best fits.

Section 22.1

ADP	ATP
autotroph	Calvin cycle
chemotroph	chlorophyll
chloroplast	NADPH
oxidation	photosynthesis
phototroph	pigment

1. Carbon dioxide and water are converted into glucose and oxygen in the process of _____.

2. Plants use _____ as an electron carrier molecule to store the energy from sunlight.

3. Adenosine triphosphate or _____ is a molecule that transfers chemical energy so that nonspontaneous chemical reactions can occur.

4. An organism that is capable of making its own food is called a(n) _____.

5. A(n) _____ uses light as its energy source.

6. A(n) _____ is an organism that makes its own food with inorganic compounds.

7. Chlorophyll is found in the membrane-bound plant cell structure called the _____.

8. Plants use _____ as the main pigment molecule to store energy from light.

9. Because sunlight is not required for the _____, it is also sometimes referred to as the "dark reactions."

10. An energy carrier molecule called _____ is regarded as "spent" and can be recharged by adding a phosphate group to the molecule.

11. _____ reactions are often used as an energy source for chemotrophs.

12. Plants are green because of the _____ chlorophyll.

Section 22.2

cellular respiration	electron transport chain
glycolysis	heterotroph
Krebs cycle	NADH

13. A(n) _____ is an organism that cannot make its own food.

14. The series of chemical reactions during cellular respiration that breaks down pyruvate into carbon dioxide is called the _____.

15. Glucose is broken down into two three-carbon molecules during _____ which is the first step in cellular respiration.

16. As a high-energy electron carrier molecule, _____ is used to transfer electrons in cellular respiration.

17. Energy for life processes is released by _____ where glucose is broken down into CO_2 and water.

18. High energy electrons are transferred through the _____ where they are used to push H^+ ions across a membrane.

Section 22.3

anabolism	catabolism
diffusion	homeostasis
metabolism	transport
work	

19. Living cells do _____ when they transfer energy between systems to produce a change.

20. _____ is the maintenance of certain conditions including pH, temperature, and oxygen-level to fall within a narrow range.

21. _____ describes metabolic reactions where molecules are broken down.

22. The movement of solutes in and out of a cell membrane is called _____.

23. _____ describes metabolic reactions where new molecules are made.

24. Cells take advantage of _____ allowing them to move small solutes in and out of a membrane based on concentration instead of using energy.

25. The collection of chemical reactions that keeps an organism alive is called its _____.

Section 22.4

buffer	electrochemical
isoenzymes	Le Châtelier's principle
optimum range	thermoregulation

26. The _____ of an enzyme is a condition, such as pH, in which it works best.

27. _____ systems help your body maintain a stable pH.

28. The action of _____ makes homeostasis possible by counteracting changes to your body's equilibrium.

29. _____ changes involve conversion between electrical and chemical energy.

30. Body temperature remains constant through _____, a physical and chemical means within the body.

31. _____ are able to catalyze the same chemical reaction while differing in amino acid sequence.

Chapter 22 review

Conceptual questions

Section 22.1

32. What happens to the energy carrier molecules ATP and NADPH once they transfer their energy in the Calvin cycle?

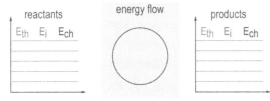

33. Using the above energy bar chart,
    a. illustrate the overall energy flow for photosynthesis; and,
    b. explain whether photosynthesis is an endothermic or exothermic process.

34. Adenosine triphosphate, ATP, is an energy carrier molecule. Where in the ATP molecule is the energy stored?

35. The pigment chlorophyll is responsible for helping a plant transfer the Sun's light energy into chemical energy. What color regions of visible light are absorbed by the chlorophyll molecule?

36. Many pigments assist chlorophyll in various plant functions and reflect different wavelengths of light. If a plant has purple leaves, what class of pigment is it likely to contain?

37. Answer the following questions about energy carrier molecules.
    a. Energy from photons is used to split water molecules during photosynthesis. The high-energy electrons gained from this oxidation reaction are held by which carrier molecule during the light-dependent reactions?
    b. What form of this carrier molecule is ready to accept more electrons?
    c. How does this energy carrier molecule "hold" energy?
    d. What happens to the energy carried by this molecule?
    e. What quality makes energy carrier molecules so useful for living systems?

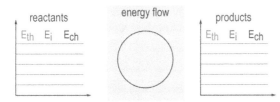

38. Using the above energy bar chart,
    a. illustrate the overall energy flow for photosynthesis; and,
    b. explain whether photosynthesis is an endothermic or exothermic process.

39. Plants work continuously to construct energy rich glucose molecules. How do plants use glucose molecules once it has been synthesized?

40. What molecule is used by plants during photosynthesis to capture and absorb the Sun's energy? How is the energy captured by this molecule?

41. Plants use water and carbon dioxide as building blocks for sugar molecules. The following questions pertain to photosynthesis.
    a. What is the balanced chemical reaction for photosynthesis?
    b. Draw the Lewis structures of the reactants of photosynthesis.
    c. What other kinds of living things are capable of using sunlight to create glucose?
    d. Would you consider photosynthesis an endothermic or exothermic reaction? Explain your answer.

42. The Calvin cycle is called a "dark reaction" because it does not require light. Summarize the main points of the Calvin cycle and state where the energy for the reactions comes from.

Section 22.2

43. Like photosynthesis, respiration occurs in a series of enzyme-catalyzed redox reactions. What three main reactions stage the breakdown of glucose molecules for cellular respiration?

44. Photosynthesis and cellular respiration work together to cycle carbon through the air, soil, water, and living things.
    a. Explain how this process works.
    b. Are matter and energy wasted during this process?

45. During which stage of cellular respiration do each of the following occur?
    a. Glucose molecules are broken down
    b. Oxygen molecules are directly used
    c. Water molecules are released
    d. Carbon dioxide molecules are released

46. Even though oxygen is not directly involved in the Krebs cycle reactions, why is it still required in order for the Krebs cycle to occur?

47. The electron transport chain requires oxygen in order to function. Why is oxygen so important for this stage? What happens with a lack of oxygen during this process?

48. Glycolysis is a term used to describe the first metabolic pathway during cellular respiration. Explain what happens to a glucose molecule during this stage.

49. During cellular respiration, glucose and other food molecules are oxidized in what appears to be a combustion reaction, but it takes place over the course of several steps. Why do cells take steps to release the energy in a glucose molecule instead of all at once?

Chapter 22 review

Section 22.2

50. Would you consider cellular respiration to be an endothermic or exothermic reaction?

    a. State your answer, and explain your reasoning.

    b. The forward direction of respiration is favorable. Does the first step of glycolysis occur spontaneously?

51. NADH carries chemical energy and transfers it down the metabolic pathway. Are the NADH molecules that are formed during glycolysis reducing agents or oxidizing agents for these later reactions? Explain.

52. Aerobic and anaerobic refer to the presence or absence of oxygen for metabolic processes. Is O_2 required for glycolysis to occur? Explain your answer.

53. What is oxygen's primary role during cellular respiration? What happens to oxygen during this process?

54. What is the balanced chemical equation for cellular respiration?

55. Compare the net amount of ATP molecules produced at the end of the three stages of cellular respiration.

    a. How much ATP is gained from the first stage?

    b. How much ATP is gained from the second stage?

    c. How much potential ATP can be yielded from the final stage of this metabolic pathway?

56. Use the energy bar chart above to:

    a. illustrate the flow of energy during a respiration reaction; and,

    b. explain whether respiration an endothermic or exothermic process.

Section 22.3

57. All cells and even some structures within cells are enclosed within a membrane to control the flow of substances through them. What is the difference between diffusion (passive transport) and facilitated diffusion?

58. Why does cellular respiration transfer only 38% of the energy in glucose molecule to ATP?

59. All living things produce ATP, the abbreviation for adenosine triphosphate. What are the 3 compounds compose an ATP molecule?

60. Specialized protein channels engage in active transport in order have solutes enter the cell through the cell membrane, requiring ATP in order to do so. Why does an active transport protein channel in a cell membrane need ATP to work?

61. Muscle cells work hard and require a lot of energy along with many other factors, including the proteins actin and myosin, to function. When a muscle contracts, these proteins do not actually change in length, why is this?

62. Adenosine triphosphate is used in biological systems to perform work. List the 3 categories of biological work and give an example of each.

63. Aside from their specific biological roles and number of phosphate groups, what is a major difference between cAMP, ADP, and ATP?

64. The body must produce ATP very rapidly and as soon as an ATP molecule is produced, it is consumed. Why is ATP used as soon as it is made?

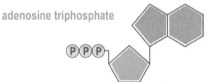
adenosine triphosphate

65. Different kinds of cells take advantage of the energy released by ATP's phosphate bond. List at least 3 examples of things that cells need ATP for.

66. Different cell types need to recognize similar or different cells along with being recognized by surrounding cells. What chemical components of cell membranes make cell signaling possible?

67. Cells communicate over short distances or when they are in direct contact with one another. Long-distance communication is generally more complex. Summarize how cells can communicate over long distances in the body.

68. Cells need to respond to changes inside and outside your body in order to maintain homeostasis. Give an example of a change that could occur in a cell due to cell messaging.

69. Insulin affects sugar, fat, and protein metabolism throughout the body. Since insulin is enclosed in vesicles within cells, how does it get into the bloodstream when glucose undergoes cellular respiration?

Section 22.4

70. The human body faces challenges to equilibrium on a regular basis. What are some examples of everyday threats to the body's homeostasis? Be sure to include some external as well as internal factors.

71. If a cell has a high pH, what would you think is the main type of work (chemical, transport, or mechanical) that the cell would do to counter the threat to homeostasis? Explain.

Chapter 22 review

Section 22.4

72. What average temperature do you believe is optimum for enzymes that work in your body? Explain your answer.

    a. Why is maintaining this temperature so important for enzymes?

    b. What happens when the temperature falls out of this range?

73. The bicarbonate buffering system is just one of a few major buffering systems in the body.

    a. What is the equilibrium expression for the bicarbonate buffering system?

    b. What property of the buffer solution in blood allows it to resist changes in pH?

74. If more bicarbonate ions (HCO_3^-) are released to the blood, will pH increase or will it decrease? Which way will equilibrium shift in this scenario? Explain your answer.

75. Use what you've learned about the equilibrium system within the human body to explain what the lungs can do to reduce blood pH.

76. The kidneys play a critical role in the body's buffering system. What can the kidneys do to reduce blood pH? Explain.

77. What are three reasons why your body temperature is kept at a constant around 98.6 °F?

    a. What role does water play in keeping your body warm?

    b. What role does water play in keeping your body cool?

78. Your nervous system uses electrochemical forces within and between atoms. What generates the electrical potential that travels down neurons?

79. What compound allows a signal to move from neuron cell to neuron cell across the gap between neurons? How is this channel opened?

80. Describe the process of how a neuron transmits a signal from the moment it receives a signal to the transmission across a synapse.

81. Chemically, homeostasis is Le Châtelier's principle in action. Explain how the body uses Le Châtelier's principle to maintain homeostasis.

82. Your body relies on thousands of enzymes to complete metabolic processes and maintain homeostasis. One such enzyme is amylase, an isoenzyme.

    a. Why does the body have more than one form of this same enzyme?

    b. Why can't only one form work for the entire digestion process?

chapter 23 | The Earth

Why is it important to recycle? How does the earth recycle matter on its own? You have probably heard the word "conservation." When thinking about the Earth, conservation typically means to preserve or protect parts of the environment. In science, we apply the word conservation a bit differently. Specifically, we talk about the laws of conservation of mass and energy. In the case of a chemical reaction, the total mass is "preserved" even though the substance changed chemically and the atoms rearranged. Total energy is "preserved" even though it is transformed from one form, like potential energy, to another, like kinetic energy.

When we look at the chemistry of the Earth, there are many examples where matter is ultimately conserved, but it is physically or chemically changed as it flows into and out of systems. For example, there is a water cycle, a carbon cycle, a nitrogen cycle, and there is even a rock cycle. These interconnected systems cycle matter through the land, water and atmosphere. In the carbon cycle, carbon could take the form of organic compounds in fuels, sugars in plants, carbon dioxide in the air, or carbonates in the ocean. The carbon atoms are conserved, but the form they take changes depending on where they are in the cycle .

Human activity can affect the balance of these systems. Since the systems are interconnected, factors that affect one of these systems will eventually affect all of them. Earth's resources need to be consumed at rates that will not alter the balance of a system. We observe the adverse effects of a chemical imbalance when conditions such as severe climate change occur. Since most of these systems occur over many years (many thousands of years!) it takes time for the Earth to equilibrate. To offset harmful human impacts, we need to plan as a global society and study the chemical balances of our planet. To accomplish this, we begin with an essential understanding of the Earth's composition and cycles.

Chapter 23 study guide

Chapter Preview

In our solar system there are other planets and even some moons that have atmospheres and water in the form of ice or oceans and yet Earth is the only one that supports life. What gives Earth the proper balance? In part it is the abundance of water that acts as a universal solvent and is cycled between the sea, the atmosphere, and the land. The tilt of the earth determines the seasons and drives the weather. The atmosphere protects life from ultraviolet light, traps heat and holds the gases needed for photosynthesis and cellular respiration. In this chapter you'll learn about Earth's chemistry from its atmosphere to its core.

Learning objectives

By the end of this chapter you should be able to:

- describe the different layers of Earth's atmosphere and events that occur at each of those levels;
- explain the greenhouse effect, why it is necessary for life, and the danger of too many greenhouse gases;
- use mole fraction to calculate ppm concentration of a gas;
- identify factors that drive the water cycle;
- explain how the carbon cycle is connected to ocean pH;
- describe ways matter cycles through Earth; and
- describe the importance of proper pH of soil for plant growth.

Investigations

23A: Greenhouse Gases
23B: The water cycle
23C: Ocean currents
23D: Ocean acidification

Pages in this chapter

748 The Atmosphere
749 The evolution of Earth's atmosphere
750 Water vapor and relative humidity
751 The dew point
752 Weather
753 Ozone
754 Sulfur and acid rain
755 Smog and air pollution
756 Carbon dioxide and the greenhouse effect
757 Section 1 Review
758 The Hydrosphere
759 The effect of ocean currents on climate
760 Chemical composition of the oceans
761 The water cycle
762 The carbon cycle
763 Acid-base reactions in the ocean
764 Ocean acidification
765 Section 2 Review
766 The Geosphere
767 Formation of the earth
768 Elemental composition of the earth
769 Minerals
770 The rock cycle
771 Rocks
772 Plate tectonics and volcanism
773 The nitrogen cycle
774 The phosphorus cycle
775 The oxygen cycle
776 Section 3 Review
777 Soil Health
778 Soil Health - 2
779 Chapter review

Vocabulary

troposphere
mesosphere
photodissociation
hydrosphere
transpiration
ocean acidification
mantle
crystalline
geologists
polymorphism
rock
lava
nitrogen fixation

stratosphere
thermosphere
greenhouse gases
salinity
stomata
geosphere
solar nebula
amorphous
sedimentary rock
igneous rock
magma
plate tectonics
phosphorus cycle

ozone layer
dew point
greenhouse effect
water cycle
carbon cycle
silicates
core
mineral
metamorphic rock
rock cycle
volcano
nitrogen cycle

23.1 - The Atmosphere

Earth's atmosphere extends from the surface of the planet to an altitude of about 100 kilometers. This blanket of air has no "end" but gets less and less dense until at 100 km, the pressure is only 0.003% of its value at sea level. For most purposes this is the border of "space." An incredible amount of life-sustaining chemistry occurs between zero and 100 kilometers up, including all our weather, the radiation-shielding ozone layer, clouds, acid rain, and much more. As our understanding of the atmosphere grows, scientists have come to realize how complex Earth's blanket of air really is.

What is the atmosphere made of?

Nitrogen and oxygen

Slightly more than 78% of dry air (by mole) is diatomic nitrogen (N_2) and 21% is diatomic oxygen, O_2. Just under 1% is argon and other trace noble gases helium, and neon. Water vapor adds between 0.1% and 4% depending on the temperature and humidity. Finally, there are trace amounts of carbon dioxide and methane, important greenhouse gases we will discuss later in this section.

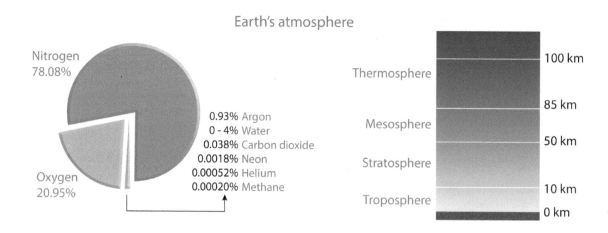

The troposphere

The atmosphere is divided into several layers. We live in the **troposphere**, the densest, warmest layer closest to Earth and extending to an altitude of about 10 kilometers. Seventy-five percent of the mass of the atmosphere is in the troposphere, which contains 99% of the water vapor in the atmosphere. Clouds that cause rain are in the troposphere. The troposphere is more turbulent than the upper layers of the atmosphere. The turbulence is created by rising air warmed by the ground and falling air cooled by the upper atmosphere. This circulation is the driver of the weather.

The stratosphere

Above the troposphere is the cooler **stratosphere** extending from 10 to 50 kilometers in altitude. The **ozone layer** is part of the stratosphere. The average temperature of the stratosphere is -3°C while the air above is slightly warmer. This creates a temperature inversion making the stratosphere relatively free of convective turbulence. Jet airliners use this "smooth air" while cruising at altitudes of 10 km.

The mesosphere and thermosphere

From 50 km to 85 km is the **mesosphere**, which has an average temperature of -65C. The pressure in the mesosphere is too low to support aircraft but great enough to be a drag on spacecraft. Extending into space from the mesosphere we reach the **thermosphere** from 85 km altitude to around 500 km. The upper boundary of the thermosphere marks the point where Earth's atmosphere collides with the Solar wind. Gases in the thermosphere are partly ionized and produce the aurora borealis in northern latitudes.

The evolution of Earth's atmosphere

Formation of Earth

Along with the rest of the Solar System, Earth formed from a cloud of molecular gas and dust 4.6 billion years ago. As the planet solidified, its nearness to the Sun meant that heat and the solar wind drove off most of the gaseous hydrogen and helium still present in the outer gas-giant planets. The first clear evidence we have of the early Earth points to an atmosphere of mostly nitrogen with volcanic carbon dioxide and other trace gases. Recent evidence suggests that much or most of the atmosphere came from comets or other space-impactors which brought water back to the planet after its surface had cooled. There was no oxygen in the early atmosphere.

The Great Oxygenation Event

About 3.5 billion years ago Earth evolved single-celled photosynthetic organisms similar to cyanobacteria. These early forms of life began producing oxygen in the oceans. For the next billion years the oxygen stayed dissolved in the ocean. Once the ocean was saturated oxygen started to escape into the atmosphere. Often referred to as the Great Oxygenation Event, the concentration of atmospheric oxygen began to rise. However, oxygen is highly reactive. Minerals such as iron chemically scavenged oxygen keeping the concentration of free oxygen low.

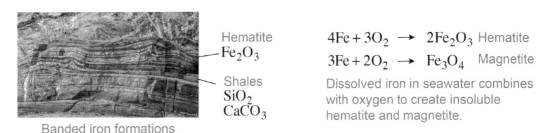

How minerals absorb free oxygen

Hematite Fe_2O_3
Shales SiO_2 $CaCO_3$

Banded iron formations

$$4Fe + 3O_2 \rightarrow 2Fe_2O_3 \text{ Hematite}$$
$$3Fe + 2O_2 \rightarrow Fe_3O_4 \text{ Magnetite}$$

Dissolved iron in seawater combines with oxygen to create insoluble hematite and magnetite.

The Cambrian explosion

An environment without sufficient free oxygen cannot provide enough chemical energy for complex animals. Once the ocean sediments and surface rocks reached saturation, oxygen began accumulating in the ocean and the atmosphere. Oxygen-fueled biochemistry sparked the furious expansion of life in the oceans of the Cambrian period starting 541 million years ago (Mya). The multicellular organisms which evolved during the Cambrian explosion included the progenitors of every family of life that exist today.

The appearance of land plants and animals

As oxygen continued to accumulate in the atmosphere, land plants and animals followed during the Devonian period (420 Mya - 359 Mya). Atmospheric oxygen reached a peak exceeding 30% during the Carboniferous period (359 Mya - 299 Mya) and the land was covered in lush jungles. The amount of decaying plant matter has never been equaled and the fossilized remains of carboniferous-era plants are the oil, gas, and coal of today.

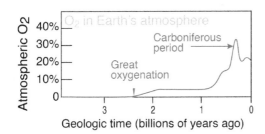

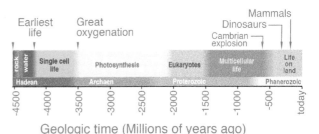

Water vapor and relative humidity

Atmospheric water

After temperature, water has the most impact on our interactions with the atmosphere. The obvious water in clouds and rain is only the most dramatic aspect of atmospheric water. The majority of water is in the form of invisible water vapor, the gas phase of H_2O. While only 0.001% of Earth's total water inventory is in the atmosphere (on average) this still amounts to 2×10^{15} kg, enough to cover the entire planet in a layer of liquid about 1" deep.

Saturation vapor pressure of H_2O

Think of the atmosphere as a gaseous solution of air and water. Like all solutes, water has a certain solubility in air. The graph below shows how the equilibrium partial pressure of water vapor in air varies with temperature. Note that at 25 °C the saturation vapor pressure is 3,200 Pa. This means air at 25°C can accept up to 3,200 Pa partial pressure of water vapor. At this temperature, any higher concentration of water vapor will partially condense into liquid until the vapor concentration is 3,200 Pa. If the air has a lower concentration (unsaturated) then liquid water can evaporate and become vapor.

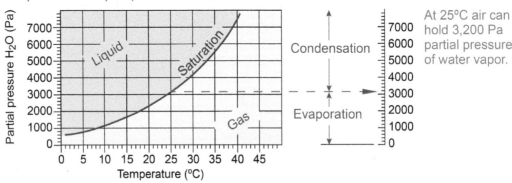

Relative humidity

Relative humidity (Rh) describes how much water vapor is in the air compared to how much water vapor the air can hold when it is completely saturated. Air at 25 °C saturates at 3,200 Pa partial pressure H_2O. A water concentration of 2,000 Pa has a relative humidity of 62.5% because 2,000 Pa is 62.5% of the saturation concentration of 3,200 Pa.

Relative humidity

The ratio of the actual partial pressure of water vapor divided by the saturation vapor pressure at that temperature, multiplied by 100.

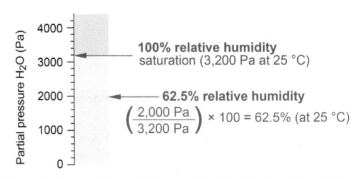

Solved problem

Calculate the relative humidity of air at 25 °C containing 2000 Pa partial pressure of water vapor.

- Asked: Relative humidity
- Given: Actual partial pressure of H_2O is 2,000 Pa at 25 °C
- Relationships: Rh = (actual partial pressure H_2O ÷ saturation partial pressure H_2O) × 100
- Solve: From the graph above at 25 °C the saturation partial pressure of H_2O in air is 3,200 Pa. The relative humidity (Rh) = (2,000 Pa ÷ 3,200 Pa) × 100 = 62.5%
- Answer: Relative humidity (Rh) = 62.5%

The dew point

Definition of the dew point

The saturation vapor pressure of water in air depends strongly on temperature. The **dew point** is the temperature at which the actual partial pressure of water vapor in the air equals the saturation partial pressure. On the liquid/gas equilibrum graph, the dew point is the saturation line (red). For convenience, the dew point is often shown on a psychrometric chart. The psychrometric chart shows the dew point as a function of temperature for different values of relative humidity.

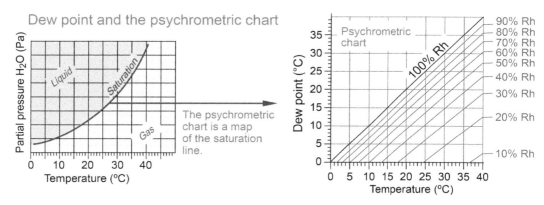

High humidity is uncomfortable

When the humidity is high, the dew point is close to the actual air temperature. That means little water can evaporate because the air is already near saturation. Humans cool by evaporation and when water can't evaporate, we experience a very uncomfortable "sticky" feeling. If the dew point is near (or above) your skin temperature, then water tends to condense rather than evaporate, and condensation gives off heat so you get hotter instead of cooler.

Condensation and the dew point

Consider air at 70% Rh and 25 °C (77 °F) on a nice day. Reading the psychrometric chart along the 70% Rh line shows that the dew point is about 18 °C. Now suppose the night temperature drops to 15 °C (59 °F). This is below the dew point! Some of the water vapor must condense into liquid. Near the ground the liquid forms droplets on every exposed surface, such as plant leaves.

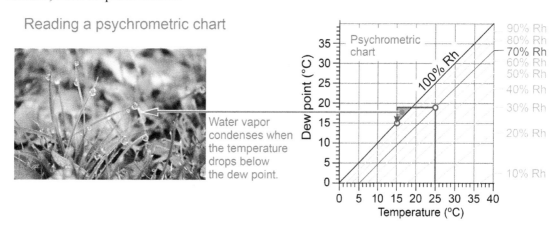

Condensation is a warming process

Condensation releases the enormous heat of vaporization of water (2,260 J/g). For comparison, cooling one gram of water from 25 °C to 15 °C releases 41.8 J of heat. Condensing one gram of vapor releases 54 times as much energy warming surfaces and the surrounding air. In humid tropical rain forests, the heat of vaporization creates an evaporative cooling effect during the day and a condensation warming effect at night. This cycling of energy helps maintain a consistent air temperature.

Weather

The energy that creates weather

The atmosphere interacts with enormous flows of energy between the Sun and Earth. The result is the complex, chaotic, and powerful interaction between wind and water that we call weather. Between wind and condensation, a single large hurricane may engage 200 times as much energy as the entire world's electrical output. Weather is driven by a number of different inter-related factors.

- Sunlight warms the ground, which warms the air near the surface. Rising warm air creates updrafts and down-drafts with falling colder air from the top of the troposphere.
- Evaporating water vapor in rising warm air carries energy from the heat of vaporization up to the upper atmosphere where it is released by condensation.
- The rotation of the Earth creates a coriolis force on the entire atmosphere which creates huge circulating patterns of air movement.

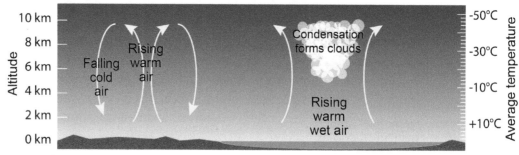

Cloud formation

Warm air absorbs water vapor over bodies of water near the ground. This raises the relative humidity and the dew point. As the warm, wet air rises, its temperature falls below the dew point and the air becomes supersaturated. Water vapor condenses out into tiny droplets to form clouds. Clouds are made up of liquid water droplets, not vapor (vapor is transparent, clouds are opaque.) The clouds often remain high because they are warmer than the surrounding atmosphere and therefore "float" on the cooler air below.

The coriolis effect and hurricanes

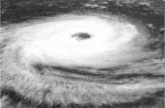
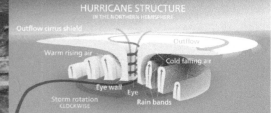

Circulation patterns caused by Earth's rotation | A hurricane | Rising warm air and falling cold air is shaped into a massive rotating storm.

Storms and hurricanes

Earth's rotation causes the atmosphere to circulate in huge gyres that spin clockwise in the northern hemisphere and counter-clockwise in the southern hemisphere. When the ocean surface temperature is warm enough, upwelling air over the oceans becomes saturated with water vapor. The energy transfer in the rising air and condensing water becomes large. The coriolis effect spins the pattern of moving air into a hurricane. The warmer the ocean surface, the more energy there is to feed and grow the hurricane.

Ozone

Absorption of UV

An important consequence of atmospheric oxygen is the absorption of ultraviolet radiation by the upper atmosphere. Short wavelength, high-energy photons from the Sun are damaging to living organisms because they can break chemical bonds. Fortunately, high energy photons are strongly absorbed by oxygen molecules which then undergo photo dissociation reactions that break diatomic oxygen into atomic oxygen. Atomic oxygen then combines with molecular oxygen to form ozone (O_3).

Ozone and the absorption of ultraviolet light

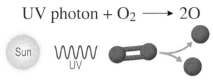
Absorption of UV photons breaks molecular oxygen into atomic oxygen

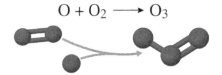
Chemical formation of ozone from atomic and molecular oxygen

O_2 and O_3

Ozone molecules also undergo **photodissociation** to absorb lower energy ultraviolet. Ozone in the stratosphere has a concentration of between 2 and 8 ppm. Together, the combination of ozone and molecular oxygen are an effective shield against ultraviolet radiation.

Ozone is toxic

While necessary for life as a UV radiation absorber in the stratosphere, ozone in the troposphere is a powerful oxidizer and is damaging to both animal and plant tissues at concentrations above 100 ppb. Ozone has a sharp smell and can be detected by most people at concentrations below 100 ppb, especially after lightning or other electrical discharges have occurred. Ground-level ozone forms from a secondary reaction among nitrogen oxides released from the burning of fossil fuels.

Chemical activity of ozone

Ozone is used in water treatment to remove dissolved manganese, iron, and hydrogen sulfide. Ozone is also used as a sterilizing agent for disinfecting water and air, and it has the benefit of decomposing to oxygen without leaving any chemical residue. In chemical manufacturing, ozone is used to break carbon-carbon bonds in the production of polymers.

CFCs and the "ozone hole"

Since the late 1970s, scientists have measured a 4% decline in the concentration of stratospheric ozone, and as much as 70% decline around the poles causing the "ozone holes." Ozone destruction is caused by photodissociation of chlorofluorocarbon compounds (CFCs) in the stratosphere. CFCs were produced by the millions of tons per year as refrigerants, most of which ultimately escaped to the atmosphere. Ozone depletion occurs because CFC molecules have an estimated 100-year lifespan in the stratosphere and each molecule can catalyze millions of ozone destroying reactions. In 1987, forty-seven nations signed the Montreal Protocol to limit the production of CFCs and by 2016 measurements confirmed that the ozone layer was slowly "healing."

Ozone depletion and chlorofluorocarbons

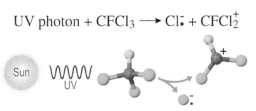
Photodissociation of chlorofluorocarbons

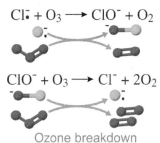
Ozone breakdown

Section 23.1: The Atmosphere

Sulfur and acid rain

Coal and the industrial revolution

The industrial revolution in the 1800's initiated large-scale human technology with the capacity to change the composition of the atmosphere. The energy source for the industrial revolution was coal. The world uses 4×10^{12} kg of coal per year to generate electricity.

Sulfur from coal

Coal contains 0.4% to 0.8% sulfur by weight. Burning coal releases up to 3×10^{10} kg per year of sulfur. Recent efforts to capture the sulfur at the power plant have made modern coal-burning plants cleaner. Since the 1970's clean-air regulation has reduced the sulfur content of gasoline; however, fuel oils used for ships are still high in sulfur. Moreover, the amount of sulfur released into the atmosphere still far exceeds natural sources.

Chemistry of atmospheric sulfur

In the atmosphere sulfur particles react with oxygen to form sulfur dioxide (SO_2) which then reacts with more oxygen to form sulfur trioxide (SO_3). The sulfur trioxide reacts with water in clouds to form sulfuric acid (H_2SO_4). The sulfuric acid precipitates out as rain. Natural (clean) rain has a pH of 5.6 due to a small concentration of carbonic acid from atmospheric CO_2. Acid rain downwind of coal-burning power plants can have a pH as low as 4.2. Since the pH scale is logarithmic, the difference of 1.4 pH units represents a factor of 25 times higher acid concentration!

Production of sulfuric acid in the atmosphere

Step 1
$$S + O_2 \longrightarrow SO_2$$
Sulfur combines with oxygen forming sulfur dioxide

Step 2
$$2SO_2 + O_2 \longrightarrow 2SO_3$$
Sulfur dioxide combines with oxygen forming sulfur trioxide

Step 3
$$SO_3 + H_2O \longrightarrow H_2SO_4$$
Sulfur trioxide combines with water forming sulfuric acid

Acid fog

In 1952, London suffered an acid fog disaster. Over a period of 5 days, Londoners burned more coal than usual to heat their homes during a period of cold weather. The foggy weather helped SO_2 form H_2SO_4 and the cold air helped to trap it. The people of London had no choice but to breathe air full of sulfuric acid. Many tried to minimize the impact by breathing through handkerchiefs. Young and old alike suffered severe respiratory damage. Thousands died of complications over several weeks and even more suffered long-term health effects.

Acid rain in the environment

Acid rain reacts with calcium carbonate in structures and artwork made of limestone and marble. Strong acids react with metals causing corrosion. These acids are also responsible for damaging ecosystems in the northeast and Great Lakes regions of the U.S. Fish and plants disappeared from hundreds of forests, lakes and ponds due to acid rain. The 1990 amendment to the Clean Air Act in the U.S. established limits on sulfur emissions from coal and other fossil fuels to reduce the risk of further acidification. Many other countries burn much more coal than the US without regulatory protection.

Many species of trees have been damaged by acid rain.

Ocean acidification

One of the most potentially damaging effects of acid rain is that it is slowly lowering the pH of the ocean. Sulfuric acid destroys the calcium carbonate in the exoskeletons of plankton, diatoms, and other life at the foundation of the ocean food chain. Even a small change in the acidity of the ocean could have an enormous effect on the marine ecosystem.

Smog and air pollution

Smog

The word "smog" first appeared in the Los Angeles Times in 1893 in which the writer described an unhealthy atmosphere of smoky fog that had become a plague in cities but was absent from rural settings. Smog is a catch-all term for visible air pollution composed of nitrogen oxides, sulfur oxides, ozone, smoke or particulates among others. Smog is caused by coal emissions, vehicular emissions, industrial emissions, forest and agricultural fires and photochemical reactions which turn these emissions into other compounds.

Smog is air pollution caused by industrial and vehicle emissions

London 1880

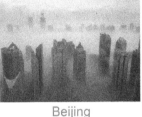
Beijing

Los Angeles

Particulates
Nitric oxide (NO)
Nitrous oxide (NO_2)
Carbon monoxide (CO)
Unburned hydrocarbons
Ammonia
Ozone
Volatile organics

The chemistry of smog

In an ideal combustion reaction hydrocarbons such as octane (C_8H_{18}) react with oxygen to produce carbon dioxide and water. However, the high heat and pressure of the combustion reaction, along with the fact that air is 78% nitrogen, allow many additional branch reactions to occur. Even in pure octane some of the oxygen reacts to produce carbon monoxide (CO) and formaldehyde (CH_2O). A fraction of the oxygen also reacts with nitrogen to form nitric oxide (NO). The nitric oxide is exhausted where it quickly reacts with oxygen and water vapor in the outside air to form nitrous oxide (NO_2) and nitric acid (HNO_3). Carbon monoxide, nitrous oxide, aldehydes, and nitric acid are major components of smog.

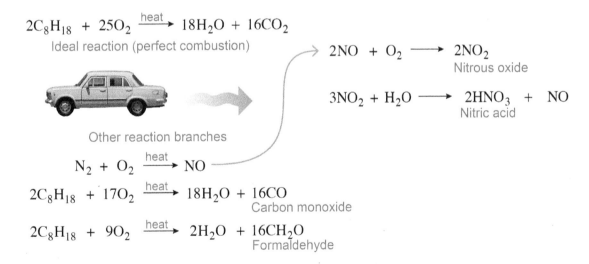

Production of nitrous oxide, nitric acid, carbon monoxide, and formaldehyde

Lead, sulfur and other impurities

Lead poisoning damages the developing brain in children and can be fatal in adults. In 1922 the compound tetraethyl lead was introduced into gasoline as a high-temperature lubricant for higher performance engines. In 1965 geochemist Clair Patterson found that lead levels in the environment were 100 times higher than natural. President Nixon signed the Clean Air act of 1970 which eliminated lead in gasoline. The Environmental Protection Agency (EPA) further reduced lead emissions in 2008.

Carbon dioxide and the greenhouse effect

Earth's energy balance

Earth's temperature is a balance between radiant energy absorbed from the Sun and energy radiated back into space. The top of the atmosphere gets hit with 1,350 W/m² of sunlight. About 25% is reflected from the atmosphere and another 5% is reflected from the surface. Seventy percent of the solar energy is absorbed by the atmosphere (25%) and the ground (45%). To maintain thermal equilibrium, all the energy absorbed from the Sun must be radiated back into space.

Thermal radiation from Sun and Earth

The atmosphere is mostly transparent to solar radiation because the Sun radiates at a temperature of 5,800 K. The Earth however, is much cooler and radiates at an average temperature of only 300 K. The energy Earth radiates into space is in the infrared. Carbon dioxide is a strong absorber of infrared. Changes in atmospheric CO_2 concentration affect the energy Earth radiates back into space.

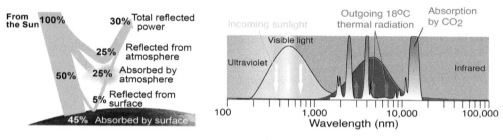

The greenhouse effect

The **greenhouse gases** are so named because the glass walls of a greenhouse pass sunlight but reflect infrared back, keeping the inside warmer than the outside. Water vapor, methane, and carbon dioxide are called greenhouse gases because they function like the greenhouse window, passing sunlight but trapping infrared. This phenomenon causes the planet to be warmer than expected, and it is known as the **greenhouse effect**. The graphs show ppm of atmospheric CO_2 average temperatures for the past 400,000 years.

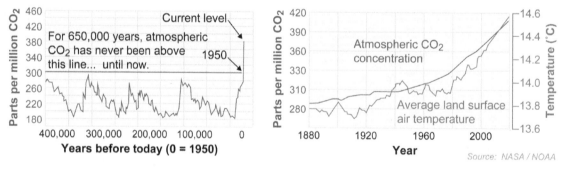

The cause of climate change

Climate change is not caused by the heat released by human activity. In one year Earth receives 5.3×10^{24} joules from the Sun, which is 10,000 times more energy that is used by all human activity on the planet. Even a 0.01% change in the absorption rate of solar energy has an enormous effect because it acts on the total energy received by the Sun.

Interactive simulation

Discover how greenhouse gases work in the interactive simulation titled Sunlight, Infrared, CO_2 and the Ground. Pay close attention to how energy moves through CO_2 and the ground. How quickly does temperature rise? When and how is heat re-radiated to Earth?

Section 1 review

Chapter 23

Typically we divide Earth's atmosphere into four progressively thinner and more distant layers: the troposphere, stratosphere, mesosphere, and thermosphere. On average most gas molecules in the atmosphere stay close to the surface of Earth due to gravity. The atmosphere is composed mainly of nitrogen and oxygen with smaller amounts of other gases. Pollutants can react with oxygen gas or water vapor with undesired results. High-energy photons from UV light can break bonds. Such a reaction leads to ozone, which protects living things from UV damage. Greenhouse gases re-radiate energy from the sun allowing Earth to be warm enough to support life. An overabundance of greenhouse gases is leading to increasing global temperatures.

| Vocabulary words | troposphere, stratosphere, ozone layer, mesosphere, thermosphere, dew point, photodissociation, greenhouse gases, greenhouse effect |

Review problems and questions

1. List the 4 layers of our atmosphere in order from space to the ground.

2. When is ozone, O_3, beneficial to life, and when is it harmful to life?

3. Identify two chemicals that can worsen acid rain when released to the atmosphere. Where do these chemicals mostly come from?

4. What does a greenhouse gas such as CO_2 do in order to classify it as a greenhouse gas?

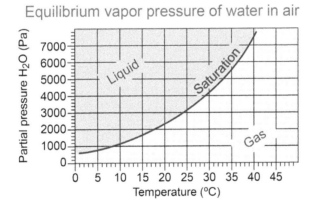

5. It's a 20 °C day and the partial pressure of water in air equals 1,920 Pa. What is the relative humidity?

Section 23.1: The Atmosphere

23.2 - The Hydrosphere

The **hydrosphere** includes all water on Earth, including rivers, oceans, the polar ice caps, icebergs, groundwater, and water in the atmosphere. If Earth were a uniform smooth sphere, the water in the hydrosphere would cover the planet in a layer 2.7 kilometers deep. Although it makes up only 0.02% of Earth's total mass, the hydrosphere is critical to life and has shaped all aspects of Earth's surface.

Distribution of water in the hydrosphere

Composition of the hydrosphere

Approximately 75% of the Earth's surface is covered by water, of which the vast majority, 97.5%, is salt water in the oceans. Only 2.5% of Earth's hydrosphere is fresh water of which two thirds is frozen in the polar ice caps and glaciers. The remaining third of the freshwater is in aquifers and other ground water. Only 0.3% of fresh water, 0.0075% of the hydrosphere, is contained in lakes and rivers.

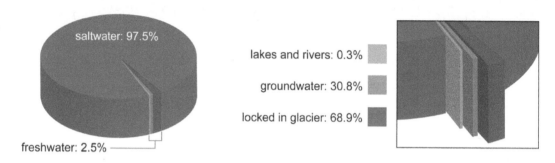

Freshwater distribution

Because of terrain and wind patterns the distribution of fresh water is very uneven around the Earth. Vast areas have little or no fresh water and are arid deserts. Other areas, such as the rainforests in equatorial Brazil have extreme rainfall, almost 7 meters per year. The figure below shows the yearly average flow of fresh water derived from computer models based on satellite observations.

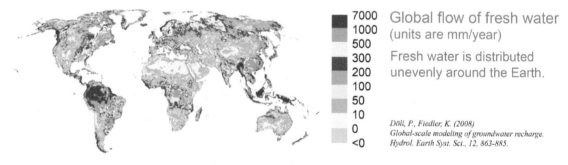

Döll, P., Fiedler, K. (2008) Global-scale modeling of groundwater recharge. Hydrol. Earth Syst. Sci., 12, 863-885.

Climate zones

Water and temperature are the primary factors that determine climate. Arid desert climates form where water is scarce. Rainforests exist where water is plentiful and the temperature is warm. Notice how little of Earth's surface has a temperate climate compared to the area which is dry desert and steppe or the opposite extreme of cold and polar.

The effect of ocean currents on climate

Ocean currents move heat

Planet-circling convection currents of ocean water transport heat from the equator to the poles. These currents have a tremendous effect on climate. The British Isles have a temperate climate even though geographically, Britain has the same latitude as Siberia (polar climate). The difference is that the Gulf Stream carries warm water into the North Atlantic Current which flows around Britain and up the western Scandinavian peninsula.

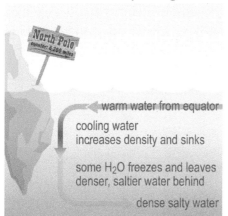
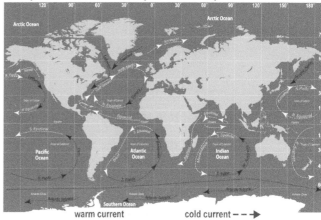

ocean-spanning currents transport heat from the equator to the poles

Cold water sinks at the poles

What enormous force can cause movement across an entire ocean? The short answer is: density! Density differences cause ocean water to rise or sink. Less dense water rises and more dense water sinks. Ocean water density depends on both temperature and salinity. These factors sometimes act in opposition, so small differences can have great effects. At the poles, cold water chilled by the ice caps is denser than warmer surface water and sinks. A subsurface river of cold dense water flows along the ocean bottom toward the equator.

Warm water rises at the equator

Warm water is less dense than cold water and the differences are greatest at the equator and the poles. The large ocean currents are the result of cold water flowing along the ocean floor from the poles pushing up warm water near the equator which then flows back along the surface toward the poles. These currents move slowly but move vast quantities of ocean water every year. Water has a high specific heat and the ocean currents transport an equally vast amount of thermal energy from the equator towards the poles.

Ocean currents circulate nutrients

The ocean currents stir up minerals and nutrients that have settled to the ocean floor. Algae and plankton that form the basis of Earth's food web thrive on the upwelling of nutrients. Migratory fish and birds follow the currents every year to take advantage of the rich feeding grounds the ocean currents create.

A warming planet will disrupt ocean currents

The ocean's surface temperature is critical to weather because surface water evaporates and eventually precipitates. Scientists are now concerned that the ocean currents are shifting, with effects on climate. As polar water becomes slightly warmer its density decreases. Also, glacial ice is melting faster than it can be replenished, adding low-density fresh water to polar ocean water. The combination of fresh water and high temperature reduces the density differences that drive the ocean currents. Any change in these currents affects the ocean's process of redistributing nutrients and the sun's energy. Many regions will experience greater weather extremes and the ocean food web could be disrupted.

Chemical composition of the oceans

Ocean water is 3.5% dissolved salt ions

Ocean water is known for its **salinity**, or saltiness. The average salinity of sea water is about 3.5% due mainly to dissolved sodium chloride. The table below shows that chloride (Cl^-) is the most abundant ionic species followed by the sodium ion (Na^+). Sulfate, magnesium, calcium, and potassium are the next most common solutes.

Saltwater is unhealthy to drink

Drinking seawater can actually cause dehydration! Your kidneys process salt water by drawing water from your body to reduce the concentration to 2% salinity. Sea water kills freshwater and land plants for a similar reason. Plants osmotically move water to dilute excess salinity. Ocean plants and animals have special adaptations that help them survive in highly saline water.

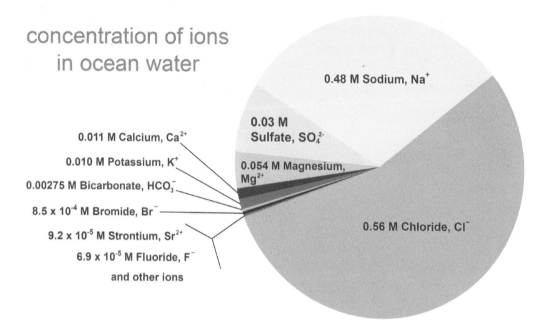

concentration of ions in ocean water

- 0.48 M Sodium, Na^+
- 0.03 M Sulfate, SO_4^{2-}
- 0.054 M Magnesium, Mg^{2+}
- 0.56 M Chloride, Cl^-
- 0.011 M Calcium, Ca^{2+}
- 0.010 M Potassium, K^+
- 0.00275 M Bicarbonate, HCO_3^-
- 8.5×10^{-4} M Bromide, Br^-
- 9.2×10^{-5} M Strontium, Sr^{2+}
- 6.9×10^{-5} M Fluoride, F^-
- and other ions

Ocean water salinity is fairly stable

Ocean water salinity remains fairly constant around the world, falling in a range between 3.1% and 3.8%. The concentration is maintained because all oceans mix and are connected through global ocean currents. Differences in salinity are due to different amounts of fresh water added from land or glaciers. Temperature also impacts salinity. In general, warm areas and tide pools experience more evaporation and higher salinity while cooler areas experience less evaporation and lower salinity, except for icy areas.

Where do dissolved substances come from?

How do minerals, ions and other dissolved substances find their way into ocean water? Many substances such as carbonates, calcium, and magnesium come from erosion and runoff from soil or from ocean sediments. Oxygen and carbon dioxide are absorbed from the air. Volcanic activity introduces some minerals such as sulfates from deep within Earth. Many minerals, such as calcium, are recycled in living organisms including diatoms, snails, and shelled creatures.

The water cycle

Where does fresh water on land come from?

Most water comes from the ocean, but it cycles from ocean to air, from air to land, and from land to ocean. When the ocean is heated by the sun, temperature differences create density differences and phase changes that drive weather and the **water cycle**. The diagram below shows paths water can take as it cycles through the ocean, air, and land.

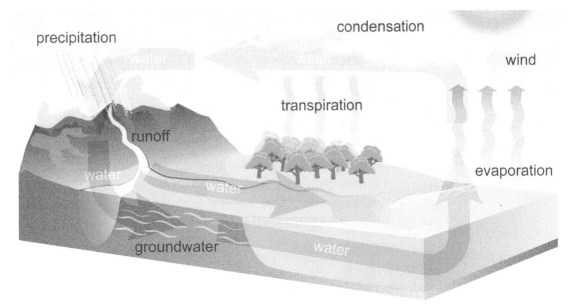

Evaporation through transpiration

As seen above, water evaporates from plants in a process known as **transpiration**. Plants have mouth-like openings on the bottom of their leaves called **stomata**. Stomata allow gas exchange for photosynthesis. When stomata are open, CO_2 diffuses in and O_2 diffuses out. Water vapor is lost as transpiration when stomata open.

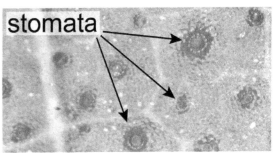

Temperature and the water cycle

The water cycle and weather are driven by the sun's energy. During the day, the sun warms the ocean and air. Solar heat causes water to evaporate from the ocean. Water vapor is taken in by the surrounding warm air. Warm air is less dense so it has space to hold water molecules. Warm, less dense air rises to a point in the atmosphere where it begins to cool. As air cools, particles come closer together. The cool air sinks as density increases. With less space available, water vapor is forced to condense into small droplets that form clouds and precipitation.

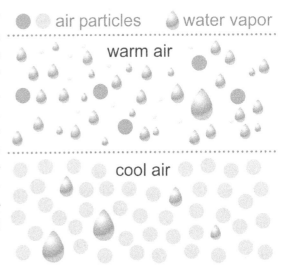

Physical changes are dominant

Except for the water molecules formed as a by-product of photosynthesis, water molecules that move through the water cycle are not chemically changed. Almost all of the parts of the water cycle involve physical instead of chemical changes.

The carbon cycle

Carbon atoms are constantly recycled

Carbon is the fourth most abundant element in the universe. The chemistry of all living organisms is based on carbon. Carbon is present in different forms throughout the atmosphere, hydrosphere, and geosphere as well as the biosphere, which includes all living things. The **carbon cycle** is a millenia-long grand circulation of carbon atoms between the oceans, land, atmosphere and living organisms. During the cycle, carbon compounds undergo many chemical changes in both living and nonliving contexts. A carbon atom that is part of a CO_2 molecule in the atmosphere is taken in by a plant and metabolized into glucose, $C_6H_{12}O_6$. It is eaten by an animal and excreted as bicarbonate, HCO_3^-, and makes its way in runoff water into the ocean where it is incorporated into the calcium carbonate shell of a clam.

Carbon recycles slowly through rocks

The four largest repositories of carbon are fossil fuels, plants, water, and sediments. Most of Earth's carbon is stored in sedimentary rocks. Shells and skeletons of marine life contain calcium carbonate, $CaCO_3$. When marine organisms die, their carbon cycles through living things or it falls to the ocean floor, locked up for ages in sedimentary rocks. The cycling of carbon takes millions of years when it is moving through sedimentary rocks that change through geologic processes such as plate tectonics or volcanism.

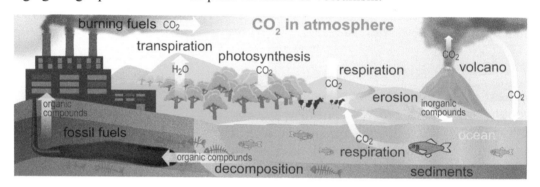

A few of the many ways carbon is recycled

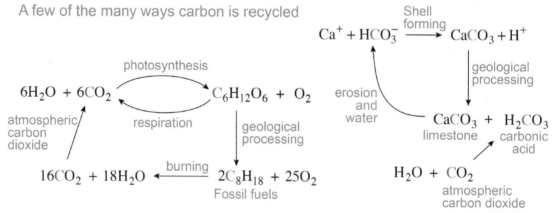

Carbon cycles through organisms

Ocean surface waters contain the largest amount of actively cycling carbon—carbon that is moving quickly through living things and the atmosphere. The cycle of photosynthesis and respiration is a good example of biological recycling of carbon. It can take from days to years for carbon to cycle through biological systems.

Fossil fuel use disrupts the cycle

Earth has had millennia to balance the carbon cycle. The amount of carbon dioxide released by living things and geologic processes is roughly equal to the amount absorbed. However, since the burning of fossil fuels became widespread, carbon is entering the atmosphere as CO_2 faster than it can be absorbed by any part of the carbon cycle.

Acid-base reactions in the ocean

In role of acidity in water

The pH of water affects both living and non-living systems.

1. Mollusks, crustaceans, and plankton slowly accrete calcium and carbonate ions in seawater to form their exoskeleton or shell. Small increases in acidity mean higher H^+ concentration which binds to carbonate (CO_3^{2-}) to make bicarbonate (HCO_3^-), decreasing the biologically available carbonate.
2. Increased acid concentration in fresh water greatly determines the rate of weathering of rocks, especially limestone (calcium carbonate).
3. Most living creatures can tolerate only a narrow range of internal blood pH. Aquatic organisms are sensitive to changes in water pH.

Ocean life that build shells or exoskeletons from carbonate

Shell-forming animals including clams, scallops, and snails

Plankton which are the base of the ocean food chain

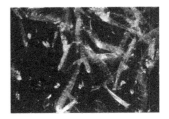
Crustaceans including these krill, lobsters, shrimp, and crabs

Ocean water is slightly basic

The pH of ocean water is determined by two factors: the concentration of carbonate ions from weathering and runoff, and the absorption of atmospheric carbon dioxide gas. Ocean water is slightly basic (pH ~ 8.1) due to dissolved carbonate and bicarbonate ions. About 88% of inorganic carbon in the ocean is in the form of the bicarbonate ion.

The CO_2 and carbonic acid equilibrium

Atmospheric carbon dioxide exists in equilibrium with dissolved CO_2 in water. As the atmospheric CO_2 concentration increases, the amount of dissolved CO_2 in the water also increases. Water reacts with dissolved CO_2 to form carbonic acid (H_2CO_3). Once formed the carbonic acid partially dissociates into bicarbonate ion (HCO_3^-) and hydrogen ion (H^+). The reactions below show the process along with the relevant equilibrium constants.

Acid equilibria in ocean water

Atmospheric carbon dioxide dissolves in water
$$CO_2\ (gas) \underset{K_H}{\rightleftharpoons} CO_2\ (aq) \qquad K_H = \frac{[CO_2\ (aq)]}{[CO_2\ (gas)]} = 29.8\ atm/mol/L$$

Carbonic acid forms in equilibrium with dissolved CO_2
$$CO_2\ (aq) + H_2O \underset{K_h}{\rightleftharpoons} H_2CO_3\ (aq) \qquad K_h = \frac{[H_2CO_3]}{[CO_2\ (aq)]} = 0.0012$$

Carbonic acid dissociates to H^+ and HCO_3^- ions
$$H_2CO_3\ (aq) \underset{K_{a1}}{\rightleftharpoons} HCO_3^-\ (aq) + H^+(aq) \qquad K_{a1} = \frac{[H^+][HCO_3^-]}{[H_2CO_3]} = 2.5 \times 10^{-4}$$

Bicarbonate ions dissociate to H^+ and CO_3^{2-} ions
$$HCO_3^-(aq) \underset{K_{a2}}{\rightleftharpoons} CO_3^{2-}\ (aq) + H^+(aq) \qquad K_{a2} = \frac{[H^+][CO_3^{2-}]}{[HCO_3^-]} = 4.7 \times 10^{-11}$$

The ocean buffer system

Because there is an excess of bicarbonate and carbonate ions, ocean water is a weak buffer. Carbonic acid is a weak, diprotic acid and the second dissociation has a very low equilibrium constant ($K_a = 4.7 \times 10^{-11}$). Excess acidity can be absorbed by converting carbonate (CO_3^{2-}) to bicarbonate (HCO_3^-).

Ocean acidification

Atmospheric CO_2 is linked to ocean pH

When the global increase in atmospheric CO_2 was first discovered in the late 1960's it was thought that the vast carbon "sink" of the oceans would help absorb excess CO_2 from the atmosphere with little ill effect. The buffering effect of the bicarbonate/carbonate system does act to remove excess H^+ ions. However, the natural concentration of carbonate ions is very low making seawater only a weak buffer. By the turn of the century scientists realized that the additional carbonic acid in seawater from rising atmospheric CO_2 was beyond the ability of the oceanic buffer system to absorb. The pH of the ocean started to fall.

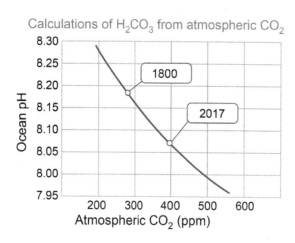
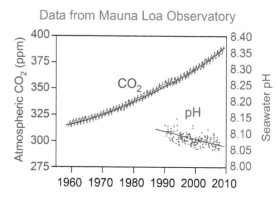

Ocean acidification

Over the past two centuries, the ocean has become more acidic with the measured pH of the ocean surface dropping about 0.1 units from 8.17 to 8.07. The graph on the left shows the predictions of the equilibrium model on the previous page. This model shows that the pH of water in equilibrium with air depends solely on the concentration of CO_2 in the air. As the atmospheric CO_2 concentration increases, the amount of disssolved CO_2 in the water increases, and the concentration of H^+ ions increases, decreasing the pH. This phenomenon is known as **ocean acidification**. Ocean acidification is sometimes referred to as the "evil twin of global warming" and "the other CO_2 problem."

Observations agree with models

Scientists estimate that 30–40% of the carbon dioxide from human activity released into the atmosphere eventually dissolves into oceans, rivers and lakes. The graph on the right above shows measurements of atmospheric CO_2 and the pH of the ocean surface near the Mauna Loa Observatory in Hawaii. The observed decrease in pH agrees very well with the calculations of the equilibrium model for the CO_2/H_2CO_3 system.

Impacts of acidification

Acidification threatens the plankton which form the basis of the ocean food chain. Plankton and other living organisms produce shells and exoskeletons from calcium carbonate ($CaCO_3$) through a process called calcification. In the calcification process, dissolved Ca^{2+} and CO_3^{2-} ions are precipitated into solid $CaCO_3$. Every molecule of carbonic acid added to the ocean converts two carbonate ions into bicarbonate ions, which cannot be used in calcification. Delicate, newly-formed calcium carbonate structures are vulnerable to dissolution unless the surrounding seawater has a saturated concentration of carbonate ions (CO_3^{2-}).

Section 2 review

Chapter 23

Water moves between the oceans, atmosphere, and land in a process known as the water cycle. Clouds are formed through evaporation of ocean water, transferred to the land by rain and snow and returned to the ocean through runoff and groundwater discharge. Plants return water to the atmosphere through transpiration. The density of water is dependent on temperature and salinity. The motion caused by more dense water sinking creates ocean currents. Ocean currents help the ocean maintain a relatively consistent level of salt ions or salinity. Similar to the water cycle, the carbon cycle shows the circular path that carbon takes in moving between the land, atmosphere, living things, and hydrosphere. The ocean relies on a bicarbonate buffer system to maintain a consistent pH. The use of fossil fuels is disturbing the balance of the carbon cycle, and excess carbon dioxide in the ocean is disturbing its pH balance.

Vocabulary words   hydrosphere, salinity, water cycle, transpiration, stomata, carbon cycle, ocean acidification

Review problems and questions

1. Explain why ocean current movements depend on salinity and temperature differences.

2. How are human activity and the carbon cycle related to ocean acidification?

3. Write a 3-step scenario that a carbon atom might follow on its way through the carbon cycle.

4. Write a 4-step scenario that a water molecule might follow in its way through the water cycle. Use the following terms in your answer: evaporation; condensation; liquid; gas.

5. The red lines numbered 1-6 in the picture are meant to be arrows that show the direction CO_2 moves through each component of the carbon cycle. For example, line #1 should be an arrow pointing up because during plant respiration, CO_2 is released from the plant to the atmosphere. Answer either "up" or "down" to indicate which direction CO_2 moves relative to the other 5 lines shown on the figure. Explain each answer.

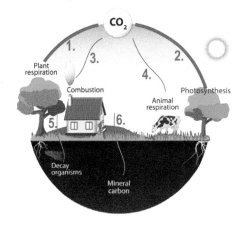

23.3 - The Geosphere

The **geosphere** includes all of the non-living matter in the planet that is not water or atmosphere. Mountains, rocks, and sand are familiar components on the surface of the geosphere. Magma and the planet's nickel-iron core are also part of the geosphere. The rich chemistry of the geosphere is driven by erosion, temperature and pressure over geological timescales lasting millions of years.

The structure of planet Earth

Age of the Earth
Our solar system formed 4.6 billion years ago from a condensing cloud of interstellar gas and dust called a **solar nebula**. From the composition of the solar system we know that the original solar nebula was mostly hydrogen and helium with trace amounts of heavier elements such as carbon, oxygen, and iron. The Sun and the outer gas-giant planets are still predominantly hydrogen and helium.

The crust
Earth's crust is the thin, solid, outer layer of the planet approximately 7 - 35 km thick. The most abundant eight elements, O, Si, Al, H, Na, Ca, Fe, and Mg, make up 98.5% of the matter in Earth's crust, with trace amounts of most other elements. The most common rocks of the crust are made of different combinations of **silicates** which are compounds of silicon and oxygen. Traces of other elements give silicates great variety in properties: granite and jade are both silicates.

The mantle
Below the crust is the **mantle** which has an average depth of 2,900 km and includes 84% of the planet's volume. Similar to the crust the mantle is primarily silicate rock but at a much higher temperature and pressure than the crust. From 500 °C near the crust the temperature in the mantle increases to near 4,000 °C at the top of the core.

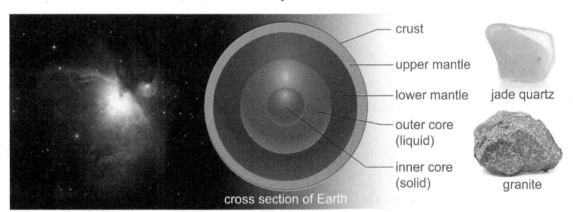

The core
The center of the planet is a nickel-iron **core** differentiated by pressure and temperature into a solid inner core and a liquid outer core. From the average density of Earth (5.5 g/cm^3) and the known density of silicate rock (2.2 g/cm^3) scientists deduced that much of the planet's heavier metals (iron, nickel) are concentrated in the core. Seismic waves pass differently through liquid and solid, and seismic data gives strong evidence that the inner core is solid and the outer core is liquid.

Pressure and temperature
Both pressure and temperature increase with depth. Both have a large effect on the chemistry of compounds. High pressure deep underground gives an energy advantage to minerals which have tighter-packed crystal structure (Le Châtelier's principle). At the surface the pressure is one atmosphere (10^5 Pa) and the average temperature is 18 °C. At the core the pressure exceeds three million atmospheres (3 x 10^{11} Pa) and the temperature is over 5,000 °C, only a little lower than the surface of the sun.

Formation of the earth

Formation by accretion

Scientists currently believe Earth formed through accretion, gravitationally attracting all matter in its orbital region until the planet was close to its final size. This hypothesis is supported by the fact that the elemental composition of Earth is almost identical to chondritic meteorites leftover from the solar nebula. The Earth's accretion phase meant continuous bombardment of objects larger than mountains and the resulting impact energy kept Earth molten for millions of years. Earth's hydrogen, helium, water, and other gases were boiled away early in the planet's history.

Density stratification

The layered structure of the planet is evidence that Earth was molten long enough for most of the iron, nickel, and dense metals to sink to the center. Lighter silicates rose to the top and formed the mantle and crust. The original composition of the crust and upper mantle are very similar except that the crust has chemically evolved through constant interaction with the atmosphere and living organisms.

disk of gas and dust spinning around a young star

dust grains → dust grains clump into planetesimals → planetesimals collide and collect into planets

A cataclysm formed the Moon

The Moon has the same chemical composition as the mantle, and an average density of only 3.3 g/cm³ compared with 5.5 g/cm³ for Earth. Astronomers believe that the Moon formed when a Mars-sized object crashed into the young Earth. The debris from the collision collected into orbit around the Earth and became the Moon. The Moon's low density comes from the fact that it was mostly the Earth's upper mantle and crust that was thrown up into space, with very little of the planet's metal core.

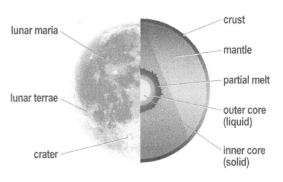

Water and atmosphere

As matter was accreted, Earth's orbit around the Sun was swept of large objects and impacts capable of melting the planet became less frequent. Earth cooled and the crust hardened. Over the next few hundred million years icy comets from the outer reaches of the Solar System were pulled Sun-ward by the gravitational tugs from the outer planets. Comets collided with the young Earth frequently enough that they are thought to have supplied most of Earth's water and atmosphere.

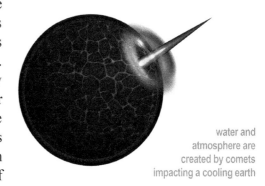

water and atmosphere are created by comets impacting a cooling earth

Elemental composition of the earth

Composition of the planets and Earth

The elemental composition of the solar system is dominated by the Sun which contains 99.99% of the mass of the system and is 75% hydrogen and 24% helium. Jupiter contains 90% of the rest of the system mass and has a similar composition to the Sun. The inner planets: Mercury, Venus, Earth, and Mars lost their hydrogen and helium in the cataclysmic heat of their early formation and rocky compositions of elements with higher melting points. The elemental composition of the planet Earth is shown below along with the composition of the crust, which is markedly different from the whole planet.

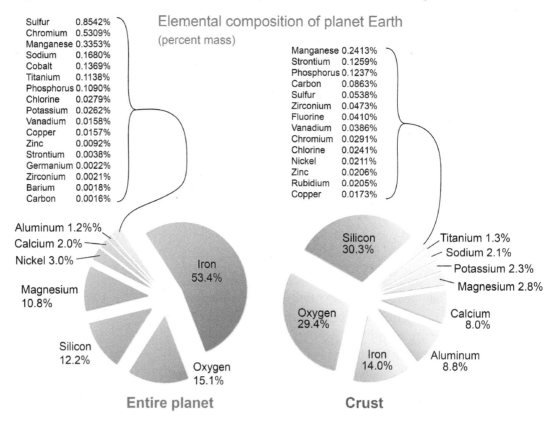

The core is Ni-Fe alloy

By mass, the Earth is more than half iron. The majority of Earth's iron however, sank to the core along with much of the other dense metals during the planet's molten phase. Scientists believe the core is a dense nickel-iron alloy that is 95% Fe and 5% Ni with 0.2% cobalt and other trace elements. We have no direct access to the core and these estimates are based on density, temperature, and pressure models as well as the known composition of meteors that are similar to the material which forms the early earth.

The crust is mainly silicates

The thin crust is mainly silicon and oxygen in mineral derivatives of silicates (SiO_2).

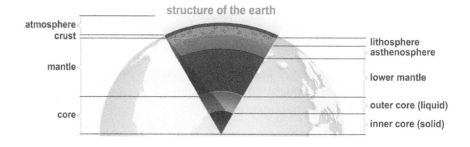

Minerals

Minerals

A **mineral** is a naturally occurring inorganic solid that has a definite chemical composition and a crystalline internal structure. A good example of definite chemical composition is quartz which has the formula, SiO_2, and is one of the most common forming minerals on Earth. The second part of the definition is peculiar to solids. Minerals are **crystalline** solids meaning the atoms have a regular and repeating pattern. The opposite of crystalline is **amorphous**, which describes a random arrangement without regular patterns. Glass is also made mostly of SiO_2 but is amorphous and therefore not technically a mineral.

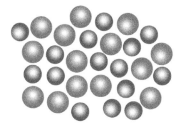
Amorphous solid

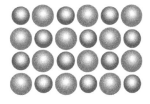
Crystalline solid

Large scale crystal

The 14 different crystal types

In 1850, French physicist Auguste Bravais proved that there were fourteen fundamentally different ways to arrange points in a three-dimensional repeating pattern. The Bravais lattices describe the fourteen different ways atoms and ions arrange themselves into crystals. The diagram below shows the fourteen crystal types. Each mineral species has both a different chemical formula and a different crystal structure. The same chemical formula in a different crystal structure is a different mineral!

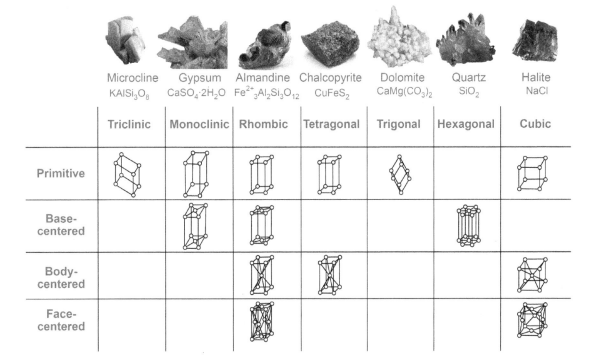

Lattice families

The Bravais lattices group into seven families. The cubic family has three variations with the simple cube, a cube with an atom in the center (body centered) and a cube with an atom at the center of each face (face-centered). All minerals belong to one of the Bravais lattice families.

Section 23.3: The Geosphere

The rock cycle

Types of rocks

The **rock cycle** is the slow recycling of Earth's crust. **Igneous rock** is worn by erosion and combined with other inorganic materials into sedimentary rock. **Sedimentary rock** is transformed by heat and pressure into **metamorphic rock**. Both sedimentary and metamorphic rock are melted and cooled into igneous rock. All three kinds of rock are exposed to the surface by plate tectonics where weathering and erosion starts the cycle again. Chemical changes occur through every stage of the rock cycle. A few dozen elements are geologically processed into more than 4,500 known mineral species which are then combined into an innumerable variety of rocks.

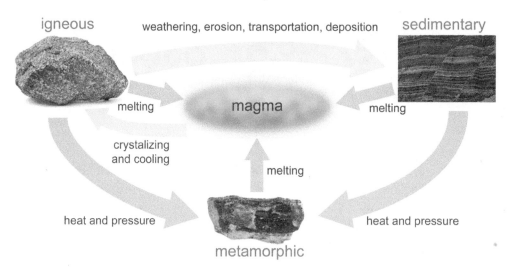

Polymorphism and equilibrium

Polymorphism describes the property of a mixture of elements to form different minerals under different conditions of temperature and pressure. Similar to solid, liquid and gas phases, the different minerals are different phases of the system which are favored under different conditions of pressure and temperature. The diagram shows the phase equilibrium for Al_2SiO_5 into kyanite, sillimanite, and andalusite. At a constant temperature of 700 °C and pressure of 4 kbar the system will form sillimanite. If slowly cooled to 300 °C at the same pressure, the mixture will change to kyanite.

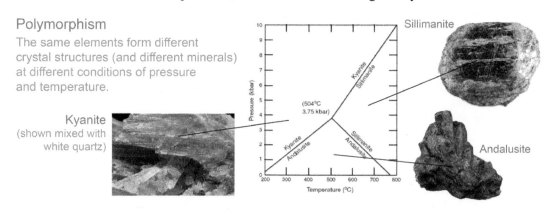

Polymorphism
The same elements form different crystal structures (and different minerals) at different conditions of pressure and temperature.

Geology

Geologists are scientists who study the Earth and the processes that change it. Geology has many applications that include, identifying areas where natural resources can be found, such as oil, coal or precious minerals. Geologists also learn about natural disasters to help us avoid building in dangerous areas. In many ways the information geologists gather is used to protect and improve our environment. By studying different types of rock, geologists discovered how matter and processes on Earth are interconnected.

Rocks

Rocks are mixtures of minerals

A **rock** is technically a solid heterogeneous mixture of different minerals. Unlike minerals, rocks have a variable chemical composition. A good example of a common rock is granite. Granite is a mixture of three or four minerals, typically quartz, feldspar and mica. If you look closely at course-grained granite you can see the different mineral crystals.

Granite (a rock) is a heterogeneous mixture of minerals

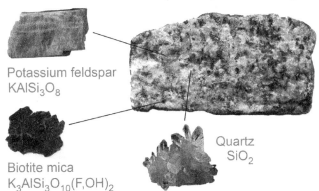

Potassium feldspar $KAlSi_3O_8$

Quartz SiO_2

Biotite mica $K_3AlSi_3O_{10}(F,OH)_2$

"Half Dome" in Yosemite National Park is a granite rock formation

The rock cycle

Very little of the rock on Earth's surface today is the same as it was when Earth was formed. Our planet has an active geology that erodes rocks into sediments, mixes and cements them into new rocks, reacts them with oxygen, salt water and acids, and then over millions of years melts, heats, and squeezes these into newer rocks.

Igneous rocks

The granite in the example is an igneous rock. Igneous rocks are formed from molten magma that has come to the surface and cooled. **Magma** is molten rock. If the magma cools slowly, relatively large crystals of the constituent minerals have time to grow out of the melt. The large crystals in the example indicate a slow-cooled magma. If the rock cools quickly the crystals are much smaller and the rock looks more uniform.

Sedimentary rocks

Earth is ancient. Over four billion years rocks on the surface are slowly worn away into sand and silt by the slow but relentless action of wind and water. There are still some places on Earth where you can find rocks as much as 2 billion years old, such as at the bottom of the Grand Canyon in Arizona. Much of the evidence of Earth's past is broken down by chemical and physical weathering. The weathered Earth is then eroded. Sand and silt eventually is carried into rivers, lakes, and the ocean where it settles to the bottom. There it mixes with the cast off calcium carbonate shells of trillions of sea creatures. Over millions of years, and pressure from layers above, the buried sediments fuse into sedimentary rock.

Fossils

Chemically, sedimentary rock is a mixture of everything, including fossilized bones. Fossils are only found in sedimentary rock. Earth's surface is in constant slow, regenerative motion. Over a hundred million years some sedimentary rocks are forced deep underground where extreme heat and pressure cause chemical changes. Fossils are destroyed under these extreme conditions.

Metamorphic rocks

If sedimentary rock forced underground becomes completely melted into magma, it becomes igneous rock when it returns to the surface. Even if the sedimentary rock does not completely melt, heat and/or pressure over time rearrange the elements into new minerals. For example, sedimentary limestone becomes metamorphic marble.

Section 23.3: The Geosphere

Plate tectonics and volcanism

Plate tectonics

Earth's crust is solid but the mantle beneath is partly molten. On a geological time scale heat from the core rises and drives huge, slow movements of the mantle. At the top of the mantle, the crust floats on a thin layer of semi-liquid molten rock called the asthenosphere. Compared to the size of the planet, the crust is thinner than an apple skin on an apple. Rigid and thin, the crust is broken up into 22 continent-sized fragments called tectonic plates. On a time scale of ten million years the movement of the mantle pushes the tectonic plates around and constantly rearranges Earth's surface. This heat-driven, global rearrangement of the Earth's crust is called **plate tectonics**.

Volcanoes and subduction zones

A subduction zone forms where plates come together. Surface rocks are driven underground and melted or partially melted into magma. A **volcano** is a site where magma forces its way back up to the surface again. Volcanoes often occur near subduction zones and are a source of igneous rocks which make up the largest fraction of Earth's crust. **Lava** is magma that has reached the surface.

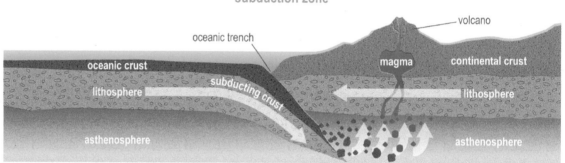

Water and volcanoes

Volcanoes tend to form around the edges of oceans, with the most famous example being the "ring of fire" of volcanoes around the Pacific ocean. This happens for two reasons: (1) subduction zones form where thin oceanic plates are driven beneath thicker continental plates, and (2) because of water. Water gets pulled down with subducted rock, and even small amounts of water reduce the melting point of granite from 1250 °C to 650 °C. The lower melting point causes granite to melt into magma at a lower temperature.

Pumice is more viscous due to its high silicate content

The melting point of rocks changes based on composition

Temperatures inside Earth can exceed 1000 °C at depths of 100 km beneath the surface. Rock may or may not melt depending on the water content, silica content, and pressure. The most common igneous rock, basalt, forms a low-silica lava with 45 to 50% SiO_2. Low silica lavas are the least viscous. Higher silicate lavas like rhyolite (70% SiO_2) have higher viscosity.

Basalt is less viscous due to its low silicate content

The nitrogen cycle

Many organisms cannot process N₂

All living organisms depend on nitrogen which moves through air, water, soil and sediments in the **nitrogen cycle**. Among the important biomolecules that contain nitrogen are amino acids (proteins), DNA, and RNA. While the atmosphere is 78% nitrogen (N_2) most living things cannot use diatomic nitrogen due to the strength of the triple bond holding the atoms together. Living organisms use nitrogen in the form of ammonia (NH_3), nitrate (NO_3^-) ions, and nitrite (NO_2^-) ions to get the nitrogen needed to synthesize biomolecules. Multicellular organisms rely on plants to get these nitrogen sources.

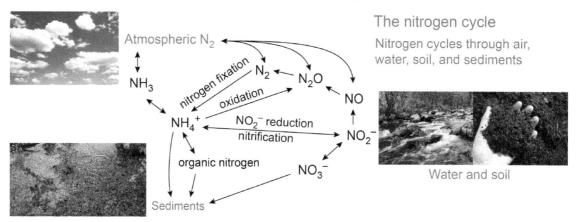

The nitrogen cycle

Nitrogen cycles through air, water, soil, and sediments

Biological nitrogen fixation

Atmospheric nitrogen is converted to biologically usable forms by the process of **nitrogen fixation**. A small amount of nitrogen is fixed by lightning which converts N_2 to NO and N_2O. However, the most important source is nitrogen-fixing soil bacteria which are symbiotic with the roots of plants. Nitrogen-fixing bacteria have a special enzyme called nitrogenase that can break the triple bond in the nitrogen molecule. Nitrogenase catalyzes the breakdown of N_2 and its incorporation into ammonia, NH_3. The presence of the H^+ ion tells us the nitrogen fixation reaction only occurs in an acidic environment and it requires energy from ATP molecules. Nitrogenase breaks down in the presence of oxygen so nitrogen-fixing bacteria live in the soil in low-oxygen environments.

Biological nitrogen fixation in plant roots

$$N_2 + 8H^+ + 8e^- + 16ATP \xrightarrow{nitrogenase} 2NH_3 + H_2 + 16ADP + 16PO_4^{3-}$$

Synthetic nitrogen fixation

As Earth's population has grown, agriculture has come to depend on synthetic fertilizer to supplement natural nitrogen fixation. Developed by Fritz Haber and Carl Bosch in the 1900s, the Haber-Bosch process produces approximately 100 million tons of nitrogen fertilizer each year. This supports agriculture for more than 30% of the world's population. The Haber-Bosch reaction produces ammonia from nitrogen and hydrogen gas but requires very high pressure conditions and the use of a catalyst. These conditions were very challenging to overcome.

The Haber-Bosch process

$$92 \text{ kj/mol} + N_2(g) + 3H_2(g) \underset{high\ pressure}{\overset{Fe\ catalyst}{\rightleftharpoons}} 2NH_3(g)$$

The phosphorus cycle

Biological use of phosphorus

Phosphorus (P) is the second most abundant mineral in the body, after calcium. About 85% of the body's phosphorus is in bones and teeth. Phosphorus atoms form the molecular backbone of DNA and RNA and are fundamental to the metabolic energy-transfer molecules: ATP (adenosine triphosphate), and ADP (adenosine diphosphate). Chemically active phosphorus in living organisms is found in the phosphate ion (PO_4^{3-}).

Phosphate minerals

Phosphorus is highly reactive and never found as a free element on Earth. With few exceptions, minerals containing phosphorus are in the maximally oxidized state, with the most common being apatite. Apatite comes in three forms, each with the same calcium-phosphorus-oxygen group, $Ca_{10}(PO_4)_6$, but different ions (OH^-, Cl^-, F^-).

Common phosphate minerals

Chlorapatite $Ca_{10}(PO_4)_6(Cl)_2$

Hydroxylapatite $Ca_{10}(PO_4)_6(OH)_2$

Fluorapatite $Ca_{10}(PO_4)_6(F)_2$

The land part of the phosphorus cycle

The **phosphorus cycle** occurs between the land and the ocean mediated by plants and animals. Phosphate minerals are weathered, dissolved, and accumulated in soil. Plants take up phosphate and bind it in organic compounds. Animals eat the plants and organic phosphate makes its way through the food chain eventually being excreted. Weathering, erosion, and runoff carry dissolved phosphates from both organic and inorganic sources into the ocean.

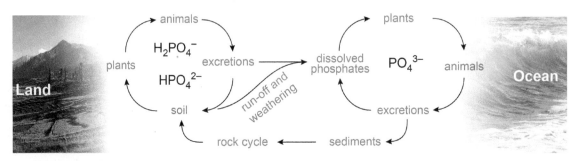

The phosphorus cycle

The ocean part of the cycle

Ocean plants and animals use phosphates in a similar way to land plants and animals. Animal excretions along with animal remains drift down to the bottom of the ocean. A large amount of inorganic phosphate salts are found in ocean floor sediments. The rock cycle gradually incorporates the phosphates into sedimentary rock, and possibly metamorphic and igneous rock. Plate tectonics brings phosphate-bearing rock back to the surface to start the cycle again.

Nitrate and phosphate pollution

Artificial fertilizers and detergents contain phosphates and are best used in moderation to avoid excessive phosphate runoff. When excess nitrates and phosphates run off into aquatic ecosystems, these nutrients cause algae to overpopulate (bloom). An algae bloom can rob plants and animals of nutrients, light, and oxygen causing them to die off.

The oxygen cycle

The source of oxygen

Almost two thirds of Earth's crust (59.7% by weight) are silicate rocks containing about half oxygen (29.4%) and half silicon (30.3%). Scientists estimate that 0.49% of the oxygen is in the crust and only 0.01% in the biosphere. Similar to phosphorus, oxygen is highly reactive and would not naturally occur in its elemental form without living organisms. The fact that 21% of Earth's atmosphere is oxygen is solely due to life and the source of the oxygen is almost wholly from photosynthesis in plants. Without living plants and other photosynthetic organisms, Earth would not have any oxygen in the atmosphere. Without oxygen there would be nothing but the most primitive life - chemosynthetic bacteria. There would be no animals, including humans. We depend on photosynthetic organisms for food as well as the air we breathe.

The oxygen cycle

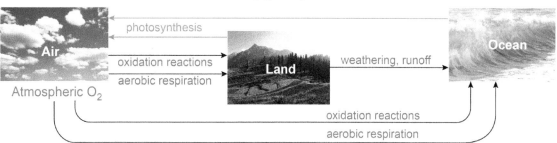

Oxygen production

In one year, the oxygen cycle accounts for the transformation of 3×10^{14} kg of oxygen. The diagram shows that production of oxygen is split about half way between ocean plants (45%) and land plants (55%). The majority of ocean photosynthesis is from algae and phytoplankton. Photosynthesis on land occurs in trees, grasses, and other plants. A very small fraction (< 0.001%) of oxygen is produced from photodissociation reactions in the atmosphere ($2N_2O + photon \rightarrow 2N_2 + O_2$).

Primary chemical reactions in the oxygen cycle

$$6CO_2 + 6H_2O \xrightarrow[\text{55\% land / 45\% ocean}]{\text{photosynthesis}} C_6H_{12}O_6 + 6O_2$$

$$6CO_2 + 6H_2O \xleftarrow[77\%]{\text{aerobic respiration}} C_6H_{12}O_6 + 6O_2$$

$$2NO_2^- + 2H_2O + 4H^+ \xleftarrow[17\%]{\text{microbial oxidation}} 2NH_4^+ + 3O_2$$

$$18H_2O + 16CO_2 \xleftarrow[4\%]{\text{fossil fuels}} 2C_8H_{18} + 25O_2$$

Oxygen use

The majority of Earth's 3×10^{14} kg per year of recirculating oxygen is used for aerobic respiration by animals (77%). Microbes such as nitrogen-fixing bacteria, decomposers, and other soil species account for another 17%. Technological uses including the combustion of fossil fuels account for about 4%. About 0.2% of surface oxygen oxidizes exposed rocks and is sequestered in the rock cycle.

Geological oxygen

The 99.5% of the oxygen trapped in Earth's crustal rock is slowly recycled by plate tectonics and the rock cycle. A significant fraction of this oxygen interacts with the biosphere through oxygen-rich calcium carbonate ($CaCO_3$) in sea shells and the exoskeleton of marine organisms. Calcium carbonate settles in sea-floor sediments which become limestone. Metamorphosis transforms limestone into marble.

Chapter 23

Section 3 review

The thin crust of the earth is made up primarily of eight elements, with silicates being the most prevalent compounds. Higher density elements such as iron sank to the Earth's core as it cooled. Between the core and crust lies the mantle containing molten magma that occasionally reaches Earth's surface through a volcano. Cooled magma forms igneous rocks. Igneous rock that is weathered over time becomes sedimentary rock. Metamorphic rock is created when igneous or sedimentary rocks are subjected to high heat and/or pressure. The nitrogen cycle shows the pathway of atmospheric nitrogen being made available to living things. Phosphorus also cycles, but only between the ocean and land without entering the atmosphere. Soil chemistry is important for agriculture as well as protecting or restoring contaminated areas. Maintaining a proper pH range is key for plant growth and the survival of other organisms that play a role in plant growth.

| Vocabulary words | geosphere, silicates, mantle, solar nebula, core, crystalline, amorphous, mineral, geologists, sedimentary rock, metamorphic rock, polymorphism, igneous rock, rock cycle, rock, magma, volcano, lava, plate tectonics, nitrogen cycle, nitrogen fixation, phosphorus cycle |

Review problems and questions

1. You learned how to classify matter as an element, compound, homogeneous mixture or heterogeneous mixture. Is a mineral an element, compound, homogeneous mixture or heterogeneous mixture? Support your answer with an explanation and an example.

2. What do the motion of Earth's plates and the motion of Earth's oceans have in common?

3. Even though 78% of the atmosphere is made of nitrogen gas, living things must rely on tiny bacteria to convert N_2 to a useable form. Why can't living things like humans convert N_2 to a useable form on their own?

4. What conditions are required for a metamorphic rock to form?

5. Pretend you are an oxygen, nitrogen, or phosphorus atom that is planning a vacation. Write a story of your 3-event vacation. The events on your vacation must follow a route within the cycle of your chosen substance.

Soil Health

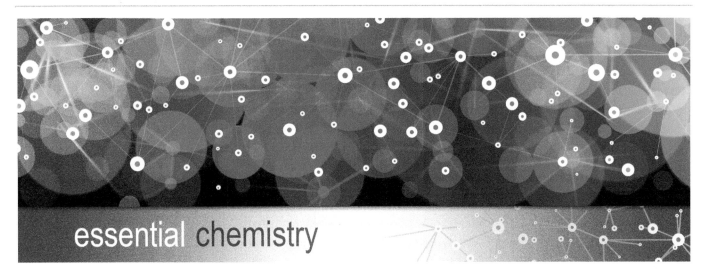

essential chemistry

An important natural resource, soil is a mixture of organic matter, clay, and rock particles. Healthy soil is essential for food production, and as you have seen, it is a component of the cycles that affect the chemical balance of the planet. Farms and farming techniques have a major impact on the health of soil. Today most farmland is prepared the same way it has been for millennia. The soil is turned over with a plow pulled by tractor or livestock. This is referred to as "tilling" the soil.

The practice of tilling turns the soil in an effort to bury weeds and the residues of the previous crops, aerate the soil, and mix in fertilizer. But is this the best way to maintain healthy soil? In conventional tillage, the field is tilled to a depth of 8-12 inches. Then it is tilled two more times to fully get the seedbed ready for planting. It seems logical that this loose earth would be ideal to allow water to drain and roots to spread so plants will grow.

The problem with using heavy machinery to plow is that it often results in broken soil laying on top of a heavily compacted layer that the plow can't reach (called the plow pan). The broken soil and plow pan make soil very susceptible to erosion. This is especially true when there are extreme weather events like droughts and floods. Tillage is the root cause of agricultural land erosion. In fact, excessive tilling was a key factor behind the Dust Bowl in the 1930s.

Some people have started advancing a new, but surprisingly old idea. In a 1943 book, Plowman's Folly, Edward H. Faulkner wrote, "The truth is that no one has ever advanced a scientific reason for plowing," and he may be right. There is a quiet growing trend towards "no-till" farming. According to the United Stated Department of Agriculture (USDA), in 2009, 35.5% of the country's cropland had at least some no-till operations and about 10% were fully no-till. No-till farming describes a way to grow crops each year without tilling the field. The residue from the previous planting remains on the field. A groove from 0.5-3 inches across is created and the seeds are planted in the groove. Overall, there is minimal soil disturbance. The previous crop residues provide considerable benefits to the soil including protection from wind and water erosion.

The benefits of no-till extend beyond soil protection. Healthy soil also impacts levels of atmospheric carbon dioxide. As you know, increases in atmospheric carbon dioxide levels shift the balance of the carbon cycle which have global consequences. No-till farming requires less use of heavy agriculture equipment so there is a decrease in fuel and in the amount of carbon dioxide that is produced.

Soil Health - 2

Tilling breaks up the organic matter in soil, and carbon from the organic matter is lost from the soil into the atmosphere during decomposition. In no-till farming, crop residue is left on the field and the soil remains intact. No-tilled fields have intact soil layers which often means a higher amount of insects, worms and microbes. Currently, the USDA considers soil to be a modest place of carbon storage only capturing about 4 million metric tons of carbon annually. This is just a tiny fraction of the US output. With the right soil practices in place, it is believed that the soil could soak up 25 times more carbon than current rates.

No-till farming practices can have a beneficial effect on another one of Earth's precious resources—water. Instead of grinding up and burying or removing previous crop residues, they remain on top of no-till soil. These crop residues can act like a moisture barrier for the top layer of the soil. This moisture barrier means there is less evaporation, so the water stays within the soil instead of going into the atmosphere. Ultimately this means these crops can be watered less.

Not only is there better retention of water with the soil, there is better infiltration of water deep into the soil. Tilled fields have less water retention and infiltration, and are more likely to have water runoff. Runoff not only carries loose soil with it, but also has the potential to carry fertilizers and pesticides to nearby waterways. Because of this, no-till farming can have benefits for the soil and for water quality.

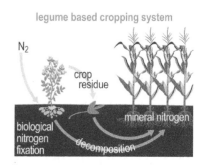

Cover crops are another technique that some farmers are using in combination with no-till farming to increase soil health. Traditionally, farmers let the land go "fallow" between plantings. Instead, some farmers are planting one, or many other crops that will turn into the crop residue for the next planting season. The cover crops help keep the soil from eroding and provide organic nutrients. For example, legume (bean) cover crops are very good at fixing nitrogen from the air. With nitrogen trapped in the roots of the legumes, less synthetic fertilizers are needed. Some farmers plant a mix of cover crops to try to mimic the diversity of Mother Nature. This can lead to a more robust biological system that requires less insecticides and fungicides.

Like anything there are pros and cons to no-till farming. One drawback of no-till farming is that it currently relies on herbicides to get rid of the weeds that tilling would have removed. Cover crops and other agro-ecological options are starting to be seen as alternatives to the reliance on these chemicals. Despite this drawback, there is definitely a growing potential for no-till, cover crop farming to increase soil health and continue to produce an abundance of food.

The chemical properties of soils such as nutrient availability are largely influenced by the soil pH. Plants obtain nutrients from the soil such as Ca^{2+}, Mg^{2+}, Zn^{2+}, Co^{2+}, Mn^{2+}, Fe^{2+}, Cu^{2+} and Al^{3+}. Elements such as phosphorous and nitrogen in a variety of ions are also important for plant growth. The pH range of 6 to 7.5 makes these elements and cations available to plants in the soil. Some plants are hardy and can survive soil pH swings, but many cannot. pH ranges outside normal also negatively affect the survival of decomposers like bacteria, fungi, and invertebrates like worms and ants.

Chapter 23 review

Vocabulary
Match each word to the sentence where it best fits.

Section 23.1

dew point	greenhouse effect
greenhouse gases	mesosphere
ozone layer	photodissociation
stratosphere	thermosphere
troposphere	

1. The _____ is where high-energy photons drive reactions that form most of Earth's ozone.
2. _____ is a process that occurs when a molecule breaks a bond due to the absorption of a photon.
3. The layer of the atmosphere that you reside in is called the _____.
4. The _____ is caused by gas molecules in the atmosphere trapping heat and reflecting it back to the Earth.
5. Substances such as water that allow sunlight to pass through the atmosphere but trap reflected infrared heat are known as _____.
6. The collection of atmospheric gas that absorbs harmful UV radiation is called the _____.
7. The _____ is located between the stratosphere and thermosphere.
8. The _____ borders space and is the outermost layer of the Earth's atmosphere.
9. The _____ is the temperature where air is saturated with water vapor.

Section 23.2

carbon cycle	hydrosphere
ocean acidification	salinity
stomata	transpiration
water cycle	

10. _____ is where water moves out of a plant into the air during photosynthesis as part of the water cycle.
11. When carbon dioxide levels dissolve into ocean water beyond buffer capacity, _____ can occur.
12. The concentration of salt in water is a measure of _____.
13. Water moves through the atmosphere, ocean, and land in a process called the _____.
14. The _____ shows how carbon moves between different areas of our Earth.
15. All forms of water on Earth, including water locked beneath Earth's crust and in the atmosphere, is called the _____.
16. Water moves from a plant to the air through _____.

Section 23.3

core	igneous rock
lava	magma
metamorphic rock	mineral
nitrogen cycle	nitrogen fixation
phosphorus cycle	sedimentary rock
solar nebula	volcano

17. A natural inorganic solid that has an organized internal structure is a(n) _____ and is the building block for a rock.
18. A(n) _____ is an opening in the Earth's crust where magma can rise to the surface.
19. Crystallized magma forms _____, covering most of the earth's upper crust.
20. Magma expelled from a volcano is called _____.
21. _____ can form from igneous rock or sedimentary rock under conditions of high heat and pressure.
22. Nitrogen becomes available to living organisms through a process called _____ when it is converted from atmospheric N_2 into a nitrogen compound.
23. The process demonstrating how nitrogen moves through the atmosphere, land, and ocean is the _____.
24. The _____ cycles between the land and ocean but never the atmosphere.
25. _____ is molten rock found below the surface of the upper mantle.
26. Earth was formed in a(n) _____, a disk-shaped region of dust and gases around a forming star.
27. The _____ of the Earth, Sun, or other planets has a very high pressure and temperature.
28. Small pieces of weathered and eroded igneous rock compress to form _____.

Conceptual questions

Section 23.1

29. What happens to an O_2 gas molecule during photodissociation? How does this benefit living things?

30. What atmospheric gases are represented by a and b in the above pie graph?
31. Rainwater is naturally acidic. Why is this?
32. The greenhouse effect reflects some infrared radiation and traps other infrared radiation. Demonstrate this with a sketch of the earth using arrows for infrared radiation.

Chapter 23 review

Section 23.1

33. How did each of the following become part of our atmosphere?
    a. CO_2
    b. O_2
    c. O_3

34. Which molecule(s) are responsible for absorbing harmful UV light?

35. The troposphere has a larger total mass than the stratosphere, but the stratosphere is much larger in size. Why?

36. The formation of two strong acids that produce acid rain can be shown in chemical reactions. What are they?

37. What happens when acid rain falls on human-made structures made of metals, limestone or marble?

38. Answer the following questions about smog.
    a. Which gases are the main contributors to smog?
    b. How does smog form?

39. What are the layers of our atmosphere in order from the ground to space?

40. The sun initiates two important chemical reactions in the upper atmosphere. Describe them.

41. During a talk show segment a medical guest recommended the use of "safe" UVB (ultraviolet B) tanning beds for getting vitamin D. Vitamin D is important for human health, yet many Americans are deficient in the vitamin. The World Health Organization recommends against using a tanning bed for cosmetic purposes and classifies the beds as "carcinogenic to humans." Research this issue and evaluate the validity of this doctor's claims. Cite specific evidence for your conclusion.

42. Find an article about humans' role in climate change in an online or print science magazine. Write a 1- to 2-paragraph summary of the article. Below your summary, write a 1- to 2-paragraph analysis of the author's purpose. Did the author aim to entertain, inform or persuade their readers? What important issues did the author leave out? If you feel the author covered all of the most important issues, identify them. Provide a copy to your teacher.

43. Which two of the most abundant gases in the atmosphere are also greenhouse gases?

44. The following pollutant gases can be naturally occurring or from human-made sources. For each pollutant, identify a source for the pollutant gas.
    a. NO
    b. SO_2
    c. O_3

45. Which type of energy on the electromagnetic radiation spectrum does carbon dioxide trap in the atmosphere?

46. One objection to scientific evidence for global warming is that the temperature measurements are unreliable. Research the topic of temperature measurements and global warming. Consult a variety of sources in a variety of formats including print and video. Note: Data can be obtained from NOAA and/or NASA, along with descriptions of the methodologies.
    a. Analyze the measurement techniques, particularly the standard method of how temperature gauges should be mounted for weather measurements.
    b. Critique whether or not siting issues create systematic errors in the global warming temperature data.
    c. Evaluate whether the warming trends are compromised by thermometer gauge siting issues.
    d. Write your most interesting thought or finding on a sticky note. The next time you go to class, place your sticky note in the area of the classroom designated by your teacher. Read another student's sticky note and write a 1-paragraph reflection on their findings or thoughts.

47. Auroras can be seen in polar regions in the thermosphere. Explain this phenomenon.

48. What are the four most abundant gases in the atmosphere?

49. Describe at least two environmental impacts of acid rain.

Section 23.2

50. What two factors primarily influence ocean water density?

51. The world's glaciers are melting at a fast rate.
    a. What impact will melting glaciers have on ocean surface water density, especially near the poles? Explain your answer.
    b. How will this density change affect deep ocean currents?

52. Why are ocean currents so important for the fish, birds, and other animals that rely of the oceans for food?

53. State the two primary mechanisms by which water vapor enters the atmosphere.

54. Approximately 75% of the Earth's surface is covered by liquid water.
    a. How much of the water on Earth is saltwater?
    b. How much of the water on Earth is contained in lakes and rivers?
    c. How much is locked in the polar ice caps and glaciers?

55. Answer the following questions about the water cycle.
    a. Sketch the water cycle; show how water moves through the geosphere, troposphere, atmosphere and biosphere (living things).
    b. Label the main processes of the cycle.
    c. Identify the location where the most evaporation occurs.
    d. Is the water cycle is a series of chemical changes, or physical changes? Explain your answer.

56. How do temperature and density differences drive the motion of ocean currents near Earth's poles?

Chapter 23 review

Section 23.2

57. Answer the following questions about the carbon cycle.
    a. Create a drawing of the carbon cycle that includes the geosphere, hydrosphere, atmosphere, and biosphere (living things).
    b. Label at least one way carbon enters the atmosphere.
    c. Label at least one way carbon returns to the geosphere.
    d. Does it appear these ways of cycling carbon through Earth's systems are in equilibrium? Why or why not?

58. Answer the following questions about the distribution of carbon on Earth.
    a. Where is the largest amount of carbon on Earth stored?
    b. Where is the largest amount of carbon contained that is actively and quickly moving through the carbon cycle?

59. The ocean is able to absorb large amounts of CO_2. What is the chemical reaction that demonstrates this?

60. Sea level has been measured for centuries using tide gauges, but more recently it has been measured using satellite data. Research these two sets of data to answer the following questions.
    a. Analyze each approach. How does each measurement technique work? Provide a written explanation of the technical details for each of the two processes.
    b. Critique each method and compare each method to the other. Would you trust one set of measurements more than another? Why?
    c. Does either data set or both show a trend over time in the change of the average sea level? Does the trend appear significant?
    d. Compare each independent data set to the other and evaluate them. Do they agree with each other?
    e. Write your most interesting thought or finding on a sticky note. The next time you go to class, place your sticky note in the area of the classroom designated by your teacher. Read another student's sticky note and write a 1-paragraph reflection on their findings or thoughts.

61. Although about 75% of Earth's surface is covered by different oceans, its chemical composition is fairly constant. How is this possible?

62. How does a higher concentration of carbon dioxide in the atmosphere affect animals with shells and exoskeletons that live in the oceans?

63. What happens to the pH of ocean water when it absorbs carbon dioxide gas from the atmosphere?

64. Why do the British Isles have a temperate climate despite them being at the same latitude as Siberia?

65. Identify the four most abundant ions in Earth's oceans.

66. Why is the arrow at Point A in the above picture sinking, and how is this related to ocean currents?

Section 23.3

67. Astronomers believe that the Moon was formed when a Mars-sized object collided with the early Earth.
    a. Why did the Moon's composition lead astronomer to this conclusion?
    b. What is the average density of the Moon?
    c. Why is the density of the Moon significantly less than the density of the Earth?

68. Current scientific theory suggests that the Earth was formed by the accretion of rocky debris orbiting within our solar system.
    a. What evidence is there to support this hypothesis for the formation of the Earth?
    b. Based on Earth's physical structure, why do scientists believe that the early Earth remained molten for a long period of time?
    c. The newly formed Earth had no almost water or atmosphere. Where did the water and atmosphere of today's planet come from?

69. Answer the following questions about the Earth's crust.
    a. Of the three rock types, which is most abundant in Earth's crust?
    b. What type of compound is the basic building block of all rocks? Describe the most common of these building blocks.

70. The Sun is by far the most massive celestial body in the solar system.
    a. What percentage of the solar system's mass is contained within the sun?
    b. What is the approximate chemical composition of the Sun?
    c. Which planet contains most of the remaining mass in the solar system?

71. State the differences between a mineral and a rock.

72. The Earth's crust is the thin solid top layer of the lithosphere, and is the home of life on Earth.
    a. What is the name of the molten rock layer that the lithosphere exists on top of?
    b. The crust is broken up in to tectonic plates that move in relation to each other. How many tectonic plates are there?
    c. What geological process drives the movement of the tectonic plates?

73. The arrangement of atoms during the rock formation process are affected by what two factors?

Chapter 23 review

Section 23.3

74. Explain why most volcanoes are found near plate boundaries.
75. Fossils are commonly found in which type of rock? Why are they uncommon in other types of rock?
76. Why is iron abundant in the Earth's core?
77. The Earth's crust is primarily composed of which two elements?
78. Answer the following questions about the layers of the Earth.
    a. Make a sketch showing the different layers of the Earth.
    b. How did these layers form?
79. What do geologists study, and how does this help to protect us and improve our environment?
80. Which types of rock can form metamorphic rock, and what conditions are necessary for the rock type metamorphosis to occur?
81. How does silica content affect the viscosity of lava?
82. What forms of nitrogen can animals and humans metabolize?
83. There are a few ways that nitrogen can become available to living organisms. What are they?
84. If nitrogen is "fixed," what does this mean?
85. Although the vast majority of the atmosphere is made of nitrogen, it is not in a form usable by most living things.
    a. Animals can't use nitrogen in the form of N_2. Why?
    b. How do nitrogen-fixing bacteria convert N_2 into a usable form?
86. Answer the following questions about the nitrogen cycle.
    a. Sketch the nitrogen cycle. Show how nitrogen moves through the atmosphere, biosphere (living things), hydrosphere and geosphere.
    b. Label places where nitrogen enters the geosphere.
87. Why do nitrogen fixing bacteria live in soil in the nodules on the roots of plants?
88. Where is phosphate primarily found in the ocean?
89. Why is the Haber-Bosch process important to you, and rest of the world's population?
90. Synthetic fertilizers are commonly used in agriculture. In future, artificial fertilizers will play an increasingly important role as the human population grows.
    a. What substance is frequently produced for use as an artificial fertilizer?
    b. How much artificial fertilizer is produced each year via the Haber-Bosch process?
    c. What percentage of the worlds population is already dependent on food produced using artificial fertilizers?
    d. What are some of the challenges of producing synthetic fertilizers?

91. What is the environmental impact of adding excess phosphorus to aquatic ecosystems by over-using artificial fertilizers and detergents?
92. Where is a large amount of inorganic phosphorus found?
93. Describe the role plants play in both the nitrogen and phosphorus cycles.
94. Identify one similarity and one difference between the nitrogen and phosphorus cycles.
95. The oxygen cycle is the movement of oxygen between the atmosphere, oceans, living organisms, and the land.
    a. How much oxygen is transformed by the oxygen cycle each year?
    b. What percentage of oxygen production is accounted for by ocean plants?
96. What percentage of oxygen is consumed by the burning of fossil fuels?
97. What percentage of Earth's atmosphere is oxygen?
98. What percent of the oxygen produced by plants is consumed via the aerobic respiration of animals?

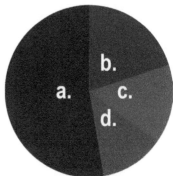

99. The above pie graph represents the abundance of elements in all Earth's layers in the entire geosphere. Which letter represents the abundance of each of the following?
    a. iron
    b. silicon
    c. oxygen

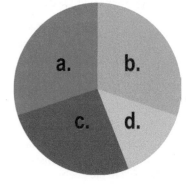

100. The pie graph above represents the abundance of elements in Earth's crust. Which letter represents the abundance of each of the following? Note: a and b are nearly equal.
    a. iron
    b. silicon
    c. oxygen

Chapter 23 review

Section 23.3

101. Phosphorus is an essential element in our diets. Answer the following questions about the role of phosphorus in the human body.
    a. Where is most the phosphorus in the human body located?
    b. Phosphorus is essential to which biological molecules?
    c. In what form is chemically active phosphorus found in living organisms?

102. Is the volcano in the picture above formed by a subducting plate? How can you tell?

104. The relative humidity of air is equal to the ratio of the partial pressure of water to how much water vapor the air can hold when it is completely saturated. The equilibrium partial pressure of water vapor in air at 25°C is 3200 Pa.
    a. If the relative humidity of the air is 51% at 25°C, what is the partial pressure of H_2O in the air?
    b. A barometric pressure is measured at 748.0 mmHg. What is the partial pressure of the rest of the air?
    c. How many molecules of water are present in a room measuring 9 x 10 x 11 feet?

105. Air at 26 °C saturates at 3400 Pa partial pressure H_2O. What is the partial pressure of water in the air if the relative humidity is 57.0 %?

Quantitative problems

Section 23.1

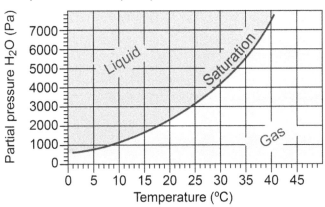

103. Use the graph above to answer the following:
What is the relative humidity when temperature is 28 °C and the partial pressure of water vapor equals 2,150 Pa?

chapter 24: The Universe

Looking up at a starry night sky can be awe-inspiring. The billions of stars prompt us to question "what resides in the heavens?" and "what is Earth's place in the universe?" Science provides both methodical and experimental ways of satisfying our curiosity. In our trek for knowledge we have developed telescopes, spacecraft and even rovers to explore objects in the heavens, including nearby planets and distant stars.

Astronomy and astrology have been used to describe the study of heavenly bodies. The difference is that the former is a science, while the latter merely resembles a science. Astronomy is the study of the position, movement and composition of celestial bodies and the universe. Astrology tries to correlate the movements of celestial bodies to peoples' lives and events on Earth. Religion, is a set of foundational beliefs. Astrology is not quite a religion and it is also not quite a science. With charts, star maps and a specialized vocabulary, astrology can seem scientific. Astrology is used by some to predict events, but astrological predictions are generally vague and untestable. Astrology is a pseudo-science because it masks itself as science and does not follow the scientific method. Scientists make predictions that are testable, collect data, share their results and ideas openly, and modify their ideas based on new evidence.

vintage astrological chart

Our knowledge of the universe is growing, but it is still incomplete. Our essential curiosity continues to push us to apply principles of science and engineering as we strive toward understanding the chemistry of space, the final frontier.

Chapter 24 study guide

Chapter Preview

Our universe is vast and expanding with evidence pointing to its origin in a massive explosion known as the Big Bang. Hydrogen and helium atoms soon formed and eventually gravity pulled clouds of gas together to form stars. In the incredibly hot interior of stars, energy is created through nuclear fusion. When larger stars die they create heavier elements that are eventually dispersed in all directions when the star explodes in a supernova. Matter is pulled together by gravity again, forming new stars and planets. In this chapter you will learn about the origin of the universe as we understand it, our solar system, the composition of the sun and planets, and explore the idea of life on other planets.

Learning objectives

By the end of this chapter you should be able to:
- describe the elemental composition of the known universe;
- explain the theory of the Big Bang and the formation of stars and galaxies;
- identify the inner and outer planets of our solar system and the differences between them;
- explain the Miller-Urey experiment and its importance to the study of early life; and
- give examples of the evidence that could point to life on other planets.

Investigations

24A: Spectroscopy

Pages in this chapter

- 786 The Universe and the Stars
- 787 Observing tools: Spectroscopy and telescopes
- 788 Redshift, the Big Bang and the expanding universe
- 789 The chemical composition of the universe
- 790 Stars
- 791 The Sun
- 792 Nuclear fusion reactions in stars
- 793 The interstellar medium
- 794 Section 1 Review
- 795 The Solar System
- 796 Chemical composition of the planets
- 797 Blackbody radiation
- 798 Temperature in the inner and outer solar system
- 799 Venus
- 800 Mars
- 801 The asteroid belt, meteors, and comets
- 802 The gas giant planets
- 803 Titan
- 804 Section 2 Review
- 805 Life Outside the Earth
- 806 The beginnings of biochemistry
- 807 Extremophiles
- 808 Oxygen is a chemical signature of life
- 809 The search for life on Mars
- 810 Is there water on Europa?
- 811 Planets in other solar systems
- 812 Section 3 Review
- 813 Spectroscopy and Celestial Bodies
- 814 Spectroscopy and Celestial Bodies - 2
- 815 Chapter review

Vocabulary

light-year
Doppler effect
plasma
photosphere
astronomical unit (AU)
blackbody
intensity
chemosynthesis

spectrometer
cosmic background radiation
ionization
interstellar medium
accretion
thermal radiation
hydrothermal vents
extremophiles

Big Bang Theory
supernova
Hertzsprung-Russell diagram
nebula
blackbody spectrum
frost line
thermophiles

24.1 - The Universe and the Stars

Our universe is incredibly vast, containing billions of galaxies each containing 50 - 500 billion stars. If only one in a thousand of those stars have planets. then there are more than 100 million planets just in our own Milky Way galaxy. People have long wondered if the universe outside of Earth was similar to what we know. Are the same elements out there? Do the same natural laws operate in other galaxies? Are the intelligent creatures out there wondering similar things about us? Science has begun to learn some of the answers, but the exciting part is that we have many more unanswered questions.

What is the universe?

Early models of astronomy

The universe is everything, including time, space, Earth, and you. Early theories of astronomy put Earth at the center of the universe with celestial objects circling around it. This theory agreed with the observations that the Sun rises in the east, sets in the west, and appears to circle around the Earth. The planets, the Moon, and the stars have similar apparent motion. Ptolemy (AD 90–AD 168) is credited with the model, illustrated in the Flammarion (below), in which an explorer reaching the edge of Earth peers out from under the first, and lowest, celestial sphere containing the sky, clouds, and air.

The sun-centered model

Nicolaus Copernicus (1473–1543) correctly deduced that the planets revolve around the Sun. Despite substantial data confirming his theory, Copernicus was not widely believed at any time during his life. Born three years after Copernicus' death, Danish nobleman Tycho Brahe (1546–1601) made the first accurate and systematic measurements of the positions of stars and planets over many years, aided by his assistant Johannes Kepler (1571–1630). Galilei Galileo's (1564-1642) discovery that there were moons circling Jupiter in 1610 offered the final refutation of Ptolemy's Earth-centered model which has stood for more than 1,500 years.

Our place in the universe

Starting from Earth and moving outward, we find that our solar system is part of the Milky Way galaxy including roughly 200 billion stars. Due to the large scale of the universe, it is helpful to use a very large unit of distance. A **light-year** is the distance that light travels through a vacuum in the time-span of one year. It is equivalent to 9.46×10^{15} m. The Milky Way is about 150,000 light-years in diameter and is part of the Virgo Supercluster which contains at least 500 additional galaxies and is 110 million light-years across. To the best of our observations the observable universe contains 10 million superclusters of galaxies similar to the Virgo Supercluster.

Observing tools: Spectroscopy and telescopes

The universe is vast

How do we know the matter and chemistry in the rest of the universe is the same as the matter on Earth? We cannot go and take samples because the distance is so vast throughout the universe it is almost beyond human comprehension. The fastest human spacecraft ever built, traveled at 158,000 mph, and even at this enormous speed it would take 18,000 years to reach the nearest star and 770 million years to cross the Milky Way galaxy.

Spectroscopy and telescopes

The answer is mostly by carefully and cleverly examining the light that reaches us at different wavelengths from high energy x-rays, through visible light, and down to the weakest infrared. To identify distant elements, we use the fact that atoms emit and absorb light in spectra that are characteristic of elements and compounds. A high resolution **spectrometer** can analyze the light from a distant galaxy, identify the elements that produced that light, and even make measurements of relative abundance and composition.

High-resolution solar spectrum

Each horizontal band scans 6 nanometers of wavelength. The bright yellow helium line is circled.

Solar absorption spectrum

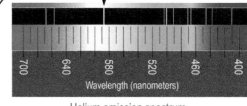
Helium emission spectrum

The solar spectrum

A good example of extraterrestrial spectroscopy is NASA's high resolution solar spectrometer. The diagram above shows the solar spectrum split into bands of visible light. The dark lines that look like stripes are absorption lines from different elements in the Sun's surface layer. The spectrum of helium is shown alongside the solar spectrum. In fact, helium was discovered this way, from spectra in the Sun and only later discovered on Earth! The spectra from distant stars and galaxies confirm that the familiar elements and compounds we know make the up the observable universe.

Other wavelengths

The matter between stars is made of diffuse clouds of gas, plasma, and dust. Optical spectroscopy reveals neutral hydrogen and ionized hydrogen as well as other trace elements. Radio telescopes observe frequencies of electromagnetic radiation characteristic of molecular absorption and emission spectra. Radio frequency "spectroscopy" reveals the presence of water, methane, formaldehyde and other compounds.

Radio telescope and a typical molecular spectra

Example spectra is of volcanic emissions on Earth

Chen, Y; Zou, C; Mastalerz, M; Hu, S; Gasaway, C; Tao, X International Journal of Molecular Sciences (2015)

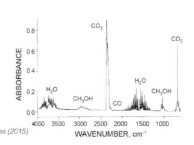

Interactive simulation

There is a lot more to electromagnetic radiation than meets the eye! While our eye detects visible light, there are many other wavelengths of light such as gamma rays and infrared radiation. Investigate frequencies and wavelengths across the electromagnetic spectrum in this interactive simulation.

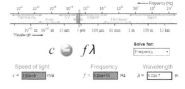

Redshift, the Big Bang and the expanding universe

The expanding universe

When astronomers compared the spectra of galaxies they discovered an odd fact: the hydrogen spectral lines were the same except shifted to the red by an amount that increased with distance. Such a shift comes from the **Doppler effect** and is due to the motion of the source away from the observer. In a Popular Astronomy article, (1915) Vesto Slipher wrote "... the great Andromeda spiral had the quite exceptional velocity of –300 km/s ..." In 1929, after analyzing many redshifts, and assistance from Edwin Hubble, he showed that the universe was expanding and galaxies were rapidly moving away from each other.

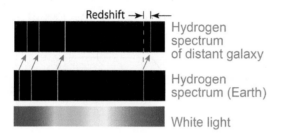

The Big Bang theory

The observed expansion implies that galaxies were closer together in the past. All the evidence points to the origin of our universe - time, space, matter, and energy - exploding outward from a massive birth cataclysm 13.7 billion years ago. This explanation for the origin of the universe is known as the **Big Bang Theory**.

Scientific development of the big bang theory

1927: Belgian mathematician Georges Lemaître proposes Big Bang theory.

1948: Ralph Alpher and Robert Herman estimate the temperature of the universe as a relic of the Big Bang.

1964: Arno Penzias and Robert Wilson use a microwave antenna to detect and measure the temperature.

Credit: NASA

1992: George Smoot and John Mather measure small variations in the microwave background, which led to present-day clusters of galaxies.

The early universe

The initial temperature of the Big Bang was so high that electrons and protons could not stay together to form atoms. As the universe expanded, its energy spread over a larger volume and the average temperature cooled. It took about three minutes for the expanding universe to cool enough for atoms to form and another 400,000 years before the first stars coalesced and started to shine. After 13.7 billion years, the universe has cooled to an average temperature of 3.7 K. The flash of energy leftover from the Big Bang is known as the **cosmic background radiation**.

Composition of the early universe

The lightest elements were the only ones to be created in any quantity during the first three minutes of the universe. As the hot matter of the Big Bang cooled, neutrons decayed into protons and electrons, then protons and electrons coalesced into hydrogen and helium. The universe cooled and expanded very quickly, ending the synthesis of atoms around the three minute point leaving 75% of the initial universe as hydrogen, 25% as helium, and less than 0.01% as lithium. There were essentially no heavier elements formed such as carbon, oxygen, and iron.

The chemical composition of the universe

Elemental abundances

The diagram shows the compositions of the early universe, the visible universe today, and the Earth's crust. The universe is still mostly hydrogen and helium. Earth's crust however, is rich in heavier elements with the most common being oxygen and silicon. How did we get from hydrogen and helium to oxygen, silicon, carbon, iron, and all the other heavier elements?

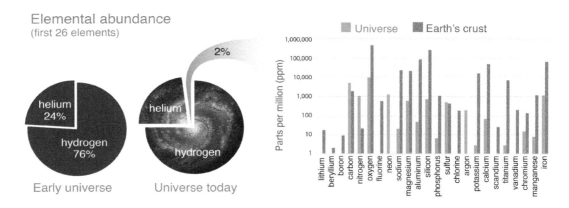

Heavy elements are created in stars

Elements heavier than hydrogen and helium were created in the nuclear furnaces at the center of stars! Stars are born when small fluctuations cause clouds of gasses and dust to collapse, under their own gravity. As the dust and gas collapses it heats up from the energy lost by the "falling" matter. At a certain point, the center of the hot, dense gas cloud reaches a temperature hot enough to cause nuclear reactions and the star begins to shine. These nuclear reactions create heavier elements such as carbon by fusing smaller nuclei.

Synthesis of carbon through nuclear fusion

Heavy elements are dispersed by exploding stars

Once a star ignites, internal nuclear reactions generate heat and pressure which stop further collapse as the core of the star gradually converts hydrogen and helium into heavier elements. Eventually, the core exhausts its hydrogen and helium. Low-mass stars then collapse into a white dwarf. Stars with more mass explode in a cataclysmic explosion called a **supernova** The explosion is so powerful one supernova can outshine an entire galaxy of 200 billion stars! Supernovae spew heavy elements into space in great clouds of plasma such as the Crab Nebula, a remnant of a supernova that exploded and humans saw in 1054.

Stars

99% of the visible universe is plasma

More than 99% of the visible matter in the universe, including all of the matter in stars, is in the form of **plasma**. Plasma is a hot gas in which the temperature is so high that electrons are stripped from atoms in a process called **ionization**. Plasma is relatively rare on Earth except in the form of lightning. A star is a hot ball of plasma held together by gravity and kept heated (and ionized) by the energy released from nuclear reactions at the core.

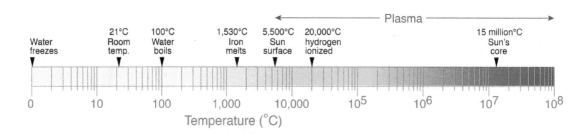

Above 20,000 °C most matter is plasma

Above a few thousand degrees molecular hydrogen (H_2) dissociates into atomic hydrogen (H). The Sun's surface is partly ionized hydrogen and helium at about 6,000 °C. Above 20,000 °C hydrogen is fully ionized. Other elements ionize at lower temperatures since their outer electrons take less energy to remove. Above 20,000 °C all of the elements are in the plasma state, although not all electrons may be stripped.

Classification of Main Sequence stars

Stellar type	M	K	G	F	A	B	O
Surface temperature	3,000K	4,400K	5,700K	6,700K	8,700K	15,000K	35,000K
Color	red	orange	yellow	yel/white	white	blue/white	blue
Mass / Sun	0.3	0.6	1.0	1.25	1.8	8	20
Luminosity / Sun	0.02	0.2	1.0	2.2	8	1,500	35,000
Population	76%	12%	8%	3%	0.6%	0.1%	0.00003%
Lifetime (× 10^9 Y)	200	36	10	6	2	0.05	0.006

Star types and the H-R diagram

Our Sun is a relatively small, yellow, G-type star. The **Hertzsprung-Russell diagram** shows the range of stars on a graph of energy output versus surface temperature derived (color). Stars spend the majority of their lives on the main sequence region of the H-R diagram. The Sun is about in the middle of the main sequence. The smallest star that can support nuclear reactions is about 10% of the Sun's mass and supergiant stars can be more than 40 solar masses. The more mass a star has, the more energy it emits and the hotter its core and surface will be. Nuclear reactions proceed at a much faster rate at higher temperatures. As a result, the more massive stars have a much shorter lifespan.

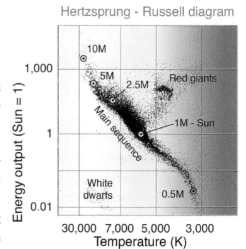

The Sun

The building blocks of the Sun

Our Sun is technically classed as a G2 yellow dwarf star and at 93 million miles it is by far the closest star to Earth. The next nearest star is Alpha Centauri, 4.37 light years away. 99.8% of the total mass of the solar system is contained in the Sun, with Jupiter containing most of the last 0.2%. The Sun is a hot ball of plasma consisting of 74% hydrogen, 24% helium, and 2% all other elements. The Sun's visible surface is the **photosphere**.

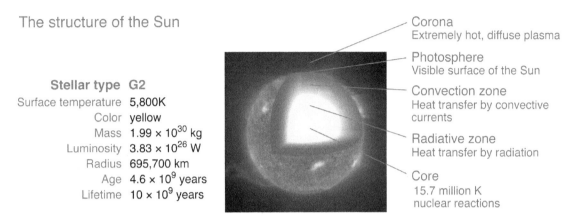

The structure of the Sun

Stellar type	G2
Surface temperature	5,800K
Color	yellow
Mass	1.99×10^{30} kg
Luminosity	3.83×10^{26} W
Radius	695,700 km
Age	4.6×10^9 years
Lifetime	10×10^9 years

- Corona: Extremely hot, diffuse plasma
- Photosphere: Visible surface of the Sun
- Convection zone: Heat transfer by convective currents
- Radiative zone: Heat transfer by radiation
- Core: 15.7 million K, nuclear reactions

The structure of the Sun

Below the photosphere, in the convection zone, there are enormous circulating currents of gas move heat outward toward the surface. Below the convection zone is the denser radiation zone in which energy from the core is carried primarily by radiation. The solar core extends to about 25% of the Sun's radius and has a density over 150 g/cm³ which is seven times more dense than platinum, the densest solid metal.

The Sun's main sequence life

The Sun ignited from a collapsing cloud of gas and dust 4.6 billion years ago, consistent with the oldest dated meteor material (4.57 billion years). The nuclear reactions fusing hydrogen into helium in the Sun's core convert four million tons of matter into energy every second, releasing 3.83×10^{23} joules every second. The Sun has enough hydrogen in its core to sustain this rate of energy production for 10 billion years as a main sequence star. During that time the core slowly shrinks and the Sun gets hotter and more energetic by about one percent every 100 million years.

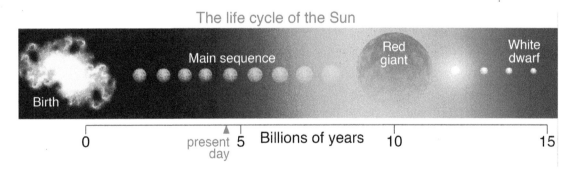

The life cycle of the Sun

Red giant and white dwarf stages

Our Sun is too small to create a supernova. When the hydrogen in the core is exhausted 5.4 billion years from now, the core will shrink, heat up, and begin fusing helium into heavier elements. A shell around the core will then become hot enough to fuse hydrogen from the radiative zone and the Sun will expand into a red giant star. As a red giant, the Sun will be larger than Earth's orbit and will consume the inner planets. A few billion years later the Sun will collapse down into a hot, compact, white dwarf star.

Nuclear fusion reactions in stars

Part one of the proton-proton chain

A complex series of nuclear reactions called the proton-proton chain fuse hydrogen to helium in the sun's core. The chance of three particles hitting each other at the same instant is negligible; therefore, each step in the proton-proton chain involves two particles. The first step is a fusion between two hydrogen nuclei (protons) to produce deuterium (^{2}H), a positron (e^+), and a neutrino (ν_e). This step is slow because the nuclear reaction that converts a proton to a neutron is mediated by the weak force. In the diagram, energy is expressed in units of megaelectron volts (MeV). One MeV is 1.6×10^{-13} J.

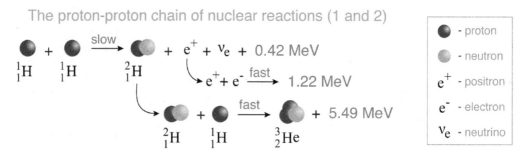

Part two of the proton-proton chain

The helium-3 produced by the first step becomes the reactant for three different branch reactions that eventually produce helium-4. The most likely branch of the reaction (86%) fuses two ^{3}He nuclei to make ^{4}He and two protons. In a second reaction sequence, about 14% of the ^{3}He nuclei fuse with ^{4}He that results in two ^{4}He nuclei and one proton. A low-probability third reaction (0.11%) fuses ^{3}He and ^{4}He nuclei together with a different intermediate stage involving boron-8.

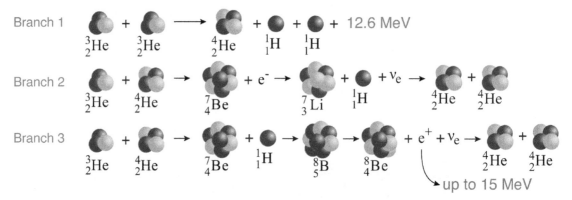

The CNO cycle

Stars more massive than 1.8 solar masses produce most of their energy through a different reaction path called the carbon-nitrogen-oxygen (CNO) cycle. The CNO cycle also fuses hydrogen into helium releasing energy but with a catalyzed sequence of nuclear reactions using isotopes of carbon, nitrogen, and oxygen. Less than 1% of the sun's energy comes from the CNO cycle but nearly 100% of the energy in a hot, high-mass star such as Sirius (close to the constellation Orion) is produced by the CNO cycle.

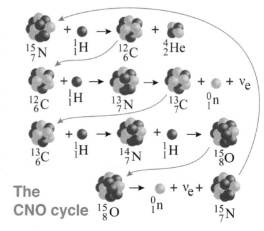

The CNO cycle

The interstellar medium

The matter between stars

Stars are separated by vast distances that are nearly, but not quite, a perfect vacuum. The **interstellar medium** is a low-density mixture of ions, atoms, molecules, and larger dust grains. Particle density ranges from thousands to hundreds of millions of particles per cubic meter. Spectroscopic evidence suggests the interstellar medium contains about 89% hydrogen and 9% helium. The remaining 2% is composed of heavier elements including oxygen, carbon, and the rest of the elements found on the periodic table. However, the density varies greatly and is much higher in nebulae and molecular clouds.

Gas and dust near the galactic center | Planetary nebula NGC 2440 | The pillars of creation in the Eagle Nebula

Nebulae

A **nebula** is an interstellar cloud of dust, hydrogen, helium and other ionized gases. Nebulae can be vast in size, up to several million light years across. The brightest nebula in the northern hemisphere, the Orion Nebula is 1,340 light years from Earth. The Orion nebula is relatively small, only 24 light years across with a total mass about 2000 times the Sun. Although denser than the space around them, nebulae are far less dense than any laboratory vacuum created on Earth. A nebula the size of the Earth would have a total mass of only a few kilograms.

Clouds of dense matter

Over billions of years, random gravitational forces from moving galaxies and stars sweeps the interstellar matter into enormous molecular clouds. Up to several thousand light years across, molecular clouds are cold enough (10 K - 50 K) that compounds can form and remain bonded together. A single molecular cloud might contain several million times as much matter as the Sun. The chemical composition of molecular clouds varies greatly and is mostly molecular hydrogen (H_2) as well as helium and a small percentage of elements and compounds including water (H_2O), carbon dioxide (CO_2), ammonia (NH_3) and methane (CH_4). Spectroscopic evidence suggests that organic compounds such as methanol (CH_3OH) and amino acids exist in molecular clouds.

Star forming regions

Molecular clouds are the birthplaces of stars and planets. This is a consequence of the combination of low temperatures and relatively high density. When any area of a cloud becomes denser than average, gravitational forces can exceed the weak gas pressure holding the diffuse matter apart. Localized areas may condense into a protostar which ignites to become a star. The photograph shows a cluster of hot, young stars that formed from a molecular cloud in the giant nebula NGC 2070. Stellar wind from these new stars has partially blown a bubble of space around the cluster. This likely triggered the collapse and formation of more stars at the bubble's edge.

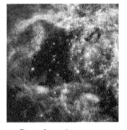

Star-forming region in NGC 2070
Hubble Space Telescope

Chapter 24

Section 1 review

Hydrogen and helium are by far the most abundant elements in the universe. They are what primarily make up stars like our Sun. The extremely high temperature and pressure of the Sun's core ionizes atoms into plasma. The energy of the Sun is generated through nuclear fusion of hydrogen nuclei into helium and occasionally into lithium or beryllium. In the early 20th century it was discovered that the universe is expanding and therefore cooling. Working backwards led scientists to the idea that 13.7 billion years ago the universe was created in the "Big Bang." As stars age they produce heavier elements. Massive stars will eventually explode in a supernova, expelling matter into the interstellar medium. Gravity can condense molecular clouds into new stars and planets.

Vocabulary words	light-year, spectrometer, Big Bang Theory, Doppler effect, cosmic background radiation, supernova, plasma, ionization, Hertzsprung-Russell diagram, photosphere, interstellar medium, nebula

Key relationships	• Distribution of elements in the universe: 75% hydrogen, 23% helium, and the remaining 2% is the rest of the elements

Review problems and questions

1. What kind of matter do you see when you look at stars?

2. What type of reaction takes place on the sun, resulting in sunlight?

3. What is the evidence for an expanding universe?

4. What happened to the very young, 3-minute-old universe to form the elements?

5. Why isn't space "truly empty?"

24.2 - The Solar System

Our solar system includes the Sun, eight major planets, more than 200 known moons, at least 5 dwarf planets (including Pluto and Charon) and innumerable asteroids, comets, and Kuiper Belt Objects. The solar system is much larger and more complex and extensive than once thought. By some estimates the solar system extends almost a light year from the Sun to the outer edge of the frozen Oort cloud.

Inside the solar system

The Oort cloud

A useful distance scale for the solar system is the **astronomical unit (AU)** which is the distance from Earth to the Sun (1 AU = 150 million km). The farthest reaches of the solar system are estimated to spread up to 50,000 AU from the Sun in the Oort cloud. The Oort cloud contains trillions of frozen objects and is the source for long-period comets. Astronomers estimate the Oort cloud to contain as much as 5 - 10 Earth masses of ices such as water, methane, ethane, carbon monoxide, and hydrogen cyanide as well as other elements including carbon compounds.

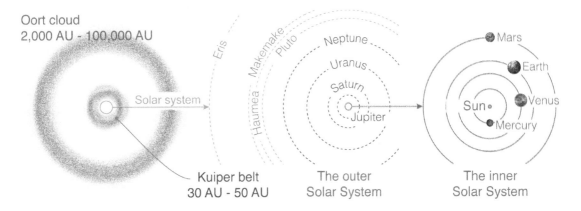

The Kuiper belt

The Kuiper belt occupies a disk-shaped volume between the orbit of Neptune (30 AU) and the inner edge of the Oort cloud (50 AU). Like the asteroid belt, it consists mainly of small bodies which are remnants from the solar system's formation. Unlike asteroids, most Kuiper belt objects are frozen ices such as methane, ammonia and water. Dwarf planets such as Pluto, Haumea, and Makemake are among the larger Kuiper belt objects (KBOs). Neptune's moon, Triton, and Saturn's moon, Phoebe, are thought to be captured KBOs.

The outer solar system

The planets naturally fall into an inner and outer group. The outer planets are gas giants composed mostly of hydrogen and helium with some methane and other gaseous compounds. The largest is Jupiter followed by Saturn, Uranus, and Neptune. All the gas planets have relatively low densities, rapid rotations, thick atmospheres, and many moons. The outer planets are much farther apart than planets in the inner system. Mars and Jupiter's orbits are about 550 million kilometers apart, nearly ten times the distance between the orbits of Earth and Mars.

The inner solar system

The inner planets are rocky and include Mercury, Venus, Earth, and Mars. All the inner planets have compositions of oxygen, silicon, iron, nickel, and other heavier elements with comparatively little hydrogen and almost no helium. The inner planets have relatively high densities, slow rotations, solid surfaces, and few moons. There are three moons in the inner system compared to Jupiter and Saturn which have more than 60 moons each!

Chemical composition of the planets

Creation of the solar system

The Sun and planets formed from the same materials in the collapse of an ancient nebula. From the abundances of heavier elements we know the Sun is a second generation star made of gases and dust that had already been exploded out of at least one previous generation of stars since the Big Bang. Matter with the smallest velocity fell inward to become the Sun. Matter with a higher velocity, coalesced into a disc, clumped, and gradually formed into into planetesimals of ~5 km in size. This process is called **accretion**. Planetesimals merged through further collisions. The composition of the planetesimals was similar to the matter that makes up the Sun itself. The differences between the inner and outer planets arose because of differences in melting point, density, and temperature of the elements that make them up.

Planets of the Solar System

	Mercury	Venus	Earth	Mars	Jupiter	Saturn	Uranus	Neptune
Type	Rocky	Rocky	Rocky	Rocky	Gas giant	Gas giant	Gas giant	Gas giant
Diameter (km)	4,878	12,102	12,756	6,794	142,796	120,660	51,200	49,500
Mass (Earth =1)	0.06	0.82	1	0.11	318	95	14.5	17.1
Orbit radius ($\times 10^6$ km)	58	108	150	228	778	1,430	2,875	4,504
Number of maj. moons	0	0	1	2	67	62	27	14
Surface gravity (Earth = 1)	0.38	0.91	1	0.38	2.5	1.1	0.89	1.1
Temperature range (°C)	-170 +400	+430 +460	-88 +48	-150 +20	-108	-139	-197	-201
P, atmosphere (Earth= 1)	0	92	1	0.01	These are gas planets with no surface			

The gas giants

Chemically, Jupiter is more like the Sun than it is like Earth! In fact, if Jupiter were about ten times larger our solar system would have two suns instead of one. Many stellar systems are binaries, dual stars orbiting each other. The chart below shows hydrogen and helium dominate the composition of all the gas giants.

Chemical composition of the Sun and gas planets

	Sun	Jupiter	Saturn	Uranus	Neptune
Hydrogen (H_2)	84%	86%	97%	83%	79%
Helium (He)	16%	14%	3%	15%	18%
Water (H_2O)	-	0.1%	-	-	-
Methane (CH_4)	-	0.2%	0.2%	2%	3%
Ammonia (NH_3)	-	0.07%	0.03%	-	-

The terrestrial planets

Once the Sun ignited and started producing energy, planetesimals close to the Sun could only accrete substances with high melting points, such as iron, nickel, aluminum, and rocky silicates. These compounds are relatively rare in the universe. Astronomers estimate they made up only 0.6% of matter that formed the solar system. This limited amount of matter is one of the reasons why inner planets are small compared to the outer planets. Another reason is that smaller objects cannot gravitationally capture material as efficiently as larger objects. Once it reached a certain mass, Jupiter efficiently swept its orbit clear of matter. Earth was too small to gravitationally sweep out a large volume of space.

Blackbody radiation

Thermal radiation

All matter has a temperature above absolute zero and gives off **thermal radiation**. Objects cooler than the Sun radiate less intense energy with longer wavelengths. At the Sun's surface, thermal radiation emitted is most intense in the visible range of wavelengths near yellow-white. An object does not have to give off visible light to give off thermal radiation. You can't see thermal radiation outside of the visible range.

The blackbody spectrum

When light strikes an object, some is reflected and some is absorbed. However, an ideal **blackbody** is an object that absorbs all the light that strikes it without reflecting any back into the environment. Transparent or shiny objects are poor blackbodies because they do not absorb as much energy as opaque objects. Only thermal radiation is emitted from a blackbody. A **blackbody spectrum** describes the thermal radiation power at different wavelengths. You can see from the diagram that room temperature (300 K) objects emit thermal radiation mostly in the infrared part of the spectrum. A stove burner at 900 K glows red but most of the energy is still in the infrared region. At 5,800 K, the Sun radiates some power in the visible range but much of the Sun's radiated energy is infrared.

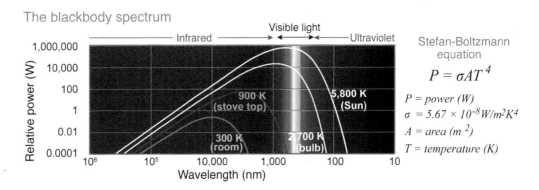

Stefan-Boltzmann equation

$$P = \sigma A T^4$$

P = power (W)
$\sigma = 5.67 \times 10^{-8} W/m^2 K^4$
A = area (m^2)
T = temperature (K)

Radiated power

The Stefan-Boltzmann equation relates the power of radiation emitted with temperature to the fourth power. If temperature doubles, the radiated power is multiplied by 16! The power radiated, P, is given by the Stefan-Boltzmann equation in the Kelvin temperature, T.

Planetary temperature and energy

A planet's surface temperature is reached when the thermal radiation emitted by the planet is exactly balanced by the energy received from the Sun! A planet must re-radiate all the energy it absorbs from the Sun or the planet will get warmer. Scientists have created a model to predict the surface temperatures of most planets, assuming they are blackbodies.

1. According to the inverse square law, the intensity of sunlight is the power emitted P_{sun} divided by the area of a sphere of radius, R, equal to the planet's orbit radius.
2. The power absorbed is the intensity of sunlight × πr^2 where r is the planet radius.
3. The thermal power emitted is the surface area of the planet ($4\pi r^2$) times σT^4

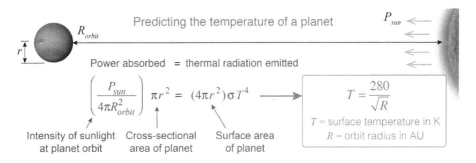

Section 24.2: The Solar System

Temperature in the inner and outer solar system

Distance and temperature

The chemistry of each planet depends strongly on the planet's surface temperature, which is determined almost completely by the distance from the Sun. The Sun radiates 3.8×10^{26} watts of energy equally in all directions. As you get farther from the Sun, that energy is spread out over a larger and larger area. The graph below shows how the intensity of sunlight changes with the distance to each planet in the solar system. With the exception of Venus, temperature decreases in line with the drop in light intensity according to the model from the previous page. The horizontal axis is in astronomical units; however, the scale is different from 0-5 AU to better separate the inner planets.

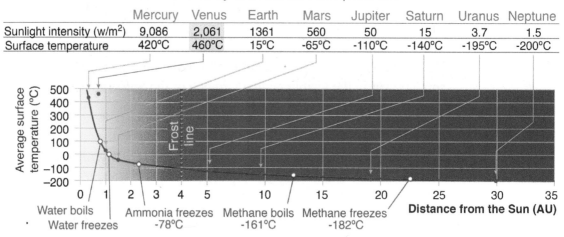

Solar intensity and surface temperature

	Mercury	Venus	Earth	Mars	Jupiter	Saturn	Uranus	Neptune
Sunlight intensity (w/m²)	9,086	2,061	1361	560	50	15	3.7	1.5
Surface temperature	420°C	460°C	15°C	-65°C	-110°C	-140°C	-195°C	-200°C

Surface temperature and light intensity

Intensity describes the power of light energy per square meter. The intensity of sunlight reaching Mercury is 9,000 watts per square meter on the day side! At the top of Earth's atmosphere sunlight has an intensity of 1,361 w/m², sufficient to maintain an average surface temperature of a comfortable 15 °C. The outer solar system is so far from the Sun that Jupiter gets 27 times less sunlight that Earth and frigid Neptune gets 907 times less.

Temperature and planetary atmosphere

The rocky planets end and the gas planets begin when the average surface temperature drops below the freezing point of volatiles such as ammonia. Astronomers refer to this as the **frost line** and it occurs around 4 AU from the Sun. The boiling points of several important compounds in planetary atmospheres are shown on the graph. At one atmosphere of pressure, water boils at 100 °C and freezes at 0 °C. Both ammonia and carbon dioxide freeze at -78 °C, methane freezes at -182 °C. Planets inside the frost line boiled away all their light gasses early in their formative years. This is the reason there is virtually no free hydrogen or helium on Earth. Water has an extraordinarily high melting point compared to other small molecular compounds, and Earth falls within a narrow range of orbits for which water is liquid at the mean surface temperature.

The problem with Venus

If you look carefully at the graph you will see that all the planets follow the same pattern of decreasing temperature with intensity except Venus. Venus is much too hot for its distance from the Sun. This has to do with the particular chemistry of Venus's atmosphere.

What makes Venus so hot?

Venus

Venus and Earth are similar planets

The Earth and Venus are very similar in size, gravity, and composition. Both planets are within the range of distance from the Sun at which water should be liquid on the surface. Up to the beginning of the 20th century people thought Venus might be a lush jungle planet. Similar to Earth, Venus has a substantial atmosphere, however Venus is perpetually covered in dense clouds. In the last fifty years space probes have revealed the truth about Venus, and how its climate and chemistry are very different from that of Earth.

Venus
Orbit radius 108 million km
Diameter 12,100 km
Composition rocky, silicates
Gravity 0.9 g
CO_2 95.6%
N_2 3.5%
O_2 0%
Pressure 92 atm

460°C

15°C

Earth
Orbit radius 150 million km
Diameter 12,756 km
Composition rocky, silicates
Gravity 1.0 g
CO_2 0.04%
N_2 78%
O_2 21%
Pressure 1 atm

The hottest planet

Far from being a tropical jungle, we now know that Venus has the highest surface temperature of any planet, over 460 °C (860 °F). Inhospitable and hot enough to melt lead, Venus is hotter than Mercury even though Venus is farther from the sun and receives only 25% as much solar intensity. Venus experiences such extreme temperatures largely because of the mass and chemical makeup of its atmosphere, especially due to the abundance of carbon dioxide and the lack of water vapor.

Greenhouse effect

Venus' atmosphere is 96% carbon dioxide with an incredible density 92 times higher than the pressure on Earth at sea level. The atmospheric pressure on the surface of Venus is comparable to being a kilometer underwater in Earth. Imagine being at the bottom of an ocean of carbon dioxide. There is almost no water, no oxygen, and only 3.5% nitrogen. Venus' atmosphere traps heat and warms the planet on average 389 °C higher than the planet would get if there was no heat trapped by the atmosphere. This is known as the "greenhouse effect" and it is occurring on Earth as well. By comparison, Earth's atmosphere makes the planet's surface on average 33 °C warmer compared to no greenhouse effect.

Venus' energy balance

Incoming radiation from the sun that reaches Venus' surface eventually finds a balance with outgoing thermal energy that leaves the planet as heat. Here is how a balance is reached:

1. Sunlight in visible wavelengths is absorbed by the atmosphere and ground.
2. Thermal radiation from the ground cannot escape into space because Venus' atmospheric CO_2 absorbs large amounts of infrared energy.
3. Venus' surface becomes hotter and hotter until a balance is reached between energy input and energy that could be radiated back into space through the thick atmosphere.

Earth reaches an energy balance in the same way. Earth's atmospheric composition allows it to reach a balance at a far lower average surface temperature than Venus.

Venus' atmosphere

The upper atmosphere of Venus is perpetually covered in dense clouds of hot sulfuric acid. It is hard to imagine any place in the solar system less hospitable than Earth's supposed sister-planet! The Russian Venera-9 probe that landed in 1975 was destroyed by the corrosive atmosphere in 127 minutes, after relaying the first pictures ever taken from Venus' surface.

Computer image of the Venusian surface from the Magellan probe. (NASA)

Section 24.2: The Solar System

Mars

Mars

Although it has always been a favorite of science fiction writers, an honest description of Mars is a frozen airless desert. With only 11 percent of Earth's mass, Mars has a surface gravity only 38% as strong as Earth. Like Venus, Mars' atmosphere is 95% carbon dioxide; however, unlike Venus, Mars' atmosphere is very thin. Mars' weak gravity is insufficient to hold a substantial atmosphere and the pressure on the surface is 100 times lower than the air pressure at sea level on Earth. The average Martian surface temperature is -65 °C, reaching above 0 °C only in summer in equatorial lattitudes. This is due to the combination of a thin atmosphere and the fact that Mars gets only 43% of the sunlight energy Earth receives.

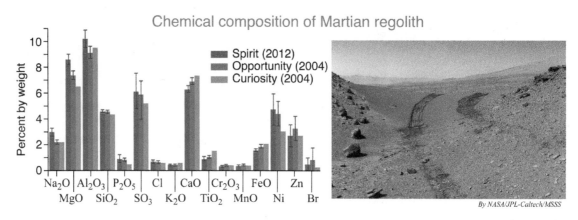

By NASA/JPL-Caltech/MSSS

The Martian regolith

Mars' surface is covered with regolith (non-biological soil) and rocks. Since 2004 there have been several robotic landers that conducted chemical assays of the Martian regolith. The graph above identifies some minerals common to Earth's crust, such as silica (SiO_2) and alumina (Al_2O_3) but very little of the more complex forms typical of most Earth soils such as feldspars ($KAlSi_3O_8 - NaAlSi_3O_8 - CaAl_2Si_2O_8$) which have volcanic origins. While calcium and magnesium oxides (CaO, MgO) are present, there is no evidence of calcium carbonate ($CaCO_3$) a significant component of Earth soils. Calcium carbonate is associated with limestone, a mineral with biological origins.

Water on Mars

Mars' red color comes from the mineral red hematite (Fe_2O_3) commonly known as rust. Hematite occurs on Earth in two forms: red and grey. Gray hematite typically forms in the presence of water and has a dark gray metallic luster due to its different crystal structure. While red hematite forms when iron is exposed to oxygen in air, grey hematite usually precipitates from iron-rich water over long periods of geological time. When gray hematite was discovered near the Martian equator, scientists had further support for the hypothesis that Mars once had deep, liquid water on the surface.

Evidence of water on Mars

Grey hematite

Red hematite

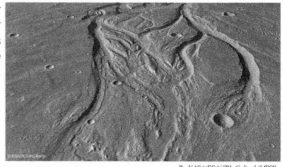

What appear to be water carved formations on the Martian surface.

By NASA/ESA/JPL-Caltech/MSSS

The asteroid belt, meteors, and comets

The asteroid belt

The asteroid belt is a region between Mars and Jupiter (1.5 AU - 5.2 AU) that contains millions of planetesimals left over from the formation of the solar system. The total mass of the main asteroid belt is about 4% of the Moon's mass, and half the total is in the four largest bodies: Vesta, Pallas, Hygiea and the dwarf planet, Ceres. Two other large groups, the Trojan asteroids, orbit with Jupiter with one group leading and the other trailing the planet in a gravitational resonance. Because many are unchanged since the solar system formed, the asteroids are important evidence to the original composition of the solar system.

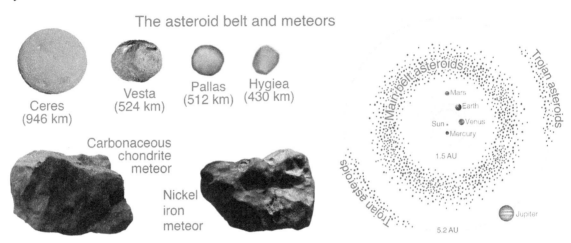

The asteroid belt and meteors

Carbonaceous chondrites

Over 75% of the visible asteroids are Carbonaceous chondrites (C-type) with chemical compositions believed to closely match the primordial early solar system. Many meteors originate in the asteroid belt and have similar chemical compositions. C-type asteroids are composed of water (up to 22%) and carbon-rich compounds including amino acids and complex aromatics.

Silicate-rich asteroids

Silicate-rich (S-type) asteroids are more common toward the inner region of the main belt and make up about 17% of the total population. Spectra reveal the silicates and some metal, but no significant carbon compounds. This indicates that these asteroids have been changed from their primordial composition, probably through melting and reforming.

Metal-rich asteroids

Metal-rich (M-type) asteroids make up about 10% of the total population and are believed to be similar to nickel-iron meteorites. M-type asteroids are most common in the middle of the main belt and may have formed from the metallic cores of once-melted asteroids that had undergone collisions with other asteroids early in the formation of the solar system.

Comets and icy bodies

Comet 67 P1 Churyumov–Gerasimenko

By ESA/Rosetta/NAVCAM

Successive images of comet PanSTARRS 2012 K1 taken from 8:14 am to 10:26 pm

Comets and icy bodies

Beyond 2 AU the asteroid belt contains many comets as the equilibrium surface temperature is below the freezing point of water. Comets are mixtures of ices, dust and rock, including water, methane and ammonia. It is thought that most of Earth's water originated in comets that collided with Earth after the planet's crust had solidified.

Section 24.2: The Solar System

The gas giant planets

Planets of gas and liquid

The common image of a "planet" with a hard, walkable surface doesn't match the enormous gas planets past the asteroid belt. The four outer planets are giant balls of cold gas and liquid without a definite surface. While they have complex atmospheres and structures, Jupiter, Saturn, Uranus and Neptune have nothing that remotely resembles features such as mountains or valleys common to the inner system planets.

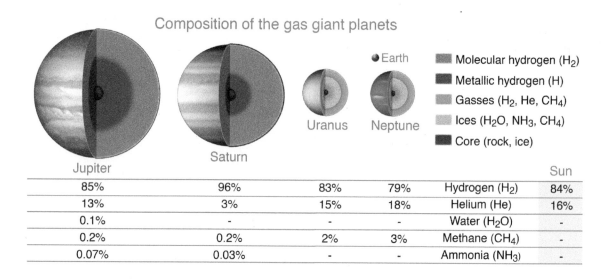

Composition of the gas giant planets

Jupiter	Saturn	Uranus	Neptune		Sun
85%	96%	83%	79%	Hydrogen (H_2)	84%
13%	3%	15%	18%	Helium (He)	16%
0.1%	-	-	-	Water (H_2O)	-
0.2%	0.2%	2%	3%	Methane (CH_4)	-
0.07%	0.03%	-	-	Ammonia (NH_3)	-

Metallic hydrogen

The pressure in Jupiter's atmosphere reaches more than a million atmospheres. At this high pressure molecular hydrogen is squeezed apart into separate protons and electrons. The combination of high pressure and low temperature creates a "cold plasma" phase, which acts more like a liquid metal than a hot gas. Hydrogen in this phase is called metallic hydrogen. At even higher pressures hydrogen becomes solid.

Structure of the gas planets

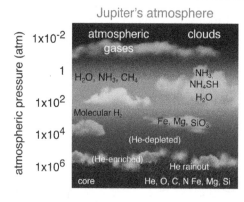
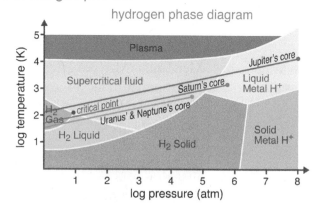

Helium sequestration

Helium is less abundant in the atmospheres of both Jupiter and Saturn because helium is soluble in metallic hydrogen. Planetary scientists believe an ultra-cold "rainout" of liquid helium depletes helium from the atmospheres of both planets and sequesters it in the metallic hydrogen region. The effect is strongest on Jupiter with its larger mass and higher proportion of metallic hydrogen compared to molecular hydrogen. Saturn has a lower core pressure and therefore less metallic hydrogen. Uranus and Neptune do not have enough core pressure to form metallic hydrogen.

Titan

Titan is unique

Circling the planet Saturn is one of the most enigmatic worlds in our solar system. The moon Titan has a diameter of 5,150 km, larger than the planet Mercury. Unique among moons, Titan has a solid surface, a dense atmosphere, oceans, rivers, and even weather! Out of a hundred or so planets and moons in our solar system, only four have solid surfaces and substantial atmospheres: Earth, Mars, Venus, and Titan. Like Earth, Titan has rainstorms, seasons, river valleys, mountains, and oceans. Unlike Earth, however, the temperature on Titan is so cold that methane runs like rivers across the surface. Titan has a methane cycle similar to Earth's water cycle.

Diameter	5,150 km
Surface gravity	0.14 g
Surface temperature	−179°C

Surface pressure	1.45 atm
Upper atmosphere	98.5% N_2, 1.4% CH_4, 0.2% H_2
Lower atmosphere	95% N_2, 4.9% CH_4

Surface conditions on Titan

The mean surface temperature on Titan is −190 °C (−270 °F). At this extremely low temperature the phases of ordinary matter become very different from what we experience on Earth. Water is as hard as granite and Titan's rocks and surface are largely water ice with hydrocarbons mixed in. Liquid water on Titan is more analogous to volcanic magma here on Earth. The volcanoes on Titan erupt with ammonia ice slush. The boiling point of methane (CH_4)—(natural gas on Earth) is −161 °C, about 30 °C above Titan's surface temperature. Titan's "water cycle" involves flowing liquid methane and ethane. Flammable natural gas and ethane flow in vast rivers and collect in deep oceans on Titan. Titan's lakes and oceans contain hundreds of times more hydrocarbon fuels than all known reserves of fossil fuels on Earth.

The Cassini-Huygens probe

On January 14, 2005, the Cassini-Huygens probe, a joint effort between NASA and the European Space Agency (ESA), landed on Titan after a journey of more than 1.3 billion kilometers, nearly 10 times the distance from Earth to the sun. Cassini-Huygens is the first and only human technology to land in the outer solar system. The car-sized probe took more than eight years to reach Titan with enough battery power to collect data for 90 minutes on the surface.

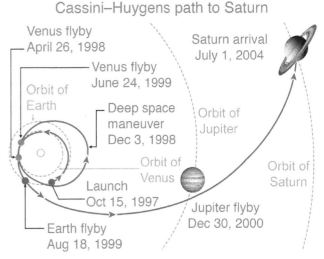

Cassini–Huygens path to Saturn

Section 24.2: The Solar System

Chapter 24

Section 2 review

The Sun was formed from a condensed molecular cloud with the leftover material forming the planets. Near the Sun metals and silicates gathered to form the inner planets while the outer planets started with water, methane and ammonia before growing large enough to capture hydrogen and helium. Distance from the Sun and the balance of absorption and radiated thermal energy determines the temperatures of the planets. Venus actually has the highest surface temperature due to a runaway greenhouse effect. Mars is essentially a frozen desert with a very thin atmosphere. Between Mars and Jupiter is the asteroid belt, some of which are silicate-rich or metal-rich. The outer planets are gas giants with "cold plasma" atmospheres and cores of rock or ice. Titan, one of Saturn's moons, actually has a solid surface and atmosphere.

Vocabulary words	astronomical unit (AU), accretion, blackbody spectrum, blackbody, thermal radiation, frost line, intensity

Review problems and questions

1. Planets after Mars are known as the "gas planets." Why are there no rocky planets after Mars?

2. Why is CO_2 such an important atmospheric gas in determining a planet's temperature?

3. How is it possible for 2 different planets to have the same average global temperature but different boiling points for substances like water and CO_2?

4. What property of the gas planets makes gases behave so differently on those planets compared to Earth?

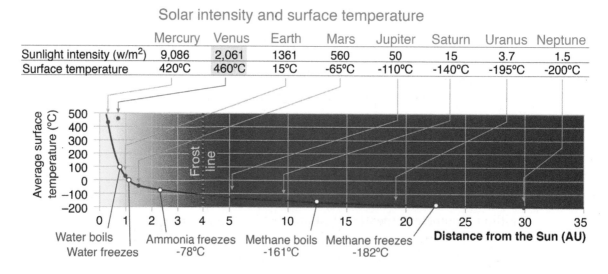

5. Use the above graph to discuss the surface conditions for a plant where the asteroid belt is today. What would the average temperature be like? Would it be likely to be a rocky planet or a gas planet?

24.3 - Life Outside the Earth

The question of the origin of life is a central mystery in science. Experiments provide tantalizing clues but definitive proof of one hypothesis or another is elusive, and may always be. We do not have the luxury of observing experiments over a hundred million years or more. Yet, there is a way to evaluate different ideas about how life began: to find life elsewhere in the universe. Given that there are 200 billion stars just in our own Milky Way, it is inconceivable to most scientists that life is unique to just one speck of a planet on the outer rim.

Extraterrestrial life

Life requires complexity

While viruses technically can reproduce, higher forms of life require complexity. Think about all the different things a single living cell must do. The cell must take in energy, maintain its integrity, reproduce, and be responsive and adaptive to its environment. These tasks cannot be accomplished by anything too simple. Interesting life, intelligent life, requires even more complexity. Since our subject is chemistry, let us further refine our discussion to chemistry-based life as opposed to more exotic ideas. All life on Earth is chemistry-based. While humans may eventually create machine-based life, this is the subject for another discussion!

Characteristics of life

It is reasonable to assume that any plausible living organism would have the following minimum characteristics.

1. A complex system of chemicals and reactions capable of fulfilling the myriad functions of life.
2. A stable body structure that organizes and maintains the living system of chemicals and reactions for extended periods of time.
3. A life-sustaining environment that contains matter and energy that can be used by the organism.

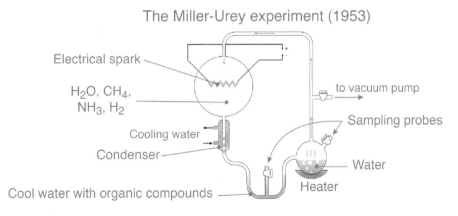

The Miller-Urey experiment (1953)

The Miller-Urey experiment

In 1953, Stanley Miller and Harold Urey put methane, ammonia, hydrogen, and water in a sealed container, then subjected the mixture to agitation, heat and lightning. The experiment simulated the simplest chemical conditions most likely to have been present in the primordial Earth. After only one week of operation Miller and Urey found five of the 20 amino acids in all proteins had been spontaneously synthesized. Fifty-four years later, (after Miller's death) scientists using more sophisticated techniques found more than 20 amino acids in sealed vials preserved from the original experiments. These included more than the 20 that naturally occur in all present-day Earth-life.

The beginnings of biochemistry

Biochemistry may be common

Five days is less than an eye-blink in geologic time. The Miller-Urey experiment proved that complex chemistries will arise almost instantly in any environment with similar conditions containing only four simple compounds known to be abundant in the molecular clouds from which planets form. In addition we know from analyzing carbonaceous chondrites, asteroids, and comets that organic compounds including amino acids are already present in interstellar molecular clouds. The Murchison meteorite that fell near Murchison, Victoria, Australia in 1969 contained over 90 different amino acids, nineteen of which are in Earth life. The same comets that brought water to the early Earth certainly carried amino acids and other complex organic molecules along with them.

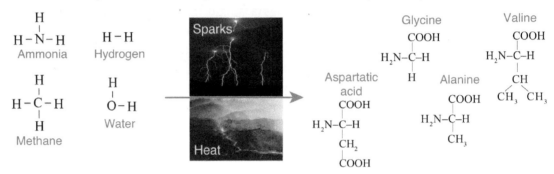
Synthesis of biological molecules from simple compounds

The oldest life on Earth

Recently discovered fossil microbes date to 4.2 billion years ago, barely 400 million years after the formation of the planet and only 100 million years after the planet cooled sufficiently to have oceans of liquid water. The microscopic bacteria were found in rock formations in Quebec, Canada and probably lived in hot vents in the 140 °F (60 °C) oceans which covered the planet. Early fossilized microbes from 3.7 billion years ago resemble the microbial mats we find in stromatolites, layered bio-chemical structures formed through the binding of sedimentary grains by microorganisms such as cyanobacteria.

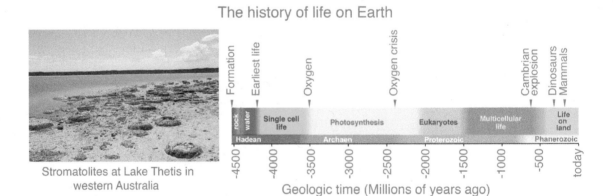
Stromatolites at Lake Thetis in western Australia

Where life may have started

It was once thought that life began in shallow, warm tide pools. However, scientists believe the early Earth was a violent, cataclysmic place with frequent meteor impacts, comet impacts, and extensive volcanism. The discovery of archaebacteria in deep-sea **hydrothermal vents** suggest that life may have begun deep underwater, shielded from the turbulence of the surface. It is probable that life got started many times, was sterilized by the next cosmic accident, then started again elsewhere.

Extremophiles

Thermophilic bacteria

In the past few decades scientists have discovered that life exists in far more places than was believed to be possible. **Extremophiles** are organisms that thrive where most other living things cannot. Because extremophiles live in harsh conditions, there is a possibility of life on other planets and moons where conditions are equally extreme. **Thermophiles**, or heat lovers, are bacteria found in very hot environments. At the time this book was written the record holder was Geogemma barossii, a single-celled microbe of the domain Archaea discovered 200 miles off Puget Sound near a hydrothermal vent. G. barossii can reproduce at 121 °C (250 °F)! Similar thermophilic microbes colonize hot springs heated by geothermal activity, such as those in Yellowstone National Park, and extremely hot geysers on the ocean floor.

Underground bacteria

A Princeton research group lead by Dr. Tullis Onstott discovered an isolated community of bacteria that derives its energy from the decay of radioactive rocks rather than from sunlight. 2.8 kilometers underground, this extraordinary bacterial community thrives in nutrient-rich groundwater near a South African gold mine. The bacteria has been isolated from the Earth's surface for several million years. The new discovery suggests deep subsurface life may be quite common.

Extremophile life

Thermophilic bacteria thrive at temperatures fatal to ordinary life.

Deep sea vent colonies form around undersea volcanic vents rich in toxic sulfides and far from sunlight

Chemosynthesis

Chemosynthesis is the biological conversion of molecules such as carbon dioxide or methane into nutrients by oxidizing inorganic compounds such as hydrogen gas and hydrogen sulfide. The oxidation reaction is a source of energy much like photosynthesis, except without light! A chemosynthetic reaction uses hydrogen sulfide (from undersea volcanic vents) to produce glucose, the same end product as photosynthesis.

Hydrothermal vents

Chemosynthetic life is the basis of the vent "food chain"

Tube worms near a hydrothermal vent in the Galapagos rift.

Hydrothermal vent life

Hydrothermal vents are volcanic geysers miles deep in the ocean that spew boiling jets of sulfuric acid and hydrogen sulfide. No sunlight ever reaches this deep abyss. These creatures use chemosynthesis to survive. No one thought life existed there until the submersible Alvin discovered the first colony of tube worms and other weird creatures around a vent in the Galapagos Rift.

Oxygen is a chemical signature of life

How could we recognize life?

How would we prove there was life on another planet if we could not physically go visit and collect specimens? How could it be proven that Earth had life if the only evidence were photographs like the ones below? Both photographs show only geologic features and atmosphere and could be taken on Mars or another world far from the Sun. If large life forms existed on Mars they would be would be obvious in the photograph but what about microscopic life similar to bacteria? Evidence shows that life begins as microscopic life. For more than three billion years of Earth's four and a half billion year history, only single celled life existed.

How would you recognize evidence of life from these photographs?

Wall Valley in Antarctica   The Sahara desert in Algiers

Oxygen

The presence of atmospheric oxygen is a sure sign of life. Oxygen is a highly reactive element and while oxygen makes up 47% of Earth's crust, without living plants virtually every molecule would be would be chemically bonded in silicates and other rocks. The primordial Earth had no free oxygen in the atmosphere. Evidence from the composition of rocks suggests oceanic cyanobacteria, which evolved about 2.45 billion years ago began photosynthetic production of oxygen.

The history of oxygen on Earth

Carboniferous period fossil of Mariopteris sauveurii

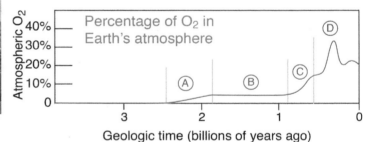

Atmospheric oxygen over time

For almost two billion years after photosynthesis evolved, the increasing amount of biologically produced oxygen stayed dissolved in the ocean, (A) in the graph above, or was captured by oxidation of minerals on land (B). It was only 750 million years ago, (C) that oxygen started to build up in the atmosphere, reaching 33% in the carboniferous era (D) when the land was covered with swamps and giant plants between 360 million and 300 million years ago. The fossilized remains of carboniferous plants are the coal and oil reserves of today. The oxygen concentration became so high that it may have contributed to a mass extinction event.

The search for life on Mars

The canals of Mars

In 1877 telescopic observations of Mars appeared to show a network of straight-line features. First described by Italian astronomer, Giovanni Schiaparelli, then confirmed by later observers, Schiaparelli called the lines canali, which translated into English as "canals." American astronomer Percival Lowell believed the canals were built for irrigation by an intelligent civilization on Mars. Lowell built an observatory in Flagstaff, Arizona to study them. The first photographs of the real Martian surface taken by Mariner 9 in 1971 finally put Mars' canals to rest. The astronomers of Schiaparelli's time spent hours staring through their telescopes waiting for a brief clear image in a moment of still air. Without cameras, they had to draw pictures of what they saw and the late 19th century was a time of popular canal building on Earth including the Suez Canal in 1869, and the first attempt at the Panama Canal in 1880.

The canals of Mars

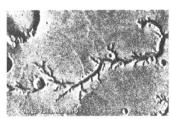

Mars through an 1870-era telescope | Schiaparelli sketches | Percival Lowell | Lowell's team sketches | Mariner 9 photograph

Viking, 1976

The NASA Viking missions (1976) searched for evidence of Martian life. Cameras on the Viking orbiters revealed geological forms typically made by large bodies of water. Huge river valleys were evidence that great floods once carved deep valleys and eroded channels in bedrock for thousands of kilometers. The Viking landers carried out chemical tests. One experiment looked for CO_2, H_2, O_2, CH_4, and N_2 given off by a soil sample treated with nutrients. Another used C-14 to test for biological reactions. A third experiment looked for photosynthesis by tracing CO_2 tagged with C-14. An on-board mass spectrometer looked for organics such as alcohols or amino acids. The results were inconclusive - evidence was not found, but nor could life be ruled out.

Pathfinder, Spirit and Opportunity

Subsequent successful NASA missions landed robotic rovers on Mars beginning with Pathfinder in 1997. Spirit and Opportunity landed on Mars in 2004 for a 92-day mission. Spirit explored Mars until 2011 and Opportunity was still running, as of the writing of this book (2017). Together the rovers have demonstrated that Mars once had oceans of water and could have supported life, but there has been no definitive evidence for life in the present or past. The tiny areas surveyed by rovers with limited instruments make finding evidence of life on Mars a matter of luck until humans can explore in person.

Three generations of Mars rovers (NASA)

Spirit and Opportunity (2004 - 2017)

Curiosity (2012)

Pathfinder (1997)

Is there water on Europa?

Europa is unusual

Europa is the smallest of Jupiter's four large moons, after Ganymede, Io, and Callisto, but is still the sixth-largest moon in the solar system. Slightly smaller than Earth's Moon, Europa has the smoothest surface of any object in the solar system. Strangely free of craters the surface is criss-crossed with strange lines bisecting flat plains. Ganymede and Callisto are covered in craters so something about Europa must be  creating active tectonics that regularly rebuild the surface. Europa's average density implies a composition of silicate rock and a nickel-iron core. The image to the right is an artists' drawing that will help you compare the size of Jupiter to the size of its moons. Notice how sunlight is far less intense compared to Earth.

Why there are few craters

Scientists believe that the surface lines are cracks in an ice crust between 10 and 30 kilometers thick. Under the crust, we believe there are the deepest liquid water oceans in the solar system, estimated to be as deep as 100 kilometers. For comparison, the deepest oceans on Earth are less than 10 kilometers deep. Craters are scarce on Europa because Jupiter's gravity causes cracks in the ice. These cracks allow liquid water onto the -160 °C surface, where it freezes to make a new surface.

The deep oceans of Europa

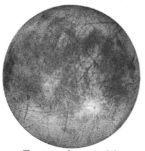

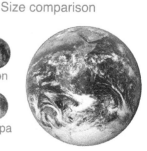

Europa has an active geology

The reason Europa is not frozen solid is the nearness of Jupiter. As Europa orbits, tidal forces from Jupiter's gravity push and pull on Europa creating enough frictional heating to maintain liquid despite the -160°C surface temperature. The entire crust "floats" on the ocean and frequent cracks in the ice permit upwelling of liquid that re-freezes and resurfaces the moon. The cracking and movement of Europa's crustal ice plates is similar to how plate tectonics on Earth constantly recycle the surface of our own planet, except water ice takes the place of magma.

Life on Europa

It is very likely that the tidal heating creates conditions at the bottom of Europa's oceans which are similar to those on Earth's sea floor. Chemosynthetic or thermophilic bacteria would presumably thrive there once they had evolved. Europa is currently considered by many scientists to be the next most likely candidate (after Mars) for extraterrestrial life in our solar system. The difficulty is in searching for evidence of it! How will a probe drill through 10 kilometers of ice 780 million miles away?

Planets in other solar systems

Exoplanets

As telescopes have become more sophisticated astronomers have identified almost 4,000 exoplanets in 2,700 stellar systems. Astronomers now believe most stars have planets and one out of every five stars may have an Earth-like planet. With 200 billion estimated stars in the Milky Way, that means it is likely that there are between 10 billion and 40 billion habitable planets just in our own galaxy.

The TRAPPIST-1 system

Astronomers recently identified another solar system with seven Earth-sized planets orbiting a small red star only slightly larger than Jupiter. Known as TRAPPIST-1, the system lies in the constellation of Aquarius, 39 light years away from Earth. Three of the planets are thought to be in the "goldilocks zone" around the star which means they likely have liquid water and are good candidates for extraterrestrial life. Notice the planets of the TRAPPIST-1 system orbit much closer to the star than Mercury and therefore have very short years compared to the inner planets of the solar system. TRAPPIST-d has a very Earth-like temperature but has a year of only 4.1 Earth-days! The initial discovery was made by TRAPPIST, the TRAnsiting Planets and PlanetesImals Small Telescope.

The TRAPPIST-1 planetary system compared to the inner Solar System

	b	c	d	e	f	g	h
Orbit radius (AU)	0.011	0.015	0.021	0.028	0.037	0.045	~0.06
Diameter (E)	1.09	1.06	0.77	0.92	1.04	1.13	0.76
Mass (E)	0.85	1.38	0.41	0.62	0.68	1.34	?
Year (days)	1.5	2.4	4.1	6.1	9.2	12.3	~20
Surf. gravity	0.72 g	1.23 g	0.69 g	0.73 g	0.63 g	1.1 g	?
Avg. temp.	130°C	72°C	19°C	-20°C	-53°C	-74°C	-100°C

	Mercury	Venus	Earth	Mars
Orbit radius (AU)	0.39	0.72	1.00	1.52
Diameter (E)	0.38	0.95	1.0	0.53
Mass (E)	0.06	0.82	1.0	0.11
Year (days)	88	225	365	687
Surf. gravity	0.38 g	0.91 g	1.0 g	0.38 g
Avg. temp.	-130°C / 400°C	450°C	16°C	-55°C

The TRAPPIST-1 star

The TRAPPIST-1 star is a small, type-M red dwarf which is much more common than G-type stars like the Sun. Type-M stars make up 76% of the Milky Way galaxy and the observation of planetary systems around a nearby star greatly increase the probability that there are many more similar planetary systems in the galaxy. Type-M stars are far cooler than the Sun but are very long-lived. Astronomers estimate that TRAPPIST-1 is at least 500 million years old and has a high concentration of heavier elements, similar to the Sun.

Chapter 24

Section 3 review

Could life exist elsewhere in the universe? The Miller-Urey experiment simulated the early earth atmosphere and subjected it to heat and electricity, producing amino acids. Complex compounds, including amino acids, have also been found on asteroids and comets, as well as in molecular clouds. It is believed that life began on Earth deep underwater, 4.2 billion years ago. The discovery of bacteria in very hot and/or very deep places has strengthened this view. It has also led scientists to look for signs of life elsewhere. Atmospheric oxygen would likely be a sure sign. The discovery that Mars once had oceans of water was significant, but no real evidence has been discovered. Europa, one of Jupiter's moons, could have conditions that might support life. Beyond our solar system, many planets have been discovered that could have liquid water and possibly support life.

Vocabulary words hydrothermal vents, thermophiles, chemosynthesis, extremophiles

Review problems and questions

1. Why is chemosynthesis a promising mechanism for life on other planets?

2. What kinds of things do scientists look for when deciding whether a planet could, or could have once supported life?

3. How does Earth act as a model for identifying life on other planets or moons?

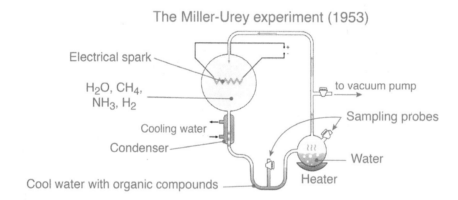

4. The Miller-Urey experiment modeled conditions on Earth that could have supported early life.
   a. What kind(s) of chemicals were believed to be present on early Earth, that were placed in the experimental reaction chamber?
   b. A spark of electricity was introduced in the chamber to model lightning. What was the likely purpose of adding energy?
   c. What was produced when the chemicals reacted after energy was added and the solution in the chamber was agitated? Why are these chemicals significant for life?

Spectroscopy and Celestial Bodies

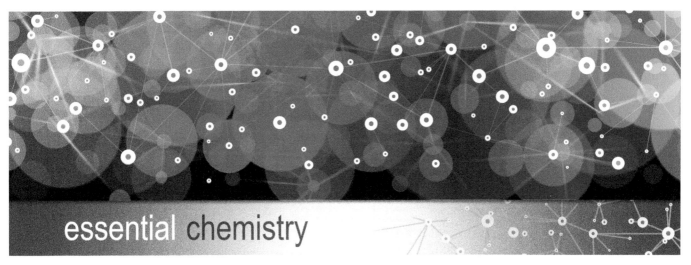

essential chemistry

How do we know the composition of the sun, stars and other heavenly bodies? Surprisingly, rainbows and prisms first pointed the way. If you have ever held a prism in the sunlight, you noticed that the light from the sun emerges as a rainbow of colors. Isaac Newton, who is generally considered the father of spectroscopy, created the foundation for our understanding of light in the book "Opticks" in 1704. With a prism, light from the sun is reflected and refracted into its constituent color components. A rainbow in the sky is produced the same way except that light is dispersed through water droplets instead of a prism.

prism

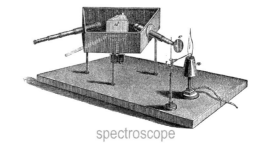
spectroscope

Through the course of the 1800s, the design of the spectroscope was improved so that the spectrum could be seen and measured in finer and finer detail. The additions included using lenses to focus the light, and replacing a prism with diffraction grating. The diffraction grating greatly improves spectral resolution and allows for the dispersed wavelengths of light to be quantified. This turned spectroscopy into a true analytical technique, and the spectra of several known metals and gases were published. Robert Bunsen created an improved flame source and experimental procedure to study the spectra of chemical compounds. (Now you know who to thank if you used a Bunsen burner to do a flame test lab!) These detailed experiments, observations, and measurements established a link between the chemicals and their spectra.

Sir Norman Lockyer took the analytical techniques of spectroscopy outside the lab and focused on celestial bodies. Using a telescope with a spectroscope he observed a prominent yellow line at about 588 nm. The line could not be explained by any known material at the time. Sir Lockyer came to the conclusion that the mystery lines must come from an unknown element. He named this element helium, from the Greek helios (ἥλιος) for sun. Amazingly, helium was not discovered on Earth for 25 more years!

The spectrum of an element is like its fingerprint and can be used for identification. In the visible solar spectrum, there are many dark absorption lines from different elements in the surface layers of the Sun. The combination of spectroscopy and telescopes together allow us to identify elements and compounds that exist outside of our Earth-bound laboratories.

Spectroscopy and Celestial Bodies - 2

The electromagnetic spectrum is much broader than the small part of visible light that our eyes detect. Infrared radiation (IR) was discovered in 1800 by William Herschel. He placed a thermometer in the different colors of light after the light passed through a prism. The temperature increase was greatest just outside the visible spectrum past the color red. The fact that there was a temperature increase indicated that there is radiation not visible to the human eye.

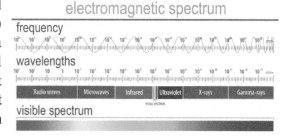

Since its discovery, efforts were made to detect IR radiation from astronomical sources. IR astronomy slowly developed throughout the 19th and 20th centuries. As detector technology increased in the 1950's and 1960's, analysis of the infrared part of the spectrum is a key tool for astronomy. One of the main benefits of using IR spectroscopy is that it peers through interstellar dust. Interstellar dust blocks visible light, but it does not block IR radiation so we peer deeper and farther with IR.

A major limitation of IR astronomy is that molecules in the Earth's atmosphere are sensitive to IR radiation. Certain molecules, like water and carbon dioxide, absorb and emit IR radiation. (This is what makes them greenhouse gases.) To get around this limitation, astronomers use IR spectrometers that are either in space, like the Hubble telescope, or at high altitudes. The Stratospheric Observatory for Infrared Astronomy (SOFIA) is a modified jumbo jet with a large reflector build into its rear section that is exposed during flight. It can fly at an altitude 39,000 to 45,000 feet, which is above 99% of the water vapor in the atmosphere.

IR astronomy is used to study and understand planet formation. NASA recently selected a new spectrometer to be used onboard the SOFIA. A team of scientists will develop a new High Resolution mid-infrared spectrometer (HIRMES) set to be used starting in 2019. The HIRMES instrument will be optimized to wavelengths between 28 and 112 nm. These wavelengths were selected because they are key in observing atomic oxygen, water, and hydrogen. The observations can lead to a better understanding of how ice and water form, and how these elements and compounds can combine with interstellar dust to eventually create planets.

The interaction between light and matter has been a powerful tool for scientists. Spectroscopy enables us to learn about everything from the structure of the atom, to the composition of celestial bodies. We have come a long way from making a rainbow with a prism, and each step along the way has led to more essential understandings about the universe.

Chapter 24 review

Vocabulary
Match each word to the sentence where it best fits.

Section 24.1

Big Bang Theory	cosmic background radiation
Doppler effect	Hertzsprung-Russell diagram
interstellar medium	ionization
light-year	nebula
photosphere	plasma
spectrometer	supernova

1. The process of gaining or losing electrons to form a charged particle is called _____.
2. Massive stars may explode in a cataclysmic explosion called a(n) _____.
3. The _____ is an extremely dilute mixture of ions, atoms, molecules, and larger dust grains that exists between stars.
4. A high energy, electrically charged state of matter is called _____.
5. The _____ is an explanation for the origin of the universe.
6. The _____ is the change in frequency of a wave due to the distance between the observer and the source changing.
7. The flash of energy leftover from the Big Bang that has since cooled is known as _____.
8. A(n) _____ is a graph showing a population of stars in which energy output is plotted against surface color temperature.
9. A(n) _____ is the distance light travels through space in the time-span of one year.
10. Scientists can use a(n) _____ to measure the composition of far away galaxies.
11. An interstellar cloud called a(n) _____ is composed of dust, hydrogen, helium, and other ionized gases.
12. The Sun's visible surface is called the _____.

Section 24.2

accretion	astronomical unit (AU)
blackbody	blackbody spectrum
frost line	intensity
thermal radiation	

13. One _____ is the average distance from the Earth to the Sun.
14. The _____ is an imaginary boundary past which it is cool enough to solidify volatile substances.
15. _____ is given off by all matter.
16. The wavelengths of light that are radiated at a specific temperatures are called _____.
17. _____ describes the power of light energy per square meter.
18. The formation of planets from smaller particles is called _____.
19. The intensity of thermal radiation emitted by a(n) _____ depends only on its temperature.

Section 24.3

chemosynthesis	extremophiles
hydrothermal vents	thermophiles

20. The process of oxidizing inorganic compounds to provide the energy to perform life functions is called _____.
21. _____ are jets of superheated hot water at the bottom of the ocean.
22. _____ are organisms that thrive in extreme conditions that would kill most life on Earth.
23. Heat-loving bacteria are called _____.

Conceptual questions

Section 24.1

24. Think about all of the matter you can see in the universe. Which of the 4 phases of matter is most common? Explain your answer.
25. Why is spectroscopy so important to understand the universe? How is it used to identify the compositions of faraway stars?
26. A by-product of the reactions taking place in the Sun's core is a massive volume of energy that is radiated outward. Are these reactions chemical reactions? Explain your reasoning.
27. The Earth may be diverse in its composition, but the universe is almost entirely composed of two elements. What is the most abundant element in the universe? The second most?
28. Why is matter rarely found naturally in the plasma state on Earth? In the rare case that plasma does appear on Earth, what form does it take?
29. In what way is the reaction that occurs during a burning campfire similar to the reaction that occurs in the sun? In what way are these reactions different? Are they more similar than different, or vice versa? Explain your answer.
30. When astronomers compared the spectra of galaxies they discovered that the spectral lines were shifted toward the red region of the spectrum. How does this "redshift" of distant galaxies provide support for the Big Bang Theory?

Chapter 24 review

Section 24.1

31. Stars form in vast nebulae when an area becomes denser than average and the gravitational forces overcome the gas pressure holding the matter apart. A star that forms in the middle of such a nebula can lead to the formation of other stars over time. Why is this?

32. Energy is produced in stars through nuclear fusion reactions that combine hydrogen into helium.

    a. Briefly describe the nuclear process that occurs in the Sun's core.

    b. How is the process different in stars with twice the mass?

33. Identify three pure substances commonly found in interstellar space, then classify each as either an element or a compound.

34. Humans have been watching the night sky for thousands of years. How did early ideas of the universe differ from the revolutionary theory of Nicolaus Copernicus?

35. In approximately 5.5 billion years, our Sun will expand into a red giant. Describe the process that will transform our Sun.

36. The early universe was composed almost entirely out of just two elements. Even though it is still primarily composed of these two elements, new ones have formed.

    a. Where did these new elements come from?

    b. How did they form?

37. Stars fuse hydrogen into helium in a series of steps. Explain why the overall nuclear fusion reaction below does not frequently occur in one step.

    $$^1H + {}^1H + {}^1H + {}^1H \rightarrow {}^4He + 2v_e + 2e^+$$

38. The Sun is the primary source of energy for life on Earth. Describe the internal structures of the Sun and explain how the energy produced within the Sun progresses through each layer.

Section 24.2

39. Research the melting points and boiling points of several common hydrocarbon compounds. Which of these hydrocarbon compounds could exist in a liquid state on the surface of Titan? Explain your reasoning.

40. The average surface temperature on Mars is -65 °C, much colder than humans are comfortable with. What feature of Mars creates this average temperature?

41. Grey hematite has a different crystal structure than the red hematite found over most of the surface of Mars.

    a. Since they both share the same chemical formula, Fe_2O_3, what does the presence of this mineral on Mars signify?

    b. How was this mineral formed?

42. The inner terrestrial planets are a great deal different than the outer gas giant planets, and are separated by a region of thousands of asteroids. Describe at least three differences between the inner and outer planets.

43. A planet's surface temperature is due to an energy balance between the thermal radiation emitted by the planet and the energy absorbed from the Sun.

    a. What would happen if a planet releases more energy than it absorbs from the Sun?

    b. Assuming that the distance between the planet and the Sun stays the same, what factors could change the thermal equilibrium between energy being absorbed and emitted by the planet?

    c. The balance between the energy being absorbed and the energy by radiated by Earth is currently changing due to increasing carbon dioxide in the atmosphere. How does atmospheric CO_2 contribute to a warming planet?

44. While there are three moons in the inner solar system, Jupiter and Saturn have more than 60 moons each. Why do the outer planets have such a large number of moons compared to the inner ones?

45. The planets of our solar system vary widely in temperature. This can be due to a single factor, or a combination of factors.

    a. Identify the planet in our solar system that has the highest average surface temperature.

    b. Why is this planet so hot?

    c. Which planet has the lowest average surface temperature?

    d. Why is this planet so cold?

    e. Earth's average surface temperature is around 15°C. Why is this ideal for supporting life?

46. Earth is the one of the few places in the solar system that has an abundance of liquid water. Where do scientists think most of Earth's water came from?

47. The asteroid belt contains millions of planetesimals left over from the formation of the solar system.

    a. Which type(s) of asteroid gives insight into the composition of the early solar system?

    b. Which type(s) do not?

    c. Explain why asteroids may not reflect the makeup of the primordial solar system.

48. The asteroid belt is located between the orbits of Mars and Jupiter. What astronomical bodies contain most of the mass of the asteroid belt?

Chapter 24 review

Section 24.2

49. Several planets are discovered to orbit a star larger and hotter than our Sun. If one of those planets was Earth-like, would it orbit closer to or farther from the star compared to Earth's distance to the sun? Explain your answer.

50. Our observations and understanding of the solar system have come a long way since Greek astronomers started on the road of scientific exploration. Since then, it has been found that the solar system formed about 4.6 billion years ago. Describe the process through which our solar system formed.

51. The surface of Venus is incredibly hot because most sunlight passes through its atmosphere while re-radiated thermal energy is absorbed. How is this possible?

52. Beyond the gas giant Neptune lies a region of space known as the Kuiper Belt, full of objects not unlike the Asteroid Belt. What is the difference between an asteroid and a Kuiper Belt object?

53. The Sun provides almost 100% of the energy found on Earth, and without it life would not be possible. We perceive this energy as sunlight and heat.
    a. What is sunlight?
    b. Why does the Sun appear to be yellow in color?

54. All known matter emits thermal radiation. What is the difference between the thermal radiation emitted by red hot iron and that of the Sun?

55. Explain how the formation of the inner and outer planets differed in relation to the frost line; how does planetary composition differ on either side of this imaginary point?

56. Mars is known for its nickname "the red planet." What is responsible for Mars' signature "red" color?

57. People are very familiar with the solid, rocky surface of the Earth and may have trouble imagining the gas giants with no surface at all. Why are the inner planets rocky while the outer planets are not?

58. On the outermost edge of our solar system exists a shell of objects known as the Oort Cloud.
    a. How massive is the Oort Cloud estimated to be when compared to Earth?
    b. What is the Oort Cloud mass made of?

59. Why is the surface of Mars cold while Venus' surface is hot even though both atmospheres are composed primarily of CO_2?

60. An ideal blackbody absorbs all electromagnetic wavelengths. For each of the following objects, state whether it would function as a good blackbody or a poor blackbody.
    a. A mirror
    b. Charcoal
    c. Glass
    d. A car tire
    e. Basalt rock
    f. A white lab coat

Section 24.3

61. When a scientist wants to look for evidence of life on other worlds, they often check for the presence of molecular oxygen in the atmosphere. Why is atmospheric oxygen a sign of life?

62. Earth is not the only object in the solar system that contains liquid water. If fact, scientists believe that several moons my contain subsurface liquid water and even vast oceans.
    a. Where in the solar system do scientists believe the deepest liquid water oceans are located?
    b. How deep are they and how does it compare to Earth's oceans?

63. Astonomers observe a moon and see that it has a surface devoid of craters but has long sections of lines bisecting flat plains. What does this observation tell us about what happens below the moon's surface?

64. Describe what the Miller-Urey experiment was trying to simulate.
    a. Describe the results of the experiment.
    b. What did the results of the Miller-Urey experiment suggest?

65. The NASA Viking missions in 1976 searched for evidence of Martian life.
    a. What did the cameras reveal about the Martian landscape?
    b. Describe three types of experiments that the Viking landers carried out to test for life on Mars.

66. Consider Jupiter's moon, Europa.
    a. How would it be possible for life as we know it to exist on Europa?
    b. By what mechanism could an organism survive on this frozen moon?

67. Give at least two examples of life forms on Earth that can make their own energy in the absence of direct sunlight.

Chapter 24 review

Quantitative problems

Section 24.1

68. Record the current time in minutes and seconds and begin reading this question; start timing yourself. Work on answering this question without stopping. If the Sun's core releases 3.83×10^{23} joules of energy every second, how much energy would it have released by the time you have finished reading this question? Stop timing yourself and use the time that has passed to calculate the answer to this question.

69. The Sun's mass is 1.989×10^{30} kg and it is composed of about 74% hydrogen, 24% helium, and 2% other elements. Based on solar energy output, the Sun fuses 3.8×10^{38} protons per second.

   a. How many years will it take the Sun to use up 25% of the hydrogen in its core? The mass of a proton is 1.67×10^{-27} kg.

   b. What is the mass of Helium that will form during this time? The overall nuclear reaction is:

   $$^1H + {}^1H + {}^1H + {}^1H \rightarrow {}^4He + 2v_e + 2e^+$$

70. Based on the following criteria, write a fusion reaction that could occur in a star.

   1. The new element must have 8 protons and 8 neutrons in its nucleus.

   2. The total mass and positive charge must balance on both sides of the reaction.

   3. The reaction can only have 2 reactants.

Section 24.2

71. The atmospheres of Mars and Venus are both primarily made of CO_2. Calculate the grams of CO_2 in one liter on Mars and Venus given the following surface pressure and temperature:

   a. The surface of Mars is $-65\ °C$ at 0.600 kPa of pressure.

   b. The surface of Venus is 460 °C at 92 atm of pressure.

72. Titan's atmosphere is mostly composed of diatomic nitrogen. The surface of the moon is $-190\ °C$ and has a pressure of 1.45 atm.

   a. What volume will 25 kilograms of N_2 occupy on the surface of Titan?

   b. How much space would the same mass of N_2 take up at the standard temperature and pressure of Earth?

   c. Which volume is larger? Why?

73. The Stefan-Boltzman law states the power of thermal radiation increases with temperature to the fourth power according to the equation $P = \sigma A T^4$ where P is the total power emitted in watts (W), A is the area of the emitting surface (m²), T is the temperature in K, and σ is the Stefan-Boltzman constant, $\sigma = 5.67 \times 10^{-8}\ Wm^2K^4$.

   a. If the power radiated by the Sun equals 3.91×10^{26} W, what must be the surface temperature of the sun? The surface area of the Sun is 6.18×10^{18} square meters.

   b. Scientists believe the sun was 30% dimmer when it first formed. How much power did the young Sun produce, and what must the surface temperature have been at that time?

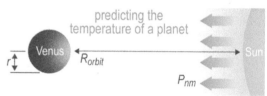

74. Most of the planets in the solar system have an average surface temperature that obeys the derived equation:

$$T = \frac{280\ K \cdot \sqrt{AU}}{\sqrt{R}}$$

T is the average temperature in Kelvin and R is the distance of the planet to the Sun in AU. The constant 280 has units of $K \cdot \sqrt{AU}$

   a. Calculate the radius of an orbit, in AU, that would result in an average surface temperature of 100 °C.

   b. Calculate the radius of an orbit, in AU, that would produce an average surface temperature of 0 °C.

   c. The region between the radii you identified in questions (a) and (b) is known as the Sun's "life zone" for potential orbits of habitable planets. Is Earth in the middle or near one of the edges of the Sun's life zone?

75. Venus is much hotter than is predicted by the equation in the previous question. Its average surface temperature is 460 °C while it is 1.08×10^8 km away from the Sun.

   a. Does Venus lie in the Sun's life zone?

   b. Calculate the average surface temperature (in Celsius) of Venus if it obeyed the derived equation.

   c. Explain whether or not liquid water could exist on the surface of a planet following the same orbital radius as Venus.

   d. What orbital radius (in AU) should Venus have so that its current temperature would follow the equation?

Chapter 24 review

Section 24.2

76. Imagine that scientists find a new planet orbiting the Sun with an orbital radius of 600 million kilometers.

    a. Based on this orbit, what would the scientists predict the temperature of the planet to be in degrees Celsius?

    b. The small rocky planet lies on the frost line and, as such, any carbon dioxide should be frozen. However, when scientists send a probe to the planet they find that volcanic activity on the planet has created a dense atmosphere of CO_2 that maintains an average surface temperature of -20 Celsius. At what distance (in km) from the Sun would a planet with this temperature be predicted to exist?

77. Recall the inverse square law: $I = P/(4\pi R^2)$. Intensity (I) of solar radiation decreases as distance from a star increases. The power (P) is the total energy output per second of the star in watts and R is the distance from the star to the planet in meters.

    a. The Sun releases 3.83×10^{26} joules every second. The distance between the Sun and Earth's thermosphere is 1.496×10^{11} meters. What is the intensity (W/m^2) of sunlight at the Earth's thermosphere?

    b. Calculate the intensity (W/m^2) of sunlight at Venus. The distance between the Sun and Venus is 1.08×10^{11} meters.

    c. Alpha Centauri A is the nearest star to Earth. This star has a luminosity of 5.815×10^{26} watts. Calculate the intensity of light at Earth's orbit if our planet was orbiting around Alpha Centauri A.

Section 24.3

78. Of the many solar systems identified by scientists, the TRAPPIST-1 system represents pehaps the best chance of finding life outside of our own solar system.

    a. The TRAPPIST-1 system is located 39 light years from Earth. What is this distance in kilometers?

    b. TRAPPIST-d is the planet with the most Earth-like conditions identified in the TRAPPIST-1 system. Its year is just 4.1 Earth-days long. How many TRAPPIST-d years are there per 1 Earth year?

79. Jupiter's moon Europa has a diameter of 3,100 km and a mass of 4.2×10^{22} kg.

    a. What is the volume of Europa?

    b. Based on volume, what is Europa's density (on average)?

25.1 - Appendix - Table of Contents

List of Reference Tables

Title	Page
Physical properties of selected elements	821
Metric prefixes and conversion factors	822
Physical constants and SI units	823
Thermodynamic quantities - Part 1	824
Thermodynamic quantities - Part 2	825
Common ions	826
Symbols used in reactions	827
Solubility rules and K_{sp}	828
Rules for assigning oxidation numbers	829
Standard reduction potentials and Activity series	830
The pH scale and indicators	831
Strong and weak acids and bases	832

Physical properties of selected elements

element name	symbol	atomic #	atomic mass (amu)	electro-negativity	density at 20 °C (g/mL)	melting point (°C) at 1 atm	boiling point (°C) at 1 atm	C_p (J/g·°C) at 23 °C	common oxidation states
hydrogen*	H	1	1.0079	2.20	0.0899	−259.34	−252.8	14.304	−1, +1
helium*	He	2	4.0028	–	0.1785	−272.2†	−268.9	5.193	0
lithium	Li	3	6.941	0.98	0.534	180.54	1,342	3.582	+1
boron	B	5	10.811	2.04	2.34	2,077	4,000	1.026	+3
carbon, diamond	C	6	12.011	2.55	3.52	3,500‡	3,930	0.516	+2, +4
carbon, graphite	C	6	12.011	2.55	2.27	sublimes at 3,642		0.717	+2, +4
nitrogen	N	7	14.007	3.04	1.2506*	−209.86	−195.8	1.042	−3
oxygen	O	8	15.999	3.44	1.429*	−218.4	−182.96	0.918	−2
fluorine	F	9	18.998	3.98	1.69**	−219.62	−188.14	0.824	−1
neon	Ne	10	20.180	–	0.9002*	−248.67	−245.9	1.030	0
sodium	Na	11	22.990	0.93	0.97	97.8	882.9	1.228	+1
magnesium	Mg	12	24.305	1.31	1.745	648.8	1,107	1.02	+2
aluminum	Al	13	26.982	1.61	2.702	660.37	2,467	0.897	+3
silicon	Si	14	28.086	1.90	2.33±0.01	1,410	2,355	0.70	+2, +4
phosphorus	P	15	30.974	2.19	1.82	44.1	280	0.769	−3
sulfur	S	16	32.065	2.58	1.96	119	444.67	0.710	−2, 0
chlorine	Cl	17	35.453	3.16	3.214*	−100.98	−34.6	0.479	−1
argon	Ar	18	39.948	–	0.00163	−189.3	−185.8	0.520	0
potassium	K	19	39.098	0.82	0.86	63.25	760	0.757	+1
calcium	Ca	20	40.078	1.00	1.54	839 ±2	1,484	0.647	+2
chromium	Cr	24	51.996	1.66	7.2028	1,857±20	2,672	0.449	+2, +3, +6
iron	Fe	26	55.845	1.83	7.86	1,535	2,750	0.449	+2, +3
cobalt	Co	27	58.933	1.88	8.9	1,495	2,870	0.421	+2, +3
nickel	Ni	28	58.693	1.91	8.90	1,455	2,730	0.444	+2, +3
copper	Cu	29	63.546	1.90	8.92	1,083±0.2	2,567	0.385	+1, +2
zinc	Zn	30	65.38	1.65	7.14	419.58	907	0.388	+2
bromine	Br	35	79.904	2.96	3.119	−7.2	58.78	0.226	−1
silver	Ag	47	107.87	1.93	10.5	961.93	2,212	0.235	+1
tin	Sn	50	118.71	1.96	7.28	231.88	2,260	0.228	+2, +4
iodine	I	53	126.90	2.66	4.93	113.5	184.35	0.145	−1
barium	Ba	56	137.33	0.89	3.51	725	1,640	0.204	+2
platinum	Pt	78	195.08	2.2	21.45	1,772	3,827±100	0.13	+2, +4
gold	Au	79	196.97	2.4	19.31	1,064.43	2,808±2	0.129	+1, +3
mercury	Hg	80	200.56	1.9	13.4562	−38.87	356.58	0.140	+1, +2
lead	Pb	82	207.2	1.8	11.3	327.462	1,748	0.129	+4, +2

*Density in g/L at STP
**Density in g/L at standard pressure and 15 °C
†Melting point at 26 atm
‡Melting point at 63.5 atm

Metric prefixes and conversion factors

metric prefixes

prefix (symbol)	factor
yotta (Y)	10^{24} = 1,000,000,000,000,000,000,000,000
zetta (Z)	10^{21} = 1,000,000,000,000,000,000,000
exa (E)	10^{18} = 1,000,000,000,000,000,000
peta (P)	10^{15} = 1,000,000,000,000,000
tera (T)	10^{12} = 1,000,000,000,000
giga (G)	10^{9} = 1,000,000,000
mega (M)	10^{6} = 1,000,000
kilo (k)	10^{3} = 1,000
hecto (h)	10^{2} = 100
deka (da)	10^{1} = 10
	Base = 1
deci (d)	10^{-1} = 0.1
centi (c)	10^{-2} = 0.01
milli (m)	10^{-3} = 0.001
micro (μ)	10^{-6} = 0.000001
nano (n)	10^{-9} = 0.000000001
pico (p)	10^{-12} = 0.000000000001
femto (f)	10^{-15} = 0.000000000000001
atto (a)	10^{-18} = 0.000000000000000001
zepto (z)	10^{-21} = 0.000000000000000000001
yocto (y)	10^{-24} = 0.000000000000000000000001

conversion factors

quantity	conversion factors		
mass	1 kg = 1,000 g	1 amu = 1.6606 x 10^{-27} kg	1 kg has a weight of 2.205 lb
length	1 in = 2.54 cm 1 ft = 12 in 1 mi = 1.609 km 1 AU = 1.50 x 10^{11} m	1 cm = 0.3048 in 1 yd = 3 ft 1 km = 0.621 mi 1 light year (ly) = 9.46 x 10^{15} m	1 m = 39.37 in = 100 cm 1 mi = 5,280 ft 1 angstrom (Å) = 10 x 10^{-10} m 1 parsec (pc) = 3.26156 ly
area	1 m^2 = 10.76 ft^2	1 in^2 6.452 cm^2	1 acre = 43,560 ft^2 = 4,048 m^2
volume	1 mL = 1 cm^3	1 gallon = 3.785 L	1 teaspoon = 5 mL
pressure	1 atm = 760 mmHg = 760 Torr = 101.325 kPa = 14.69 psi		
temperature	K = °C + 273	°C = (°F − 32) ÷ 1.8	°F = °C × 1.8 ÷ 32
energy	1 calorie = 4.184 J 1 eV = 1.602 x 10^{-19} J	1 Calorie = 1,000 calories 1 rad = 0.010 J/kg tissue	1 V = 1 J/C 1 J = 1 kg·m^2/s^2
power	1 watt (W) = 1 J/s = 0.001341 horsepower (hP) = 550 ft lb/s		

Physical constants and SI units

physical constants

quantity (symbol)	value	unit
absolute zero	0 −273	K °C
atomic mass unit	1.66×10^{-27}	kg
Avogadro's number	6.02×10^{23}	kg
charge of a proton	1.602×10^{-19}	C
charge of an electron	1.602×10^{-19}	C
Faraday constant (F)	96,000	C/mol
ideal gas constant (R)	0.0821 8.31	L·atm/mol·K L·kPa/mol·K
mass of a neutron	1.6749×10^{-27} 1.0087	kg amu
mass of a proton	1.6726×10^{-27} 1.0073	kg amu
mass of an electron	9.11×10^{-31} 5.49×10^{-4}	kg amu
molar volume of a gas at STP	22.4	L/mol
Planck's constant (h)	6.626×10^{-34}	J·s
speed of light in a vacuum (c)	2.998×10^{8}	m/s
standard pressure	101.325	kPa
standard temperature	0	°C

SI units

measurement	unit	symbol
amount of a substance	mole, mol	n
electric current	ampere, A	I
length	meter, m	l
mass	kilogram, kg	m
temperature	kelvin, K	T
time	second, s	t

units derived from SI units

quantity	unit	symbol	SI equivalent	derived equivalent
energy	joule, J	q	$kg·m^2/s^2$	N·m, C·V, or W·s
force	newton, N	F	$kg·m/s^2$	−
frequency	hertz, Hz	v	1/s	−
mass density	kg/m^3	d	kg/m^3	kg/L
power	watt, W	P	$kg·m^2/s^3$	J/s or V·A
pressure	pascal, Pa	p or P	$kg/m·s^2$	N/m^2
solution concentration	molarity, M	M	mol/m^3	mol/L
volume	liter, L	V	m^3	−

Section 25.1: Appendix

Thermodynamic quantities - Part 1

specific heat (C_p) of selected substances*

substance	C_p (J/g·°C)
air	1.006
ammonia (gas)	2.090
benzene (liquid)	1.750
carbon (diamond)	0.509
carbon (graphite)	0.708
concrete	0.880
cork	2.000
ethanol (gas)	1.420
ethanol (liquid)	2.440
glass	0.800
marble	0.858
methane	3.49
oil	1.900
polyethylene (plastic #4, LDPE)	1.676
polypropylene (plastic #5, PP)	1.571
polystyrene (plastic #6, PS)	0.927
polyvinylchloride (plastic #3, PVC)	0.90
steel	0.470
water (solid)	2.027
water (liquid)	4.184
water (vapor)	2.015
wax	2.10
wood	2.500

*See the Appendix page titled Physical properties of selected elements to find specific heat values for pure elements

heat of fusion (ΔH_{fus}) and heat of vaporization (ΔH_{vap}) for selected substances

substance	ΔH_{fus} (J/g)	ΔH_{vap} (J/g)
ammonia	33	134
ethanol	104	854
nitrogen (liquid)	25.5	201
iron	267	6,265
water	335	2,256
silver	88	2,336

enthalpy of formation ($\Delta H°_f$) of selected substances (at 25 °C)

formula	name	$\Delta H°_f$ (kJ/mol)
C_3H_6O (l)	acetone	−249.4
NH_3 (g)	ammonia, gas	−45.90
NH_3 (aq)	ammonia, aqueous	−80.8
C_6H_6	benzene	49.0
C (s)	carbon (diamond)	1.9
C	carbon, graphite	0
CO_2	carbon dioxide	−393.5
CO	carbon monoxide	−110.5
CS_2 (g)	carbon disulfide	116.7
$CuSO_4$	copper(II) sulfate	−771.4
C_2H_5OH (l)	ethanol	−277.0
H_2 (g)	hydrogen, gas	0
Fe_2O_3	iron(III) oxide	−824.2
NO_2	nitrogen dioxide	33.2
O_3	ozone	142.7
O_2	oxygen	0
SO_2 (g)	sulfur dioxide	−296.84
H_2O (l)	water, liquid	−285.83
H_2O (s)	water, solid	−291.83
H_2O (g)	water, vapor	−241.82

Thermodynamic quantities - Part 2

enthalpy, free energy change and entropy for selected substances

substance	formula and state	ΔH°$_f$ (kJ/mol)	ΔG° (kJ/mol)	S° (J/mol·K)
bromine	Br$_2$ (g)	31	3	245
bromine	Br$_2$ (l)	0	0	152
calcium oxide	CaO (s)	-635	-604	40
calcium silicate	CaSiO$_3$ (s)	-1630	-1550	84
carbon (diamond)	C (s)	2	3	2
carbon (graphite)	C (s)	0	0	6
carbon dioxide	CO$_2$ (g)	-393.5	-394	214
carbon monoxide	CO (g)	-110.5	-137	198
chlorine	Cl$_2$ (g)	0	0	223
copper	Cu (s)	0	0	33
copper(II) oxide	CuO (s)	-156	-128	43
hydrochloric acid	HCl (g)	-92	-95	187
hydrogen	H$_2$ (g)	0	0	131
iron	Fe (s)	0	0	27
iron(III) oxide	Fe$_2$O$_3$ (s)	-824.2	-740	90
methane	CH$_4$ (g)	-75	-51	186
methanol	CH$_3$OH (g)	-201	-163	240
methanol	CH$_3$OH (l)	-239	-166	127
nickel	Ni (s)	0	0	30
nickel(II) chloride	NiCl$_2$ (s)	-316	-272	107
oxygen	O$_2$ (g)	0	0	205
silicon dioxide (quartz)	SiO$_2$ (s)	-911	-856	42
sodium chloride	NaCl (s)	-411	-384	72
sodium hydroxide	NaOH (aq)	-470	-419	50
water	H$_2$O (g)	-241.82	-229	189
water	H$_2$O (l)	-285.83	-237	70

average bond energies (kJ/mol)

C-H, 413	C=C, 614	N-H, 391	O-H, 463	H-H, 436	S-H, 339	I-Cl, 208
C-C, 348	C≡C, 839	N-N, 163	O-O, 146	H-F, 567	S-F, 327	I-Br, 175
C-N, 293	C=N, 615	N-O, 201	O-F, 190	H-Cl, 431	S-Cl, 253	I-I, 151
C-O, 358	C≡N, 891	N-F, 272	O-Cl, 203	H-Br, 366	S-Br, 218	
C-F, 485	C=O, 799	N-Cl, 200	O-I, 234	H-I, 299	S-S, 266	
C-Cl, 328	C≡O, 1,072	N-Br, 243	O=O, 495			
C-Br, 276		N=N, 418		Br-F, 237	S=O, 523	
C-I, 240	Cl-F, 253	N≡N, 941	F-F, 155	Br-Cl, 218	S=S, 418	
C-S, 259	Cl-Cl, 242	N=O, 607		Br-Br, 193		

Section 25.1: Appendix

Common ions

common positive ions

1+	ammonium	NH_4^+	**2+**	chromium(II)	Cr^{2+}	**2+**	tin(II)	Sn^{2+}
	copper(I)	Cu^+		cobalt(II)	Co^{2+}		zinc	Zn^{2+}
	hydrogen	H^+		copper(II)	Cu^{2+}	**3+**	aluminum	Al^{3+}
	lithium	Li^+		iron(II)	Fe^{2+}		boron	B^{3+}
	potassium	K^+		lead(II)	Pb^{2+}		chromium(III)	Cr^{3+}
	silver	Ag^+		magnesium	Mg^{2+}		cobalt(III)	Co^{3+}
	sodium	Na^+		manganese(II)	Mn^{2+}		iron(III)	Fe^{3+}
2+	beryllium	Be^{2+}		mercury(I)*	Hg_2^{2+}		nickel(III)	Ni^{3+}
	cadmium	Cd^{2+}		mercury(II)	Hg^{2+}	**4+**	lead(IV)	Pb^{4+}
	calcium	Ca^{2+}		nickel(II)	Ni^{2+}		tin(IV)	Sn^{4+}

*The mercury(I) ion is a diatomic ion made of two Hg^+ ions to form Hg_2^{2+}.

common negative ions

1−	acetate	CH_3COO^- or $C_2H_3O_2^-$	**1−**	hydrogen sulfate**	HSO_4^-	**2−**	oxalate	$C_2O_4^{2-}$
	bicarbonate*	HCO_3^-		hydroxide	OH^-		oxide	O^{2-}
	bisulfate**	HSO_4^-		hypochlorite	ClO^-		peroxide	O_2^{2-}
	bromate	BrO_3^-		iodide	I^-		silicate	SiO_3^{2-}
	bromide	Br^-		nitrate	NO_3^-		sulfate	SO_4^{2-}
	chlorate	ClO_3^-		nitrite	NO_2^-		sulfite	SO_3^{2-}
	chlorite	ClO_2^-		perchlorate	ClO_4^-		sulfide	S^{2-}
	chloride	Cl^-		permanganate	MnO_4^-		thiosulfate	$S_2O_3^{2-}$
	cyanide	CN^-	**2−**	carbonate	CO_3^{2-}	**3−**	nitride	N^{3-}
	dihydrogen phosphate	$H_2PO_4^-$		chromate	CrO_4^{2-}		phosphate	PO_4^{3-}
	fluoride	F^-		dichromate	$Cr_2O_7^{2-}$		phosphate	PO_4^{3-}
	hydrogen carbonate*	HCO_3^-		***hydrogen phosphate	HPO_4^{2-}		phosphide	P^{3-}

* = bicarbonate is the common name for the hydrogen carbonate ion
** = bisulfate is the common name for the hydrogen sulfate ion
*** = hydrogen phosphate is the common name for the monohydrogen phosphate ion

Symbols used in reactions

chemical reactions

symbol	meaning
+	reacts with
(g)	gas (reactant or product)
↑	gas evolved as a product
(l)	liquid (reactant or product)
(s)	solid (reactant or product)
↓	precipitate formed as a product
(aq)	aqueous reactant or product (dissolved in water)
→	forms, produces, yields, etc. to completion
⇌ ↔ ⇌	A two-direction arrow could indicate any of the following: resonance, equilibrium, reversibly forms, reversibly produces, reversibly yields, etc.
$\xrightarrow{\Delta}$	reacts when heat is added
$\xrightarrow{cat}$	reacts with an added catalyst
$\xrightarrow{pressure}$	reacts above normal atmospheric pressure
+ energy + ____ kJ + heat	with reactants: endothermic; with products: exothermic
−ΔH	negative change in enthalpy; exothermic reaction
+ΔH	positive change in enthalpy, endothermic reaction
NR	no reaction takes place

nuclear reactions

symbol	meaning
A	mass number placeholder
Z	mass number placeholder
X	element symbol placeholder
X-A	isotope of element X
$^{A}_{Z}X$	isotope of element X
α	alpha particle
β	beta particle
$^{0}_{-1}e$	beta particle or electron
γ	gamma radiation
$^{1}_{0}n$	neutron
$^{1}_{+1}p$	proton
$^{1}_{+1}e$	positron

Solubility rules and K_{sp}

solubility rules for common compounds

soluble compounds	insoluble compounds (except with group 1 metal ions and NH_4^+)
group 1 metal ions Li^+, Na^+, K^+, Rb^+, Cs^+	carbonates, CO_3^{2-}
ammonium, NH_4^+	hydroxides, OH^-, except Ba^{2+}
acetate, $C_2H_3O_2^-$ or CH_3COO^-	chlorides of Cu, Pb, Ag and Hg
nitrates, NO_3^-	bromides of Cu, Pb, Ag and Hg
sulfates, SO_4^{2-}, except Ba, Pb, and Sr	iodides of Cu, Pb, Ag and Hg
chlorides, except with Cu, Pb, Ag and Hg	sulfides, S^{2-}
bromides, except with Cu, Pb, Ag and Hg	phosphates, PO_4^{3-}
iodides, except with Cu, Pb, Ag and Hg	

solubility product constants at 25 °C

salt	name	ion product	K_{sp}
$Al(OH)_3$	aluminum hydroxide	$[Al^{3+}][OH^-]^3$	1.3×10^{-33}
$BaCO_3$	barium carbonate	$[Ba^{2+}][CO_3^{2-}]$	5.1×10^{-9}
$BaCrO_4$	barium chromate	$[Ba^{2+}][CrO_4^{2-}]$	1.2×10^{-10}
BaF_2	barium fluoride	$[Ba^{2+}][F^-]^2$	1.0×10^{-6}
$BaSO_4$	barium sulfate	$[Ba^{2+}][SO_4^{2-}]$	1.1×10^{-10}
$CaCO_3$	calcium carbonate	$[Ca^{2+}][CO_3^{2-}]$	2.8×10^{-9}
CaF_2	calcium fluoride	$[Ca^{2+}][F^-]^2$	1.0×10^{-6}
$Ca(OH)_2$	calcium hydroxide	$[Ca^{2+}][OH^-]^2$	5.5×10^{-6}
$Ca_3(PO_4)_2$	calcium phosphate	$[Ca^{2+}]^3[PO_4^{3-}]^2$	1.0×10^{-26}
$CaSO_4$	calcium sulfate	$[Ca^{2+}][SO_4]^{2-}$	9.1×10^{-6}
$CoCO_3$	cobalt(II) carbonate	$[Co^{2+}][CO_3^{2-}]$	1.4×10^{-13}
CoS	cobalt(II) sulfide	$[Co^{2+}][S^{2+}]$	4.0×10^{-21}
$CuCl$	copper(I) chloride	$[Cu^{+1}][Cl^{-1}]$	1.2×10^{-6}
$Cu(OH)_2$	copper(II) hydroxide	$[Cu^{2+},OH^-]^2$	2.6×10^{-19}
CuS	copper(II) sulfide	$[Cu^{2+}][S^{2-}]$	6.3×10^{-6}
$Fe(OH)_2$	iron(II) hydroxide	$[Fe^{2+}][OH^-]^2$	8.0×10^{-16}
FeS	iron(II) sulfide	$[Fe^{2+}][S^{2-}]$	6.3×10^{-18}
$Fe(OH)_3$	iron(III) hydroxide	$[Fe^{3+}][OH^-]^3$	4.0×10^{-38}
$PbCl_2$	lead(II) chloride	$[Pb^{2+}][Cl^-]^2$	1.6×10^{-5}
$PbCrO_4$	lead(II) chromate	$[Pb^{2+}][CrO_4^{2-}]$	2.8×10^{-13}
$PbSO_4$	lead(II) sulfate	$[Pb^{2+}][SO_4^{2-}]$	1.6×10^{-8}
PbS	lead(II) sulfide	$[Pb^{2+}][S^{2-}]$	8.0×10^{-28}
$MgCO_3$	magnesium carbonate	$[Mg^{2+}][CO_3^{2-}]$	3.5×10^{-8}
$Mg(OH)_2$	magnesium hydroxide	$[Mg^{2+}][OH^-]^2$	1.8×10^{-11}
MnS	manganese(II) sulfide	$[Mn^{2+}][S^{2-}]$	2.5×10^{-13}
HgS	mercury(II) sulfide	$[Hg^{2+}][Hg^{2-}]$	1.6×10^{-52}
$AgC_2H_3O_2$	silver acetate	$[Ag^+][CH_3COO^-]$	2.0×10^{-3}
$AgBr$	silver bromide	$[Ag^+][Br^-]$	5.0×10^{-13}
Ag_2CO_3	silver carbonate	$[Ag^+]^2[CO_3^{2-}]$	8.1×10^{-12}
$AgCl$	silver chloride	$[Ag^+][Cl^-]$	1.8×10^{-10}
AgI	silver iodide	$[Ag^+][I^-]$	8.3×10^{-17}
Ag_2S	silver sulfide	$[Ag^+]^2[S^{2-}]$	6.3×10^{-50}
$SrSO_4$	strontium sulfate	$[Sr^{2+}][SO_4^{2-}]$	3.2×10^{-7}
SnS	tin(II) sulfide	$[Sn^{2+}][S^{2-}]$	1.0×10^{-25}
ZnS	zinc sulfide	$[Zn^{2+}][S^{2-}]$	1.6×10^{-24}

Rules for assigning oxidation numbers

rules for assigning oxidation numbers

Rule	Example	Oxidation #
1. The oxidation number of an atom in a pure element is zero.	Fe Cl Cl_2	0 0 0
2. The sum of the oxidation numbers in a neutral molecule is zero. Subscripts become coefficients in the equation: $0 = n_{element} + n_{element}$	CCl_4 $0 = n_C + 4(-1)$	Cl: -1 (Rule 5) C: +4 (Rule 2)
3. The sum of all atoms' oxidation numbers for an ion is equal to the charge of the ion. For polyatomic ions, Charge $= n_{element} + n_{element}$	Cu^{2+} SO_4^{2-} $-2 = n_S + 4(-2)$	+2 O: -2 (Rule 5) S: +6 (Rule 3)
4. Group 1 and 2 metals in compounds have oxidation numbers according to group number. Group 1 metals = +1. Group 2 metals = +2.	$CaCl_2$ $0 = 2 + 2(n_{Cl})$	Ca: +2 (Rule 4) Cl: -1 (Rule 2)
5. The oxidation number of a nonmetal in a compound is as follows; if the compound is made of two different nonmetals, assign the oxidation number to the nonmetal that appears first in this list and calculate the other number with the appropriate equation. • Fluorine (F): -1 • Hydrogen (H): +1 except when combined with a metal • Oxygen (O): -2 • Group 17 (Cl, Br, I): -1 • Group 16 (S, Se, Te): -2 • Group 15 (N, P, As): -3	F_2O $0 = 2(-1) + n_O$ Pb_2O_3 $0 = 2n_{Pb} + 3(-2)$ CuS $0 = n_{Cu} + (-2)$ LiH $0 = 1 + (n_H)$	F: -1 O: +2 (Rule 2) O: -2 Pb: +3 (Rule 2) S: -2 (Rule 5) Cu: +2 (Rule 2) Li: +1 (Rule 4) H: -1 (Rule 2)

Standard reduction potentials and Activity series

standard reduction potentials

reduction half-reaction	E^0 (V)	
$Li^+(aq) + e^- \rightarrow Li(s)$	-3.04	
$Na^+(aq) + e^- \rightarrow Na(s)$	-2.71	
$Mg^{2+}(aq) + 2e^- \rightarrow Mg(s)$	-2.38	Strong reducing agent, strong tendency to be oxidized, strong tendency to give up electrons, weak oxidizing agent, tendency to occur in the reverse direction.
$Al^{3+}(aq) + 3e^- \rightarrow Al(s)$	-1.66	
$2H_2O(l) + 2e^- \rightarrow H_2(g) + 2OH^-(aq)$	-0.83	
$Zn^{2+}(aq) + 2e^- \rightarrow Zn(s)$	-0.76	
$Cr^{2+}(aq) + 2e^- \rightarrow Cr(s)$	-0.74	
$Fe^{2+}(aq) + 2e^- \rightarrow Fe(s)$	-0.41	
$Ni^{2+}(aq) + 2e^- \rightarrow Ni(s)$	-0.23	
$Sn^{2+}(aq) + 2e^- \rightarrow Sn(s)$	-0.14	
$Pb^{2+}(aq) + 2e^- \rightarrow Pb(s)$	-0.13	
$Fe^{3+}(aq) + 3e^- \rightarrow Fe(s)$	-0.04	
$2H^+(aq) + 2e^- \rightarrow H_2(g)$	0.00	Reference half-cell
$Cu^{2+}(aq) + 2e^- \rightarrow Cu(s)$	0.34	
$O_2(g) + 2H_2O(l) + 4e^- \rightarrow 4OH^-$	0.40	Strong oxidizing agent, strong tendency to be reduced, strong tendency to attract electrons, weak reducing agent, tendency to occur in the forward direction.
$Cu^+(aq) + e^- \rightarrow Cu(s)$	0.52	
$Ag^+(aq) + e^- \rightarrow Ag(s)$	0.80	
$ClO_2(g) + e^- \rightarrow ClO_2^-(aq)$	0.95	
$O_2(g) + 4H^+(aq) + 4e^- \rightarrow 2H_2O(l)$	1.23	
$Cl_2(g) + 2e^- \rightarrow 2Cl^-O(aq)$	1.36	
$PbO_2(s) + 4H^+(aq) + 4e^- \rightarrow Pb^{2+} + 2H_2O(l)$	1.46	
$Au^{3+}(aq) + 3e^- \rightarrow Au(s)$	1.50	
$H_2O_2(aq) + 2H^+(aq) + 2e^- \rightarrow 2H_2O(l)$	1.78	
$F_2(g) + 2e^- \rightarrow 2F^-(aq)$	2.87	

activity series of metals in aqueous solution

metal	oxidation reaction
lithium	$Li(s) \rightarrow Li^+(aq) + e^-$
potassium	$K(s) \rightarrow K^+(aq) + e^-$
barium	$Ba(s) \rightarrow Ba^{2+}(aq) + 2e^-$
calcium	$Ca(s) \rightarrow Ca^{2+}(aq) + 2e^-$
sodium	$Na(s) \rightarrow Na^+(aq) + e^-$
magnesium	$Mg(s) \rightarrow Mg^{2+}(aq) + 2e^-$
aluminum	$Al(s) \rightarrow Al^{3+}(aq) + 3e^-$
manganese(II)	$Mn(s) \rightarrow Mn^{2+}(aq) + 2e^-$
zinc	$Zn(s) \rightarrow Zn^{2+}(aq) + 2e^-$
chromium(III)	$Cr(s) \rightarrow Cr^{3+}(aq) + 3e^-$
iron(II)	$Fe(s) \rightarrow Fe^{2+}(aq) + 2e^-$
cobalt(II)	$Co(s) \rightarrow Co^{2+}(aq) + 2e^-$
nickel(II)	$Ni(s) \rightarrow Ni^{2+}(aq) + 2e^-$
tin(II)	$Sn(s) \rightarrow Sn^{2+}(aq) + 2e^-$
lead(II)	$Pb(s) \rightarrow Pb^{2+}(aq) + 2e^-$
hydrogen gas	$H_2(g) \rightarrow 2H^+(aq) + 2e^-$
copper(II)	$Cu(s) \rightarrow Cu^{2+}(aq) + 2e^-$
silver	$Ag(s) \rightarrow Ag^+(aq) + e^-$
mercury(II)	$Hg(l) \rightarrow Hg^{2+}(aq) + 2e^-$
platinum	$Pt(s) \rightarrow Pt^{2+}(aq) + 2e^-$
gold	$Au(s) \rightarrow Au^{3+}(aq) + 3e^-$

The pH scale and indicators

What is pH?

- pH is defined as the negative common logarithm of the concentration of hydrogen ions in a solution in mol/L. A common logarithm is used because [H$^+$] can be very large or very small in an acidic solution. A change in one pH unit represents a difference of 10 times. "p" is an abbreviation for the mathematical relationship of "$-\log 10$".
- pH lower than 7 is acidic
- pH lower than 7 is acidic
- pH above 7 is basic
- [H$^+$] > [OH$^-$] in an acidic solution
- [H$^+$] < [OH$^-$] in a basic solution
- [H$^+$] = [OH$^-$] in a neutral solution
- The pH where [H$^+$] = [OH$^-$] is the equivalence point in a titration

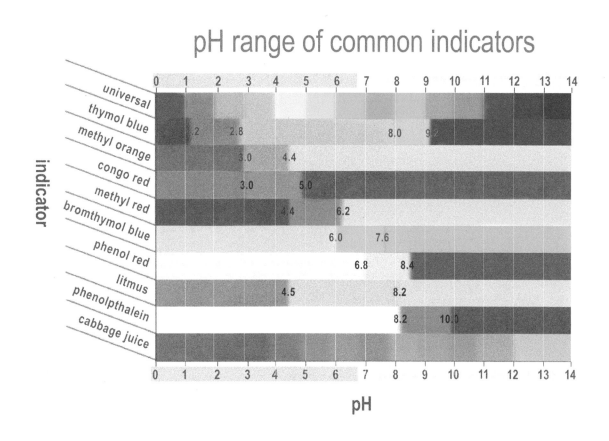

Strong and weak acids and bases

Strong acids are completely ionized in water to yield one or more H^+ ion per molecule:

- HBr, hydrobromic acid
- HCl, hydrochloric acid
- HI, hydroiodic acid
- HNO_3, nitric acid
- $HClO_4$, perchloric acid
- H_2SO_4, sulfuric acid (only first proton completely ionizes)

Strong bases are completely ionized in water to yield one or more OH^- ion per molecule:

- $Ba(OH)_2$, barium hydroxide
- $Ca(OH)_2$, calcium hydroxide
- LiOH, lithium hydroxide
- KOH, potassium hydroxide
- NaOH, sodium hydroxide
- $Sr(OH)_2$, strontium hydroxide

interpreting K and pK values

strength	K	pK
very strong	> 0.1	< 1
moderately strong	$10^{-3} - 0.1$	1 – 3
weak	$10^{-5} - 10^{-3}$	3 – 5
very weak	$10^{-15} - 10^{-5}$	5 – 15
extremely weak	$< 10^{-15}$	> 15

K and pK values for selected acids and bases

acid	name	K_a	pK_a
$H_2C_2O_4$	oxalic acid*	5.9×10^{-2}	1.23
H_3PO_4	phosphoric acid*	7.52×10^{-3}	2.12
HF	hydrofluoric acid	7.2×10^{-4}	3.14
HCH_3COO	acetic acid	1.76×10^{-5}	4.75
H_2CO_3	carbonic acid*	4.3×10^{-7}	6.37
HCN	hydrocyanic acid	6.17×10^{-10}	9.21
base	**name**	K_b	pK_b
NH_3	ammonia	1.8×10^{-5}	7.74
*value for first hydrogen only			

Periodic Table

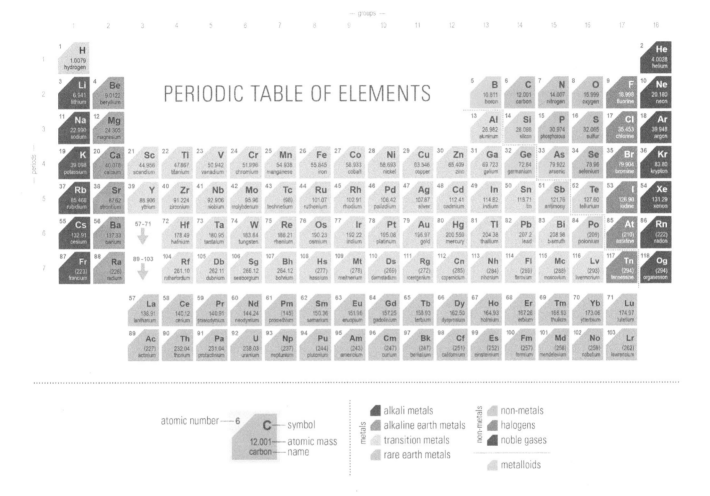

Chemistry Equation Builder

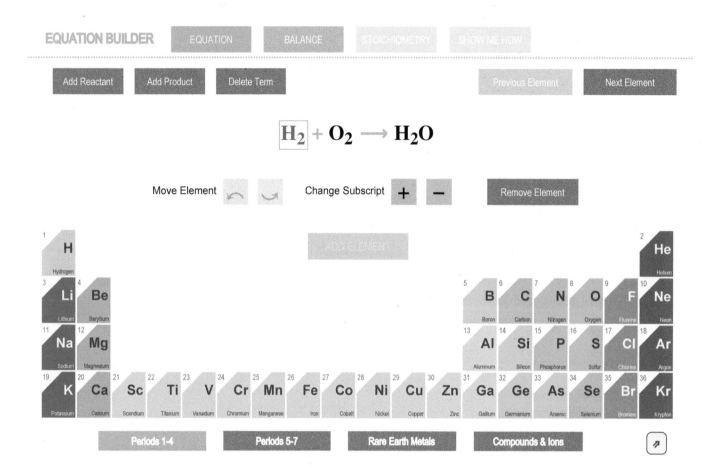

Chapter 25: Appendix

Glossary

A

absolute zero — the lowest possible temperature at which the energy of molecular motion is essentially zero, or as close to zero as allowed by quantum theory

accretion — the process of coming together under the influence of gravity to form a larger body.

accuracy — a term that describes how close a measurement is to the true value

activated complex — a high-energy state where bonds are being broken and reformed; also referred to as the transition state

activation energy — the energy necessary to break bonds in the reactants so a chemical reaction can occur, measured in units of joules per mole (J/mole) or kilojoules per mole (kJ/mole); the minimum amount of energy required for molecules to react

active site — a small region of an enzyme molecule that is responsible for binding or holding on to the substrate molecules

activity series — a list of metals arranged in order of reactivity by easiest to oxidize to hardest to oxidize

actual yield — the amount obtained in the lab by actual experiment

addition polymerization — the process of forming a polymer by adding monomers together using their double bonds

addition reaction — an organic chemical change that adds atoms to parent carbon atoms by removing multiple bonds

adiabatic — are theoretical conditions in which no energy is lost to the surroundings.

adipose tissue — a type of connective tissue in living things that stores fat for energy storage, insulation, and cushioning

ADP — an energy carrier molecule, adenosine diphosphate, regarded as the spent form of ATP that can be recharged when a phosphate group is added to the molecule

alcohol — a hydrocarbon with a hydroxyl group, –OH, attached

aldehyde — a functional group in which the central carbon makes a double bond with oxygen and is then bonded to hydrogen on one side and to another hydrocarbon

alkane — a hydrocarbon containing only single bonds; a saturated hydrocarbon

alkene — a hydrocarbon containing one or more double bonds

alkyl halide — a functional group also known as haloalkanes that contain a halogen such as fluorine, chlorine, bromine, or iodine bonded to an R-group

alkyne — a hydrocarbon containing one or more triple bonds

alloy — a solid material made up of two or more elements (usually metals) that is evenly mixed on the microscopic level

alpha decay — decay that occurs when a nucleus emits a helium nucleus

alpha helix — a common form of the secondary structure of a protein that forms a spiral or twist similar to a Slinky® or spiral staircase, in which a single protein chain is held together by hydrogen bonding attractions

amine — an organic molecule containing an amino group, NH_2

amino acid — a chiral molecule that has an amino group on one end and a carboxyl group on the other; there are 20 different naturally occurring amino acids

Glossary

amorphous — a microstructure that does not have any ordered pattern

ampere — the SI unit of electrical current

amphipathic — a molecule that has dual polarity with a polar, hydrophilic region and nonpolar, hydrophobic region

amphoteric — a substance that can act as either an acid or a base under different circumstances; water is amphoteric

anabolism — The process of chemical reactions that constructs or synthesizes molecules from smaller units, usually requiring input of energy (ATP) in the process.

anhydride — A compound that contains no water particles.

anion — a negatively charged ion formed when an atom gains valence electrons; ions of non-metals and most polyatomic ions are anions

anode — the electrode at which oxidation occurs

antimatter — particles with the opposite electrical charge of normal matter particles

aqueous (aq) — dissolved in water

aqueous equilibrium — condition when the amount of dissolved solutes remains constant over time

aromatic hydrocarbon — a hydrocarbon with a distinct aroma containing one or more benzene ring structures

Arrhenius acid — a substance that produces hydrogen ions, H+ in water

Arrhenius base — a substance that produces hydroxide ions, OH- in water

astronomical unit (AU) — a unit of measurement used for distance measures of the solar system. One AU is equal to 1.50×10^{11} m, the approximate distance from Earth to the Sun

atmospheric pressure — the pressure we feel from the air around us, 101,325 Pa or 14.7 psi (average)

atomic mass — the mass of an atom's protons and neutrons expressed in atomic mass units (amu) on the periodic table; on the periodic table, calculated as a weighted average that reflects isotope abundance

atomic mass unit (amu) — a mass unit equal to 1.66×10^{-24} g or 1.661×10^{-27} kg

atomic number (Z) — the number of protons in a nucleus, unique to each element

atomic radius — the distance from the center of an atom to its outer edge

ATP — adenosine triphosphate, a molecule that carries chemical energy from spontaneous chemical reactions to nonspontaneous reactions

Aufbau principle — states that electrons enter the lowest energy orbitals before entering higher energy level orbitals.

autotroph — an organism that creates food or energy molecules from inorganic chemicals in the environment

average — value obtained from a set of measurements by adding up all their measurement values and dividing by the total number of measurements

Avogadro's law — Volume is directly proportional to moles when pressure and temperature are constant, stated as $V/n = k$.

Avogadro's number — the number of atoms contained in one mole, 6.02×10^{23}

B

balanced chemical equation — a chemical equation that satisfies the law of conservation of mass, when the total numbers of atoms of each element are the same on both the reactant side and the product side of a chemical equation

Glossary

bar graph — a graph in which the height or length of columns or bars represent the values of data

barometer — an instrument that measures atmospheric pressure

base — a chemical that dissolves in water to create fewer H^+ ions than there are in neutral water (or equivalently, more OH^- ions)

Beer's law — states the absorbance of a solution is directly proportional to its concentration

beta decay — decay that occurs when an unstable nucleus releases an electron

beta pleated sheet — a secondary structure of protein that forms a folded or pleated pattern similar to an old-fashioned fan; formed by two or more side-by-side protein chains held together by hydrogen bonding

beta radiation — the radiation resulting from beta decay

Big Bang Theory — The theory that illustrates the universe's origin via its expansion from a cluster of small, very hot, particles

bimolecular — an elementary process involving the collision of two molecules

binary compound — a compound made of only two atoms or ions

binding energy — the energy that keeps nucleons inside a nucleus together

blackbody — an idealized object that absorbs all the light that strikes it without reflecting any of it back into the environment, emitting thermal radiation

blackbody spectrum — the range of electromagnetic radiation from infrared to ultraviolet that is emitted from a blackbody

boiling point — the temperature at which a substance changes phase from liquid to gas; for example, the boiling point of water is 100°C

Boyle's law — states that the pressure and volume of a gas have an inverse relationship, when temperature is held constant.

brittle — a property of materials that are hard and require little energy to fracture, used to describe materials that break or shatter relatively easily

Brownian motion — the erratic, jerky movement of tiny particles suspended in a fluid caused by the random impacts of individual molecules in thermal motion

Brønsted-Lowry — a different way to look at what defines acids and bases, in which acids are compounds that donate protons (H^+) and bases are compounds that accept protons

buffer — a solution that resists small changes in pH by chemical action

buffer capacity — the amount of excess acid or base a buffer can neutralize without changing the pH of the solution

C

calorimeter — is a well-insulated device in which a reaction takes place.

calorimetry — The study of heat flow and its measurement

Calvin cycle — a sequence of chemical reactions that makes energy-rich sugars from CO_2 by using the high-energy molecules ATP and NADPH formed in light-dependent reactions

carbohydrates — organic molecules that occur in food and living tissue which contain a hydrogen to oxygen ratio of 2:1 and are a source of energy for living organisms

Glossary

carbon cycle — the process showing how carbon moves among the atmosphere, the land, and the ocean in its different chemical forms, such as CO_2, $C_6H_{12}O_6$, and H_2CO_3

carbon dating — a technique that uses the ratio of C-12 to C-14 isotopes to determine the age of organic materials

carbon-14 — a radioactive isotope of carbon

carbonyl group — common organic reference, formed when a carbon atom is double bonded to an oxygen atom (C=O).

carboxylic acid — an organic molecule containing a carboxyl group, COOH

catabolism — is the process of chemical reactions that break down complex molecules into smaller units, usually releasing energy in the process.

catalyst — a substance that speeds up the rate of a chemical reaction, by providing a pathway with a lower activation energy

cathode — the electrode at which reduction occurs

cation — a positively charged ion formed by losing valence electrons, including metals and the ammonium ion

cell emf — abbreviation for the electromotive force of a cell

cellular respiration — a chemical reaction that breaks down glucose and other food molecules in the presence of oxygen, releasing energy that is used to carry out life processes

Celsius scale — a temperature scale with 100 degrees between the freezing point and the boiling point of water; water freezes at 0°C and boils at 100°C

chain reaction — term used to describe when the neutrons produced in one fission reaction cause more fission reactions to take place, releasing yet more neutrons

Charles' law — states that the volume of a gas is directly proportional to the temperature at constant pressure.

chemical bond — a relatively strong attraction between two atoms that forms when electrons are shared or transferred between the two atoms

chemical change — a change that affects the structure or composition of the molecules that make up a substance, typically turning one substance into another substance with different physical and chemical properties

chemical equation — an expression that describes the changes that happen in a chemical reaction, typically of the form reactants → products

chemical formula — a combination of element symbols and subscripts that tells you the ratio of elements in a compound

chemical property — a property that can only be observed when one substance changes into a different substance, such as iron's tendency to rust

chemical reaction — a process that rearranges the atoms in any substance(s) to produce one or more different substances; the process that creates chemical changes

chemical species — a chemical entity such as a particular atom, molecule or ion that is taking part in a chemical process or is being measured

chemosynthesis — the formation of organic matter via the oxidation of inorganic compounds

chemotroph — an organism that can synthesize its own organic food molecules from inorganic substances in the absence of light

chiral carbon — carbon atom in a ring or chain that has four different groups attached to it

chlorophyll — a pigment molecule that absorbs sunlight and uses it to excite electrons in its molecular structure, thereby storing energy to use in photosynthesis

Glossary

chloroplast — a membrane-bound structure found inside plant cells where pigments like chlorophyll are found and where photosynthesis occurs

chromatography — solution separation technique where solutes pass through a medium at different rates

claim — a statement asserting something to be true

closed system — a system that is not allowed to exchange matter or energy with its surroundings; in thermodynamics, a closed system can exchange energy, but not matter with its surroundings

colligative property — a physical property of a solution that depends only on the number of dissolved solute particles, not on the type (or nature) of the particle itself

colorimeter — a device that uses the absorbance or transmittance of light shined through a colored solution to determine its concentration

combined gas law — mathematical relationship that merges Boyle's law and Charles' law

combustion — a chemical reaction that involves the rapid combination of a fuel with oxygen

common ion — an ion produced by two or more different chemicals in the same solution

complex carbohydrates — larger molecule made of three or more monosaccharides generally used to store energy

compound — a substance containing more than one element in which atoms of different elements are chemically bonded together

concentrated — term used to describe a solution containing a lot of solute compared to solvent

concentration — the amount of each solute compared to the total solution

conclusion — a stated decision whether or not the results of experiments or observations confirm an idea or hypothesis

condensation — a phase change from gas to liquid; a substance in its gas phase may condense at a temperature below its boiling point

condensation polymerization — type of reaction that creates a polymer by linking monomers through the loss of a small molecule such as water

condensed matter — matter in which atoms or molecules are closely packed together and strongly interacting, includes both liquids and solids

conduction — the flow of heat energy through the direct contact of matter

conductivity — a measure of the degree to which a substance can or cannot conduct electricity

conjugate acid-base pairs — a pair of molecules or ions where one molecule or ion acts like an acid by donating a proton, and the other molecule or ion acts like a base by accepting the proton

control variables — variables that are kept constant during an experiment

conversion factor — a ratio of two different units that has a value of 1 (for example, 3.785 liters/1 gallon), where the numbers are different but the actual physical quantity is the same.

correlation coefficient (r) — a number between +1 and -1 which describes the goodness of fit for a regression

corrosion — the process where a metal deteriorates during a redox reaction

cosmic background radiation — radiation formed as a byproduct of the heat released from the Big Bang

coulomb (C) — the SI unit of electrical charge

Glossary

covalent bond — a chemical bond that consists of two or more shared electrons; also called a molecular bond

cracking — a process used to break long-chain hydrocarbons down into smaller hydrocarbon fragments, requiring high temperatures and a catalyst

critical mass — the minimal mass required to sustain a nuclear chain reaction

crystal — a piece of solid, nonamorphous matter in which the microstructure is uniform and continuous over the entire piece

crystalline — a microstructure with a repeating, ordered pattern

D

Dalton's law — states that in a mixture of non-reacting gases, the total pressure exerted is equal to the sum of the partial pressures of the individual gases.

decomposition — a reaction in which a single substance breaks apart to form two or more new substances

dehydrogenation reaction — chemical process in which hydrogen (H_2) is removed from the alkane, causing formation of double bonds and resulting in the formation of unsaturated alkenes

denatured — amino acid or protein that has suffered damage to its 3D structure resulting in compromised function

density — a property of a substance that describes how much matter the substance contains per unit volume; typical units are grams per cubic centimeter (g/cm^3)

dependent variable — in an experiment a variable that is assumed to respond to changes in another variable or which evolves over time; on a graph, the dependent variable is plotted on the vertical (y) axis

deposition — the conversion of a gas to a solid without ever passing through the liquid phase

desalination — use of reverse osmosis to convert saltwater into freshwater

dew point — the temperature at which air is saturated with H_2O vapor (Rh = 100%)

diatomic molecule — a molecule composed of only two of the same or two different atoms

diffusion — the spreading of molecules through their surroundings through constant collisions with neighboring molecules

dilute — term used to describe a solution containing relatively little solute compared to solvent

dimensional analysis — technique of using conversion factors to convert between units

dipole — a structure such as a molecule in which there is some separation of positive and negative charges

dipole–dipole attraction — the attractions between the positive part of one polar molecule and the negative part of another polar molecule

diprotic acid — an acid in which one mole of acid yields two moles of protons in water

disaccharides — oligosaccharides composed of two monosaccharides

dissociation — the separation of a substance into its ions

dissolution — the process where a solvent dissolves a solute to form a solution

dissolved — term used to describe when particles of solute are completely separated from each other and dispersed into a solution

distillation — a process of separating, concentrating, or purifying liquids by boiling and condensing them

Glossary

distilled water — water purified by heating water to steam and condensing the steam into a clean or sterilized container

DNA — a long deoxyribonucleic acid polymer chain of nucleotides that stores and transmits genetic information

Doppler effect — the change in frequency detected as the distance between the observer and the source changes

double bond — a chemical bound in which two pairs of electrons are shared

double replacement — reactions in which positive ions are exchanged between compounds; also known as double displacement

ductile — a property of materials that describes ease of stretching or molding and lack of brittleness

E

electric current — the directed movement of electric charges

electrochemical — describes chemical change brought about through an electrical change or vice versa; conversion between chemical and electrical energy

electrochemical cell — a device in which redox reactions take place

electrochemistry — the study of change in oxidation state usually due to electron transfer during chemical processes

electrode — the part of a cell where oxidation and reduction reactions occur

electrolysis — the result of a chemical reaction in an electrolytic cell

electrolyte — a solute capable of conducting electricity when dissolved in an aqueous solution

electrolytic cell — an electrochemical cell where current is generated when electrical energy is converted to chemical energy in a nonspontaneous reaction

electromotive force (emf) — the difference in the electrical potential between the anode and the cathode of an electrochemical cell

electron configuration — a description of which orbitals contain electrons for a particular atom

electron density — the number of electrons in a specific volume of space.

electron transport chain — a sequence of biologically important chemical reactions that uses high-energy electrons from the Krebs cycle to make ATP molecules

electronegativity — a value between 0 and 4 that describes the relative pull of an element for electrons from other atoms, with higher numbers meaning stronger attraction for electrons; the ability of an atom to attract another atom's electrons when bonded to that other atom

electroneutrality — the property of charges to balance each other, resulting in a charge neutral system

electroplating — nonspontaneous redox process where an electrolytic cell is used to deposit a thin metal coating on a material by oxidizing anions or reducing cations

element — substance that cannot be separated into simpler substances by chemical means

element symbol — a one- or two-letter abbreviation for each element

elementary steps — a series of simple reactions that represent the overall progress of the chemical reaction at the molecular level

empirical formula — the simplest ratio of atoms in a substance

Glossary

endothermic — a physical or chemical process that absorbs energy

energy — a measure of a system's ability to change or create change in other systems, typically measured in joules

energy level — the set of quantum states for an electron in an atom that have approximately the same energy

enthalpy — the energy potential of a chemical reaction at standard temperature and pressure, measured in joules per mole (J/mole) or kilojoules per mole (kJ/mole)

enthalpy of formation — the change in energy when one mole of a compound is assembled from pure elements

entropy — is the way energy is distributed in a system or substance, symbolized by S

enzymes — biological catalysts, responsible for speeding up the chemical reactions inside our bodies

equilibrium — a balance in a chemical system, at which the rate of the forward reaction is equal to the rate of the reverse reaction, and the concentrations of reactants and products remain constant over time

equilibrium constant — the numeric value of the equilibrium expression

equilibrium expression — a special ratio of product concentrations to the reactant concentrations, in which concentration is raised to the power of the corresponding coefficient in the balanced equation

equilibrium position — the favored direction of a reversible reaction, determined by each set of concentrations for the reactant(s) and product(s) at equilibrium

equivalence point — in a titration, the point at which the moles of H^+ from the acid and OH^- from the base are exactly equal

equivalent — is the mass of acid that produces one mole of protons in water.

error — the unavoidable difference between a real measurement and the unknown true value of the quantity being measured

error bars — vertical and horizontal lines added to the plotting of a data point to show the uncertainty in the x and y values

ester — an organic molecule containing a –COOR group, responsible for many smells and tastes in fruits and flowers

ether — a molecule where oxygen is bonded to two carbon groups, often hydrocarbon chains

evaporation — a phase change from liquid to gas at a temperature below the boiling point

excess reactant — the substance that remains after the reaction is complete

excited state — condition of an atom when one or more electrons occupies a higher energy orbital than in the ground state

exothermic — a physical or chemical process that releases energy

experiment — a situation specially set up to observe how something happens or to test a hypothesis

experimental variable — the single variable that is changed to test its effect

exponent — the power of 10 for a number written in scientific notation, such as the 2 in the number $5 \times 10^2 = 500$

extremophiles — organisms that have the ability to survive in conditions that would kill most other life forms on Earth

F

Fahrenheit scale — a temperature scale with 180 degrees between the freezing point and the boiling point of water; water freezes at 32°F and boils at 212°F

Glossary

family — another name for a group or vertical column on the periodic table

fat — nonpolar molecules made from fatty acids and a glycerol molecule, efficient for storing energy for many types of animals and not soluble in water

first law of thermodynamics — states that energy can neither be created nor destroyed, and thus the total energy in an isolated system remains constant; all the energy lost by one system must be gained by the surroundings of another system

fission reaction — a nuclear change in which a large nucleus is split into smaller nuclei

formula mass — the sum of the atomic masses of all atoms in a chemical formula expressed in atomic mass units (amu); also known as *molecular mass* for covalent substances

formula unit — the chemical formula of an ionic compound in its lowest ratio

free radical — a molecule or atom that is highly reactive owing to its having one or more unpaired valence electrons

frequency table — a table in which a large data set data has been sorted into a smaller number of categories with each category representing a range of values or a count of items - useful for seeing large scale patterns in data with high point-to-point variability

frost line — the point in the solar system, approximately 4 AU from the Sun, where compounds like ammonia have the ability to condense into solids

functional group — an atom or group of atoms that defines chemical and physical properties of an organic compound

fusion reaction — a nuclear change where two small nuclei are fused under conditions of high temperature and pressure, creating a large nucleus and vast amounts of energy

G

galvanic cell — an electrochemical cell in which spontaneous chemical reactions generate electricity; also known as a voltaic cell

gamma decay — decay that occurs when a nucleus releases electromagnetic energy

gene — a segment of DNA that contains the code for a specific protein

geologists — scientists who study the Earth

geometric isomer — a molecule with the same sequence of atoms, but different placement of groups around a double bond, as another molecule

geosphere — the total of all non-living matter on Earth - such as rocks and mountains - excluding water and the atmosphere

Gibbs free energy — energy that is free to do work

Gibbs–Helmholtz equation — a thermodynamic equation used for calculating changes in the Gibbs energy of a system as a function of temperature

glycolysis — the first step in cellular respiration, in which one molecule of glucose is broken in half, yielding two three-carbon molecules of pyruvic acid

graduated cylinder — an instrument used to measure volume

gram — unit of mass equal to 1/1000 kg

greenhouse effect — occurs when heat-trapping gases in the atmosphere redirect some heat back to Earth, warming the planet

greenhouse gases — gases in Earth's atmosphere such as water vapor, carbon dioxide and methane that re-radiate heat and trap it near Earth's surface

Glossary

ground state — condition of an atom when every electron occupies the lowest energy orbital allowed by the exclusion principle

H

half-life — the time it takes for half of the atoms in a sample to decay

half-reactions — the oxidation and the reduction parts of a redox reaction

halogens — group 17 elements, including chlorine, fluorine, and bromine

heat — energy that is transferred due to differences in temperature

heat of fusion — the energy required to change the phase of one gram of a material from liquid to solid or solid to liquid at constant temperature and constant pressure

heat of solution — the energy absorbed or released when a solute dissolves in a particular solvent

heat of vaporization — the energy required to change the phase of one gram of a material from liquid to gas or gas to liquid at constant temperature and constant pressure

Heisenberg's uncertainty principle — a foundation of quantum mechanics that states that the values of position and velocity, or energy and time, cannot be known to better than $h/2\pi$

Henry's law — a given temperature, gas solubility in a liquid is proportional to the pressure and concentration of the gas above the liquid

Hertzsprung-Russell diagram — a graph showing a population of stars in which energy output (luminosity) is plotted on the vertical axis and surface color temperature on the horizontal axis

Hess's law — the total enthalpy change for the reaction is the sum of all changes regardless of the multiple stages or steps of a reaction

heterogeneous mixture — a mixture that is not uniform, in which different samples may have different compositions

heterotroph — an organism that consumes other living things to obtain energy

homeostasis — a condition where the body maintains a stable equilibrium between independent internal systems

homogeneous mixture — a mixture that is uniform throughout, in which any sample has the same composition as any other sample

Hund's rule — states that one electron will occupy each available sub-orbital before any sub-orbital accepts two electrons

hybridization — a quantum behavior in which the s and p orbitals combine to form sp-hybrid orbitals; provides a foundation to explain chemical bonding

hydrate — a compound that has water attached to its structure

hydration — the process of molecules with any charge separation to collect water molecules around them; not chemically bonded, but holding tightly to its private collection of water molecules

hydrocarbon — a molecule made entirely from carbon and hydrogen atoms

hydrogen bond — an intermolecular attraction that forms between a hydrogen atom in one molecule and the negatively charged portion of another molecule (or another part of the same molecule); in water, the attractions between the partially positive hydrogen from one molecule to the partially negative oxygen on an adjacent molecule

hydrogenation reaction — a type of addition reaction where hydrogen gas, H_2, is added to a double or triple bond in a hydrocarbon, causing the structure to become saturated

Glossary

hydronium ion — H_3O^+ ion that forms when an H^+ ion bonds to a complete water molecule, giving acids their unique properties

hydrophilic — water-loving, polar portion of a molecule

hydrophobic — water-hating, nonpolar end of a molecule

hydrosphere — the total mass of water on Earth in all forms - including the oceans, lakes, rivers, clouds, rain, and groundwater

hydrothermal vents — an opening in the Earth's surface from which water, heated within the Earth, is released

hypothesis — a tentative explanation for something, or a tentative answer to a question

I

ideal gas — a gas that is made up from molecules that have no volume or interactions with each other and that obeys the various gas laws; gases with high temperatures and low pressures are good approximations this kind of gas

immiscible — condition where a solute and solvent are unable to mix; a solution cannot form

independent variable — in an experiment, a variable which may be purposefully changed and which; (a) is assumed to affect the values of other variables or (b) is time in an experiment in which things change over time; this variable is plotted on the horizontal (x) axis of a graph

indicator — a chemical that turns different colors at different values of pH, used to determine pH directly or in reactions with solutions of known pH

inhibitor — a substance that slows down or prevents a particular chemical reaction or other process, or that reduces the activity of a particular reactant, catalyst, or enzyme.

inquiry — the process of learning through asking questions

insoluble — not dissolvable in a particular solvent

intensity — the amount of energy that flows per unit area per unit time

intermediate — a chemical species that is formed during the elementary steps but is not present in the overall balanced equation

intermolecular forces — forces that act between molecules, typically much weaker than the forces acting within molecules

interstellar medium — a mixture of ions, atoms, molecules, and dust grains that occupies the space between stars

intramolecular forces — are forces between atoms within a molecule

inverse square law — states that the intensity of radiation emitted from a point source decreases inversely over the square of the distance

ion — an atom or small molecule with an overall positive or negative charge as a result of an imbalance of protons and electrons

ion product constant — the product of $[H^+] \times [OH^-] = 1.0 \times 10^{-14}$ in pure water at 25 °C, symbolized as K_w

ionic bond — an attraction between oppositely charged ions, occurring with all nearby ions of opposite charge

ionic compound — a compound such as salt (NaCl), in which positive and negative ions attract each other to keep matter together

ionic radius — the approximate size of an atom's charged ion measured in a variety of different ways

Glossary

ionization — the process involving the gain or removal of electrons from an atom which can create a positive or negative ion

ionization energy — the energy required to completely remove an electron from an atom

ionizing radiation — radiation with high enough energy to remove electrons from their atoms

isoelectronic — two or more atoms or ions with the same electron configuration.

isoenzymes — are enzymes that differ in amino acid sequence but catalyze the same chemical reaction.

isomers — are specific structures of a molecule; a term used only when a chemical formula could represent more than one molecule

isotopes — atoms or elements that have the same number of protons in the nucleus but different number of neutrons

J

joule — the fundamental SI unit of energy (and heat)

K

Kelvin scale — a temperature scale that starts at absolute zero and has the same unit intervals as the Celsius scale. $T_{Kelvin} = T_{Celsius} + 273$

ketone — a functional group in which the central carbon makes a double bond with oxygen and is then bonded to two hydrocarbons one on each side

kilogram — SI unit of mass

kinetic energy — the energy of motion

kinetic molecular theory — the theory that explains the observed thermal and physical properties of matter in terms of the average behavior of a collection of atoms and molecules

Krebs cycle — a series of energy-extracting chemical reactions that break pyruvic acid down into $CO_2(g)$, during cellular respiration

L

lattice — a regular repeating pattern of ions in a salt crystal

lava — molten rock or magma released from the vent(s) of a volcano

law of conservation of energy — states that energy can never be created or destroyed, just converted from one form into another

law of conservation of matter — states that the total mass of reactants (starting materials) and the total mass of products (materials produced by the reaction) is the same

law of mass action — a general description of any equilibrium reaction that illustrates the relationship of chemicals and their molar coefficients

Le Châtelier's principle — states that when a change is made to a system at equilibrium, the system will shift in a direction that partially offsets the change; the change can be defined as a change in temperature, concentration, volume, or pressure

Lewis dot diagram — shows one dot for each valence electron an atom has; these dots surround the element symbol for the atom

light-year — the distance that light travels through a vacuum in the time-span of one year. It is equivalent to 9.46×10^{15} m.

limiting reactant — the substance that is used up first during a reaction

linear regression — a mathematical technique for calculating the best straight line fit to a set of x-y data

Glossary

lipid — a molecule that typically falls into the category of fat or steroid

lipoproteins — small spherical assemblies with hydrophilic heads lining the exterior with hydrophobic tails inside where a lipid is carried

liter — an SI unit of volume equal to a cube 10 centimeters on a side, or 1,000 cm^3

logarithm — in base 10, a number A derived from another number B such that $10^B = A$

London dispersion attraction — the attraction that occurs between nonpolar molecules owing to temporary slight polarizations that occur when the normally equal distribution of electrons is shifted

M

macronutrients — elements needed in large quantities by your body

macroscopic scale — on the scale that can be directly seen and measured, from a colony of bacteria up to the size of a planet

magma — molten rock found below the surface of the upper mantle

malleable — able to be hammered into thin sheets without cracking

mantissa — a decimal number that multiplies the power of 10 in scientific notation

mantle — the 2,900 km thick layer of silicate rock beneath the Earth's crust and above the Earth's core

mass — the amount of matter, measured in units of grams or kilograms (SI)

mass defect — the difference in mass between the reactants and products of a nuclear reaction is converted to energy released during the reaction

mass number (A) — the total number of protons and neutrons in a nucleus

matter — material that has mass and takes up space

measurement — information that describes a physical quantity with both a number and a unit

melting point — the temperature at which a substance changes phase from solid to liquid; for example, the melting point of water is 0°C

mesosphere — the layer of Earth's atmosphere where meteors burn up, located between the stratosphere and thermosphere

metabolism — the collection of chemical reactions that convert food into building blocks for biomolecules that sustain life, such as proteins

metal — any of a class of elementary substances, as gold, silver, or copper, all of which are crystalline when solid and which are good conductors of electricity and heat.

metallic bond — an attraction between metal atoms that loosely involves many electrons

metalloids — a class of elements that have electrical properties between those of metals and non-metals. This group includes boron, silicon, germanium, arsenic, antimony, tellurium, and polonium.

metamorphic rock — created from igneous and sedimentary rock under conditions of high heat and pressure

microscopic scale — on the scale of atoms and molecules, of order 10^{-10} meters and smaller

milliliter — a unit of volume equal to a cube 1 cm on a side, or 1 cm^3

mineral — a natural nonorganic solid with an organized internal structure where the atoms have a definite pattern

Glossary

miscible — capable of mixing to form a solution

mixture — matter that contains more than one pure substance

molality (m) — a unit of concentration used when the temperature varies, expressed in moles of solute per kilogram of solvent

molar mass — the mass of one mole of a substance; the term formula mass is used for ionic substances and molecular mass for covalent substances

molar volume — the amount of space occupied by a mole of gas at standard temperature and pressure, equal to 22.4 L per mole at 0°C and 1 atm

molarity (M) — the number of moles of solute per liter of solution

mole ratio — a ratio comparison between substances in a balanced equation, obtained from the coefficients in the balanced equation; the ratio allows for the conversion of one substance to another substance by using molar equivalent amounts

molecular formula — the exact number and types of atoms in a molecule

molecular mass — the formula mass of a covalent compound

molecular weight — the mass in grams of one mole of an element or compound - identical to formula mass and molecular mass

monomer — a small molecule that is a building block of larger molecules called polymers

monosaccharide — a single sugar unit

N

NADH — nicotinamide adenine dinucleotide, a molecule capable of carrying high-energy electrons and transferring them to another pathway

NADPH — nicotinamide adenine dinucleotide phosphate, a molecule that carries two high-energy electrons and stores sunlight as chemical energy

nanotechnology — Technology that relies on manipulating individual atoms or molecules, at the nanometer scale

natural laws — the unwritten rules that govern everything in the universe

nebula — a sizable cloud composed of gas and dust that forms in the space between stars

Nernst equation — the mathematical equation that relates the electrical potential of a cell to the standard reduction potential and the state of the reaction under nonstandard conditions as given by the concentration of reactants and products

network covalent — a type of large molecule, usually made from hundreds to billions of atoms, in which each atom is covalently bonded to multiple neighboring atoms, forming a web of connections

neutral — an atom or molecule with zero total electric charge; in the context of acids and bases, a solution with the pH = 7.0, which also means the concentrations of H^+ and OH^- ions are equal

neutralization — a reaction in which the pH of an acid is raised by combining with a base, or the pH of a base is lowered by combining with an acid; complete neutralization results in a pH of 7, the same as neutral water

nitrogen cycle — the process that shows how nitrogen moves from the atmosphere to the ocean, land, and living systems

nitrogen fixation — the ability to use nitrogen gas from the atmosphere and convert it into a nitrogen compound such as ammonia or nitrate

nitrogenous base — a nitrogen-containing base; one component of nucleotides

Glossary

nomenclature — the system and set of rules used to name chemical compounds

non-ionizing radiation — radiation that cannot cause ionization of atoms

non-metals — a class of elements which typically are electrical insulators, including oxygen, carbon, nitrogen, neon, fluorine and elements to the right of the periodic table.

non-polar covalent bond — a bond formed between two atoms in which electrons are shared equally or almost equally between the two atoms

normality — a unit of concentration measured in equivalents per liter, symbolized as N

nucleons — all of the particles inside a nucleus, including protons and neutrons

nucleotide — a molecular unit composed of a five-carbon sugar, a phosphate group, and a nitrogenous base

O

objective — describes only what actually occurs without explanation, opinion or bias

ocean acidification — the decrease in pH in the oceans due to increasing concentration of atmospheric carbon dioxide

octet rule — rule that states that elements transfer or share electrons in chemical bonds to reach a stable configuration of eight valence electrons; the light elements H, Li, Be, and B have He as the closest noble gas, so the preferred state is two valence electrons instead of eight

ohm (Ω) — the unit of electrical resistance

open system — a system in which matter and energy can be exchanged with the surroundings

optical isomers — mirror image compounds that form when the carbon atom has four different groups attached to it; they rotate light differently with respect to each other

optimum range — a set of conditions such as pH and temperature range within which an enzyme works best

orbital — group of quantum states that have similar spatial shapes, labeled s, p, d, and f

organic chemistry — the study of carbon-containing molecular compounds

osmosis — natural tendency of a solution to keep its solute particles evenly distributed

oxidation — a chemical reaction that increases the charge of an atom or ion by giving up electrons

oxidation number — a number given to an element that indicates the number of electrons lost or gained by that element in a particular compound

oxidizing agent — element that is reduced and thus oxidizes another element

ozone layer — a layer of atmospheric ozone, O_3 molecules found in the stratosphere that absorbs harmful UV radiation

P

parent compound — the longest continuous chain of carbons in an organic compound, which tells us the base alkane name

partial hydrogenation — a chemical reaction where only some of the unsaturated fat is hydrogenated; in this case some double bonds remain in the structure, causing the formation of transfats in place of the naturally occurring *cis* configuration

pascal (Pa) — SI unit of pressure, equal to one newton of force per square meter of area (1 N/m^2)

Pauli exclusion principle — states that two electrons in the same atom may never be in the same quantum state

Glossary

peptide bond — a bond between the amino group and the carboxyl group of two amino acids; formed through condensation polymerization or the loss of water molecules

percent yield — the ratio of the amount of product actually obtained by experiment (actual yield) as compared to the amount of product calculated theoretically (theoretical yield) multiplied by 100

period — a row of the periodic table

periodic — repeating at regular intervals

periodic table — a graphical chart of information on the elements that groups the elements in rows and columns according to their chemical properties

petroleum — a blend of hydrocarbons, formed from prehistoric organic matter, containing mainly straight- and branched-chain alkanes

pH scale — a measurement of the H^+ ion concentration that indicates whether a solution is acidic or basic; pure water has a pH of 7; solutions with pH < 7 are acidic.; solutions with pH > 7 are basic

phase change — conversion of the organization of molecules in a substance without changing the individual molecules themselves, such as changing from solid to liquid or liquid to gas

phase diagram — is a plot of pressure and temperature that shows the state of matter for a particular substance.

phospholipid — a lipid that has a phosphate group attached, giving it a polar head and a nonpolar tail; the phosphate group is the polar region of the molecule and the hydrocarbon chain is the nonpolar region

phosphorus cycle — the process that shows how phosphorus moves among the land, the ocean, and living systems

photodissociation — the breaking of a chemical bond in a molecule caused by the absorption of a photon

photon — the smallest possible quantity (or quantum) of light

photosphere — the visible surface of a star from which its light radiates

photosynthesis — the endothermic chemical reaction, driven by sunlight, that combines CO_2 and H_2O to form glucose ($C_6H_{12}O_6$) and oxygen

phototroph — an organism that requires light to make its own organic food molecules

physical equilibrium — a balance in a physical system where there is an equilibrium between two or more phases of the same substance

physical property — property such as mass, density, or color that you can measure or see through direct observation

pie graph — a circular diagram resembling slices of a pie in which each slice represents the relative size of one part of a group in relation to the whole group

pigment — a colored compound found in nature

Planck's constant (h) — the scale of energy at which quantum effects must be considered, equal to 6.626×10^{-34} joule-seconds (J·s)

plasma — a hot, energetic phase of matter in which the atoms are broken apart into positive ions and negative electrons that move independently of each other

plate tectonics — the slow rearrangement of the Earth's crust driven by flows of heat from the core outward

Glossary

polar covalent bond — a bond formed between two atoms in which electrons are unequally shared

polarization — an uneven distribution of positive and negative charge

polyatomic ion — a small molecule with an overall positive or negative charge

polymer — a long-chain molecule formed by connecting small repeating units with covalent bonds; a molecule built up from many repeating units of a smaller molecular fragment

polymerization — a reaction that assembles a polymer through repeated additions of smaller molecular fragments

polymorphism — the property of the same composition of elements to form different minerals under different conditions of temperature and pressure

polyprotic acid — contains more than one hydrogen ion, or proton

positron — a particle that has the same mass as the electron and positive charge

positron emission — decay that occurs when a nucleus releases a positron

precipitate — an insoluble compound that forms in a chemical reaction in aqueous solution

precision — term that describes how close measured values are to each other

pressure — force per unit area, with units of Pa (N/m^2) or psi (lb/in^2) or atm; acts equally in all directions within a liquid or a gas

primary structure — the sequence of amino acids in a protein chain

procedure — detailed instructions on how to do an experiment or make an observation

product — a substance that is created or released in a chemical reaction

protein — biologically functional polymers of amino acids, typically very large molecules with 100 or more amino acids

pseudoscience — describes ideas that are often presented as scientific but which are not supported by scientific evidence - examples include astrology, phrenology, alchemy, crystal healing, creationism, and parapsychology

pure substance — an element or compound; a kind of matter that cannot be separated into other substances by physical means such as heating, cooling, filtering, drying, sorting, or dissolving

Q

quantum state — a specific combination of values of variables such as energy and position that is allowed by quantum theory

quaternary compound — a compound that contains four different elements

quaternary structure — 3D structure formed by complete proteins when they bond or interact with one another

R

R group — abbreviation for any atom or group of atoms attached to the main organic molecule

radiation — the transmission of energy through space

radiation dose — the amount of radiation absorbed by the body

radioactivity — a process by which the nucleus of an atom spontaneously changes itself by emitting particles or energy

rate constant — a number that describes how fast a radioactive nuclide decays

Glossary

rate determining step — the slowest elementary step in the reaction mechanism that determines the overall rate (or speed) of the chemical reaction

rate law — an equation that links the reaction rate with the concentrations or pressures of the reactants and constant parameters, usually rate coefficients and partial reaction orders

rate of decay — the number of radioactive decays per unit time

reactant — a substance that is used or changed in a chemical reaction; the starting material or substance in a chemical reaction

reaction mechanism — a proposed sequence of elementary steps that leads to product formation and must be determined using experimental evidence

reaction rate — the speed at which a chemical reaction occurs; the change in concentration, of a reactant or product, per unit time

reactivity — the tendency of elements to form chemical bonds

redox — the abbreviation for oxidation–reduction

reducing agent — substance that is oxidized and thus reduces another substance

reduction — a chemical reaction that decreases the charge of an atom or ion by accepting electrons

repeatable — when identical results of an experiment or observation can be obtained by anyone performing the same experiment or observation in the same way

resistance — the measure of the degree to which a conductor opposes an electric current through that conductor.

resolution — the smallest quantity that can be measured or displayed

resonance — when delocalized electrons in molecules or polyatomic ions are shared equally, more than one Lewis structure is possible

reverse osmosis — water purification process that uses high pressure and a semi-permeable membrane to force separation between solute and solvent particles

RNA — a ribonucleic acid polymer composed of nucleotides that contain a five-carbon sugar called ribose, a phosphate group, and a nitrogenous base; transmits the information stored in DNA to cell machinery

rock — a solid composed of two or more minerals with a variable composition

S

salinity — a measure of the concentration of salt in water

salt — an ionic compound in which the positive ion comes from an acid and the negative ion comes from a base; an ionic compound that dissolves in water to produce ions

salt bridge — an electrical connection between the oxidation and the reduction half-cells of an electrochemical cell

saturated — situation that occurs when the amount of dissolved solute in a solution gets high enough that the rate of undissolving matches the rate of dissolving

saturated hydrocarbon — organic compound containing only single carbon-to-carbon bonds

scale — a level or unit of measurement that shows a certain level of detail

scientific method — a process of learning that constantly poses questions and tentative answers that can be evaluated by comparison with objective evidence

Glossary

scientific notation — a method of writing numbers as a base, times a power of 10

second law of thermodynamics — states that the entropy of a spontaneous occurrence in an isolated system always increases.

secondary structure — refers to how a section of a protein is oriented, generally as an alpha helix or beta sheet

sedimentary rock — formed from the compression of small pieces of igneous rock

significant figures — a way of recording data that tells the reader how precise a measurement is

silicates — a class of compounds that contains silicon and oxygen.

simple carbohydrates — small sugar molecules made of one or two monosaccharides

single bond — a chemical bond in which one pair of electrons are shared

single replacement — a reaction in which a single element replaces another element in a compound, also known as displacement

slope — on a graph - the ratio of increase in the vertical (y) variable divided by increase in the horizontal (x) variable

solar nebula — a disk-shaped region of dust and gases around a forming star, which can lead to planetary formation

solubility — the amount of a solute that will dissolve in a particular solvent at a particular temperature and pressure

solubility product constant — the equilibrium constant for a solid substance dissolving in an aqueous solution that represents the level at which a solute dissolves in solution; abbreviated as K_{sp}

soluble — a substance that is able to dissolve in water.

solute — any substance in a solution other than the solvent

solution — a homogeneous mixture composed of a solute and a solvent

solvation — the process where a solvent surrounds a solute to dissolve it and form a solution; called hydration when water is the solvent

solvent — the substance that makes up the biggest percentage of the mixture, usually a liquid

specific heat — the quantity of energy, usually measured in $J/(g \cdot °C)$, it takes per gram of a certain material to raise the temperature by one degree Celsius

spectator ion — an ion that does not change its oxidation state and exists in the same form on both the reactant and product sides of a reaction

spectral line — a single wavelength or color of light appearing in a spectrum that is characteristic of particular energy level transitions between atomic electrons for any given element

spectrometer — a device that measures the spectrum of light

spectroscopy — the science of analyzing matter using electromagnetic emission or absorption spectra

spectrum — a representation of a sample of light into its component energies or colors, in the form of a picture, a graph, or a table of data

spin — a quantum property that can have two states referred to as up and down ($+1/2$ and $-1/2$); electrons, protons, and similar particles called fermions have this property

spontaneous — refers to a reaction that occurs in the indicated direction without any energy input

Glossary

standard deviation — the average spread of all the data points away from the average value of the whole set of data points

standard error — the error in the average (or mean) of a set of measurements - mathematically equal to the standard deviation divided by the square root of the number of measurements

standard molar volume — the amount of space taken up by 1 mole of any gas at standard temperature (0°C) and pressure (101.325 kPa); equal to 22.4 L

standard reduction potential — the potential of a cell measured under standard conditions of temperature, pressure, and concentration

standard temperature and pressure (STP) — conditions of one atmosphere of pressure and 0°C

state function — is a property whose value does not depend on the path taken to reach that specific value.

stoichiometry — the study of the amounts of substances involved in a chemical reaction

stomata — mouth-like openings on the underside of leaves through which gases are exchanged

stratosphere — the second layer of gases after the troposphere, extending from 10 to 50 km above the Earth

strong acid — a compound that dissociates completely (or almost completely), usually yielding 1 mole of H^+ for every mole of acid dissolved

strong base — a compound that dissociates completely (or almost completely), usually yielding 1 mole of OH^- for every mole of base

strong nuclear force — a short range attractive force that acts between protons and neutrons, but does not affect electrons

structural isomer — a molecule with the same number and type of atoms as another molecule but a different bonding pattern

sublimation — the conversion of a solid to a gas without ever passing through the liquid state.

substitution reaction — a chemical process in which one or more hydrogen atoms on an alkane are removed and replaced by different atoms

substrate — a reactant molecule that is able to be bound by a particular enzyme

supernova — a star that suddenly increases greatly in brightness because of a catastrophic explosion that ejects most of its mass.

supersaturated — term used to describe when a solution contains more dissolved solute than it can hold; these solutions are always unstable and the excess solute recrystallizes, often rapidly

surface tension — a force created by intermolecular attraction in liquids, such as hydrogen bonding in water, acting to pull a liquid surface into the smallest possible area, for example, pulling a droplet of water into a sphere, a property of liquids to resist having their surface broken, usually measured in J/m^2

surroundings — everything outside the defined system

synthesis — a reaction in which two substances combine to form a third substance, also known as a combination reaction

system — a group of interacting objects and effects that are selected for investigation

System International — a complete, coherent system of units used for scientific work, in which the fundamental quantities are length, time, electric current, temperature, luminous intensity, amount of substance, and mass

T

tap water — a solution of H_2O and other dissolved substances obtained from a faucet

Glossary

temperature — a measure of the average kinetic energy of atoms or molecules, measured in units of degrees Fahrenheit (°F), degrees Celsius (°C), or kelvin (K)

ternary compound — a compound that contains three different elements

tertiary structure — a level of protein organization held together by intermolecular attractions, hydrophobicity of nonpolar regions, and ion–dipole interactions, that forms from the bending or folding of the secondary alpha helix

theoretical yield — expected amount produced if everything reacts completely

theory — an explanation that is supported by evidence

thermal conductor — a material that conducts heat easily

thermal energy — the total energy in random molecular motion contained in matter due to its temperature

thermal equilibrium — a condition where the temperatures are the same and heat no longer flows from one material to another

thermal insulator — a material that resists the flow of heat

thermal radiation — the electromagnetic energy radiated by matter with a temperature above absolute zero

thermistor — an electronic sensor for measuring temperature by changes in resistance

thermocouple — an electronic sensor for measuring temperature that produces a temperature-dependent voltage

thermometer — an instrument that measures temperature

thermophiles — a classification of extremophiles that have undergone adaptations allowing them to live in high or low temperature extremes

thermoregulation — maintenance of a constant body temperature through physical and chemical means within the body

titration — a laboratory process to determine the precise volume of acid or base of known concentration that exactly neutralizes a solution of unknown pH

trace amount — a very small quantity

trace elements — elements that are needed in very small quantities to maintain optimum health; too little or too much of any one can be toxic or cause disease

transcription — the stage of gene expression where specific DNA sequences are copied and edited in the nucleus

transition metals — groups 3 - 12 of the periodic table which include elements with 1 to 10 electrons in d orbitals

translation — phase of gene expression that occurs in the cytoplasm where amino acid sequences in mRNA molecules are decoded and built into proteins by tRNA

transpiration — the release of water from the leaves of plants as they carry out photosynthesis

transport — is the movement of materials in and out of a cell.

triglyceride — a fat or lipid molecule used for energy storage, made of a glycerol molecule and three long-chain fatty acids

triple bond — a chemical bond in which three pairs of electrons are shared

triple point — a combination of temperature and pressure in which a substance is a solid, liquid, and gas simultaneously.

Glossary

triprotic acid — compound in which one mole of acid yields three moles of hydrogen ions or protons in water

troposphere — the layer of gases closest to the Earth, extending up to approximately 10 km above the Earth's surface

U

unbalanced chemical equation — does not satisfy the law of conservation of mass because the number of each type of atom on the reactant side does not equal the same number for each atom on the product side

unimolecular — an elementary step where only one reactant is involved

universal gas constant — the value for any gas under any conditions in which conditions are unchanging, represented by the letter R

unsaturated hydrocarbon — a compound that contains carbon and hydrogen with multiple bonds between the carbon atoms in the parent chain

V

vacuum — a space that does not contain any particles of matter

valence electrons — electrons in the highest unfilled energy level, responsible for making chemical bonds

Van der Waals attractions — a term used to describe the attractions between molecules, typically referred to as intermolecular attractions, but associated with the London dispersion type of intermolecular attraction

volcano — opening in the Earth where magma, gases, and other materials can escape from inside the Earth

volt — the unit of voltage

voltage — the electrical potential difference between two objects

voltaic cell — an electrochemical cell in which spontaneous chemical reactions generate electricity; also known as a galvanic cell

volume — an amount of space having length, width, and height

VSEPR (Valence Shell Electron Pair Repulsion) — a theory that states that the shapes of molecules are dictated, in part, by the repulsion of the shared electrons and the unshared pairs of electrons

W

water cycle — the process showing how water molecules move among the atmosphere, the land, and the ocean

water displacement — a gas collection technique that uses the gas to force water out of a container

watt — the SI unit of power or energy flow, equal to 1 J/s

weak acid — a compound that only partially dissociates in solution, typically with only a few percent (or less) of the substance yielding H^+ ions

weak base — a compound that only partially dissociates in solution, typically with only a few percent (or less) of the substance producing OH^- ions

weight — a force (push or a pull) that results from gravity acting on mass, measured in newtons in the SI system of units and pounds or ounces in the English system of units

work — the transfer of energy from one system to another to produce a change

Index

A

absolute zero	99
absorbance	431
accretion	796
accuracy	30
acidic solution	509
naming and writing formulas	512
activated complex	450
activation energy	450, 458
active site	699
activity series	590
addition reaction	669
adenosine diphosphate	717
adenosine triphosphate	717
adiabatic	110
adipose tissue	686
ADP	717
alcohol	658, 658
aldehyde	659
alkane	643
branched	645
alkene	650
alkyne	650
alloy	336
alloys	414
alpha (α-) helix	702
alpha- (α-) helix	697
amine	662
amino acids	65, 691
Amontons, Guillaume	390
amorphus	769
ampere	574
amphipathic	688
amphoteric	514
amplitude	265
anabolism	726
anhydride	177
anions	136
anode	578
anthocyanin	716
antifreeze	437
antioxidant	544
aromatic hydrocarbon	327
aromatic hydrocarbons	655
Arrhenius, Svante	471
Arrhenius	
acid	509
base	509
asteroid	801
astronomical unit (AU)	795

atom	
size of	32
atomic mass	64, 164, 259
atomic mass unit	164, 169
atomic number	62
atomic number (Z)	64, 258, 603
atomic radius	285
ATP	717
Aufbau principle	271
autotroph	714
average	43
Avogadro's law	393
Avogadro's number	164
Avogadro, Amadeo	49, 164

B

Baekeland, Leo	670
balancing an equation	200
ball and stick model	70
Balmer, Johann	267
bar graph	45
barometer	383
base	510
Beer's law	431
bent molecule	323
benzene	327, 655
best fit line	48
beta (β-) sheet	702
beta- (β-) pleated sheet	697
beta-carotene	2
beta-decay	612
beta-radiation	612
Big Bang theory	788
bimolecular	466
binding energy	606
blackbody spectrum	797
Bohr's hypothesis	269
Bohr, Neils	267
boiling	118, 479
boiling point	115
boiling point	
and particle size	333
and polarity	332
bond angle	322
bond energy	356
of carbon-carbon bonds	650
bond triangle	316
box and whiskers plot	46
Boyle's law	387
Brahe	786
branched alkane	645

Index

brittle 78
Brownian motion 96, 97
Brønsted-Lowry 532
Brønsted-Lowry definition 510
buckyball 335
buffer 532, 735
buffer capacity 533

C

calorie (cal) 101, 110
Calorie (Cal)
   kilocalorie 101, 110
calorimeter 110, 110
calorimetry 101
Calvin Cycle 718
capsanthin 2
carbohydrate 66
carbohydrates 660, 680
carbon allotropes 335
carbon cycle 762
carbon dating 616, 616
carbon dioxide
   dry ice 350
carbonyl 659
carboxylic acid 658
catabolism 726
catalyst 458
catalysts 194
cathode 578
cathode rays 263
cation 136
cell emf 585
cell membrane 731
cellular respiration 199, 721
cellulose 680
Celsius (°C)
   conversion 95
   scale 95
chain reaction 622
change of state graph 351
changes that impact pressure 382
charge separation 309, 310
Charles' law 391
Charles, Jacques 390
chemical 2
chemical bond 68, 298, 299
   equilibrium distance 299
chemical equation 192, 193
   product 192
   reactant 192
chemical equations
   balancing 197

chemical equations (cont.)
   method for balancing 199
chemical forensics 175
chemical formula 68
   of acids and bases 512
   of covalent compounds 153
   of simple ionic compounds 142, 314
   of ternary and quaternary ionic
compounds   145
chemical property 81
chemical reaction 82, 192
   combination 203
   combustion 207
   decomposition 204
   double replacement 206
   polymerization 328
   single replacement 205
   synthesis 203
chemical species 489
chemosynthesis 807
chemotroph 714
chiral carbon 683
chlorophyll 715
CHON 65
chromatography 411
cis- and trans- 653
citric acid 2
closed system 102, 493
colligative effect
   boiling point elevation 436
colligative property 435
   freezing point depression 435
colorimeter 431
combined gas law 392
common ion 532
complex carbohydrates 681
compound 68
compounds 68
concentration
   concentrated 417
   dilute 417
condensation 115, 115, 116, 118
condensation polymerization 670, 682
condensation reaction 687, 696
condensed matter 348
conduction 104
conductivity 415
   of ionic substances 334
conductivity sensor 415
conductor 315
   thermal 104
conjugate acid-base pairs 514

Index

control variable 12
convection zone 791
conversion factors 38, 39
Copernicus 786
correlation coefficient 48
corrosion 583
cosmic background radiation 788
coulomb (C) 574
covalent bond 149, 306
   sigma and pi bonds 324
cracking 668
critical mass 622
critical point 350
Crookes, William 263
crystal lattice 334
crystal structure 306, 314
crystalline 769
Curie 263
cyclo- 654
cytochromes 723

D

D isomer
   dextrorotation 683
Dalton 254
Dalton's Law of Partial Pressure . . . 384
de Broglie 269
dehydrogenation reaction 666
delocalized electrons 316
Democritus 254
denatured 692
density 18
   water displacement 40
dependent variable 46
deposition 350
desalination 411
dew point 751
diamond 335
diatomic molecule 150, 299
diffusion 414, 731
dimensional analysis 39
   and the mole 166, 169, 171, 172, 173
dipole-dipole attraction 331
dipole-dipole forces 333, 334
disaccharide 681
dissociation 412
dissolution 412
dissolved 410
distillation 76
   of hydrocarbons 648
distilled water 76

DNA 5, 66, 183, 701
doppler effect 788
double bond 304
   and hybrid orbitals 324
double bonds
   bond energy of 356
double replacement reaction 206
dry ice 350
ductile 78
ductility 336
dwarf planet 801

E

electric current 574
electrochemical 738
electrochemical cell 578, 578
electrochemistry 544
electrode 578
electrolysis 118, 580
electrolyte 415, 578
electrolytic 579
electromagnetic attraction 298
electromagnetic force 259
electromotive force (emf) 585
electron configuration 276
   and ions 312
   bonding 300
   noble gas 300, 312
electron density 319
electron mobility 316
electron transport chain 724
electron transport chain (ETC) 723
electronegativity 306, 308
   and bond type 309
   and predicting bond type 308, 311
electroneutrality 544
electrons
   localized and delocalized 316
electroplating 581
electrostatic attraction 299
electrostatic repulsion 299
element 62
element symbol 62
elementary steps 463
elements
   in the human body 65
empirical formula 178, 179, 180, 181
endothermic 118
   definition of 355
   phase change 113
endothermic reaction
   photosynthesis 363

Index

energy
    and chemical bonds 307
    and enthalpy 354
    as mediator of change 19
    Earth's energy balance 117
    in chemical bonds 356
    mechanical energy 19
    nuclear energy 19
    of a system 106
    radiant energy 19
    thermal 106, 107, 109
energy bar diagram 103
    adiabatic 110
    exothermic 362
    glycolysis 724
    photosynthesis 719
    physical change 103
energy level 271
enthalpy 354
    change in reaction ΔH 355
    Hess's Law 359
enthalpy of formation 358
enthalpy of reaction
    photosynthesis 363
entropy 366
enzyme 66, 318, 691
enzymes 458
equilibrium 478
equilibrium constant 485, 512
equilibrium exression 485
equilibrium postion 483
equivalence point 528
error 30, 43
    error bars 44, 46
ester 661
    polyester 661
ethene 650
ether 661
ethyne 650
evaporation 116, 118
excited state 273
exothermic 118
    definition of 355
    phase change 113
experiment 28
    reliability 28
    results 28
exponent 31
extremeophiles 807

F

factors affecting reaction rate 451

Fahrenheit (°F)
    conversion 95
    scale 95
Faraday (F) 585
fat 66, 680, 687
femtochemistry 51
Ferigna, Bernard 183
first law of thermodynamics 106, 109
first order reaction graph 455
Flammarion 786
forces of attraction
    and polymers 339
    intermolecular 118
    intramolecular 118
forensic chemistry 176
formula mass 168
formula unit 137
four primary forces
    electromagnetic 257
    gravity 257
    strong nuclear 257
    weak nuclear 257
free radical 544, 667
freezing 118
freezing point depression constant . . . . . .
(K_f)   437
frequency 265
frequency table 45
frost line 798
fructose 660
fullerenes 335
functional group 657

G

G-type star 790
galactose 660
galvanic cell 578
galvonic cell 579
gamma-decay 613
gas laws 387
    and STP 394
    Avogadro's law 393
    Boyle's law 387
    Charles' law 391
    combined gas law 392
    Dalton's law 384
    ideal gas law 397
gases 98
    kinetic energy 98
Geiger 264
gene 704
geologists 770

Index

geometric isomer 653
geosphere 766
Gibbs free energy 369
Gibbs-Helmholtz equation 369
glucose 660
glycolysis 722, 724
graduated cylinder 17
gram 16
graph
   change of state 351
graphite 335
graphs
   line 46
gravity 7
gray 629
greenhouse effect 756, 799
greenhouse gases 756
ground state 273
group 64

H

half life
   uranium 622
half-life 614
   of carbon-14 . . . . 616, 616
half-reactions 559
Haumea 795
heat 101
   specific 107
heat of fusion 114
heat of solution 433
heat of vaporization 115
heating curve
   water 114
Heisenberg's uncertainty principle . . 268
hematite 800
Henry's law 421
Hertzsprung-Russell diagram 790
Hess's Law 359
heterogeneous mixture 75
heterotrophs 721
Hillary, Sir Edmund 402
homeostasis 729
homogeneous mixture 74
Hund's Rule 272
hybrid orbitals 324
hybridization 272
hydrate 177
hydration 412
hydrocarbon 207, 642
hydrocarbons
   naming 643, 663

hydrocarbons (cont.)
   hydrogen bond 331
   hydrogenation reaction 669
hydronium ion 511
hydrophillic 688
hydrophobic 688
hydrothermal vent 806
hydroxyl 658
hypothesis 8

I

ideal gas 396
ideal gas law 397
   fixed conditions 398
   variable conditions 399
igneous rock 770
immiscible 413
independent variable 46
inhibitor 458, 691
inquiry 7
insoluble 194, 417
inspection method 198
insulator 335
   thermal 104
intensity 798
intensity of radiation 628
intermediate 464
intermolecular forces 84, 118, 330, 333, 422
intersellar medium 793
intramolecular forces 84, 118, 330
inverse square law 628
ion 135, 260, 300
ion, polyatomic 312
ionic bond 137, 306, 311
   and conductivity 334
   properties of ionic substances . . . 334
ionic radius 286
ionization 287, 627, 790
ionization energy 287, 315
ionizing radiation 627
ions
   anions 136
   cations 136
isoelectronic 300
isoenzymes 735
isolated system 102
isomer
   structural, geometric, and optical . . 652, 653
isomers 325
isooctane 648
isotope 259

Index

isotopes 258

J

joule 20, 101

K

K_a 512
K_b 512
Kelvin 391
kelvin
   °C to K formula 390
Kelvin (K)
   conversion 99
   scale 99
Kepler 786
ketone 659
kilogram 16
kinetic energy 96, 97
   average 98, 98
kinetic molecular theory 97, 380, 396
kinetic theory of matter
   heat transfer 104
Krebs cycle 723, 724
Kuiper Belt Object 795
K_w 511

L

L isomer
   levorotation 683
lactic acid 653
lattice structure 334
lava 772
Lavoisier, Antoine 42, 49
law of conservation of matter . . . 195
law of mass action 485
Le Châtelier's Principle 492
Lewis dot diagram 302
   and molecular shape 323
   and resonance structures . . . 326
   for a molecule 303
Lewis, Gilbert 262
linear molecule 320
linear regression 48
lipid 686
lipoprotein 687
lipoproteins 328
liquid crystal 339
liters 17
localized electrons 316
logarithm 517
logarithmic scale 516

London dispersion attraction . . . 333
London dispersion forces 334
lone pairs of electrons
   and molecular shape 320, 323
Lowell 809

M

macronutrient 66
macroscopic scale 6
Makemake 795
malleable 78
manometer 387
mantissa 31
mantle 766
mass 16
   formula mass 168
   molecular mass 168
mass defect 608
mass number 258
mass number (A) 258, 603
mass spectrometer 175
matter 4
Maxwell 266
measurement 14
melting 118
   melting point 114
Mendeleev, Dmitri 262, 284
meniscus 17
mesosphere 748
metabolism 726
metal 315
metallic bond 306, 315
   properties of metallic substances . . . 336
metalloids 315
micelle 688
microscopic 6
microscopic scale 6
Milky Way galaxy 786
Miller-Urey experiment 805, 806
milliliters 17
mineral 769
miscible 413
mixture 73, 73
mobile electrons 316, 336
molal 437
molality 436
molar mass 164, 169, 170
molar volume 394
molarity 425
   brackets 425
mole 164

Index

mole (cont.)
molecular clouds 793
molecular formula 178, 180, 181
molecular geometry 318, 319
   and polarity 310
molecular ion 260
molecular mass 168
molecular surface model 70
molecular weight 169
molecule 71
   representations of 70
monatomic 260
monomer 212, 328, 682
monosaccharide 681

N

NADH 722
NADPH 717
nanotechnology 151, 183
nanotube
   carbon 335
natural laws 7
nebulae 793
Nernst Equation 591
network covalent 151
network covalent structure 335
neutral 509
neutralization 526
neutralized 527
neutrons 257
nitrogen cycle 773
nitrogen fixation 773
nitrogenase 773
nitrogenous base 701
noble gas
   electron configuration 300
noble gas configuration 312
non-catalyzed 458
non-ionizing radiation 627
non-polar bond 310
non-polar covalent bond 309
nonmetals 315
Norgay, Tenzing 402
Normality
   equivalents 530
nuclear reaction 602
nucleon 602, 606
nucleotide 701
nucleus
   size of 257

O

objective 8

ocean acidification 536, 764
ocillating electromagnetic force 266
octane 648
octet rule 301
ogliosaccharide 681
ohm (Ω) 576
Ohm's law 576
Oort cloud 795
open system 102
Opportunity 809
optical isomer 653
optimum range 735
orbital 270
orbitals
   d 270
   f 270
   p 270
   s 270
organic chemistry 642
osmosis 411
outliers 43
oxidation 544, 714
oxidation number 546
oxidizing agent 555
ozone layer 748

P

parent compound 645
partial hydrogenation 669
Pascal (Pa) 381
Pauli exclusion principle 272
Pauling scale, the 308
Pauling, Linus 262
pentane 643
peptide bond 696
percent composition 175, 176, 177, 178
period 64
periodic table 63, 64
   and ionic charge 311
   family 131
   periodic 128
periodic trends
   electronegativity 308
Perrin, Jean 49
petroleum 668
pH 517
pH scale 508
phase change 113
   boiling 479
   evaporation 118
   and intermolecular forces 330

Index

phase change (cont.)
    boiling 115, 118
    condensation 115, 115, 116, 118
    endothermic 118
    equilibrium 117
    evaporation 116
    exothermic 118
    freezing 114, 118
    melting 114, 118
phase diagram 348
phenolphthalein 528
phenylthiocarbamide (PTC) 45
phlogiston theory 42
phospholipid 688
phosphorus cycle 774
photodissociation 753
photon 269
photoshphere 791
photosynthesis 212, 717, 719
    enthalpy of reaction 363
phototroph 714
photsynthesis 715
physical change 78, 118
physical property 78
pi bond 324
pie graph 45
pigment 715
Planck's constant 268
Planck, Max 51
plantetesimal 796
plasma 790
plate tectonics 772
Pluto 795
polar bond 310, 310
polar covalent bond 309
polarity
    and dissolving 332
polarization 299
polyatomic ion 140, 312
polyatomic ions 313
polycrystalline structure 336
polymer 212, 328, 682
    and attractive forces 339
polymerization 212, 328, 670, 682
polymorphism 770
polyprotic
    diprotic 529
    triprotic 529
polysaccharide 681
positron 613
positron-emission 613

precipitate 83, 206, 209
precipitate reaction 210
precision 30
prefixes for hydrocarbons 643, 663
pressure 381
    from vapor 479
    in space 383
primary structure 696
properties of acids 508
properties of bases 508
properties of gases 380
proportion 37, 47
protein 66, 680, 691
protein structure
    secondary 697
proteins 65
proton-proton chain 792
protons 64
pseudoscience 9
psychrometric chart 751
Ptolemy 786
pure substance 73, 428

Q

quanta 269
quantity 30
quantum effects 259
quantum mechanics 262
quantum state 270
quartz 62
quaternary compound 145

R

R-group 645
rad 629
radiation 610
    gray, unit of 629
    sievert, unit of 629
    rad, unit of 629
    rem, unit of 629
radiation dose 629
Radio telescopes 787
radioacivity 610
radioactive 259
radon 611
rate 37, 46, 47, 448
rate constant 614
rate determining step 466
rate law 466
    first order 452
    rate constant 452

Index

rate law (cont.)
   zero order 452
ratio 37, 47
reaction mechanism 463, 463
reaction profile 465
reaction rate 449
reactivity 82
redox 545
reduction 545
regression 48
relative humidity
   dew point 751
   psychrometric chart 751
rem 629
reolith 800
repeatable 8
repulsive force 319
resist 526
resistance 576
resolution 30
resonance 326, 327, 655
reverse osmosis 411
ring structure 327
RNA 704
rock 771
rules for balancing an equation . . . 197
Rutherford's Gold Foil Experiment . . 264
Rutherford, Ernest 263

S

salinity 760
salt 509
salt bridge 578, 583
saturated bonds 650
saturated hydrocarbon 642
saturation 418, 418
Sauvage, Jean-Pierre 183
scale 6, 60
scanning tunneling microscope . . . . . .
(STM) 51
Schrödinger, Erwin 270
scientific method
   conclusion 49
scientific notation 31
   powers of ten 31, 32
   sigificant figures and, ... 34
sea of electrons 306, 315
second law of thermodynamics . . 102, 366
second order reaction graph 455
secondary protein structure 697
sedimentary rock 770
semiconductors 315

semiperiable membrane 411
SI unit
   energy 101
   heat 101
sievert 629
sigma bond 324
significant figures 33
   zeros 35, 36
silicates 62, 766
simple carbohydrates 681
single bond 304
size of the nucleus 257
slope 37, 46
   equation 47
solar nebula 766
solubility 332, 417
solubility product constant 496
solubility rules 209
soluble 194, 209, 417
solution 74
   of solids 336
   solute 410
   solvent 410
solvation 412
space-filling model 70
specific heat 107
spectator ions 211
spectral line 267
spectrometer 274, 787
spectroscopy 274
spectrum 274
speed 448
spin 272
Spirit 809
spontaneous 366, 579
standard deviation 43
standard error 43
standard molar volume 173
Standard Reduction Potentials . . . 587
standard temperature and pressure . . 173
standing wave 270
starch 680
state function 359
states of matter 113
Steps for Balancing an Equation . . 201
stereoisomer 653
stock solution 430
Stoddart, Sir J. Fraser 183
stomata 761
story problem 208
stratosphere 748

Index

strong acid (cont.)
strong acids 531
strong base 512
strong nuclear force 257
structural diagram 70, 303
    for ionic compounds 311
structural isomer 652
sublimation 350
substitution 667
substrate 699
suction 389
supernova 789
supersaturated 419
surface tension 331
symbols 194
synthesis reaction 203
system
    closed 102
    energy flow 103
    isolated 102
    open 102
system and surroundings 102, 103, 366

T

tap water 76
temperature 94, 97, 98, 101
    final, of a mixture 106
    scales 95, 99
ternary compound 145
tertiary structure 698
tetrahedral molecule 318, 319, 324
the conservation of energy 20
the first law of thermodynamics . . . 354
the scientific method 10
theory 8
Theory of Atoms 254
thermal conductor 104
thermal energy 101, 106
thermal equilibrium 105
thermal insulator 104
thermistor 94
thermocouple 94
thermometer 94
thermophiles 807
thermoregulation 737
thermosphere 748
Thompson, J.J. 263
Titan 803
titration 528
trace amounts 65
trace element 66

transcription 704
transition metals 282
translation 704
transmittance 431
transpiration 761
transport 731
TRAPPIST-1 811
triglyceride 687
trigonal planar molecule 320
trigonal planar shape 322
trigonal pyramidal 321
trigonal pyramidal molecule 320
triple bond 304
    and hybrid orbitals 324
triple bonds
    bond energy of 356
triple point 350
tritium 602
troposphere 748

U

unbalanced equation 196
uncertainty 46
unimolecular 466
unsaturated bonds 650
unsaturated hydrocarbon 642
uranium
    half life 622
uranium-235 622

V

vacuum 383
valence 301
    bonding 300
van der Waals attractions 333
van't Hoff factor 437
vapor pressure 479
vaporization 115
variable 11, 12
    control 12
    dependent 12, 46
    independent 12, 46
    relationships between 48
Viking 809
volcano 772
volt (V) 576
voltage 576
voltaic cell 579, 579
volume 17, 40
VSEPR
    bent shape 321

Index

VSEPR (cont.)
   molecular shapes 322
   polarity 322
   tetrahedral shape 321
   trigonal planar shape 322
   trigonal pyramidal shape 321
VSEPR theory 319, 324
   lone pairs 320

W

water
   triple point of 350
water cycle 117, 761
water displacement 40, 384
water quality 52
wave-particle duality 269
wavelength 265
weak acid 512
weak base 512
weight 16

X

xanthophyll 716

Z

zero order reaction graph 455
Zewail, Ahmed 51

Fundamental constants

Quantity	Symbol	Value	Units
Speed of light in a vacuum	c	3.00×10^8	m/s
Gravitational constant	G	6.67×10^{-11}	N m²/kg²
Boltzmann's constant	k_B	1.38×10^{-23}	J/K
Stefan–Boltzmann constant	σ	5.67×10^{-8}	W/(m² K⁴)
Planck's constant	h	6.63×10^{-34}	J s
Avogadro's number	N_A	6.022×10^{23}	mol⁻¹ (or particles/mol)
Gas constant	R	8.315	J/(mol K)
Permittivity of free space	ε_0	8.85×10^{-12}	C²/(N m²)
Permeability of free space	μ_0	$4\pi \times 10^{-7}$	T m/A
Mass of the electron	m_e	9.11×10^{-31}	kg
Mass of the proton	m_p	1.6726×10^{-27}	kg
Mass of the neutron	m_n	1.6749×10^{-27}	kg
Charge of the electron	e	-1.60×10^{-19}	C
Coulomb's law constant	k_e	8.99×10^9	N m²/C²
Absolute zero	0 K	−273.15	°C

Useful values constants

Quantity	Symbol	Value	Units
Strength of gravity at Earth's surface	g	9.81	m/s² or N/kg
Speed of sound at 20°C and 1 atm	c_s	343	m/s
Standard atmospheric pressure	atm	1.01×10^5	Pa or N/m²

Scientific notation and powers of ten

Description	Scientific notation	Equivalent value	SI prefix	Abbreviation
One quadrillion	1×10^{15}	1,000,000,000,000,000	peta	P
One trillion	1×10^{12}	1,000,000,000,000	tera	T
One billion	1×10^9	1,000,000,000	giga	G
One hundred million	1×10^8	100,000,000		
Ten million	1×10^7	10,000,000		
One million	1×10^6	1,000,000	mega	M
One hundred thousand	1×10^5	100,000		
Ten thousand	1×10^4	10,000		
One thousand	1×10^3	1,000	kilo	k
One hundred	1×10^2	100	hecto	h
Ten	1×10^1	10		da
One	1×10^0	1		
One tenth	1×10^{-1}	0.1	deci	d
One hundredth	1×10^{-2}	0.01	centi	c
One thousandth	1×10^{-3}	0.001	milli	m
One ten thousandth	1×10^{-4}	0.0001		
One hundred thousandth	1×10^{-5}	0.00001		
One millionth	1×10^{-6}	0.000001	micro	μ
One ten millionth	1×10^{-7}	0.0000001		
One hundred millionth	1×10^{-8}	0.00000001		
One billionth	1×10^{-9}	0.000000001	nano	n
One trillionth	1×10^{-12}	0.000000000001	pico	p
One quadrillionth	1×10^{-15}	0.000000000000001	femto	f

Astronomical data

Quantity	Symbol	Value	Units
Astronomical unit (average Earth–Sun distance)	AU	1.50×10^{11}	m
Average Earth–Moon distance		3.84×10^{8}	m
Radius of the Sun	$r_\odot$	6.96×10^{8}	m
Radius of the Earth	$r_\oplus$	6.37×10^{6}	m
Radius of the Moon		1.74×10^{6}	m
Mass of the Sun	$m_\odot$	1.99×10^{30}	kg
Mass of the Earth	$m_\oplus$	5.98×10^{24}	kg
Mass of the Moon		7.35×10^{22}	kg

Conversion factors

Quantity	Some conversions		
Length	1 in = 2.54 cm	1 cm = 0.3048 in	1 m = 39.37 in
	1 foot (ft) = 12 in	1 yard (yd) = 3 ft	1 mile (mi) = 5,280 ft
	1 mi = 1.609 km	1 km = 0.621 mi	1 angstrom (Å) = 10^{-10} m
	1 AU = 1.50×10^{11} m	1 light year (ly) = 9.46×10^{15} m	
Area	1 m^2 = 10.76 ft^2	1 in^2 = 6.452 cm^2	1 acre = 43,560 ft^2 = 4,048 m^2
Volume	1 m^3 = 1,000 liters (L)	1 L = 1,000 cm^3	1 gallon = 3.785 L
Time	1 year (yr) = 365.24 day	1 day = 24 hr	1 hr = 60 min
	1 min = 60 s	1 yr = 3.156×10^{7} s	1 day = 86,400 s
Speed and velocity	1 m/s = 2.237 mi/hr (mph)	1 mph = 1.609 km/hr	1 m/s = 3.600 km/hr
Force	1 N = 0.2248 pounds (lb)	1 lb = 4.448 N	1 ton = 2,000 lb
Pressure	1 Pa = 1 N/m^2	1 Pa = 1.45×10^{-4} lb/in^2 (psi)	1 atm = 1.013×10^{5} Pa
Energy	1 J = 10^{7} erg	1 calorie = 4.184 J	1 Calorie = 1,000 calorie
	1 eV = 1.602×10^{-19} J	1 kWh = 3.6×10^{6} J	1 Btu = 1,055 J
Power	1 horsepower (hp) = 746 W	1 hp = 550 ft lb/s	
Mass	1 kg = 1,000 g	1 amu = 1.6605×10^{-27} kg	1 kg has a weight of 2.205 lb
Angular measures	1 radian (rad) = 57.3°	π rad = 180°	1 rad/s = 9.55 rev/min (rpm)